9th International Conference on Wind Turbine Noise (WTN 2021)

Online
18-21 May 2021

ISBN: 978-1-7138-3184-6

Printed from e-media with permission by:

Curran Associates, Inc.
57 Morehouse Lane
Red Hook, NY 12571

Some format issues inherent in the e-media version may also appear in this print version.

Printed with permission by Curran Associates, Inc. (2021)

For permission requests, please contact Institute of Noise Control Engineering/Europe (INCE/Europe) at the address below.

Institute of Noise Control Engineering/Europe (INCE/Europe)
Riverside House
4 Oakland Vale
New Brighton, Merseyside CH45 1LQ
United Kingdom

Phone: +44 (0)151 638 0181
Fax: +44 (0)151 639 5212

Mail@inceeurope.org

Additional copies of this publication are available from:

Curran Associates, Inc.
57 Morehouse Lane
Red Hook, NY 12571 USA
Phone: 845-758-0400
Fax: 845-758-2633
Email: curran@proceedings.com
Web: www.proceedings.com

TABLE OF CONTENTS

9th International Conference

Wind Turbine Noise 2021

Remote from Europe

Tuesday 18th - Friday 21st May 2021
www.windturbinenoise.eu

Conference Proceedings

v2.1 10th May 2021

INCEEUROPE

Wind Turbine Noise 2021

These Proceedings are those for the Ninth International Conference on Wind Turbine Noise organized by INCE/Europe. This was our first wind turbine e-conference, but not the first one for INCE/Europe as we hosted Quiet Drones 2020 last October in collaboration with CidB in Paris.

One of the biggest problems with an e-conference is scheduling for time zones. Inevitable there will be some people who can only watch live sessions at thoroughly inhospitable times. At the time of writing it looks like 75% of our delegates will be from Europe, about 5% from Asia/Pacific and 22% from the Americas. Reflecting these numbers we operated the conference in the afternoons European time. But almost everything was recorded and delegates were able to watch during the conference and for two weeks after.

You can see an overall view of the papers in the programme below and you should go to a particular paper by clicking the paper in the programme. The papers are identified by lead author family name and appear in alphabetical order.

Organisation

Chair

- Dick Bowdler, UK

INCE-Europe

- Jean Tourret, France, President
- Geoff Leventhall, UK

Administrator

- Cathy MacKenzie, UK

Committee members

- Mark Bastasch, USA
- Franck Bertagnolio, Denmark
- Norm Broner, Australia
- Sylvia Broneske, UK
- Matthew Cand, UK
- Kristina Conrady, Germany
- Christophe Delaire, Australia
- Malcolm Hayes, UK
- Michaela Herr, Germany
- Cordula Hornung, Germany
- Brian Howe, Canada
- Simon Jennings, Ireland
- Damian Kelly, Ireland
- Soogab Lee, Korea
- David Michaud, Canada
- Bo Søndergard, Denmark
- Frits van den Berg, Netherlands
- Wouter van der Velden,
- Sabine von Hunerbein, UK

Co-Hosts: Technical operation of the conference

- Payam Ashtiani
- Matthew Cand
- Krispian Lowe
- Andy McKenzie

Tuesday 18th May

13:00 - Introductory Conversation

Dick Bowdler and Jean Tourret

Meet the other delegates and the organisers.

Find out how the conference operates.

14:00 - Session 1 - Source Noise Analysis and Prediction

Session Chairs	**Cordula Hornung and Pietro Bortolotti**
Martuscelli Faria	A review on the development of airfoils for wind turbine blades
Hornung	Turbulence inflow noise prediction of wind turbine rotors: The physically correct representations of the Simplified Amiet and Lowson Model
Hasheminasab	Effect of grid resolution on airfoil self-noise prediction by large eddy simulation
Bortolotti	Validation efforts of an open-source aeroacoustics model for wind turbines
Rodriguez	The Quasi-3D blade and rotor noise prediction methodology for the PNoise code and preliminary results
Bertagnolio	A tower wake model for Low-Frequency Noise of downwind turbine rotors
	Extended Discussion

16:00 - Break and Networking

16:30 Session 2 Large Blades - Do They Pose Special Acoustic Problems?

Session Chairs	**Franck Bertagnolio and Alexandre Suryadi**
Seel	Numerical study of the impact of vortex generators on trailing edge noise
Suryadi	Identifying the flap side-edge noise contribution of a wind turbine blade section with an adaptive trailing-edge
Saab	Developing new airfoils for larger wind turbine blades
	Extended Discussion

18:00 - Meet in the Wonder Room

Go to the Wonder Room and meet the other delegates

INCE EUROPE

International Year of Sound 2020+

Wednesday 19th May 2021

13:00 - Conversation Task 39. Update on International Energy Agency progress

14:00 Influences	**Kristina Conrady on the Wind**
	Low Level Jets and Low Level Wind Maxima - what they are and how they affect wind turbine noise.

14:45 - Break and Networking

15:05 Session 3	**Propagation and Modelling**
Session Chairs	**Matthew Cand and Susanne Martens**
Kayser	Calculation of wind turbine noise uncertainty for downwind conditions
Søndergaard LS	Long distance noise propagation over water for an elevated height-adjustable sound source
Clark	Comparison between modelled and measured noise impact with varying ground factors
Halstead	A study of the relationship between wind direction and sound level for wind turbines measured in the far-field
Elsen	Different sound source setups in the simulation of wind turbine sound propagation
Dutilleux	Meteorological effects on wind turbine noise at the receptor location
Bolin	Wind sector management using Beilis Tappert Parabolic Equation
	Extended Discussion

17:05 - Break and Networking

17:25 - Session 4	**Impact Studies and Regulations**
Session Chairs	**Mark Bastasch and Andy McKenzie**
Søndergaard B	On the need for improved prediction models and updated noise regulations to utilize the advanced controls strategies that are available for modern wind turbines
Palmer	Stymied by Standards? Arguments for wind turbine noise standards that actually measure irritant drivers
Tam	A characterization of wind turbine and background noise distributions in far-field receptor testing of wind turbine facilities
	Extended Discussion

18:30 - Meet in the Wonder Room

Thursday 20th May 2021

13:00 Conversation

Frits van den Berg talks to Mirjam Davidson and Anne Struijs of RWE about working with neighbours

13:45 - Break and Networking

14:05 Session 5 — **Noise Measurement and Assessment**

Session Chairs — **Payam Ashtiani and Sabine von Hunerbein**

Ashtiani	A review of different methodologies for the measurement of sound pressure level from wind farms
Finez	Robust noise indicators using Gaussian Processes
Gloaguen	Estimation of the sound emergence of wind turbines by semi-supervised learning technique
Broneske	Development of IEC/TS 61400-11-2: Measurement of wind turbine noise characteristics in receptor position
Summers	Further experience of reviewing noise assessments for wind farms in Scotland and the implementation of the IOA Good Practice Guide to the Application of ETSU-R-97 for the Assessment and Rating of Wind Turbine Noise
Bartolazzi	A model to calculate the delta between internal noise with open windows vs external noise
	Extended Discussion

16:05 - Break and Networking

16:25 - Session 6 - Tonality

Session Chairs — **Sylvia Broneske and Robin Woodward**

Old	Wind turbine sound quality rating
Busse	The (psychoacoustic) basics of tonality perception
Søndergaard LS	Tonality content analysed with both 1/3 octave band and narrowband methods with comparison to listening test
Woodward	An implementation of ISO/PAS 20065:2016 for the analysis of wind turbine sound at receptor distances
Munro	Comparison of tonality analysis methods for wind turbine receptor based long-term monitoring data sets
	Extended Discussion

17:55 - Short Break

Thursday 20th May 2021

18:05 Session 7	**Amplitude Modulation**
Session Chairs	**Sarah Large and Duncan Halstead**
Mascarenhas	Physics-based auralization of wind turbine noise
Pies	Assessment and rating of wind turbine noise immission at dwellings - the influence of amplitude modulation, aerodynamic noise sources and the Doppler effect. **POSTER - no presentation see poster on poster link on website
Lotinga	Subjective responses to wind turbine noise amplitude modulation: pooled analysis of laboratory listening studies and synthesis of an AM character rating penalty
	Extended Discussion
19:05 - End	

Friday 21st May 2021

13:00 Conversation

Frits van den Berg talks to Aileen Jackson and Rosemary Milne about what it is like to live near wind turbines

13:45 Session 8	**Infrasound**
Session Chairs	**Norm Broner and Bruce Walker**
van den Berg	Audibility and health effects of infrasound
D'Amico	Prediction of wind turbines infrasound from meteorological parameters
Leventhall	If they are not being made ill by infrasound, then what is it?
	Extended Discussion

14:50 - Break and Networking

15:10 - Session 9	**Perception and Health**
Session Chairs	**Irene van Kamp and David Michaud**
Liebich	A meta-analysis on the impact of wind turbine noise on sleep using validated objective sleep assessments
Søndergaard LS	Wind farm neighbourship investigated by a daily app questionnaire combining weather, noise and annoyance
Preihs	Assessing wind turbine noise perception by means of contextual laboratory and online studies
Ollson	Establishing sound limits for wind energy: What is the role of annoyance?
	Extended Discussion

16:45 Conversation and Closing - We need your feedback

Jean Tourret and Dick Bowdler discuss the advantages and disadvantages of e-conferences with the delegates - and what plans are for 2023.
Give us your feedback.

17:30 Conference Closes

| Conference Proceedings

INCEEUROPE

9th International Conference on Wind Turbine Noise
Remote from Europe – 18th to 21st May 2021

A review of different methodologies for the measurement of Sound Pressure Level from Wind Farms

Payam Ashtiani, Aercoustics Engineering Limited PayamA@aercoustics.com

Summary

The measurement of noise immission from wind turbines is often carried out to determine the acoustic contribution of a wind farm in the far-field, often at the dwelling location, or an equivalent point of environmental compliance. In the case of pre-construction noise modelling, the methodologies employed to determine the expected sound levels at these locations are largely similar. They are typically based on one of two noise modelling algorithms, and the difference between employed methods lies in the assumptions for the modelling parameters. However, in the realm of post construction noise immission measurements, there is a relatively wide array of methodologies employed to determine the noise contribution from the wind farm. This paper outlines the major methodologies being employed in the field and evaluates their benefits and drawbacks.

1. Introduction

There have been many different measurement methods employed when attempting to determine noise immission from wind farms. One of the likely reasons for this wide variation is the inherent complexity of measuring noise from wind farms in a consistent, repeatable, and accurate way. The difficulties of conducting far field measurements of wind turbines is outlined by many previous studies [1][2][3][4][5][6][7]. The primary complications are:

- The noise source is intermittent and controlled by the weather
- The noise source is often loudest during windy conditions which are not ideal for acoustic measurements
- Contaminating background noise from vegetation is also wind speed dependant
- The noise source is generally broadband in nature as is the background vegetation noise making source separation more difficult
- Regulatory limits often place turbines at distances where the sound levels are close to ambient sound levels making it more difficult to obtain adequate signal-to-noise ratio
- All of the above is also dependent and affected by wind direction, seasonality and instantaneous atmospheric conditions.

Due to the complex nature that arises from the combination of the above facotrs, noise measurement techniques have evolved to try and overcome these challenges. This paper outlines the major methodologies being employed in the field, and evaluates their benefits and drawbacks. The goal is to aid any future measurement campaign in deciding on methodology before undertaking such a process which is often long and expensive.

2. Measure or Predict?

Methods employed by various measurement institutes and regulatory authorities broadly fall into two categories

1. Direct measurement at the point of interest and;
2. Indirect measurement at more favourable position

In the case of the direct measurement, microphones and other instrumentation is deployed at the location of interest ("receiver position") and measurements are carried out in order to quantify the noise levels attributable to the wind farm directly. In this case, the measurement campaign is faced with all the complexities outlined in the previous section.

In the case of indirect measurements, measurements are carried out on individual turbines in accordance with IEC 61400-11[8] to determine its sound power and the results of those measurements are used in an acoustic model to infer what the sound pressure level would be in the farfield at the receiver positions under the desired conditions.

2.1 Benefits of a direct measurement approach

Some regulators like the direct measurement approach as it provides more direct evidence of the compliance status of a project. Because the measurements are carried out at the location of interest, in cases of dispute all parties can point to the results of measurements as the direct evidence of the sound levels that occur at the point on interest.

Such measurements can also identify factors outside of the sound pressure level that may be of concern such as the prevalence of Amplitude Modulation, Low Frequency Noise, and Tonality. Many of these quantities are either not predictable or come with high margins of uncertainty when modelling. In cases where regulators use post construction measurements as a means of verifying that modelling was conducted appropriately, some regulators have been uncomfortable with the concept of "using the model to verify the model" and would prefer direct measurements to achieve that outcome.

2.2 Benefits of an in-direct measurement approach

Some regulators prefer the in-direct measurement approach. This method will typically have a lower measurement uncertainty of the sound power portion as the microphone will be placed relatively close to an individual turbine. Whilst there will be an additional uncertainty introduced by using a noise model to predict what the sound pressure level would be at the receiver position, many regulators consider that the models used have been sufficiently validated and their results can be used reliably.

The benefit of using this approach is that from a regulatory perspective there will be more consistency, and the line of determination of compliance status can be thinner. i.e. the determination of compliance is not affected by atmospheric conditions, time of day, wind direction, etc as would be during measurements at the far field. The result would be that the determination of compliance would be more consistent between different measurement institutes, different measurement campaigns, and at different sites. The variations would largely be due to the behaviour of the turbine(s) during the measurements, and individual site conditions for the measurement of sound power in accordance with IEC 61400-11.

3. Direct measurement methods

Within the branch of direct measurement approaches, there are a host of different methods being employed. All of them in some form are influenced by the question they are trying to answer. The question of compliance usually comes from a regulatory agency and has aspects that are baked into it that influence the way measurement methods have evolved.

As an example, some regulations define compliance as being determined as a function of ambient noise. This means that the method in determination of ambient noise will influence the method for determination of a change in the noise levels as a result of the wind farm. In the UK and Australia, for example, ambient conditions are determined using measurements conducted in 10 minute intervals, and using the statistical indicator of LA90. As it is acknowledged that ambient noise conditions vary with wind speed, this data is gathered before and after the wind farm is built and binned according to wind speed (measured as standardized 10m wind speed).

3.1 Organization of data by condition vs time

In the direct measurement category, there is a further breakdown in methods that can be either characterised by the organization of data by measurement condition vs by time. In the measurement of noise from wind farms, one of the chief challenges is the separation of the wind farm noise from the ambient noise. Because both quantities are ever changing, and the frequency content is often overlapping, separation is often conducted by quantifying the ambient noise by temporarily parking the nearby turbines. Once this is done, the operational noise can be adjusted by logarithmically subtracting the measured ambient noise, and arriving at the estimated wind farm contribution. This is often referred to as the 'shutdown method'

The instantaneous ambient levels can never be fully known when the turbines are operational. The data during shutdowns can be used in two ways:

1. To infer what the ambient conditions were during times that are just before / after the shutdown; looking at data as a function of time.
2. To build a relationship between ambient sound levels and wind speed; looking at the data as a function of measurement condition.

In the first method, referred to here as a "time-based method", ambient data is collected at regular intervals, and establishes the ambient level for a short duration before and after the shutdown period. For example, a shutdown would be conducted between 1-2 am, and the results would represent the ambient sound level from 10pm the earlier that night to 7am the coming morning. For those times then, the operational data would give the total sound level, and the ambient sound level would be logarithmically subtracted, and the wind farm contribution would be the result for each hour between 10pm and 7am.

In the second method, referred to here as the "binning method", shutdown data from multiple nights is included into a data-set that is binned as a function of measured wind speed. The ambient sound level is then computed for each wind bin. Subsequently, data gathered during wind farm operation is similarly binned, and an operational sound pressure level is then computed for each wind bin. Then, for each wind bin, the wind farm contribution is derived by logarithmic subtraction occurring at each individual wind bin.

Both these methods have benefits and drawbacks.

3.1.1 Benefits of Time-based methods

Some practitioners and regulators favour a time-based method as it can result in an estimated sound contribution from a wind farm on an hourly (or even more granular) basis. This allows a regulator to point to a specific hour as evidence of a violation or of compliance. The base assumption is that ambient conditions haven't substantially changed within the timeframe of the comparison.

Factors that would affect whether this assumption holds should be analysed in detail, as ambient conditions have the potential to change significantly over very short periods of time. The main determiner of ambient sound levels is wind speed, and thus, if this method is employed special considerations should be given to the wind speeds occurring during the shutdown and relevant operational times to ensure they are, in fact, comparable. A secondary source of variation could also be wind direction. If the source of ambient noise is vegetation, then the wind direction with relation to that vegetation could make a difference to the ambient sound levels measured. And lastly, change in environmental activity in the area could have a significant effect on the ambient levels. For example, if the time range includes a dawn chorus (birds singing) or early farmer activity, those times will not be comparable to other parts of the early morning, and have a potential to introduce significant error in the estimation of the wind farm noise. The same would be expected of the change in sound levels between daytime and night time.

3.1.2 Benefits of binning methods

Binning methods strive to define the ambient noise levels as a function of wind speed, wind direction, or other factor. The philosophy behind binning methods is that if the variables effecting the ambient sound levels are measured and controlled for, the difference attributable to the wind farm would become easier to tease out. The benefit of this method is that if there are larger changes in weather conditions on a given night, those data would be binned accordingly and used to build a trend that could be used when enough data is gathered over a sufficiently large dataset.

If this method is employed, care should be taken that long term changes in ambient conditions are taken into account, and that seasonal changes that would effect the ambient sound levels are not mischaracterised.

3.2 Acoustic descriptors and time intervals

The choice of acoustic descriptors to measure are usually laid out by regulators, giving practitioners limited ability to make decisions about what is most appropriate. However, this section outlines the benefits and drawbacks of the most commonly used configurations. Acoustic descriptors and measurement intervals go hand-in-hand. The main choice is between an Equivalent Energy Level (such as Leq), or a statistical index (such as LA90 or LA50). Leq levels fit in better with Health Based regulatory frameworks and annoyance studies, and as such are often the default noise limit descriptors in jurisdictions. However, due to the logarithmic nature of Leq levels, the measure is biased toward the higher sound levels in a given interval, and has the potential to be effected by transient contamination. Due to this sensitivity, many jurisdictions opt to measure using a statistical descriptor of LA90. The LA90 of an interval will automatically remove the levels occurring 90% of the time, and results in the level that represents the lowest 10 percentile of the interval. This is assumed to be the stead-state sound immission from the wind farm during representative conditions.

In practice both LA90 levels and LAeq levels can be used with relatively high success rate. For both conditions, the important metric that has to go with them is the time interval. LAeq levels can be measured in short intervals (such as 1 minute) which can be used to identify and remove transient contamination from the data. LA90 levels can be used with intervals as long as 10 minutes, and somewhat simplify the data contamination aspect, at the cost of losing time granularity in the data, as there will be only one value every 10 minutes.

3.3 Binning variable

In various binning methods, the acoustic data is binned by wind speed, and sometimes by wind direction. However, there are differences in the way the values can be derived, and this too, has an effect on the results.

3.3.1 Binning vs. filtering

In the context of wind turbine sound immission, it should be acknowledged that the "wind speed" of a given measurement is not one quantity. During one instance, the wind speed at different heights could best describe different noise sources at play. For example:

- The wind speed at hub height would best correlate to the sound being emitted from the turbines
- The wind speed at microphone height could best correlate to the self-noise of the measurement system (as wind speed impinging on the wind screen)
- The wind speed at low elevation (for e.g. 2m) could correlate to vegetation noise from low level shrubbery
- The wind speed at higher elevation (for e.g. 10m) could correlate to the vegetation noise from taller trees
- The wind speed at near hub height or other height could correlate to vegetation noise from forested areas at different elevations in a complex topography

It is important, then, that when presented with the opportunity to conduct measurements, that a reasonable approach is taken to quantify the variables contributing to these potential noises so that appropriate data handling can occur.

When there is wind speed data for multiple heights, a binning approach should pick one wind speed value to bin the data with. All other wind speed values would then be used as a filter on the data to ensure that both ON and OFF conditions are within an acceptable or consistent range.

For example, a measurement campaign may opt to bin data by 'standardized 10m wind speed'. This will have an effect of correlating with the sound emission of the turbines. However, ground level or 10m height measured wind speed could be used to filter out intervals that would be expected to contaminate the measured levels, or to keep the ground level wind speeds within a narrow range such that both ON/OFF conditions would have similar ambient noise conditions.

Similarly, a measurement campaign may opt to bin data by 'measured 10m height wind speed'. This will have an effect of correlating with the ambient sound levels in the region. However, hub height wind speed (either measured, or by power output proxy) could be used to filter out intervals where the noise source would not be operating in the desired range.

The same process would apply for binning by wind direction.

4. Conclusions

Due to the complex nature of measurement of noise immission from wind farms, many different noise measurement techniques have been developed to quantify the noise contribution of wind farms in the far field. This paper outlines the major methodologies being employed in the field, and evaluates their benefits and drawbacks. The goal is to aid any future measurement campaign in deciding on methodology before undertaking such a process which is often long and expensive.

This paper does not cover other choices required to be made during the measurement of wind turbine noise such as the determination of representative conditions, the appropriate choice of operational wind speeds, electrical power range, wind direction conditions, appropriate ground level wind speed ranges, time of day, meteorological conditions such as wind shear , wind weer, turbulence intensity, atmospheric stability, Scada data synchronization and calibration, or methods for removing data contamination due to human and natural activity (insects, car pass-bys, distant airplane pass-bys, variable water noise from streams, roads after rainfall, etc)

To be sure, successful environmental noise measurements have many hurdles to cross, and the experience for wind turbine noise measurements shows that there are many ways to peel an orange. However, care should be taken and appropriate consideration must be given to the myriad of factors that can influence the resulting sound levels measured. When comparing measured levels between different acoustic campaigns, in different regulatory jurisdictions, or using different methodologies, the results may not be comparable if the measurement methods used to arrive at the numbers are structurally different.

References

[1] Egedal, R. et. al "*Danish experiences with measuring wind turbine noise at neighbor dwellings*", 8th International Conference on Wind Turbine Noise, Lisbon 2019

[2] Christensen, N. et. al "*Noise propagation at short range farfield position investigated by* in *simultaneous measurements the nearfield and the farfield*", 8th International Conference on Wind Turbine Noise, Lisbon 2019

[3] Fredianelli, L. et al "*Assessment of WTN by separating residual noise without the farm shutdown: validation of the Italian procedure*." 7th International Conference on Wind Turbine Noise, Rotterdam 2017

[4] De Beer, E. "Using long term monitoring for noise assessment of wind farms", 7th International Conference on Wind Turbine Noise, Rotterdam 2017

[5] Brush, E. "The Challenges and Benefits of Long-Term Sound Monitoring of Wind Farm Sites", 7th International Conference on Wind Turbine Noise, Rotterdam 2017

[6] Ashtiani, P. et al "Analysis of Noise Immission levels measured from Wind Turbines". 4th International Conference on Wind Turbine Noise, Rome 2011

[7] Ashtiani, P. et al "Spectral discrete probability density function of measured wind turbine noise in the far field", 6th International Conference on Wind Turbine Noise, Glasgow 2015

[8] IEC 61400-11 - Wind turbines – Part 11: Acoustic noise measurement techniques Ed 3.1, International Electrotechnical Commission, 2018

INCEEUROPE

9th International Conference
on
Wind Turbine Noise
Remote from Europe – 18th to 21st May 2021

A model to calculate the delta between internal noise with open windows vs external noise

Andrea Bartolazzi, Studio Rinnovabili, a.bartolazzi@studiorinnovabili.it
Cecilia Pompili, Studio Rinnovabili, c.pompili@studiorinnovabili.it

Summary

To evaluate if a receptor is hit by a wind farm it is important that the acoustician measures inside the dwelling. In some countries this is requested by regulation. At least for Italian and French regulations the concept of inside noise with open windows is well known. There is also a secondary concept, inside noise with closed windows. Secondary in the sense that it is easier to handle by increasing the noise absorbing structures.
The difference between external noise and internal noise is present in the literature, but it is subject to large variations.
The possibility to measure inside a building while developing a wind park farm is usually very limited. For this reason, we tend to measure outside the dwelling. Because of that it is interesting to understand what is the difference between the outside noise of the dwelling and inside the dwelling before we have the possibility to measure it. To have an idea of this noise difference between external noise and internal noise we propose to model the building in CadnaA. This calculates what is the noise reduction from outside to inside with given relative positions on the turbines versus the position of the windows.
The model shows the very high importance of the relative position of the window respect to the position of the turbine.
After calculating these numbers, we have performed some tests in order to understand if the model is reliable. Tests show the importance of the position of the sound level meter inside the building and relative to the window. Tests have shown a good reliability of the model.

1. Introduction

The reason why we are interested in calculating the difference between external and internal noise with the open windows is that the regulation utilizes the internal noise with open windows to verify noise limits. A usual way to obtain this number is to make a contemporary measurement of noise inside and outside of the house measuring at the same time the wind and comparing the noise curve over the wind classes. This experiment it is complicated by many typical facts. Primarily, houses in the countryside – a typical location where wind farm are installed - are not always accessible, because they are used only in some seasons or

during some weekdays. Additional reason is a large number of windows that could be hit by the noise. If someone would want to understand the situation for every window it would need to make many different measurements. Another fundamental difficulty in these measurements is the fact that to replicate the situation we would need the presence of the turbine. This situation is clearly impossible in the preliminary stage of the project. Finally, once the wind farm has been built, we would typically turn the turbines on and off in order to understand the difference between external background and total noise vs internal background and total noise in similar wind conditions. As everybody understands this is a very costly experiment.
Because of all these reasons, it is useful for us to have a model of the noise difference between outside and inside the dwelling with open windows. With such a model, we can anticipate the analysis of the impact on the environment in a very preliminary stage of the project. Moreover, as we will see key parameters for understanding the impact are already clear from aerial photos. This will increase the possibilities to tackle the potential problem in the early stage of the project.
Although there is a large bibliography analyzing the noise difference between inside and outside of a building, literature on the open window situation is not very precise. Literature on differential noise seems general purpose. In particular usually noise is coming from many different points and therefore the difference between inside and outside noise is the statistical average. M. Cosa [3] for example proposes delta noise of around 10 dB.
In the case of a wind park the noise sources positions are fixed. The UNI11143-7 norm [2], that refers specifically to wind turbine noise, proposes 6 dB, but this level compared with experience is very high for situations where the turbine is just in front of the window.
Literature frequently considers closed windows situations. With the open windows the situation changes a lot because the influence of the walls and other absorbing structures is much lower.
The interest to analyze the problem with closed windows is minor. In this case the possible counteractions that are possibly usable are at low costs. As a matter of fact, if a plant is disturbing with closed windows we can think at reinforcing the noise absorption on the walls and windows. This is much easier and less costly than solving a noise problem with the open windows.

For this reason, the main problem is to analyze the open windows situation. The first step we have done in order to analyze the situation has been to build a CadnaA model of the wind farm and of the terrain including a model of the disturbed room in which we have represented the room with the internal and external walls. Simulating the noise emitted by the turbine we have seen immediately how important the relative position between the turbine and the direction of the window was. As a matter of fact, if the virtual sound level meter is inside the room there is almost a geometric situation of impact. In case the sound level meter sees the turbine noise it is highly impacted. If the connection line between the center of the turbine and the sound level meter is obstructed by a wall there is a strong noise reduction.
For this reason, we have analyzed the geometrical area where we are allowed to measure in order to comply with the legislation. Referring to Italian rules [4-5] we are obliged to measure at 1 m from every reflecting surface and at 1.5 m from the floor. This reduces the positions for the sound level meter, as you can see in the Figure 5 and Figure 6.

2. Noise model

The first step we have performed has been the preparation of a noise model that represents the wind turbines situation and the receptor situation comprehensive of a description of the receptors room. The noise model comprehends topography of the site and the presence of the turbines.

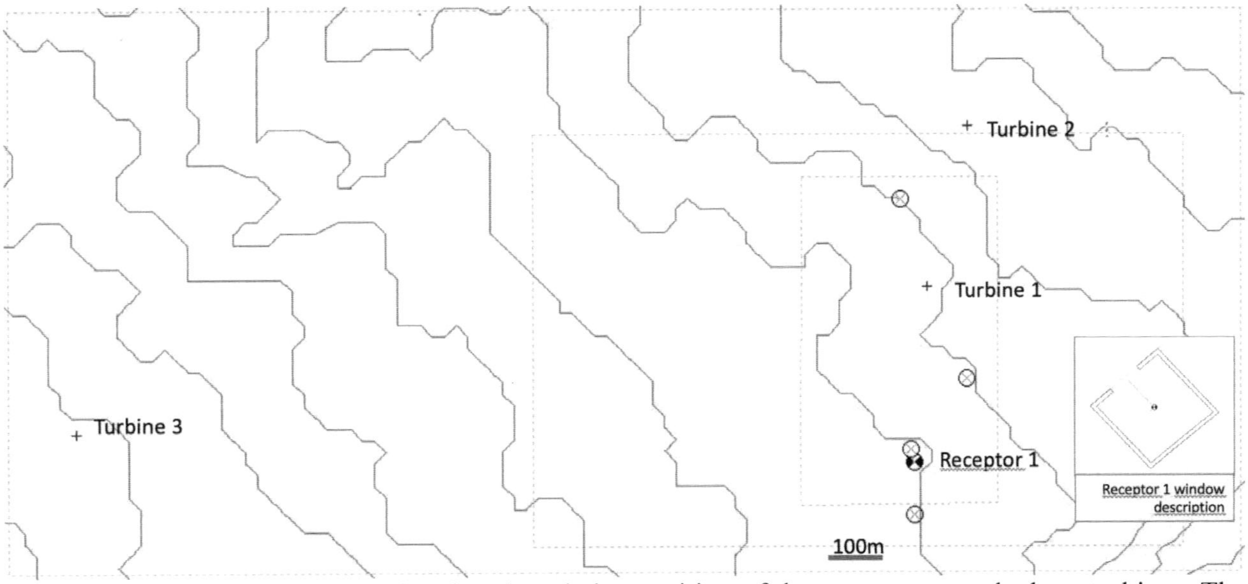

Figure 1 – Map of the situation showing the relative position of the receptor towards three turbines. The land is almost flat degrading from west to east from 260 m asl to 210 m asl.

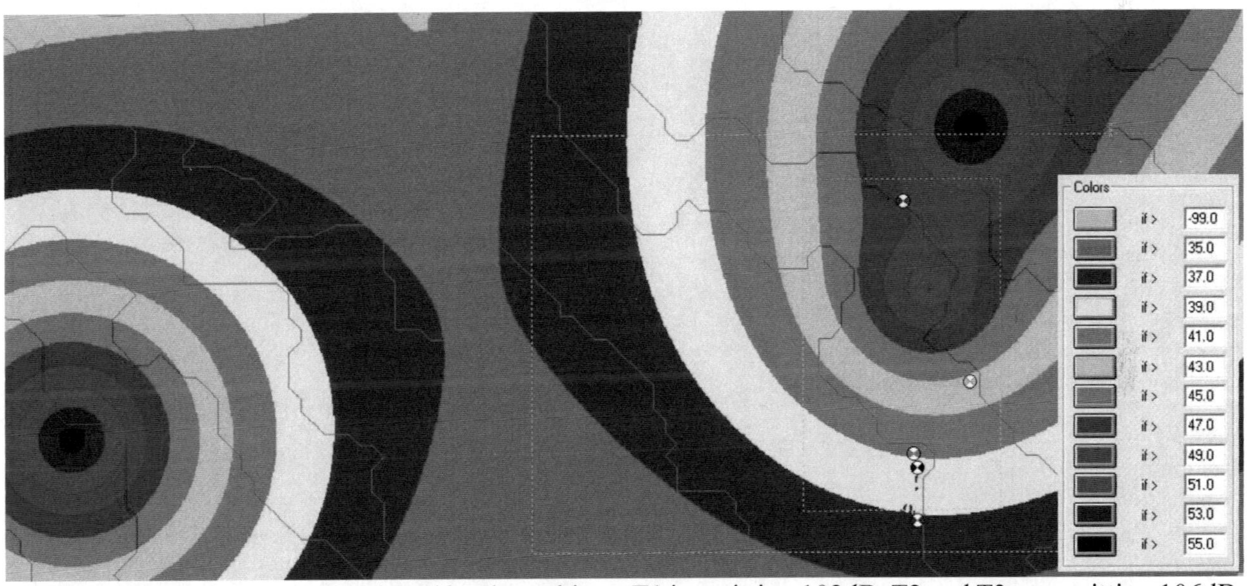

Figure 2 – Map of the noise produced by the turbines. T1 is emitting 102dB, T2 and T3 are emitting 106dB.

The program CadnaA has the possibility to include walls and other type of obstacles inside the description of the dwelling. Since the model has the objective to analyze the noise inside the dwelling with open windows, the noise absorption of the walls has a smaller effect compared to a situation with closed windows. The parameters for the analysis are described below.

Parameter	Value
Norm	ISO 9613
Receiver height	1.5 m
Turbine height	91 m
DTM Model	Grid 50-90 m
Barrier	Not present

Parameter	Value
Terrain absorption	0.8
Reflection max order	3
Temperature	10°C
Humidity	70%
Terrain attenuation	Spectral, all sources
Wind calculation method	Cmet, C0=1,5
Stability class	D

Table 1 – Configuration parameter for calculation

The situation of the furniture inside the dwelling plays an important role. Anyway, it is almost impossible to describe it and it is also usually unknown – considering the difficulty to have access to the building – so it should be taken as an additional error.

In the Figure 3 you can see the results of the differential noise inside the room. What appears clear from this simulation is that the most important parameter that the software considers is the relative position of the turbine towards the window direction. As a matter of fact, even the turbine T3 on the west side which is more than 1,5 km away has a direct influence inside the house which is easily recognizable.

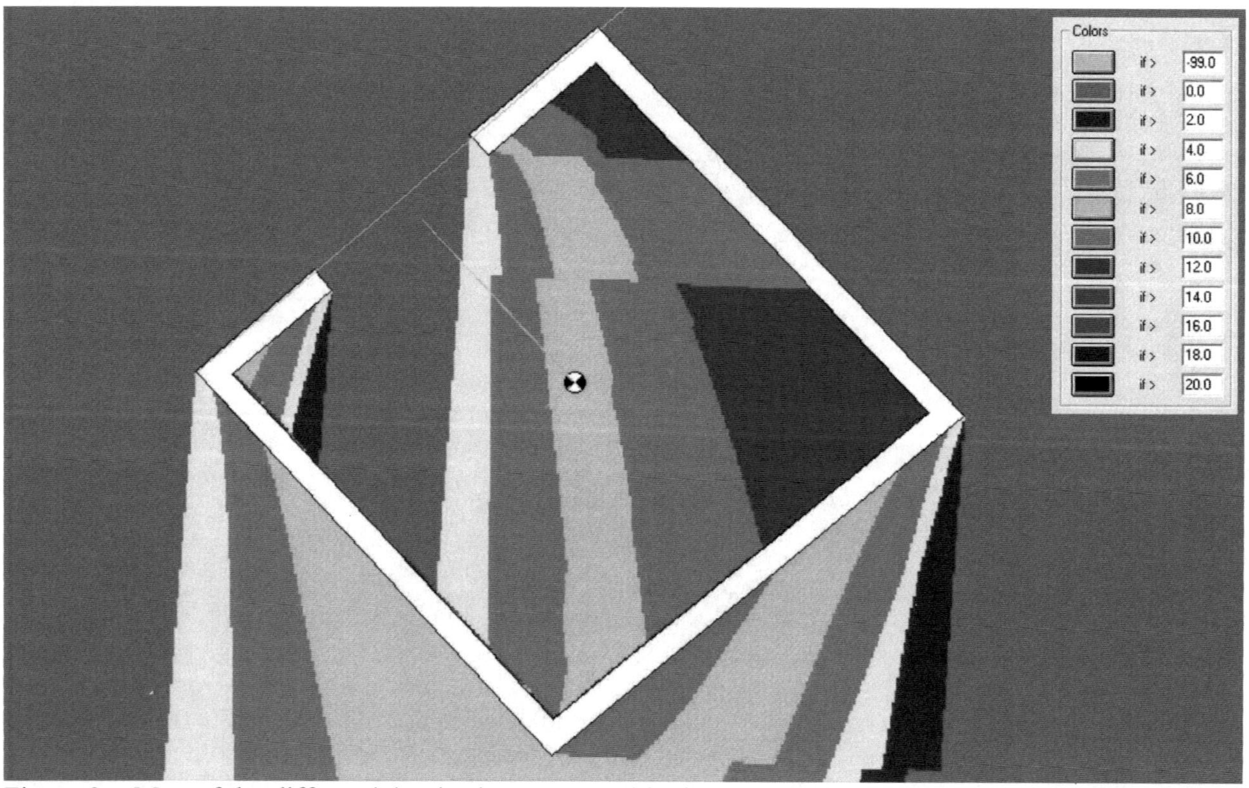

Figure 3 – Map of the differential noise between outside the room in front of the window and inside, and position of the microphone

The model of the room is focused on the necessity to understand the role of the walls, and to distinguish it from the role of the ceiling of the room. The model composed of the walls and open ceiling up all the way long the height of the house. If we limit the height of the house the real height we have a strong three-dimensional effect, and the results are unreadable. The height of the walls are increased to 10 m height, in order to avoid the noise comes from the upper part of the structure complicating the readability of the results (see Figure 4).

This result may be depending on the functioning system of the software and on the various uncertainties that are included in the model. For this reason, it seems logical to perform verifications, a first one of which is in the last paragraph.
Anyway, if the results are confirmed, we can say that influence inside the house with open windows is mostly a geometrical assessment.

Figure 4 – In order to avoid the influence of the ceiling we have modelled a 10 meters high structure. Inside the structure the measuring point at 1,5 meter from the ground

2.1 Geometrical analysis

The possible impact of the wind turbine noise inside the building seems to be very well related to the geometry of the room and by the distances that the sound level meter need to have following regulation. The main parameters of this geometry are the dimensions of the room, the length of the window and the distance between the side wall and the first corner of the wall opening. The technical regulation we use prescribes that the sound level meter is placed at least at 1 m distance from every reflecting surface and at 1,5 m from the floor. Referring to the Figure 5 underneath we can calculate what is the minimum angle from which the turbine is able to send its noise inside the measurable area. Over this angle the noise of the turbine will be highly impacting.

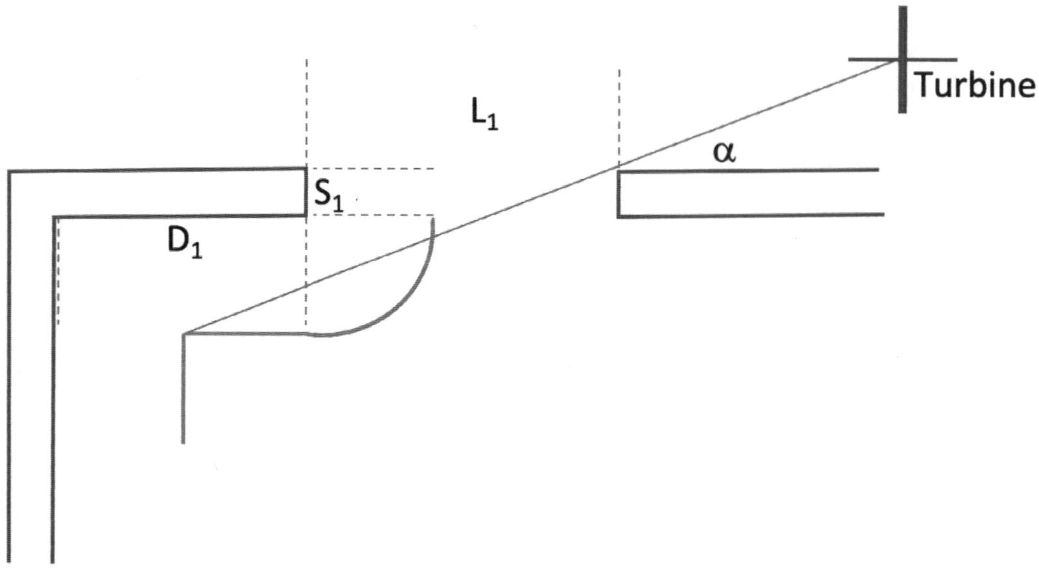

Figure 5 – Angle from which the noise arrives from the turbine on the plane

$$\alpha = \arctan\frac{S_1 + 1}{(L_1 + D_1 + 1)} \qquad [1]$$

For example, as in our CadnaA model, if the length of the window L_1 is 1 meter, the distance from the wall D_1 is 0,6 meters, and the size of the wall S_1 is 0,2 meters, the angle α, formed by the ray, between the turbine and the limit of the measurable position inside the room, and the external plane of the building, will be 63°. This means that for angles above 63° degrees the turbine is visible from the sound level meter, otherwise it is not as in our example in which the angle is 45°. If the turbine is invisible from the sound level meter the noise differential will be significantly lower.
If, for example, we parameterize some of these dimensions, the following table comes out:

	Window width								
Wall/ Window distance	1	1,1	1,2	1,3	1,4	1,5	1,6	1,7	1,8
0,5	67	63	60	56	53	50	47	45	43
0,6	63	60	56	53	50	47	45	43	41
0,7	60	56	53	50	47	45	43	41	39
0,8	56	53	50	47	45	43	41	39	37
0,9	53	50	47	45	43	41	39	37	35
1	50	47	45	43	41	39	37	35	34
1,1	47	45	43	41	39	37	35	34	32
1,2	45	43	41	39	37	35	34	32	31
1,3	43	41	39	37	35	34	32	31	30
1,4	41	39	37	35	34	32	31	30	29
1,5	39	37	35	34	32	31	30	29	28

Table 2 – Minimum angle from which the sound level meter sees the turbine centre in function of window length (L_1) and wall/window distance (D_1) with a thickness of the wall of S_1 = 0,2 m

Another level of difficulty comes if we want to consider the effect of the ceiling of the room, which was not acting in this case. In this case the turbine maybe invisible because on the ceiling.

2.2 Neglected effects

There are some effects that are neglected. The first one as we said is the three-dimensional effect of the walls and the ceilings together. Then there is the reflection of the internal walls. When the noise enters into the room there is a reflection from the windows and from the floor that is also depending on the specific furniture of the room. Another aspect that is neglected is the noise absorption power of the walls.

3. Experiment

The set-up of the experiment we have performed contains a noise measurement point inside the dwelling and another measurement appointment outside of it. There is a wind measurement point at 10 m height on a mast that measures wind speed and direction and it is placed 100 m northern than turbine T1.

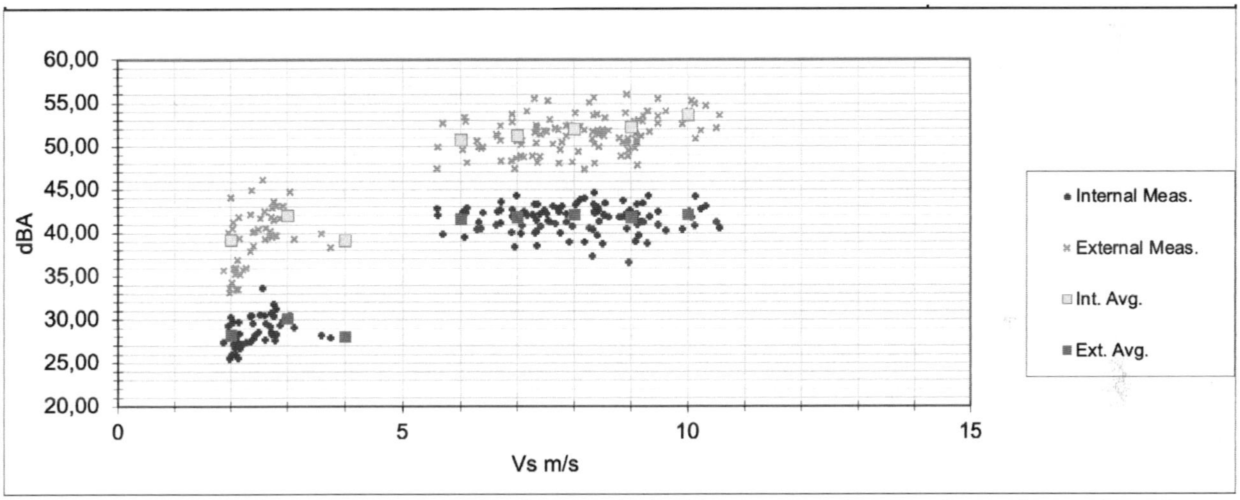

Figure 6 – Results in the measurement of noise vs wind with an external and an internal sound level meter.

Values	Points	External	Internal	$\Delta L_{WA,k}$
Noise 2 m/s	26	39,3	28,2	11,0
Noise 3 m/s	19	42,1	30,2	11,9
Noise 4 m/s	2	39,2	28,1	11,1
Noise 5 m/s				
Noise 6 m/s	11	50,8	41,6	9,1
Noise 7 m/s	26	51,2	41,8	9,4
Noise 8 m/s	29	52,0	42,1	9,9
Noise 9 m/s	32	52,2	41,8	10,4
Noise 10 m/s	9	53,6	42,1	11,5
Average				*10,6*

Table 3 – Results in the measurement of noise vs wind with an external and an internal sound level meter.

The results show a constant differential noise between inside and outside of the dwelling. The levels of noise in the external sound level meter are more dispersed. The level of differential noise is in average 10,6 dB, which can be compared to 8,5 to be given by the model.

4. Conclusions

We analyse differential noise between outside and inside a dwelling at open windows in order to understand if there is a possible method of modelling and calculate it. The results of a simple model indicate that the main parameter that influences the differential noise is the position of the turbine towards the direction of the window.
For this reason we analyse the geometric situation in order to table blind angles from which the turbine is out of the influence of the sound level meter.
Supporting these geometrical reflections we performed a test that partially confirms the calculation structure.

References

[1] Intern. Standard IEC 61400-11 (2012) *Wind turbine systems. Part 11: acoustic noise measurement techniques*

[2] UNI / TS 11143 (2013) *Metodo per la stima dell'impatto e del clima acustico per tipologia di sorgenti*

[3] Cosa M., Nicoli M. (1991) *Valutazione e controllo del rumore e vibrazioni, edizioni ESA*

[4] D.P.C.M. 01.03.1991 (1991) *Limiti massimi di esposizione al rumore negli ambienti abitativi e nell'ambiente esterno*

[5] Bartolazzi A. Mariani M. (2011) *Italian regulations on wind farm noise: A state of the art*

9th International Conference
on
Wind Turbine Noise
Remote from Europe – 18th to 21th May 2021

A Tower Wake Model for Low-Frequency Noise of Downwind Turbine Rotors

Franck Bertagnolio
DTU Wind Energy, Roskilde, DK-4000, Denmark

Helge Aa. Madsen
DTU Wind Energy, Roskilde, DK-4000, Denmark

Summary

In this contribution, a simplified model for the turbulent wake of a wind turbine tower interacting with the rotor blades, in the case of a downwind rotor concept, is proposed. The goal is to model the effect of the vortex shedding and associated turbulent flow vortices on the noise emissions.

The proposed model consists of 2 parts. The first one is a CFD simulation of the flow field around a circular cylinder accounting for the vortex shedding and turbulence structures. This model is limited in term of resolution for obvious computational requirements. In addition, a temporal and spatial sampling is used to store the flow data to manageable file sizes.

The second part of the model consists of using the above data to recreate the flow features to a realistic level so that it can be used for the prediction of Low-Frequency Noise emissions from a wind turbine rotor located downstream of the tower.

1. Introduction

For various technical and economical reasons, there exists currently a renewed interest for industrial scale downwind turbines. The latter have been discarded in the late 80s, partly because it was recognized that this design concept may emit higher noise levels, in particular in the low-frequency range [6]. These Low-Frequency Noise (LFN) and Infra-Sound (IS) emissions are mainly the result of the blade-tower wake interaction.

A few models have been proposed in the literature to predict the above phenomenon [9, 12, 17–19]. The authors developed earlier a model that is capable of predicting LFN emissions from, in addition to atmospheric wind turbulence and turbulence from the wake of a potential upstream turbine, the blade-tower interaction. Its methodology is based on the simulations of the wind turbine flow and acoustic response in the time domain [2]. In this model, the blades are the noise sources and the effect of the tower can be accounted for as a wind velocity deficit in the tower wake for a downwind turbine rotor concept, or a flow disturbance in front of the tower for an upwind turbine. In both cases, these flow patterns are modelled as a time-average of the actual wind flow field. However, it is well known that the Reynolds number of the flow around a MW-size wind turbine tower (even at a low wind speed) is large enough so that the tower wake is turbulent, including vortex shedding and multiple turbulent vortex structures of all sizes. These structures do also interact with the passing blades and contribute to the noise generation.

In the present study, a new methodology is proposed to simulate the turbulent vortex shedding in the wake of a wind turbine tower and calculate the resulting noise emission from the blades interacting with this wake. The overall computational framework is described in Section 2, and the existing tower wake model and its contribution to noise in Section3. The main contribution of the present work is presented in Section 4 and the results in Section 5. Conclusions are drawn in Section 6.

2. LFN rotor noise model including blade-tower wake interaction noise

The basic modeling approach for predicting LFN and IS from a wind turbine rotor was originally proposed by the authors in an earlier paper [2]. It consists of two main components:

- The first component is the HAWC2 aero-elastic code for the simulation of wind turbines. It provides the necessary aerodynamic information to calculate aerodynamic noise from a turbine. For the present purpose, these data boil down to the time-series of the lift forces along the blades' span.
- The above numerical output of the HAWC2 calculations are used as input for the noise model module implemented in HAWC2. In the present case, time-series of the acoustic pressure in the far-field are calculated to evaluate LFN and IS.

These two parts of the wind turbine noise model are discussed in more details separately below.

2.1. HAWC2 aeroelastic code

The HAWC2 code [10] is a multibody aeroelastic code designed to simulate wind turbine structural loads and power extraction in the time domain. The Blade Element Momentum theory by Glauert [5] is applied in order to calculate the aerodynamic loading resulting from the incoming wind and the rotation of the blades. Various wind flow patterns can be accounted for in this model. The most interesting features in the present context are the possibility to model the atmospheric turbulence, as well as the influence of the tower on the mean wind flowfield, which are main sources of aerodynamic noise in the low-frequency range. Note here that recent studies suggest that the tower itself might be a strong LFN source in connection with the passage of the blades [9, 20].

The HAWC2 code has been extended in order to simulate the aeroacoustic of wind turbines by coupling additional models for turbulent inflow noise, trailing edge noise and stall noise [1] in the spectral domain. But in the present context, the acoustic predictions are conducted in the time domain [2] by calculating acoustic pressure resulting from the unsteady loads on the blades as explained below.

2.2. LFN noise prediction

Two main noise generation mechanisms are considered here. The first one is the atmospheric turbulence impacting the blade, generating a dynamic loading which radiates as noise. The second one is the effect of the tower flow deflection (upstream of the tower for an upwind turbine concept) or wake flow deficit (in the wake of the tower for a downwind concept) on the blades. When the blades pass in front or behind the tower, the flow disturbances similarly generate dynamic loading on the blades radiating as noise.

From the computed time-series of the aerodynamic loads exerted on the blades, the method developed by Farassat [3] is used to calculate the acoustic response. In brief, this method is an advanced expression of Ffowcs Williams-Hawking solution [4] of the original Lighthill's acoustic analogy [11]. In our context, since the aerodynamic loading is the main noise generation mechanism, the Farassat formulation is nearly identical to the Curle's analogy. Although, all terms of this formulation are included in the present model, it is found (as expected) that the main contribution originates from the unsteady aerodynamic loading, which can be summarized with this equation for the acoustic noise in the far-field from a varying loading:

$$p_a(t) = \frac{1}{4\pi C_0} \int_S \left[\frac{\dot{\vec{l}} \cdot \vec{r}}{r\,(1 - M_r)^2} \right]_{\text{ret}} \mathrm{d}S$$

where C_0 is the speed of sound, $\vec{l}$ is the lift force vector applied on the elementary surface $\mathrm{d}S$ and the upper dot denotes a time derivative, $\vec{r}$ is the vector between the elementary surface where the lift is applied and the listener location in the far-field where the acoustic pressure is calculated pointing toward the receiver, r is its norm, and M_r is equal to $\vec{M} \cdot \vec{r}$ where $\vec{M} = \vec{v}/C_0$ with $\vec{v}$ being the velocity of the elementary surface $\mathrm{d}S$ relative to the listener. The integration is conducted over the whole blade surface S, for all blades.

3. Steady-state approach for the influence of the tower

In the HAWC2 aeroelastic calculation framework, the influence of the tower on the blades can be considered with simple, but realisitic, steady-state flow models. Both the situations of an upwind and downdwind rotor concept can be considered as illustrated in Fig. 1.

In the case of an upwind turbine, the atmospheric inflow onto the turbine rotor (including possible shear and veer) is modified by enforcing an inviscid potential flow around the tower cylinder. When the blades pass in front of the tower (with respect to the wind direction), then the influence of the tower is felt as a deceleration of the wind speed together with a deflection of the wind speed toward the sides of the tower.

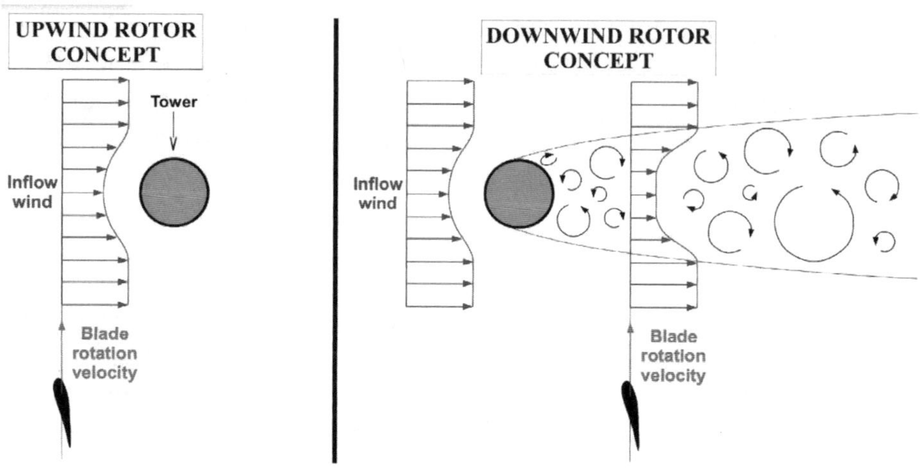

Fig. 1 Slice of a wind turbine tower and a passing blade for an upwind or downwind rotor design concept (Left: upwind passing blade experiencing a wind flow slow-down caused by the tower; Right: downwind passing blade experiencing a wake flow deficit and associated shedding vortices caused by the tower).

In the case of a downwind turbine, the flow deficit in the wake of the tower is calculated with a model based on the boundary layer of a jet flow. In this model, the flow deficit becomes wider and less intense away from the turbine which is a reasonable description of reality, if considering the time-averaged flow field. The intensity of this flow deficit can be scaled with a single parameter, which is the drag coefficient exerted on the tower cylinder. A value of $C_D = 0.3$ that is typical for a high Reynolds number flow around a circular cylinder is usually preferred. It should be noted here that the actual turbulent structures that are expected to occur in the wake of the tower (see the turbulent vortices illustrated as black circles with an arrow inside the tower wake on the left side of Fig. 1) are neglected in this steady-state flow model.

The upwind and downwind flow models above are based on the definition of an average flowfield, i.e. the tower effect of the wind flowfield onto the blade is considered constant in time. However, the blades move in and out of these pertubated flow regions, thereby creating dynamic loading on the blades, and subsequently LFN and IS.

4. Development of a new tower wake model in HAWC2

The tower wake flow models considered in the above section are stationary in time. However, for high Reynolds numbers typical of flows around large turbine towers, the wake flow is characterized by turbulent vortex shedding with a multitude of vortical scales, the largest being often coherent along larger distances over the tower span (as illustrated in Fig. 2). The blades traversing the tower wake as they rotate will interact with these vortices and these interactions may be quite different from the simpler interaction with a stationary wake flow deficit using a jet model as described in the previous section.

The objective of the present work is to develop a more realistic model for the tower wake flowfield and its interaction with the blade, including the above unsteady turbulent flow features.

The main idea is to use a 3D CFD solver (here, the in-house code EllipSys3D [14, 15]) in a configuration for which the main vortical structures of the turbulent tower wake flow can be captured. This requires to conduct an unsteady calculation with an appropriate turbulence modelling strategy. Indeed, Direct Numerical Simulation for such a configuration is out of reach even with the most powerful computers existing to date. The chosen approach is Detached Eddy Simulation [16] which combines an unsteady RANS model in the vicinity of the (tower) wall to calculate the boundary layer flow, and a Large Eddy Simulation model away from the surface that simulates the largest vortical structures to smaller ones down to the local computational grid cell size. The LES approach accounts for the turbulent sub-grid scales (i.e. the vortices that are smaller that the local cell size) with a turbulent viscosity model. A snapshot of the computed flowfield in Fig. 2 illustrates the vortex shedding and the resolved vortex structures in the tower wake.

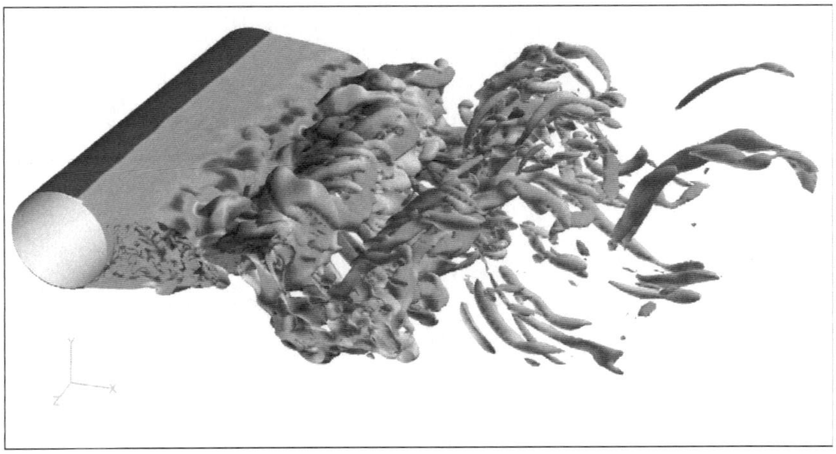

Fig. 2 Iso-vorticity in the flow behind a circular cylinder at $Re = 3$M.

The first step of the model consists in recording a time-series of the flow field calculated with the CFD solver described above. However, the present 3D CFD calculations involve very large spatial grids (in the order of hundreds of millions of computational cells) and very small time-steps, due to the fine discretization requirements in order to capture both the boundary layer and the turbulent vortices. Hence, it is decided to:

1) record the flow inside a limited 3D rectangular box located in the wake region of the tower representative of the flow of interest,
2) define a 3D cartesian mesh inside this box with a much coarser spatial resolution than the original CFD calculation itself in order to reduce the size of the recorded file, and the CFD flow field is interpolated at the nodes of this coarse mesh,
3) conduct the above recording interpolation procedure at regular times during the CFD simulations, but at a much lower sampling rate than the one corresponding to the small time-steps required for the CFD calculation.

A 2D slice of a snapshot of this recording at a given time is displayed in Fig. 3. In this snapshot and as for the calculation results presented in the next section, the coarse cartesian interpolation 3D box mesh contains 120 (in the flow direction-X) $\times$ 48 (transverse-Y) $\times$ 34 (spanwise-Z as in the snapshot perspective of Fig. 2) square cells. In addition, 400 time-steps are stored corresponding to the release of a dozen of large scale vortices of the von Kármán alley street.

The time-series of dynamic wake flow deficit recorded on this coarse mesh can be enforced during a HAWC2 simulation to model the effect of the tower wake, much like in the same way as it is done for the Dynamic Wake Meandering model in HAWC2 [13], although its effect is here limited to a specific region of space defined by the above 3D rectangular box located in the wake of the tower.

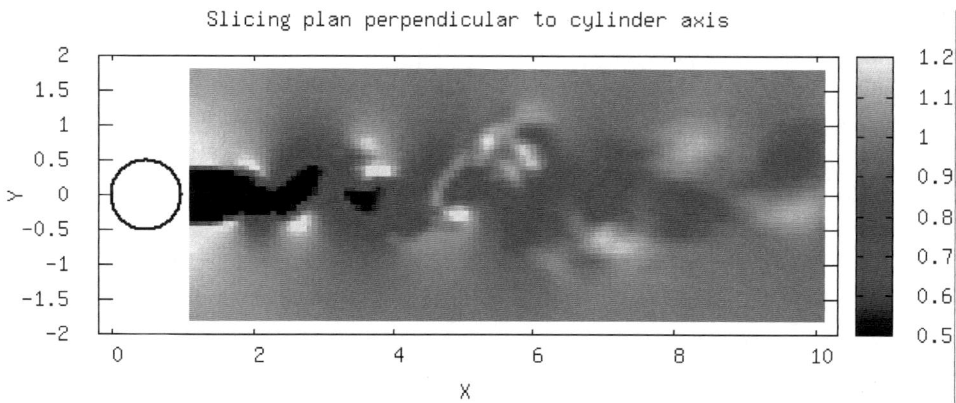

Fig. 3 Snapshot of the time-series of the tower wake flow recorded on a coarse grid in a rectangular box located in the tower wake. The present snapshot displays a slice perpendicular to the tower axis in the 3D box interpolated from the original flow as displayed in Fig. 2 (The colors indicate the intensity of the wake flow deficit along the main flow direction with black indicating the largest deficit to yellow indicating an accelerated flow compared to the inflow velocity far upstream of the cylinder, which corresponds to a value of 1 in the color scale).

As a second step in the model, some additional aspects of the original turbulent flowfield are considered. In the above interpolation and recording procedure, some small turbulent scales (smaller than the cell size of the coarse cartesian mesh defined above) present in the original CFD calculation have been filtered out, in addition to the even smaller turbulent scales that had already been filtered out by the LES model in the DES calculation itself. However, as mentioned in the beginning of this section, small turbulence scales have an impact on LFN emissions as explained in Section 2.2. Hence, it is proposed to re-inject some small scale turbulence in the recorded filtered flow field. This is done using a Mann-box-type approach as normally used in HAWC2, i.e. a single 3D turbulent periodic flow field is computed and stored. Then, it is convected in the HAWC2 calculation domain at a given convection velocity in a frozen state. For the purpose of the present study, this small-scale turbulence is assumed isotropic. Furthermore, the size of the turbulence box is twice the size of the coarse mesh cells in order to only add turbulent scales smaller than these cells. The turbulence integral length scale is chosen large enough so that the generated turbulence is characteristic of turbulence scales in the well-known turbulent energy cascade. More importantly, the added turbulent kinetic energy from the turbulence box is calculated so that it matches the turbulent kinetic energy of the sub-grid scales as in a LES model, using a cell size as defined by the coarse mesh from the interpolation defined above.

It is important to note that the above approach is very crude as turbulence is known to be a very highly non-linear phenomenon. It is assumed here that the contributions of the different turbulent scales can be added up linearly. Nevertheless, it is also expected, and therefore assumed in the present model, that these aspects have a small influence on the noise generation mechanisms involved.

5. Noise results for the different blade-tower interaction models

The results displayed in this section are based on time-series of the acoustic pressure calculated at a given observer/listener position away from the turbine using the computational framework described in the previous sections. These time-series are processed by Fast-Fourier Transform to obtain acoustic power spectra using the classical Welch method. Furthermore, these spectra can be integrated and two types of filtering are used to obtain the noise metrics as defined in the Danish regulations [7] for LFN and IS:

- The first one, denoted as L_{pG}, is an integration of the overall noise spectra applying a G-weighting frequency filter. This weighting progressively, but rapidly, filters out frequencies below 10 Hz and above 20 Hz. It is therefore more representative of IS immission levels.
- The second one, denoted as $L_{pA,LF}$, is based on a spectral integration using the more standard A-weighting filter, but the integration is limited between 10 to 160 Hz. Note that the A-weighting is progressively filtering out frequency below 1 kHz, although much less rapidly than the G-weighting. Therefore, the contribution of the frequencies close to 160 Hz will be larger than those around 10 Hz (provided that physical noise levels are equal in both frequency ranges). This filtering approach is therefore more representative of LFN (i.e. in the frequency range from 20 Hz to 160-200 Hz, while IS is commonly referring to frequencies below 20 Hz).

The NREL 5MW reference turbine [8] is used as a test-case for the present model. The noise emissions are calculated for several wind speeds and the optimal operational conditions for the turbine (rotational speed and blade pitch) are modified accordingly. A wind shear is enforced using a wind velocity as function of height corresponding to a power law with a power coefficient equal to $1/7$. The atmospheric turbulence is included in all the calculations using a Mann box, as it is the standard for HAWC2 simulations (see earlier considerations). However, it is important to note that the same turbulence box is used in the calculations at all wind speeds. Hence, the actual turbulence intensity of the generated turbulence is decreasing with wind speed (roughly from 14% at 6 m/s to 4% at 18 m/s). It may be argued that turbulence intensity should remain more or less constant in reality. However, the purpose here is to avoid to many parameters varying at the same time for the sake of comparison.

The $L_{pA,LF}$ and L_{pG} noise levels as a function of wind speed are displayed for a listener located on the ground at 500 m and 1000 m from the turbine in Fig. 4. It is observed that from the upwind, downwind without a tower, and downwind with a jet model configurations, the tower has virtually no impact on the $L_{pA,LF}$. A larger impact is observed on the L_{pG} at higher wind speeds which increases with the drag coefficient of the jet model (from $C_D = 0.3$ to $C_D = 0.6$). The latter observations also permit to conclude that in all the above cases, it is the atmospheric turbulence that is mainly driving the LFN and IS emissions.

If looking at the downwind configurations and including the newer tower wake model, the trends are different. As for the $L_{pA,LF}$ emissions, the new model predicts slighly larger noise levels compared to the jet model (even for $C_D = 0.6$). However, if considering the L_{pG} emissions, large noise level increases are observed at higher wind speeds (up to 10 dB at 16 m/s). By removing the subgrid scale turbulence, it can be concluded that the increases observed above are caused by the larger scales of the vortex shedding in the tower wake flow. Note however that the subgrid scale turbulence model is very preliminary and need further investigations in order to be validated.

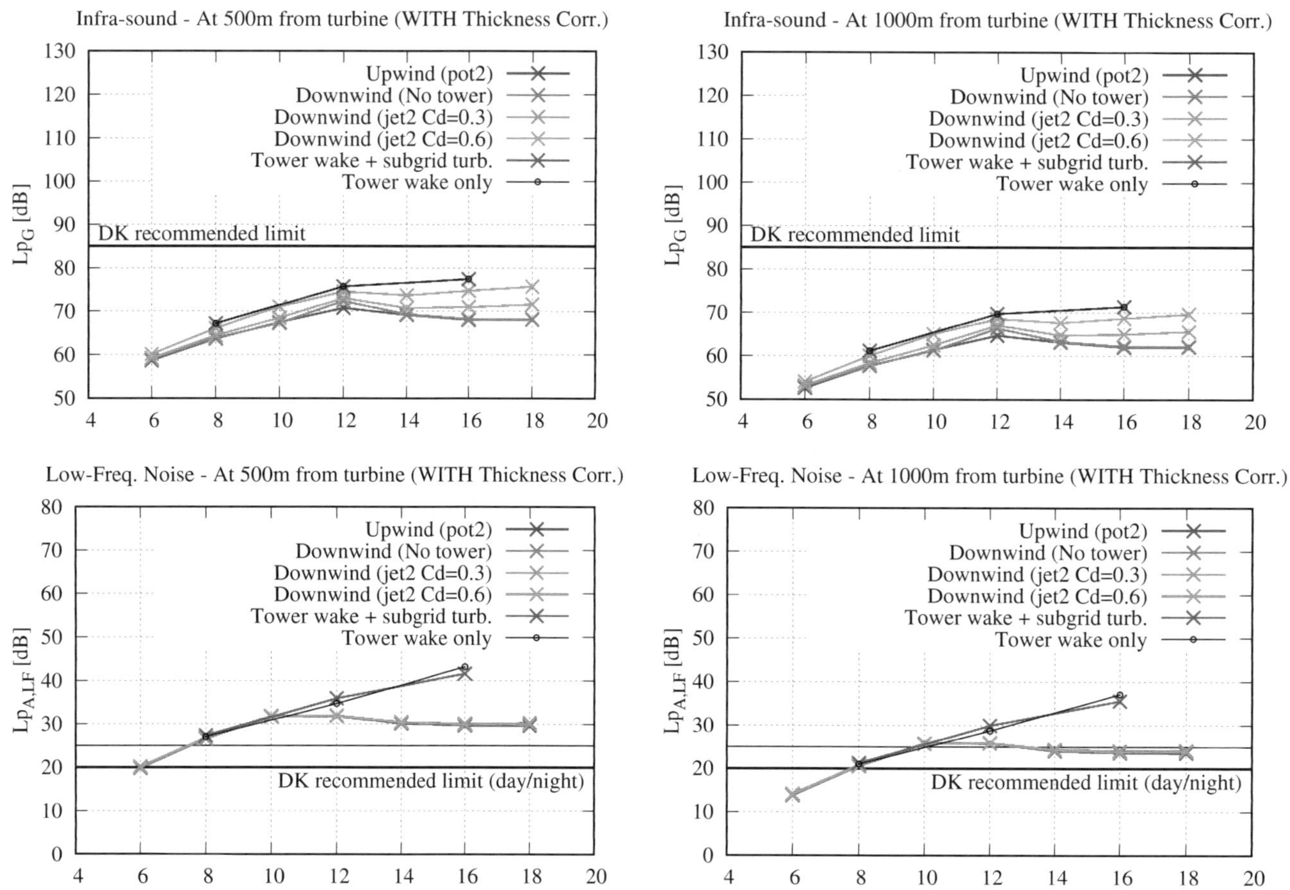

Fig. 4 $L_{p_{A,LF}}$ **(Bottom) and** L_{p_G} **(Top) as a function of wind speed for a listener located at 500 m (Left) and 1000 m (Right) from the turbine. The black horizontal lines correspond to the Danish recommended limits.**

6. Conclusions and perspectives

A new model is proposed for predicting LFN and IS from the interaction of the wind turbine wake and blades passing into it. Some preliminary qualitative results are presented in this paper.

The main findings from the present study are that downwind turbines do produce more IS (as characterized by L_{p_G} levels) as it was expected from earlier studies. But, the validity of the tower wake model is critical to predict these differences correcly, as some large differences in the present results indicate. If the present new wind tower wake model and the associated noise generation model are correct, or at least provide the correct trends, significant increases in LFN (as characterized by $L_{p_{A,LF}}$ levels) can be expected at high wind speeds.

It is important to remind that the results presented in this paper are qualitative and preliminary. The present different models, both tower wake and noise generation, have not been thoroughly tested and validated. For validating such a model, there is a the lack of publicly available data because of the sensitive issue of LFN and IS for the industry.

Therefore, the next step is to get access to reliable wind turbine LFN noise data. Indeed, in the first place LFN and IS from wind turbines are not easy to measure since the noise levels are often comparable to the background ambient noise. In the second place, there exist very few operating downwind MW-size turbines in the world, restricting the possibly available data. Finally, manufacturers and operators are usually reluctant to divulge the necessary operational parameters (e.g. wind speed, wind turbine geometry, rotational speed, etc) that are necessary as inputs for the models in order to validate them.

Nevertheless, as suggested in the introduction, if downwind turbines have to be considered in the future as viable technological alternatives in the energy system and landscape, the issues of LFN and IS will most probably have to be investigated further.

Acknowledgments

This work was partly supported by the EUDP project LowWind (Jr.nr. 64019-0074) funded by the Danish Energy Agency (Energistyrelsen).

References

[1] Bertagnolio, F., Madsen, H. A., and Fischer, A. (2017a). A Combined Aeroelastic-Aeroacoustic Model for Wind Turbine Noise: Verification and Analysis of Field Measurements. *Wind Energy*, 20(8):1331–1348.

[2] Bertagnolio, F., Madsen, H. A., and Fischer, A. (2017b). A Temporal Wind Turbine Model for Low-Frequency Noise. In *InterNoise17*, INTER-NOISE and NOISE-CON Congress and Conference Proceedings, pages 5025–5033, Hong Kong, China. INCE (publisher).

[3] Farassat, F. and Succi, G. P. (1980). A Review of Propeller Discrete Frequency Noise Prediction Technology with Emphasis on Two Current Methods for Time Domain Calculations. *J. Sound Vib.*, 71(3):399–419.

[4] Ffowcs Williams, J. E. and Hawkings, D. L. (1969). Sound Generated by Turbulence and Surfaces in Arbitrary Motion. *Philosophical Transactions of the Royal Society*, A264(1151):321–342.

[5] Glauert, H. (1935). *Airplane Propellers*, volume In: Aerodynamic Theory Volume IV. Springer, Berlin, Heidelberg.

[6] Hubbard, H. H. and Shepherd, K. P. (1990). Wind Turbine Acoustics. Technical Paper 3057 DOE/NASA/20320-77, NASA. Retrieved from http://citeseerx.ist.psu.edu/viewdoc/download?doi=10.1.1.460.5771&rep=rep1&type=pdf.

[7] Jakobsen, J. (2001). Danish Guidelines on Environmental Low Frequency Noise, Infrasound and Vibration. *Journal of Low Frequency Noise, Vibration and Active Control*, 20(3):141–148.

[8] Jonkman, J., Butterfield, S., Musial, W., and Scott, G. (2009). Definition of a 5-MW Reference Wind Turbine for Offshore System Development. Tech. Rep. NREL/TP-500-38060, National Renewable Energy Laboratory, Golden (CO). (Contract No. DE-AC36-08-GO28308).

[9] Klein, L., Gude, J., Wenz, F., Lutz, T., and Krämer, E. (2018). Advanced CFD-MBS coupling to Assess Low-Frequency Emissions from Wind Turbines. *Wind Energy Science*, (3):713–728.

[10] Larsen, T. J. and Hansen, A. M. (2007). How 2 HAWC2, The User's Manual. Tech. Rep. RISØ-R-1597(ver.3-1), Risø-DTU, Roskilde, Denmark. Available at `http://www.hawc2.dkdownload/hawc2-manual` (Accessed: 2021-03-11).

[11] Lighthill, M. J. (1952). On Sound Generated Aerodynamically. I. General Theory. *Proceedings of the Royal Society London A*, 211(1107):564–587.

[12] Madsen, H. A., Johansen, J., Sørensen, N. N., Larsen, G. C., and Hansen, M. H. (2007). Simulation of Low Frequency Noise from a Downwind Wind Turbine Rotor. In *45th AIAA Aerospace Sciences Meeting and Exhibit*, AIAA 2007-623, Reno (NV).

[13] Madsen, H. A., Larsen, G. C., Larsen, T. J., Troldborg, N., and Mikkelsen, R. (2010). Calibration and validation of the dynamic wake meandering model implemented in the aeroelastic code HAWC2. *Journal of Solar Energy Engineering*, 132(4):(14 pages).

[14] Michelsen, J. A. (1994). Block Structured Multigrid Solution of 2D and 3D Elliptic PDE's. Tech. Rep. AFM 94-06, Technical University of Denmark.

[15] Sørensen, N. N. (1995). General Purpose Flow Solver Applied to Flow over Hills. Risø-R-827-(EN), Risø National Laboratory, Roskilde, Denmark.

[16] Travin, A., Shur, M., Strelets, M., and Spalart, P. R. (2000). Detached-eddy simulations past a circular cylinder. *Flow Turbul. Combust.*, 63:293–313.

[17] Viterna, L. A. (1981a). Method for Predicting Impulsive Noise Generated by Wind Turbine Rotors. In *Second DOE/NASA "Wind Turbine Dynanmics" Workshop*, Conference Proceedings, Cleveland (OH).

[18] Viterna, L. A. (1981b). The NASA-LeRC Wind Turbine Sound Prediction Code. In *Wind Turbine Dyn.*, Conference Paper, pages 411–418. (Available online: https://ntrs.nasa.gov/search.jsp?R=19820015854).

[19] Viterna, L. A. (1982). Method for Predicting Impulsive Noise Generated by Wind Turbine Rotors. In *The 1982 International Conference on Noise Control Engineering*, Conference Proceedings, San Francisco (CA). (DOE/NASA/20320-36, NASA TM-82794).

[20] Yauwenas, Y., Zajamšek, B., Reizes, J., Timchenko, V., and Doolan, C. J. (2017). Numerical Simulation of Blade-Passage Noise. *Journal of the Acoustical Society of America*, 142(3):1575–1586.

9th International Conference
on
Wind Turbine Noise
Remote from Europe – 18th to 21st May 2021

Wind sector management using Beilis Tappert Parabolic Equation

Karl Bolin, Kungliga Tekniska högskolan, Stockholm, Sweden, kbolin@kth.se

Summary

This paper investigates the possibility to optimise the operation of wind turbine by allowing increased power output when favourable sound propagation occurs and decrease the power when unfavourable sound propagation can be expected, so called wind sector management. The study is performed by using sound propagation calculations using a terrain dependent sound propagation algorithm, the Beilis Tappert Parabolic Equation and meteorological input from weather forecast data. The site is a wind farm in Sweden that already implement Wind sector management with a ray tracing method (Nord2000) and the Parabolic Equation results are compared to the ray tracing to investigate similarities and differences.

1. Introduction

Wind power is rapidly growing around the world and can today be viewed as one of our most important available sources of renewable energy. As pitch regulation is normal and wind farm surveillance around the clock is common for larger today possibilities to actively adapt wind turbine operation is utilized at wind sites nowadays.

Wind sector management is when a wind farm is changing operational modes of the wind turbines depending on prevailing meteorological conditions to maximize the electric power output when favourable sound propagation conditions occur and restrict the electric power and thus the sound power when downward refracting sound propagation occur. Examples of downward and upward refraction can be seen in Figure 1 (a) and 1 (b) left figures show the sound speed profile (c(z)) and the right figures show the Transmission Loss relative to 1 m distance in dB.

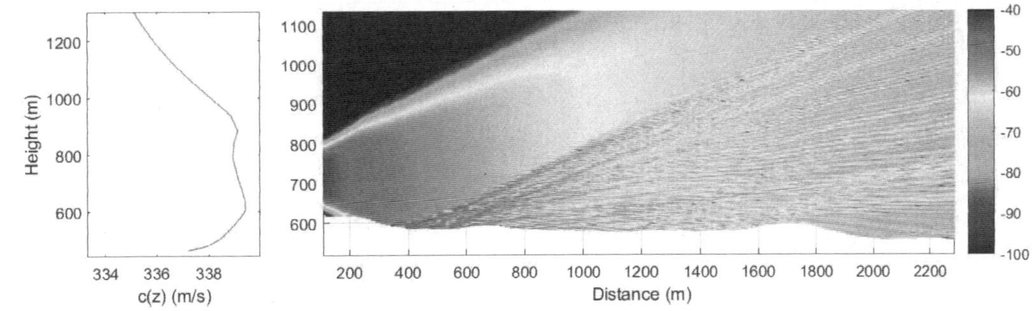

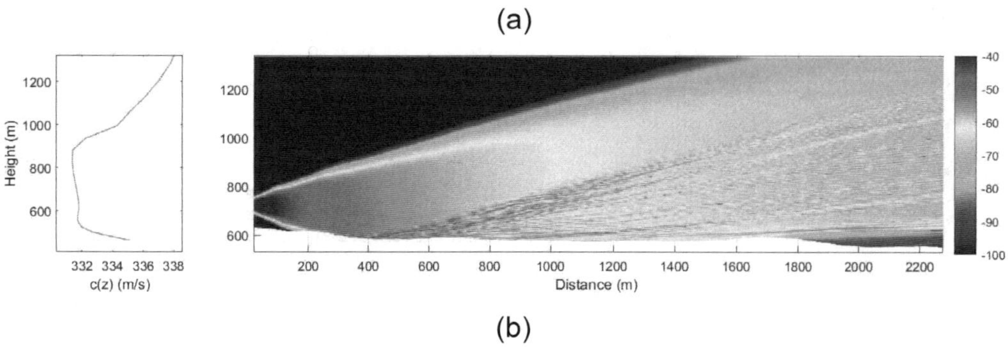

Figure 1: Cross section of acoustic transmission from a source showing (a) downward and (b) upward sound refraction.

2. Site description

The site contains three different wind farms and is situated in Southern Sweden. In total 34 wind turbines of different types with varying 2-3.3 MW electric power are installed in the area and the hub-heights vary between 92.5-105 m. A map of the site is seen in figure 2. There are dwellings circumvented by wind turbines, by the road and South of the largest lake, and thus simultaneous downwind sound propagation from all turbines will not be probable at this site. Therefore, the 12 Southwestern windfarms have been applying wind sector management of the turbines to decrease the sound levels for the roadside dwellings but the other (22) turbines are not moded. This study uses two reception points, one at the Southern border of the wind farm and one point Northeast of the farm as these have been categorized as interesting reception points in earlier sound exposure calculations (Fredriksson, 2017).

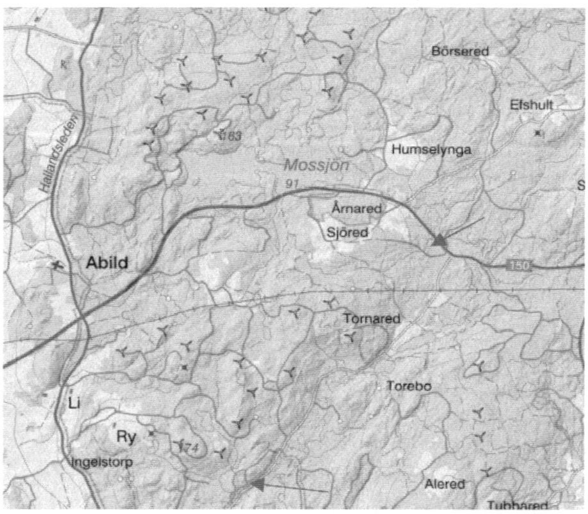

Figure 2: Map of wind turbine site (from https://minkarta.lantmateriet.se/). The turbines are shown as propellers and the reception positions are shown as red arrows (Southern point, reception point 1 and North-eastern reception point 2)

3. Sound propagation calculations

3.1 Beilis Tappert Parabolic Equation

The Beilis Tappert Parabolic Equation method has been developed for underwater sound propagation but adjusted to atmospheric conditions by Parakkal et al (2012). The algorithm has three main advantages compared to many other sound propagation algorithms, firstly it is numerically fast compared to other parabolic equation models, secondly it can incorporate arbitrary sound speed profiles as input and thirdly it considers ambulating terrain conditions- both changing ground height and changing acoustic properties of the ground. Therefore, the Beilis Tappert Parabolic Equation has been implemented in MATLAB (version 2019a) and validated to benchmark cases in Parakkal et al (2012).

For the current implementation 1/3 octave frequencies between 25-1000 Hz were calculated between the 5:th and 30:th November of 2018. To decrease the number of calculations the results include wind turbines closer than 3 km for 1/3 octave frequencies below 315 Hz and 2.5 km above. The numerical parameters are chosen to vertical discretization of 0.1λ (where λ is the wavelength) and horizontal discretization of 10 λ for frequencies below 315 Hz and 15 λ for higher frequencies which is in accordance with recommendations of the Beilis Tappert Parabolic Equation method.

3.2 Environmental parameters

3.2.1 Sound speed calculation

The effective sound speed calculations are commonly approximated as logarithmic sound speed profiles for sound propagation calculations but as the meteorologic conditions are constantly varying and usually far too complex, especially at higher heights than the atmospheric boundary layer this paper has used the Scandinavian weather forecast data from the software Arome-HIRMLAM which is publicly available from the Norwegian Meteorological Institute. Hourly forecast data are available for a 2.5x2.5 km grid covering Scandinavia and the sound speed profiles are based in wind velocity, temperature, and humidity.

3.2.2 Terrain alteration

Information of the terrain ambulation has been retrieved from the Swedish Metria database and a resolution of 50x50 m are used for the calculations. As can be seen in figure 2 is that the terrain is ambulating with wind turbines on the hills and dwellings situated in the valleys.

3.2.3 Ground conditions

The Corine database pan-European classification of ground type data divide the terrain into approximately 80 types of ground types ranging from water surfaces, forest floors to glaciers etc. This database is available for a 25x25 m grid resolution in Sweden and these data have been obtained in the propagation paths to decide flow resistivity according to Sohlman et al (2004) and then using the corresponding flow resistivity as suggested by Embleton et al (1983) and implemented into the Delany and Bazley (1970) ground impedance model.

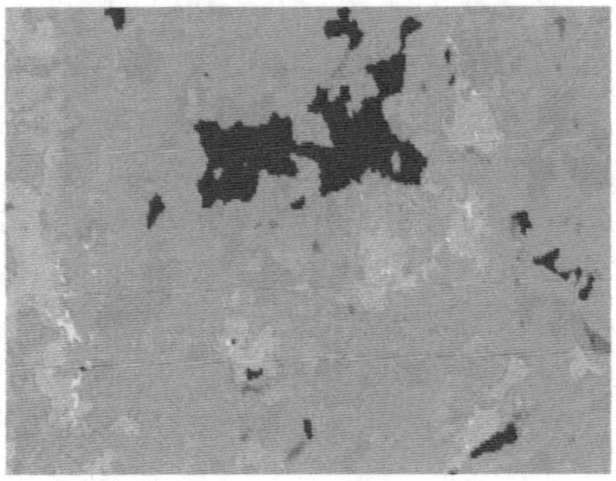

Figure 3: Terrain type of the site, blue areas show water, green represent forest terrain and yellow farmland.

4. Results

4.1 Sound pressure levels

If assuming no modes are activated, that is no wind sector management the total A-weighted sound levels of the two reception points are shown in figure 4. As can clearly be seen the general levels for the first position is lower than at the second reception point which is partly due to the longer distances from the turbines to position 2 and due to the prevailing wind direction, which is South-westerly. It is observed a shift in the data at the 18:th of November when position 1 have less sound exposure than position 2. Overall, the dynamics of the present data are most probably exaggerated as the shadow regions are not accurate due to omitting turbulence in the propagation as shown in Bolin, et al (2009) to have a levelling effect.

Comparing the results to the Nord2000 calculations from Fredriksson J, (2017) the levels are 35 dBA for the first reception position and 40 dBA for the second reception position. As can be seen from figures 4 and 5 the parabolic equation seems to give results that are similar as the Ray tracing algorithm but sometimes exceeds the 35 dBA levels for the first reception point but never exceeds 40 dBA at the second reception point.

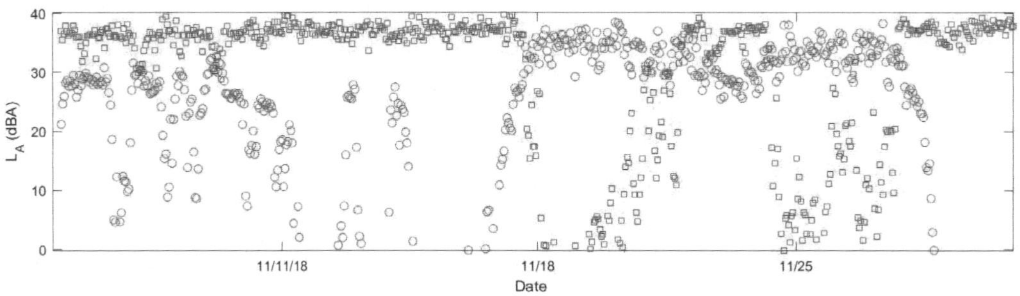

Figure 4: Calculated sound pressure levels for (o) receiver position 1 (Eastward propagation) and (□) position 2 (North-westward propagation). The data show "normal" operations- mode regulation is included.

If the wind sector management are considered and compared to the normal operations the results are shown in figure 5. As can be seen the second receiver position have decreased levels but the modes do not capture all the highest sound levels, for instance at the 11:th November full operations are noted while the sound levels sometimes are almost 40 dBA. From the comparison of the two sound propagation models, it is suggested from the parabolic equation results that the wind sector management could be improved as there are several occasions when turbines have restricted operations while the sound level is well below 40 dBA which is the guideline for wind turbine noise in Sweden (Naturvårdsverket, 2020).

For the Parabolic Equation method the dynamics of the varying sound propagation is as noted before exaggerated by the inaccurate modelling of acoustic shadow zones that would be remedied by including turbulence in the method. However, when disregarding the sound levels for the shadow zones, the variations of the sound exposures are in the same order of magnitude as the mode dynamics which indicate that for these two points the sound propagation is as important as the source to the emitted sound level.

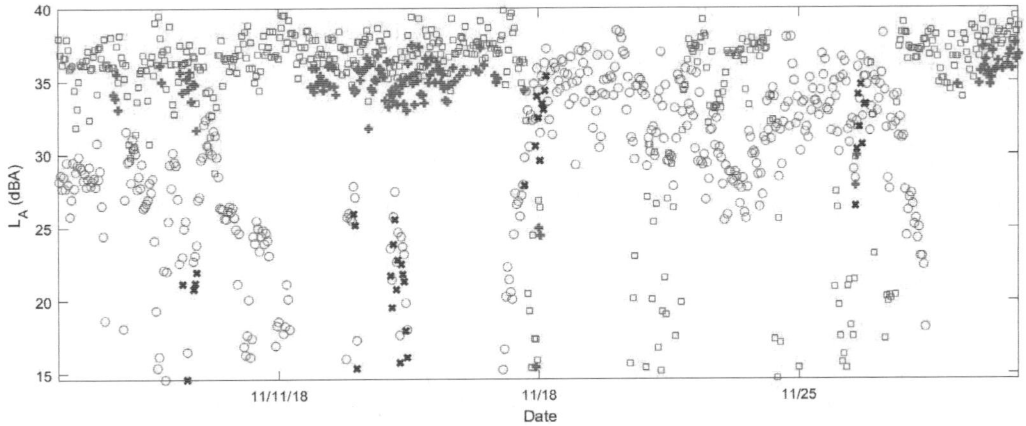

Figure 5: Sound level at receiver position 1(blue, o for normal operations and x for moded operations) and receiver position 2 (red, □ for normal operations and + for moded operations)

4.2 Wind data

The wind forecast data at hub-height are shown in Figure 7. As can be seen the wind speeds are generally above 4 m/s which is approximately the cut-on wind speed for the turbines. As can be seen in the lower graph showing the wind direction is observed that a dominating Westerly wind is prevalent during the period but at around the 18:th and 25:th November there are shifts to Southerly winds and then a gradual swing towards Westerly winds again. The Southerly winds are associated with the shifting sound levels in figure 4 at these same times, increasing for position 1 and decreasing for position 2.

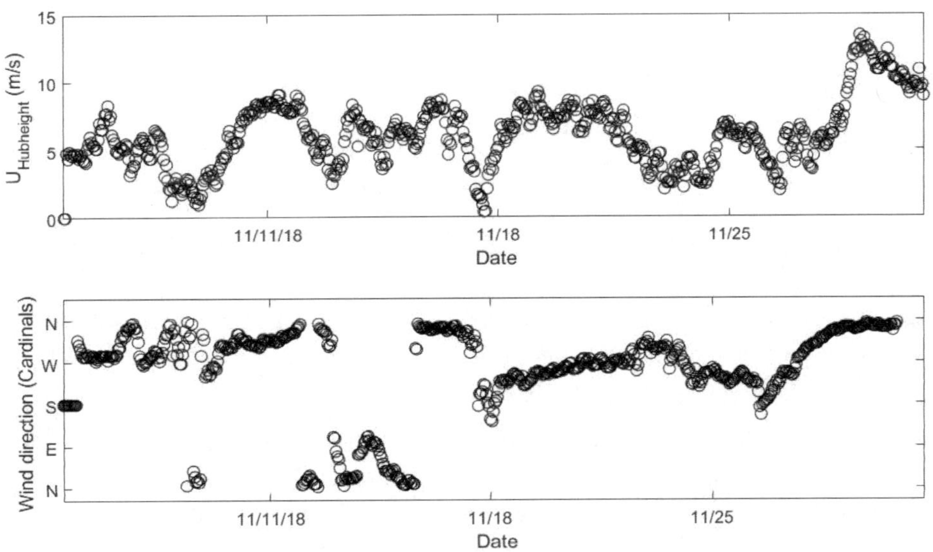

Figure 7: Wind speed at hub-height (upper figure) and wind direction (lower figure).

4.3 Computational times

In total 25 days hourly propagation codes from 12 turbines are computed for 17 third-octave bands, thus adding up to approximately 60 000 propagation calculations per reception point. A laptop computer (Lenovo Thinkpad E480 with an Intel Core 5i-8250U CPU 1.60 GHz processor and 8 GB RAM memory) has been used in the calculation. The computational times for the propagation codes are spanning from around 10 minutes for the lower frequencies to around one hour for the 1 kHz third octave band. As the propagation code is possible to parallelize the computational times could be reduced significantly on a computer with multiple processors which would be beneficial, especially if turbulence effects should be included using Monte Carlo-simulations as in Bolin et al (2009).

5. Conclusions

The computational times of the Beilis Tappert Parabolic equation are fast enough to compute a daily sound propagation forecast of the wind farm although turbulent effects have not been included in the present article which will increase the duration of the computation. The shadow zones are severely exaggerated by the parabolic equation when turbulence is not included. However, for the current application this is not of outmost importance as the occasions when high sound levels occur are the most interesting as these would be the occasions when guidelines could be exceeded and therefore reducing the source is necessary.

Overall, the results show interesting results as local meteorological profiles seems to have an impact on the sound levels at reception points that are comparable in magnitude of the moding of the wind turbines. This would suggest that there are opportunities for optimizing electric power which is not captured when using logarithmic sound speed profiles and ray tracing algorithms. Thus, fast parabolic equation models could possibly increase the electrical output while still complying to guidelines. Further studies at more wind farms and validation to measurements are of course needed before using the described methodology in reality.

References

Bolin, K., Boué, M., & Karasalo, I. (2009). Long range sound propagation over a sea surface. *The Journal of the Acoustical Society of America*, *126*(5), 2191–2197. https://doi.org/10.1121/1.3238236

Delany M. E. and Bazley E. N (1970). *Acoustical properties of fibrous absorbent materials*, Appl. Acoust. **3**, p. 105–116

Embleton T. F. W., Piercy J. E., and Daigle G. A. (1983), *Effective flow resistivity of ground surfaces determined by acoustical measurements*, J. Acoust. Soc. Am. **74**, p. 1239–1244

Fredriksson J. (2017) 10-16057 A01-A02 Ljudberäkning vindpark Högabjär-Kärsås 171030, Technical report (in Swedish)

Naturvårdeverket (2020), Vägledning om buller från vindkraftverk, Naturvårdsverket, https://www.naturvardsverket.se/upload/stod-i-miljoarbetet/vagledning/buller/buller-vindkraft/vagledning-om-buller-fran-vindkraftverk.pdf (in Swedish)

Parakkal, S., Gilbert, K. E., & Di, X. (2012). Application of the Beilis-Tappert parabolic equation method to sound propagation over irregular terrain. *The Journal of the Acoustical Society of America, 131*(2), 1039–1046. https://doi.org/10.1121/1.3676693

Sohlman M.,Jonasson H., and Gustavsson A. (2004) *Using Satellite data for the determination of the acoustic impedance of ground,* Technical report Metria/SP, Kiruna, Sweden

INCE EUROPE

9th International Conference
on
Wind Turbine Noise
Remote from Europe – 18th to 21st May 2021

Validation Efforts of an Open-Source Aeroacoustics Model for Wind Turbines

Pietro Bortolotti, National Renewable Energy Laboratory, USA: pietro.bortolotti@nrel.gov
Yi Guo, National Renewable Energy Laboratory, USA
Eric Simley, National Renewable Energy Laboratory, USA
Jason Roadman, National Renewable Energy Laboratory, USA
Nicholas Hamilton, National Renewable Energy Laboratory, USA
Patrick J. Moriarty, National Renewable Energy Laboratory, USA
Carlo R. Sucameli, Technical University of Munich, Germany
Franck Bertagnolio, DTU Wind Energy, Denmark

Summary

The open source aeroservoelastic wind turbine solver, OpenFAST, now includes an aeroacoustics model, which is described here and validated against experimental measurements recorded on a GE 1.5-MW wind turbine installed at the Flatirons Campus of the National Renewable Energy Laboratory in Boulder, Colorado. The validation demonstrates satisfactory agreement between numerical predictions and experimental recordings, with discrepancies up to 7 dB in the overall sound pressure levels at low wind speeds and a better agreement around the rated wind speed of the turbine.

1. Introduction

A higher penetration of wind into the energy mix implies a higher number of wind turbines installed both offshore and on land. Land-based wind farms are often already located in close proximity to residential areas, and the installation of larger turbines in areas subjected to noise generation limits is expected to continue. To estimate compliance with such limits, numerical models that estimate the aeroacoustics emissions of wind turbines are crucial to assist turbine manufacturers in designing quieter turbines and operators with noise-mitigation strategies while minimizing energy production and revenue losses. The National Renewable Energy Laboratory (NREL) has years of experience in developing numerical models to estimate airfoil noise (Moriarty, 2005) and wind turbine noise (Moriarty and Migliore, 2003, Moriarty et al., 2005). While the confidence in the airfoil models has been satisfactory, thanks to decades-old validation campaigns (Amiet, 1975, Brooks et al., 1989), the models to estimate the aeroacoustics noise of wind turbines have been less reliable and have had less documented validation.

Recently, researchers moved the NREL wind turbine aeroacoustics code into the latest release of the open source aeroservoelastic framework, OpenFAST (Bortolotti et al., 2020). While reimplementing the models, a code-to-code comparison against a similar code at the Technical University of Munich was executed (Sucameli et al., 2018). The comparison highlighted some discrepancies, opened new questions, and did not provide any indication about the accuracy of the models in predicting the noise of modern turbines. In this work, those discrepancies are investigated, and a validation effort based on experimental measurements obtained from a GE 1.5-MW wind turbine installed at NREL is presented.

The paper is organized as follows: Section 2 introduces the framework used to run the numerical simulations. The experimental setup and the recorded measurements are presented in Section 3; followed by the results of the validation process, which are discussed in Section 4. Section 5 closes the paper with key takeaways and an overview of the ongoing work.

2. Numerical Models and Code-to-Code Verifications

OpenFAST includes a variety of models to estimate the aeroacoustics noise of wind turbines (Bortolotti et al., 2020). The next two subsections elaborate on the two models that were verified and validated in past work. The first model aims to capture the noise generated along the leading edges (LEs) of the blades, the second model along the trailing edges (TEs). All other noise sources are excluded from the analysis and are assumed to make minor contributions to the total noise levels.

2.1 Leading-edge noise

The turbulent inflow noise represents the noise generated by an arbitrary body immersed in a turbulent flow. For a wind turbine, we assume that this noise radiates from the LE of the rotor blades. To predict turbulent inflow (TI) noise, OpenFAST implements the Amiet model (Amiet, 1975), which adopts a flat-plate approximation. A correction for finite blade thicknesses has also been developed and implemented (Moriarty et al., 2005), and is added to the Amiet sound pressure levels (SPLs). In this work we restricted the verification and validation processes to the Amiet model, which computes the SPL of an airfoil as

$$SPL_{TI} = 10\log_{10}\left(\rho^2 c^4 \frac{L_t d}{2r_e^2} M^5 I_1^2 \frac{\hat{k}_1^3}{\left(1+\hat{k}_1^2\right)^{\frac{7}{3}}} \overline{D}\right) + 78.4 \tag{1}$$

where ρ is the air density, c the speed of sound, L_t the turbulent length scale, d the blade element span, r_e the effective distance between LE and observer, M the Mach number, I_1 the turbulence intensity of the airfoil inflow, $\hat{k}_1$ a wave number function of frequency f, and $\overline{D}$ the directivity term. The directivity term is different above and below a threshold frequency that depends on chord and local inflow velocity.

The model was verified against the results presented in Figure 4 of Amiet, 1975, and code-to-code compared against the results generated at DTU Wind Energy in Denmark with the code HAWC2 (Bertagnolio et al., 2017), and at the Technical University of Munich (TUM) with the code Cp-Max (Sucameli et al., 2018). Note that both DTU and TUM implement the formulation of the Amiet model described in Paterson and Amiet, 1976, with TUM also implementing the model from Amiet, 1975. Only the results from the former model are reported here for TUM because those from Amiet, 1975 overlap those from OpenFAST.

The inputs to the models used for the comparison are listed in Table 1, and the results of the comparison for two sets of chordwise and spanwise directivity angles are shown in Figure 1.

Table 1. Inputs to the Turbulent Inflow Noise Models Used in the Verification Studies. Inputs correspond to those of the Lowest Spectrum from Figure 4 in Amiet, 1975.

Input	**Unit**	**Value**
Air density ρ	kg m^{-3}	1.225
Speed of sound c	m s^{-1}	340.270
Turbulent length scale L_t	m	0.032
Span length d	m	0.533
Observer distance r_e	m	2.134
Incident turbulence intensity I_1	-	0.044
Incident wind speed U_1	m s^{-1}	30.965
Mach number M	-	0.091
Chord c	m	0.457

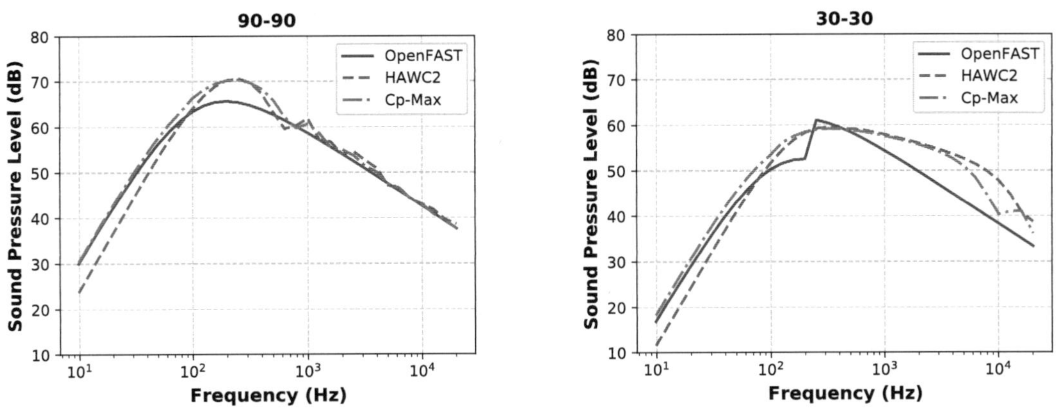

Figure 1. Results of the code-to-code verification of the turbulent inflow noise models applied at the airfoil level implemented in OpenFAST (NREL), HAWC2 (DTU), and Cp-Max (TUM). The left plot corresponds to an observer sitting on top of the airfoil LE, whereas the right plot corresponds to a case where chordwise and spanwise angles are both equal to 30 deg, see Figure 1 in Bortolotti et al., 2020 for a graphical representation of the angles.

At the airfoil level, the models are found to match fairly well. At chordwise and spanwise angles different than 90 degrees (i.e., for the observer not sitting on top of the airfoil LE), the model implemented in OpenFAST shows a discontinuity at the cut-off frequency because of the directivity term. Note that the directivity term at low frequency for the LE was originally missed in Bortolotti et al., 2020, which only reported directivity for the TE noise contribution. The documentation of OpenFAST v3.0.0 reports all formulas correctly (NREL, 2021).

2.2 Trailing edge noise

The second major source of aeroacoustics noise of wind turbines is noise that radiates from the TE of the blades. This noise source is usually referred to as turbulent boundary layer TE noise. OpenFAST implements two models to simulate this noise mechanism, namely the Brooks-Pope-Marcolini model (Brooks et al., 1989) and a more recent model implemented at the Dutch research institute TNO (Parchen, 1998). This work adopts the former model, which was code-to-code compared to the one available at TUM reporting minor discrepancies (Bortolotti et al., 2020).

Table 2. Key Characteristics of the GE 1.5-MW Wind Turbine.

Model	GE 1.5-MW SLE
Serial number	N000780-N/TB059-3
Configuration	Horizontal axis, upwind, three bladed
Control strategy	Pitch control, variable speed
Generator	Winergy, doubly fed induction, JFEC-500SS-06A
Gearbox	Winergy multistage planetary / helical model PEAB 4410.4, serial number NFR-W-111620
Blades	GE37c, fiberglass, S00028, S00029, S00030
Rated power (kW)	1,500
Rotor diameter (m)	77
Hub height (m)	80
Rated wind speed (m s^{-1})	14

3. Experimental Setup

An experimental campaign was conducted at the Flatirons Campus at NREL using an instrumented GE 1.5-MW wind turbine owned by the U.S. Department of Energy. The next two subsections briefly describe the equipment used and the measurement processing. Readers interested in the details of either topic should refer to an upcoming NREL technical report (Hamilton, 2021).

3.1 Equipment

The key characteristics of the GE 1.5-MW wind turbine are reported in Table 2, and a photo of the machine is shown on the left of Figure 2. This wind turbine model is representative of a large number of wind turbines operating in the United States, with more than 18,000 units installed.

The turbine installed at the Flatirons Campus has a permanent set of sensors measuring various quantities across the components. During this project, the area surrounding the wind turbine was further instrumented with 11 sound boards located as illustrated on the right of Figure 2.

Researchers chose the locations approximately symmetrical on either side of an axis aligned with the prevailing wind direction which, during the winter months, is equal to 285 degrees (Hamilton and Debnath, 2019). The international standards for wind turbine noise IEC 61400-11, for sound-level metering IEC 61672, and measurement microphones IEC 61094-4, were followed. Eight of the 11 microphones have a standard measurement range between 20 Hz and 11.2 kHz. The microphone marked as Mic 4C is one of these eight and is located at the IEC-prescribed location, downwind on the ground at a distance from tower base equal to the turbine height. The last three of the 11 microphones, located at the three data acquisition system (DAS) locations because of extra requirements in terms of power supply, are low-frequency microphones capable of measuring in the subaudible frequency range, as low as 1 Hz. The measurements of these microphones were not part of this study but will be used for future validations of the numerical framework to estimate the low frequency noise emissions of wind turbines.

Each microphone communicated with a DAS subsystem over coaxial connection. Signal degradation was mitigated in the instrumentation plan by limiting coaxial cable lengths to a maximum of 50 m (green lines in Figure 2). Communication from each DAS location to a central data storage server occurred over fiber-optic cables, which have less stringent constraints in terms of maximum distance.

Figure 2. (Left) The GE 1.5-MW wind turbine located at the Flatirons Campus of NREL in Boulder, Colorado. (Right) Bird's-eye view of microphone locations and cabling. Mic 4C is at the IEC-prescribed location for a wind direction of 285 degrees.

3.2 Measurements and data processing

The measurement campaign took place between December 2020 and February 2021, with measurements obtained on eight different days. Because winter is the windy season at the Flatirons Campus of NREL, measurements could be obtained for a diverse set of wind speeds and atmospheric conditions for wind directions near 285 degrees. All microphones were calibrated at the beginning and at the end of each measurement period, and measurements were recorded during both day and night. Background noise was recorded by parking the wind turbine during a wide range of atmospheric conditions. Although not shown in Figure 2, reference wind speed measurements were obtained from a cup anemometer at 80-m height on a met tower located 153 m (2 rotor diameters) west of the wind turbine at a heading of 280 degrees. In addition, the turbulence intensity of the wind at different heights and at different average wind speeds was also reconstructed, and it is reported in Figure 3. Notably, all data streams were synchronized using Global Positioning System (GPS), which provided a way to not only synchronize the microphone data with the turbine and met tower data, but also to ensure that all recording windows for the microphones started at the same time and remained synchronized, greatly simplifying processing.

The rest of this section explains the data processing steps used to derive wind turbine SPLs binned by wind speed. First, for each microphone recording, the following steps are used to calculate unweighted and A-weighted SPLs in 10-second samples. Ten-second samples are used based on guidance in the IEC 61400-11 standard (International Electrotechnical Commission, 2020). For each 10-second sample, the unweighted 1/3-octave band SPL spectrum is calculated for center frequencies from 1 Hz to 20 kHz using the noiseLAB software (DELTA Acoustics and Vibration, 2014). The equivalent overall SPL is formed by energy summing the individual 1/3-octave bands (International Electrotechnical Commission, 2020). Similarly, the A-weighted 1/3-octave band SPL spectrum is calculated by applying the A-weighting transfer function to the unweighted 1/3-octave bands. The individual A-weighted 1/3-octave bands are then summed to determine the equivalent A-weighted SPL.

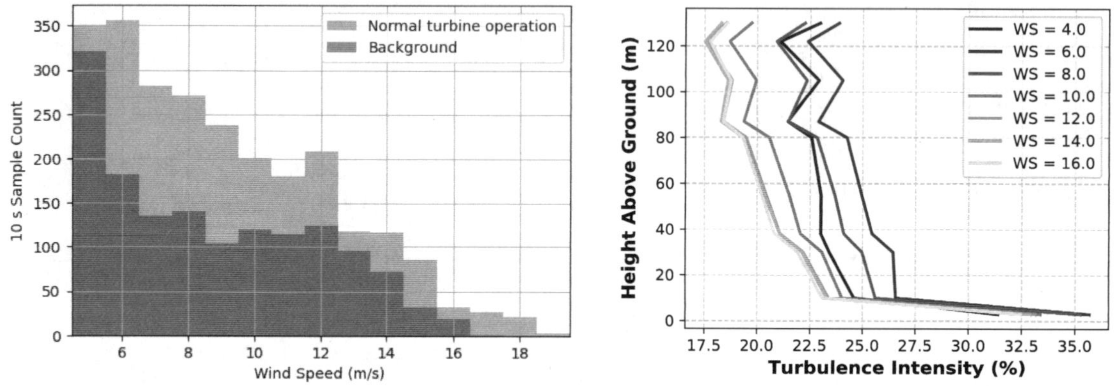

Figure 3. (Left) Number of valid 10-second SPL samples obtained for each 1 m s^{-1} wind speed bin for background and normal operation periods. (Right) Turbulence intensity versus height above ground at various average wind speeds at hub height.

Supervisory control and data acquisition (SCADA) data from the wind turbine and meteorological measurements from the met tower, sampled at a rate of 1 Hz, are processed into 10-second mean values and synchronized with the SPL samples. Together with the reference wind speed measurements at 80-m height from the met tower, the following two variables are used in the remainder of the data processing. Wind direction is estimated using the wind turbine's nacelle orientation, following guidance provided by the IEC 61400-11 standard (International Electrotechnical Commission, 2020). A turbine status signal indicating the operating state of the wind turbine is used to separate the data into background and normal operation categories.

After separating the synchronized noise, SCADA, and met tower data into background and normal operation periods, 10-second samples in which the nacelle position is more than 15 degrees from the intended wind direction of 285 degrees are removed, as recommended by the IEC 61400-11 standard. However, because of a lack of background measurement data, we do not apply any wind direction filtering to the background noise periods.

Lastly, noise data from the valid 10-second samples are binned by wind speed, using a bin width of 1 m s^{-1}. The number of 10-s samples for background and normal operation periods, binned by wind speed, are shown in Figure 3. Note that the IEC 61400-11 standard recommends a minimum of ten 10-s samples for background and normal operation periods within a wind speed bin. Based on this guidance, we analyze SPLs for wind speed bins up to 16 m s^{-1}.

Prior to using the noise measurements for model validation, the SPLs within a particular wind speed bin are averaged and the wind turbine noise is distinguished from background noise (e.g., wind-induced noise). For the SPL variable of interest, the data for both the background and normal operation periods are energy averaged using the following formula:

$$SPL_{\text{avg}} = 10\log_{10}\left(\frac{1}{N}\sum_{i=1}^{N} 10^{\frac{SPL_i}{10}}\right), \qquad (2)$$

where SPL_i represents the SPL for the ith 10-second sample in the wind speed bin and N indicates the number of samples in the bin. The average background SPL is then subtracted from the average SPL during normal operation, yielding the SPL attributed to wind turbine noise. Examples of the unweighted and A-weighted 1/3-octave band SPL spectra for normal operation and background periods, along with the spectra representing wind turbine noise, for the 10 m s^{-1} wind speed bin are provided in Figure 4. Note that for frequencies below 315 Hz in this wind speed bin, the background noise is greater than the total noise when the turbine is operating.

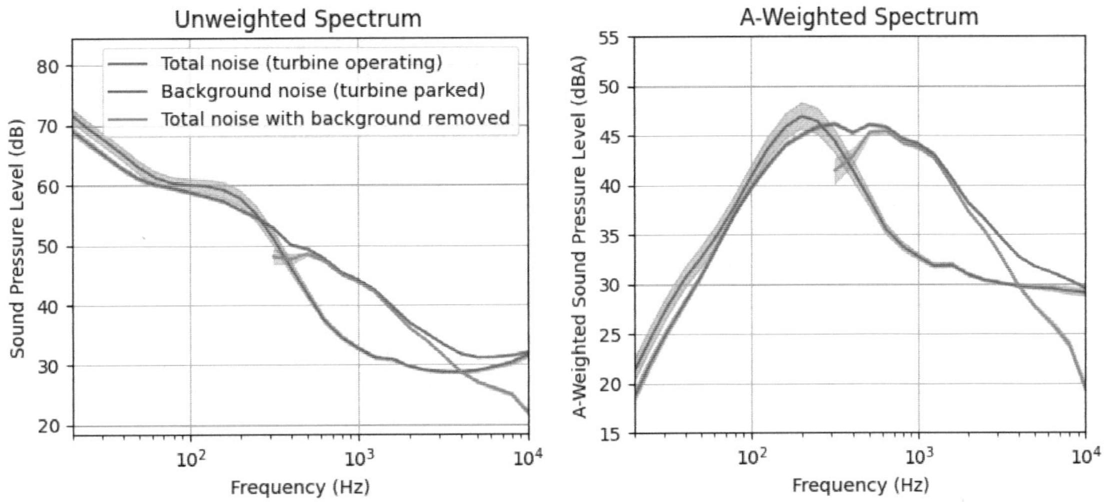

Figure 4. Unweighted (left) and A-weighted (right) 1/3-octave band SPL spectra for the 10 m s^{-1} wind speed bin. Energy averaged spectra for normal operation and background periods are shown together with the spectra for normal operation with background noise removed (i.e., the spectra attributed to wind turbine noise). The shaded regions indicate the standard error associated with the energy averaged SPL estimates.

Therefore, the SPL attributed to wind turbine noise cannot be calculated for these frequencies. Ideally, background noise should not exceed the combined background and wind turbine noise. But in this data set, the background noise measurements were likely obtained in different atmospheric conditions (e.g., with higher turbulence intensity) than the normal operation periods, leading to higher low-frequency wind-induced noise levels.

4. Validation

This section presents the comparison between the numerical predictions of OpenFAST and the field measurements of the noise emissions of the GE 1.5-MW turbine. The overall SPLs between 10 Hz and 20 kHz are compared at various wind speeds for the eight standard microphones described in Section 3.1. The OpenFAST simulations are run with the same rotor speed and pitch angle of the real turbine. The simulations are run at steady wind speed, and time-invariant turbulence intensities are assumed for the LE noise model. The turbulence intensity is assumed to follow the lines shown in Figure 3. The numerical simulations also assume a turbulent length scale of 42 m, which is the value prescribed by the IEC 61400-11 standard.

Figure 5 shows the SPL spectrum predicted by OpenFAST in comparison with the experimental results for average wind speeds at hub heights of 6 m s^{-1}, 8 m s^{-1}, 10 m s^{-1}, and 12 m s^{-1} for the microphone 4C, which is located at the IEC-prescribed location. Overall, fairly good agreement between OpenFAST and experimental results is reported at 8 m s^{-1} and 10 m s^{-1}, whereas OpenFAST underpredicts SPL at 6 m s^{-1} and overpredicts at 12 m s^{-1}. At frequencies less than 0.2 kHz, experimental data is unavailable because the background noise matches the intensity of turbine noise. When frequencies move beyond 0.2 kHz to 0.3 kHz, the TE noise starts dominating the spectrum and both numerical and experimental lines show a bump. Notably, at 10 m s^{-1} and 12 m s^{-1}, and at frequencies higher than 2 kHz, the OpenFAST predictions start to deviate from experimental data. The discrepancy may originate from the turbulent boundary layer TE noise component of the Brooks-Pope-Marcolini model. Another source of error could be the high background noise, which above 4 kHz is higher than the turbine noise with background removed, see Figure 4.

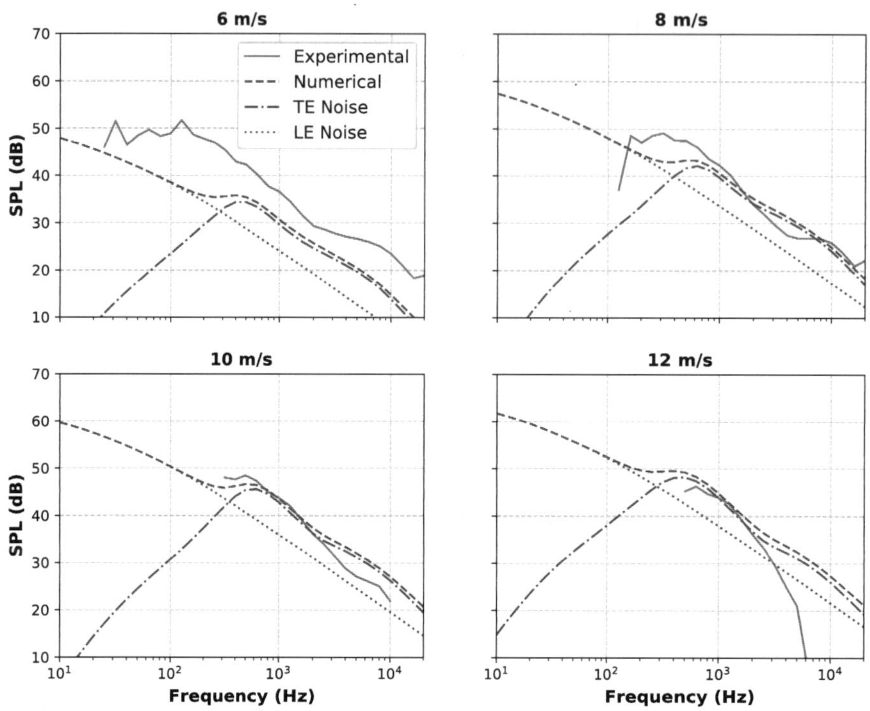

Figure 5. Numerical and experimental frequency spectra of the sound pressure levels (SPLs) at four wind speeds at IEC-prescribed location Mic 4C. The numerical predictions are the sum of LE and TE noise.

Future work will investigate if the TNO model, combined with accurate descriptions of the boundary layer characteristics along blade span, can reduce the gap.

Figure 6 compares the measured and numerical overall SPLs at various wind speeds at all eight microphone locations. The validation shows a somewhat surprising different slope with respect to wind speed. OpenFAST predicts a more marked dependency of the overall A-weighted SPL (OASPL) to wind speed primarily caused by variations in rotor speed, whereas the experimental recordings show a milder dependency. The differences are highest at 6 m s^{-1} and are between 5 dB(A) and 7 dB(A), depending on the microphone. Mic 2N also shows a discrepancy of 10 dB(A) at 14 m s^{-1}, but the experimental recording that decreases with wind speed suggests that the experimental data point may be flawed. Finally, Mic 6S reports the poorest comparison with the experimental line above the numerical one by 7 dB(A) at 6 m s^{-1} and by 4 dB(A) at 10 m s^{-1}. The fact that Mic 6S is located near a hill might be making the comparison more challenging than for the other microphones.

5. Conclusions and Ongoing Activities

This work presents the latest update of the aeroacoustics model in the wind turbine aeroservoelastic solver, OpenFAST, and its validation against measurements recorded on a GE 1.5-MW machine installed at the Flatirons Campus of NREL. The validation returns differences up to 3 dB(A) between numerical and experimental lines in the range of 8 m s^{-1} to 12 m s^{-1} of wind speed at hub height, but errors up to 7 dB(A) outside this range. The discrepancies are generated by the different slope of the overall SPLs in respect to wind speed, with OpenFAST predicting a steeper curve, whereas the experimental recordings are found less sensitive to wind speed.

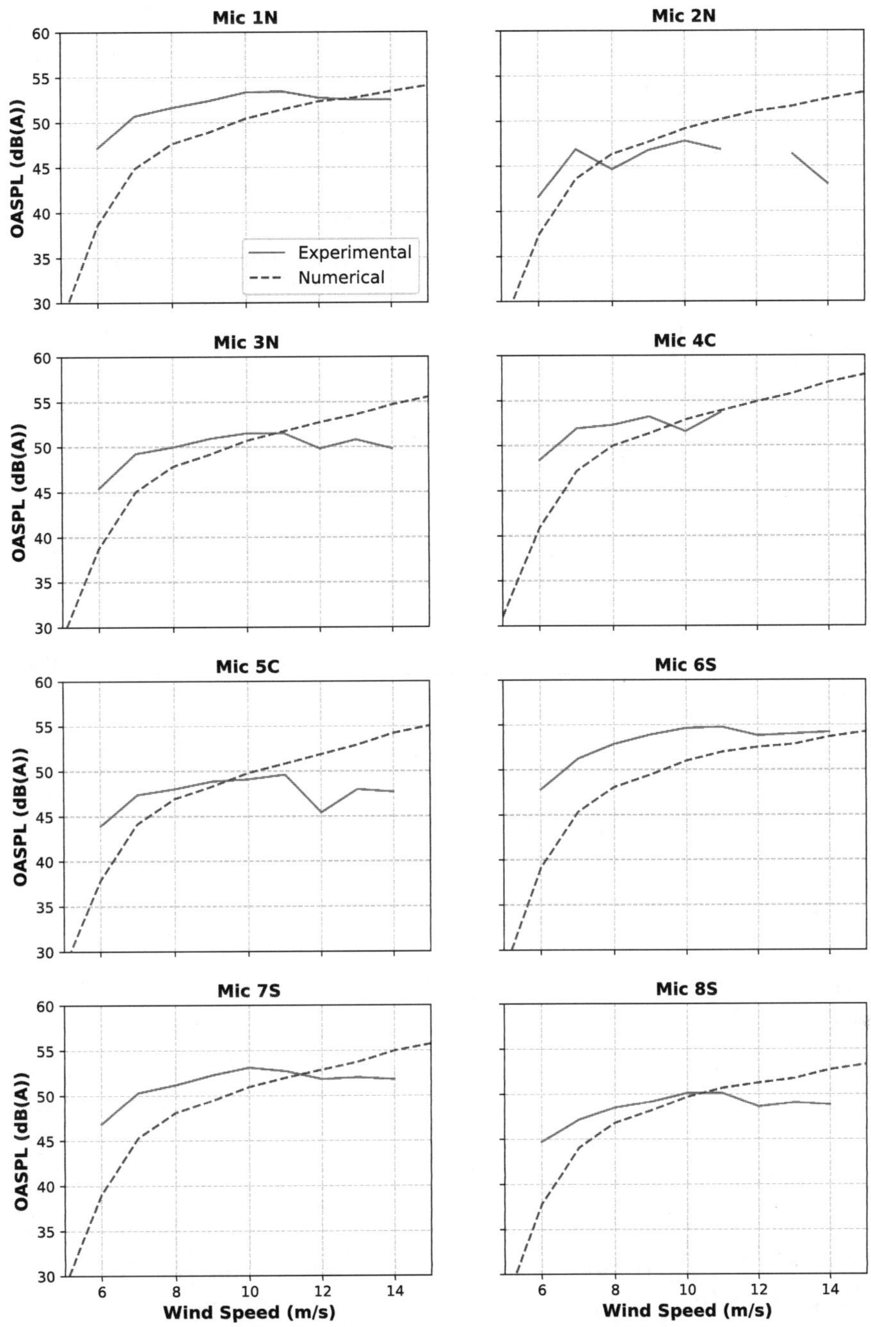

Figure 6. Numerical and experimental overall A-weighted sound pressure levels (OASPL) at the eight microphone locations at varying wind speeds.

Several activities are ongoing. The verification step presented in Section 2 was restricted to the airfoil level and only to three numerical frameworks. Studies within the IEA Wind Task 39 on Quiet Wind Turbine Technology are comparing the outputs of multiple codes at the turbine level. The comparison is based on the NM80 wind turbine located at the Danish research center Risø. A dedicated technical report is being prepared. In addition, the validation process should include the other noise models implemented in OpenFAST, such as the LE model described in Moriarty et al., 2005, and the TE noise model described in Parchen, 1998. Finally, more work is required on processing of low-frequency noise measurements, first focusing on distinguishing the turbine noise from the background and later validating the models available at DTU and TUM, among others.

Acknowledgments

The authors would like to acknowledge the precious work of the field technicians of NREL. This work was authored by the National Renewable Energy Laboratory, operated by Alliance for Sustainable Energy, LLC, for the U.S. Department of Energy (DOE) under Contract No. DE-AC36-08GO28308. Funding provided by the U.S. Department of Energy Office of Energy Efficiency and Renewable Energy Wind Energy Technologies Office. The views expressed in the article do not necessarily represent the views of the DOE or the U.S. Government. The U.S. Government retains and the publisher, by accepting the article for publication, acknowledges that the U.S. Government retains a nonexclusive, paid-up, irrevocable, worldwide license to publish or reproduce the published form of this work, or allow others to do so, for U.S. Government purposes.

References

Amiet RK (1975) *Acoustic Radiation from an Airfoil in a Turbulent Stream* Journal of Sound and Vibration 41(4), 407-420 doi: 10.1016/S0022-460X(75)80105-2

Bertagnolio F, Madsen HAa, and Fischer A (2017) *A Combined Aeroelastic-Aeroacoustic Model for Wind Turbine Noise: Verification and Analysis of Field Measurements* 20(8), 1331-1348 doi: 10.1002/we.2096

Bortolotti P, Branlard E, Platt A, Moriarty PJ, Sucameli C, and Bottasso CL (2020) *Aeroacoustics Noise Model of OpenFAST* NREL/TP-5000-75731

Brooks TF, Pope DS, and Marcolini MA (1989) *Airfoil self-noise and prediction* NASA Reference Publication 1218

DELTA Acoustics and Vibration (2014) "noiseLAB User's Guide for Version 4.0". https://noiselabdk.files.wordpress.com/2011/10/noiselab-4-0-users-guide.pdf

Hamilton N, and Debnath MC (2019) *National Wind Technology Center-Characterization of Atmospheric Conditions* NREL/TP-5000-72091

Hamilton N, Bortolotti P, Simley E, Guo Y, and Roadman J (2021) *Aeroacoustic Assessment of Wind Plant Controls*. NREL Technical Report. In Preparation

International Electrotechnical Commission (2020) "International Electrotechnical Commission (IEC) standard, Wind Turbines Part 11: Acoustic Noise Measurement Techniques". IEC 61400-11, Edition3.1

Moriarty PJ (2005) *NAFNoise User's Guide* https://www.nrel.gov/wind/nwtc/assets/pdfs/nafnoise.pdf

Moriarty PJ, and Migliore P (2003) *Semi-Empirical Aeroacoustic Noise Prediction Code for Wind Turbines* NREL/TP-500-34478

Moriarty PJ, Guidati G, and Migliore P (2005) *Prediction of Turbulent Inflow and Trailing-Edge Noise for Wind Turbines* 11th AIAA/CEAS Aeroacoustics Conference

NREL (2021) *OpenFAST Documentation - Read the Docs v3.0.0*

Parchen RR (1998) *Progress Report DRAW: A Prediction Scheme for Trailing Edge Noise Based on Detailed Boundary Layer Characteristics* TNO Institute of Applied Physics

Paterson RW, and Amiet RK (1976) *Acoustic radiation and surface pressure characteristics of an airfoil due to incident turbulence* NASA Contractor Report 2733

Sucameli CR, Bortolotti P, Croce A, and CL Bottasso (2018) *Comparison of some wind turbine noise emission models coupled to BEM aerodynamics* J. Phys.: Conf. Ser. 1037 022038

9th International Conference
on
Wind Turbine Noise
Remote from Europe – 18th to 21st May 2021

Development of IEC/TS 61400-11-2: Measurement of wind turbine noise characteristics in receptor position

Sylvia Broneske, RWE Renewables UK: sylvia.broneske@rwe.com
Bo Søndergaard, Sweco Danmark A/S: bo.sondergaard@sweco.dk

Summary

Following an intensive period of weekly online meetings, the Project Team PT61400-11-2 is very close to submitting a committee draft to the Technical Committee TC 88 of the new Technical Specification IEC TS 6140-11-2 *Measurement of wind turbine noise characteristics in receptor position*. At the time of writing this paper, it is thought to only require a few more meetings until a first draft can be submitted to the member National Committees of IEC for review by their experts. It is hoped that the review process in the National Committees is under way at the time of the Wind Turbine Noise conference in May 2021.

This paper gives an overview of the work of the Project Team and the topics included in the Technical Specification to describe the sound characteristics of wind turbine noise through measurements and assessment at the receptor position. This includes physical characteristics like amplitude modulation, tonality and sound pressure levels amongst others and their correlation with wind data where applicable.

The methods are presented as ‘operating instructions’ for conducting measurements for authorities and measurement institutes, tailored to the specific intentions of the measurements. The TS will be of particular interest in countries where no guidance on wind farm noise measurements is available but also in countries where guidance and legislation of environmental noise measurements may not be sufficiently adapted for wind turbine noise specifics.

Acknowledgments

Writing the Technical Specification has been a challenging task not only due to the comprehensive topics to cover and the challenging times since the outbreak of Covid-19 but also to bring all the different experts with a high level of expertise, experience and motivation together in one place (in person and virtually) to discuss and understand each other’s views. It has been a successful process which has resulted in a workable document and also lead to more investigations e.g. on tonality on wind turbine noise. A number of papers and presentations from that work are being presented at this conference. Many ideas have been taken from the discussions for further research work and much knowledge has been gained during sometimes intensive discussions.

The authors would like to thank all members of the project team for their huge effort in bringing together the required information, expertise and experience to write this document. It has truly been a world-wide cooperation and exchange of experiences. Owing to the high frequency of online meetings in the last months, it has been difficult for many members to attend all the meetings, however, we appreciate any input and all the productive discussions we have had since 2018.

As next step, the Project Team is now looking forward to the comments and suggestions they are expecting to receive in the next phase, which will be the review period by the National Committee's acoustic experts, but also to good discussions with many questions and comments at this conference.

1. Introduction

At the Wind Turbine Noise Conference 2019 in Lisbon/Portugal, Broneske and Søndergaard (2019) introduced the proposed Technical Specification IEC TS 61400-11-2 *Measurement of wind turbine noise characteristics in receptor position* (TS) to the audience and gave an overview of the process of writing documents like standards or technical specifications for the IEC[1] or ISO[2]. More details about the work up to the WTN 2019 can be found in said paper.

Since then, the way of working in IEC committees has completely changed following the worldwide outbreak of COVID-19 in early 2020, from several days in-person meetings to online meetings only. In addition to the general challenges of writing a document with an international expert group and achieving consensus by bringing many opinions and varying methods together, it brought the additional challenge of trying to make online meetings work across the time zones with active members working in the west of the USA, across Europe and all the way to India and China. The members of PT11-2 met this challenge by joining at 5 in the morning, fitting the meeting in between breakfast and school runs, staying past their office hours in the afternoon, arranging child care and home-schooling around it or even attending late in the evening from 8 to 10 at night. Other obstacles were the sheer amount of content to be delivered and the time members were able to spend on working on the document in addition to a high workload in their companies and institutions, so that deadlines for the committee draft had to be postponed again and again. However, the Project Team pushed through all those challenges and is finally on the finish line for a Committee Draft (CD) which is planned to be submitted to IEC/TC88[3] in spring 2021.

In general, the document can be used by regulators and authorities, measurement laboratories, developers, operators and manufacturers for the comparison with local regulation, the comparison with guarantee values and the assessment of sound characteristics in wind turbine noise at the receptor position. It is aimed at countries and regions without a long-standing tradition and knowledge of wind turbine acoustics and PT11-2 invites those countries to draw on the experience of this expert group instead of having to build up the knowledge from scratch.

It needs to be pointed out, that it is not the intention of the TS to replace existing wind-turbine-specific national regulations and expertise, but to be used as guide in such experienced

[1] International Electrotechnical Commission: International Standards and Conformity Assessment for all electrical, electronic and related technologies

[2] International Organization for Standardization

[3] Technical Committee TC88 "Wind energy generation systems"; https://www.iec.ch/dyn/www/f?p=103:7:5599856330528::::FSP_ORG_ID,FSP_LANG_ID:1282,25 (last accessed 22/03/2021)

countries to improve their understanding and develop a state of the industry legislation. We invite all experts and policy makers to learn from the vast experience and considerations put into this document by the world's leading wind turbine acoustic experts.

Since this paper follows on from the WTN 2019 paper by Broneske and Søndergaard, it will not discuss the basics of standardisation but will give an overview of the proposed chapter structure and recommended methods described in the TS.

2. History of the IEC/TS 61400-11-2 with the working title "Wind energy generation systems - Part 11-2: Measurement of wind turbine noise characteristics in receptor position"

Standards dealing with sound measurements of sources and propagation calculations of industrial noise have traditionally been written and maintained by ISO. An exception is the standard for the determination of sound power levels of wind turbines, which was included in the IEC series because it is strongly linked to the operation of this particular plant and therefore fits into the IEC series which describes this product. The IEC 61400-11 is an expansion of the ISO standards for the determination of sound power levels (especially ISO 3744), but which includes as a special feature the correlation of the sound power level with (hub height) wind speed and in particular describes measurements at higher wind speeds.

It is particularly the measurement at higher wind speeds, the correlation with the wind turbines' operational data, measurement and assessment of long periods (up to several weeks or months) that prompted the new work item for a document detailing the determination of wind turbine noise characteristics in receptor position, expanding the measurement position from near the wind turbine to the far-field and thus directly at the noise receptor positions. This new document will be Part 2 of the 'acoustics series' within the IEC 61400 standards, dealing with wind energy generation systems. It had been decided to write a Technical Specification (TS) instead of an International Standard (IS), as it was felt that e.g. the topic of amplitude modulation still requires further research and therefore standardization as an IS was considered to be premature. A TS refers to international standards with respect to detail and completeness. Furthermore, a TS is similar to an IS in that it is normative and developed by consensus. A TS is approved by a two-thirds majority of the participating members of a technical committee of the IEC (TC) or a subcommittee (SC). The approval process for the TS is similar to that of the IS, but saves one round of voting and therefore goes through the IEC process faster for publication.

2.1 The Project Team and Expert Groups

The PT61400-11-2 (PT11-2) project team currently consists of 35 experts, representing 11 National Committees (shadow committees to TC88). Only countries where their national standards committee is a member of TC88, can send experts on their behalf. Experts of other countries can only take part in meetings of the project team in an advisory manner at the discretion of the Convenor. Project management for PT11-2 had been assigned to Mr. Bo Søndergaard of Sweco Danmark A / S, who is already in charge of the MT11 maintenance team. A project manager can be assisted by a secretary which in this case has been taken on by Ms Sylvia Broneske of RWE Renewables UK.

It became apparent that in order to progress the document, some of the more difficult technical topics could not be written in the regular meetings of PT11-2. Subsequently some smaller expert groups were set up in order to work in parallel with the larger project team. Owing to their informal nature, an expert group can consist of specialists not associated with the IEC project team, but particularly experienced in a specific area, with the task to answer or support the project team on those specific topics. The expert groups on Wind Turbine Amplitude Modulation and Tonality Assessment were organised by Mr Robin Woodward of Hayes

McKenzie Partnership Ltd./United Kingdom supported by input from several leading wind turbine acoustics consultancies across the world who will also give presentations at this conference. Mr Payam Ashtiani of Aercoustics/Canada organised the expert group on sound pressure level determination with members of PT11-2 and Mr René Haevernick and his colleagues of Nordex/Germany provided valuable input for the wind speed and direction chapter. Many other members wrote individual sections outside of the PT11-2 meetings that were then discussed in the larger group.

2.2 Progress

Since the first meeting in June 2018, PT11-2 has met five times in various locations. Each of those meetings was hosted by the company of a different member of the group as is custom in IEC and ISO standardisation work. innogy (now RWE Renewables), NREL, Sweco, Aercoustics and ACCIONA Energía S.A. kindly provided meeting facilities and refreshments during those meetings.

PT11-2 has a core of 10 to 15 members regularly attending the meetings who are also very active in writing or commenting on the text of the TS. The PT11-2 found a meeting rhythm of initially every two weeks from May 2020 which was increased to weekly meetings from January until April 2021.

When proposing the new work item for this TS in 2018, the initial date given for submission of the Committee Draft to TC88 was January 2019. Generally, IEC expects a TS to be written within 18 months. However, extensions of this deadline had been requested and granted twice, owing to the work being much more complex than initially anticipated. When looking at the topics included it is clear where some of the delay came from: there are at least three individual documents combined into the one TS since there were no existing standards which could be referred to without further amendments, with the exception of the impulsivity section. Consensus had to be reached on complex topics which resulted in some discussions lasting several months.

2.3 Further Steps to Publication

The current preparatory stage, where the document has the status of a Working draft (WD), ends with the submission of the Committee Draft (CD) to all IEC members of TC88. National Committees (NC) that actively participate in TC88 (P-members) and those that have observer status (O-members) can comment on and approve the CD. National Committees have between 8 and 16 weeks to submit their comments. This is the most important commenting stage and the aim is to reach consensus on the technical content. The members of PT11-2 will come together after the commenting period and address the received comments. If there are topics to be revised based on comments or important aspects have been missed or requested by a National Committee, the PT11-2 will further discuss and address this at that time.

Following implementation and response to the received comments, the TS will be published as a Committee Draft for Vote (CDV). The CDV is considered approved if 2/3 of all votes by P-members are in favour of the publication. At this stage, only editorial comments can be incorporated. This enquiry stage lasts 12 weeks. Following approval of the TC88 member committees, the TS will be published by IEC central office in Geneva within 6 weeks of approval.

Given some of the external time constraints, a publication of the TS is not expected before the end of the year. A great unknown is also the required further work once the comments of the NCs come in. Provided consensus can be reached not only within PT11-2 but also with the NCs in a timely manner, a publication of the TS is now anticipated for the first half of 2022.

3. Structure of the Document

It was initially thought that the TS would cover lots of different 'traditional' locally embedded methods for all aspects of the sound characterisation, however, in the end PT11-2 have reduced the choice to a few recommended methods. When working on the document, it was found that existing methods had to be adapted to the specifics of wind turbine sound. Most important, however, it is to give guidance on how to carry out measurements the best possible and most reliable way for a given task through detailed planning and creating a test plan for the measurements.

An overview of the contents of the TS is given in Figure 1 below. The document has a main section with topics that apply to all measurements and assessments, and sections that are specific to a defined task for the receptor measurement, such as sound pressure level, tonality, amplitude modulation or impulsivity. An emphasis is put on the creation of a test plan and it should be noted that not all measurements have to be carried out at the same time but have to be selected in accordance to the task given to the expert in the field. The TS is intended to be used as a 'toolbox' and gives multiple options that can be put together for the measurement, depending on the purpose of the assessment. Furthermore, there are eleven Annexes which give additional information (informative, not normative) but are not mandatory for any of the methods described in the main section.

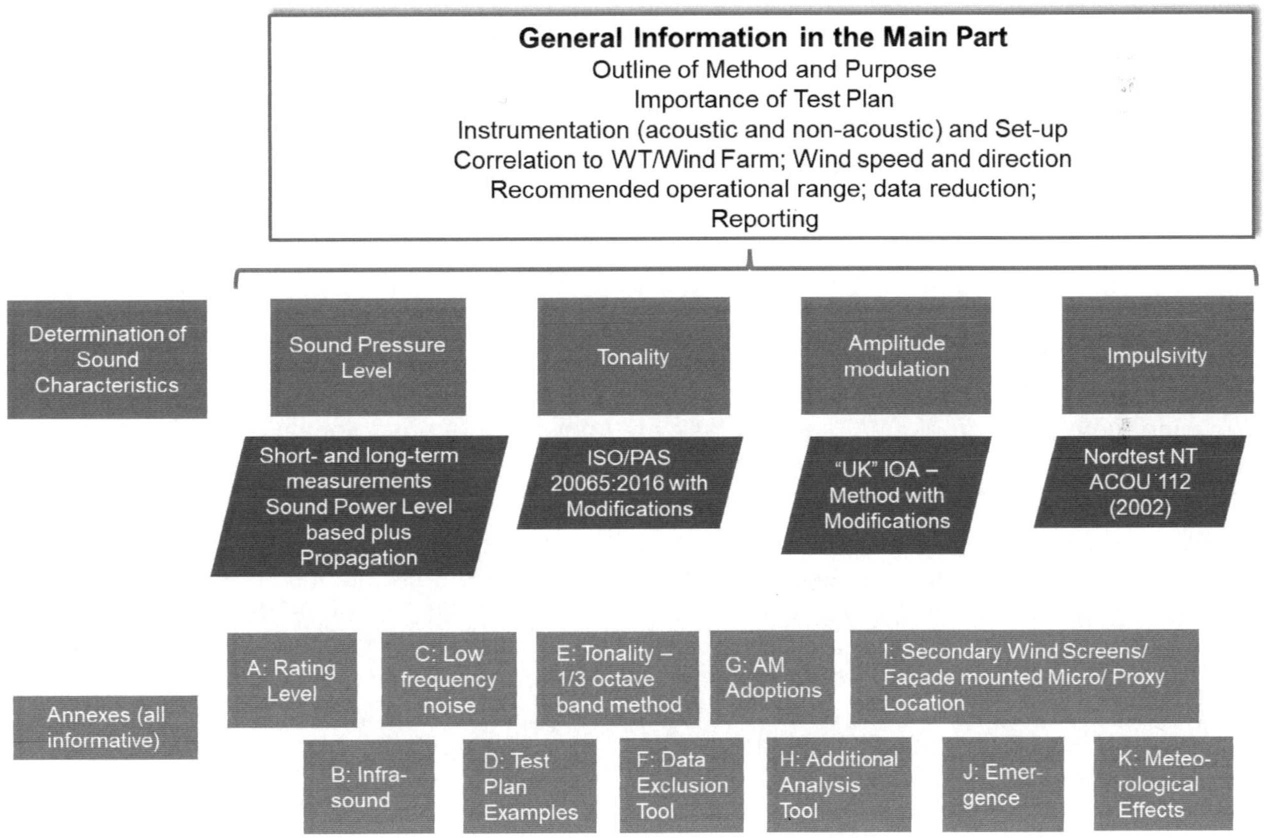

Figure 1: Overview of IEC/TS61400-11-2 structure

The list of contents is provided in Appendix 1 of this paper and will give the reader an idea of the structure and sections discussed in the TS. This also shows the variety of topics that were discussed, agreed upon and written about.

3.1 Defining the Scope of the Test – The Test Plan

The purpose of the TS is to give guidance for the measurements to secure robust results. Before every measurement, the scope of the test should be clear and preferably agreed with the appropriate authorities or involved parties. Reasons for testing at receptor position can include compliance measurements (compare measured data with noise limits given for the wind farm) or the investigation of certain complaints, amongst others. Complaints can for example include that the wind farm is generally perceived as too loud (sound pressure level testing), that a whining or humming can be heard (tonality assessment) or that the wind farm sound has been described as having a thumping character (AM investigation).

Setting up a test plan is the key to a successful measurement campaign as it should address the measurement conditions and the environment in great detail before the measurements are initiated. The following topics are a guide on the decisions to be made in advance of the actual measurement period:

- Purpose of the measurements: Sound pressure level, AM, Tonality, Low Frequency Noise
- Quality of the measurement site: Ambient noise, representative measurement conditions, other noise sources in the vicinity
- Receptor locations and microphone setup
- Instrumentation requirements for overall sound pressure levels, 1/3-octave band, FFT
- Wind speed determination methods (at the wind farm, at the receptor locations, what is the wind speed reference in the planning conditions etc.)
- Other atmospheric conditions with consequences for the sound emission, sound characteristics and propagation

An example for a test plan from the TS is given here:

Situation: Compliance of the wind farm noise contribution with a fixed sound pressure level limit shall be proven, as well as non-occurrence of tonal audibility noise above a certain threshold, while masking of tonal noise by ambient noise created near the receptor locations (e.g. wind rustling in trees around the house) shall be acceptable.
The wind turbines that make up the wind farm are known to have a potential risk of tonal noise emission around 8 m/s at hub height, mostly occurring in cross-wind conditions. The maximum sound power levels of the aforesaid wind turbine type are in the range of 6–12 m/s at hub height.

Proposed measurement setup: Measurements suitable for the determination of sound pressure levels shall be executed at a proxy location to the receptor under investigation as the signal-to-noise ratio (SNR) is not sufficient to measure directly at the receptor location. The proxy location is chosen to be on the same isophone as the receptor location (if feasible and if sufficient to reach desired SNR) and in the free field at least 100 m distance from any trees. If sufficient SNR cannot obtained at that location, the measurement should be moved closer to the wind turbines.
Alternatively: The microphone should be set up on the façade of the residential property to further increase SNR to a sufficient level.
This measurement shall be a long-term measurement to cover a wind speed range of at least 6-12 m/s (at hub height), filtered for down-wind conditions, and needs to cover the minimum amount of data points per wind speed bin under favourable noise propagation conditions (temperature inversion during night time). In order to distinguish between wind farm noise contribution and background noise levels, wind turbine on-/off-testing is required, where measuring with the wind farm being offline needs to cover at least xxx % of the overall testing time per wind speed bin (alternatively: reference to formerly measured background noise levels

or background noise levels determined by any other method). The percentage may be prescribed by or agreed with the Authorities/Regulator.

A second noise measurement shall be executed as a short-term measurement in order to investigate tonal audibility, covering 7-9 m/s (at HH) in cross-wind conditions. The microphone shall be located at the receptor location (exact position tbd, e.g. representative for typical / worst-case listener positions). This measurement shall be executed during the fall/winter period, while the trees have no leaves, to represent worst case (i.e. least) masking conditions for tonal noise from the wind farm. No on-/off-testing is required except if a potential tone shall be proven to be not originating from the wind farm.

3.2 Generally applicable Topics in the Main Part of the TS

Several chapters in the TS apply more or less to all of the sound characteristic assessment methods. Where further refinement is required to suit a particular task, this is described in the individual measurement chapters. The TS also has mandatory chapters that can be found in all IEC standards/technical specifications.

The technical part starts with the outline of the method, followed by the requirements for instrumentation, for both acoustic and non-acoustic measurements, including calibration and time synchronisation. As described before, test planning plays an important part in the assessment of sound characteristics at receptor positions, therefore this topic has its own chapter. The TS gives furthermore minimum requirements for the setup of the acoustic measurements at the measurement location, a proxy location and the mounting of the microphone. Since wind turbine operation depends on wind conditions and sound emission are correlated with wind speed, the knowledge of such wind data is required for the assessment. A chapter is dedicated to the different methods of wind speed and direction determination, wind speed binning for turbine on-periods and binning wind speed by background noise (with wind speed measurements near the measurement location). And finally the last chapter of the TS gives a descriptions of the information to be included in a measurement report.

3.3 Sound Pressure Level

There is not just one single method to be used in the determination of the sound pressure level at receptor position but several methods that can be chosen depending on the task at hand. Some of the methods are integrated in the legislation of the countries where the PT11-2 members come from. This chapter describes long-term, short-term and modelling based approaches to determine the sound pressure level at receptor position. It is anticipated that this chapter could be used either by authorities in outlining appropriate methods for a given jurisdiction, or by measurement institutes in interpretation of local regulations and requirements, or in research efforts where quantifying the sound pressure level in the far field is desired. Only one of the detailed methods should be chosen for the assessment. The preferred method to determine the sound pressure level at a far field location is with long term measurements.

The main acoustical quantities to be measured are the equivalent sound pressure level, L_{Aeq}, or statistical levels, L_{Axx}. This data is obtained by determining the appropriate quantities in short measurement intervals such as between 10 seconds and 10 minutes to be able to determine measurement validity, and subsequently aggregated into longer periods to correspond with the intended analysis. L_{Aeq} intervals should be chosen to be shorter intervals, such as 1 minute, but not shorter than 10 s, in order to isolate contaminating transient events within an interval that can then be discarded. Statistical indices such as L_{Axx} sound levels are typically analysed in 10-minute intervals with a recommended maximum interval length of 10 minutes given potential variability in wind speeds. Furthermore, wind data should be collected at or near the acoustic receptor location and meteorological data at a representative location for the measurement area (of each receptor location). The wind speed and wind direction data intervals should preferably match the intervals of the A-weighted sound pressure levels.

Three measurement methods are described that vary in detail and duration of the operation cycling (short-term and long-term on and off periods). Data gathered in this fashion can be binned by wind speed and wind direction for a binned analysis methodology and to determine the wind farm acoustic distribution by background noise correction (logarithmically).

Where legislation requires or where a small signal-to-noise ratio doesn't allow for a meaningful result from measurements (weak signal and/or contamination by ambient noise levels) the sound pressure level existing at a specific point far from the wind turbines can be estimated using a combination of near-field measurements for determination of the apparent sound power level and sound model propagation. In such a scenario, the apparent sound power level of the wind turbines could be determined using the methods outlined in IEC 61400-11 and used as inputs to a noise model. A recommendation for parameters used in noise modelling with ISO 9613-2 and Nord2000 is given as well.

3.4 Tonality

The PT11-2 had selected the ISO/PAS 20065 method as the appropriate tonality analysis method for the assessment of sound characteristics at receptor position. It became apparent that the method required adaptation to the specific conditions and requirements of wind turbine sound since this method was primarily used, and presumably intended, to quantify the tonality of individual samples or short-term data sets to confirm the subjective assessment of the acoustic expert at the receptor position. It is not known to be used as either a primary assessment tool or on long-term and thus larger data sets in a more automated fashion. Tones from wind turbines and masking noise can vary with wind speed so that a range of conditions needs to be assessed and correlated to the tone assessment as well.

In September 2020 a 'Tonality Expert Group' was formed in an attempt to allow a wider group of experts to dedicate time specifically to this aspect of the TS and to advise the PT11-2 on this matter. Further information on this expert group and their findings can be found in Woodward's paper *An Implementation of ISO/PAS 20065:2016 for the Analysis of Wind Turbine Sound at Receptor Distances* also being presented at this conference.

3.5 Amplitude Modulation

One sound characteristic of wind turbines, that is related to the rotational speed of the rotor, can be described as a regular pattern of a rising and falling amplitude of the sound emission and is referred to as amplitude modulation. Quantifying the amplitude modulation allows insight into the overall character of the sound and allows an estimation of the extent to which the noise impact might be greater than expected for steady noise with similar averaged broadband noise level.

The method detailed within the AM section is an implementation of the UK Institute of Acoustics Amplitude Modulation Working Group (IOAAMWG) Final Report *A Method for Rating Amplitude Modulation in Wind Turbine Noise*. An expert group has been working on the content of this section, suggesting several improvements and a more concise wording than in the original report. The use of the 'IOA method' has been approved by the IOA prior to the work on the TS section. The primary changes from the method detailed within the Final Report relate to more specific output requirements, and the use of worst case results in each bin. The method provides the average modulation depths for bins of 1 m/s wind speed and 30° wide wind direction sectors. It is applicable for upwind three-bladed wind turbines with a rotor speed of 6 to 32 rpm. The measurement shall be carried out in free-field conditions externally. An Annex is provided with suggestions to adapt the method to wind turbines not meeting the aforementioned criteria.

3.6 Impulsivity

Even though wind turbine sound is not typically considered to be impulsive, PT11-2 decided to give guidance on how to assess impulsivity in case there are impulsive sound sources, most commonly mechanical sources. Examples are banging noise like loose cables, yaw motors, loose glue in the blades etc. However, audible impulses must not be confused with aerodynamic blade swish and thumping sounds which are better described by the 'Amplitude Modulation' method outlined in section 13 of the TS. An impulse is defined as a sudden onset of sound – having a gradient that exceeds 10 dB/s for the total A-weighted noise. The prominence of the impulse also depends on the corresponding increase in the noise.

PT11-2 recommends the use of the Nordtest method *Acoustics: Prominence of impulsive sounds and for adjustment of* L_{Aeq} (NT ACOU 112:2002-05) which is also the basis for the future ISO/PAS 1996-3 *Acoustics – Description, measurement and assessment of environmental noise – Part 3: Objective method for the measurement of prominence of impulsive sounds and for adjustment of* L_{Aeq}. The ISO/PAS 1996-3 is currently in its Draft stage and circulated for voting to the member committees of ISO/TC/SC 1 'Noise'. Provided a successful outcome of the voting, the ISO/PAS 1996-3 is expected to be published in the second half of 2021. Depending on timing, this may mean, that PT11-2 can refer to the ISO/PAS in its final version of the TS instead of the Nordtest document.

The TS deviates from the Nordtest method in the following way: a shorter recording interval of 10 minutes instead of 30 minutes, verification of the signal coming from a wind turbine by audio recording for unattended measurements and a wind direction sector for analysis of ±45° from the wind farm. All recorded 10 minute periods are analysed for impulsivity without averaging.

3.7 Annexes

The PT11-2 members have a wealth of information and experience to share, but not all of it was considered of a normative nature and are therefore provided in one of the eleven Annexes. Those give additional information to existing chapters in the TS or discusses new topics that the group felt should be commented on but did not consider to be an appropriate method for the TS.

Annex A provides suggestions to adjust the equivalent continuous sound pressure level for different sound characteristics, which, for wind turbines, are mainly tones, amplitude modulation and impulses. It should be noted that only one type of level adjustment applies at a time and that the level adjustments are typically defined in national regulations. Annex B gives references to recent infrasound research which did not find evidence that infrasound levels from wind turbines are loud enough to have a measurable effect on humans. Annex C suggests a method to determine low frequency noise evaluation by measuring the respective 1/3 octave band levels near the wind turbine in accordance to IEC 61400-11, carry out transmission loss calculations with the prediction method Nord2000 to derive the external levels and finally apply façade insulation levels for indoor noise level assessment. Examples for creating a test plan appropriate to the cases to be investigated are given in Annex D. Annex E provides some reference to existing standards that describe the 1/3 octave band method for tonality determination, however, it is generally not recommended to use this method for the determination of tonality from wind turbines as this method will not give detailed and very accurate results. For the less experience reader, Annex F provides examples of different types of data exclusion tools and methods for time based, meteorology based or irregular sound occurrences. For cases where the method described in the main part of the TS does not characterise AM well, some adaptions are given in Annex G which are most commonly appropriate for smaller scale wind turbines. Some additional analysis tools for special events are described in Annex H and special situations like secondary wind screens, façade mounted microphones and proxy locations are included in Annex I. Annex J gives a translation and description of the French term 'emergence' in wind turbine noise evaluation. Since the propagation of sound is influenced by the micrometeorology at the wind turbine(s) site and at the receptor position, Annex K gives an explanation on the effects and measurements to describes those aspects. And lastly, areas of further consideration for the implementation of ISO/PAS 20065:2016 for wind turbine sound are given in Appendix L, which the expert group identified but could not further investigate in the given time frame.

References

Broneske, S and Søndergaard, B (2019) *Measurement of wind turbine noise characteristics in receptor position – A new IEC Technical Specification*, 8th International Conference on Wind Turbines Noise (Lisbon, Portugal), 2019

Institute of Acoustics (2016) *AMWG Final Report: A Method for Rating Amplitude Modulation in Wind Turbine Noise.* IOA, Milton Keynes/United Kingdom of GB and NI.

ISO 1996:2017 *Acoustics - Description, measurement and assessment of environmental noise -- Part 2: Determination of environmental noise levels.* ISO, Geneva/Switzerland.

NT ACOU112 Acoustics: Prominence of impulsive sounds and for adjustment of L_{AEQ}

ISO/PAS 1996-3 *Acoustics - Description, measurement and assessment of environmental noise — Part 3: Objective method for the measurement of prominence of impulsive sounds and for adjustment of L_{Aeq}.* To be published within the next year.

ISO 9613-2:1996 *Acoustics - Attenuation of sound during propagation outdoors - Part 2: General method of calculation.* ISO, Geneva/Switzerland.

ISO/PAS 20065:2016 *Acoustics – Objective method for assessing the audibility of tones in noise – Engineering method.* ISO, Geneva/Switzerland.

Woodward, R (2021) *An Implementation of ISO/PAS 20065:2016 for the Analysis of Wind Turbine Sound at Receptor Distances*, 9th International Conference on Wind Turbines Noise (remote from Europe), 2021

Appendix 1: List of Contents of IEC/TS61400-11-2 as of March 2021

CONTENTS

9th International Conference
on
Wind Turbine Noise
Remote from Europe – 18th to 21st May 2021

The (psychoacoustic) Basics of Tonality Perception

Markus Buße, RWE Renewables GmbH: markus.busse@rwe.com

Summary

This paper gives an introduction to the psychoacoustics of tonality and provides a basis for subsequent papers on *Further investigations into a suitable implementation of ISO/PAS 20065:2016 for the analysis of wind turbine sound at receptor distances* (Woodward) and *An implementation of ISO/PAS 20065:2016 for the analysis of wind turbine sound at receptor distances* (Woodward).

As a standardized term, tonality describes the prominence of a tonal component within a noise background. The cochlea within the inner ear of the peripheral auditory system is responsible for a frequency place transformation of any stimulus and can be seen as a less than perfect Fourier analyser of the auditory system. The ability to detect a tone within a complex noise background is determined by frequency selectivity. Frequency selectivity can be demonstrated with masking experiments and shows that the detection threshold of a sinusoidal tone within white noise is dependent on the noise energy within a critical band centred around that tone and the energy of the tone itself. The magnitude of the perceived tonality is also dependent on various aspects of pitch perception elicited by different characteristics of the tonal component.

1. Introduction

The prominence of discrete frequency components or narrow frequency bands within a noise background can be described as tonality. While wind turbine noise is known to be dominated by aero-acoustic broadband noise from the blades' movement through the air, tonal components, for example from mechanical sources, might also occur. According to the German standard DIN 45681:2005-03 (2005) with the title *Acoustics - the detection of tonal components of noise and determination of tone adjustment for the assessment of noise immissions*, noise immissions are generally more irritating, if they contain tonal parts. Thus, it is important to quantify the subjective perception of tonality for a comprehensive noise immission assessment. For this reason, several industry standards such as the DIN 45681, the almost identical ISO/PAS 20065:2016 (2016), the ISO 1996-2 (2007) or the NZS6808:2010 (2010) are based on psychoacoustic findings to a varying degree. The standards usually follow a practical approach but the perception that underlies these metrics is often oversimplified (Vormann 2011). As Vormann explains, the term *tonality* is not part of common parlance. It is a term for specialists, generally describing a tonal noise component in a noise background. Therefore, it is difficult to measure and quantify the perception of tonality by individual untrained listeners.

This paper is intended to give an introduction to the psychoacoustics of tonality and to be a brief overview of the relevant basic principles and characteristics of the human auditory system so that the papers *Further investigations into a suitable implementation of ISO/PAS 20065:2016 for the analysis of wind turbine sound at receptor distances* (Woodward)*, An implementation of ISO/PAS 20065:2016 for the analysis of wind turbine sound at receptor distances* (Woodward) and this session in general can be understood in this context.

2. Peripheral Auditory System

The human peripheral auditory system consists of the outer ear with the pinna, the ear canal and the ear drum at the transition to the middle ear. From there, the acoustic vibrations are transferred into mechanical vibrations and progress via the bones of the ossicle chain, consisting of the malleus, the incus and the stapes (also known as hammer, anvil and stirrup), to the inner ear with the cochlea, where the incoming signal is transferred into electrical action potentials for further neuronal processing. The path is depicted in Figure 1.

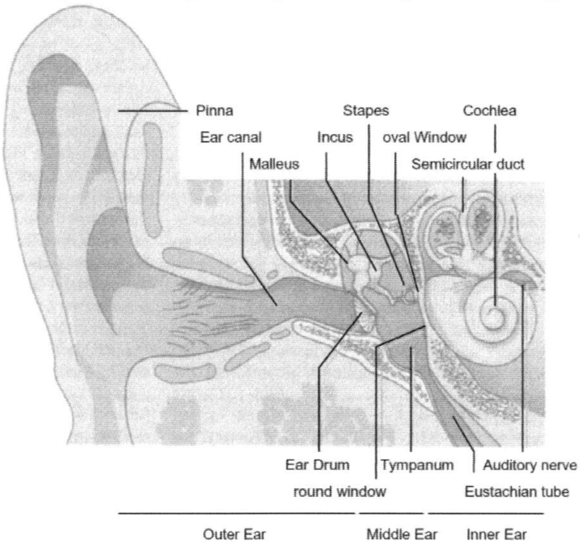

Figure 1, adapted from Kollmeier (2008): Overview of the anatomy of the outer ear, middle ear and inner ear.

Any incoming acoustic signal is subject to the nonlinear characteristics of the peripheral auditory system. The ear canal acts like a pipe of around 2 cm length and is responsible for an increased sensitivity in the region of 4 kHz. The middle ear has an impedance matching function from the air-filled outer and middle ear with small forces and large displacements to the fluid-filled inner ear with large force and small displacements. This is mechanically achieved with a lever system by the ossicle chain and the difference in effective area of the ear drum and the oval window at the transition to the cochlea (Moore 2004). The impedance is matched almost perfectly in the mid-frequency range around 1 kHz but varies for other frequency regions (Fastl & Zwicker 2007).

In order to understand the auditory perception of tonality, it is important to understand the basic functioning of the cochlea. However, it is beyond the scope of this paper to provide a description of all sections and physiological processes. The cochlea is shaped like the spiral snail shell and is located in the extremely hard temporal bone of the skull (Fastl & Zwicker 2007). Inside the cochlea, there are three fluid-filled channels from the base at the oval window to the apex at the other end. Along the length, the channels are divided by two membranes, the Reissner's membrane and the basilar membrane. The latter is of particular interest for the perceptual response to different sounds. With the oscillation at the oval window, induced by the stapes, a pressure difference inside the cochlea excites the basilar membrane and a travelling wave moves from the base to the apex causing a stimulus-dependent vertical displacement of the basilar membrane. Owing to differences in mechanical properties, like mass and stiffness along the basilar membrane, different locations along its length are tuned to different frequencies, where the magnitude of the displacement reaches its maximum for a characteristic frequency. At the base, the basilar membrane is relatively stiff and narrow, while wider and less stiff at the apex. For this reason, high frequencies cause a maximum displacement near the base and low frequencies closer to the apex (Moore 2004). This mapping of characteristic frequencies is called the *place-frequency transformation* or *place principle* and the basilar membrane can be seen as a Fourier analyser with a less than perfect accuracy (Fastl & Zwicker 2007 and Moore 2004). The frequency place transformation is visualised in the schematic representation of this principle in figure 2.

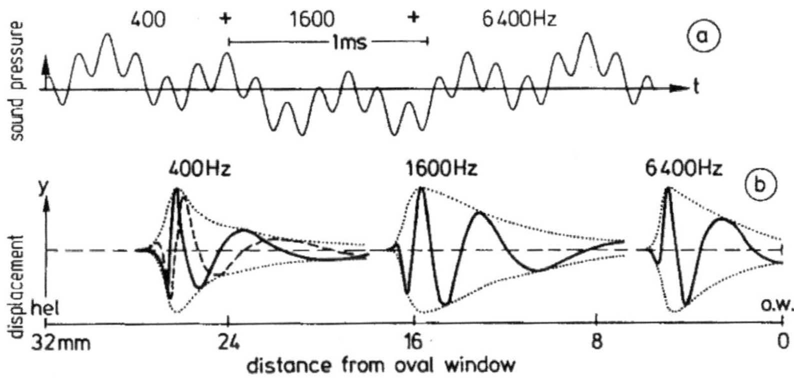

Figure 2, adapted from Fastl & Zwicker (2007): Schematic drawing of the transformation of frequency into place along the Basilar membrane. In (a) three simultaneously presented tones of different frequencies expressed as compound time function produce travelling waves (b), that reach their maximum at three different places corresponding to the characteristic frequencies.

3. Frequency Selectivity and Masking

For the perception of tonality, the ability of the auditory system to detect single sinusoidal components from a complex stimulus plays an important role. This ability is referred to as *frequency selectivity* (Moore 1995). Frequency selectivity can be demonstrated and measured by masking, where masking is either the *process* by which the threshold of audibility of the frequency increases relative to the threshold in silence owing to the presence of another masking stimulus, or the *metric* of the difference between the raised audibility threshold and the corresponding threshold in silence at that frequency. Generally speaking, masking limits the frequency selectivity of the auditory system. Masking phenomena are often classified by their timing of the test tone and the masking noise: *simultaneous masking* occurs, when both signals are presented at the same time; *post-masking* is observed, when the test tone to be detected is presented after the masking noise; and *pre-masking* describes the process, when the masking noise is presented before the test tone. For the following sections, only simultaneous masking will be considered.

3.1 Power Spectrum Model and Critical Bands

In 1940 the US-American physicist Harvey Fletcher conducted an experiment that contributed to the understanding of the auditory system's frequency selectivity (Fletcher 1940). Fletcher used a detection experiment, where a test tone at a fixed frequency had to be detected within a noise of variable bandwidth centred around the tone. The noise had a constant power density, so that the total energy of the noise increased proportionally with the bandwidth of the noise. The level of the test tone was varied, and the hearing threshold of the test tone was measured depending on the bandwidth of the masking noise. The results of the experiment showed that the threshold increased with an increasing bandwidth of the masking noise until a certain bandwidth was reached after which the threshold stayed approximately constant, even with further increasing bandwidth of the noise. The experiment has been repeated multiple times since then, for example by Schooneveldt and Moore (1989) as shown in Figure 3.

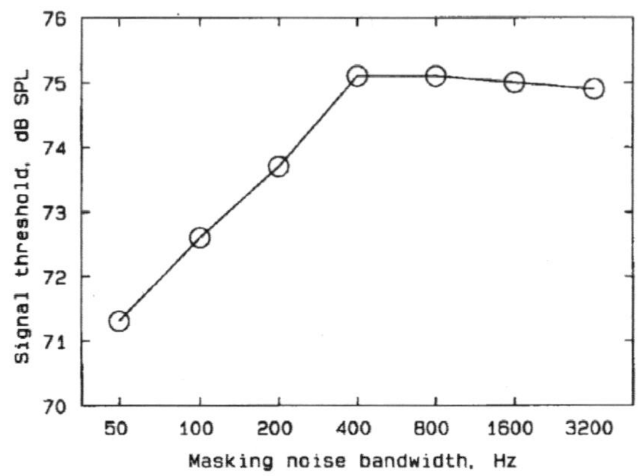

FIGURE 3.1 The threshold of a 2000-Hz sinusoidal signal plotted as a function of the bandwidth of a noise masker centered at 2000 Hz. Notice that the threshold of the signal at first increases with increasing masker bandwidth and then remains constant. From Schooneveldt and Moore (1989), by permission of the authors.

Figure 3, adapted from Moore (2004): Threshold of a 2 kHz sinusoidal signal plotted as a function of the bandwidth of a masking noise centred around that tone with variable bandwidth.

Fletcher postulated that the basilar membrane works as a bank of band-pass filters, where each filter has a centre frequency and a certain bandwidth called *critical bandwidth,* according to the location on the basilar membrane (Fletcher 1940). According to this, the auditory system uses a filter with a centre frequency near the test tone. The filter removes some energy from the masking noise but lets the test tone pass unhindered. The detection threshold of the tone is thus determined by the energy within this filter and not by the total energy of the masking noise. The filters are called critical filter or auditory filter. These findings are summarized under the term *Power Spectrum Model* (Moore 2004).

Fletcher was able to define his theory by assuming the critical band filters have an ideal rectangular transfer function. The actual level-dependent filter form can be determined with various methods that are beyond the scope of this work. A typical auditory filter has a rounded top and sloping edges as depicted in figure 4, determined with the so-called *Notched Noise Method* (Patterson 1976).

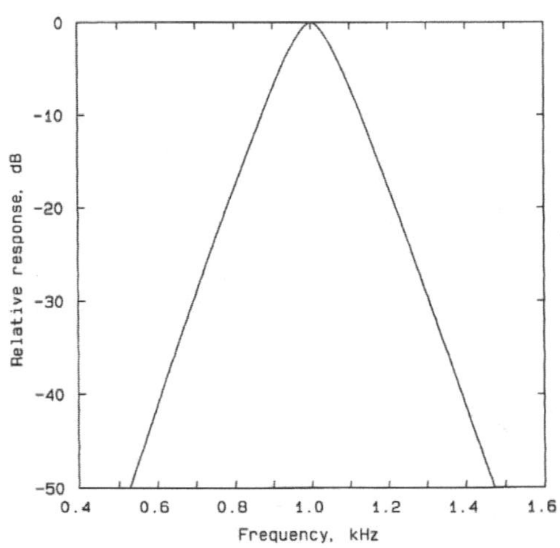

Figure 4, adapted from Moore (2004): A typical filter shape determined using Patterson's method. The filter is centred at 1 kHz. The relative response of the filter (in dB) is plotted as a function of frequency.

It is not possible to specify the exact shape of the auditory filters with just one parameter, but a useful parameter of filters is the 3 dB bandwidth. For the filters derived with Patterson's method, the 3 dB bandwidth is typically between 10 % and 15 % of the centre frequency (Moore 2004). Moore also proposes an alternative measure as a description for an equivalent rectangular bandwidth (ERB). This is a description for a perfect rectangular filter as assumed by Fletcher, where the transmission in its passband is equal to the maximum transmission of the specified filter and transmits the same power of white noise as the specified filter. The ERB of the auditory filter has a bandwidth of 11 % to 17 % of the centre frequency and can be derived as a function of frequency F in kHz by:

$$\mathrm{ERB} = 24.7(4.37F + 1).$$

According to Fastl and Zwicker (2007) the bandwidth of the auditory filters for centre frequencies up to 500 Hz is 100 Hz, and 20% of the centre frequency for higher frequencies. An analytical expression of this relationship as a function of frequency f in kHz, which is also used in the standards DIN 45681 or ISO/PAS 20065:2016, is

$$\Delta f_G = 25 + 27[1 + 1.4(f)^2]^{0.69}.$$

3.2 Critical Ratio

Fletcher explained the data from his experiment by assuming a constant power ratio between the sinusoidal tone at the level of the hearing threshold (P_S^{thres}) and the noise within the corresponding critical filter (P_N), independent of the centre frequency and the bandwidth of the noise masker:

$$\frac{P_S^{thres}}{P_N} = C.$$

In case the noise has a constant power density N_0 within its bandwidth W and the power of the tone at the threshold is known, the constant C can be derived by:

$$\frac{P_S^{thres}}{N_0 \cdot W} = C.$$

If C is known, the bandwidth W could be determined to:

$$W = \frac{1}{C}\frac{P_S^{thres}}{N_0}.$$

Fletcher estimated the constant to be $C = 1$ and called the expression *critical ratio* which can be interpreted as a bandwidth.

$$\text{critical ratio} = \frac{P_S^{thres}}{N_0}$$

Thus, according to Fletcher, a tone within a masking noise can just be heard, if the power of the tone and the power of the noise within the critical band is equal. However, in later experiments by B. Scharf, C was determined to be $C = 0.4$ and also to be frequency dependent (Scharf 1970, cited in Moore 2004). According to this, a tone within a masking noise can just be detected, if the power of the tone is roughly 4 dB lower than the power of the masking noise within the corresponding auditory filter.

3.3 Masking Index

As Scharf has determined, the ratio of the tone signal's power at the threshold and the noise power within a critical band is not the same for all frequencies, as Zwicker postulated. This means, the frequency selectivity and the ability of the auditory system to detect tones in masking noise is not the same for all frequencies. Fastl and Zwicker (2007) clarify this relationship with data derived from an experiment, where instead of white noise, a so-called *uniform excitation noise* was used. This noise is defined to give a uniform excitation over the complete basilar membrane. Thus, each critical band contains the same energy for each centre frequency. Figure 5 depicts the results of the experiment. It shows the thresholds where the tone was just audible are approximately 2 dB lower than the noise level of the critical band for low frequencies but 6 dB lower for high frequencies with a slope in between. This is also represented by the so-called *Masking Index* in the upper part of the figure, which represents the difference between the tone level and the critical band level.

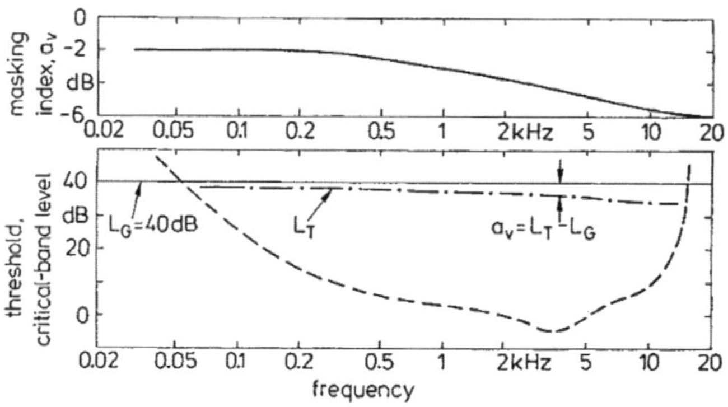

Figure 5, adapted from Fastl & Zwicker (2007): The upper part shows the Masking Index as the difference between the Tone level at threshold and the critical band level as a function of frequency. The lower part shows the constant level of the uniform excitation noise L_G (solid line) and the tone level at threshold L_T (dashed-dotted line). Also depicted is the hearing threshold in quiet (dashed line).

Fastl and Zwicker explain the reduced frequency selectivity of the hearing system at lower frequencies with the width of the critical bands. According to them, for frequencies < 500 Hz, the bandwidth is 100 Hz, whereas the bandwidth is much larger for frequencies > 500 Hz. This means the fluctuations of the noise at high frequencies are not audible and the Masking Index is -6 dB. For low frequencies these fluctuations reduce the frequency selectivity of the auditory system with the result of a higher Masking Index of -2 dB. This is taken into account by the standard DIN 45681, for example.

4. Pitch

The perception of tonality is not just dependent on the frequency selectivity of the auditory system but also related to the subjective perception of pitch. In fact, a prerequisite for the assessment of tonality is the existence of a component that has a tonal character with an associated perception of pitch height and pitch strength (Hansen, Verhey & Weber 2011). Pitch is a relatively controversial topic because it is a perceptual, rather than a physical and easily measurable metric (Plack and Oxenham 2005). Definitions of pitch usually are musical references and associations between pitch and the musical scale but also operational definitions. For example: "A sound can be said to have a certain pitch if it can be reliably matched by adjusting the frequency of a pure tone of arbitrary amplitude" (Hartmann 1997, cited in Plack and Oxenham 2005). In other words, a pitch can be ordered on a scale from high to low (Fastl & Zwicker 2007). This is described by pitch height.
Not only pure tones induce a sensation of pitch, but also narrow-band noises or high- and low-pass noises. However, the perceived strength of pitch is different. Fastl and Zwicker (2007) define pitch strength as a sensation that can be labelled from faint to strong and ordered on a scale. Fastl & Zwicker assessed different signals for quantitative pitch strength in a magnitude estimation experiment and ordered them on a scale, where a pure tone was set to a relative pitch strength of 100 %. The pitch strength of narrow-band noise decreases with increasing bandwidth. They concluded that sounds with line spectra generally elicit relatively large pitch strength and sounds with a flat, continuous spectrum elicit only small values of relative pitch strength. The DIN 45681 takes this into account by excluding tonal components which have a relative pitch strength of less than 70 % (Hansen 2010).

In general, pitch strength is a complex topic and depends on various attributes. Fastl & Zwicker (2007) have shown in further experiments that the pitch strength of a pure tone increases with its duration up to around 300 ms. The strength also increases with an increasing sound pressure level. Furthermore, the relative pitch strength is frequency dependent with the largest values at mid frequencies as depicted in figure 6.

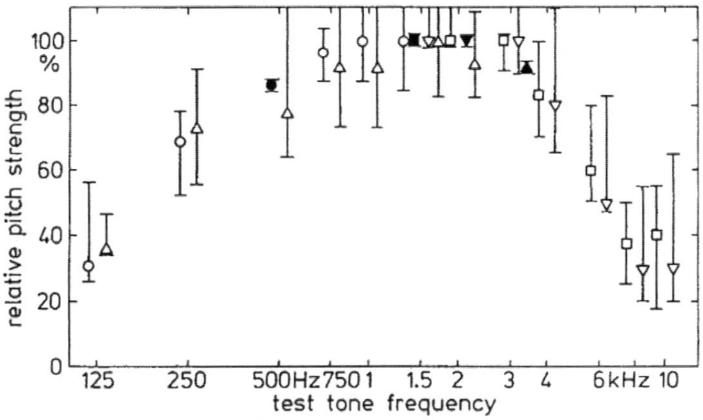

Figure 6, adapted from Fastl & Zwicker (2007): Relative pitch strength of pure tones at 80 dB SPL with 500 ms duration as a function of test tone frequency.

5. Magnitude of Tonality

In the context of annoyance and immission assessment, the perceived magnitude of a tonal components as an acoustical foreground with an acoustical background of masking noise is of great interest.

There are several factors that have an influence on the magnitude. Hansen (2010) describes this as a two-fold tone-noise dichotomy. On the one hand, the tonal part and the noise are observed separately as two auditory objects and compared in terms of the magnitude. On the other hand, the noise is observed in terms of the quality, where an auditory object can be perceived as noisier or more tonal with a gradient in-between. Neither concept can be observed in isolation for the determination of the magnitude of tonality. In order for a tone to be heard in a complex noise signal, a prerequisite is a dominating tonal part over a noise as explained by the frequency selectivity of the auditory system and the signal to noise ratio in the previous sections. If this is given, the magnitude of perceived tonality could further be influenced by the perceived relative pitch strength of the tonal part. As introduced above, the frequency of the tonal part could be taken into account. A tonal part at a relatively low or high frequency but at the same level would elicit a smaller relative pitch strength than at mid frequencies. In standards such as the DIN 45681, this is currently not implemented.

The influence of several tone parameters, such as the frequency and the bandwidth of the tonal part on the tonality derived by the DIN 45681(2005) compared to the perceived tonality from listening tests has been investigated by Vormann (2011). He found a significant effect on tonality by the frequency of sinusoidal tones of constant level in uniform excitation noise. The perceived tonality was overestimated by the DIN 45681 below frequencies of approximately 500 Hz while the calculated tonality matched the perceived tonality from the listening tests above 500 Hz relatively precisely.

The results of the influence of the bandwidth of small band noise on tonality showed a decreasing perceived tonality for an increasing noise bandwidth, as expected. This is taken into consideration by the DIN 45681, while Vormann concluded, that the tonal prominence of at least 70 %, as used in the standard, is too strict and small band noise with a smaller relative pitch strength should not be underestimated.

6. Future Work

Current standards and methods for an objective assessment of tonality are generally a compromise between complexity of the algorithms and the accuracy of describing the subjective perception. It is obvious that this means the output of the standards is biased in some aspects. Most of the standards such as the DIN 45681 and ISO/PAS 20065 are based on basic psychoacoustic findings that were established by the time the standards were written. However, as with a lot of topics in psychoacoustics, there is still ongoing research about the perception of tonality in general and related topics, such as pitch perception, in particular. In research about perceived annoyance, it was established that the loudness seems to be a more suitable metric, rather than the spectral energy from an FFT analysis. These findings, and for example the frequency dependency of pitch strength, need to be implemented in the algorithms and investigated with studies and listening tests.

In addition, standards are established to work with a variety of signals for environmental noise assessment and noise quality in product development alike. Wind turbine noise includes tones of lower frequency, amongst others. However, as Vormann has shown in his dissertation, the DIN 45681 is not as suitable for low frequency tones as it is for higher frequency tones above approximately 500 Hz. Future work should take this into account.

In the context of noise-related environmental health aspects, and as standards and methods are part of penalty schemes in regulations, it is especially important to adapt the current methods for the assessment of tones in wind turbine noise and improve the objective description of subjective tonality perception.

7. Conclusions

The psychoacoustics of tonality perception is a complex topic that depends on various factors, as this brief summary of some basic concepts has shown. It begins at the peripheral auditory system with the physiology of the cochlea, including the mechanisms of frequency selectivity. The findings of Fletcher and the determination of the critical filters as part of a bandpass filter bank on the cochlea was an important step to the understanding of how tonality is perceived and how it could be described quantitatively by the signal to noise ratio of the tonal component and the noise within the corresponding critical filter. Another important topic is the relative pitch strength of the tonal part within the noise to be assessed. It shows that there are additional psychoacoustic mechanisms taking place at various stages of the auditory system. As pitch in itself is still a topic of investigation, the interaction between pitch and the perception of tonality still needs more research.

The standards that are currently available as an objective metric, such as the DIN 45681, rely mostly on the signal to noise ratio by assessing the energy of the spectral components of an FFT spectrum but take the frequency-dependent frequency selectivity and some aspects of pitch strength into account. While this is a practical approach that works sufficiently, there seem to be some frequency-dependent effects that are not represented, yet. A promising metric for the understanding of several psychoacoustic effects such as the noise-induced annoyance could be the loudness (Fastl & Zwicker 2007). In this context, Hansen, Verhey and Weber (2011) suggested to assess the partial loudness of the tonal part of a noise for an assessment of tonality.

A description of current work for an adaption of the ISO/PAS 20065 for wind farm tonality is given in Woodwards' papers *Further investigations into a suitable implementation of ISO/PAS 20065:2016 for the analysis of wind turbine sound at receptor distances* and *An implementation of ISO/PAS 20065:2016 for the analysis of wind turbine sound at receptor distances*.

References

DIN 45681:2005-03 (2005) *Acoustics - Detection of tonal components of noise and determination of tone adjustment for the assessment of noise immissions,* Beuth Verlag, Berlin, Germany. German industry standard.

Fastl, H. and Zwicker. E. (2007). *Psychoacoustics: facts and models.* 3rd ed. Springer series in information sciences 22. Berlin; New York: Springer.

Fletcher, H. (1940). *Auditory Patterns* In: *Reviews of modern physics* 12, pp. 47–65.

Hansen, H., Verhey J. L. and Weber, R. (2011) *The Magnitude of Tonal Content. A Review* Acta Acustica United with Acustica, 97(3), 355-363

Hansen, H. (2010) *Tone-Noise Dichotomy: Investigating Tonal Content Magnitude & Pitch Strength.* University of Oldenburg, Dissertation

ISO/PAS:20065: 2016 (2016) *Acoustics – Objective method for assessing the audibility of tones in noise – Engineering method,* BSI Standards Limited, London.

Kollmeier, B (2008) *Anatomy, Physiology and Function of the Auditory System* In: *Handbook of Signal Processing in Acoustics.* Ed. by Havelock, D., Kuwano, S., and Vorländer, M. New York, eSpringer.

Moore, B. C. (2004) *An introduction to the Psychology of Hearing.* 5th ed. London: Academic Press

Moore, B. C (1995) *Hearing – Handbook of perception and cognition.* 2nd ed. San Diego: Academic Press.

Patterson, R. (1976) *Auditory filter shapes derived with noise stimuli.* J. Acoust. Soc. Am., 59:640-654.

Plack, C. J. and Oxenham, A. J. (2005) *Overview: The Present and Future of Pitch in Pitch: Neural Coding and Perception. Ed. by* Plack, C. J., Oxenham, A. J., Fay, R. R and Popper, A. N. Springer Handbook of Auditory Research, Springer New York.

Schooneveldt, G. P. and Moore, B. C. (1989) *Comodulation masking release (CMR) as a function of masker bandwidth, modulator bandwidth and signal duration* J. Acoust. Soc. Am. 85, 273-281.

Vormann, M. (2011) *Untersuchungen zu psychoakustischen Mess- und Berechnungsverfahren der Tonhaltigkeit.* University of Oldenburg, Dissertation.

INCEEUROPE

9th International Conference
on
Wind Turbine Noise
Remote from Europe – 18th to 21st May 2021

Comparison of Measured and Predicted Turbine Immission Noise Levels using Known Inputs to Model

Kohl Clark, B. Eng., Aercoustics Engineering Limited: KohlC@Aercoustics.com,
Payam Ashtiani, B.A.Sc., P.Eng, Aercoustics Engineering Limited:
PayamA@Aercoustics.com

Summary

In order to minimize the risk of adverse effect from a proposed wind facility, predictive modelling is often used as a pre-construction assessment method. In many jurisdictions around the world, ISO 9613-2 is used to predict the noise impact from industrial noise sources, including wind facilities.

ISO 9613-2 makes use of several modelling parameters which affect the predicted attenuation over distance. The ground factor, G, dictates the degree of attenuation provided by the interaction of the wave front and the ground between a source and a receiver. Values between 0 and 1, representing perfectly reflective and perfectly absorptive ground, respectively, have been applied in the context of industrial noise modelling. There remains a lack of consensus within the industry on the appropriate ground factor to be used during the modelling for the permitting process for a wind farm. Studies have been performed assessing the applicability of various ground factors by comparing the sound levels predicted with ISO 9613-2 and those measured on site.

This study seeks to add further context to the comparison of these levels by using the sound levels measured according to IEC 61400-11 for inputs as noise source power levels in the model and making predictions at 15 locations across seven wind farms. The cumulative impact from all relevant turbines has been assessed at each of the 15 locations using ground factors between 0.0 and 0.8. These predicted sound levels are then compared to average measured sound pressure levels from high-quality datasets collected at the same locations. These noise measurements are conducted such that the worst-case impact from the wind facility is assessed.

Comparison of the measured and predicted sound levels across all sites shows a trend indicating that lower values of G, between G=0.2 and G=0.4, may be most appropriate to accurately predict the sound levels on site, providing that the actual turbine emission levels and spectra closely match those of the specification. Further, in situations where a specific ground factor must be used, a statistical analysis on the dataset indicates that adjustment factors ranging from 0.3 dB for G=0 to 3.9 dB for G=0.7 may be used to better ensure that the measured immission sound level falls at or below the predicted sound level.

To further reduce the degree of uncertainty in these calculations steps will be taken to more accurately quantify the uncertainty inherent to the immission measurements.

1. Introduction

In order to minimize the risk of adverse effect from a proposed wind facility, predictive modelling is often used as a pre-construction assessment method. In Ontario, and elsewhere in the world, ISO 9613-2 [1] is used to predict the noise impact from industrial noise sources, including wind facilities.

Although it has been shown that more complex modelling methods such as Parametric Equation Modelling can make predictions with a higher accuracy, the ISO 9613-2 standard is still applied widely for the permitting of wind facilities. The applicability of the ISO 9613-2 standard for wind turbine noise propagation prediction has been the subject of much research. Many studies have compared sound levels measured from wind farms to predicted noise levels in modelling, and specifically to those employing ISO 9613-2 [2] [3] [4] [5].

When applying ISO 9613-2 in this context, there has been some variation in the literature regarding the most appropriate ground factor ("G") to ensure that the noise modelling does not under-predict the noise impact from the facility.

Previous research that was conducted in order to determine the most appropriate ground factor to use have largely been based on comparing the result of a measurement campaign with the predicted level from the model [2] [3] [4] [5] . However, a major shortcoming of these studies is the assumption that the turbines were emitting sound levels and spectra equal to that of the specifications used in the model. From experience, the sound emission output of the turbines have an inherent variation based on the turbine model, and may be subject to differences when compared to the manufacturer specified maximum sound emission. Consequently, any conclusions from those previous studies may have been biased by whether the turbines in the study were louder or quieter than their manufacturer specification used in the model at the time of measurement, as well as the associated differences in spectral levels.

As part of this study, far field measurement datasets have been identified where the signal-to-noise ratio is sufficiently high, and the sound emission levels have been measured in-situ at one or more turbine from the site. In some cases, this is the closest turbine to the measurement location. To assess the most appropriate ground factor, a comparison has been made between the noise levels predicted using measured sound power levels according to ISO 9613-2 and the noise levels measured during immission measurements at the same locations.

2. Background

The following section describes the nature of the acoustical immission measurements as well as the selection criteria for which sites were assessed in the study. Further, additional context is provided regarding the ISO 9613-2 modelling parameters. Finally, the methodology for the comparison of the measured and predicted sound levels is described.

2.1 Immission Measurements

Many of the datasets chosen for this study were from measurements carried out in Ontario. The measurements are typically conducted at a selection of the worst-affected points of reception around a wind farm or proxy locations in cases where access was restricted. Details of measurement requirements are outlined in the Provincial guidelines [6]. The primary requisites for a receptor being classified as worst-case are as follows:

- High predicted noise impact

- Situated downwind of the farm in the prevailing wind direction

The measurements target the worst-case noise impact conditions, only allowing the assessment of data satisfying the following criteria:

- Nearest turbine is producing an electrical power output above a level corresponding to either 85% of its rated electrical power, or to 90% of its rated sound power.
- All turbines within 1.5 km are operating according to the scenario:
  - Total Noise: All turbines are spinning and producing power
  - Background: All turbines are parked and not producing power
- Measurement location is downwind of nearest turbine – within +/- 45° of the nearest turbine's measured yaw angle.
- Significant transient contamination is not present.

Night-time (22:00 – 05:00) noise measurements are carried out in two phases, with one taking place in the fall/winter (typically September-December), and the other taking place in the spring (typically March-May). Measurement locations are often between 550m and 850m from the nearest turbine and are sufficiently far removed from any reflective surfaces or foliage. Using a microphone with a primary and secondary wind screen at a typical height of 4.5 meters, audio data is recorded in one-minute intervals between frequencies of 20Hz to 10,000Hz. The insertion loss of the wind screen has been tested and accounted for in the processing. One-minute LA_{eq} and L_{90} values are recorded for determination of average sound levels and for transient event filtering, respectively. Weather data including wind speed and direction, as well as atmospheric conditions, are recorded simultaneously at a height of 10 meters above ground level (10m-AGL). Measurement datasets from two of the assessed sites are included below and generally demonstrate the quality of the data being used in this assessment.

Figure 1: Site A Measurement Data - Spring Measurement Period

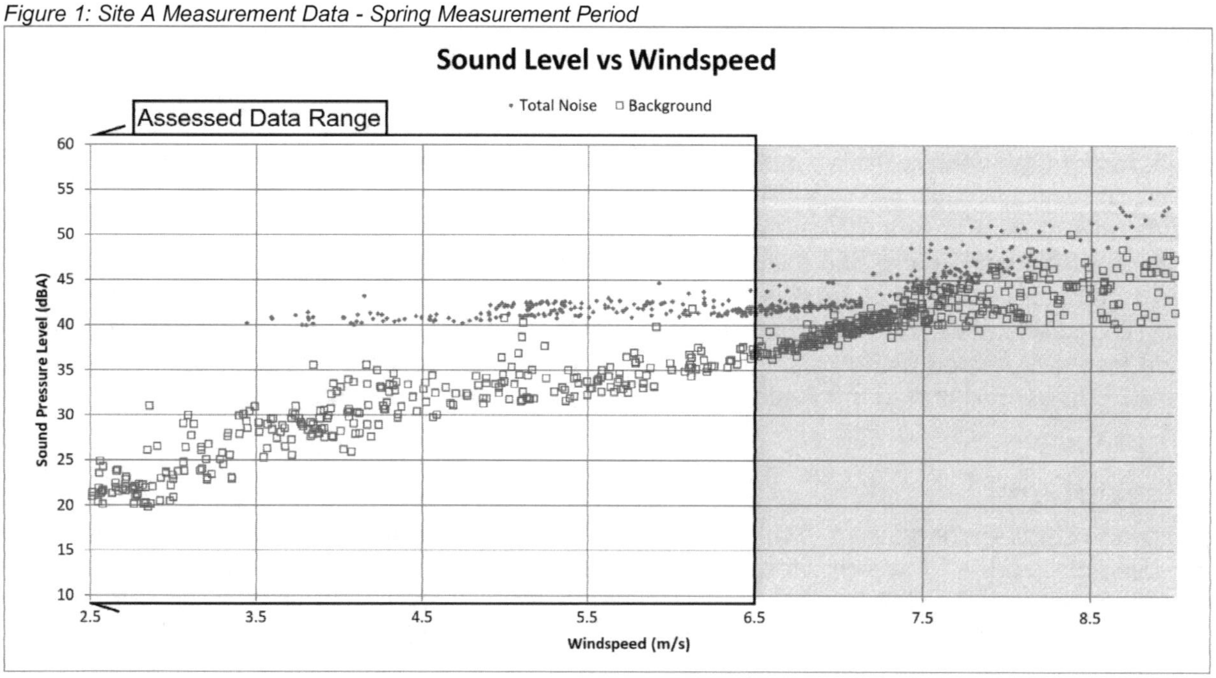

Figure 2: Site D2 Measurement Data - Fall Measurement Campaign

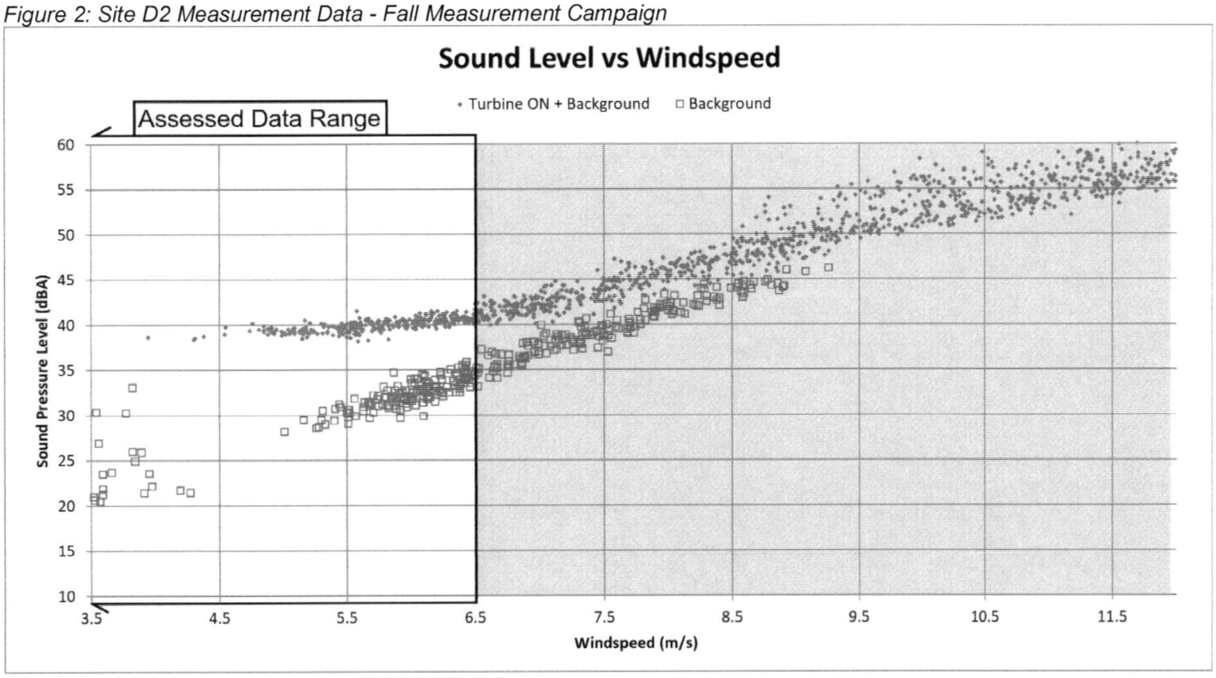

To account for the influence of wind speed on the acoustical measurements, data are sorted into integer wind speed bins ranging from 1-7 m/s (0.5-1.49 m/s, 1.5-2.49 m/s, etc).
The assessed project noise impact level is an average Turbine-Only sound level, which is taken to be the result of logarithmically subtracting the average measured Background sound level from the average measured Total Noise sound level, for each wind speed bin. The standard imposes data filtering to ensure representative conditions between these Total Noise and Background periods, such that the subtraction of windbin-averaged values for these periods results in the noise impact from the turbine operation only. In cases where noise from the ambient environment was observed to have a negligible impact on the Total Noise sound levels and where fewer than 30 valid Background intervals were collected, an ambient sound level of 30 dBA was assumed. This assumed background sound level was used for at least one wind bin at four of the 15 sites.

2.2 Sites Selected for Comparison

Aercoustics has conducted immission measurements at numerous sites for which comparisons could be made. However, in order to remove as much uncertainty as possible from the noise level comparisons, the following considerations were made regarding the selection of sites for this study:

- Emission test data is available for turbines near the monitor; ideally the closest turbine.
- High data quality was demonstrated, demonstrating separation between the Total Noise and Background data across multiple wind speed bins.

15 measurement locations across seven wind farms in Ontario were selected for the assessment. For five of the sites, both a spring and a fall/winter dataset was assessed. For seven of the sites, only a fall/winter measurement campaign was assessed, and for two only a spring measurement campaign. Finally, fall/winter measurements were taken at site E1 before and after de-rating the two closest turbines to the measurement location. In this scenario, IEC measurements were also conducted at the de-rated turbines after the modifications were implemented.

Approximately half of the measurement locations in this assessment featured wind turbines in the 2 MW range with hub heights of roughly 100 m. The turbines at the other eight sites featured nameplate capacities in the range of 2.9 – 3.5 MW with hub-heights of 120 to 130 m. The assessed measurement locations are described in Table 1.

Table 1: Measurement Location Summary

<table>
<tr><th colspan="3">Site</th><th>A</th><th>B</th><th>C</th><th>D1</th><th>D2</th><th colspan="2">E1</th><th>E2</th></tr>
<tr><td rowspan="2">Measurement Period</td><td>Spring</td><td># Days</td><td>8</td><td>58</td><td>-</td><td>-</td><td>-</td><td colspan="2">-</td><td>-</td></tr>
<tr><td>Fall / Winter</td><td># Days</td><td>29</td><td>36</td><td>23</td><td>72</td><td>73</td><td>166</td><td>32</td><td>124</td></tr>
<tr><td>Wind Farm Size</td><td colspan="2">Nameplate Capacity Class [MW]</td><td>2-10</td><td>2-10</td><td>2-10</td><td colspan="2">100+</td><td colspan="3">100+</td></tr>
<tr><td rowspan="2">Measurement Location Details</td><td colspan="2">Distance from Nearest Turbine [m]</td><td>570</td><td>385</td><td>675</td><td>724</td><td>630</td><td colspan="2">683</td><td>613</td></tr>
<tr><td colspan="2">Ground Cover between Turbine and Measurement Location</td><td>Crop / Grassland</td><td>Forest / Crop</td><td>Crop / Grassland</td><td>Crop</td><td>Crop</td><td colspan="2">Crop</td><td>Crop</td></tr>
</table>

<table>
<tr><th colspan="3">Site</th><th>F1</th><th>F2</th><th>F3</th><th>F4</th><th>F5</th><th>G1</th><th>G2</th><th>G3</th></tr>
<tr><td rowspan="2">Measurement Period</td><td>Spring</td><td># Days</td><td>59</td><td>118</td><td>80</td><td>56</td><td>-</td><td>-</td><td>-</td><td>-</td></tr>
<tr><td>Fall / Winter</td><td># Days</td><td>-</td><td>24</td><td>17</td><td>89</td><td>17</td><td>118</td><td>165</td><td>166</td></tr>
<tr><td>Wind Farm Size</td><td colspan="2">Nameplate Capacity Class [MW]</td><td colspan="4">10-100</td><td colspan="4">100+</td></tr>
<tr><td rowspan="2">Measurement Location Details</td><td colspan="2">Distance from Nearest Turbine [m]</td><td>601</td><td>787</td><td>717</td><td>610</td><td>670</td><td>428</td><td>592</td><td>540</td></tr>
<tr><td colspan="2">Ground Cover between Turbine and Measurement Location</td><td>Crop / Open</td><td>Crop / Open</td><td>Crop / Open</td><td>Crop / Open</td><td>Crop / Open</td><td>Crop / Open</td><td>Crop / Open</td><td>Crop / Open</td></tr>
</table>

2.3 Noise Level Prediction Using ISO 9613-2

Sound propagation according to ISO 9613-2 is calculated according to several attenuation parameters. In the context of a typical wind facility, the dominant factors are usually A_{div}, (geometric divergence), A_{gr} (ground effect), and A_{atm} (atmospheric absorption). The divergence term is based on spherical spreading of the wave front and varies solely with distance from the source. The ground attenuation term varies in frequency over distance and is dictated predominantly by the ground effect constant, G, which ranges between values of 0 and 1. These extremes are intended to represent perfectly reflective (G=0) and completely porous ground (G=1), with the value G=0.5 representing ground of 'mixed porosity'. The atmospheric term varies spectrally with the attenuation-per-distance depending on temperature and humidity values. The propagation calculations accounted for a temperature of 10 degrees Celsius and a relative humidity value of 70%. Average meteorological values for each measurement campaign at each site are included in the table below. These values often shift throughout these measurement campaigns, however the average for the measurement periods generally agree with the modelled conditions.

In this data subset, 14 of the 21 datasets were collected in the fall/winter months, and seven of the 21 were collected in the spring months. The average measured atmospheric conditions are presented for each measurement location and period in Table 2.

Table 2: Measured Atmospheric Conditions at Each Measurement Location

Site	Period	Average Relative Humidity [%]	Average Temperature [°C]
A	Spring	44	9
	Fall	88	18
B	Spring	71	1
	Fall	70	3
C	Spring	60	1
D1	Fall	73	-3
D2	Fall	73	0
E1	Fall	80	5
E2	Fall	76	8
F1	Spring	69	19
F2	Spring	72	20
	Fall	71	10
F3	Spring	74	25
	Fall	68	3
F4	Spring	64	18
	Fall	78	4
F5	Fall	58	16
G1	Fall	75	1
G2	Fall	73	1
G3	Fall	75	2
Fall Average		**74**	**5**
Spring Average		**65**	**13**

The acoustical modelling software CadnaA, which utilizes ISO 9613-2, was used to make worst-case noise predictions at the five sites considered in the study. All wind turbines considered in the permitting of the wind farm were accounted for in the noise modelling, including those from adjacent facilities. Predictions of the cumulative noise impact from all relevant turbines were made at noise receptors corresponding to the real-world measurement locations at ground factors ranging from 0.0 - 0.8 in increments of 0.1.

Wind turbine noise sources were modelled as point sources at heights corresponding to the hub-height of the turbine. The worst-case sound power spectra measured according to IEC 61400-11 [7] [8] were used for each turbine where measurement data was available. Measurement data was available for some but not all turbines on a given site; for turbine noise sources that were not measured, turbine sound power spectra for were taken to be the energy-average of the tested values from other turbines on-site, for all turbines having the same model and rated sound output as those tested. Where no suitable test data was available for a given turbine model or rated sound power output, sound power spectra from the turbine specification were used. This was only the case for one model of turbine at Site D1/D2, and, at all sites, for the turbines from adjacent facilities. The table below compares the manufacturer specified sound power levels with those measured on site and provides context regarding the predicted acoustical contribution from measured sources compared to the overall predicted noise level.

Table 3: Measurement Location Noise Impact Comparison

Site	Nearest Turbine Rated Sound Power [dBA]	Measured Turbine Max IEC Sound Power [dBA]	Predicted Sound Level[1] from All Noise Sources [dBA]	Predicted Sound Level[1] from Measured Noise Sources [dBA]
A	103.2	102.6, 103.2	39.5	34.3
B	103.2	103.2, 103.2	42.7	42.3
C	103.2	102.9, 103.7	38.0	35.2

Site	Nearest Turbine Rated Sound Power [dBA]	Measured Turbine Max IEC Sound Power [dBA]	Predicted Sound Level[1] from All Noise Sources [dBA]	Predicted Sound Level[1] from Measured Noise Sources [dBA]
D1	102.0	102.0	39.5	35.0
D2	104.0	103.7	37.8	31.4
E1	104.0	103.3	36.9	15.5
E1[2]	102.0	102.0, 102.1	35.9	34.4
E2	106.0	106.1	38.9	22.0
F1	104.0	105.0	41.4	37.7
F2	106.0	104.6	40.0	33.4
F3	106.0	104.6	38.9	34.1
F4	106.0	104.6	38.8	29.8
F5	106.0	104.6	38.8	22.3
G1	105.0	104.8	39.8	23.6
G2	105.0	104.8	39.5	12.8
G3	105.0	104.6	39.4	37.8

1 – Sound level Predicted at Measurement Location using Measured Values and G=0.7.
2 – Listed values reflect the application of a -2dB derate to the two nearest turbines at site E1.

2.4 Comparison of Measured and Predicted Noise Levels

Data had been collected at a range of wind speeds at the sites mentioned above. A high-quality subset of this data was selected for assessment based on a number of criteria. Specifically, the windbin-average Turbine-Only sound pressure level for a given measurement location was only considered where:

- A minimum 1-minute data count of 20 was achieved for Total Noise intervals
- A minimum 1-minute data count of 20 was achieved for Background Noise intervals, OR the background was observed to have a negligible influence on Total Noise sound levels.
- A minimum signal-to-noise ratio between average Total Noise and Background sound levels of 6 dBA was achieved
- The wind speed bin was below 7 m/s, in order to reduce the influence of wind-related noise on the measurement data.

Applying these filters resulted in a dataset with a population of N=53. By virtue of the signal-to-noise threshold, minimum turbine power threshold, and downwind criterion for all data, these final Turbine-Only average sound levels represent the most accurate assessment of a worst-case impact from the nearby wind farm in the far field.

3. Results

The filtered selection of data has been grouped into wind speed bins for assessment of the average difference between measured and predicted noise levels for different values of G across multiple sites. Figure 3 below shows, for various wind speeds, the difference between measured levels and modelled levels with different ground factors. The grading is in 0.25 dBA increments, and the white area represents deviations of -0.25 to +0.25 dBA, i.e. most accurately reflecting the measured levels. Areas below the 'white belt' represent conditions where the measurements were less than the modelled levels (i.e. the model overpredicted) while the area above the white line represents times where the measured levels where higher than the modelled level (i.e. the model underpredicted).

Figure 3: Difference between Measured and Predicted Sound Level for Different values of G at Different 10m-AGL Wind Speeds

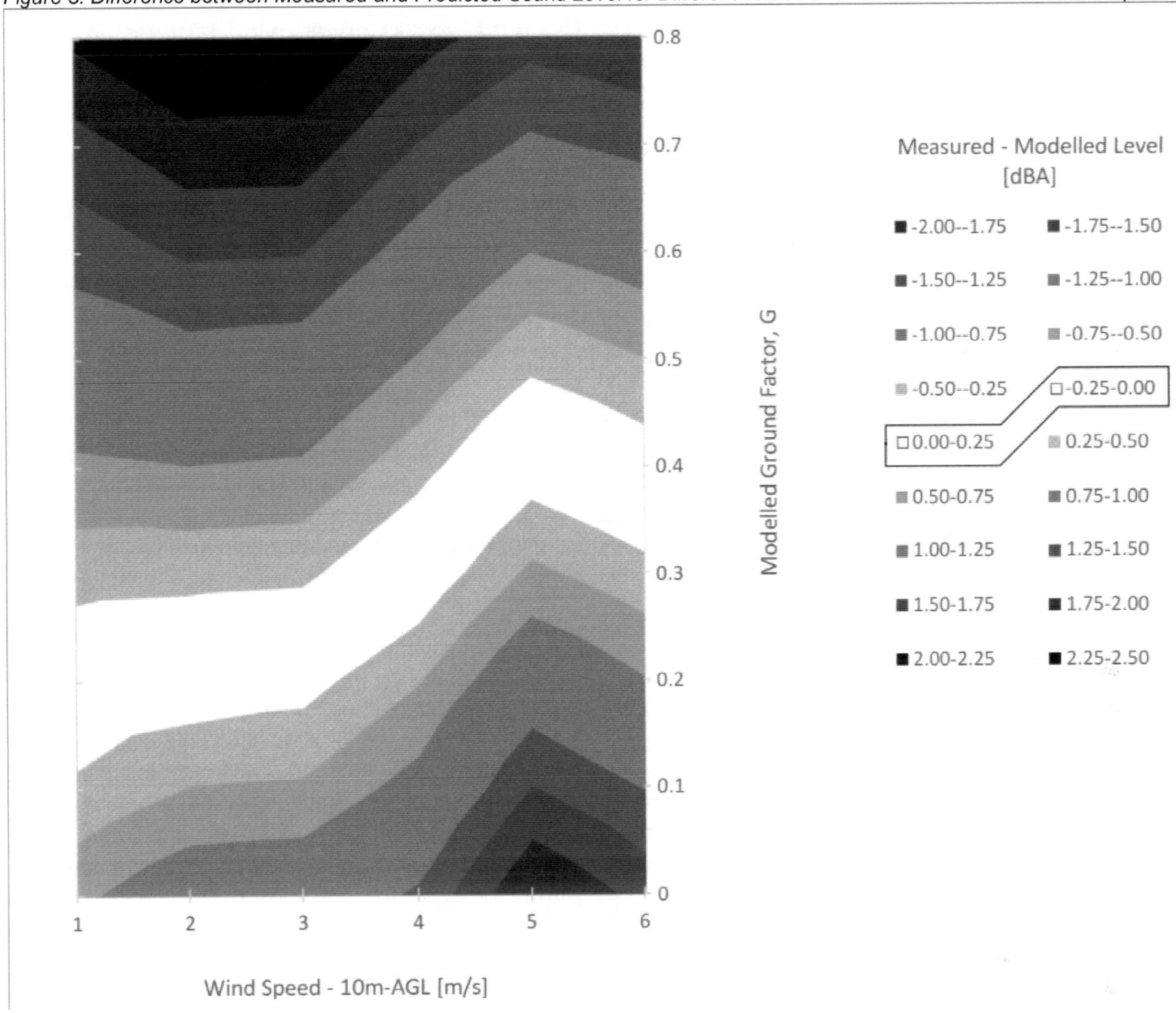

This same data is also presented in a more granular format in Figure 4. Values falling around the y-axis represent those most accurately reflecting the measured levels. The average wind shear value for each wind bin at a given measurement location was calculated, and Figure 4 includes the wind shear value for each wind bin averaged among all sites with available data.

Figure 4: Difference between Measured and Predicted Sound Level for Different values of G at Different 10m-AGL Wind Speeds

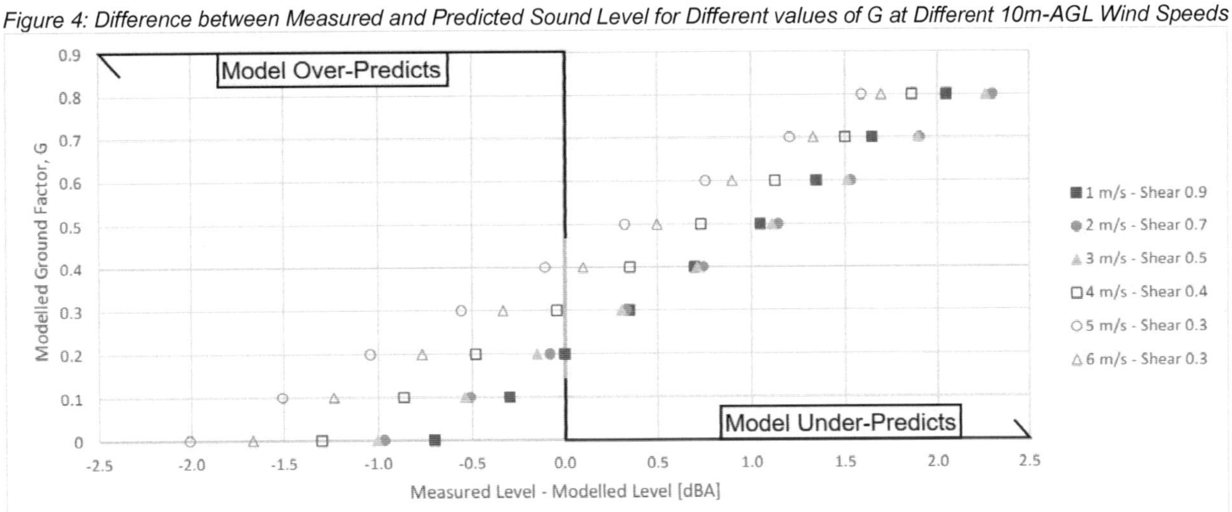

With a sample size of 53 wind bins, collected at locations across seven wind farms of varying sizes, the findings of this study represent a consistent trend. On average it was found that application of a ground factor of about G=0.3 was required in order to align the model with the measured value. This value is lower than findings from other studies [3]; potential reasons for these findings are presented in Section 5 of this study.

Furthermore, the data was also analysed by the measured ground level wind speed. It should be noted that the analysis only considers datapoints where the turbine is producing at least 90% of its maximum sound output, typically occurring at power outputs of at least 85% of rated output. Accordingly, lower ground level wind speeds generally represent higher wind shear events, and higher wind speeds represent comparatively lower wind shear events.

In some cases, a turbine reaches its maximum sound power output at windspeeds corresponding to an electrical power output below 85% of the turbine's rated value. This is the case for three of the sites assessed. In these datasets, a reduced electrical power threshold corresponding to 90% of the turbine's maximum sound output was applied – in some cases as low as 47% of the rated power. Regardless, the use of a reduced power threshold was found not to materially affect the average shear value for data in a given wind bin when compared to the average shear values using an 85% power threshold.

An interesting trend appears to be that at lower wind speeds (i.e. higher shear conditions) the ground factor to match the model to the measurements would have to be considerably lower than originally anticipated – roughly G=0.2. As the ground level wind speed increases and the associated wind shear decreases, the ground factor that best matches the measurements increases to G=0.4. This could be due to more extreme downward refraction conditions when the ground level wind speed is below 3 m/s and the hub height wind speed is ~10 m/s. However, the difference in measured level between extreme and moderate level shear was found to be relatively minor, in the order of 0.5 – 1 dBA.

The average wind shear for each wind speed bin at a given measurement location has been calculated based on the measured 10m-AGL and hub-height windspeeds. Figure 5 presents these average shear values plotted against the difference between measured and predicted sound levels for the location using a ground factor of G=0.5.

Figure 5: Average Wind Shear vs. Difference between Measured and Modelled Sound Level

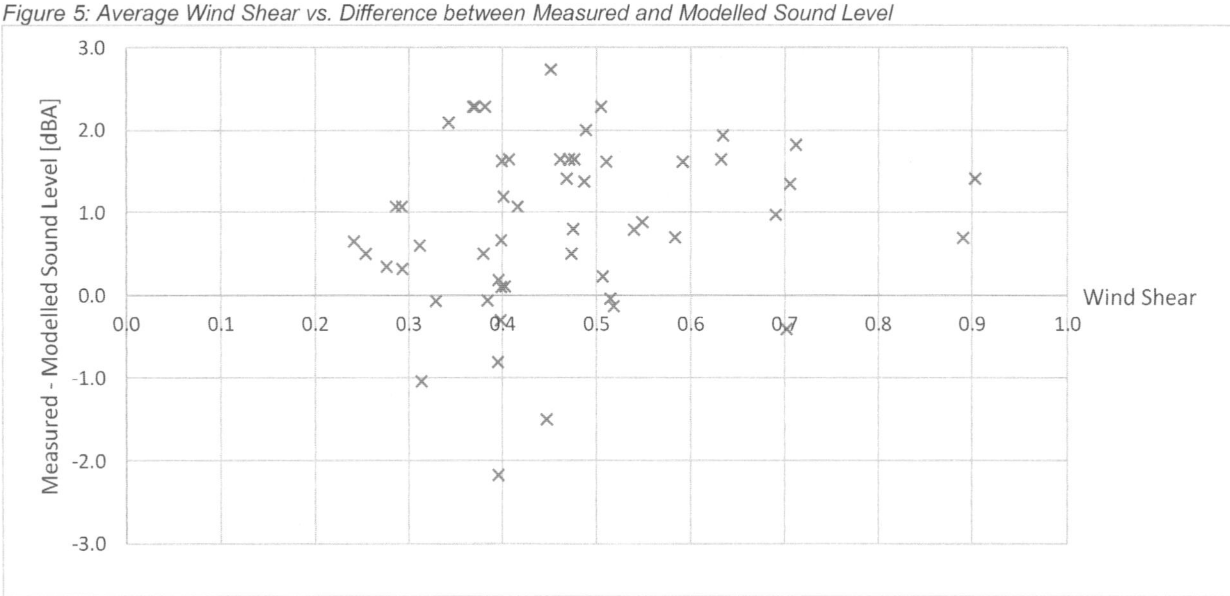

The above figure illustrates a weak tendency for sound levels measured under conditions of higher shear to exceed the sound levels predicted at the same locations using a ground factor of G=0.5. Due to the uneven distribution of wind shear values in the available data, this trend is not readily observable.

A statistical analysis of the differences between measured and modelled values has been carried out for a subset of ground factors commonly used for environmental noise predictions. The frequency plot for G=0.5 is presented below. While the dataset is limited, the distribution of values appears to follow a normal distribution, which is shown overlaid on the figures below. It is to be noted that no goodness-of-fit test has been carried out, and accordingly the results of this statistical analysis should be considered preliminary.

Figure 6: Statistical Analysis of Measured vs Modelled SPL for G=0.5

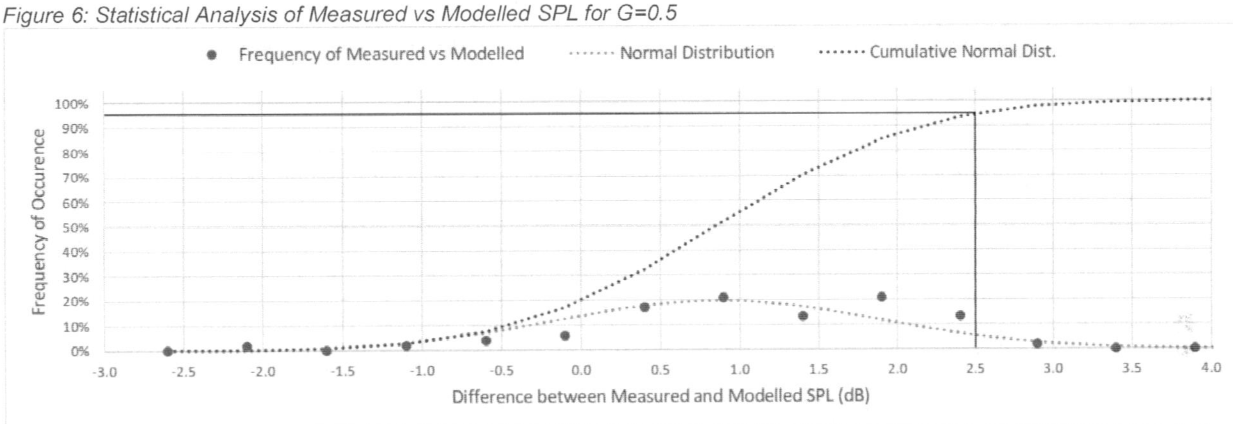

Assuming a normal distribution on the data above, it can be shown that, for G=0.5, an exceedance over the predicted level of up to 2.5 dBA accounts for 95% of the results. Put differently, including 2.5 dBA of conservatism in the modelling process for G=0.5 should result in a 5% or lower chance of the measured sound level exceeding the predicted value at the same location.

4. Limitations of Study

Differences of fractional decibels are not practically relevant in terms of environmental acoustics but become important in the context of wind farm permitting. Even so, it is important to understand that the degree of uncertainty in this context often exceeds the magnitude of the value being assessed.

Compared to other studies regarding the applicability of various ground factors in wind turbine noise modelling, this study represents a lower degree of uncertainty in the results by virtue of using measured sound power levels combined with high quality immission data.

Regardless, there remain numerous sources of uncertainty involved in this assessment which arise from multiple sources. Specifically;

1. Wind Turbine Sound Emission Levels used for Modelling
    a. *IEC measurement uncertainty*
    b. *Variation in Turbine sound emission for turbines of the same model within a wind farm*
    c. *Stability of Turbine sound emission over time*

2. ISO 9613-2 Modelling Uncertainty
    a. *General uncertainty of model*
    b. *Downwind condition from all sources simultaneously*
3. Calculated Turbine-Only Sound Pressure Levels (Immission) - Unknown
    a. *Variation in environmental weather conditions across campaign*
    b. *Uncertainty related to the signal-to-noise ratio*
    c. *Uncertainty related to the measurement chain*
    d. *Variation in conditions between the worst-case sound power spectrum used and those present during immission measurements*

Items 1a, b, c, 2a, and 3a, b, c, d may bias values either above or below the real value, where item 2b is expected to have the effect of over-predicting the sound pressure level in situations where the receptor is surrounded by wind turbines and therefore cannot be downwind of all turbines simultaneously. This situation is most relevant at sites D, E, F, and G, and less relevant at the other sites, where the measurement location was downwind of the majority of the closest turbines.

In this study, the maximum measured IEC sound power spectrum was used as the input to the model. The immission measurement methodology ensures that only the intervals corresponding to turbine operation within 10% (0.5 dBA) of its maximum overall sound power are assessed. Nonetheless, for data in a given wind speed bin, it is possible that the average turbine operational conditions correspond a sound power output up to 10% below the maximum. Consequently, it may be ideal to select an IEC sound power spectrum in line with the environmental conditions that were most prevalent in the data at a given site.

Additionally, it is to be noted that for some sites, the impact from directly-measured sources represent a comparatively small proportion of the cumulative noise impact. As mentioned earlier, where measurement of the nearest turbine(s) was not possible, the sound power level measured at another turbine of the same specification at the same site would be used. While this approach is expected to remove uncertainty from the comparison of modelled and predicted values, it may be ideal to ascribe higher significance to the results from sites where the directly-measured turbines represent the dominant contribution to the overall impact.

The uncertainty associated with the ISO 9613-2 methodology is stated to be +/- 3.0 dB in situations where the average height between source and receiver is below 30 meters, and where noise sources are within 1,000 meters of the receptor. In the context of most wind farms, this is not the case, and so the uncertainty may exceed the stated value.

Additionally, all of the measurement scenarios in this study generally correspond to flat ground. Variations with respect to hilly and complex terrain was not part of the scope of this study; the appropriateness of various values of G has been assessed to vary based on topography [4].

A detailed assessment of uncertainty was not possible at the time of this study. A simplified uncertainty for the results was calculated based on the various discrete uncertainties mentioned above and was found to be in the order of at least +/- 1.6 dB.

As mentioned previously, the results presented here represent a set of samples with high quality far-field data, and with field tested turbine emission data. While the same size achieved in this study is substantial, future work on this subject could include the addition of further sample sets in order to further improve the quality of the conclusions, and potentially to extend the findings to other topographical conditions.

5. Implications of Study - Appropriateness of G<0.5 Ground Factors

The results of the study indicate that a value of G=0.3 or lower is appropriate for conservatively and accurately predicting the sound level at a point of reception for a wind farm. Higher values of G, up to G=0.5, may be appropriate for lower-shear conditions, and may be valid for sites where high-shear conditions occur rarely. Conversely, for worst-case assessment of sites with a greater prevalence of high-shear conditions, values of G down to G=0.2 may be applicable. As stated earlier in this study, the applicability of these lower ground factor values is valid for situations where the sound power spectra used in modelling are known or expected to closely match the real-world turbine outputs.

For most wind facilities, the ISO 9613-2 methodology is inherently conservative in that it considers the point of reception to be down wind of all noise sources. This will not be the case for any receptor that is surrounding by turbines and therefore will only be down-wind of a selection of those. Conservatism can be added in different areas during the noise impact assessment, including in the manufacturer's turbine noise specifications, in the selection of ground factors, and in the application of explicit safety factors on the noise output level.

In scenarios where the ground factor value to be used in modelling is stipulated by regulatory or other governing documents, and where a post-construction verification audit is required, it may be desirable to apply additional conservatism in the model. For example, if regulations stipulate that G=0.5 must be used, a 'adjustment factor' of between 2 - 2.5 dB could be added to the turbine sound power level to greatly reduce the likelihood that the measured sound levels exceed the applicable sound level limit.

Higher values of G may be applicable in situations where the representative or average noise impact of the facility is being assessed, rather than the absolute worst-case.

In consideration of the findings of this study, it is important to be sufficiently conservative in the modelling process for wind facilities, particularly so considering the limitations of the ISO 9613-2 methodology. Modelling using ground factors as high as G=0.7 has somewhat consistently resulted in compliant immission sound levels in Ontario, providing that the turbine sound power output is within specification; this effect may not be the result of the applicability of that ground factor and may be due to other factors regarding sound output spectrum of the turbine(s) in-situ as compared to the assumed values in the pre-construction noise studies. However, this should not be an indication of the validity of using ground factor of G=0.7 in the future.

References

[1] ISO, "ISO 9613-2:1996 Acoustics - Attenuation of sound during propogation outdoors - Part 2: General method of calculation," International Organization for Standardization, Geneva, Switzerland, 2017.

[2] J. A. M. J. M. C. A.J. Bullmore, "Wind Farm Noise Predictions and Comparisons with Measurements," in *Third International MEeting on Wind Turbine Noise*, Aalborg, Denmark, 2009.

[3] D. G. D. W. C. Delaire, "Comparison of predicted wind farm noise emission and measured post-construction noise levels at the Portland Wind Energy Project in Victoria, Australia," in *Fourth International Meeting on Wind Turbine Noise*, Rome, Italy, 2011.

[4] J. C. T. Evans, "Comparison of predicted and measured wind farm noise levels and implications for assessments of new wind farms," in *Proceedings of Acoustics, 2011*, 2011.

[5] Kaliski et al., "Regulating and Predicting Wind Turbine Sound in the U.S.," in *NOISE CONTROL ENGINEERING. INTERNATIONAL CONGRESS AND EXPOSITION. 47TH 2018. (INTERNOISE 2018)*, Chicago, USA, 2018.

[6] Ministry of the Environment and Climate Change, "Compliance Protocol for Wind Turbine Noise," Government of Ontario, Toronto, 2017.

[7] IEC 61400-11 Edition 2.1, "Wind Turbine Generator Systems, Part 11: Acoustic Noise Measurement Techniques," International Electrotechnical Commission, Geneva, Switzerland, 2002.

[8] IEC 61400-11 Edition 3.0, "Wind Turbine Generator Systems, Part 11: Acoustic Noise Measurement Techniques," International Electrotechnical Commission, Geneva, Switzerland, 2016.

INCEEUROPE

9th International Conference
on
Wind Turbine Noise
Remote from Europe – 18th to 21st May 2021

Prediction of wind turbines infrasound from meteorological parameters

Sarah D'Amico, Information Technology Department, Ghent University: Sarah.DAmico@UGent.be
Timothy Van Renterghem, Information Technology Department, Ghent University: Timothy.VanRenterghem@UGent.be
Dick Botteldooren, Information Technology Department, Ghent University: Dick.Botteldooren@UGent.be

Summary

Infrasound exposure has become a main cause of opposition towards new wind turbine projects. In order to assess its relevance, the data from a long-term measurement campaign was used to develop a prediction scheme for wind turbine infrasound using regression models and neural network techniques. Therefore, a measurement campaign at close distance from a wind turbine was conducted using a low-frequency microphone. An ad-hoc designed and validated wind-shielding dome was used to improve the signal-to-noise ratio in the infrasonic frequency range. The acoustic measurements were accompanied by meteorological data at three heights, high-frequency wind measurements and wind turbine operational data. The blade passing frequency was accurately predicted using the operational data. In the on-going study, the prediction of the blade passing frequency levels yielded a determination coefficient of 79% using the regression approach and of 89% using the neural network model, allowing to identify the most significant atmospheric variables on the infrasound immission.

1. Introduction

The need for renewable sources of energy is resulting in an increment in the construction of wind farms. While wind turbines have a good reputation as efficient devices for the generation of green energy, their noise emissions are often considered as the main negative impact on the environment. New generation wind turbines are also growing in size, renewing the interest towards their low frequency and infrasound emissions.

Infrasound from wind turbines consists of broadband noise and tonal components. Blade-tower interaction is mainly responsible for the tonal components of the infrasound (van den Berg, 2005). During its rotation, the blade encounters a region of velocity deficit in front of the tower, resulting in temporal change of the angle of attack. This generates acoustic emissions at the blade passing frequency. Broadband infrasound from wind turbines originates from the interaction between the turbulent eddies advected with the wind flow and the blades (Wagner et el., 1996). Especially the large eddies lead to sudden changes in the aerodynamic loading of the blade.

Infrasound immissions from wind power generators are measured simultaneously with wind-induced microphone noise, potentially masking wind turbine infrasound. Therefore, correctly measuring infrasound from wind turbines presents difficulties, leaving uncertainties on the determining parameters generating their infrasound.

This paper studies infrasound from wind turbines in correlation with different inflow and atmospheric conditions. A wind shielding dome was used to improve the signal-to-noise ratio in the infrasonic frequency range whose development and assessment was discussed in a previous work (D'Amico et al., 2021). A long-term measurement campaign allowed developing regression models and neural network routines in order to study the influence of different atmospheric parameters on the generation of infrasound from wind turbines.

2. Methodology

2.1 Measurement campaign

The measurement campaign took place on the wind farm at the Melle Campus of Ghent University from September 2019 to November 2020. The wind farm is composed of three Enercon wind turbines of 2 MW, each placed at 450 m from each other. The wind turbines are pitch regulated and have a hub height of 108 m and a rotor diameter of 82 m.

The area is a fully flat agricultural terrain and some sparse buildings are located at a distance of 200 m from the microphone. A main motorway is at approximately 250 m from the microphone. Figure 1 illustrates the wind farm.

Figure 1. Measurement location

Measurements were taken at 120 m from the central wind turbine. Acoustic pressure fluctuations were measured using a GRAS 46 AZ low frequency microphone and preamplifier that ensures linear response from 0.5 Hz. The microphone was used together with a Alix System single board computer for continuous recordings at 48 kHz. 57s continuous recordings were collected every minute. A GPS clock was used to synchronize the recordings every minute.

The microphone was equipped with rain protection and a 10-cm diameter wind screen. On top of that, the microphone was covered by a previously designed wind-shielding dome to reduce the wind-induced microphone noise at infrasound in order to improve the signal-to-noise ratio (D'Amico et al., 2021). The Svantek 30 A Class 1 acoustic calibrator was used to record a 1 kHz 94 dB calibration signal on a weekly basis.

The meteorological data were acquired at 2, 10 and 108 m height. A Gill WindMaster 3D sonic anemometer was used at 10 m height to acquire wind data at 20 Hz acquisition frequency, at a resolution of 0.01 m/s and 0.01°C.

Meteorological data at 2 m height was provided by the Department of Physics of Ghent University and were acquired at a station at 40 m from the microphone. The data at 108 m height (hub height) were provided by the wind farm operator.

Figure 2 illustrates a schematic of the set up while Figure 3 presents some pictures of the measurement instrumentation in the field.

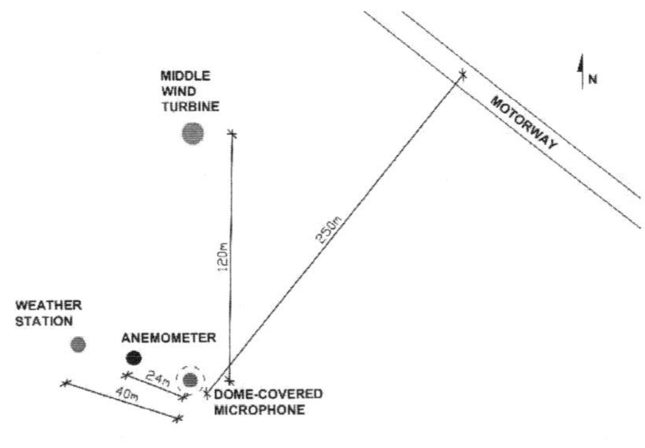

Figure 2. Schematic of the field measurement set up

Figure 3. Field measurement set up: sonic anemometer at 10 m height (left), wind-shielding dome covering the microphone (right, bottom) and the meteorological station (right, top)

The meteorological data, as well as the wind turbine data were averaged over 10 minute-periods. The TI was calculated at the reference height (10 m). From the instantaneous data of the anemometer, the mean and fluctuating values of the three components of the instantaneous wind vector and temperature scalar were calculating using Reynolds decomposition, on a 10-minute base.

From the recorded signals, rainy period were excluded, as well as reported period with disturbing noises in the proximity of the microphones. For each of the 57-s signals, the wave file recording was first calibrated using the 94 dB signal at 1kHz of a calibrator. The signal was then compensated for the internal high-pass filter of the Alix System at infrasound. The PSDs were then calculated for each 57s signal. In a final step, the PSDs were linearly averaged over 10-minute periods and synchronized with the meteorological data.

2.2 Modelling

The data collected during the measurement campaign were used to fit both a regression model and a neural network routine, with the tonal sound pressure levels at the blade passing frequency (BPF) as target.

The operational data from the wind turbines were used to identify the BPF (Fig.4) from the rotational speed. Only the 10-minute data points that presented a clear peak at the BPF were considered. The BPF peaks were identified using a 5 dB emergence criteria above broadband noise. The input variables included the 10-minutes statistics of the meteorological data gathered at the three heights and the nacelle position data.

For the regression analysis, a stepwise linear regression model with parameters interactions was used while for the neural network routine, the Levenberg-Marquardt algorithm was used with 10 hidden neurons. In a next step, the neural network that was generated was then used to analyse the influence of specific input meteorological parameters on the BPF levels.

3. Results

Figure 1 presents a spectrogram of one day. The BPF levels and its harmonics are clearly visible and can be predicted from the rotation speed data with a root-mean-squared-error (RMSE) of only 0.04 Hz.

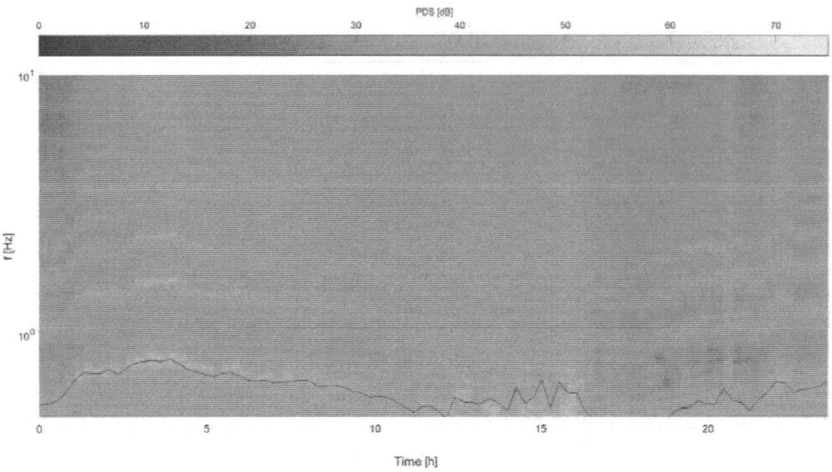

Figure 2. Acoustic spectrogram with predicted BPF (black line)

A time series of the data fitting using the two different models is presented in Figure 5. The errors are also presented in Figure 5 for both modelling approaches. Table 1 shows the determination coefficient (R^2) and the RMSE for both models. The regression model shows a R^2 of 79% while the neural network routine explains 89% of the observed variance. The RMSE, instead, remains the same, circa 3 dB, for both models.

Table 1: Prediction coefficient and RMSE for neural network and regression model

	R^2	$RMSE$
Neural network	**0.89**	**3.0 dB**
Regression model	**0.79**	**3.1 dB**

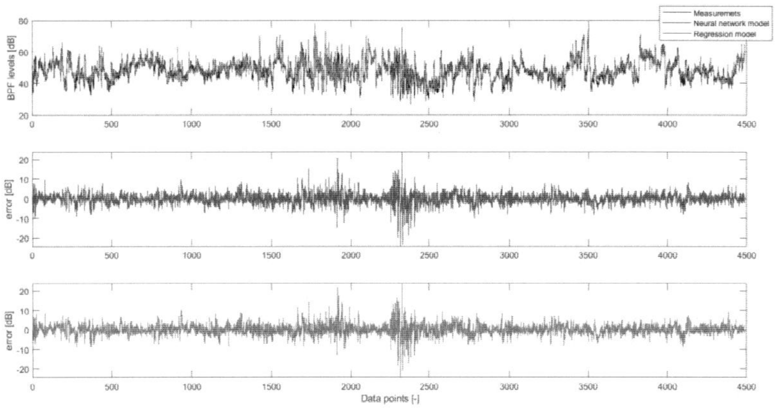

Figure 5. BPF levels in dB measured and predicted by the regression and neural network model together with the errors

The regression model indicates the following parameters as relevant predictors; the mean wind speed (MWS) and temperature at the three heights, as well as the turbulence intensity (TI) at 10 m height and nacelle position. The mean v and w component of the wind vector calculated from the anemometer data, were also identified as relevant predictors together with the vertical heat flux and the friction velocity. The latter two were estimated respectively as the covariance of the fluctuating vertical component of the wind velocity vector and of the temperature, $\sigma(w'T')$ and as $\sqrt[4]{\overline{u'w'}^2 + \overline{v'w'}^2}$ (Klipp, 2018).

The neural network developed was used to analyse what meteorological parameters have a relevant influence on the BPF levels. Figure 6 presents the variation of the BPF levels with MWS and TI; both factors lead to a significant increase in BPF levels. The vertical heat flux and the vertical temperature gradient, $\Delta T/\Delta h$, only resulted in variations of less than 1 dB. This shows that the BPF infrasound is only influenced to a limited extent by the stability of the atmosphere, while strongly sensitive to the wind speed and intensity of velocity fluctuations in the wind inflow. Further analysis is needed to corroborate these initial findings.

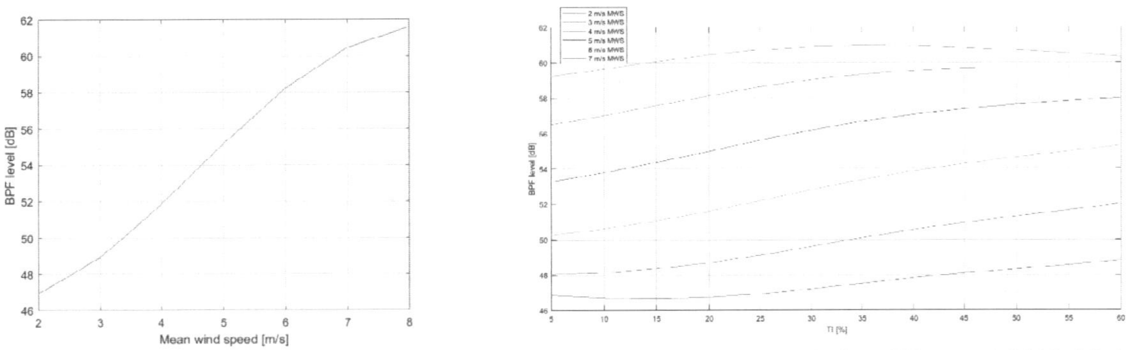

Figure 7: dB BPF levels variations with MWS (left) and with TI for different MWS (right)

4. Conclusions

The study presented allowed to define a regression model and a neural network routine for BPF infrasound level predictions with a high determination coefficient, using a vast meteorological and acoustical data set collected in a long-term measurement campaign. The preliminary results showed that the BPF levels increase significantly both with increasing MWS and TI.

References

van den Berg, G., (2005). *The beat is getting stronger: the effect of atmospheric stability on low frequency modulated sound of wind turbines*. Journal of Low Frequency Noise, Vibration and Active Control, 24(1), 1–24.

Wagner, S., Bareiss, R., Guidati, G., (1996). *Wind turbine noise*, Springer, Berlin.

D'Amico, S., Van Renterghem, T., Botteldooren, D., (2021) *Measuring infrasound outdoors with a focus on wind turbines: the benefits of a wind-shielding dome.* Applied Acoustics, accepted.

Klipp, C., (2018). T*urbulence friction velocity calculated from the Reynolds stress tensor.* Journal of Atmospheric Science, 75(4), 1029-1043.

9th International Conference
on
Wind Turbine Noise
Remote from Europe – 18th to 21st May 2021

Meteorological effects on wind turbine noise at the receptor location

Pierre Dutilleux, Evergy Engineering GmbH, pierre.dutilleux@evergy.de
Astrid Ziemann, TU Dresden, Chair of Meteorology, astrid.ziemann@tu-dresden.de

Summary

During the development phase of a windfarm project, assumptions are made as to the acoustic propagation conditions of the wind farm noise. During the operation phase, the results of a wind farm noise measurement can only be compared to the prediction as far as the effective operating conditions match the assumptions made for the acoustic propagation calculation.

The micrometeorology at the wind farm site and at the dwellers locations has a large impact on the acoustic propagation.

Practical and effective methods of collecting information on the micrometeorology have yet to be developed for the specific needs of wind farm acoustics. Hints on possible methods can be derived from the experience gained in the realm of the acoustic assessment of ground-based noise sources.

The height profile of the speed of sound must be considered. It is related to the vertical wind profile and the temperature gradient, which depends on the cloud cover and the solar radiation but the question arises how to collect detailed site-specific information.

After many years using for wind farms, and their noise sources at high height, the same methods which have been developed for ground-based noise sources, some discrepancies have appeared which called for justifications or for more specific assessment methods. In several countries, detailed investigations have been performed which show how the results of an acoustic propagation calculation depends on the assumptions on the properties of the atmosphere as a propagation medium.

Collecting detailed information especially about the vertical temperature gradients and the wind shear is recommended during wind farm noise measurements.

1. Introduction

The prediction and the measurement of the acoustic noise can only be accurate and reliable as far as the effective operating conditions match the assumptions made during the acoustic propagation calculation of the wind farm project.

The micrometeorology at the wind farm site and at the dwellers locations has a large impact on the acoustic propagation of the noise from the wind turbines. Depending on the sound source and the distance between the sound source and the dwellers locations, different effects can dominate. In particular, the effect of sound refraction is not realistically considered in standardized models of sound propagation in terms of its spatial and temporal variability, which can lead to increased uncertainty in the prognoses for immission control and noise pollution for affected residents. This paper addresses the aspects of which meteorological variables should be monitored during a measurement campaign and of which practical models can be used to analyse and classify the data.

Practical and effective methods of collecting information on the micrometeorology have yet to be developed or adapted for the specific needs of wind farm acoustics. Hints on possible methods can be derived from the experience gained in the realm of the acoustic assessment of noise sources, which are close to the ground. A challenge is however the measurement within the wake of the wind farms.

2. Meteorological measurements to assess ground-based noise sources in comparison to wind farm noise

The acoustic assessment methods for ground-based noise sources such as road traffic noise stress the importance of taking into account the height profile of the speed of sound [5], ([10] Section 8 Meteorological conditions).

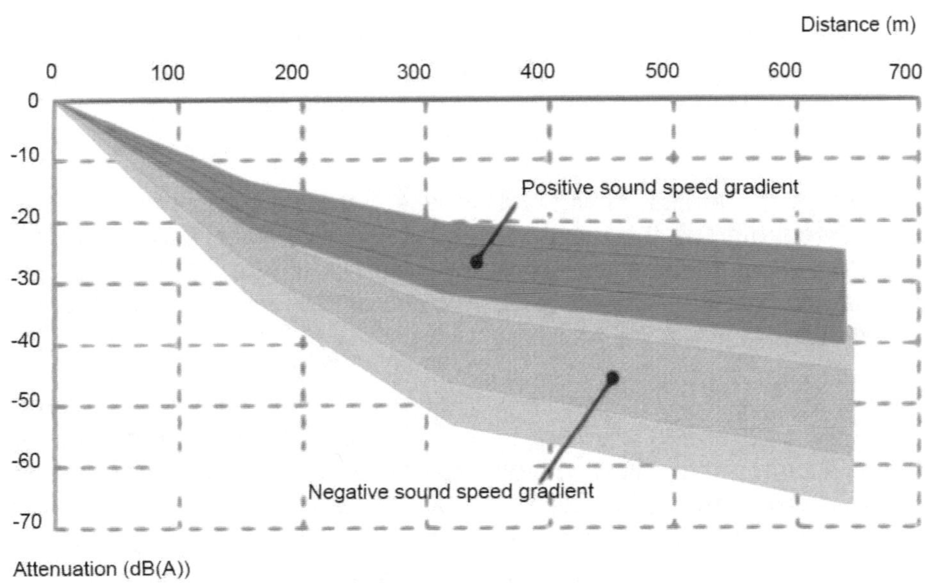

Figure 1: Attenuation and dispersion of sound levels based on the distance from the source for downward (positive gradient) and upward (negative gradient) refraction conditions (Source: [5], [2])

Whereas the sound speed profile is seldom measured directly, it is often modelled on the basis of the temperature and wind speed profiles. A detailed identification of these profiles is strongly recommended but is sometimes difficult to implement in the field. In practice, a classification of the time periods according to the averaged wind speed component in the direction of sound propagation at one height and the solar radiation during the day as well as cloud cover especially during the night has shown to be sufficient for an initial information about the sound propagation situation. Vertical gradients of wind speed and of temperature can be qualitatively associated to each of the resulting classes.

The thermal stratification of the atmosphere and therewith the temperature gradient can be qualitatively related to the cloud cover and the solar radiation ([10] Tables A.2 and A.3).

The cloud cover can be estimated in octas by the personal at the site during the measurement campaign but this method cannot be applied for longer measurement periods. An automatic procedure should be preferred instead. An alternative can be to use the cloud cover information collected by a nearby weather station. Depending on the region on the globe, this weather station might be too remote and/or the available information too coarse. The distance to the weather station might also induce a large time delay, which would impede the coherence with the other measured data sets. The time-averaging and time-sampling at the weather station might also be inappropriate for the needs of acoustic measurements around a wind farm where a measurement is typically needed every 10 minutes.

The incoming solar radiation quantifies the energy reaching the ground and is expressed in W/m^2. The road noise prediction method NMPB2008 [5] accepts two conventional values for strong radiations and average radiations but reminds that it might not allow for a full characterisation of all the classes of acoustic propagation conditions. According to the French standard NF S 31-120 [12], the propagation conditions can be characterised by five wind-speed/wind-direction classes named U1 to U5 and five thermal classes named T1 to T5. This classification is usually referred to as UiTi ([5] Table D.3). An extended approach might consider the energy budget of the surface described by the sum of all short- and long-wave radiation fluxes to and outgoing from the surface as well as the turbulent sensible and latent heat transfers, possible soil heat transfer and storage in vegetation layers. There are only a few stations, which are measuring the full spectrum of radiation fluxes as well as turbulent heat fluxes, because especially for the latter one a special equipment and data analysis algorithms are needed. In short of local data, collecting this information from a nearby weather station might be affected by the same drawbacks as for the cloud cover.

A challenge in evaluating the sound propagation of wind turbines is that the sound source is at a great height above the ground. Depending on the terrain and land use as well as on the time of the day/year and the meteorological conditions, temperature gradient differences can occur between these altitudes and near the ground. A direct measurement of the temperature gradient at these heights is therefore desirable.

The vertical wind speed gradient is usually determined based on wind speed measurements at several heights, at least two. This wind speed gradient is typically valid for the measurement location. When no further detailed information is available, the same wind speed gradient is usually applied to the whole projected area. Care should however be given to the fact that the wind speed gradient might vary significantly across this area. The wind speed gradient close to a dwelling might significantly differ from that at the wind farm site, especially if the wind farm is located on top of a hill and if the dwellings are located in valleys. Furthermore, there are phenomena like the nocturnal low-level jet (heights between 50-500 m), which lead to deviations from the known daytime logarithmic wind profile [13]. In order to measure such phenomena that are relevant for wind turbines, high meteorological measuring masts up to the hub height of the wind turbines are necessary to measure the wind and temperature profile adequately.

3. Classification of the meteorological influence on the noise immission

After many years using for wind farms the same methods which have been developed for noise sources close to the ground, some discrepancies have appeared which called for justifications or for more specific assessment methods. In several countries, detailed investigations have been performed which show how the results of an acoustic propagation calculation depends on the assumptions on the properties of the atmosphere as a propagation medium [7][16]. As an example, we give here a few details about the current German framework.

In order to take into account the variability of the meteorological influence in improved standardized sound propagation models, an acoustically motivated classification of meteorological input variables, in particular the wind and temperature profile, is required. A reference to standardized models as well as stability and wind classes in the dispersion calculation is recommended in order to examine different environmental stressors with suitable and homogenized approaches [3]. A draft German guideline VDI 4101-1 [15] addresses the specific issues of the acoustic propagation at long distances of sound sources at typical heights for traffic noise and wind turbine noise.

For this reason, the one-dimensional, two-layer model of the atmospheric boundary layer [1] from the German guideline VDI 3783-8 [11] was used to simulate the meteorological data basis. The Monin-Obukhov Similarity Theory (MOST, overview by Stull, 1989 [1] and Foken, 2006 [4]) forms the basis for this model. Simplifications include, among other things, stationarity, homogeneous and flat terrain, no mesoscale circulations or weather fronts, and an averaging time of the meteorological variables between 1 min and 1 h. The model enables, e. g., a description of the vertical profiles of horizontal wind speed and direction as well as of air temperature.

The sound propagation classes are derived from the same six dispersion classes, which are used for the assessment of air quality [3]. To distinguish between these classes (O1: very stable, O2: stable, O3: neutral-stable, O4: neutral-unstable, O5: unstable, O6: very unstable), the Obukhov length was used as a measure for strength of thermal stratification. The Obukhov length is that height at which turbulence is generated more by buoyancy than by wind shear. Thermal stratification, or atmospheric stability, is a relevant measure of the support or the suppression of the vertical motion of air parcels. The kind of stability (stable, neutral, unstable) is directly related to the vertical temperature profile. During a clear night, the surface and near-ground air layers cool down due to the long-wave radiation from the ground followed by the formation of a temperature inversion with increasing temperature by height (i. e. stable stratification). During a sunny day, a profile with decreasing air temperature values develops (i. e. unstable stratification).

The wind speed at 10 m height is binned in nine dispersion classes (W1: 1 m/s...W9: 12 m/s). Purely arithmetically, 54 different wind and temperature profiles for one type of surface roughness result from the class combinations of wind speed close to the ground and the stability of the stratification. In order to be able to calculate one set of vertical profiles in a numerical example, a geographical latitude of 51° North and a roughness length of 0.1 m (grassland) were assumed ([15] Attachment B). However, it must be taken into account that not all possible class combinations occur in nature. Stable stratification occurs, for example, only at relatively low wind speeds close to the ground. Accordingly, there are 38 realistic cases with different wind and stability classes.

In a further step, vertical profiles of the effective speed of sound were calculated from the wind and temperature profiles. Sound propagation in the direction of the wind (downwind) or in the direction of headwind (upwind) was assumed. The height of 100 m was used as the reference wind direction, i. e. the height of the sound source. Figure 2 shows an example with vertical profiles of the effective speed of sound for very stable stratification.

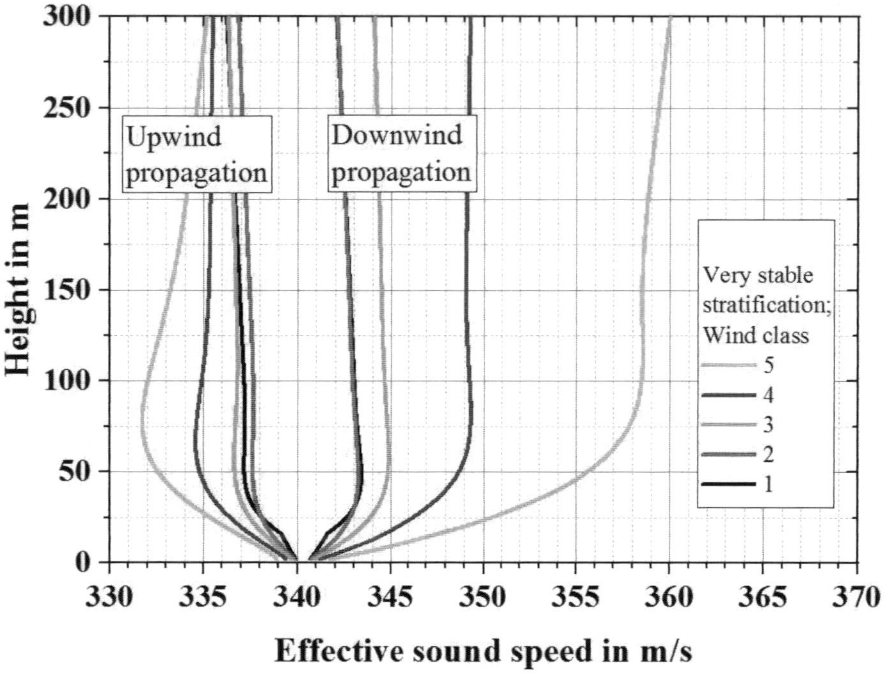

Figure 2: Vertical profiles of the effective sound speed for class O1 (very stable stratification) and five wind classes (reference wind speed: W1: 1; W2: 1.5; W3: 2; W4: 3; W5: 4.5 m/s) in upwind and downwind sound propagation direction (wind direction → sound source at 100 m height).

The draft guideline details a sample calculation, which is part of the algorithmic confirmation of the method (VDI 4101-1 Attachment B [15]). The vertical profiles of horizontal wind speed and direction as well as air temperature served as input data for sound propagation modelling with a two-dimensional CNPE (Crank-Nicholson Parabolic Equation) wave model [8]. In addition to sound wave divergence, the phenomena of sound wave reflection on the ground and sound refraction due to vertically variable temperature and wind fields were taken into account. In the horizontal direction, the atmosphere and the grassy soil surface are assumed to be homogeneous. The results of the calculations at the centre frequencies of the 1/3rd octave bands 16 to 2000 Hz, for sound sources at 1 m and at 100 m height, for downwind and upwind, at distances ranging from 500 to 1000 m show among others that (Figure 3):

- for downwind conditions, the variations of the acoustic level deviations from the free-field propagation levels are larger when the boundary layer of the atmosphere is very stable (O1)

- for upwind conditions in particular, the level variations are higher for neutral dispersion classes (O3, O4) especially for the higher sound source

- the acoustic level deviations from the free-field propagation levels are typically smaller for the source at 100 m height than for the source at 1 m height

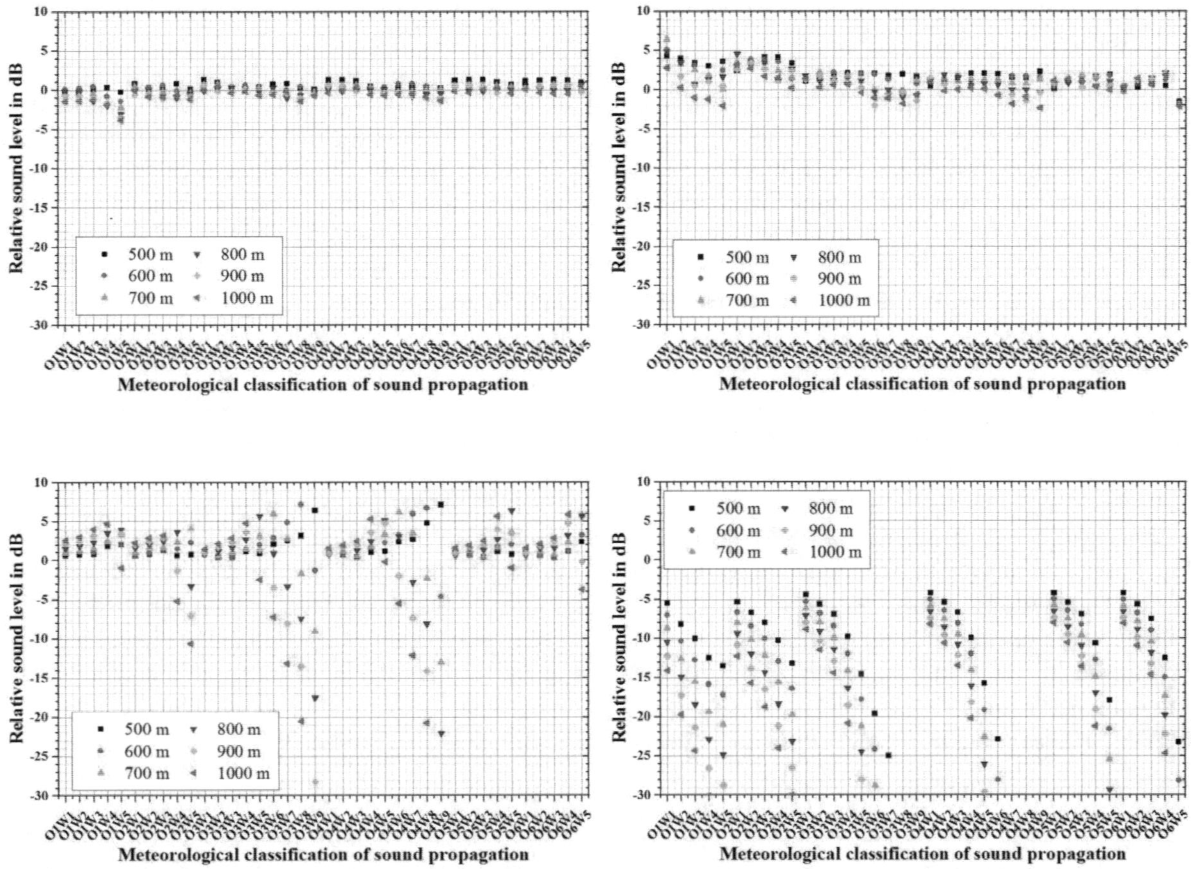

Figure 3: Relative sound level (difference between sound level with ground surface and sound level with free propagation without surface, sum of the sound level over frequencies 16…2000 Hz) as a measure for the ground influence modified by the meteorologically induced sound refraction for 38 meteorological classes (O…stability, W…wind) and 6 distances between sound source and receiver (500…1000 m). Left: wind turbine noise (sound source at 100 m height), right: traffic noise (sound source at 1 m height), above: downwind, below: upwind

For upwind conditions, it should be noted that comparatively higher sound level values (relative sound level greater than 0 dB) can occur for higher sound sources. These situations must be taken into account for noise protection of dwellers.

In summary it can be said that the properties of the acoustic propagation medium must be carefully characterised if accurate calculation results are needed. In order to transfer the results of the above simulation to a specific location, climatological studies on the occurrence of certain wind and stability classes at the specific location are required.

4. Meteorological conditions for acoustic measurements

If no detailed meteorological variables have been collected during the acoustic measurement campaign, the optionally available sources of reanalysis weather data (e. g. ERA5 [17]) can help assess the typical average meteorological conditions for the projected area. The grid of the reanalysis data is however usually much larger than the distances between wind turbines and dwellings. They can provide information about the overall meteorological conditions in the area but do not yet provide micrometeorological information. In addition to the coarse spatial (horizontal and vertical) and temporal resolution of these data, the used meteorological models are not suitable for the incorporation of microscale effects and variability of vegetation or terrain on the wind and temperature field.

According to ISO 1996-2 ([10] Section 8), the meteorological variables shall be measured and information about the atmospheric stability shall be provided. This standard gives an example of atmospheric stability classes and even recommends a more detailed description of the atmosphere, especially for very high sound sources such as wind turbines but neither describes nor specifies how to do it.

4.1 Temperature gradient and vertical temperature difference

According to Gauvreau et al. 2009 [6], the temperature gradient close to the ground can be characterised (for ground-based sources) using temperature measurements at 1 m, 3 m and 10 m height. For wind turbines, a temperature measurement at hub height is recommended and should be planned when a high met mast is installed at the measurement location.

The potential temperature is an important variable for the characterisation of the boundary-layer. Its vertical gradient, or potential temperature difference, can be used to classify the atmospheric stability. A typical device is an electrical temperature sensor, which measures the air temperature via temperature-dependent resistors (e. g. Pt100). These sensors are often coupled with capacitive sensors for humidity measurements. If these sensors are ventilated and shielded against radiation, then an adequate measurement of the air temperature and humidity is possible. This low-cost method can potentially be deployed at several or at all sound level meter locations, allowing a comparison of the atmospheric stability at these various locations.

In some wind turbine types, the ambient temperature outside the nacelle is measured for the purpose of air temperature regulation inside the nacelle. These measurements are usually not suitable for the purpose of acoustic calculations because the heat emanating from the machinery can affect the measured values. For temperature regulation purpose, a resolution of 1°C is usually sufficient. Such measurements are unfortunately neither accurate nor reliable enough for the purpose of acoustic propagation calculation. A specific temperature measurement of the ambient temperature with a resolution and uncertainty better than 0.1°C would however allow a better determination of the temperature gradient between ground and hub height.

Another possibility to measure the required temperature profile is the ultrasonic anemometer/thermometer, which is a standard device in micrometeorology since many years now. It serves for the 3-dimensional capture of the wind speed and the wind direction as well as the acoustic virtual temperature without influences of radiation or inertia. The last-mentioned effects cause a (positive during daytime or negative during nighttime) radiation error also for shielded temperature sensors (like a Pt100) and lead to a start-up threshold for cup anemometers. The ultrasonic anemometer measures the acoustic travel time of a sound signal which is sent and received by ultrasonic converters within a small measuring volume (about 20 cm edge length). Because of the high measuring rate, this device is very well suited to capture gusts and peak value measurement. Due to the measured acoustic virtual temperature it is possible to derive the effective sound speed in a correct way, i. e. including the effect of air humidity on the sound speed.

4.2 Wind shear

In a small operating wind farm, the wind shear is typically determined from a wind-speed measurement at 10 m height and at hub height.

At larger wind farms, a high met mast might be available at the site and the wind shear could also be derived from the anemometers of this met mast. A lidar (Light detecting and ranging) or a sodar (Sonic detecting and ranging) could also be used. Thereby, it has to be kept in mind that a sodar is working with sound waves within the audible spectrum, which may disturb the wind farm noise measurement at the dwellers location. A lidar system is able to provide a large amount of detailed information about the wind profile typically up to 200 m and every 20 m. The advantage of a lidar is that the measurement is easy to set up and take down. In addition, spatially averaged values for wind speed and direction are provided. This enables statements to be made about the wind field over the entire rotor area. Care should be given at the minimum height of the measurement. If the acoustic propagation close to the ground and around the dwellings has to be calculated accurately, detailed information about the atmosphere close to the ground (<= 10 m) is also needed. The typical and commercial lidar systems are designed to provide information at such a low height. Special care should however be given to the periods with precipitation and fog because they affect the lidar measurements.

An example of the use of the wind shear information during low frequency noise measurement campaign is provided in Blumendeller et al. 2020 [14]. In this example, the low frequency noise can be more easily identified in the frequency spectrum when considering only the slightly-stable atmospheric conditions (m = 0,35, where m is the logarithmic shear exponent between 10 m and hub heights).

4.3 Cloud cover

Although the documentation of the cloud cover during the measurement period is required by the standards on acoustic measurements in the environment [10][12], suitable data acquisition systems of such an information are still seldom found in wind energy applications.

Systems are however available which analyse the infrared emissions from the sky to determine the cloud cover. Such systems have been primarily developed for the operation of PV plants but they could potentially also find application in wind energy. The ceilometer which is used at airports can provide automated continuous information about the cloud height and secondary the cloud cover in octas.

Primarily a temperature profile is needed to discuss the meteorological influence on sound propagation. Therefore, direct measurement of the temperature profile should be the first choice in comparison to using the cloud cover for a rather qualitative estimation of the stratification stability.

5. Outlook

Collecting detailed information about the wind speed and the air temperature is the matter of several standards of the IEC 61400 series. Whereas we cannot assume that a lidar based weather station is always available for the purpose of the acoustic assessment, nacelle mounted lidars [18] might become available in the near future. The acoustic assessment of a wind farm project could benefit from the data collected by these lidars as well as from meteorological masts equipped by ultrasonic anemometers measuring the required variables directly.

Other effects of the meteorology on the overall outcome of an acoustic measurement campaign, such as turbulent eddies on the propagation path, the background noise and the signal-to-noise ratio of the measured data are also important but outside the scope of this paper.

References

[1] Stull, R.B. (1989) *An introduction to boundary layer meteorology* Dordrecht: Kluwer Academic Publishers

[2] Zouboff, V., Laporte, J.C. and Brunet, Y. (1998) *Effets des conditions météorologiques sur la propagation du bruit, Prise en compte pratique* LCPC

[3] Federal Ministry for Environment, Nature Conservation and Nuclear Safety, Germany (2002) *Technical Instructions on Air Quality Control – TA Luft, 8 Meteorological Data*

[4] Foken, T. (2006) *50 Years of the Monin-Obukhov similarity theory* Boundary-Layer Meteorology 119, pp. 431–447

[5] Sétra (2008) *Road noise prediction – 2 Noise propagation computation method including meteorological effects NMPB2008*

[6] Gauvreau, B., Écotière, D., Lefèvre, H. and Bonhomme, B. (2009) *Propagation acoustique en milieu complexe, Méthodes et métrologie* LCPC

[7] Heutschi, K., (2010) *Lärmermittlung und Massnahmen zur Emissionsbegrenzung bei Windkraftanlagen* EMPA

[8] Wilson, D. K., Vecherin, S. N. and Ostashev, V. E. (2012) *Modeling sound fields from directive noise sources in the presence of atmospheric refraction and ground effects* In: INTER-NOISE and NOISE-CON Congress and Conference Proceedings. Institute of Noise Control Engineering, 1094-1106

[9] Bigot, A., Slaviero, D., Mirabel, Ch. and Dutilleux, P. (2015) *Influence of vertical temperature gradient on background noise and on long-range noise propagation from wind turbines* Wind Turbine Noise, Glasgow

[10] ISO 1996-2 (2017) *Determination of sound pressure levels*

[11] VDI 3783-8 (2017) *Environmental meteorology. Turbulence parameters for dispersion models supported by measurement data*

[12] NF S 31-120 (2018) *Characterization and measurement of environmental noise - Soil and weather conditions influence* AFNOR

[13] Ziemann, A., Starke, M. and Leiding, T. (2019) *Sensitivity of nocturnal low-level jets to land-use parameters and meteorological quantities* Adv. Sci. Res. 16, 85–93, https://doi.org/10.5194/asr-16-85-2019

[14] Blumendeller, E., Kimmig, I., Huber, E., Rettler, Ph. and Cheng, P.W. (2020) *Investigations on Low Frequency Noises of On-Shore Wind Turbines* Acoustics 2020, 2, 343–365, doi:10.3390/acoustics2020020

[15] VDI 4101-1 (2020) *Sound propagation outdoors in consideration of meteorological and topographical conditions – Part 1: Phenomena and Procedures*

[16] VDI 4101-2 (2020) *Sound propagation outdoors in consideration of meteorological and topographical conditions – Part 2: Wind turbines*

[17] ERA5 *Hourly data on single levels from 1979 to present* https://cds.climate.copernicus.eu/cdsapp#!/dataset/reanalysis-era5-single-levels?tab=overview

[18] IEC PT 61400-50-3 *Use of nacelle mounted lidars for wind measurements*

INCE EUROPE

9th International Conference
on
Wind Turbine Noise
Remote from Europe – 18th to 21st May 2021

Different sound source setups in the simulation of wind turbine sound propagation

Katharina Elsen, German Aerospace Center Oberpfaffenhofen, Germany:
katharina.elsen@dlr.de
Arthur Schady, German Aerospace Center Oberpfaffenhofen, Germany:
arthur.schady@dlr.de

Summary

The sound propagation of a wind turbine in complex terrain was simulated using a ray-tracing approach in order to study the effects of different sound source setups on the sound pressure level close to the ground. Apart from the meteorology also the wake of the wind turbine itself was considered within the simulations. Sound source setups were defined with different complexity in terms of spatial resolution and geometry and statistically analysed where the spatial distribution of the sound pressure level was considered in relation to the different wind directions. Whilst, according to theory, in case of flat topography and negligible meteorological effects no differences in the far field sound pressure levels can be found, significant differences in the sound pressure level distributions can be observed between different sound source setups in a more complex simulation. Our simulations suggest, that at least in case of complex terrain and under consideration of meteorology, the sound source setup must be chosen with care.

1. Introduction

Modelling the sound propagation of a wind turbine is a key component within the planning process of wind farms. In simulations, the initialization of the source have a decisive influence on the sound immission. Not only the frequency range and directivity, but also the location of every single sound source causes different ray trajectories and refraction patterns. However, detailed simulations, also including the far field of the wind turbine, can easily become very complex and, depending on the approach, be computationally expensive. In order to obtain accurate results, simulations do not only need to consider external factors as topography and meteorology, but also the sound source itself. On the other side, engineering models used for the prediction of sound pressure levels generated by wind turbines are required to obtain reproducible and accurate results within a reasonable computational time while being as user friendly as possible.

The aim of the project *Schall_KoGe* is to develop an engineering model for the sound propagation of wind turbines that is applicable in flat but also complex terrain and under consideration of different meteorological conditions. A key point within the project is to study the effect of differently detailed setups of the sound source on the sound propagation. Here, we primarily consider only the geometrical arrangement of the source points. Starting from the simplest possible (abstract source description) towards realistic initializations we investigate the relevance of a detailed description of the source geometry.

While in simple terrain and without the effects of meteorology the sound propagation in the far field does not depend on the exact sound source geometry, this is not necessarily the case for complex topography, in particular in combination with special meteorological effects. In this paper we want to demonstrate the effects of different sound source setups on the simulation of the sound propagation in complex terrain within a ray-tracing approach. Hereby, also the effects of meteorology as well as the wake of the turbine itself were taken into account, where the according flow field was derived using a WRF-simulation.
The present study completely relies on numerical simulations. As the simulations presented in the following were performed using a frequency spectrum that does not exactly correspond with the actual spectrum of the wind turbine in the field, nor was the sound source setup adjusted to the prevailing meteorological and topographic field conditions, the results were not compared to any measurement results. The goal of the present study is however to study the influence of different sound source setups on the sound propagation in complex terrain and to quantify its effects. The main goal is to find a sound source setup that offers a balance between accuracy and usability within an engineering model framework.
The paper is organized in the following manner: in section 2 an overview on the terrain in the Vale de Cobrao (Perdigão, Portugal) and the meteorological situation used for the simulations is given. Section 3 concentrates on the sound propagation model and the simulation setups itself. The simulation results as well as a statistical analysis of the results are then described in section 4. Finally, conclusions are drawn in section 5.

2. Input Data

2.1 Input Data – Terrain and Meteorology

Within the context of the New European Wind Atlas (NEWA)-Project, long-term high-resolution atmospheric and sound data were collected in the Vale de Cobrao (Perdigão, Portugal) in a large measurement campaign with European and US-American participation (Fernando, 2019). As a part of the Schall_KoGe project these data have been analyzed in order to study the sound propagation in vicinity of a wind turbine over complex terrain and will later on be used to validate our simulation results. The terrain at Perdigão is characterized by two parallel ridges, about 4500 m long, approximately the same height and 1400 m apart. Figure 1 shows a north headed view on the terrain at Perdigão, where the wind turbine (WT), located on the southwestern ridge, is indicated by a red circle.

The ridges are typically inflown perpendicular, i.e. from the southwest or northeast. Figure 2 shows the u-component of the wind velocity including the wake of the WT, coming from WRF-simulations for May 22, 2017 around 3:45 am, where a typical southwest inflow situation with a Low-Level-Jet had prevailed. One can clearly see the increased wind speed over the ridge of the hill and the decreased wind speed (wake) behind the wind turbine. Whilst in front of the WT the u-component of the wind velocity is around 6 m/s, it is in the range of 0 m/s to 1 m/s in the valley.

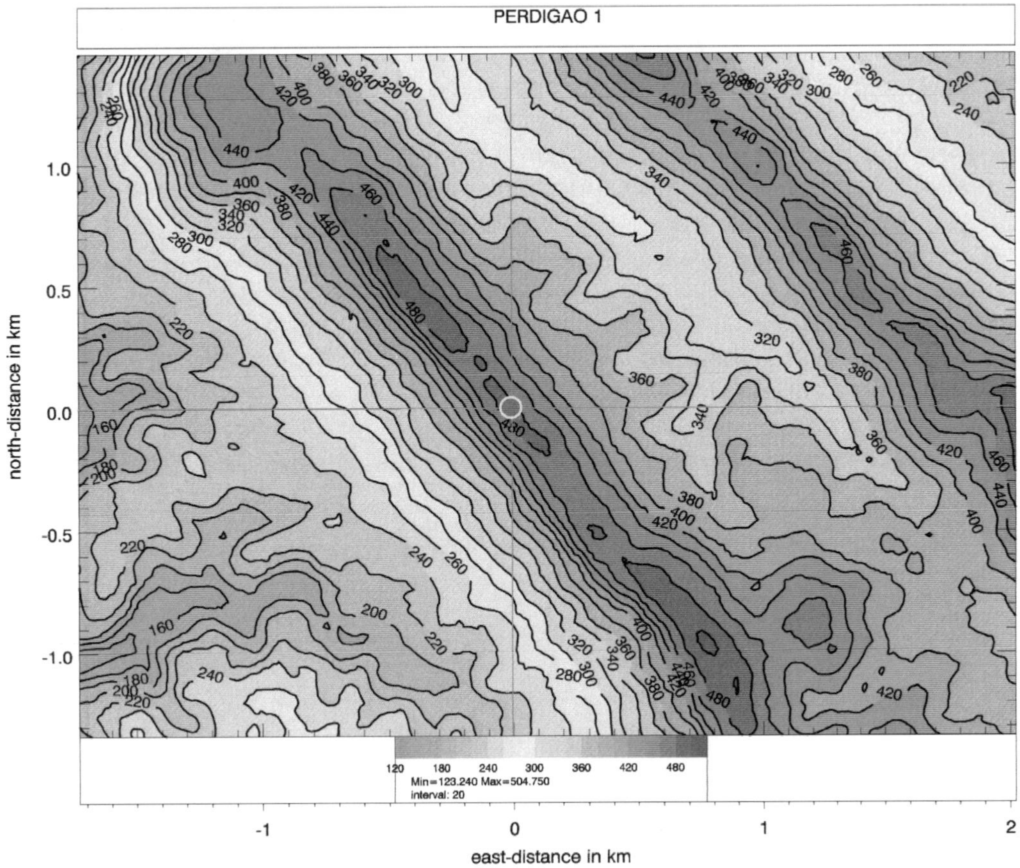

Figure 1: North headed view on the terrain at Perdigão, the WT is indicated by the red circle.

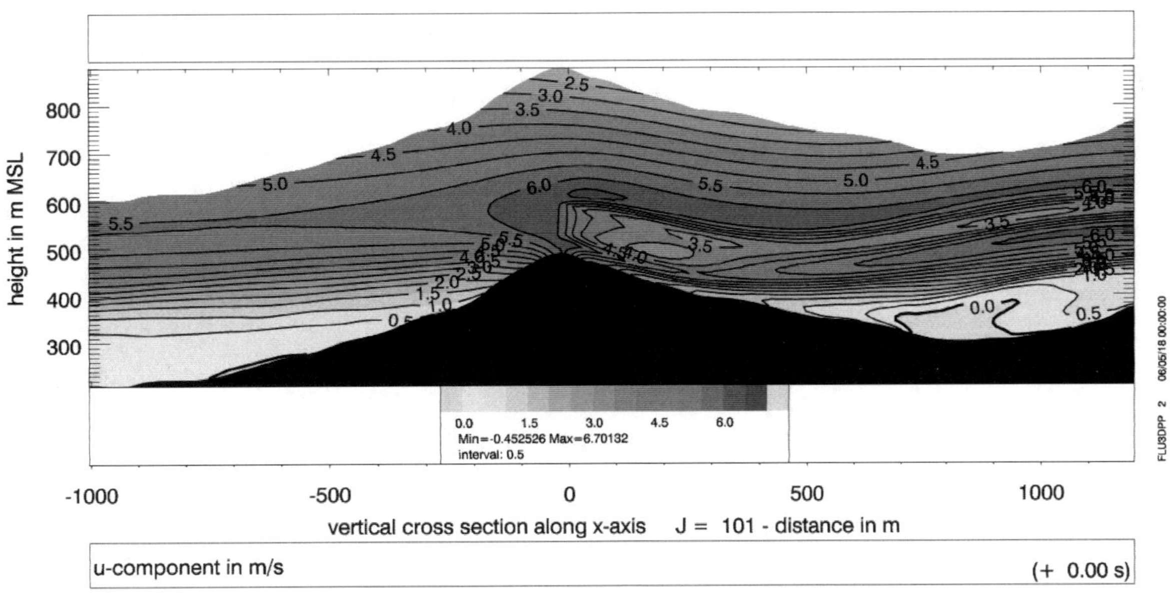

Figure 2: Typical southwest inflow situation including a Low-Level-Jet. The figure shows the u-component of the velocity.

2.2 Input Data – Sound Source Setups

Six different sound source setups with different spatial resolution and geometry are presented. Each sound source is made up from a number of point sources, where in the simplest case a single point-source at hub height is used whilst in the most complex case a total of 36 point-sources are used. Hereby the sound power of each single point source is emitted over the whole sphere where the directivity of the source is taken into account following (Ffowcs Williams & Hall, 1970). The frequency spectrum is set according to (Deutsche WindGuard Consulting, 2015). The total sound power level was set to $L_w = 107\ dB$ in all the cases.
In the following a quick overview on the six different sound source setups that are used in the following simulations shall be given. They are also depicted in Figure 3.. The aim of the 6 different cases was to be able to analyze the influence of different levels of setup complexities on the sound propagation. Whilst we cannot simulate the time-dependency of the source, i.e. the moving blades, within our ray-tracing approach model we rather chose to emulate the characteristics of the rotating blades by adapting the geometry of the sound source setup. In the simplest case, only one source at hub position is used to represent the wind turbine.

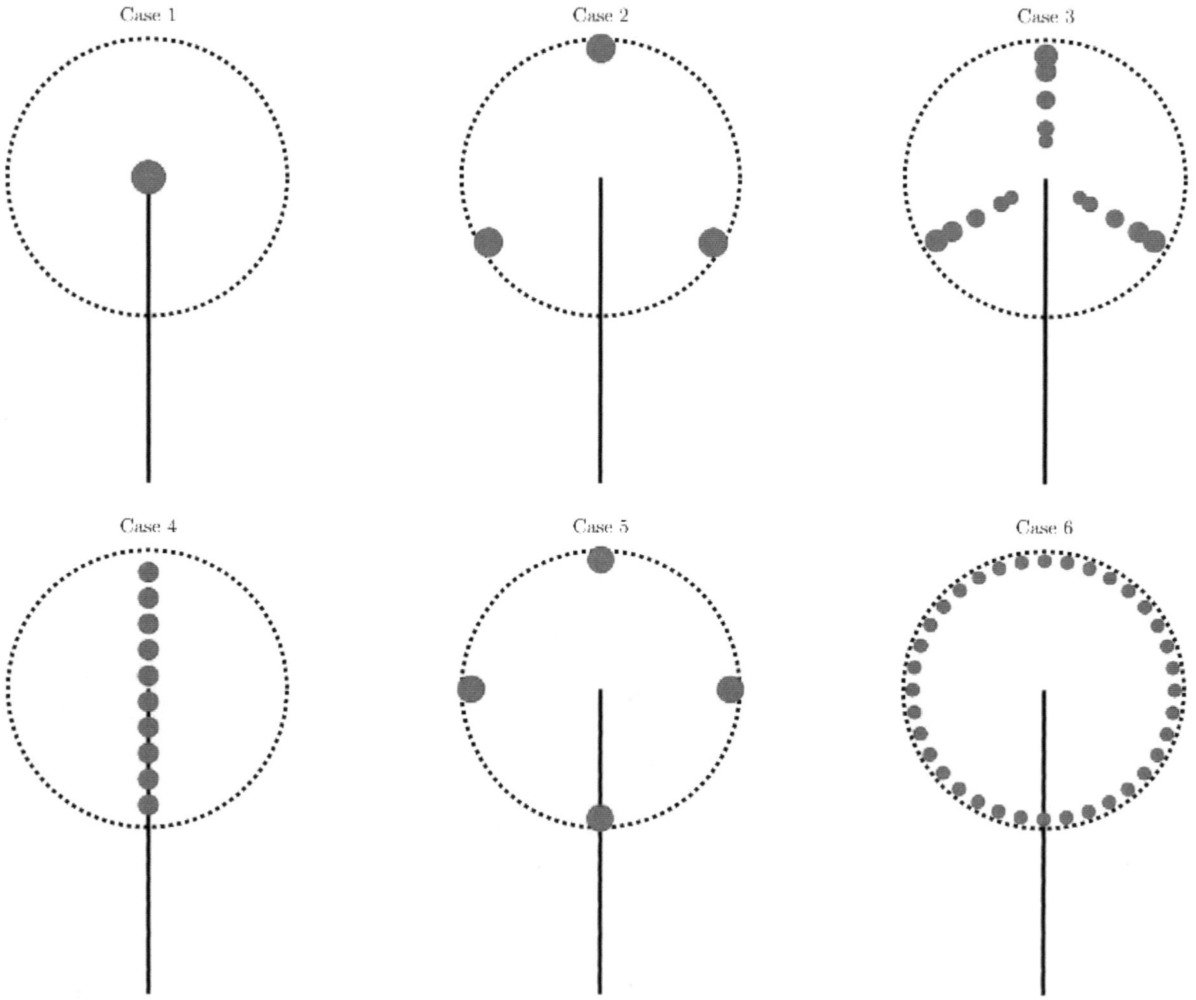

Figure 3: Overview on the different sound source setups.

Case 2 is made up of 3 point sources that are positioned at the 3 blade tips, where the blade positions are assumed to be at 90°, 210° and 330° with the angles being defined according to the unit circle. Case 3 is similar to case 2 but also considers the sound emission along the blade. Therefore 5 point sources are used to represent the sound emission at each blade where the blade positions are set again to be at 90°, 210° and 330°. Case 4 was defined to also allow for the consideration of the wind speed profile and at the same time distribute the sound sources over different heights. Case 6 considers the rotation of the blades by equally distributing 36 point sources along the circle that is described by the blade tip motion. This means that one point source is placed at every 10° of angle at a blade tip position. Case 5 is essentially a simplification of case 6 where only 4 instead of 36 point sources are used and the blade positions are assumed to be at 0°, 90°, 180° and 270°.

Table 1: Overview on the different sound source setups.

Case	# Sources	Description
1	1	At hub position
2	3	One at every of the three blade tips, blade positions at 90°, 210° and 330°
3	15	5 per blade, blade positions at 90°, 210° and 330°
4	10	In different heights over ground
5	4	1 each for four different possible blade tip positions, at 0°, 90°, 180°, 270°
6	36	Equally distributed over 36 possible blade tip positions, 1 source at blade tip position at every 10°

A summary of these setups is also given in Table 1. As case 6 has the highest spatial resolution, it will be considered the reference case in the following.

3. Description of Propagation Model and Simulation Setups

3.1 Model description

An efficient three-dimensional sound propagation model is needed which accounts for three-dimensional inhomogeneities in the air, the effects of the uneven ground and a variable definition of source setup. In this study a three-dimensional ray-based sound particle model is used (Heimann and Gross; 1999). This type of model is much less expensive than a 3-dimensional time-domain Euler model, for instance. Therefore, it can be applied to a high number of situations and a rather large range. The model was previously also applied to wind turbine noise by Heimann et al. (Heimann D K. Y., 2011).

In the model a greater number of so-called sound particles are uniformly released at the position of one or several source points from where they propagate into the surrounding airspace. Each particle carries a certain amount of sound energy depending on the strength of the source and the starting direction to account for the source directivity. The particles travel with the effective speed of sound along rays which can be curved according to the local vertical and horizontal gradients of the sound and wind speed (refraction). The ray coordinates are calculated by a time integration of the following equations which determine the position and the direction of travel of the particles (Pierce, 1981):

$$\frac{dx_i}{dt} = u_i + cn_i \tag{6}$$

$$\frac{dn_i}{dt} = -\frac{\partial c}{\partial x_i} - \sum_{j=1}^{3}\left(n_j \frac{\partial u_j}{\partial x_i}\right) \tag{7}$$

x_i (with $i = 1, 2, 3$ referring to the coordinates x, y, z, respectively) are the components of the position vector and n_i are the components of the wave vector which is perpendicular to the wave front and determines the direction of the particle movement. The sound speed c is a function of the local temperature T

$$c = \sqrt{\frac{c_p}{c_v} R_d T}$$

where c_p and c_v are the specific heat capacities of air at constant pressure and volume, respectively, and R_d is the gas constant of dry air.

Particles which hit the ground are reflected, i.e. they change their direction (specular reflection), lose a part of their energy and change their attributed phase according to the complex impedance of the ground Z_G, the grazing angle ϕ, and the plane-wave reflection coefficient

$$R = \frac{\sin\phi - \beta}{\sin\phi + \beta} \quad \text{with} \quad \beta = \frac{\rho c}{Z_G} \tag{8}$$

where $\rho = p_0/(R_d T)$ is the density of air. The impedance ratio β is parameterized by the flow resistivity of the ground σ according to Delany and Bazley (Delany ME, 1970), with $\sigma = 300$ kPa s m^{-2} in the whole domain and for all cases.

The sound energy of passing particles is added within defined receiver volumes considering their actual phase. The chosen total number of particles must be large enough to ensure that sufficient particles pass even remote receiver volumes to achieve statistically robust results.

The computational efficiency of the ray-based sound particle model is an important advantage if a set of several three-dimensional simulations over irregular terrain and in an inhomogeneous atmosphere or different source configurations are envisaged. Another advantage of the Lagrangian concept is the easy implementation of the source directivity.

However, there are also shortcomings. The ray theory is a high-frequency approach so that the results for low-frequency sound can be inaccurate if strong refraction plays a role. The treatment of diffraction is another weakness of ray-based models. In the present study diffraction in shadows behind topographical features or refractive shadows seem to be of relative importance.

3.2 Simulation setup

The terrain was rotated such that the wind direction at the wind turbine is parallel to the x-axis (comp. Figure 5) for the simulations. Hereby, the computational domain ranges from -1004.0m to 1204.0m in x-direction and from -804.0, to 804.0m in y-direction. The computational domain was chosen such that also a part of the hillside of the opposite ridge is covered by the domain, thus allowing for reflections from that hillside back into the valley. A grid width of (dx,dy,dz) = (8m,8m,8m) with 32 terrain-following z-levels i.e. with a total height of 256m over ground was used for the acoustic simulations. The meteorological grid has the same resolution in x- and y-direction as the acoustical grid but used 20 terrain-following levels in z-direction with a height-dependent grid-resolution (i.e. resolution is decreasing with height) ranging from 0 m to 500 m. The meteorological data are then interpolated onto the acoustical grid. The wind turbine is located at (x,y,z)=(0m,0m,97m). In all seven cases, a total amount of about 10 million sound particles was used for the simulations.

Turbulence is considered as far as turbulence data are an output of the WRF-simulations. Both, refraction and air absorption are taken into account. The time step was set to dt=0.004 s while the total simulation time was 5 s.

Two sets of 6 simulations each, i.e. one for each sound source setup, and therefore a total of 12 simulations were performed. One set contains simulations with flat topography and a neutral atmosphere from an acoustics point of view, i.e. zero wind speed and isothermal temperature profile, while the other set contains the simulations in complex terrain including meteorology and wake. In the first set of simulations we could confirm that between the different sound source setups there are no significant differences in the sound pressure levels in the far field. This matches with what is expected from the theory. Therefore, in the following we will concentrate on the second set of simulations only.

4. Simulation Results and Statistical Analysis

The main goal behind the statistical analysis of the sound sources is to determine if there are significant differences between the different source setups and whether or not these differences might be of importance in the context of sound propagation models. This is an important distinction as finding statistically significant differences between different source setups does not necessarily imply that these differences actually play a role throughout the propagation process and, even more important, for the recipient of the sound as differences in the sound pressure level of below 1 dB are generally barely hearable (Ortscheid & Wende, 2004). As we do not only want to find the best representation of the sound source but also have an

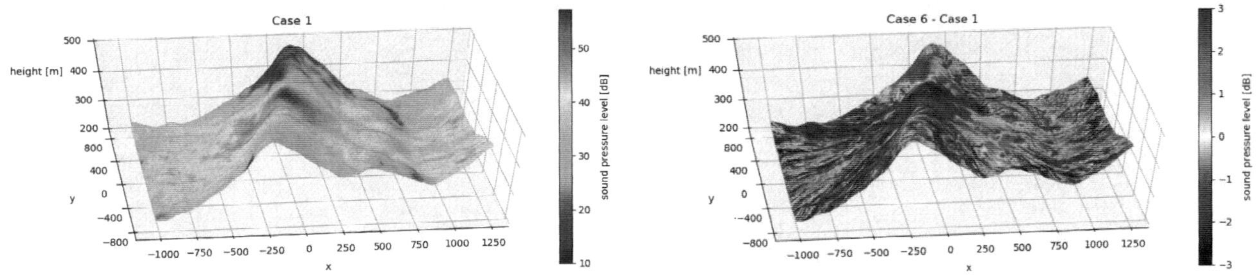

Figure 4: Left: Simulated sound pressure level 4 m above ground (terrain following coordinated) for case 1. Right: Sound pressure level difference with respect to the reference (case 6) in 4 m over ground for case 1.

engineering approach in mind, we are especially interested in finding a good balance between the necessary accuracy on one hand and a possibly low computational cost as well as usability on the other.

Whilst, generally speaking, increasing the number of point sources within the model does not directly affect the computations time of our model it still has an indirect effect as increasing the number of point sources also implies increasing more knowledge on each single point source (emission intensity, directivity, sound generation dependent on flow conditions).

4.1 Probability distribution of the sound pressure level differences

In order to better understand the differences between different sound source setups, the probability distribution of the sound pressure level differences between different cases shall be analysed. By $L_p^i(x,y,z)$ the sound pressure level field of Case *I* is denoted. As an example, the field of the sound pressure level in 4 m over ground for Case 1 is shown in Figure 4 on the left. The difference in sound pressure levels between two cases *i* and *j*, $\Delta L_p^{i,j}(x,y,z)$ is then defined as

$$\Delta L_p^{i,j}(x,y,z) = \Delta L_p^i(x,y,z) - \Delta L_p^j(x,y,z).$$

As reference, case number 6 is used, as it is one of the two cases with the highest resolution. Figure 4 on the right shows – as an example – the sound pressure level differences between Case 6 and Case 1. In the following, Case 6 will also be denoted by a superscript *R*, i.e. the sound pressure level differences between the reference case and the case 1 to 6 are denoted by

$$\Delta L_p^{R,j}(x,y,z), j = 1, \dots, 6.$$

Just by looking at the figures, one can barely notice any differences between the sound pressure levels of the different cases though. This also holds for the sound pressure level differences, where the differences mainly range from -3 dB to +3 dB and the largest differences can be found at the border of the computational domain. Therefore, the probability density

functions of the sound pressure level differences between cases 1 to 5 and the reference case were computed. Those functions are shown in Figure 5.
Even though the total number of point sources in case 5 differs strongly from the reference case 6 (4 vs. 36 point sources) this is the most similar case to the reference. The probability density function of case 5 resembles a gaussian distribution with its maximum being located around x=0 (i.e. on average case 5 and 6 are equally loud at the ground level) and being significantly higher than those of the remaining cases 1 to 4. The largest differences between case 5 and the reference is in the range of +/- 2 dB.
The maxima of the probability density functions of cases 2 to 4 are all equally high but not centered around x=0. Whilst in case 2 the maximum is shifted to the right (i.e. on average case 6 (at ground level) is louder than case 2) it is the contrary for cases 3 and 4.
As the sound propagation of a wind turbine also strongly depends on the wind direction – here

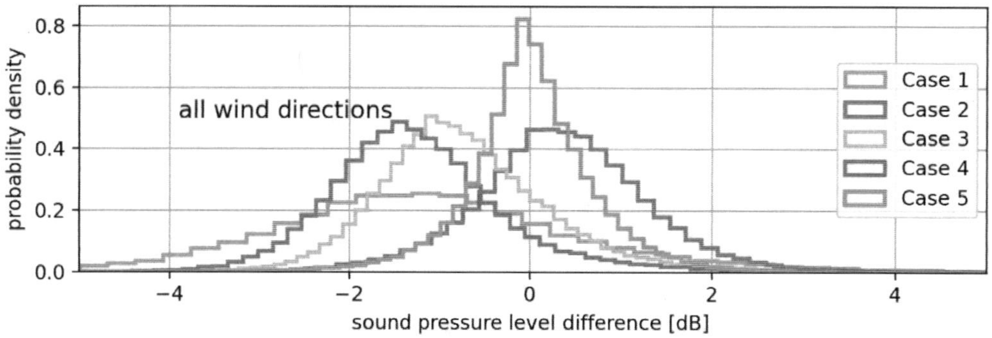

Figure 6: Probability distributions of the sound pressure level differences between cases 1 to 5 and the reference (case 6).

we need to consider effects like the differences in the vertical gradient of the effective sound speed, the wake of the wind turbine in downwind direction, or the directivity of the sound source – the domain was split into three subdomains, with regard to the wind direction. Hereby, downwind, cross wind and upwind were distinguished as is shown in Figure 6, where the respective domains are indicated in red, grey and blue.

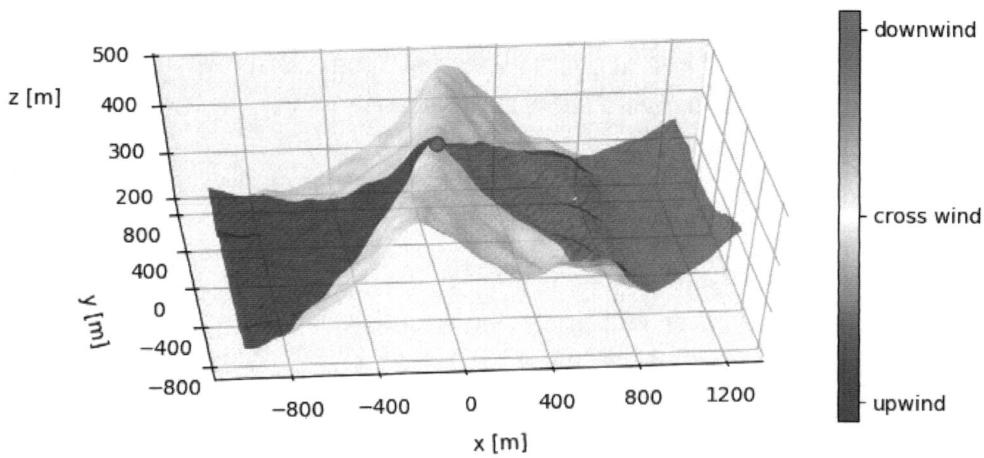

Figure 5: View on the Perdigão terrain. The three main wind directions used in the statistical analyses, downwind, cross wind and upwind, are indicated in red, grey and blue, respectively.

The wind turbine is marked by the red and blue dot located roughly in the middle of the domain. As the wind direction is not constant throughout the computational domain, the main wind

directions were defined using the wind direction at the wind turbine. As the wind direction at the wind turbine is roughly perpendicular to the ridge, the distinction of the three main wind direction goes along with a distinction of the
three main topographic aspects. In the downwind domain, the terrain is characterized by a hillslide leading into the valley between the two ridges, followed by a rise which would be leading to the second ridge. The latter however is not part of the computational domain. The upwind domain is characterized by a hillside only, whereas the two cross wind domains contain the ridge as well as the remaining area of two flanks of the hill. The probability density functions of the sound pressure level differences between cases 1 to 5 and the reference (case 6) for the three domains described above are shown in Figure 7.

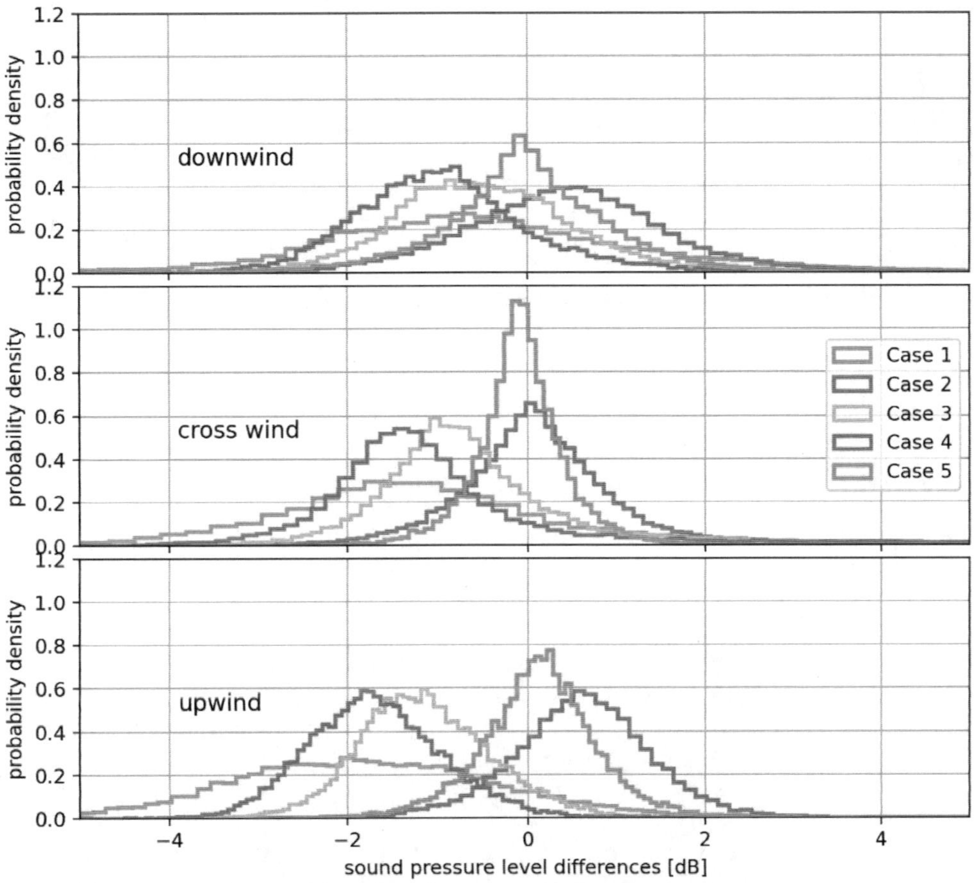

Figure 7: Probability distributions of the sound pressure level differences between cases 1 to 5 and the reference case (6) for the three main wind directions, downwind, cross wind and upwind.

From top to bottom, the probability density functions for the downwind, cross wind and upwind case are illustrated. It is easily seen that in the downwind case the curves are significantly flatter than in the cross-wind case and still clearly flatter than in the upwind case. Also here we find that for all wind directions case 5 is most similar to the reference (case 6) and on average the difference between those two cases is nearly 0.

In the downwind case the maximum of the pdf of case 2 is slightly shifted to the right, being located around x=0.5 dB whilst the maxima of cases 3 and 4 are shifted to the left by about 1 dB. The maximum of case 1 is shifted to the left by about 0.75 dB but shows the largest variation of all cases.

The upwind case shows a very similar picture to the downwind case, however, the maxima of all functions differ stronger from x=0 and the standard deviation of all PDFs is larger than in the downwind case.

In *Table 2*, selected percentiles from the probability density function of the sound pressure level differences (in dB) between the reference case and the cases 1 to 6. Values less or equal to 1 are indicated in green. Values larger than 1 but less or equal to 2 are indicated in yellow. Those values larger than 2 but less or equal to 3 are indicated in orange, whilst those larger than 3 are marked in red. This distinction was made, as differences in sound pressure levels of below 1 dB in the sound pressure level range of a wind turbine were reported to be barely noticeable *(Ortscheid & Wende, 2004)*, whilst those sound pressure level differences of more than 3 dB are usually very well audible.

Table 2: Selected percentiles from the probability density function of the sound pressure level differences in dB between the reference case (6) and the cases 1 to 5. Values less or equal to 1 are indicated in green. Values larger than 1 but less or equal to 2 are indicated in yellow. Those values larger than 2 but less or equal to 3 are indicated in orange, whilst those larger than 3 are marked in red

Percentile	Case 1	Case 2	Case 3	Case 4	Case 5
2.5 %	-4.46	-1.69	-2.42	-2.99	-1.57
5 %	-3.86	-1.27	-2.14	-2.70	-1.19
25 %	-2.24	-0.23	-1.35	-1.87	-0.36
50 %	-1.20	0.34	-0.82	-1.32	-0.01
75 %	-0.12	0.94	-0.18	-0.71	0.39
95 %	2.16	1.97	1.04	0.79	1.22
97.5 %	3.15	2.40	1.60	1.68	1.57

4.2 Dependency of the sound pressure level from the distance from the source

In order to better understand the distribution of the sound pressure level differences, also the dependency of those differences from the distance from the source were analysed.

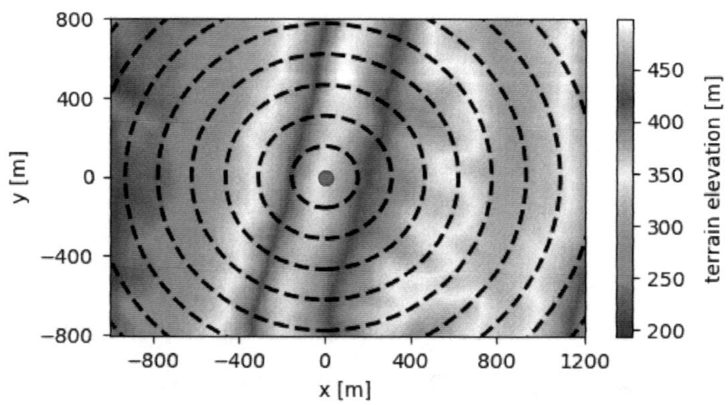

Figure 8: View on the terrain. The wind turbine is marked with a red dot. The black dashed rings indicate the rings within which the averaged sound pressure levels are computed.

Therefore, the computational domain was split into rings with a width of 16 m, which corresponds to the double grid cell width. This is indicated in Figure 8, where the wind turbine is marked by a red dot and the rings are indicated by the black dashed line.
Within these rings, the sound pressure level differences of case 6 against the other studied cases at the ground level were averaged using an energetic mean. In Figure 9 the dependency of the sound pressure level differences from the distance of the source are indicated for the cases 1 to 5, where case 6 is the reference and averaged over all wind directions. The largest differences are found around the wind turbine at x=0, over the following 600 m the sound

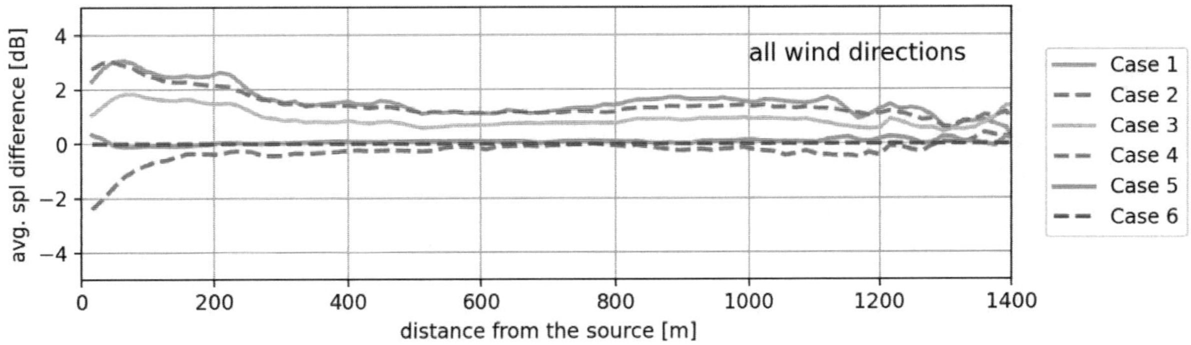

Figure 9: Dependency of the sound pressure level differences from the distance from the source for the cases 1 to 5, where case 6 is the reference again, and for all wind directions.

pressure levels slowly converge towards y=0 as can be expected from the theory. However, at around 800 m we can observe that the differences in sound pressure level become larger. The difference between case 5 and the reference case is close to zero independently from the distance to the source.

Aditionally we analysed the influence of the different wind directions and therefore split the domain into the three above mentioned sub-domains. The dependency of the sound pressure level differences from the distance of the source for the single main wind directions is shown in Figure 10. In the downwind case we can now observe a much faster convergence towards y=0 where at around x=600 m the difference in sound pressure level is below 0.5 dB for almost all cases. Whilst cases 2 stays close to 0 for the next 400 m, cases 1, 3 and 4 start to deviate compared to case 6. At around x=1000 m also case 2 finally starts to diverge from y=0. The above described behaviour, i.e. diverging from y=0 again correlates with the change of the orography in that area, where it changes from a downhill slope into a valley and further into an uphill slope. A very similar behaviour can be observed in the upwind domain, where between 600 and 800 m of distance from the source the orography changes from a downhill slope into a flatter terrain. In the crosswind case however, the differences between cases 1 to 5 and the reference case stay relatively constant, with the cases 2 and 5 showing significantly smaller differences from the reference than cases 1, 3 and 4.

5. Conclusions

The aim of this study was to show the influence of different sound source setups on the sound propagation of a wind turbine in complex terrain. It shall be mentioned that the reference case is not necessarily the best or most realistic representation of the sound source, but it is the case with the highest spatial resolution utilized here, representing the full rotation of a blade. We could show that in combination with complex terrain and meteorology there can be significant differences between different sound source setups that might be of importance when simulation the sound propagation of a wind turbine. Even though the effects of different sound source setups are less strong than those originating from meteorology – variations due to meteorology can make up to 25 dB in 1000 m of distance from the source (D., 2018) – depending on the wind direction they might still in the range of the effects caused by the orography or the wake of the wind turbine itself and therefore should be considered to be included in sound propagation simulations in complex terrain. Further research will be dedicated to compare sound propagation simulations performed using more realistic sound source setups originating from CFD-simulations with measurements from the Perdigão field experiment.

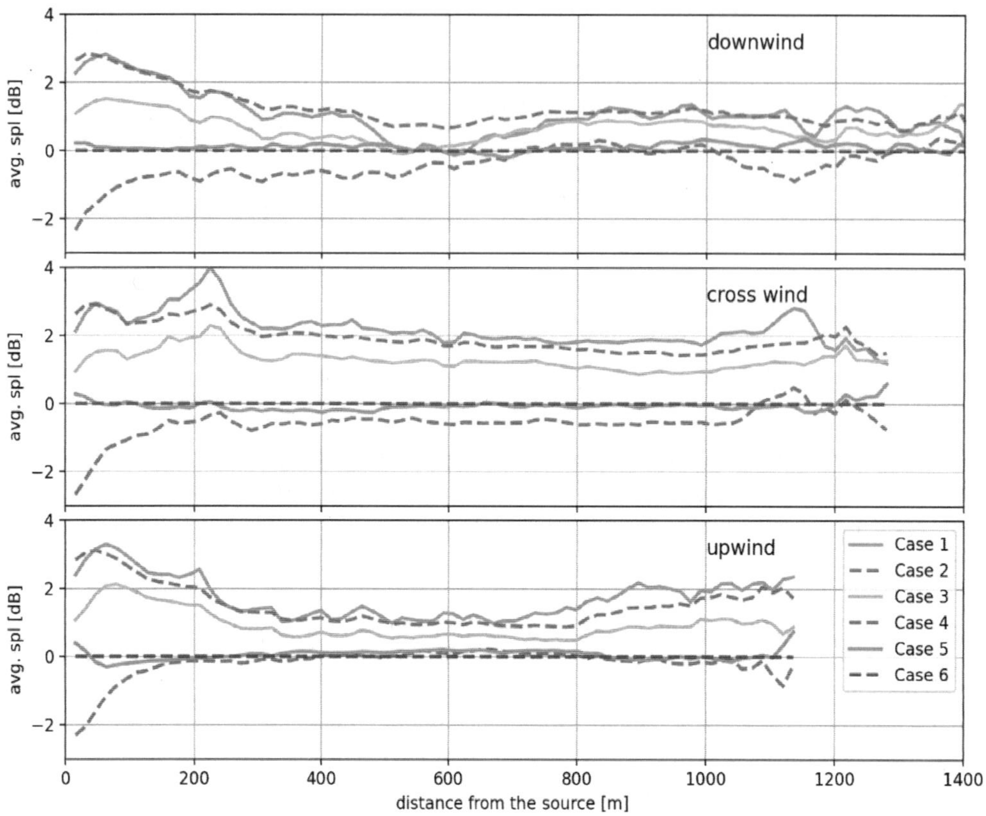

Figure 10: Dependency of the sound pressure level differences from the distance from the source for the cases 1 to 5, where case 6 is the reference, seperated by wind direction.

Acknowledgement

The studies for this project are supported by the Federal Ministry of Economics and Energy within the framework of the Schall_KoGe project with the funding code: 0324337B.

References

A.D., P. (1981). *Acoustics - An introduction to its physical principles and applications.* New York: McGraw-Hill Inc.

D., H. (2018). Sound propagation from a wind turbine under various atmospheric conditions. *Meteorologische Zeitschrift* , pp. 27: 265-275.

Delany ME, B. E. (1970). Acoustical properties of fibrous absorbent materials. *Applied Acoustics* , pp. 3 : 105-116.

Deutsche WindGuard Consulting, G. (2015). *Auszug aus dem Prüfbericht MN15016 zur Schallemission der Windenergieanlage vom Typ Enercon E-115 Betriebsmodus BM0s.*

Fernando, H. J. (2019). *The Perdigão: Peering into Microscale Details of Mountain Winds. .* Bull. Amer. Meteor. Soc., 100, 799–819.

Ffowcs Williams, J. E., & Hall, L. H. (1970). Aerodynamic sound generation by turbulent flow in the vicinity of a scattering half plane. *Journal of Fluid Mechanics*, pp. 40: 657-670.

Heimann D, G. G. (1999). Coupled simulation of meteorological parameters and sound intensity in a narrow valley. . *Applied Acoustics* , pp. 56: 73-100.

Heimann D, K. Y. (2011). The wake of a wind turbine and its influence on sound propagation. . *Meteorologische Zeitschrift* , pp. 20 : 449-460.
Ortscheid, J., & Wende, H. (2004). Retrieved from https://www.iserlohn.de/fileadmin/user_upload/Dokumente/Umwelt/Laermaktionsplan_Stufe_3/lmm-geringer-akustischer-wirkung.pdf

9th International Conference
on
Wind Turbine Noise
Remote from Europe – 18th to 21st May 2021

A Review on the Development of Airfoils for Wind Turbine Blades

Alexandre Martuscelli Faria (University of São Paulo): martuscellifaria@usp.br
Joseph Youssif Saab Jr. (Mauá Institute of Technology): saab@maua.br
Sara Rodriguez (University of São Paulo)
João Paulo Barreto (University of São Paulo)
Marcos de Mattos Pimenta (University of São Paulo)

Summary

Airfoil geometry is key to fulfill two important requirements while designing utility-size wind turbine equipment: (i) aerodynamic performance for high power coefficients during energy conversion and (ii) low noise generation for compliance with sound pressure level limits. Contrary to the classic structural trade-off, a low noise requirement does not seem to pose an antagonistic design objective with aerodynamic efficiency, at least in the range of audible noise frequencies and yet, airfoils designed for low noise emissions seem to be an exception to the rule, at least within the published literature. This review paper reports on the results of several airfoil development research for large-size wind turbine blade application, covering geometries developed solely with aerodynamic performance in mind and those few also concerned with noise emission from the inception. A reference table with data of the airfoils discussed is provided along with the associated design Reynolds numbers, whenever available. Finally, a selected sample of design criteria for large-size wind turbine airfoils is briefly discussed.

Nomenclature

a = speed of sound [$\frac{m}{s}$]

c = Airfoil chord [m]

C = coefficient

$Cl\,x\,AoA$ = 2D (airfoil) coefficient of lift versus angle of attack diagram

I = noise or sound intensity [$\frac{W}{m^2}$]

L = span of the airfoil section considered [m]

l = length scale of the turbulent region [m]

M = U/c_0, the eddy convection velocity Mach number

P = mechanical power converted from the available wind power [W]

r = distance from source to observer [m]

t = airfoil thickness [m]

U = typical eddy convection velocity [$\frac{m}{s}$]

V = average wind velocity [$\frac{m}{s}$]

x = length over the airfoil chord [m]

α = normalized turbulence intensity

ρ = air density [$\frac{kg}{m^3}$]

Subscripts

$_l$ = lift

$_d$ = drag

$_{design}$ = @ design point

$_{max}$ = @ maximum value

$_0$ = reference fluid state

$_P$ = power

1. Introduction

A consistent trend found in the manufacturing process of utility-size, horizontal-axis wind turbines (HAWTs) throughout the years is the steady growth of the rotor diameter, which has led to a decreasing cost of energy conversion [1]. However, while the power conversion increases with the third power of the wind velocity, the acoustic power of the trailing-edge noise (TEN) source, which is the main airfoil self-noise source for large size HAWT equipment scales with the fifth or sixth power of the flow velocity, depending whether the noise source size is compact or not in relation to the wavelength under consideration [2]. Thus, the aeroacoustic noise associated with the ever-larger rotor diameter and higher local speeds due to increasing tower heights might impose limitations to the size of inshore-bound equipment in the future. This environmental restriction could compose the structural and material technical challenges already faced by the wind energy industry, which must meet challenging clean energy goals set by Countries throughout the World in order to fulfill greenhouse gas emissions (GHG) limitations and global warming temperature limits proposed by the PARIS agreement [3] and other subsequent Conferences of the Partiess (COP) of the United Nations.

For the modern HAWT, the optimum range of Tip Speed Ratios (TSR, λ) is between 6 and 8 [4], with 7 being the most frequently reported value. For a 12 m/s wind and a TSR of 7, a modern HAWT operates at a tangential velocity of 84 m/s at the tip of the blades. Because of the diameter growth trend, blades sized exclusively based on the aerodynamic performance will produce ever stronger sound power levels (SPW) and will demand ever larger deployment sites and associated capital expenditure (CapEx) for operating within acceptable sound pressure levels (SPL) as measured at the dwellings closest to the wind farms. These considerations suggest that a modern set of requirements for designing airfoils for HAWTs should include, apart from the robust aerodynamic behavior expected, restrictions on aerodynamic noise generation [5]. The power converted by a wind turbine may be calculated, once the coefficient of power is known, with the aid of Eq. 1,

$$P = 0.5 \cdot \rho \cdot V^3 \cdot S \cdot C_p \tag{1}$$

Equation 1 asserts the dependence of the power production on the wind speed cubed, curbed by the theoretical Betz limit for the Coefficient of Power (C P) of 59.25% for unshrouded rotors [6]. On the other hand, the scaling of the sound intensity, I, resulting from the interaction of the turbulent boundary layer (TBL) with the airfoil trailing-edge (TE), for a compact source, may be calculated with the aid of Eq. 2 [7],

$$I \propto \rho_0 c_0^3 M^5 \alpha^2 \frac{L\, l}{r^2} \cos^3(\bar{\theta}) \tag{2}$$

The noise intensity is then influenced by four main flow parameters, i.e., the eddy convection velocity, M ; a length scale of the turbulent region, l; the nor- malized turbulence intensity, α; and the angle between the eddy convective velocity and the edge, $\bar{\theta}$. Except for the latter, which may be manipulated by 3the addition of serrations in the TE [2] for instance, all other parameters are influenced by the native geometry of the airfoil. The state of the turbulence in the vicinity of the TE is determined by the development history of the boundary layer (BL) and, hence, by the shape of the pressure distribution and gradient along the blade section [8]. As a suitable length scale of the turbulent region, the boundary layer displacement thickness is often adopted by both semi-empirical and theoretical noise emission models [9], with direct implication that attached flows will lead to less noise emission at the trailing- edge of the airfoil. The remaining part of this review is organized in two sections: the review of selected researches aiming to develop wind turbine (WT) airfoil geometries through various approaches, with and without concern regarding noise sources for the last twenty years, including a summary data table of major airfoils developed for HAWTs and their associated theoretical design characteristics and some selected examples of criteria for designing HAWT airfoils.

2. An overview of airfoil development research for large-size HAWT equipment

A summary of the theoretical performance data for the main airfoils described in this section may be found in Table 1, whenever available from the original sources, along with design Reynolds number range and blade-length recommendation for applicability.

A three-step noise prediction method was set up by Guidati and Wagner [10] in order to study the influence of the turbulent boundary layer (TBL) structure on the TE noise and to investigate airfoil shape influence on noise emission. The first step of the method employed the X-FOIL code [11, 12, 13] to compute the BL flow around the airfoil. The second step was to reconstruct the TE velocity profile by using the Cole law of the wall [14], and the last step was to obtain the far field noise employing the theoretical TNO TE noise model [8, 15, 16]. One important outcome of that research was the comparison of the X-FOIL e n transition model [11] against data measured at the IAG Laminar Wind Tunnel, in the form of drag-polar curves for the different airfoils tested, with favorable results. The main conclusion of the research [10] was that a thin, attached BL is favorable for reducing TE noise and that a moderate noise reduction was achievable by airfoil shape manipulation. The airfoils developed in this fashion displayed a reduction in sound pressure level (SPL) between 2 and 4 dB(A) in the frequencies between 1 kHz and 2 kHz for the untripped flow case only, while for tripped flow they emitted more noise than the NACA 64418 airfoil used as a benchmark. No trade-off study was reported regarding noise and performance for the newly developed airfoils, the profiles of which are not provided in the research paper. During the development of the RISO family of airfoils by Fuglsang and Bak [17], the authors employed a direct design method detailed in Fuglsang et al. [18], where a traditional Simplex optimizer was used with a finite difference sensitivity analysis method and coupled with the X-FOIL code. The design variables were the control points of the curves that described the airfoil shape. The solution method was considered robust, even if computationally expensive. The optimization problem did not consider minimizing TE noise as part of the objective function or as a constraint. The available design data for the RISO family of airfoils is shown in Table 1. The RISO and the CU family of airfoils are among the few high Reynolds number airfoils detected in this review.

In USA, most of the published WT airfoil development work was accomplished under the initiatives of the National Renewable Energy Laboratory (NREL) and it was organized in the form of public reports released from 2005 on, although the research work had been ongoing for many years. Most airfoil families developed have been designed under a partnership with Airfoils Inc. [19, 20, 21] and were recommended for blade lengths ranging from 3 m to 50 m. Their development was carried out on theoretical grounds alone [19], based on results from simulations run with the Eppler airfoil code [22]. The resulting airfoils generally display thick sections with laminar flow development (retarded transition) in the "clean" condition. A list of Somers and Tangle's airfoil and patents may be found in the NREL website [23]. The EU initiative called SIROCCO [24] entailed developing silent rotors by acoustic optimization as a primary target. The idea was to modify the BL state at the TE, by adjusting the pressure recovery at the rear part of the air- foil. In accordance with the authors, in this region there is a significant fluid deceleration which violates the equilibrium boundary layer approach and would render previous noise prediction schemes based on the TNO-TPD TE noise method, inappropriate. Thus, a modified-TNO-TPD TE noise method was proposed, now considering the anisotropy effects via a Reynolds Stresses turbulence model. The combined aerodynamic and aeroacoustic models were implemented in a numerical optimization environment that generated airfoil shapes with "minimal noise production", in an automated fashion. The air- foil shapes produced by the SIROCCO initiative are not available to the public and the wind tunnel noise measurements reported indicated from 1 to 3 dB(A) less noise emitted at the design-lift range. Field measurements, however, showed the noise reduction of these airfoils to be scant (of the order of ≈ 0.5 dB(A)), with reasons not fully understood but assigned by the authors to possible unsteady inflow conditions. Bertanoglio et al. [16] concluded that the TNO model for TE noise displayed an offset of about 5 to 10 dB SPL under prediction when compared to

NACA0012 noise data measured by NASA [25] and at the Aeroacoustic Wind tunnel Braunschweig (AWB). However, the authors considered the model still useful for relative performance comparison of airfoils. They next coupled the model to an optimization code developed at RISO in order to reduce a cost function - made up of aerodynamic plus TE noise components - subjected to many constraints. Basically, the code calculated the local gradient of the cost function associated with each parameter, adopting values that would minimize the cost function. The treatment to avoid local minimum values is not described. The RISO-B1-18 airfoil produced by the method, was subjected to optimization at the single angle of attack (AoA) of 6°. The chord-based Reynolds number for the simulation is not specified, although previous validations attempts in the same research were accomplished at Re = 1.6E + 06 and 2.9E + 06. The geometry and boundary layer thickness of the modified airfoil were reportedly little different from the benchmark values, however the authors concluded that the Turbulent Kinetic Energy (TKE) had been reduced in the optimized design, particularly near the TE, with a resulting reduction of noise in the range of 1-2 dB, which they considered relevant for the purpose of WT design.

Another optimization technique for wind turbine airfoils was presented by Li et al. [26], where a combination of methods was used to optimize the lift-to-drag ratio of an airfoil at the design angle of attack. The upper and lower surfaces of a reference airfoil were replaced by two B-spline curves with four control points each. The optimization goal was to reach the maximum C l /C d ratio at the airfoil design AoA, but there was no declared concern with airfoil self-noise. Four examples of geometrical optimizations are shown in the paper with up to 24% improvement in (C l /C d) max at the airfoil design point. However, the design requirements for WT blades usually involve many simultaneous requirements other than the isolated aerodynamic efficiency at the design point, as will be seen in section 3. In Petrone et al. [27], airfoil self-noise plus inflow noise sources were superimposed in order to calculate the full acoustic signature of a complete, operating WT. A design optimization process based on a genetic algorithm was then applied to the geometry in order to seek for noise reduction while considering many uncertainties involved in the real process, e.g. surface contamination by insects, geometrical imperfections and meteorological conditions. A new WT blade was designed and constructed from 3 different airfoils (root, middle and tip), with interpolated geometries in between them. The X-FOIL was used for 2D flow estimation over each section and a finite-span effect was also integrated. The resulting blades were applied to an AOC 15/50 WT (15 m-span, 50 kW rated power), with the aim to maximize the mean power coefficient while reducing the SPL at a microphone located 20 m downwind of the turbine, at the ground level. The authors claimed they were able to find a trade-off blade geometry with significant less noise emission and with negligible power coefficient reduction with respect to the baseline, factory optimized geometry. However, no geometric data is provided for the blade sections of the final blade configuration. The authors also claimed to have demonstrated how the construction and operating uncertainties considered resulted in a general decrease in performance with respect to the design scenario, suggesting that designers should not rely on hydraulically-smooth and geometrically-perfect blade surfaces in order to reach the necessary performance.

Xiaomim [28] described the use of a genetic algorithm to optimize WT airfoils. The airfoil upper and lower curves were parameterized and the thick- ness was allowed to vary within a prescribed percentage of the chord. For each parametric combination a CFD-RANS simulation was run in order to evaluate the cl and cd of the airfoil configuration. The "quality" of each variety was then assessed in order to breed another generation and so on, successively, until a maximization goal was achieved. However, no acoustic restraints or objective functions were included in this research and the resulting airfoils or performance were not published. Cheng et al. [29] have coupled numerical optimization to the flow solver RFOIL [30], and also employed a genetic algorithm method in order to perform the optimization of a WT blade design by adapting and morphing a "shape function", based on Taylor high-order polynomial series. This study has resulted in a family of three WT airfoils, which were described as "WT" series, with relative thicknesses in the range of 15 to 21%. The work shows the comparison of the newly developed WT-180 airfoil with some selected bench- marks (NREL-S810 and the NACA-63-418) in the RFOIL environment, with the WT-180 displaying higher theoretical C L /C D ratio, as well as higher C L values when compared to the benchmarks, at Re = 3E + 06. A similar RFOIL analysis was made for WT-150 and WT-201 airfoils, but only the WT-180 is later subjected to a wind tunnel testing for three Reynolds numbers (Re = 2E + 06, Re = 3E + 06, Re = 4E + 06), which basically agreed with the RFOIL analysis. No acoustic design constraints were considered on the algorithm development, however, the authors prescribed a sharp trailing-edge in order "to prevent excessive boundary layer noise". Grasso [31] also reported on the development and wind-tunnel testing of a family of airfoils developed for very large WT. The development was accomplished via multidisciplinary design optimization (MDO) founded on a gradient-based algorithm. The airfoil was represented by Bezier curves, with 13 control points in this specific case, and the aerodynamic behavior is evaluated via the RFOIL code. The objective function is a linear, weighted combination of the aerodynamic efficiency and sectional moment of inertia. A family of airfoils designated ECN-G1-XX was designed with relative thicknesses ranging from 15% to 21% and the 21% thick ECN-G1-21 model was wind tunnel tested at very low turbulence intensity and Reynolds numbers of 1E + 06 and 3E + 06, in free and fixed transition, plus with the addition of vortex generators (VG). In free transition at the highest Reynolds number, the stall behavior was more sudden than predicted but the pressure distributions were correctly predicted by the RFOIL. Drag-polar curves from RFOIL were also well predicted for both Reynolds numbers after the RFOIL drag values were adjusted with an uniform increase of 10%. In order to try to improve the stall characteristics, a post-stall lift drop constraint was added, resulting in the ECN-G3-XX family of airfoils, with relative thicknesses in the same range as the original family. The resulting family showed smoother stall characteristics at the sake of a decreased C l,max . Liu [32] have presented a review on WT noise mechanism, as well as "de-noising techniques", characterizing the dominant noise sources for the frequency spectrum ranging from zero to 10 kHz. The de-noising techniques presented referred both to aerodynamic noise suppression by blade optimization combining numerical and experimental methods, as to mechanical de-noising methods through signal processing diagnosis and decomposition techniques. This work, however, does not introduce any new WT airfoil geometry, dealing only with the classical NACA0012 benchmark geometry. Miller et al. [33] presented a new family of flatback airfoils created by optimization through a genetic algorithm in which the airfoil shapes were modified using Bezier curve parameterization. The aerodynamic performance of the family of airfoils was evaluated with the aid of X-FOIL code and structural analysis was also considered in this case. Flatback airfoils may be found in the inner-to-mid region of the wind turbine blades and they usually have a high resistance to flapwise bending and may attain high maximum lift coefficients. Thus, the CU-W1 XX family of airfoils was designed and numerically tested at Re = 6E + 06 and M = 0.1 with the aid of X-FOIL, resulting in C l,max and C l /C d ratio values larger than the similar existing airfoils, such as NREL S8 series, and Delft University DU series. A single mention on airfoil noise is made related to the operating conditions, but not as a design constraint. The family has relative airfoil thicknesses varying in 3% increments, from 21% to 36% chord length. Zhu et al. [34] have made an overview of the activities of the Wind Turbine Department of the Technical University of Denmark (DTU) in regards to coupling wind turbine noise generation and propagation models. The activities are reported to include the development of high fidelity Navier-Stokes-based computational aeroacoustics methods suitable for understanding the noise generation mechanisms and yielding engineering oriented models suitable for airfoil and rotor design applications.

Li et al. [35] introduced a complex methodology for design and optimization of large HAWT airfoil for application in sites with prevalent low wind speed with high inflow turbulence. The parametric study considered aerodynamic and structural aspects of the problem, but also included a noise emission assessment of the geometries by application of the NREL Airfoil Noise code (NAFNoise) [36, 37] based on the NASA BPM self-noise code [38, 39]. Noteworthy are the special design considerations regarding sensitivity to inflow turbulence, with special attention to the small-scale turbulence, which has intensities recorded through anemometry in excess of 20%, although classic inflow turbulence intensity modeling in blade airfoil design is usually set to very low values due to historical influence of aviation airfoil design and low turbulence wind tunnels. Unfortunately, an inflow noise model was not included as part of the optimization methods, where the authors have employed an overall design optimization (ODO) framework and an extended design optimization (EDO) framework. A case design was then introduced, having the NACA 63421 outboard airfoil as the departure point for the design optimization. According to the authors, the optimized airfoil geometries developed, designated CAS-W1-210 and CAS-Ti-210 have presented better theoretical results on aerodynamic and aeroacoustic efficiency as evaluated by the X-FOIL/RFOIL and NAFNoise analysis. Table 1 shows available, approximate design point data for selected HAWT airfoils. It is interesting to notice that the maximum design Reynolds numbers point found in the selected discussed papers is limited to 6E + 06, while the more characteristic operating conditions estimated via the Blade Element Momentum (BEM) method [6] suggest Reynolds numbers is in the range of 7E + 06 to 9E + 06 and Mach numbers in the range 0.14 to 0.21 for a 100 m-diameter-rotor designed with the Somers S830 airfoil from 55% to 90% blade span, subjected to a wind speed of 12 m/s and TSR of 7 [40]. A summary of WT airfoil data resulting from the researches within this review may be seen in Table 1.

Table 1 - A summary of major airfoils developed for large-size horizontal axis wind turbines and their theoretical aerodynamic design characteristics.

Airfoil	t/c		α (°)		Reynolds #	Clean Airfoil		Tripped Airfoil			Blade size	Ref.
	max.	@ chord %	zero lift	design	(design)	$C_{lmax.}$	$(Cl/Cd)_{max.}$	trip loc.	$C_{lmax.}$	$(Cl/Cd)_{max.}$	[m]	
DU 91-W2-250	0.25	0.3	-3.2	6.6	3.0E+06	1.37	128	5% u.s.	1.16	62	N/A	[31], [42]
DU 93-W-210	0.21	N/A	-4.2	N/A	3.0E+06	1.35	143	5% u.s.	1.17	65	N/A	[31], [42]
DU 95-W-180	0.18	0.3	-2.0	N/A	3.0E+06	1.21	143	5% u.s.	1.14	70	N/A	[31], [42]
DU 96-W-180	0.18	0.3	-2.7	6.6	3.0E+06	1.26	145	5% u.s.	1.17	73	N/A	[31], [42]
DU 97-W-300	0.30	0.3	-2.2	9.3	3.0E+06	1.56	98	5% u.s.	1.17	53	N/A	[31], [42]
DU 00-W-212	0.212	0.3	-2.7	6.5 @ Cl=1.06	3.0E+06	1.29	132	N/A	N/A	N/A	N/A	[31], [42]
DU 00-W-350	0.35	0.3	-2.0	7.0 @ Cl=1.13	3.0E+06	1.39	81	N/A	N/A	N/A	N/A	[31], [42]
DU 00-W-401	0.401	0.3	-3.0	5.0 @ Cl=0.82	3.0E+06	1.04	54	N/A	N/A	N/A	N/A	[31], [42]
Risø-B1-15	0.15	0.278	-4.1	6.0 @ Cl=1.21	6.0E+06	1.92	157	N/A	N/A	N/A	N/A	[42]
Risø-B1-18	0.18	0.279	-4.0	6.0 @ Cl=1.19	6.0E+06	1.87	166	N/A	N/A	N/A	N/A	[42]
Risø-B1-21	0.21	0.278	-3.6	6.0 @ Cl=1.16	6.0E+06	1.83	139	N/A	N/A	N/A	N/A	[42]
Risø-B1-24	0.24	0.27	-3.1	6.0 @ Cl=1.15	6.0E+06	1.76	120	N/A	N/A	N/A	N/A	[42]
Risø-B1-30	0.30	0.27	-2.1	5.0 @ Cl=0.90	6.0E+06	1.61	N/A	N/A	N/A	N/A	N/A	[42]

Risø-B1-36	0.36	0.27	-1.3	5.0 @ Cl=0.90	6.0E+06	1.15	N/A	N/A	N/A	N/A	N/A	[42]
Risø-P-15	0.15	0.328	-3.5	6.0 @ Cl=1.12	3.0E+06	1.49	173	N/A	N/A	N/A	N/A	[42]
Risø-P-18	0.18	0.328	-3.7	6.0 @ Cl=1.15	3.0E+06	1.50	170	N/A	N/A	N/A	N/A	[42]
Risø-P-21	0.21	0.323	-3.5	6.0 @ Cl=1.14	3.0E+06	1.48	159	N/A	N/A	N/A	N/A	[42]
Risø-P-24	0.24	0.32	-3.7	6.0 @ Cl=1.17	3.0E+06	1.48	156	N/A	N/A	N/A	N/A	[42]
NREL S830	0.21	0.3	-5.8	5.0 @ Cl=1.25	4.0E+06	>1.6	170	2% u.s.; 5% l.s.	>1.6	74	40-50	[20], +XFLR5, V6_40, M=0.2 run [13]
WT150	0.15	0.254	N/A	N/A	3.0E+06	1.77	158,079	N/A	N/A	N/A	N/A	[30]
WT180	0.18	0.294	N/A	N/A	3.0E+06	1.63	151.6	N/A	N/A	N/A	N/A	[30]
WT201	0.201	0.305	N/A	N/A	3.0E+06	1.57	152.65	N/A	N/A	N/A	N/A	[30]
ECN-G1-21	0.21	N/A	N/A	N/A	6.0E+06	≈2.1	≈165	1%	≈1.75	≈105	100	[32]
ECN-G3-21	0.21	N/A	N/A	N/A	6.0E+06	≈1.6	≈160	1%	≈1.75	≈105	100	[32]
CU-W1-21	0.21	0.27<x/c <0.29	N/A	N/A	6.0E+06	1.55	156	N/A	N/A	N/A	N/A	[34]
CU-W1-24	0.24	0.27<x/c <0.29	N/A	N/A	6.0E+06	1.55	140	N/A	N/A	N/A	N/A	[34]
CU-W1-27	0.27	0.27<x/c <0.29	N/A	N/A	6.0E+06	1.55	138	N/A	N/A	N/A	N/A	[34]
CU-W1-30	0.3	0.27<x/c	N/A	N/A	6.0E+06	1.65	131	N/A	N/A	N/A	N/A	[34]

	0	<0.29										
CU-W1-33	0.33	0.27<x/c <0.29	N/A	N/A	6.0E+06	1.80	122	N/A	N/A	N/A	N/A	[34]
CU-W1-36	0.36	0.27<x/c <0.29	N/A	N/A	6.0E+06	1.90	116	N/A	N/A	N/A	N/A	[34]
CAS-Ti-210	0.21	0.305<x/c< 0.35	N/A	5	4.0E+06	1.39	162.56	N/A	1.12	N/A	N/A	[36]
CAS-W1-210	0.21	0.32	N/A	5	4.0E+06	1,673	160,307	N/A	1.19	N/A	N/A	[36]

Notices:
1) While no coordinates of the airfoils described in Table 1 are in public domain, the coordinates of the NREL "S" family of airfoils is available for research purposes.
2) N/A=Not available.

3. Design criteria for airfoils bound for large-size HAWT

A high lift-to-drag ratio and a low sensitivity 1 to airfoil surface contamination and manufacturing process tolerances have been generally the primary design drivers for WT airfoils, [17, 19, 20, 21, 30, 42]. In fact, despite the many approaches available for airfoil geometric optimization in order to delay transition, manufacturing tolerances and limited maintenance during operation contribute to WT blade surfaces to be inevitably far from optimal shape and surface finish [43]. Apart from the just described primary requirements, there are other specific design constraints prescribed depending upon the type of WT or on the radial position that the airfoil will assume along the span of the blade. Airfoil sections close to the hub, for instance, are strongly affected by structural requirements and also by the 3D flow effects, such as the centrifugal pumping [6]. According to Burton et al. [41] the most general trade-off faced while designing a WT airfoil is the aerodynamic-structural one: a low relative thickness results in lower drag but higher thickness allows for higher stiffness and strength with low weight due to the increased moment of inertia of the section. On the other hand, all significant self-noise sources are empirically confirmed to be located at the external (large radial distance from the hub) portion of the blade due to increased tangential speed and hence, higher local Mach numbers [2, 44]. However, despite the trend of increase in wind turbine diameter since the years 1990s, the noise emission started to appear as a new requirement to be complied with only by 2005 [20], when NREL explicitly listed low noise emission as a design requirement for the first time and that requirement translated into limitations to the lift coefficient towards the outboard quarter of the blade. This does not translate into aerodynamic efficiency losses due to the fact that the highest (C_l/C_d) usually occurs at C_l values quite below its peak as may be seen in Table 1. The development of WT airfoils at Delft University of Technology (DU), initially reported by Timmer and van Rooji [30], had to meet additional requirements, e. g., they should be suitable for both fixed and variable pitch blades, which demanded some compromise regarding the stall and post stall behavior. A summary of design features embodied in the family of DU WT airfoils is shown in Table 2: Thinner airfoils of the DU family, like the DU 95-W-180 and DU 96-W-180, are suitable for blade sections with larger radii outboard) and, as mentioned in section 2, were designed with the aid of the X-FOIL tool, while thicker ones, more suitable for sections with smaller radii (inboard), where the 3D effects are more pronounced, were designed with the help of the RFOIL software, a code derived from X-FOIL but modified in-house in order to include the 3D rotational flow effects.

Table 2 – Some aerodynamic performance criteria employed in the development of the DU family of HAWT airfoils. Source: [31]

Parameter	Requirement
$C_{l,design}$	For the outboard airfoils, the peak value of the c_l at the design point (usually $\frac{Cl}{Cd}$ maximum) is relatively unimportant in a pitch-controlled WT, from the performance perspective.
$C_{l,max}$	Moderate values of $C_{l,max}(1.4 1.5)$ are recommended. High values of this parameter associated with high AoA are not desirable, since the degradation of surface finish and corresponding increased roughness will reduce the $c_{l,max}$ corresponding angle by several degrees, resulting in large differences in lift between the clean and soiled airfoil surface conditions.
$(C_{l,design} - C_{l,max})$	A small (0.2) difference between $C_{l,design}$ and $C_{l,max}$ is considered desirable in order to prevent excessive loads from gusts and also stall due to slow control actuation.

Table 3 compares design requirements and wind tunnel results for the DU 95-W-180 and DU 96-W-180 outboard airfoils.

Table 3 - Airfoil design requirements for DU 95-W-180 / DU 96-W-180 outboard airfoil and wind tunnel test results. Source: [31].

Parameter	Design Requirements for DU 95-W-180 and DU 96-W-180, 18% thickness airfoils	Wind Tunnel Test results for earlier version (1995)	Wind Tunnel Test results for latter version (1996)
Cl_{max}	1,25 @ $3E+06$	1,20 @ $3E+06$	1,32 @ $3E+06$
Cl/Cd	>140@ $3E+06$	-	149
low sensitivity to Leading-Edge roughness	-	-	cl_{max} loss is 0.09

The NREL reports [19, 20, 21, 42] also stressed the need to boost efficiency and maintain it in the presence of LE debris as the primary design drivers for the "S" family of airfoils.

The requirements translated into (i) high maximum lift; (ii) relative insensitivity to roughness and; (iii) low profile drag. The resulting airfoils generally display thick sections with laminar flow development, i.e. retarded transition when in "clean". The NREL S series of airfoils were inversely designed from a prescribed pressure distribution manipulated to comply with a set of requirements, including noise emission control, as exemplified by the 4-million Reynolds number S830 primary airfoil, whose requirements are listed in Table 4.

Table 4 – NREL S830 primary airfoil design requirements and constraints.

General requirements and constraints	**Quantitative requirement for S830 airfoil** [20]
High maximum lift coefficients.	1.60
The maximum lift coefficient may not decrease significantly with transition fixed near the leading edge in both surfaces.	1.60 (smooth)-1.57(rough)
The airfoil should exhibit docile stall characteristics.	Ok (theoretical)
Low profile-drag coefficients over the specified ranges of lift coefficients.	0.80-1.40
Thicker than the S825 and S826 airfoils and appropriate for higher Reynolds numbers (larger machines).	21%
Lift (lift coefficient times blade chord) produced by the outboard portion of the blade should be constrained to alleviate the noise (this is clearly a trade-off with the first requirement).	Lift limited to 1.6
The zero-lift pitching-moment coefficient must be equal to or larger than a specified amount.	≥ -0.15
The airfoil thicknesses must equal the one specified.	21%

Other families of airfoils were designed by NREL with other priority drivers [41], like improving the post-stall control (for stall-regulated rotors), high maximum lift coefficient (for variable-pitch, variable-speed rotors), etc.. The thicker NREL WT airfoils are not always equipped with a large leading-edge (LE) radius [45], and, as pointed out by Migliore and Oerlemans [46], sharp LE radius may lead to higher inflow noise, which could neutralize any efforts to reduce airfoil trailing-edge noise.

4. Conclusions

Selected researches on airfoil development devoted towards large-size horizontal axis wind turbines in the last twenty years were described. One summary table was assembled

with theoretical aerodynamic design data and operating parameters for the main families of airfoils encountered. The table reveals that most airfoils designed and published to date were designed for operating points Reynolds numbers of up to 6 million. However, existing calculation show that the current 100 m-diameter equipment already reaches Reynolds numbers in the range of 7 - 9 million in high wind conditions and 150 to 220 m-class HAWT equipment that will be commissioned to cope with the increasing pressure for replacement of intensive CO_2 emitting systems will need airfoils developed for even higher local Reynolds and Mach numbers. This reveals a lack of development or dissemination of high Reynolds WT airfoils designed both under aerodynamic and noise requirements. While the typical utility-size WT diameter has been growing steadily in order to drop the wind energy conversion cost, the first requirements for designing quieter airfoils did not become public until 2005, and many developments have not yet adopted the noise emission requirement as a design driver to date. Fortunately, there seems to be no design trade-off between aerodynamic efficiency and low TE noise emission requirements, since attached flow through all operational envelope AoA is a necessity from both the stand point of aerodynamic torque generation and low TEN emission.
The primary aerodynamic design requirements for HAWT airfoils include moderate to high coefficients of lift, little sensitivity to surface roughness, a flat top C_l x AoA curve before the stall angle and a docile stall. Other requirements depend upon the WT type and/or the airfoil position, such as the airfoil thickness requirements for structural soundness near the hub and noise emission constraints for airfoils located in the outer part of the blades.

Acknowledgements

The authors would like to acknowledge the Coordination for the Improvement of Higher Education Personnel of the Brazilian government (CAPES) for the financial support.

References

[1] D. Griffith, T. Ashwill and B. Resor, "Large Offshore Rotor Development: Design and Analysis of the SANDIA 100 m Wind Turbine Blade", 53rd AIAA/ASME/ASCE/AHS/ASC Structures, Structural Dynamics and Materials Conference, Honolulu, 2012.

[2] S. Oerlemans, "Wind Turbine Noise: Primary Noise Sources", Amsterdam, 2011.

[3] United Nations - Climate Change, "The Paris Agreement", 05 dec 2019. [Online]. Available: https://unfccc.int/process-and-meetings/the-paris-agreement/the-paris-agreement.

[4] M. Ragheb and A. M. Ragheb, "Wind Turbines Theory - The Betz Equation Equation and Optimal Rotor Tip Speed Ratio", Urbana-Champain, Il., 2011.

[5] G. Pechlivanoglou, "Passive and Active Flow Control Solutions for Wind Turbine Blades", Berlin, 2012.

[6] M. Hansen, "Aerodynamics of Wind Turbines", Second Edition, London: Earthscan, 2008.

[7] S. Wagner, R. Bareiß and G. Guidati, "Wind Turbine Noise", Berlin: Springer, 1996.

[8] T. Lutz, A. Herrig, W. Würz, M. Kamruzzaman and E. Krämer, ”Design and Wind Tunnel Verification of Low-Noise Airfoils for Wind Turbines”, AIAA Journal V.45 N.4, pp. 779-785, April 2007. https://doi.org/10.2514/1.27658.

[9] J. Y. Saab Jr. and M. Pimenta, ”Displacement Thickness Evaluation for Semi Empirical Airfoil TE Noise Prediction Model”, Journal of the Brazilian Society of Mechanical Sciences and Engineering, V.38, N.2, pp. 385-394, Feb 2016. http://dx.doi.org/10.1007/s40430-015-0341-5.

[10] G. Guidati and S. Wagner, ”Design of Reduced Noise Airfoils for Wind Turbines”, in ECCOMAS 2000: European Congress on Computational Methods in Applied Sciences and Engineering, Barcelona, 2000.

[11] M. Drela and M. B. Giles, ”Viscous-Inviscid Analysis of Transonic and Low Reynolds Number Airfoil”, AIAA Journal V25 (10), pp. 1347-1355, October 1987. https://doi.org/10.2514/3.9789.

[12] M. Drela, ”X-FOIL”. MIT, 2000. http://web.mit.edu/drela/Public/web/xfoil/. [Online].

[13] M. Drela, H. Youngren and A. Deperrois, ”XFOIL/XFLR5”, Cambridge, Ma., 2009.

[14] H. Schlichting and K. Gersten, ”Boundary Layer Theory”, 8th ed., Berlin: Springer, 2003.

[15] R. Parchen, ”A Prediction Scheme for Trailing-Edge Noise Based on Detailed Boundary Layer Characteristics. Progress Report”, The Netherlands, 1998.

[16] F. Bertagnolio, H. A. Madsen and C. Bak, ”Experimental Validation of the TNO Trailing Edge Noise Model and Application to Airfoil Optimization”, Roskilde, Denmark, 2009.

[17] P. Fuglsang and C. Bak, ”Development of the Risø Wind Turbine Airfoils”, Roskilde, 2004.

[18] P. Fuglsang, C. Bak, M. Gaunaa and I. Antoniou, ”Design and Verification of the Riso-B1 Airfoil Family for Wind Turbines”, Transactions of the ASME, Vol. 126, pp. 1002-1010, November 2004. http://doi.org/10.1115/1.1766024.

[19] D. Somers and J. Tangler, ”The S822 and S823 Airfoils,” Golden, Co., 2005a.

[20] D. Somers and J. Tangler, ”The Airfoils S830, S831 and S832,” Golden, 2005b.

[21] D. Somers and J. Tangler, ”The Airfoils S833, S834 and S835,” Golden, Co., 2005c.

[22] R. Eppler, ”EPPLER AIRFOIL DESIGN AND ANALYSIS CODE,” Stuttgart, C.2000.

[23] National Renewable Energy Laboratory - NREL 3, ”Patent Report by Invention”, October 2nd, 2015. [Online]. Available: https : //wind.nrel.gov/airf oils/Documents/AirfoilP atentReport 2 009 − 03 − 18.pdf .

[24] J. Schepers, A. Curves, S. Oerlemans, K. Braun, T. Lutz, A. Herrig, W. Wuerz, A. Mantesanz, L. Garcillán, M. Fischer and K. M. T. Koegler, "SIRROCO: Silent Rotors By Acoustic Optimisation," in Second International Meeting on Wind Turbine Noise, Lyon, Fr., 2007.

[25] T. Brooks and T. H. Hodgson, "Trailing Edge Noise Prediction From Measured Surface Pressures", Journal of Sound and Vibration 78(1), pp. 69-117, 1981. https://doi.org/10.1016/S0022-460X(81)80158-7.

[26] J.-Y. Li, R. Li, Y. Gao and J. Huang, "Aerodynamic optimization of wind turbine", Proc. IMechE Vol. 224 Part A: J. Power and Energy, pp. 827-838, 2010. https://doi.org/10.1243/09576509JPE888.

[27] G. Petrone, C. de Nicola, D. Quagliarella, J. Witteveen and J. Iaccarino, "Analysis and Optimization of Wind Turbine Noise Under Uncertainty", in 4th International Meeting on Wind Turbine Noise , Rome, 2011.

[28] C. Xiaomim, Optimization of Wind Turbine Airfoils/Blades and, Saint Louis, Missouri: Washington University in Saint Louis, 2014.

[29] J. Chen, Q. Wang, S. Zhang, P. Eecen and F. Grasso, "A new direct design method of wind turbine airfoils and wind tunnel experiment", Applied Mathematical Modeling, pp. 2002-2014, 2016. https://doi.org/10.1016/j.apm.2015.09.051.

[30] W. Timmer and R. van Rooji, "Summanry of the Delft University Wind Turbine Dedicated Airfoils", Transactions of the ASME, Vol. 125, pp. 488-496, Nov. 2003. https://doi.org/10.1115/WIND2003-352.

[31] F. Grasso, "ECN-G1-21 Airfoil: Tunnel Testing", Journal of Aircraft, https://doi.org/10.2514/1.C032744. Design and Wind. pp. 1478-1484, 2016.

[32] W. Liu, "A review on wind turbine noise mechanism and de-noising techniques", Renewable Energy, pp. 311-320, 2017. https://doi.org/10.1016/j.renene.2017.02.034.

[33] M. Miller, K. Slew and E. Matida, "The development of a flat- back wind turbine airfoil family", Wind Energy, pp. 1372-1382, 2018. https://doi.org/10.1002/we.2260.

[34] W. Zhu, W. Shen, E. Barlas, F. Bertagnolio and J. Sorensen, "Wind turbine noise generation and propagation modeling at DTU Wind Energy: A review", Renewable and Sustainable Energy Reviews, pp. 133-150, 2018. https://doi.org/10.1016/j.rser.2018.02.029.

[35] X. Li, L. Zhang, J. Song, F. Bian and K. Yang, "Airfoil design for large horizontal axis wind turbines in low wind speed regions", Renewable Energy, pp. 2345-2357, 2020. https://doi.org/10.1016/j.renene.2019.07.163.

[36] P. Moriarty, "NAFNoise User's Guide", NREL, Golden, CO, 2005.

[37] P. Moriarty, "NAFNoise and FAST noise modules", in NREL Design Codes Workshop, Golden, 2008.

[38] T. Brooks and M. Marcolini, "Scaling of Airfoil Self-Noise Using Measured Flow Parameters", AIAA Journal, pp. 207-213, February 1985. https://doi.org/10.2514/3.8896

[39] T. F. Brooks, S. Pope and M. A. Marcolini, "Airfoil Self-Noise and Prediction", NASA, Langley, 1989.

[40] J. Y. Saab Jr, M. M. Pimenta, A. M. Faria and S. Rodriguez, "Determination of local flow conditions and preliminary investigation on the validity of the PNoise code for large size wind turbine", Journal of the Brazilian Society of Mechanical Sciences and Engineering, pp. 493-504, September 28th 2018. http://dx.doi.org/10.1007/s40430-018-1412-1.

[41] T. Burton, N. Jenkins, D. Sharpe and E. Bossanyi, Wind Energy Handbook, 2nd ed., Chicester: J. Wiley & Sons, 2011.

[42] D. M. Somers, "The S825 and S826 Airfoils", NREL/SR-500-36344, Golden, Co, 2005d.

[43] O. Eisele, G. Pechlivanoglou and C. ,. P. C. Nayeri, "Experimental & numerical investigation of inflow turbulence on the performance of wind turbine airfoils", Proceedings of ASME Turbo Expo 2013: Turbine Technical Conference and Exposition, San Antonio, Tx, 2013.

[44] P. Moriarty and P. Migliore, "Semi-Empirical Aeroacoustic Noise Prediction Code for Wind Turbines", Golden, CO, 2003.

[45] W. Devenport, R. Burdisso, H. Camargo, E. Crede, M. Remilieux, M. Rasnick and P. Van Seeters, "Aeroacoustic Testing of Wind Turbine Airfoils", NREL/SR-500-43471, 2010.

[46] P. Migliore and S. Oerlemans, "Wind Tunnel Aeoracoustic Tests of Six Airfoils for Use in Small Wind Turbines.", NREL/CP-500-35090, Golden, CO., 2003. https://doi.org/10.1007/s40430-018-1412-1.

INCEEUROPE

9th International Conference
on
Wind Turbine Noise
Remote from Europe – 18th to 21st May 2021

Robust noise indicators using Gaussian Processes

Finez Arthur, Engie Green, arthur.finez@engie.com,
Favrot Nicolas, INSA Lyon, n.favrot@gmail.com,
Petit Arthur, Engie Green, arthur.petit@engie.com,
Le Bourdat Colin, Engie Green, colin.lebourdat@engie.com,
Antoni Jérôme, INSA Lyon, jerome.antoni@insa-lyon.fr

Summary

In some countries, acoustic regulations regarding wind turbines require to compare the noise in "on" & "off" conditions at nearby dwellings. As all wind conditions are hardly met in a few weeks of measurement campaign, it is often necessary to estimate the noise in rare conditions from the measurements. In other cases, measured noise indicators are obviously misleading : for instance in cases of negative emergence (the noise index with the "windfarm on" is lower than with the "windfarm off"), or if the background noise indicator at low speed is much higher that the noise at higher windspeed. In such cases and in the process of building a curtailment plan, it is necessary to modify measured acoustic indicators to produce more realistic noise levels. Today, this modification is operator dependent and thus relies on his experience.

This contribution deals with establishing a more robust and unbiased procedure. Use is made of Gaussian Processes (GP) and multivariate normal (MVN) distributions as they provide a suitable framework to mix measured data with the "expert feeling". This vague but meaningful notion is replaced by a database of noise indices measured in past measurement campaigns. Similar measurement are selected to constitute a prior distribution which models "in such conditions what noise curves usually look like". Conditional laws then allows to deduce the posterior : the site specific distribution knowing the small amount of measured data. Moreover the statistical framework gives access to confidence intervals of this estimates.

The application on wind farm noise assessment show that this technique provides comparable results to the manual expertise. Moreover computed indicators are much more robust than classical binned medians evaluation against the lack of data.

1. Context and motivations

Assessing the 'true' and complete noise situation around wind farms is a matter of prime interest for wind farm operators, in particular when it comes to designing noise curtailment plans. They consist in defining in which conditions (wind direction, wind speed, period of the day) and in what extent turbine rotor speeds must be limited so as to get a noise reduction compared to a full power operation. Such curtailment plans are usually associated with consequent production losses and must be carefully designed. Their optimisation results firstly from the absolute necessity to comply with local noise regulation, and secondly from a balance between an additional limitation of neighbourhood annoyance and electrical production losses. The evaluation of representative and complete noise levels is all the more complex in countries where the notion of *noise emergence* at nearby housings is controlled (Italy and France), which shall be understood as the difference in dB between the *background noise* level (when the wind farm is not in operation) and the *Total noise* level (when the windfarm is in operation). The background noise evaluation is tricky since it is usually based on a limited number of samples : cyclic stops of the windfarm are required which are also linked to lost revenue. As a matter of fact the operator usually limits the duration of the acoustic measurement campaigns to a minimum allowing a satisfactory description of the noise situation. Moreover short time environmental noise levels are by nature not repeatable and can be thought as random variables. A statistical treatment of noise sample is thus required. In France for instance noise indicators are evaluated by a median calculation of the sample levels contained in a windspeed bin for specific homogeneous noise conditions. In cases where only a few measurement samples are available, it is thus not surprising to see non intuitive situations arising, where such noise indicators are obviously misleading :

- Noise levels lower in “on” phases than in “off” phases (apparent negative emergences)
- Noise levels lower at high windspeeds than at low windspeeds (decreasing noise curve[1])
- Lack of measurement in certain rare conditions (absence of noise indicator)

To correctly design curtailment plans, it is mandatory to have realistic noise curves, which are as valid as possible for all possible conditions. Because of aforementioned discrepancies the acoustic expert is forced to manually correct, interpolate and/or extrapolate computed noise indicators to produce more realistic and complete noise curves based on his/her experience. Today this process is operator dependant and a more robust and unbiased procedure would be welcome to rationalize this step in wind farm noise management.

Aspects of robust wind farm noise assessment have of course been addressed in the scientific community, if only to propose relevant compliance testing procedures to regulatory authorities. In a recent review, Hansen [1] pointed out the difficulty to provide such representative and complete analysis both for background noise and total noise. Apart from the statistical treatment of windspeed bins, Hansen mentioned that a best fit regression technique using polynomial curve with order 1 (linear regression) to 3 is today widely used [2]. The best fit high order polynomial (up to 4) was mentioned in 1996 by the ESTU report [3] to interpolate collected noise samples, although it is stressed out that extrapolated data at very low and high wind speed must be used with care. To better describe the different regimes in the noise curve, namely a plateau in the lower speed range and a polynomial increase in the higher speed range, a log based regression is also proposed. These techniques provide the advantage of completeness and can sometimes be extrapolated to cases where no experimental data has been obtained. However the choice of the model complexity is crucial to fully render the typical “elbow shape” of background noise curves and is always prone to over or underfitting. It still depends on the nature on the noise environment (according to best practice guides [4]) and the operator expectations on the characteristics of these curves.

The key point in the manual expert intervention is the use of *a priori* knowledge of “what noise curve usually look like in these conditions” to deduce “what this specific curve probably looks like” given a set of collected samples. Among various machine learning techniques, Gaussian Processes (GP) provide an interesting and rigorous framework for a less subjective use of prior

---

[1] “Noise curve” means the function of the noise indicator versus wind speed in certain homogeneous noise conditions and farm operation.

information. In this perspective the role of the expert knowledge is played by a database containing all noise indicators computed in past acoustic campaigns, possibly with labels of the acoustic environment (traffic, nature presence) and conditions (period of the day, wind direction sector, period of the year). Figuratively it can be used to "steer" the noise curve, on the basis of partial measurements in a sensible direction when data are missing. The aim of this work is to propose an adaptation of this powerful technique to the specific case of wind farm noise measurements. A brief description of GP is presented in section 2, illustrative applications are proposed in section 3. In section 4 is provided a more statistical assessment of the robustness of GP compared to the classical analysis provided by current regulatory methodology in France.

2. Gaussian Process method

Gaussian Process (GP) [5] is a generic denomination for a stochastic process of several random variables, every subset of which follows a multivariate normal distribution (MVN). It is used for example in biology (heart rate estimates, gene expression studies) [6], finance [7], spatial models in meteorology and geology and the analysis of computer experiments [8], as it is a handy and flexible tool for comprehension models applied to noisy data. In our case, the acoustic level is supposed to be issued from a GP and the random variables are the noise levels at a given windspeed. This stochastic formulation allows us to apply Bayes' rule and to compute properly a posterior probability density function (pdf) given a prior pdf and a statistical model for the observed data.

The proposed procedure for estimating robust noise indicators using Gaussian Processes is based on:

- **Observations:** a set of measured windspeed and noise level (V, L) sample pairs. A new pair is typically available each 10 minutes. V is the mean of windspeed during this period, as measured by a trusted device (LIDAR, mast or mean of anemometer data from a selection of wind turbines). The noise level L is typically the median of all the A weighted overall short term (1 sec) equivalent noise levels. This set must be coherent in itself as it is assumed to be produced by a single random process. In particular, measurements of a single measurement point and in the same homogeneous noise condition [2] must be present,
- **A database:** a set of noise curves coming from previous studies in similar conditions (weather, environment, period, farm operation). This information will serve as a support for filling out missing data or correcting unexpected samples from the above mentioned observation set.

In the following, the notion of noise curve means the statistically representative noise level $L(V)$ observed in a single observation point for a specific homogeneous noise condition and wind farm operation setting, as a function of the wind speed V. This variable is not used as continuous variable but as a discrete variable $\boldsymbol{V} = \{V_i, i = 1..n\}$, which is either a set of integers, a fraction of integers or the set of measured values, as the case may be.

2.1 Multivariate normal (MVN) distributions

The basic idea of the proposed procedure is to consider all noise curves $L(V)$ as produced by a global GP. As all quantities depend on the windspeed variable V, use is made of multivariate normal distribution, which is an extension of normal distribution to multi dimensional objects [3]. Noise curves are seen as random draws of an MVN entirely defined by :

- A mean curve $\mu(V)$ which is the extension of the concept of mean value

[2] For example "at night, with the wind coming from the south sector"
[3] The number of dimension being n

- A covariance function $\Omega(V_i, V_j)$ which is the extension of the concept of variance (squared standard deviation).

The mean curve describes the evolution of the mean noise level as a function of the discretised windspeed. The covariance function relates the covariance of the two dependent random variables which are the noise levels at speeds V_i and V_j. Thus the general random process producing noise curves is regressed to an MVN process $L(V) \sim MVN(\mu, \Omega)$ called **the prior distribution.** The noise curve to be deduced from the specific site measurement is considered a realisation of this general gaussian process. Both function μ and Ω are applied a discrete sets of winds speeds, in which case the are translated to vectors and matrices. $\boldsymbol{\mu_1}$ will be used to denote the vector of the mean curve $\mu(V)$ computed on the speed set $\boldsymbol{V_1}$ and $\boldsymbol{\Omega_{12}}$ the matrix of function Ω computed on $\boldsymbol{V_1}$ and $\boldsymbol{V_2}$.

2.2 Prior model

The specificity of the problem addressed lies in the distinction between the status of context variables. While it is clear that for a given measurement site sound levels at adjacent windspeed are correlated, it is not the case between measurements at two different sites. A specific geographic location may indeed be relatively noisy at all wind speeds while another one may *independently* stand considerably below the "overall mean noise curve". As a consequence the proposed model should account for this distinction between the measurement site defined by a categorial variable S and the speed variable V. It is proposed here to model the site specific behaviour as a linear trend deviating from the overall mean curve:

$$L(V,S) \sim \mu(V) + \boldsymbol{g}(V)^T \boldsymbol{a}(S) + \varepsilon \qquad \text{(eq. 1)}$$

where

- $\mu(V)$ is the overall noise mean curve extracted from the database. It can be empirically computed as the mean of the noise curve present in the database.
- $\boldsymbol{g}(V) = [1, V]^T$ is a vector of monomial functions of the wind speed. In this study, a linear trend is chosen to relate the site specific deviation from the overall mean curve, so there are only two components in $\boldsymbol{g}$,
- $\boldsymbol{a}(S) = (a_0, a_1)$ is the coefficient vector of the linear deviation from the mean curve for a specific measurement site S. Themselves follow an MVN defined by the mean values $\boldsymbol{\mu_A} = (\overline{a_0}, \overline{a_1})$ and covariance matrix $\boldsymbol{\Omega}_A[2 \times 2]$. They can be estimated on each site by a linear regression, which provides coefficients (a_0, a_1). Then statistics can be computed on the all set of sites and measurement points to provide $\boldsymbol{\mu_A}$ and $\boldsymbol{\Omega_A}$.
- ε is the residue MVN and relates the noise curve variations around the site trend $\mu(V) + a_0 + a_1 V$. By construction this MVN has zero mean but its covariance function $\Omega_\varepsilon$is central to the modelling process. For instance, it encodes the typical wave length of those variations. It can be estimated in several ways : directly from measurements or by further modelling. For this exploratory work, a classical model for the covariance function is used:

$$\Omega_\varepsilon(V_i, V_j) = \frac{e^{(V_i - V_j)^2}}{\theta} + \eta\, \delta_{ij} \qquad \text{(eq. 2)}$$

  where θ and η are hyperparameters. θ is linked with the correlation length in the noise curve and η describes the amplitude of the uncorrelated noise. $\delta_{ij} = 1$ whenever $i = j$, else 0.

In the following subsections, it will be assumed that the whole MVN has been characterised from the database : quantities $\boldsymbol{\mu_A}, \boldsymbol{\Omega_A}$ have been computed from the prior database and μ and Ω_ε can be expressed on various windspeed set V thanks to an important interpolation step. Details of their computation will be given in the application section **Erreur ! Source du renvoi introuvable.**.

2.3 Conditional probabilities

On site measurements formally noted D_n are seen as realizations of the general MVN in the context of a specific site and for various wind speeds. The goal is to predict the likelihood of future measurements on this same site from the general prior description *and* given the partial past observations of D_n. The realisation of the noise curve cannot be computed in an deterministic manner but it is possible to characterise its distribution which is also an MVN called the **posterior distribution** :

$$L(\boldsymbol{V})|D_n \sim MVN(\widetilde{\boldsymbol{\mu}}, \widetilde{\boldsymbol{\Omega}}). \qquad \text{(eq. 3)}$$

The power of GP is to provide a direct expression for the desired quantities $\widetilde{\boldsymbol{\mu}}$ and $\widetilde{\boldsymbol{\Omega}}$. If D_n consists in a vertical vector of noise levels $\boldsymbol{L_2}$ measured for a set of $n_2$windspeed $\boldsymbol{V_2}$ and it is desired to estimate the probability of noise level $\boldsymbol{L_1}$ on a new set of n_1 windspeeds $\boldsymbol{V_1}$, then conditional probabilities (Bayes rule) read

$$\widetilde{\boldsymbol{\mu}} = \boldsymbol{m_1} + \boldsymbol{\Omega_{12}}\boldsymbol{\Omega_{22}^{-1}}(\boldsymbol{L_2} - \boldsymbol{m_2}) \qquad \text{(eq. 4)}$$

$$\widetilde{\boldsymbol{\Omega}} = \boldsymbol{\Omega_{11}} - \boldsymbol{\Omega_{12}}\boldsymbol{\Omega_{22}^{-1}}\boldsymbol{\Omega_{21}} \qquad \text{(eq. 5)}$$

Covariances matrices $\boldsymbol{\Omega_{11}}, \boldsymbol{\Omega_{12}}, \boldsymbol{\Omega_{21}}, \boldsymbol{\Omega_{22}}$ all come from the prior model of the covariance function applied to either $\boldsymbol{V_1}$ or $\boldsymbol{V_2}$:

- $\boldsymbol{\Omega_{11}}$ is Ω_ε computed on the basis of the new windspeed $\boldsymbol{V}_1$,
- $\boldsymbol{\Omega_{22}}$ is Ω_ε computed on the basis of measured windspeed $\boldsymbol{V}_2$,
- $\boldsymbol{\Omega_{12}}$ is the rectangular cross-covariance matrix issued from the computation of Ω_ε on $\boldsymbol{V}_1$ and $\boldsymbol{V}_2$
- $\boldsymbol{\Omega_{21}}$ is the transpose of $\boldsymbol{\Omega_{12}}$.

It can be noticed in equation 5 that the input of additional data on $\boldsymbol{V_2}$ leads to variance reduction compared to the prior model given only on $\boldsymbol{V_1}$; it is a core aspect of GP regression as highlighted in [5] (eq. 5.1). However $\widetilde{\boldsymbol{\Omega}}$ being not influenced by measured values in L_2 is certainly surprising. It actually stems from the assumption of identity between the prior Gaussian process and the one producing the data. In our case this assumption is actually violated since the prior is constituted of processed acoustic indicators (binned medians) while $\boldsymbol{L_2}$ data are 10 minutes samples. The former is associated to much lower variance than the latter. As a consequence, the notion of variance in the proposed method can not be directly interpreted as the acoustic indicator uncertainty. This point is a good candidate for further investigations.

Concerning equation 4, site dependent trend vectors $\boldsymbol{m_1}$ and $\boldsymbol{m_2}$ are an adaptation of the prior model to the measurement site including the linear regression adjustment. More precisely

- $\boldsymbol{m}_1 = \boldsymbol{\mu}_1 + \boldsymbol{G}_1{}^T\widehat{\boldsymbol{a}}$ is evaluated on the new basis of windspeeds $\boldsymbol{V_1}$, with $\boldsymbol{\mu_1} = \mu(\boldsymbol{V_1})$ and $\boldsymbol{G}_1 = [\boldsymbol{g}\left(V_1^{(1)}\right), \boldsymbol{g}\left(V_1^{(2)}\right), \dots, \boldsymbol{g}(V_1^{(n_1)})]$,
- $\boldsymbol{m}_2 = \boldsymbol{\mu}_2 + \boldsymbol{G}_2{}^T\widehat{\boldsymbol{a}}$ is similarly evaluated on the basis of measured windspeeds $\boldsymbol{V}_2$, with $\boldsymbol{\mu}_2 = \mu(\boldsymbol{V}_2)$ and $\boldsymbol{G}_2 = [\boldsymbol{g}\left(V_2^{(1)}\right), \boldsymbol{g}\left(V_2^{(2)}\right), \dots, \boldsymbol{g}(V_2^{(n_2)})]$.

The linear trend coefficient vector $\widehat{\boldsymbol{a}} = (\widehat{a_0}, \widehat{a_1})$ is estimated for the specific site thanks to a regularized inverse procedure using the direct measurements $\boldsymbol{L_2}$:

$$\widehat{\boldsymbol{a}} = (\boldsymbol{G}_2\boldsymbol{\Omega}_{22}^{-1}\boldsymbol{G}_2^T + \lambda\boldsymbol{\Omega}_A^{-1})^{-1}(\boldsymbol{G}_2\boldsymbol{\Omega}_{22}^{-1}(\boldsymbol{L}_2 - \boldsymbol{\mu}_2) + \lambda\boldsymbol{\Omega}_A^{-1}\boldsymbol{\mu}_A) \qquad \text{(eq. 6)}$$

where λ is a regularization parameter. It is set to 10 in the present study, but varying its value appeared to have very little influence on the results.

2.4 Deduced information

Once the posterior mean $\tilde{\boldsymbol{\mu}}$ and covariance matrix $\tilde{\boldsymbol{\Omega}}$ are estimated on an arbitrary set of windspeeds $\boldsymbol{V}_1$, we have access to valuable information, for instance :

- The posterior mean curve $\tilde{\mu}$ is the required noise indicator which is also the most probable curve knowing the onsite observations and the information of the database
- Confidence intervals of 5% and 95% around this mean curve,
- More generally it is possible to draw candidate noise curves using the posterior statistical law computed $L(V)|D_n$ and to compute their likelihood.

3. Application

As an illustrative example, the method described in the previous section is applied on the data of a wind farm compliance study in France. The wind farm consists of 3 wind turbines of 70 m diameter operating in standard mode in the daytime and in a curtailed mode in the night time. The acoustic measurement point is located approximately 500 m from the closest wind turbine. The noise of the wind turbine could be heard from this position in several (windspeed/wind direction configuration). The windfarm was subject to a reinforcement of the curtailment plan consequently to the acoustic study, making this example an interesting application for the GP process.

3.1 Prior mean curve

The first step in the GP procedure is to select in the database of past campaigns the noise curves in similar situations. In this example, we are first interested in total noise during the daytime with the wind coming from the west sector. So we select of the noise curve in this situation. Note that we could easily add geographic and/or environment criteria in the selection. The 184 selected curves are shown in Figure 1.

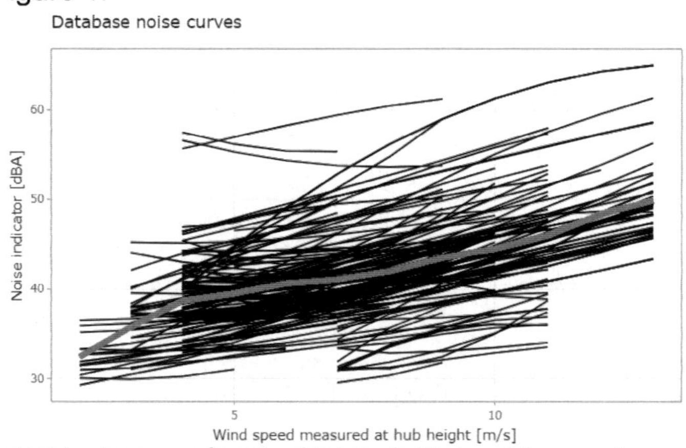

Figure 1: Selected total noise curves from the database in black and the overall mean curve $\mu(V)$ in red

For all these curves, the measured windspeeds have been translated to the hub height assuming standard shear coefficients (0.1 for the day, 0.2 for the evening and 0.3 for the night). The noise levels have been interpolated to integer values of the windspeed at hub height.

The empirical mean of all these curve correspond to $\mu(V)$ and is shown in a think red line in Figure 1. Note that extreme values of the windspeed (2 m/s and 13 m/s) have rarely been characterised, be it because wind farm are most often stopped at very low wind speed (so total noise is not really measured), because high windspeeds correspond to rare conditions or because expert offices do not provide information at such low speed which are rarely acoustically challenging. Observed cases where bin 2 m/s has been measured were generally quieter that others, which is translated by drop of the mean curve at low speeds. This effect may have an detrimental impact on the noise evaluation, in which case a manual completion of missing values could be considered.

3.2 Prior linear deviations

The difference of each noise curve with the mean curve $L(V) - \mu(V)$ is then computed and presented in Figure 2. It is clear that some site are considerably quieter than the average (-10dB) and some of them are considerably louder (+10dB).

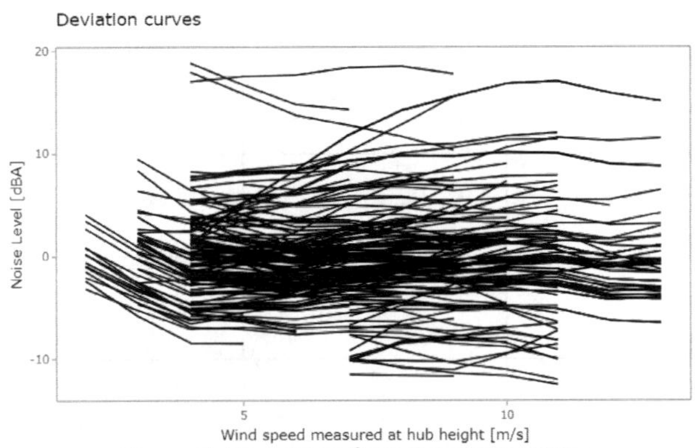

Figure 2: Deviations from the mean curves

For most of the noise curves, a linear trend can be drawn which relates more or less accurately the deviation from the curve. A linear regression model is the applied separately to each curve, coefficients (a_0, a_1) are computed and corresponding histograms are presented in Figure 3. In can be seen that the computed offsets a_0 are in the vast majority between -10 and +10 dBA, while most trend coefficients a_1 lie between -1 and +1 dBA/(m/s).

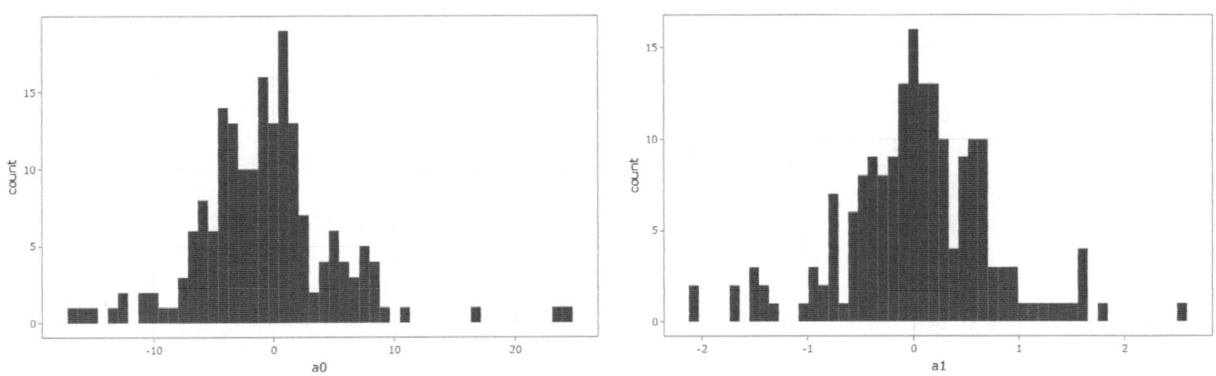

Figure 3 : Histrogram of linear regression coefficients a_0 and a_1.

All these coefficients allow to compute the mean coefficients $\overline{a_0} = -0.7\ dBA$ and $\overline{a_1} = 0.035$ dBA/(m/s) and the covariance matrix depicted in Figure 4. It can be seen that very low correlation is found between both coefficients (off the main diagonal).

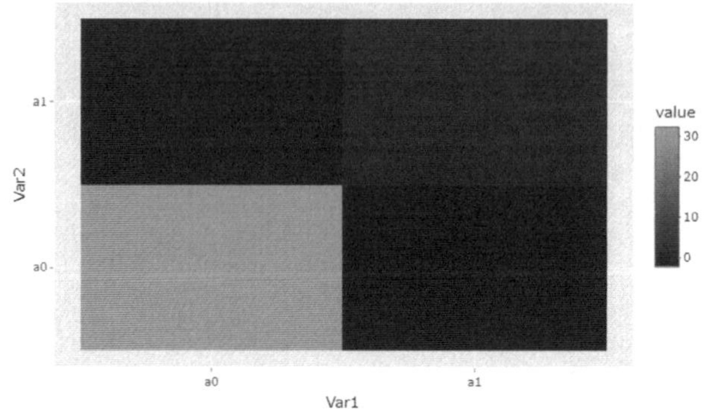

Figure 4: Covariance matrix of the linear regression coefficients

3.3 Prior residual curves

The subtraction of the linear trend $a_0 + a_1 V$ to each curve of Figure 4 leads to the residuals depicted in **Erreur ! Source du renvoi introuvable.** which can be viewed as random draws of the $\varepsilon(V)$ MVN. A particular behaviour can be seen for one curve reaching 2.5 dBA at 8 m/s (which was already noticeable in Figure 1 and Figure 2), but it will not ruin the modelling attempts as it is an isolated case.

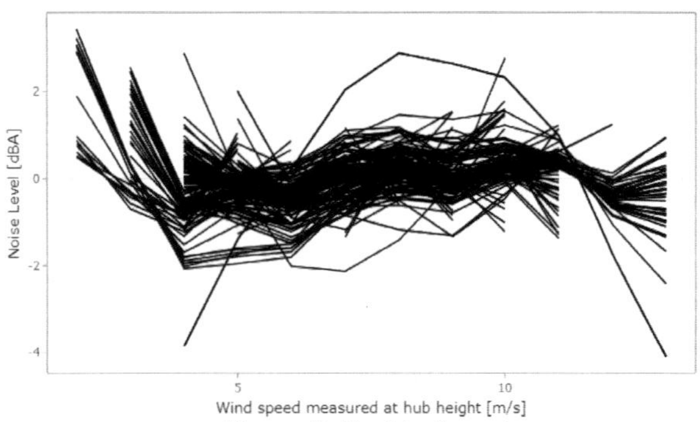

Figure 5: Residual curves

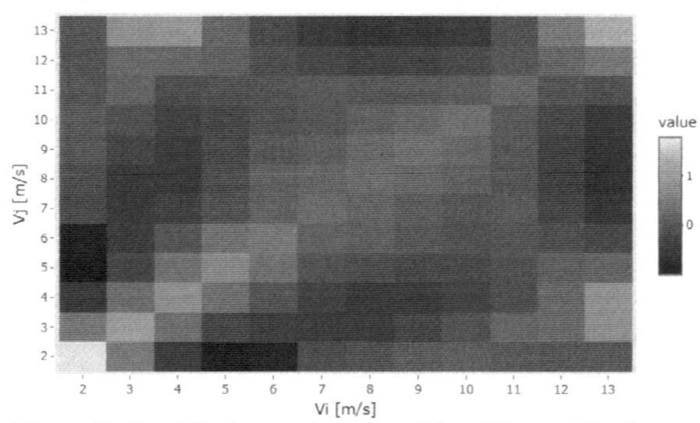

Figure 6: Empirical covariance matrix of the residual curves

The empirical covariance matrix of the residual curves is plotted in Figure 6. It can be seen that the empirical variance is maximal on the diagonal (whenever V_i and V_j) which supports the analytical model proposed in equation 2. However after a first decrease of the variance with the distance $|V_i - V_j|$, Figure 6 shows an increase of the correlation. For instance, $V = 4$ m/s and $V =$ 13 m/s appear to be significantly correlated in terms of residue value according to the prior database. It is possible that this is an effect of the chosen model adopted in eq. 1. As the linear trend as been removed for each site, the residue curve fluctuates around this trend ; the fluctuating pattern depends on the position of the elbow in the noise curve. It is thus likely that low speed and high speed residues present some repetitive behaviour. A more usual aspect of the covariance matrix may be found if higher order terms were used in vector $\boldsymbol{g}(V)$, which is a perspective of this work for a better modelling of the covariance matrix.

To apply equations 4 and 5, it is mandatory to estimate the covariance function Ω_ε at non integers speed values $\boldsymbol{V}_1$ and $\boldsymbol{V}_2$. In this goal a first idea was to interpolate and extrapolate the empirical covariance matrix of Figure 6. This important step necessitates a finely tuned covariance model and could not be achieved in an acceptable way for the time being. For this reason the more classical covariance model of equation 2 was chosen with hyperparameters empirically set to obtain visually acceptable results in the following section.

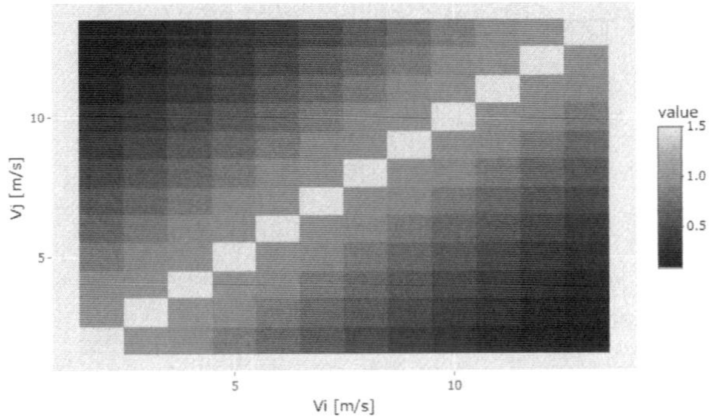

Figure 7: Modelled residue covariance matrix with $\theta = 50$ and $\eta = 0.5$

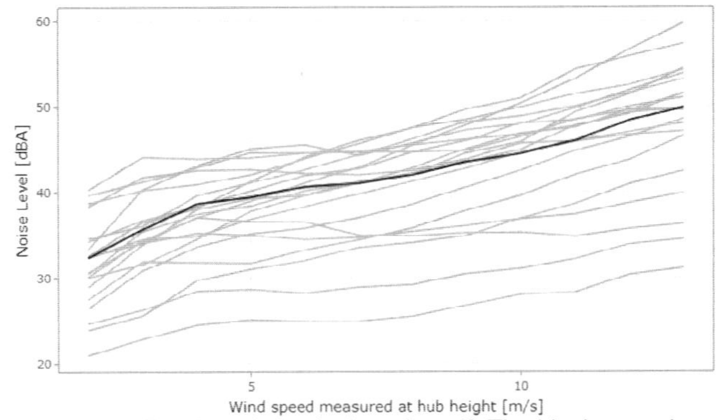

Figure 8: Random draws in the prior pdf. The black curve is the mean curve of the prior; grey curves are 20 realisations of the identified GP.

3.4 Random draws in the prior probability density functions (pdf)

Once the general prior GP is fully described by means of $\boldsymbol{\mu}_A, \boldsymbol{\Omega}_A$, and $\boldsymbol{\mu}_1, \boldsymbol{\Omega}_{11}$, for $\boldsymbol{V}_1 = \{2, 3, .., 12\}$ m/s, it is possible to randomly draw feasible noise curves according to this pdf using a random number generator for multivariate normal distributions. This is proposed in Figure 8 with 20 realisations. It can be seen that even with the simplified covariance matrix model, all produced curves are feasible candidates for noise indicator curves, when compared to the database curves of Figure 1, though in a more complete way (no windspeed indicator missing).

3.5 Posterior pdf computation

Mixing the knowledge encoded in the prior pdf and samples collected from the example windfarm is now possible using equations 4, 5 & 6. The mean curve plotted in Figure 8 is fitted using cubic splines ; it allows to compute estimates of $\mu(V)$ for arbitrary values of V without generating overly large values outside the characterised speed range. The main output of the GP computation is presented in Figure 9 for the illustrative example used here above. Measurements $L_{50,10min}$ in these conditions (daytime, farm in operation) are shown in yellow points ; they exhibit a large variance although invalid and rainy data have been filtered out as the local guidelines recommend. The French regulatory computation process [10] has been applied on these data ("Binned Medians") and is presented also on Figure 9. The corresponding curve shows large variations especially at low wind speeds where the number of samples is rather low (only 10 samples in the 3 m/s bin for instance). As a consequence it has been chosen by the acoustic expert to modify these values by hand to draw the smoother green curve at 3, 4 and 6 m/s.

The GP procedure has been applied on these data. The prior mean curve is presented in dashed symbol for comparison with Figure 1 and Figure 8. It is significantly above the measured point cloud designated this site as rather quiet compared to passed campaigns. The GP posterior mean curve (PMC) $\tilde{\mu}(V)$is also shown on this graph. It is well below the prior mean showing the noticeable adaptability of the GP technique when measured data differ significantly from the prior features. It can be seen that the PMC passes homogeneously through the middle of the measurement point cloud. In particular, at low speed it appears that values at 3 ,4 and 6 m/s are closer to the PMC than values at 5 and 7 m/s. As a consequence, they can be viewed as more likely indicator values; the relevance of the expertise office choice could be questioned in this case. Another interesting feature of the GP give access to the variance of the indicators. A transparent ribbon in Figure 9 shows the 5%- 95% confidence interval of the posterior mean curve, allowing to state that values below 5 m/s are associated with much higher uncertainty values than at 8 m/s, and this unlikelihood can be quantified, with the limitations exhibited in section 2.3 . At last, it can be noticed that the PMC exhibit a hump around 4 m/s which appears not to be physical. The constraint on the mean curve to be monotonically and smoothly increasing has indeed not been included in Ω_ε but it could be a future development of this tool since theorical tools appear to be well established [9].

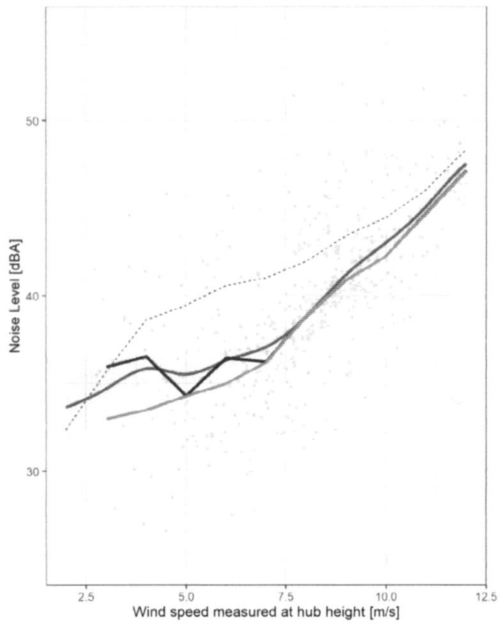

Figure 9: Day, total noise

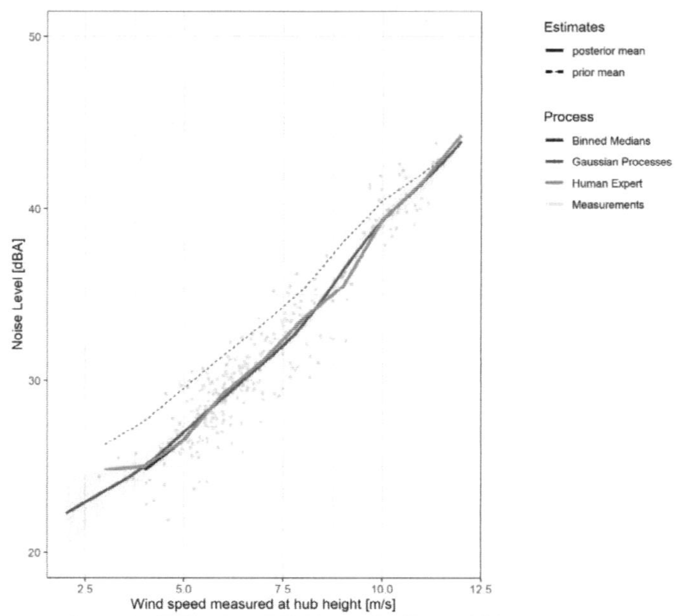

Figure 10: Night, total noise

Comparaison of posterior mean curve ("Gaussian Processes") with the French regulatory computation ("Binned Medians"), the manual modification of this curve by an operator ("Human Expert'). Also shown are measurement samples $L_{50,10min}$ in yellow, and the prior mean curbe (dashed).

Another example of the application of GP is illustrated in Figure 10, corresponding to total noise recordings during the night. No additional tuning of the method has been made, except of course the curve selection in the prior database. Chosen hyperparameter values appear to be rather adapted to all tested cases. In Figure 10 measurements samples are less spread than in the previous case, showing less difference between all the methods. However it can be noticed that the binned median method provides indicators at 8 m/s which are a bit off centred. This may be due to the uneven availability of the samples in this particular windspeed bin. Also it can be noticed that at low windspeeds, the PMC is linearly increasing while the expected behaviour of the noise curve is to reach a plateau around $V = 0$ m/s. As for the monotonicity, the constraint $\frac{d\mu}{dV}(V = 0) = 0$ has not been added in the presented implementation.

Figure 11 shows the trickier case of background residual noise at night, which is actually the most interesting since

- the regulatory constraints in France are more restrictive at night than during the day,
- these measurements require stops of the windfarm and may be subject to a lack of statistical convergence as explained in the introduction.

In this case, the binned median at 5 m/s value has been corrected by the acoustic consultant. The GP PMC is found to be very close to the corrected value. However at higher speed, it seems that the PMC does not perfectly fit to the abrupt increase of noise with windspeed, which is particular to this site compared to the database curves. This results in an overestimate of the noise level between 6 and 8 m/s and an underestimate of noise between 8 and 11 m/s. This may be solved in the future by a more flexible model of the database curves including higher order terms in the vector $\boldsymbol{g}(V)$ or by a better covariance matrix fitting as mentioned previously.

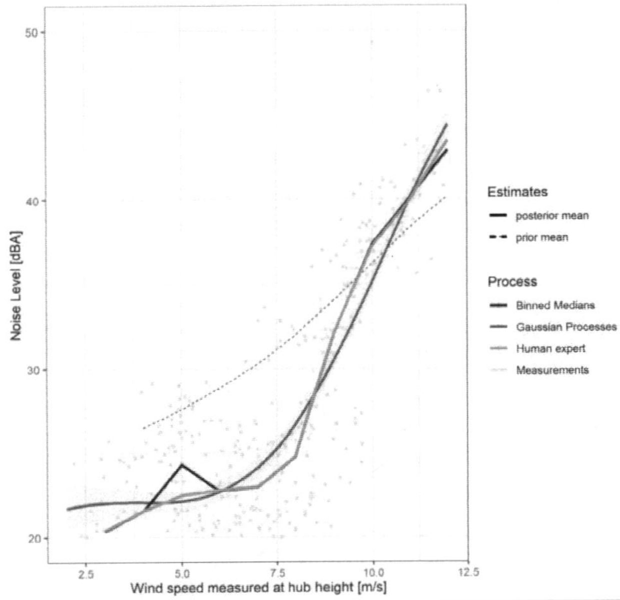

Figure 11: Night background noise

4. Robustness of noise indicators

The term 'robust' in the title of this paper refers to the ability of an automatic method to provide stable results even in cases of missing data. A comparison between the French regulation median based method [10] and the Gaussian Process method is carried out on the case of the total noise measurement in daytime (data of Figure 9). The full dataset is composed of 545 measurements points. From this table, a number of samples is randomly selected and removed. Both processes (GP and Binned Medians) are followed to produce an estimated curve of noise indicators as already presented in Figure 9, Figure 10 and Figure 11. The noise curves based on the partial dataset are compared to the one obtained using the full dataset and root mean square errors (in dBA) are computed as an indicator of the robustness.

This evaluation is repeated with 100 different completeness rates, scaling down from 100% (545 samples) to 1% (only 5 samples) and with 10 random draws for each completeness rate. It must be stressed out that the binned median technique can usually be applied in a wind speed bin only if at least 10 measurement samples are available in this bin. In this conditions, it is not possible to compute a single indicator below the extreme rate of 2%. For the sake of method comparability, this sanity rule has been dropped in the following.

The result of this robustness study is presented in Figure 12. Both method reach of course the zero error line at the rate of 100% since it is the baseline. But as the completeness rate is progressively decrease, it becomes clear that the Gaussian Process method raises on average half the error of the binned medians technique.

As a consequence, assuming that there is no bias error between both methods (and thus discrepancies outlined in section 3 have been treated), it can be interpreted that for a given acceptable error, the GP method requires half the number of stop hours of the binned median technique.

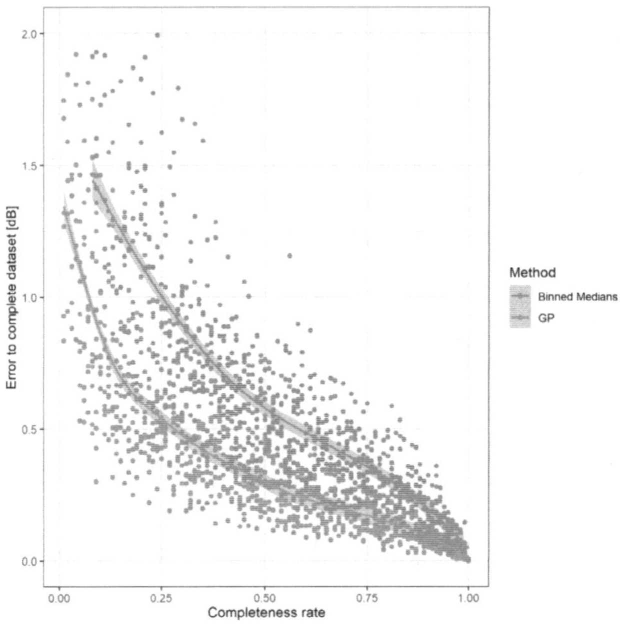

Figure 12: Error to complete dataset results for various rates of missing data, both for the French regulation method ("binned medians"), but also for the present method ("GP").

5. Conclusions

In this study, the well developed statistical technique of Bayesian inference using Gaussian Process is adapted to the applicative case of wind farm noise indicator deduction based on 10 min measurement samples. It make use of a database of previous measurements in the neighbourhood of different windfarms, keeping track of the conditions for the indicator computation (wind speed, wind direction, period of the day, acoustic environment). This database mimics the experience of the operator which is forced to manually adjust the noise levels when the classical binned medians procedures leads to unlikely results.

The Bayesian framework offers a rigorous yet elegant manner to mix prior information of what a noise curve should like to direct measurements on a specific site. A first application on terrain data showed that the noise levels computed are close to the manually adjusted indicators. Moreover it has been shown that it presents much more stable results in the case of rare data than the classical technique.

However some patterns of the noise curve could not yet be fully reproduced, especially when rapid slope variations of the noise curve occurs as it is the case for background noise during night time. In this regards, the method could benefit from further improvements like a better fitting of the covariance matrix of the residue MVN, adding constraints like the monotonicity of the noise curve, or a more rigorous evaluation of the noise indicator variance.

A direct application of the proposed method concerns the computation of adjusted wind turbine curtailment plans for which complete and robust noise indicators are needed, even in conditions that have seldom been characterised.

References

[1] Hansen C. and Hansen J. (2020) *Recent advances in wind turbine noise research* Acoustics 2020(2), 171-206

[2] EPA. *Wind Farms Environmental Noise Guidelines*; Technical Report; South Australian Environmental Protection Agency: Adelaide, SA, Australia, 2009.

[3] ETSU, *The assessment & rating of noise from wind farms,* ESTU-R-97, 1996

[4] Cand M., Davis R., Jordan C., Hayes M., Perkins R., *A good practice guide to the application of the ETSU-R-97 for the assessment and rating of wind turbine noise*, Institute of Acoustics, May 2013 accessible through http://www.ioa.org.uk/about-us/newsarticle.asp?id=260

[5] Gramacy R. B. *Surrogates : Gaussian Process Modeling, Design, and Optimization for the applied sciences* ISBN 9780367415426, Chapman and Hall/CRC 2020

[6] Stegle O., Borgwardt K., *GP Applications* accessible via https://ethz.ch/content/dam/ethz/special-interest/bsse/borgwardt-lab/documents/slides/CA10_GPApplications.pdf 2010

[7] Gonzalvez J., Lezmi E., Roncalli T. Xu J. *Financial Applications of Gaussian Processes and Bayesian Optimization* arXiv 1903.04841 q-fin.PM 2019

[8] Rasmussen C. E., Williams C. K. I., *Gaussian Processes for Machine Learning*, the MIT Press, 2006

[9] Da Veiga S., Marrel A. *Gaussian process modeling with inequality constraints*, Annales de la faculté des sciences de Toulouse – Mathématiques, 21(3) 2012

[10] Projet de norme française NFS 31-114, v9 ; 22/10/2014

9th International Conference
on
Wind Turbine Noise
Remote from Europe – 18th to 21st May 2021

Estimation of the sound emergence of wind turbines by semi-supervised learning technique

Jean-Rémy Gloaguen, UMRAE, Cerema, Univ. Gustave Eiffel, Ifsttar, Strasbourg, 67035, France: jean-remy.gloaguen@cerema.fr
David Ecotière, UMRAE, Cerema, Univ. Gustave Eiffel, Ifsttar, Strasbourg, 67035, France
Benoit Gauvreau, UMRAE, Univ. Gustave Eiffel, Ifsttar, Cerema, Bouguenais, 44344, France
Arthur Finez, ENGIE Green, Nantes, 44000 France
Arthur Petit, ENGIE Green, Nantes, 44000 France
Colin Lebourdat, ENGIE Green, Nantes, 44000 France.

Summary

The sound emergence is the main regulatory estimator for wind turbine noise in France. This criterion aims to limit their noise impact on local residents and is highly dependent on the variation of residual noise over time. Therefore, initially defined curtailment plans can sometimes become inadequate, in which case they cannot easily be updated without leading to significant production losses. Machine learning techniques allow today to consider the continuous estimation by measurements of the sound contribution of wind turbine noise in the ambient noise and thus its noise emergence, without needing to stop the wind farm. This operation makes it possible not only to regularly adapt these reduction plans, thus optimizing electricity production, but also limiting the possible noise annoyance for local residents.
For this purpose, semi-supervised Non-negative Matrix Factorization method is considered, enhanced by a temporal regularity constraint. This approach combines a wind turbine dictionary designed on a learning basis and a free dictionary that allows the adaptation of the method to the variability of residual noise. Tests conducted on simulated measurements reveal satisfactory performances with mean estimation errors lower than 2 dBA for wind noise emergences lower than 5 dBA. Finally, the presence of these two types of dictionaries makes it possible to estimate the wind noise emergence according to one or the other depending on the predominance of the estimated wind turbine noise.

1. Introduction

In some European countries, the sound impact of wind farms is assessed by their sound emergences. This indicator is estimated by the difference between the A-statistical sound pressure level (SPL) of the ambient noise at the receiving point $L_{A,50,amb.}$ (i.e. when the wind farm is operating) and the residual noise $L_{A,50,res.}$ (i.e. when the wind turbine is shutdown):

$$E = L_{A,50,amb.} - L_{A,50,res.} \quad (1)$$

The statistical SPL presents the advantage to be more suited for the wind turbine (WT) noise, which remains quite constant during time, while equivalent SPL is more sensitive to extreme sound pressure levels that might result from emergences of residual noises.
These sound emergences are then limited to regulation thresholds to not exceed. To respect these criteria, the most common solution is the application of a curtailment plan on WTs to limit their operation and thus their noise emission (Rogers, 2020). However, this plan affects strongly the production capacity of the machines.
Furthermore, this solution has the main drawback to depend on the variability of the residual noise, which is a time varying component that changes over the meteorological conditions, the ground characteristics and the surrounding infrastructures. Consequently, this may result in two situations where the curtailment plan might be too strong (i.e. the electrical production is unnecessarily limited) or too low (i.e. the sound emergences exceed the regulation thresholds).
Currently, to update this plan, *on/off* cycle measurements are made. These measurements consist in assessing alternatively the SPLs of the ambient noise and the residual noise by stopping the wind farm periodically. If needed, a new curtailment plan can then be formulated.

However, these measurements require the shutdown of the machines, which also significantly affects the electrical production. Furthermore, they are only carried out during a couple of weeks and are not sufficiently representative over a long period. Consequently, this updated plan does not guarantee to be still valid few months later. To repeat this process regularly will then increase again the loss of electrical production and is not a suitable solution.
Consequently, there is an interest to be able to estimate the sound emergence of wind farms continuously without stopping them. Thus, it would be possible to control regularly the noise annoyance for neighboring inhabitants, while optimizing the electrical production.
Instead of seeking to improve the sound source emission of WT and the models for outdoor sound propagation (Cotté, 2019), this emergence could be estimated by measurements made *in situ*. Recently, (Gloaguen et al., 2020) propose a first tool to estimate the sound emergence of WTs. This tool is based on the Non-negative Matrix Factorization (NMF) used as a source separation method to first extract the WT component from simulated ambient scenes and then to estimate the sound emergences. This preliminary study has been extended in (Gloaguen et al., 2021) on a larger number of situations where the WT component has been propagated at different distances in different propagation conditions. Supervised learning associated with a temporal constraint reached satisfactory results with a mean error inferior to 2 dBA on many cases. In this study, only WT noise was considered in NMF. It might be then helpful to better consider the residual noise in NMF to bring more flexibility but also robustness to this method.
Thus, this paper extends these works and considers a semi-supervised learning, which allows NMF to consider labelled data (i.e. WT noise) and unknown data that can be adapted to the diversity of the residual noise. Section 2 introduces Semi-supervised NMF, Section 3 presents the different corpora used in this experiment and Section 4 summarizes the main results and details the behavior of the proposed method.

2. Semi-supervised Non-negative Matrix Factorization

In audio field, Non-negative Matrix Factorization (NMF) is usually introduced as a linear approximation of an audio spectrogram $V_{F\times N}$ by the product of two matrices, W and H, such as $V \approx \tilde{V} = WH$. $W_{F\times K}$, called *dictionary*, includes audio spectra and $H_{K\times N}$, called the *activation matrix*, corresponds to the temporal evolution of each of the spectra such as $\hat{V} = WH$ (Dikmen & Mesaros, 2013; Lee & Seung, 1999). As only amplitude (or power) spectra are considered in V and W, only additive combinations are possible. The dimensions of the matrices are chosen

such as $K < \min(F, N)$ to avoid overcomplete representation. In supervised learning, W is learned on labelled samples and H is the unknown to estimate. In the present case, W includes WT noise spectra.
For semi-supervised learning (Kitamura et al., 2014), a second dictionary and activation matrix are considered, respectively $Y_{F\times J}$ and $Z_{J\times N}$. The dimension J corresponds to the number of considered spectra, which is often set such as $J < K$ in order to allow NMF to stay focused on the data learned in W.
These two matrices are learned on each spectrogram to represent, the best as possible, the residual noise. This second part makes it possible to increase the degree of freedom of NMF and to include residual noise directly learn on the spectrogram $\tilde{V}$. In supervised learning, it would have been necessary to learn a residual dictionary on labelled data, which is more complex to do, as it is a noise composed of multiple different kind of sound sources not easy to define precisely.
Furthermore, instead of a linear sum between $[WH]$ and $[YZ]$, the energetic sum between the two components is considered here:

$$V \approx \tilde{V} = \sqrt{[WH]^2 + [YZ]^2}. \tag{2}$$

This modification of the NMF problem is proposed because the available data are expressed in Pa, thus the energetic sum is more appropriate (see Sec. 3). The approximation between V and $\tilde{V}$ is performed by minimizing the cost function $D_\beta(V||\tilde{V})$:

$$\min D_\beta(V||\tilde{V}) = \min\left(\sum_{f=1,n=1}^{F,N} d_\beta(V_{fn}||\tilde{V}_{fn}) + \alpha_{sm}\sum_{n=1}^{N}\left(h_{k,n} - h_{k,n-1}\right)^2\right) \tag{3}$$

$D_\beta(V|\,|\tilde{V})$ belongs to the β-divergence family where only the Euclidean distance is considered such as $d_2(x|y) = \frac{1}{2}(x-y)^2$. The added regularization term, weighted by the coefficient α_{sm}, corresponds to a temporal regularity constraint applied only on H (Févotte et al., 2018). This constraint forces the shape of the activations in H to adapt a smoother behaviour and thus to better corresponds to the time signature of WT noise, which is quite constant in practice. As the NMF formulation has been modified on Eq. 2, the update algorithms of H, Y and Z are changed. The new update algorithms are estimated as in (Lee & Seung, 2000) with a descent gradient. Equation 4 presents the update algorithms for the Euclidean distance for this new formulation:

$$Y \rightarrow Y.\frac{\left[[YZ].V\right]Z^T}{\left[[YZ].\tilde{V}\right]Z^T} \tag{4.1}$$

$$Z \rightarrow Z.\frac{Y^T\left[[YZ].V\right]}{Y^T\left[[YZ].\tilde{V}\right]} \tag{4.2}$$

$$H \rightarrow H.\frac{W^T\left[[WH].V\right] + 2\alpha[H_{+1} + H_{-1}]}{W^T\left[[WH].\tilde{V}\right] + 2\alpha\,[H + H_{12}]} \tag{4.3}$$

with $H_{-1} = [\mathbf{0}\; H_{1:N-1}]$, $H_{+1} = [H_{2:N}\; \mathbf{0}]$ and $H_{12} = [\mathbf{0}\; H_{2:N-1}\; \mathbf{0}]$ where $H_{n:m}$ refers to the selection of columns n to m of H and $\mathbf{0}$ a column of K zeros. More details on NMF and some applications can be found in (Févotte & Idier, 2011) , (Heittola et al., 2011) and (Ludeña-Choez et al., 2017).

In summary, among different parameters, each taking several possible values, one tries to find an optimal NMF formulation: the number of basis in W ($K \in \{5, 10, 20\}$), the number of basis in Y

($J \in \{2, 5, 10\}$) and the weight of the temporal regulation ($\alpha_{sm} \in \{0, 0.5, 1, 2\}$). Finally, NMF is performed for 1000 iterations where H, Y and Z are initialized randomly.
Sound emergences can be estimated in two different ways with semi-supervised NMF (SS NMF). First, one estimates from $[WH]$ the WT noise component in V, deduces the residual component, its estimated statistical SPL $\tilde{L}_{A,50,res.}$ and then the estimated emergence $\tilde{E}_{WH}$,

$$\tilde{E}_{WH} = L_{A,50,amb.} - \tilde{L}_{A,50,res,WH}. \quad (5)$$

In a second way, one considers $[YZ]$ as the exclusively residual noise and directly use it to estimate the sound emergence, called $\tilde{E}_{YZ}$. Figure 1 summarizes this process. In Section 4.1.1, the influence of this choice on the setting on the estimation of the sound emergence is exposed.

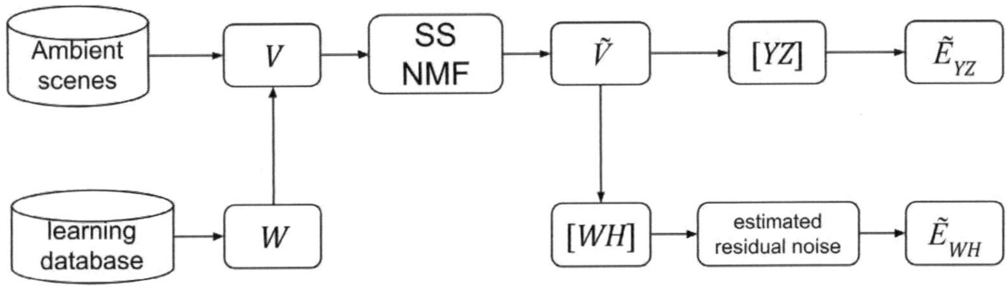

Figure 1: block diagram of the estimation of the sound emergence by NMF.

3. Corpora of environmental sounds

To perform the calculation, two different corpora are used. The first one is a corpus of ambient scenes, built in order to simulate the method based on *on/off* cycle measurements (see Section 3.1). However, instead of considering long periods, one considers short periods of 10 minutes of ambient noise (i.e. *on* cycles when the WTs operate) and only residual noise (i.e. *off* cycles when the WTs are shutdown). This second period simulates the current process used *in situ* that estimates the emergence $E_{on/off}$. This indicator is a baseline method, which will help to compare the performances of NMF. A second corpus composed of exclusively of WT noise spectra is dedicated to the learning of the dictionary W and is presented in section 3.2.
All the following data are expressed in Pa for each third octave band between 20 Hz and 10 kHz.

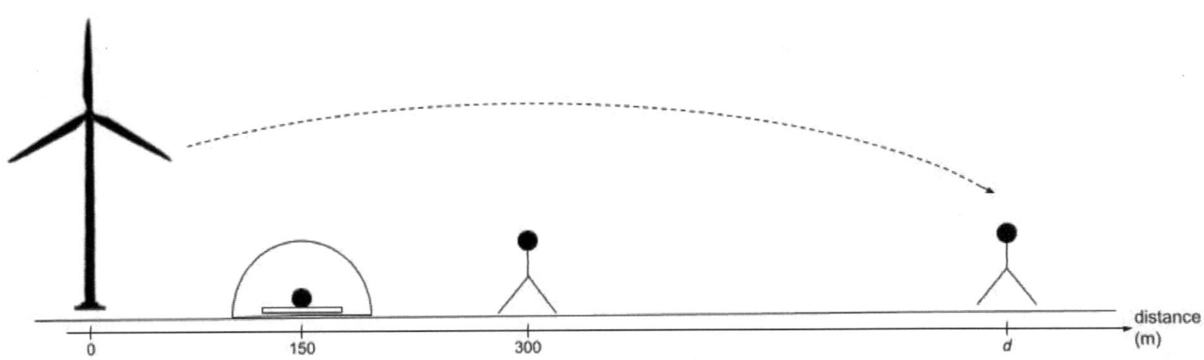

Figure 2: schema of the different measurement points used for the dictionary corpus (150 m) and the corpus of simulated ambient scenes (300 m). The distance d corresponds to the distance where the ambient scene are located (500 m, 1000 m, 1500 m).

3.1 Corpus of simulated ambient scenes

A corpus of 30 simulated ambient scenes is built for this experiment. This simulation process consists in the sum of a WT noise component with a residual noise component. This process makes it possible to know the 'exact' sound emergence $E_{ex.}$(see Eq. 1) and can then be compared to the one estimated by NMF. By using *in situ* measurements, this 'exact' value would have been unknown.

3.1.1. Wind turbine noise samples

The WT noise components are collected during *on/off* cycle measurements at 300 m from the front line of a wind farm at 1.5 m height (Kayser et al., 2018) (see Figure 2). To consider the most accurate samples, the residual noise spectra, estimated with the 30 minutes preceding and following the *off* measurements, filter the *on* measurements. From these cleaned samples, 30 samples of 10 minutes are extracted (see Figure 3). To generate ambient scenes accurately, an attenuation filter must filter the WT noise in order to simulate its propagation at a certain distance through an inhomogeneous media.

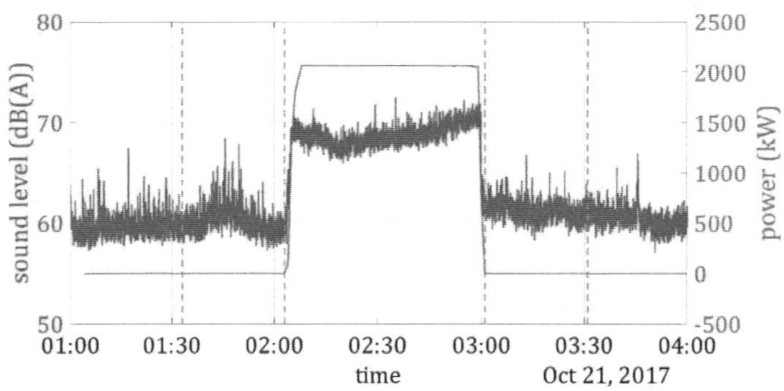

Figure 3: example of the extraction of wind turbine spectra. The emergence on *cycle is considered and filtered by the mean residual spectra estimated from the measurements carried out between the dashed lines.*

3 distances between the WTs and a receiving point are considered ($d \in \{500, 1000, 1500\}$ m) through one inhomogeneous media, which is considered here as a downward propagation condition. Propagation conditions are characterized although the vertical profile of the effective sound celerity, defined as:

$$c_{eff}(z) = \sqrt{\gamma RT(z)} + U(z)\cos(\theta) \tag{6}$$

with γ the heat capacity ratio of dry air at constant pressure and volume (γ = 1.4), R, the specific gas constant for dry air (R = 287 J.kg^{-1}.K^{-1}), θ the angle between the wind direction and the propagation direction and $T(z)$ and $U(z)$ the mean vertical profiles of air and temperature and wind speed respectively. As turbulence is not considered, these profiles are only expressed according to their mean part:

$$T(z) = T_0 + a_t \log\left(\frac{z-d}{z_0}\right), \tag{7.1}$$

$$U(z) = a_u \log\left(\frac{z-d}{z_0}\right) \tag{7.2}$$

where T_0 (K) is the ground surface temperature, d = 0.66h_v is the displacement height accounting for the influence of the vegetation height h_v (m), z_0 = 0.13h_v is the roughness height of flux profiles and a_t and a_u are the coefficients that determine the shape of the temperature and wind vertical profiles, respectively. The relative humidity is set to 70 %. The settings of the downward refraction condition used in the study are T_0 = 15°C, a_u = 100 m.s^{-1}, a_t = 0.15 K.m^{-1}, θ = 45.
Sound propagation is modelled through the use of Parabolic Equation (PE) (Kayser et al., 2020) and resolved by the split-step Padé approach with an average ground floor impedance (airflow

resistivity σ = 500 kN.s.m^{-4}). To improve the estimation of this filter, the sound emission of WT noise is defined by the extended source model proposed by Cotté (Cotté, 2019).
PE is performed for all the third octave bands from 20 Hz to 3150 Hz. This process makes it possible to obtain an accurate estimation of the SPL of WT to ensure the validity of the results of NMF.
From this resolution, the attenuation filters at the distances d are estimated with a reference microphone at 300 m (corresponding to the position where the WT samples in the ambient scenes have been collected) (see Figure 4). Due to computation limit, the attenuations for the third octave bands from 4 kHz to 10 kHz are modelled only with the geometrical divergence and the atmospheric absorption. This limitation is not prohibitive as the WT noise contribution in these bands are very low.

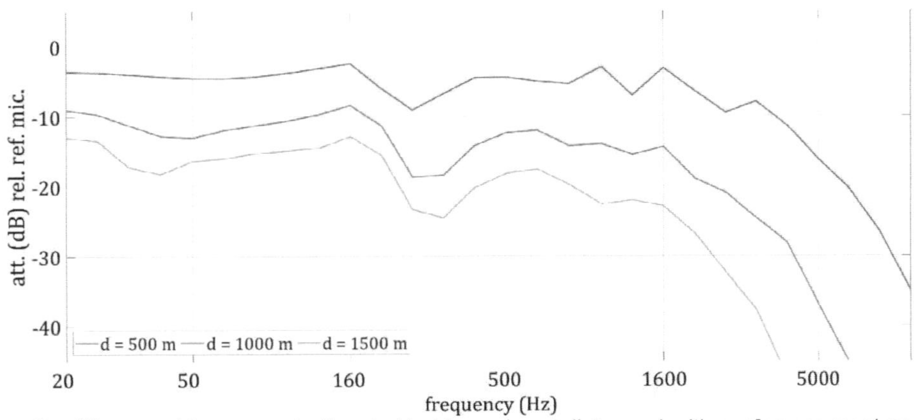

Figure 4: attenuation filters used to propagate the wind turbine noise at distance d with a reference receiver located at 300 m from the front line of the wind farm and at 1.5 m height above the ground.

3.1.2. Residual noise samples

The residual noise samples come from two different locations (called WF1 and WF2) and have been collected during *on/off* cycle measurements, but this time during the *off* periods. For each location, 30 samples are collected. The residual noises from WF1 account for weak residual SPL as the wind farm is located far from any anthropogenic noise sources. In the opposite, WF2 is located close to a highly frequently road. The residual noise samples related to this place are then much more dynamic.
Each sample lasts 25 minutes. The first 10 minutes are used to build the ambient noise sample with the WT noise sample. The last 10 minutes are used as only residual noise to simulate the shutdown of the wind farm and to estimate the baseline emergence $E_{on/off}$,

$$E_{on/off} = L_{A,50,amb.} - L_{A,50,res.,off} \tag{8}$$

The last 5 minutes (between the 10th minutes and the 15th minutes) can be seen as the transition time before the WT are shut down and are then discarded.

3.1.3. Building of the ambient scenes

From the WT and residual noise samples, the 30 simulated ambient scenes are built. These scenes length 10 minutes with a 1 second step. The WT noise is first propagated to 3 different distances (d ∈ $\{500, 1000, 1500\}$ m) for one propagation condition (see Section 3.1.1). The WT and the residual noise samples are then summed following the Signal-to-Residual-Ratio (SRR) defined as:

$$SRR = L_{A50,WT} - L_{A50,res.,on}, \tag{9}$$

where $L_{A,50,WT}$ is the A-statistical SPL of the WT component and $L_{A,50,res.,on}$ the A-statistical SPL of the residual component used for the ambient scene. The SRR is defined for {-9; -6; -3; 0; 3; 6; 9} dBA. When $SRR < 0$ dBA, the residual noise is predominant over the WT noise and inversely when $SRR > 0$ dBA. These SRR values correspond to 'exact' sound emergences $E_{ex.}$ of {0.5; 1.0; 1.8; 3; 4.8; 7.0; 9.5} dBA respectively. Finally, the simulated ambient scenes are the spectrograms V of NMF with dimensions F = 28 (i.e. number of third octave bands between 20 Hz and 10 kHz) and N = 600. The adjustments made to set the residual noise at a SPL corresponding to the SRR are also made on the last 10 minutes of the residual noise to assure its continuity. The Figure 5 displays for 2 SRR values the different parts of an ambient scene and Figure 7 summarizes the steps to build the ambient scenes.

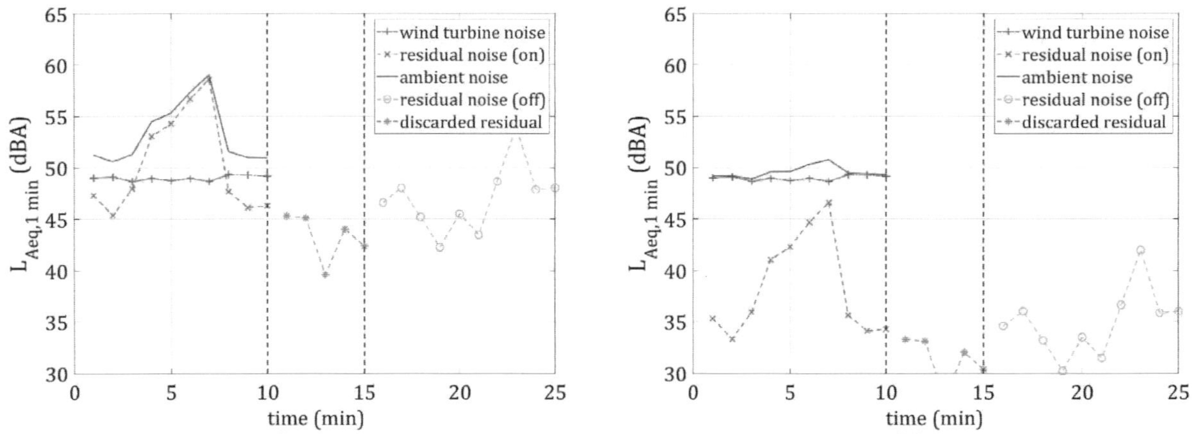

5-a 5-b

Figure 5: details on the building of the simulated ambient scene for SRR *= -6 dBA (5-a) and* SRR *= 6 dBA (5-b) (scene 6, WF1, d = 500 m).*

Finally, a pre-processing step is performed on the ambient scenes before considering it in NMF. This pre-process consists in the attenuation of the most energetic frames. One considers here that these frames are due to high emergences of residual noises. Thus, to help the approximation made by NMF and limit the influence of the residual noise, it might be beneficial to remove them. To do so, the temporal frames where the equivalent SPLs are superior to the $L_{A,10}$ are discarded and replaced by interpolated values to ensure coherent continuity in the spectrogram. Figure 6 displays an example. To know its impact, NMF is also performed without this operation.

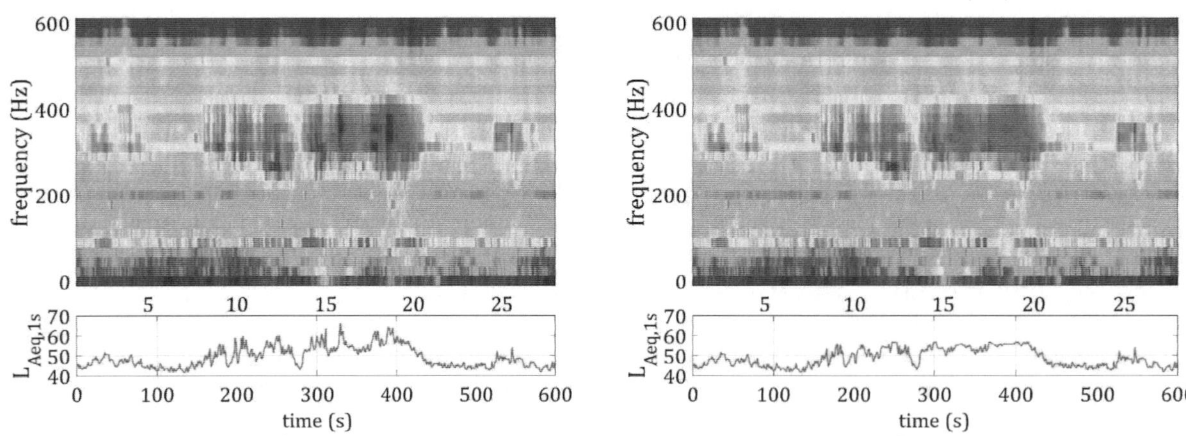

6-a 6-b

Figure 6: influence of the pre-processing step on the spectrogram and on the A-SPL. On the initial spectrogram (6-a), the residual noise provokes high emergences. On the pre-processed spectrogram (6-b), these emergences are reduced, for instance in the intervals {327; 334} s and {386; 396} s

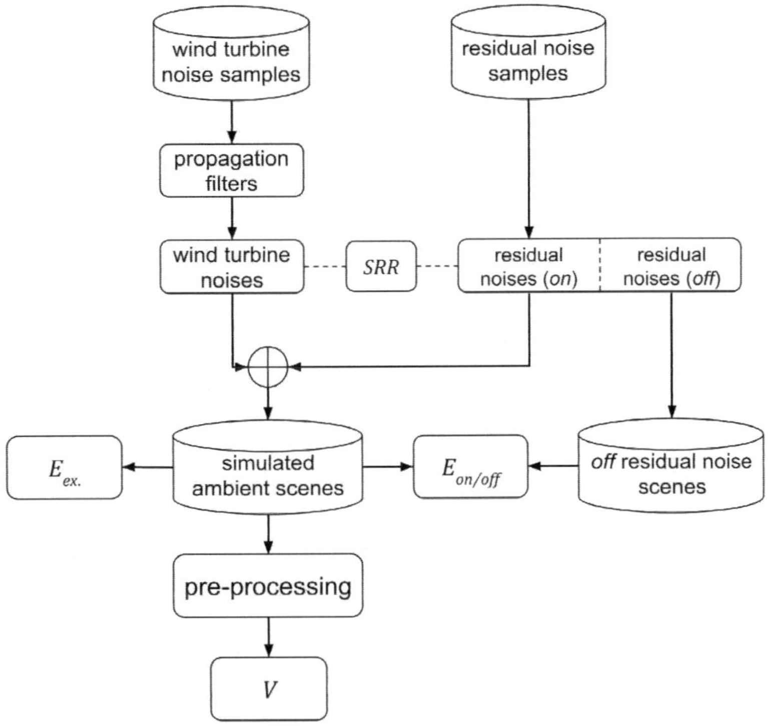

Figure 7: block diagram of the ambient scene corpus building.

3.2 Corpus for the dictionary *W*

To build a dictionary of WT spectra, measurements made during *on/off* cycle measurements on the same wind farm than in Section 3.1 are collected. To ensure to have spectra the less polluted by surrounding residual noise samples, only measurements made following the IEC standard protocol are considered as they are closed to WTs (see Figure 2). In addition, only the most emerging measurements are considered. They are then filtered by the estimated residual noise spectra measured during the *off* cycle as in Section 3.1.1, to ensure the most accurate WT spectra.

In all, 14200 third octave band spectra (ranking from 20 to 10 kHz) are extracted. This large number of spectra cannot be directly in the multiplicative update algorithms (Eqs. 4) as it will highly increase the computation cost. Furthermore, these spectra present redundant information. Consequently, a K-mean algorithm reduces these 14200 spectra to $K \in \{5, 10, 20\}$.

As the ambient scenes simulate the WT noise contribution at 3 distances from the front line of a wind farm, an attenuation 'transfer function' filters is also applied on the initial dictionary at the considered distances. Instead of using PE, one chooses a simpler approach to facilitate the implementation of the proposed method in an industrial process. The Weyl's Van der Pool equation is then considered and solved by the sound rays model (Salomons, 2012). The sound emission models the WT noise by 3 moving monopoles (hub height = 80 m, speed rotation = 20 rpm). A downward meteorological condition is chosen with a linear vertical profile of the effective celerity: $c_{eff}(z) = c_0 + b_{lin}z$ with c_0 the sound celerity (m/s) with an air temperature of 10°C, b_{lin} = 0.1 m^{-1} and z the height (m). The ground impedance is defined by the Miki's model (Miki, 1990) with an air flow resistivity representative of countryside (σ = 500 kN.s.m^{-4}). The corresponding attenuation filters are displayed on Figure 8. They are generated for the 3 distances with a reference microphone corresponding to the one on the ground located at 150 m from the front line of the wind farm (see Figure 2). Finally, each filtered spectra of the dictionary is normalized such as $||W_k||$ = 1.

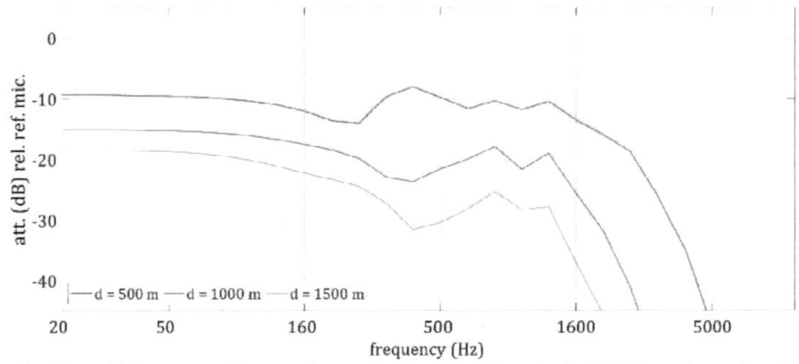

Figure 8: attenuation filters for the dictionary with a reference receiver located at 150 m from the front line of the wind farm and on the floor.

4. Computation and discussions

The objective of the experiment is to find an optimal setting of NMF (β, K, J, α_{sm}, number of iteration, estimation by $[WH]$ or $[YZ]$, pre-processing step) on 30 simulated ambient scenes with the WT noise propagated at different distances ($d \in \{500, 1000, 1500\}$ m), for one downward propagation condition and for two locations of residual noises (WF1 and WF2). For each ambient scene, one gets the 'exact' sound emergence $E_{ex.}$, thanks to the simulation process, the estimated emergence by the simulated *on*/*off* cycle scenes $E_{on/off}$ and the estimated one by NMF $\tilde{E}_{WH/YZ.}$.

To compare these indicators, the Mean Absolute Error MAE is considered. It is expressed such as:

$$MAE_{WH/YZ} = \frac{\sum_{m=1}^{M} \left| \tilde{E}_{WH/YZ,m} - E_{ex.,m} \right|}{M} \tag{10}$$

This metric is also computed with $E_{on/off}$ instead of $\tilde{E}$. M is the number of considered scenes. It can be computed for the total number of scenes (M= 180, see Section 4.1.1) and for each distance and residual noise (see Section 4.1.2).

4.1.1. Global results on all the corpus of simulated ambient scenes

The MAE error is computed for each association of setting (384 associations in all) on all the corpus (M = 180 scenes). Instead of summarized all of these, only the ones with the lowest errors are displayed in Table 1 according to the dimension J, the use of the pre-processing step (*no* or $L_{A,10}$) *and* the choice of the component in $\tilde{V}$ used to estimate the sound emergence ($[WH]$ or $[YZ]$). Then, the association of settings reaching the lowest error is the optimal SS NMF formulation the most efficient that is considered.

Table 1: lower errors of SS NMF among all the tested association of settings according to the dimension J, the choice of the component in $\tilde{V}$ estimation and the pre-processing step. In bold letters, the best results.

Method	K	J	α_{sm}	Iteration	Pre-processing step	Component in $\tilde{V}$	MAE (dBA)
Baseline	-	-	-	-	-	-	2.7 (± 2.8)
SS NMF	5	2	0.5	50	*no*	[WH]	2.1 (± 1.8)
	5	2	0.5	50	$L_{A,10}$	[WH]	**1.9 (± 1.7)**
	5	5	0.5	50	*no*	[WH]	2.1 (± 1.8)
	5	5	0.5	50	$L_{A,10}$	[WH]	**1.9 (± 1.7)**
	5	2	1	5	*no*	[YZ]	6.2 (± 5.8)
	5	2	0.5	1000	$L_{A,10}$	[YZ]	3.1 (± 2.0)

5	5	1	1000	*no*	[YZ]	4.0 (± 4.5)
5	5	0.5	1000	$L_{A,10}$	[YZ]	2.5 (± 1.8)

The lowest error is reached when emergences are estimated from $[WH]$, with the pre-processing step and for $J \in \{2;5\}$. More generally, the lowest MAE errors are obtained when the emergence is estimated from $[WH]$ instead of $[YZ]$. This choice happens to be the most influential as the NMF settings are for the most part similar (i.e. the dimensions K and J, the number of iteration, the smoothness weight α_{sm} stay the same).
The addition of the pre-processing step improves the results, even if its impact is more reduced when emergences are estimated from $[WH]$. This result can be understandable as it limits on the spectrogram V the presence of the residual noise, which can naturally help the approximation of the WT component by NMF.
Furthermore, the increase of basis in Y only improves the results when $[YZ]$ is considered. The approximate WT component is then constant for J = 2 or J = 5. This behavior reveals that the estimation of the $[WH]$ component is quite robust and the increase of the dimension J does not result in more confusion. For the next, we consider J = 5 as it is also where the estimations of the sound emergence with $[YZ]$ are the best.
Finally, the proposed method shows lower errors with reduced standard deviations than the baseline method on this corpus, which reveals that the estimation by SS NMF is more accurate. The baseline method, based on *on/off* measurement process, makes the hypothesis that residual noise is constant over time, which might not be the case. This is particularly the case for the samples that come from the WF2 wind farm. Applied directly to the ambient scenes, this versatility does not affect SS NMF, in addition to not having to shut down the wind farm.

4.1.2. Behaviors of the optimal SS NMF detailed

From this main results, the behavior of the optimal SS NMF (K = 5, J = 5, α_{sm} = 0.5, 50 iterations, with the pre-processing step and with the emergence deduced from $[WH]$ component) is now according to the distance and the location of the residual noises in Figure 9.

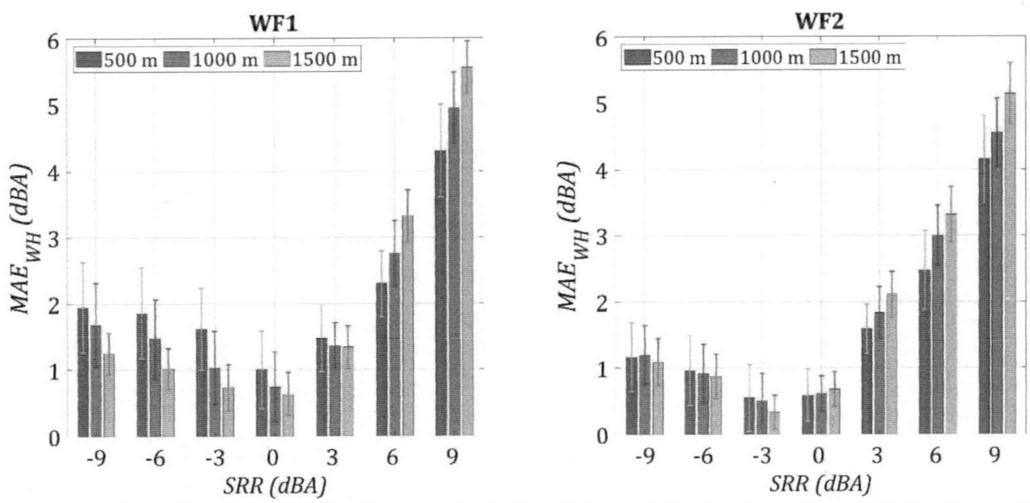

Figure 9: errors of the optimal SS NMF details according to the distance d, the location of the residual noises and the SRR.

First, one notices the strong dependence of the results to the SRR values. The evolution of the errors according to the residual noise location or the distance follows the same pattern. The MAE_{WH} tends to decrease when SRR increases and is inferior to 0 dBA. Then, it increases strongly with the SRR as the WT become more and more predominant over the residual noise.
To apprehend these results, the 1 second equivalent SPL are displayed in Figure 10 for SRR = -6 dBA and for SRR = 6 dBA in Figure 11.

In these representative examples, when SRR is negative, the estimated WT noise is overestimated, which results in the underestimation of the estimated residual noise and so the overestimation of the sound emergence.
Conversely, when SRR = 6 dBA, the estimated WT noise is more equivalent to its 'exact' value. Consequently, the strong errors observed in Figure 9 are not due to a bad estimation of the WT component, but to a higher sensitivity of the emergence indicator in these cases when SSR is negative. For instance, an underestimation estimation of the WT SPL of 1 dBA for SRR = 9 dBA results of an error on the residual noise SPL of 4 dBA. This behavior happens specifically when the sound emergence is high (above 5 dBA) and can still be detected.

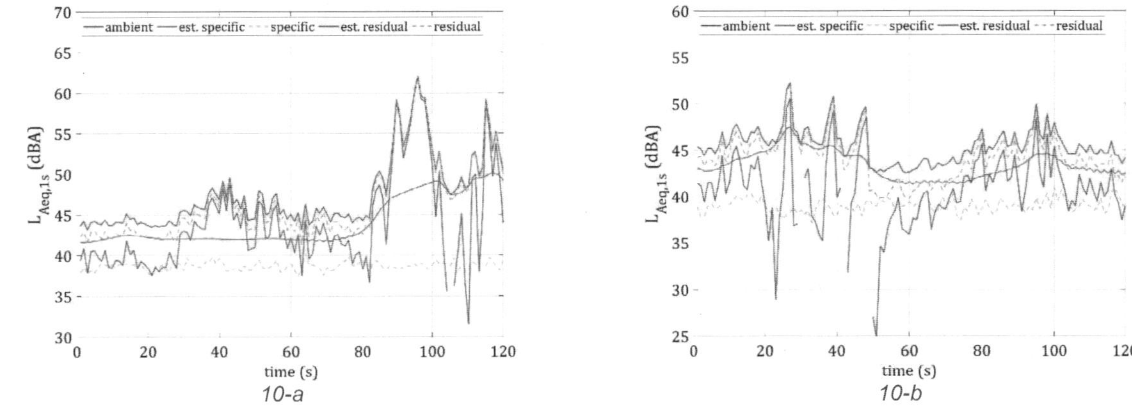

Figure 10: a 120 seconds extract of the 1 second equivalent SPL for SRR *= -6 dBA with residual noise from WF1 (scene 4, d = 500 m) (10-**Erreur ! Source du renvoi introuvable.**) and WF2 (scene 4, d = 500 m) (10-**Erreur ! Source du renvoi introuvable.**).*

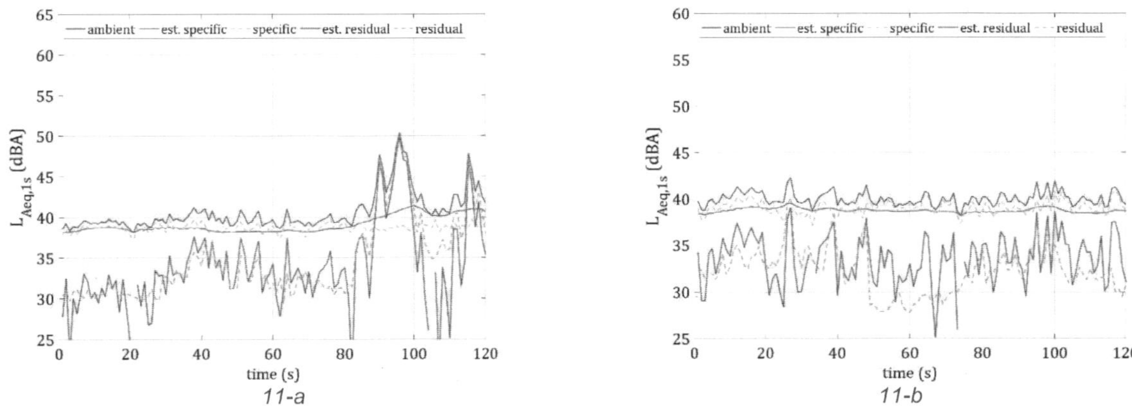

Figure 11: a 120 seconds extract of the 1 second equivalent SPL for SRR = +6 dBA with residual noise from WF1 (scene 4, d = 500 m) (11-a) and WF2 (scene 2, d = 500 m) (11-b).

If the SRR is the main influence parameter, the location and the distance have also a significant impact (but reduced). According to their evolution on Figure 9, it might be some interaction phenomena between these parameters. At 500 m for SRR< 0 dBA, for WF1 residual samples, the errors are higher than at 1500 m. They decrease with WF2 residual noise samples. When, SRR becomes positive, these patterns reverse. At d = 500 m, residual samples from WF1 are less noisy, which results in more similar spectral shapes to the WT noise and then generates confusion. With the distance, the attenuation filters altered the spectra, which results in a better fit of the estimated and the 'exact' WT noise spectra. This observation can be seen in Figure 12 where the mean spectra of each component are displayed. Finally, when SRR increases, the error becomes higher with the distance. This might be due to the difference in the propagation filter applied on the dictionary W (see Section 3.2) emphasized with the distance and the high value of the SRR where the sound emergence is more sensitive.

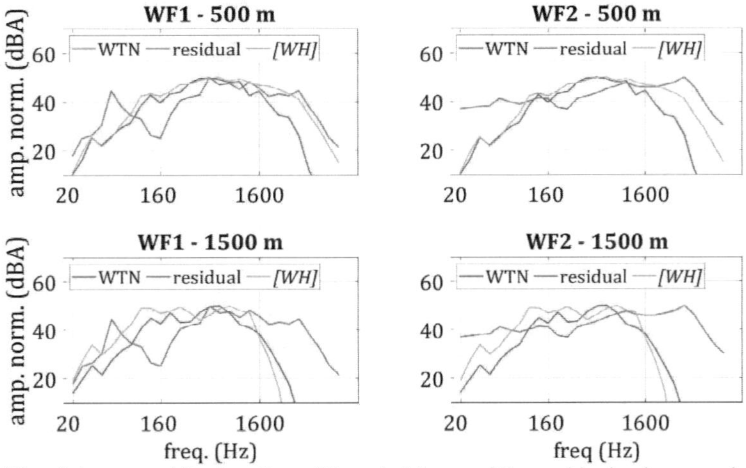

Figure 12: influence of the distance and the location of the wind farms of the residual noise samples on the mean spectra of wind turbine noise, residual noise and estimated WTN ($[WH]$) (scene 2, SRR = -6 dBA).

4.1.3. Comparison of the estimated residual noises

In this last part, the residual noise deduced from the estimated WT noise and the residual noise directly computed with $[YZ]$ are compared. To do so, the best setting with $[WH]$ and $[YZ]$ (see Table 1) are selected and their errors according to SRR are displayed in Figure 13.

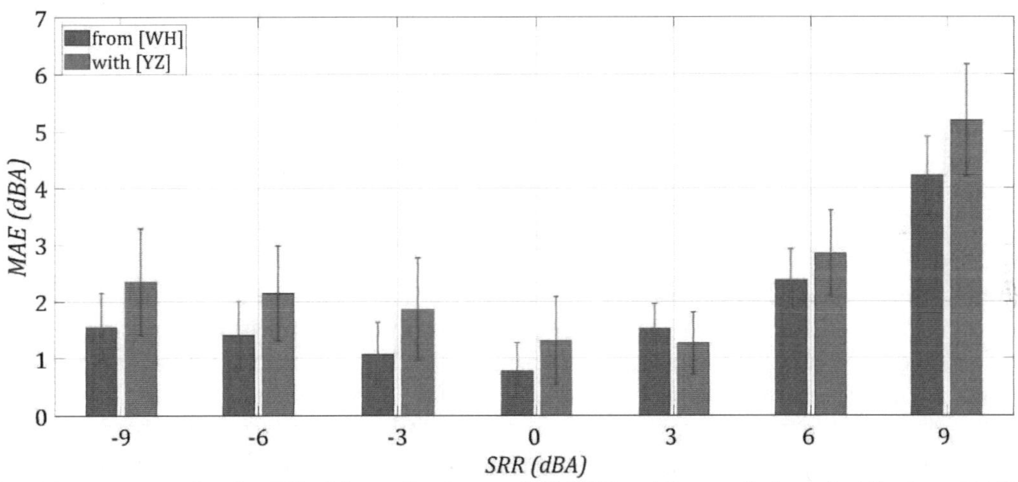

Figure 13: MAE errors according the SRR of the optimal error of SS NMF and the one that reached the lowest MAE error with the residual noise estimated with $[YZ]$ (see Table 1).

For both settings, the errors follow a similar respond according to the SRR, meaning that the estimation of the residual component is similar too. To illustrate this observation, Figure 14 summarizes 120 seconds of the residual components of one scene.

The behavior of the two residual components are similar in this example. The one deduced from $[WH]$ does not present the effects of the pre-processing step (between 46 s and 66 s, see Figure 14) as it is deduced from the ambient noise and from $[WH]$. Conversely, $[YZ]$ is learned on processed spectrograms. Finally, both approaches reveal to be very similar and to be mostly influenced by the approximation of the WT component by $[WH]$, which provokes the overestimation of the residual component (between 74 s and 83 s for instance, see Figure 14).

By using $[YZ]$, the main limit is the reduced number of spectra to approximate this sound source. The results in Table 1 reveal that the increase of the dimension J is beneficial (MAE_{YZ} = 3.1

($\pm$ 2.0) dBA with J = 2 and MAE_{YZ} = 2.5 ($\pm$ 1.8) dBA with J = 5). However, increasing this dimension also increases the risk that the cost function (Eq. 3) might be minimized mainly with this component at the expanse of $[WH]$, which is moreover subject to a temporal constraint. Consequently, the dimension J must stay low in order to not include in $[YZ]$ some WT noise component.
Furthermore, the reduced number of spectra in Y might be limited in some cases to model a component that includes many kinds of sound sources and might be insufficient to model it properly. From $[WH]$, the WT noise is more constant over time, if the dimension K is also low, it is still sufficient to approximate this component. Consequently, deduced directly from the ambient signal, the sound diversity of the residual component is better considered even if the estimated WT component presents some mistakes. This way limits the error and then improves the estimation.

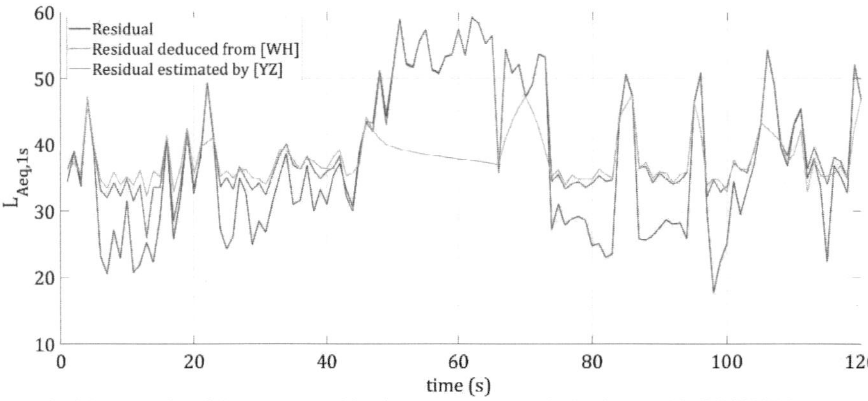

Figure 14: extract of 120 seconds of the exact residual components and obtained with SS NMF from $[WH]$ and from $[YZ]$ (scene 10, WF2, d = 500 m).

5. Conclusions

Semi-Supervised Non-negative Matrix Factorization (SS NMF) has been considered to estimate the sound emergence of wind turbine noise. To estimate the optimal settings, simulated ambient scenes have been built. This process makes it possible to compare the estimated emergences with their 'exact' values, which is not possible with *in situ* measurements without stopping the machines. The semi-supervised learning consists in adding in the dictionary a free part that can be adapted on each ambient scene. In addition, a regulation term is added through a temporal continuity constraint applied only on H (i.e. the temporal activations of the wind turbine spectra). This constraint makes it possible to have more continuous shapes of the activations and then more similar to the wind turbine noise behavior.
The common formulation of SS NMF has been adapted to correspond to the form of the input data through an energetic sum instead of a linear sum. This change required generating new update algorithms.
After computation, the settings of the optimal SS NMF have been found (K = 5, J = 5, α_{sm} = 0.5, 50 iterations, with the pre-processing step and with the emergence deduced from $[WH]$ component), which result in a global MAE error of less than 2 dBA. The simulated ambient scenes enables the study of this result according to the SRR, the distance and the location of the residual noise. The SRR reveals to be the more influential setting as it impacts directly the sound emergence. For negative SRR, SS NMF performs appropriately, but the estimation errors increase significantly as the wind turbine noise becomes more predominant. This behavior is more due to a high sensitivity of the emergence indicator than to a poor estimation of wind turbine noise. The different locations of residual noise in the scene have also an impact on the results

according to its similarity of its audio spectra with the wind turbine noise: the more similar they are, the higher the risk of confusion.
Finally, this proposed method, despite its limitations, still generates estimation errors lower than the ones based on *on/off* cycle measurements, which is a great improvement for wind farm operators. Now, these results and observations make it possible to consider this method using real *in situ* measurements and allow a better apprehension of the estimations according to the dynamic of the ambient noise or the distance between the wind farm and the measurement point.

References

Cotté, B. (2019). Extended source models for wind turbine noise propagation. *The Journal of the Acoustical Society of America*, *145*(3), 1363-1371. https://doi.org/10.1121/1.5093307

Dikmen, O., & Mesaros, A. (2013). Sound event detection using non-negative dictionaries learned from annotated overlapping events. *2013 IEEE Workshop on Applications of Signal Processing to Audio and Acoustics (WASPAA)*, 1-4. https://doi.org/10.1109/WASPAA.2013.6701861

Févotte, C., & Idier, J. (2011). Algorithms for nonnegative matrix factorization with the β-divergence. *Neural Computation*, *23*(9), 2421-2456.

Févotte, C., Vincent, E., & Ozerov, A. (2018). Single-Channel Audio Source Separation with NMF : Divergences, Constraints and Algorithms. In S. Makino (Éd.), *Audio Source Separation* (p. 1-24). Springer International Publishing. https://doi.org/10.1007/978-3-319-73031-8_1

Gloaguen, J.-R., Ecotière, D., Gauvreau, B., Finez, A., Petit, A., & Lebourdat, C. (2020). *Automatic estimation of the sound emergence of wind turbines using Non-negative Matrix Factorization : A preliminary study.*

Gloaguen, J.-R., Ecotière, D., Gauvreau, B., Finez, A., Petit, A., & Lebourdat, C. (2021). Automatic estimation of the sound emergence of wind turbine noise with Non-negative Matrix Factorization. *In review.*

Heittola, T., Mesaros, A., Virtanen, T., & Eronen, A. (2011). Sound event detection in multisource environments using source separation. *in Workshop on Machine Listening in Multisource Environments, CHiME2011.*

Kayser, B., Cotté, B., Ecotière, D., & Gauvreau, B. (2020). Environmental parameters sensitivity analysis for the modeling of wind turbine noise in downwind conditions. *The Journal of the Acoustical Society of America*, *148*(6), 3623-3632. https://doi.org/10.1121/10.0002872

Kayser, B., Ecotière, D., Bourdat, C., & Gauvreau, B. (2018, juin 12). *A new experimental database for wind turbine noise propagation in an outdoor inhomogeneous medium*. 17th International Symposium on Long Range Sound Propagation, Lyon, France.

Kitamura, D., Saruwatari, H., Yagi, K., Shikano, K., Takahashi, Y., & Kondo, K. (2014). Music Signal Separation Based on Supervised Nonnegative Matrix Factorization with Orthogonality and Maximum-Divergence Penalties. *IEICE Transactions on Fundamentals of Electronics, Communications and Computer Sciences*, *E97.A*(5), 1113-1118. https://doi.org/10.1587/transfun.E97.A.1113

Lee, D. D., & Seung, H. S. (1999). Learning the parts of objects by non-negative matrix factorization. *Nature, 401*(6755), 788-791. https://doi.org/10.1038/44565

Lee, D. D., & Seung, H. S. (2000). Algorithms for Non-negative Matrix Factorization. *In NIPS*, 556-562.

Ludeña-Choez, J., Quispe-Soncco, R., & Gallardo-Antolín, A. (2017). Bird sound spectrogram decomposition through Non-Negative Matrix Factorization for the acoustic classification of bird species. *PLOS ONE, 12*(6), e0179403. https://doi.org/10.1371/journal.pone.0179403

Miki, Y. (1990). Acoustical properties of porous materials-Modifications of Delany-Bazley models-. *Journal of the Acoustical Society of Japan (E)*, *11*(1), 19-24. https://doi.org/10.1250/ast.11.19

Rogers, J. (2020). Optimal strategies for wind turbine environmental curtailment. *Wind Energy*, *23*(5), 1331-1350. https://doi.org/10.1002/we.2489

Salomons, E. M. (2012). *Computational Atmospheric Acoustics*. Springer Science & Business Media.

INCEEUROPE

9th International Conference on Wind Turbine Noise
Remote from Europe – 18th to 21st May 2021

A study of the relationship between wind direction and sound level for wind turbines measured in the far-field

Duncan Halstead, Aercoustics Engineering Limited
Hillary Fung, Aercoustics Engineering Limited

Summary

This study examines the relationship between wind direction and far-field sound pressure level measured near one or more wind turbine generators. Sound propagation through the atmosphere is a well-studied concept [1] [2] [3] [4], and downwind sound propagation is typically found to be the most efficient. Current standards [1] [2] also suggest that a downwind[1] position from a wind turbine will experience the highest sound pressure levels. Far-field sound levels from wind turbines, however, are seldom the subject of robust study, due largely to the lack of available datasets having appropriate weather conditions or supporting data (wind speed/direction, turbine outputs, etc.) with which to study.

A wind turbine is directive based on the profile of the trailing edge (or other) noise source [3], and the distance to a receiver over which wind turbine noise regularly propagates is typically much farther than the distances at which we have robust verification measurements. Further, the hub height of modern megawatt-scale wind turbines now regularly exceeds 100 m, which is a source height much higher than what is usually considered in empirical research. As such, the effect of wind direction on far-field sound pressure levels warrants further study. This paper presents a review of measurement data collected in different wind directions across a selection of sites in southern Ontario.

Six measurement datasets collected at five different wind facilities are examined, each having a different layout of nearby turbines. All locations chosen for study had close turbines in only one direction, making for a good basis to study the effect of wind direction. The data was collected in 1-minute averages of sound pressure level, wind speed and direction, temperature, humidity, and atmospheric pressure. Data provided by each wind farm are also used in the study, including the outputs of the nearby turbines.

It was found that the influence of wind direction did not affect the sound levels as expected in the sites under study. Notable findings in the data are that the sound levels in the upwind direction were among the louder sound levels measured in five of the six datasets. Further, the direction

[1] Wind blowing from the nearest wind turbine to the measurement position.

counterclockwise to the downwind angle of the nearest turbine was louder than the clockwise direction (i.e. the 315-360/0° angle was louder than 0-45° angle, when 0° = downwind). These findings illustrate some interesting areas for further research, and subsequent studies will benefit from an expanded analysis dataset to determine if the findings here are indicative of a general trend.

1. Introduction

The effect of wind direction on far-field sound propagation is well studied for other sound sources. However, fewer studies exist that examine this relationship specifically for wind turbines. Wind turbines are a unique sound source, they are elevated, directive and the motion of their blades creates a moving source. As such, wind turbines warrant specific examination and validation. Few such studies exist, likely due to a lack of available data with which to conduct such a study.

This study takes data from several measurement campaigns spanning different wind farms, layouts, and turbine models, and examines the relationships between wind direction and sound level. Differences in both overall noise level and 1/3rd octave band frequency are examined, and the uncertainty associated with the measured levels is also be provided.

Far-field noise testing of wind farms is the predominant method for assessing acoustic compliance of wind farms in North America. Measurement campaigns are difficult, and at the mercy of weather during the available measurement time. An understanding of the relationship between wind direction and wind turbine noise level is important to determine the likely changes in noise level a receptor may experience with different wind conditions. If a reliable and consistent relationship between wind direction and measured level can be found, this information may be used to guide future work using measured or modelled wind turbine sound levels.

2. Measurement Methodology

The measurement data used for this study was collected during long-term measurement campaigns at a variety of locations across wind farms in southern Ontario, Canada. Measurement equipment was installed in the far-field, near residences or in vacant lots surrounding a wind facility. The measurement campaigns typically lasted anywhere between 4 to 16 weeks long and had to follow a measurement protocol established by the Ontario Ministry of the Environment [8].

2.1 Measurement Equipment

The data used in this study was collected using measurement equipment installed near wind facilities and configured to remotely record sound levels and meteorological ("MET") information each night. The equipment synchronously captures acoustic and MET data at heights of 4.5 metres and 10 metres, respectively. The microphone is fitted with a primary and secondary windscreen to reduce the effect of wind generated self-noise. The 10-metre weather station measures a range of weather parameters for each interval. A picture of a typical measurement apparatus is shown in Figure 1.

Figure 1: Measurement apparatus example

These measurement stations are typically installed at distances of 500 to 800 metres from the nearest wind turbines. All measurements are unattended and measurement equipment is calibrated at the beginning, end, and periodically throughout each measurement campaign. A total of 4,480 hours of measurement data was assessed in this study.

2.2 Measurement Parameters

The datasets presented here are comprised of 1-minute averages taken from 9 pm to 6 am every night during the measurement period. Sound and 10-metre MET data are acquired synchronously, and turbine operational data is received separately via the facility SCADA system. Acoustic parameters include overall LA_{eq}[2], 1/3rd octave band level, and L90[3] level for each interval. The 1/3rd octave band and L90 data are used in this study to remove acoustic contamination (insects and car pass-bys, primarily). The 10-metre MET data used includes wind speed, wind direction, and precipitation. In this study, the wind speed and precipitation data are relied upon to remove periods that are likely influenced by rain or wind-induced noise.

From the facility SCADA system, the electrical power is used to constrain the periods under study to high output conditions of the nearest turbine. This was applied because turbine power is highly correlated to sound emission for most pitch-controlled[4] wind turbines. The yaw position of the turbine is used to determine the wind direction, as the wind direction at hub height is considered more stable and important to the propagation path of the sound wave. Other analysis using wind turbine information, including wind shear and wind veer, are outside the scope of this study.

2.3 Measurement Environments

The measurement locations used in this study were chosen from a larger available measurement database. Locations were chosen based on the layout of the surrounding turbines and the range of wind directions available in each dataset. Locations having one or more close turbines in only one direction from the measurement position were prioritized to ensure that differences in sound level at each wind direction would not be influenced by turbines from other directions. This way, measurement locations having only one downwind direction were selected for further analysis.

A total of 14 measurement locations were initially evaluated and reduced to six locations having a suitable turbine layout and range of observed wind directions. Sites having one or more turbines in a single direction were prioritized, with secondary turbines in other directions allowed if they are at least twice as far away as the nearest turbine. A summary of the locations and their approximate surrounding turbine layout is presented in Table 1.

Table 1: Description of Monitoring Locations and Surrounding Environment

Location ID	Distance to Nearest Turbine	Downwind Direction	Turbine Layout
Site A	610 m	150°	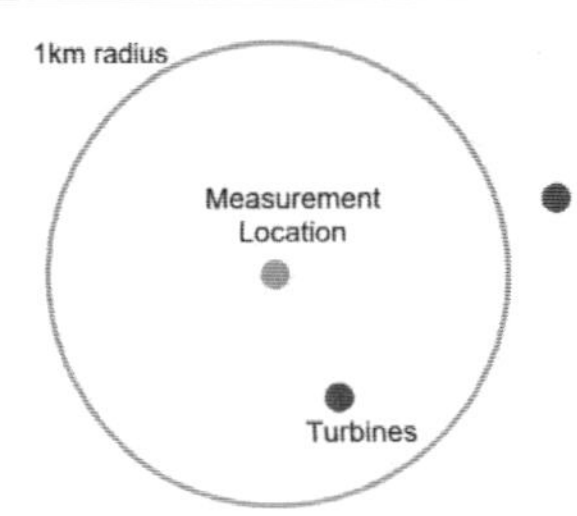

² Energy-equivalent sound level, A-weighted.
³ Sound level exceeded 90% of the time for a given measurement interval.
⁴ Also known as variable speed turbines.

Site B	680 m	148°	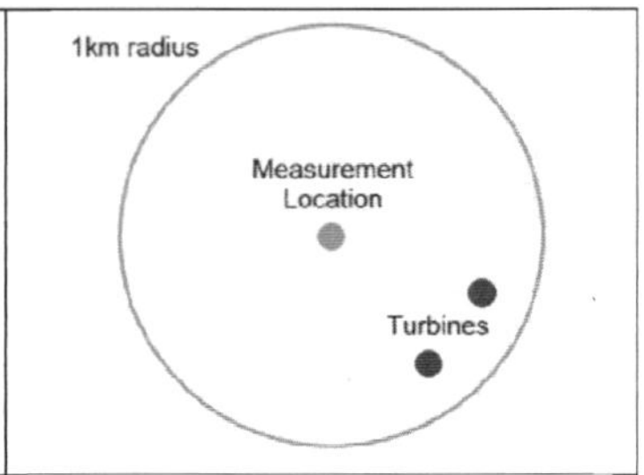
Site C	630 m	272°	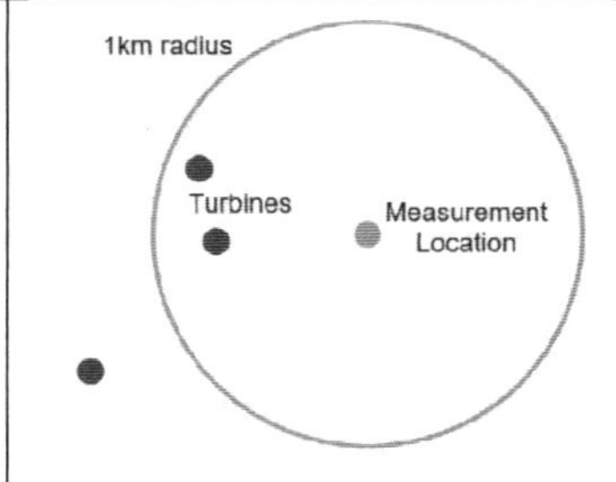
Site D	770 m	286°	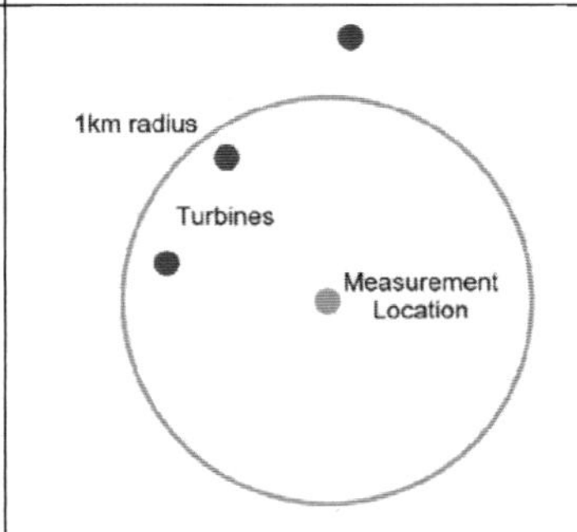
Site E	580 m	275°	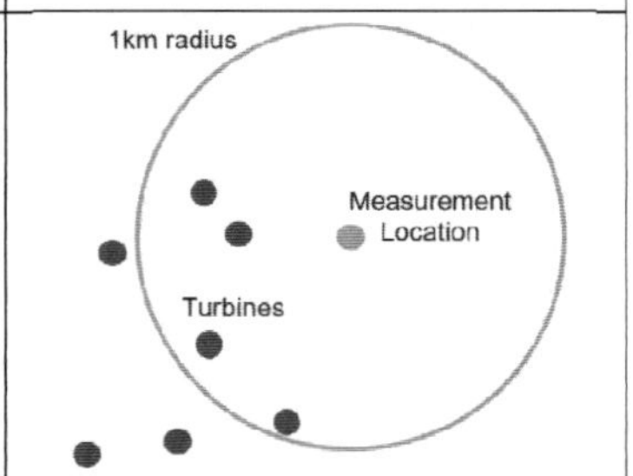
Site F	560 m	197°	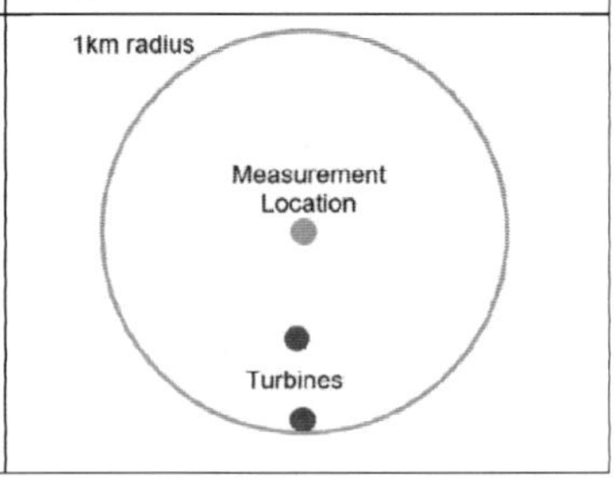

2.4 Data Filtering

Each measurement dataset is filtered to remove excessive ambient sound levels. Periods of high 10-metre wind speeds are removed from the analysis, as high wind speeds will increase the ambient sound level above the level of the wind turbine [9]. The threshold for 10-metre wind speed is chosen based on a visual inspection of each dataset plotted as sound level vs. wind speed (see Figure 2 for an example).

Ambient contamination was further reduced in the datasets by removing high frequency noise above 1250 Hz[5] via the 1/3rd octave band levels, and removing intervals having transient contamination identified by examining the difference between average (LAeq) and minimum (L90) sound levels for each interval.

The measurement datasets are also filtered for high turbine output, above 60% of nominal power for the nearest turbine. This ensures the wind turbines are emitting high sound levels while maximizing the amount of low 10-metre wind speed data available to compare.

The periods that remain after filtering include only conditions when the turbine has a high output, and the ambient influence is low. The intent is to maximize the signal to noise between turbine and ambient for each dataset so a better comparison of the turbine sound levels at different wind directions can be made.

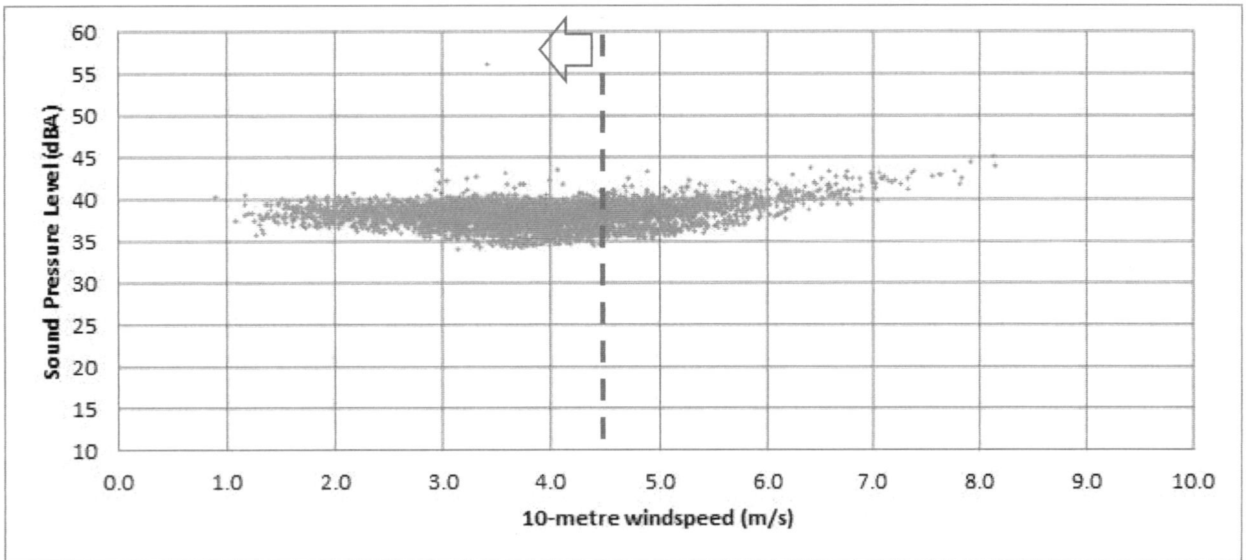

Figure 2: Example dataset of sound levels plotted against 10-metre wind speed; the cut-off point for wind direction is shown as a red line, and only the data left of that line is kept for further analysis. Data in this plot is already filtered for 60% power and includes all wind directions.

The remaining data is filtered and binned by wind direction, in 10° increments. The datasets after filtering are presented in the following section in several histogram and spectrogram plots. Data is finally compiled in tabular format and a radial plot showing the average difference in sound level with respect to downwind direction across all sites is provided.

3. Results and Discussion

Each of the six measurement datasets are presented below, after the filtering described in the previous section was applied. The different plots illustrate how the sound levels vary with wind direction, both in overall level as well as by frequency. The differences in measured sound level by wind direction, compared to the downwind direction, are also presented in tabular format at the end of this section. The results of the data are then averaged by wind direction to develop a single wind rose with the average difference in measured sound level provided at each wind speed.

[5] Due to the distance from the turbines of each measurement point (500m and greater), acoustic energy above these frequencies is not expected to originate from the wind facility due to higher levels of atmospheric attenuation at these frequencies.

3.1 Variation in Overall Sound Level with Wind Direction

The overall sound pressure level (filtered) is plotted against the measured wind direction for each dataset under study. The data is plotted in both a histogram and spectrogram format. The histogram plots in Figure 3, below, show the distribution of measured sound levels as the wind direction changes. The histogram plots are normalized to 0°=downwind for easy visual comparison between different sites.

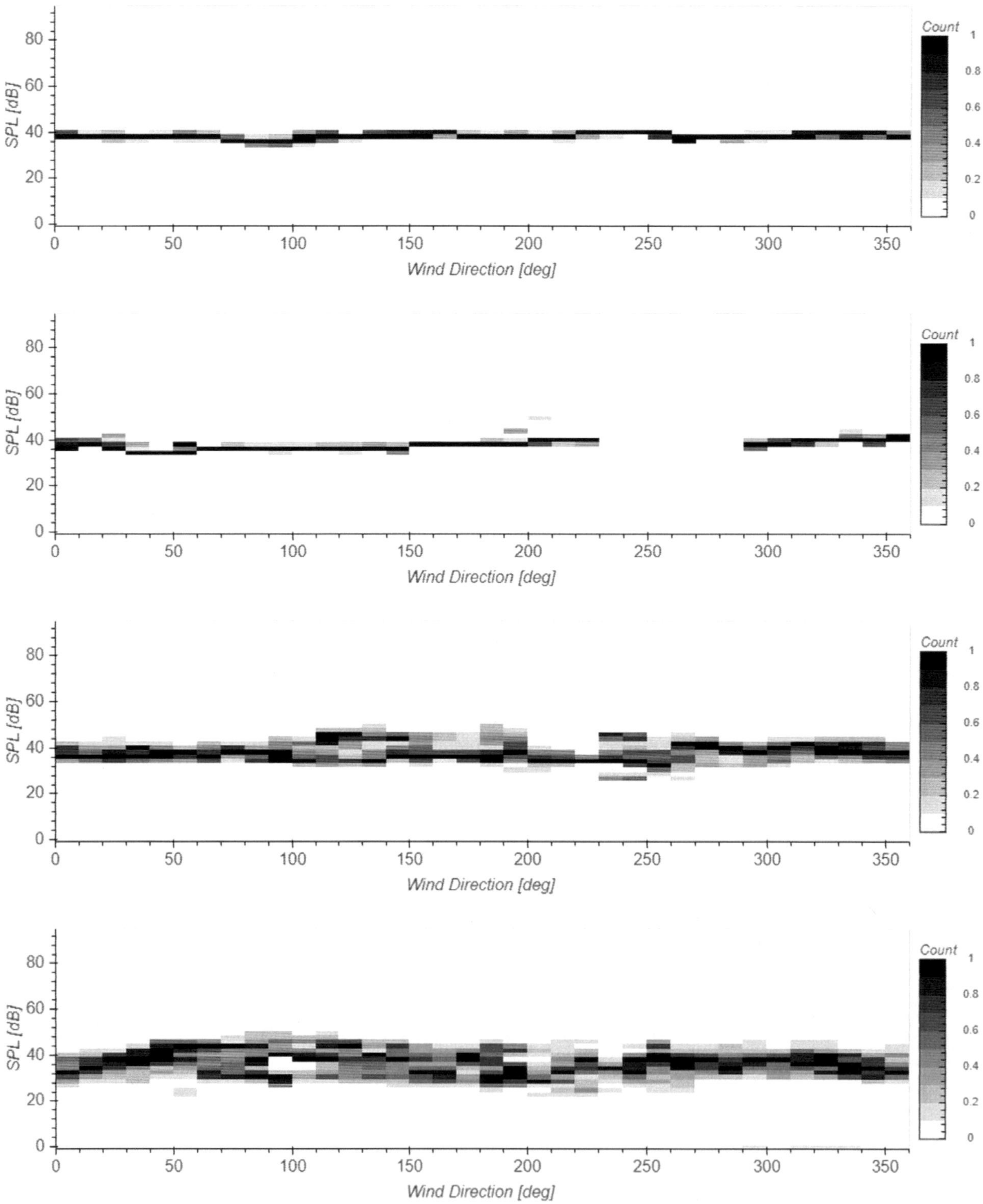

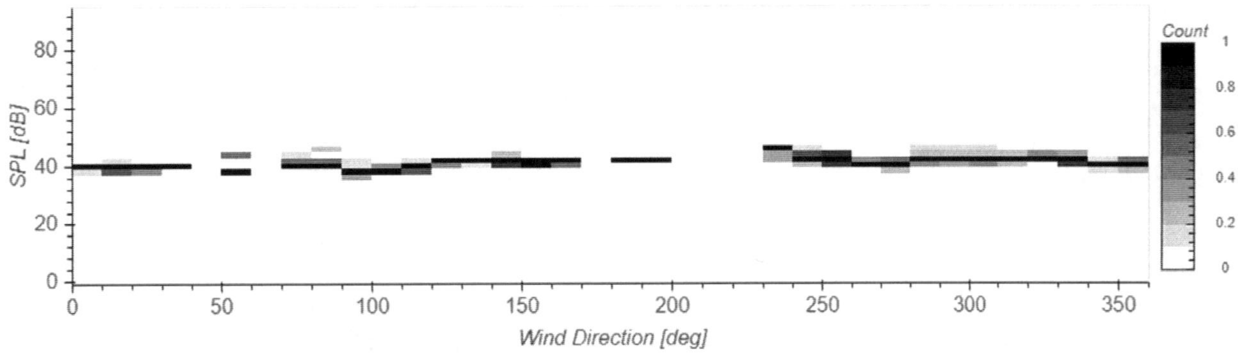

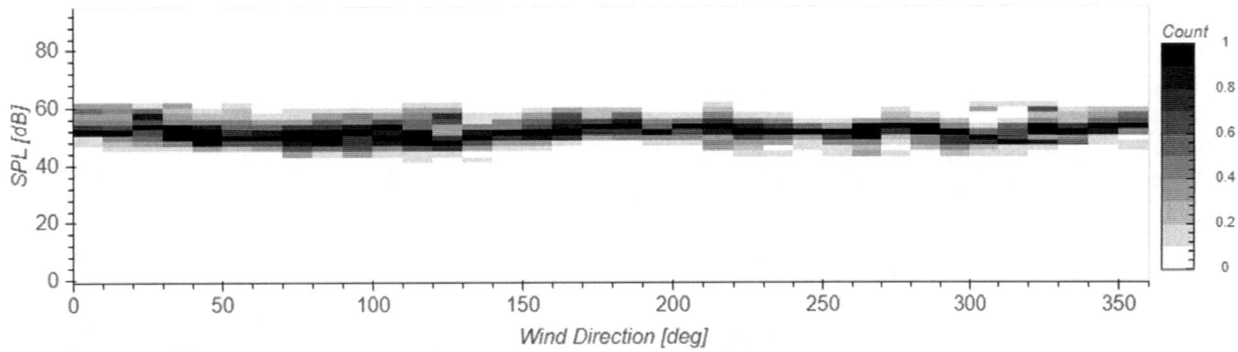

Figure 3: Measured sound pressure level vs. wind direction (Site A-F). Darker colour indicates higher density of measurement points. 0° = Downwind from nearest turbine for all plots.

From the plots above, a relationship between wind direction and measured sound level does not appear to be clearly defined, and the measured sound level does not appear to fluctuate or reduce as much as would be expected, particularly in the upwind direction (around 180°). In fact, the measured sound levels appear consistent across all wind directions, with relatively small fluctuations in level with wind direction.

3.2 Average Sound Spectrum by Wind Direction

As an aside, the sound level by 1/3rd octave frequency is plotted against the measured wind direction. The sound level at each frequency is represented by a colour, or "heat", with yellow for higher and blue for lower relative levels. The colour range is normalized relative to each frequency across the wind direction range, and so a yellow colour represents the wind direction at which the sound level for each 1/3rd octave frequency is highest. Plots for each site are presented below in Figure 4.

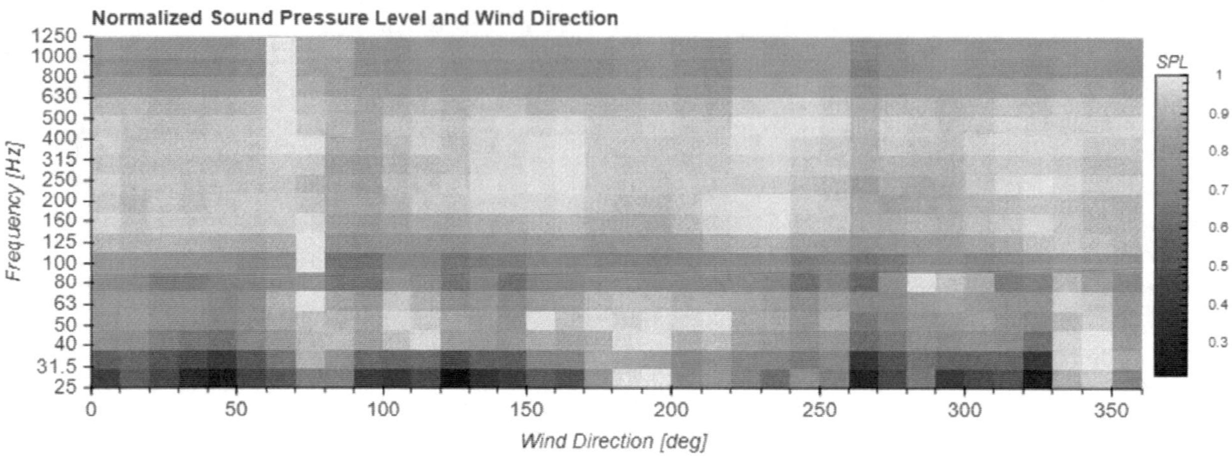
Normalized Sound Pressure Level and Wind Direction
Frequency [Hz]
1250
1000
800
630
500
400
315
250
200
160
125
100
80
63
50
40
31.5
25
0
50
100
150
200
250
300
350
Wind Direction [deg]
SPL
1
0.9
0.8
0.7
0.6
0.5
0.4
0.3

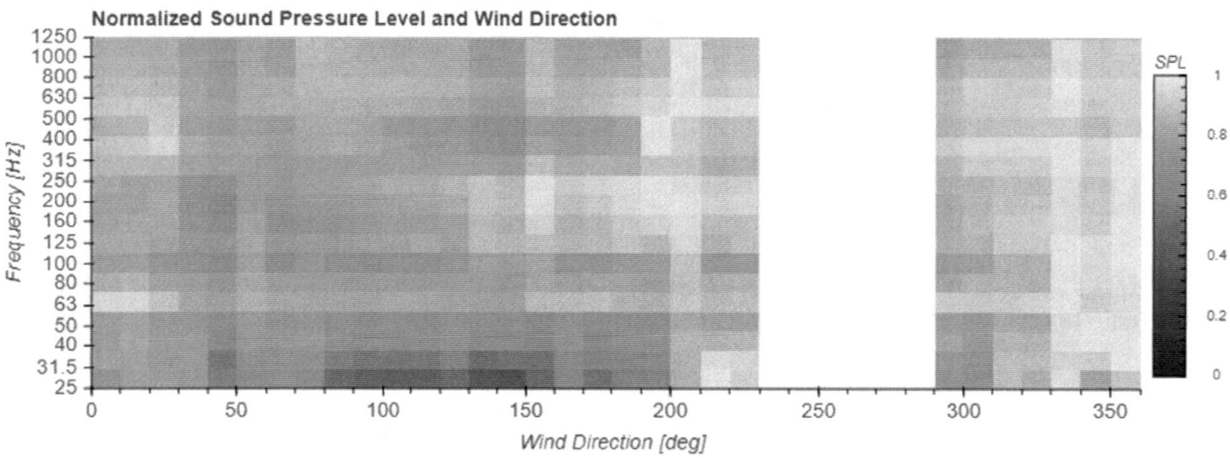
Normalized Sound Pressure Level and Wind Direction
Frequency [Hz]
1250
1000
800
630
500
400
315
250
200
160
125
100
80
63
50
40
31.5
25
0
50
100
150
200
250
300
350
Wind Direction [deg]
SPL
1
0.8
0.6
0.4
0.2
0

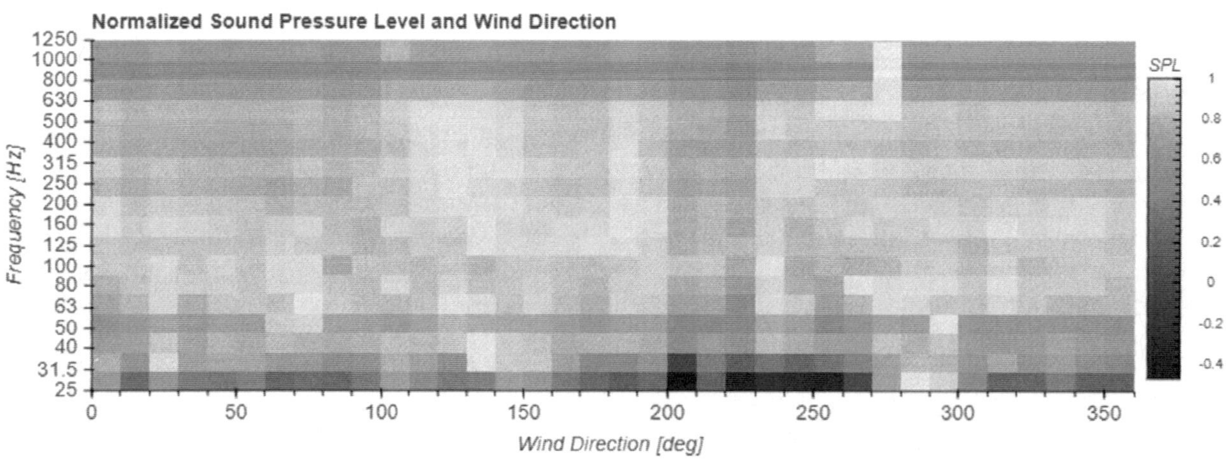
Normalized Sound Pressure Level and Wind Direction
Frequency [Hz]
1250
1000
800
630
500
400
315
250
200
160
125
100
80
63
50
40
31.5
25
0
50
100
150
200
250
300
350
Wind Direction [deg]
SPL
1
0.8
0.6
0.4
0.2
0
-0.2
-0.4

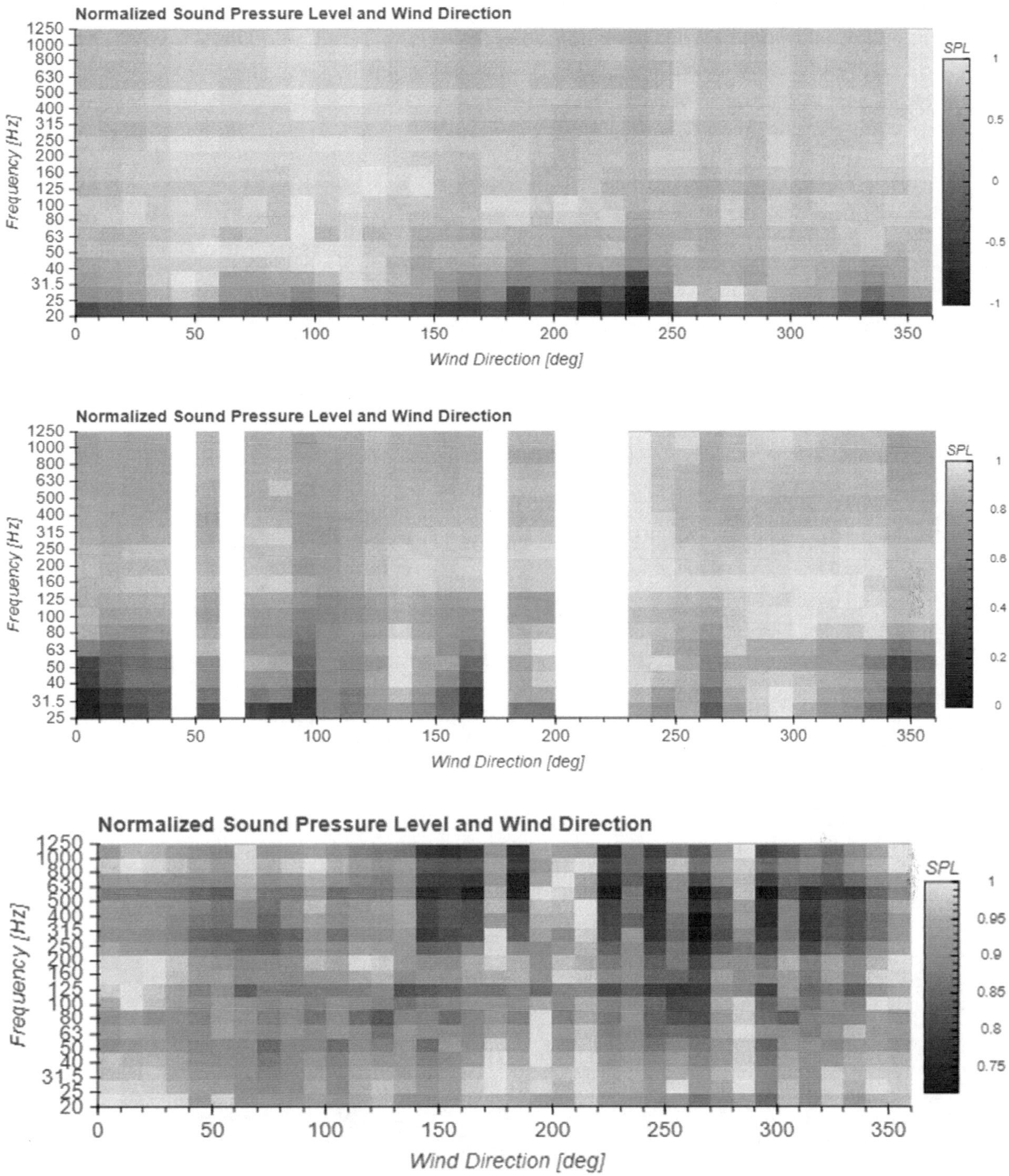

Figure 4: Average 1/3rd octave band sound level for each wind direction bin, plotted as a colour, or "heat", from the lowest (blue) to the highest (yellow) normalized for each 1/3rd octave band frequency.

Based on the above plots, there is a possibility of frequency-specific trends in the high and low sounds levels as wind direction changes. However, further investigation of the relationship by 1/3rd octave frequency has been limited by time constraints and as such is outside the scope of this study.

3.3 Average Overall Sound Level by Wind Direction

The average overall sound level in each wind direction bin is presented in tabular format below in Table 2. The labels for each wind direction bin indicate the highest angle in the bin (i.e. 50° represents data from 41-50°). Each 45° quadrant is shaded representing a typical separation of downwind (green), crosswind (peach), and upwind (blue) directions. Blank cells indicate no data available in that bin.

Table 2: Average measured SPL by wind direction

Wind Dir	Site A	Site B	Site C	Site D	Site E	Site F
10	39.1	38.1	39.7	30.5	39.8	51.8
20	38.4	38.3	38.1	36.1	39.7	49.7
30	38.2	38.6	38.5	35.4	39.7	55.8
40	38.1	35.6	40.6	40.2	39.8	53.4
50	38.3	34.5	39.2	40.0		53.7
60	38.6	36.8	37.8	39.6	40.7	52.0
70	43.2	35.7	38.9	37.5		52.3
80	40.3	37.1	40.0	36.1	41.3	53.3
90	38.1	37.0	37.3	37.6	42.0	54.7
100	37.2	36.8	38.5	42.7	38.4	54.5
110	37.6	36.6	41.2	40.6	38.8	52.9
120	38.7	36.6	45.1	42.0	39.8	56.7
130	38.1	36.6	45.2	46.3	41.5	54.0
140	39.0	36.5	47.8	48.5	41.7	52.7
150	38.9	36.1	46.7	43.6	41.7	53.3
160	39.0		44.5	40.2	41.3	56.1
170	39.3	37.4	44.2	38.4	41.6	52.4
180	38.5	37.9	45.9	39.9		
190	38.5	38.6	47.1	42.0		45.6
200	38.8	40.6	44.6	35.8		
210	38.5	41.5	36.3	37.6		49.4
220	38.6	39.3	37.5	31.2		43.8
230	39.8	39.7	35.9	34.1		45.8
240	39.9		35.6	34.8	46.2	51.7
250	39.5		37.9	38.5	43.5	50.4
260	39.3		49.1	36.5	42.5	48.6
270	37.3		49.9	39.5	40.8	46.4
280	37.9		43.7	39.9	41.8	47.4
290	38.3		40.7	37.2	43.1	54.4
300	38.2	38.1	40.8	36.9	42.8	49.8
310	38.6	39.3	40.2	37.2	42.6	52.5
320	39.2	39.6	41.8	35.8	42.1	53.0
330	39.2	39.8	41.4	34.7	42.3	55.6
340	39.3	41.8	41.7	36.1	41.0	56.8
350	39.2	40.2	39.5	44.2	40.2	53.0
360	38.7	41.3	37.8	53.7	40.5	54.8

The data presented in Table 2 is used to evaluate the difference in measured sound level compared to the 10° (i.e. downwind) wind direction bin. These results are presented in Section 3.5.

3.4 Discussion on Uncertainty

As part of the analysis, consideration must be given to the uncertainty of the calculated average sound levels used for comparison. To quantify this, the Type A[6] uncertainty, or standard error, is calculated for the data in each wind direction bin. Any difference in sound level at a particular wind direction must be at least greater than the combined uncertainty of the two average sound levels being compared. The standard error of the average overall sound level in each wind direction bin is presented in Table 3. Note that Type B[7] uncertainty has not been calculated due to time constraints in conducting the study.

Table 3: Type A Standard Error of average sound level by wind direction

Wind Dir	Site A	Site B	Site C	Site D	Site E	Site F
10	0.11	0.24	0.10	0.06	0.09	0.07
20	0.04	0.19	0.10	0.08	0.12	0.06
30	0.05	0.25	0.10	0.06	0.12	0.12
40	0.06	0.33	0.04	0.08	0.10	0.10
50	0.04	0.16	0.08	0.12		0.08
60	0.07	0.28	0.06	0.16	2.20	0.06
70	0.38	0.08	0.12	0.17		0.07
80	0.21	0.18	0.12	0.23	0.36	0.06
90	0.16	0.11	0.17	0.23	0.84	0.10
100	0.13	0.10	0.08	0.86	0.27	0.12
110	0.12	0.08	0.15	0.17	0.14	0.13
120	0.09	0.05	0.09	0.20	0.10	0.24
130	0.04	0.22	0.12	0.12	0.07	0.17
140	0.06	0.10	0.21	0.31	0.06	0.11
150	0.05	0.31	0.15	0.10	0.12	0.11
160	0.08		0.04	0.06	0.07	0.19
170	0.06	0.00	0.04	0.13	0.11	0.12
180	0.17	0.11	0.09	0.10		
190	0.11	0.08	0.13	0.18		0.04
200	0.17	0.57	0.07	0.07		
210	0.09	0.60	0.06	0.16		0.12
220	0.23	0.19	0.11	0.15		0.01
230	0.18	0.17	0.03	0.08		0.07
240	0.08		0.01	0.06	0.60	0.11
250	0.06		0.06	0.12	0.21	0.19
260	0.19		0.38	0.11	0.15	0.18
270	0.40		0.32	0.08	0.07	0.13
280	0.08		0.12	0.11	0.06	0.13
290	0.15		0.05	0.27	0.09	0.30
300	0.14	0.26	0.04	0.47	0.06	0.16
310	0.06	0.19	0.06	0.39	0.05	0.24
320	0.06	0.24	0.05	0.77	0.04	0.30
330	0.05	0.22	0.08	0.68	0.04	0.25
340	0.16	0.32	0.06	0.54	0.08	0.34
350	0.08	0.67	0.13	0.34	0.29	0.06
360	0.12	0.47	0.09	0.45	0.31	0.05

[6] Type A uncertainty is the standard error computed on a dataset and represents the likely interval around sample the average inside which the true mean of the population exists.

[7] Type B uncertainty is the uncertainty associated with the performance of the measurement equipment.

Due to the extended measurement duration and unattended nature of the measurement campaign, effects of ambient may still have an impact on uncertainty calculations, as an ambient sound source present on one night when a particular wind direction was measured may increase the sound level and could potentially decrease the calculated uncertainty if the source is relatively steady. While a great deal of ambient noise is eliminated by the 10-metre wind speed and transient filters, there could still be an influence from ambient that skews the results. A great deal of verification, including listening tests, is conducted on the measured data located in the downwind quadrant, as this is the wind direction range under study for each facility due to regulatory requirements. However, this verification is typically not performed on other wind directions. As such, a more robust investigation on the effects of ambient noise on the individual datasets may be warranted in future iterations of this work.

3.5 Difference in Sound Level by Wind Direction

To evaluate on aggregate the influence of wind direction on sound level, the difference in level compared to the downwind angle in each wind bin is provided in Table 4. The downwind position is taken to be the 10° wind bin, which represents wind directions from 0-10°. The differences in sound level for each site are then averaged together to determine the total average difference from downwind across all site in each wind direction. A linear average used to to mitigate the effect of outliers on the calculation, as logarithmic averages will give more weight to higher numbers. As a visual aide, the sound levels in the average are colour-coded to show the relative difference in sound level, with red indicating higher levels and green indicating lower levels.

The shaded grey cells indicate a difference that is less than the combined uncertainty of the two levels used to compute the difference. If the difference in level is less than the combined uncertainty of the two levels, then it is not included in the calculation of the average difference for that wind bin (i.e. cells shaded in grey are excluded from the average). Note that Site D has been removed from the average. Site D was found to have an abnormally low sound level in the downwind angle (30.6 dBA), which skewed the differences in all other wind bins upward. Removing Site D reduced the levels but did not change the relative difference in level across all directions.

Table 4: Difference in sound level compared to Downwind direction (10° bin). Positive numbers indicate a higher sound level than downwind direction. Sound levels of the two averages are coloured red for higher and green for lower levels.

Wind Dir	Site A	Site B	Site C	Site D	Site E	Site F	Average Difference*
10	0.0	0.0	0.0	0.0	0.0	0.0	0.0
20	-0.7	0.2	-1.7	5.6	-0.2	-2.2	-1.5
30	-0.9	0.5	-1.2	4.9	-0.1	4.0	0.6
40	-1.0	-2.5	0.9	9.7	0.0	1.6	-0.2
50	-0.8	-3.6	-0.5	9.5		1.9	-0.8
60	-0.4	-1.3	-1.9	9.1	0.8	0.2	-0.9
70	4.1	-2.3	-0.8	7.0		0.5	0.4
80	1.2	-0.9	0.3	5.6	1.5	1.5	0.7
90	-1.0	-1.1	-2.4	7.1	2.2	2.9	0.1
100	-1.9	-1.2	-1.2	12.2	-1.4	2.7	-0.6
110	-1.5	-1.5	1.4	10.1	-1.0	1.0	-0.3
120	-0.4	-1.5	5.3	11.5	0.0	4.9	2.1
130	-0.9	-1.4	5.4	15.8	1.7	2.1	1.4
140	-0.1	-1.6	8.0	18.0	1.9	0.8	2.3
150	-0.2	-1.9	7.0	13.1	1.9	1.5	2.1
160	-0.1		4.8	9.7	1.5	4.3	3.5
170	0.2	-0.7	4.5	7.9	1.8	0.6	1.3
180	-0.6	-0.1	6.1	9.4			2.8
190	-0.6	0.5	7.4	11.5		-6.2	0.3
200	-0.3	2.5	4.9	5.3			3.7
210	-0.6	3.4	-3.5	7.1		-2.4	-0.8
220	-0.5	1.2	-2.2	0.7		-8.0	-2.4
230	0.7	1.6	-3.9	3.6		-6.0	-1.9
240	0.8		-4.2	4.3	6.4	-0.1	1.0
250	0.4		-1.8	8.0	3.7	-1.4	0.2
260	0.2		9.3	6.0	2.6	-3.2	2.9
270	-1.8		10.2	9.0	0.9	-5.4	1.0
280	-1.2		4.0	9.4	1.9	-4.4	0.1
290	-0.8		0.9	6.7	3.2	2.6	1.5
300	-0.9	0.0	1.0	6.4	3.0	-2.0	0.3
310	-0.5	1.2	0.5	6.7	2.8	0.7	0.9
320	0.1	1.5	2.1	5.3	2.3	1.1	1.7
330	0.1	1.7	1.7	4.2	2.5	3.8	2.4
340	0.2	3.7	1.9	5.6	1.1	5.0	2.9
350	0.1	2.1	-0.2	13.7	0.4	1.2	1.0
360	-0.3	3.2	-1.9	23.2	0.7	3.0	0.9

* average 2 excludes Site D.

From Table 4, above, it is apparent that the higher relative sound levels appear to occur in the 120-200° direction. This is essentially the upwind direction from the turbine, indicating that the sound levels in the upwind direction may be on par and potentially louder than the sound levels in the downwind direction. However, the angle from 310-360/0° is also relatively louder than the downwind angle, which may indicate that the downwind range of +/-45° may still result in high overall levels compared to other directions. A radial plot of the differences is provided in Figure 5 for another representation of the measured differences.

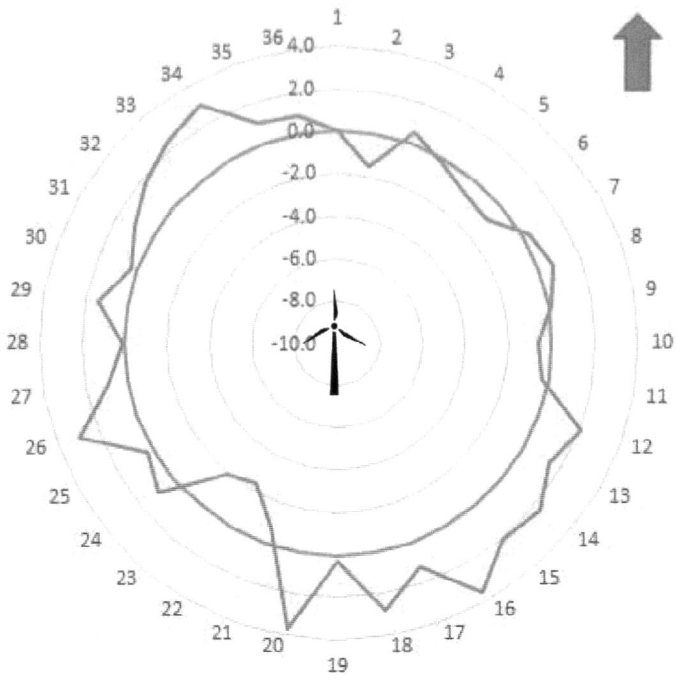

Figure 5: Sound level difference from downwind vs. wind direction (x10°s). Arrow represents wind downwind direction.

4. Conclusions

This study examined the variation in measured sound levels by wind direction at six locations near wind facilities in southern Ontario. It was found that the sound level did not fluctuate or attenuate as expected in different wind directions, particularly in the upwind condition. In fact, both the upwind direction (120-200°) and the direction just counterclockwise of the downwind vector (310-360/0°) had the highest relative sound levels.

4.1 Further Study

Due to time constraints, the scope of the study covered in this paper was limited. However, the findings of this study would greatly benefit from further research before more general conclusions may be drawn. Specific areas for investigation include:

1. More robust evaluation of the variation in noise level between assessment datasets.
2. Examination of trends between different turbine models.
3. Correlation of these results with the expected directivity of a wind turbine sound source.
4. Additional datasets for investigation.

References

[1] F. M. Wiener and D. N. Keast, "Experimental Study of the Propagation of Sound over Ground," *Journal of the Acoustical Society of America,* vol. 31, no. 6, 1959.

[2] L. Hannah, "Wind and Temperature Effects on Sound Propagation," *Factors Affecting Outdoor Sound Propagation, Massey University,* 2006.

[3] L. Nijs and C. P. A. Wapenaar, "The influence of wind and temperature gradients on sound propagation, calculated wiht the two-way wave equation," *Journal of the Acoustical Society of America,* vol. 87, no. 5, pp. 1987-1998, 1990.

[4] M. Diamond, "Cross Wind Effect on Sound Propagation," *Journal of Applied Meteorology,* vol. 3, no. 2, pp. 208-210, 1963.

[5] IEC TC88, "61400-11 Wind turbines - Part 11: Acoustic noise measurement techniques," International Electrotechnical Commission, Geneva, 2012.

[6] ISO, "1996-2 - Description, measurement and assessment of environmental noise," International Organization for Standardization, Geneva, 2017.

[7] D. Bowdler and G. Leventhall, Wind Turbine Noise, Multi-Science Publishing, 2012.

[8] Ministry of the Environment and Climate Change, "Compliance Protocol for Wind Turbine Noise," Government of Ontario, Toronto, 2017.

[9] D. Halstead and N. Tam, "A study of background noise levels measured during far-field receptor testing of wind turbine facilities," in *8th International Conference on Wind Turbine Noise*, Lisbon, 2019.

INCEEUROPE

9th International Conference
on
Wind Turbine Noise
Remote from Europe – 18th to 21st May 2021

Effect of grid resolution on airfoil self-noise prediction by large eddy simulation

S.M. Hasheminasab: smhasheminasab@aut.ac.ir
S.M.Hossein Karimian, Sahar Noori
Department of Aerospace Engineering, Amirkabir University of Technology, Tehran, Iran

Summary

Airfoil self-noise is one of the dominant sources of wind turbine noise. This paper presents the results of numerical investigations exploring the airfoil self-noise prediction employing large eddy simulation (LES). In the first step, an incompressible large eddy simulation of fluid flow around a NACA0012 airfoil at zero angle of attack with a chord-based Reynolds number of 6.4×10^5 is performed. In the second step, the far-field noise is predicted by Ffowcs-Williams & Hawkings (FW-H) model using the LES solution. Two different grid resolutions are used to investigate the effect of grid resolution on the accuracy of acoustic results. The results are compared with experimental data of wind tunnel tests and noise measurements through microphones. The comparison shows that the grid resolution has a significant effect on the acoustic predictions and by increasing the number of grid points in the spanwise and streamwise direction, numerical noise predictions approach the experimental results. In some frequencies, by increasing the number of the grid points, the differences between numerical and experimental results are increased to less than 5 dB.

1. Introduction

Over the past decade, wind turbine use has increased more than 25 percent per year, however, noise and the visual impact caused by wind turbines give rise to problems concerning the public acceptance of wind energy especially in densely populated countries (Wagner, et al., 1996). Wind turbine noise consists of mechanical and aerodynamic noise. Aerodynamic noise is emitted from the blades and is mainly due to the interaction of turbulence with the blade surface. The dominant source of aerodynamic noise at low Mach numbers is the interaction of the turbulent boundary layer and the trailing edge of the airfoil called Turbulent Boundary Layer-Trailing Edge (TBL-TE) noise.

Numerous empirical models for TBL-TE noise prediction are available in the literature. Yet, the required input parameters for accurate noise predictions still reside in data that is best accessed from numerical simulations (Winkler, et al., 2010). An alternative is using Computational aeroacoustics (CAA) methods based on Lighthill's theory (Lighthill, 1952) and followed by various aeroacoustic analogies (Ffowcs Williams, et al., 1969) (Lilly, 1974). One prediction strategy for aerodynamic sound is using the near-field solutions of unsteady fluid flow for the far-field solution of sound propagation. This strategy decouples the propagation of sound from its generation and separates the flow solution process from the acoustic analysis. The near-field solution of unsteady fluid flow by computational fluid dynamics (CFD) must identify the noise sources due to turbulence, the interaction of flow structures and solid surfaces, etc. An accurate prediction of far-field noise requires well-resolved transient flow field data, especially the flow field near the solid surfaces (Arakawa, et al., 2005). Hence by using Direct Numerical Simulation (DNS) and Large Eddy Simulation (LES) of fluid flow more accurate data will be achieved in comparison to unsteady Reynolds Average Navier-Stokes (RANS) solutions. But DNS is computationally expensive requiring a fine computational grid, so it is limited to simple geometries and low Reynolds number flows.

In recent years a variety of large-eddy simulations were performed to generate the source terms for wave propagation equations (Wagner, et al., 1996). (Wang, et al., 2000) performed LES computations of the flow past an asymmetrically beveled trailing edge of a flat strut at a chord Reynolds number of $Re = 2.15 \times 10^6$. Their computed mean and fluctuating velocity profiles compare reasonably well with experimental measurements. They concluded that, for accurately predicted noise radiation using the LES solution, the size of the spanwise domain must be larger than the coherence length of the source field in this direction. An incompressible LES was performed by (Winkler, 2009) over a NACA 6512-63 airfoil at zero angle of attack with Reynolds number of $Re = 1.9 \times 10^5$ for near-field noise source identification. Then three different CAA methods are used for far-field noise measurement and results are compared to experimental results. One year later (Winkler, et al., 2010) they investigate the effect of a grid refinement in LES on the same geometry and flow conditions for trailing edge noise prediction. Results are compared with experimental and theoretical predictions for far-field noise. The comparison shows the necessity of accurate boundary conditions in the LES to arrive at comparable results with wind tunnel measurements. (Wasala, et al., 2015) used Ansys Fluent (ANSYS Inc.) for LES and CAA analogies for the prediction of noise generated by the outer part of a CART-2 wind turbine blade. Results show a good agreement with experimental data.

In this work, an incompressible large eddy simulation of flow over a NACA0012 airfoil at zero angle of attack is performed by Ansys Fluent (Ansys Inc.), and far-field noise is predicted using FW-H acoustic analogy. The grid is refined in two steps and results are compared with experimental data for both aerodynamic parameters and far-field noise measured by (Brooks, et al., 1989) in a wind tunnel.

2. Simulation

2.1 Geometry and grid generation

Flow around a NACA0012 is simulated in a C-type domain with the dimensions as shown in Figure 1. Distance from the inlet to the leading edge is 12.5C and the spanwise length of the airfoil is 0.1C. the chord length is assumed to be C=0.3048 to be comparable with experimental acoustic data of (Brooks, et al., 1989). Velocity inlet and pressure outlet conditions are used as boundary conditions. At velocity inlet, velocity magnitude of 31.7 is prescribed in the *x*-direction and the gauge pressure at pressure outlet is set to zero. Hence the Reynolds number based on the chord length is 6.4×10^5. The upper and lower surfaces of the airfoil are defined as stationary walls with the no-slip condition. Symmetry condition is prescribed on sides surfaces. There is no convective flux across the symmetry plane, the normal velocity component at the symmetry plane is thus zero. Also, there is no diffusion flux across the symmetry plane, the normal gradients of all flow variables are thus zero at the symmetry plane. Therefore, normal velocities and gradients of all variables are assumed to be zero at symmetry planes.

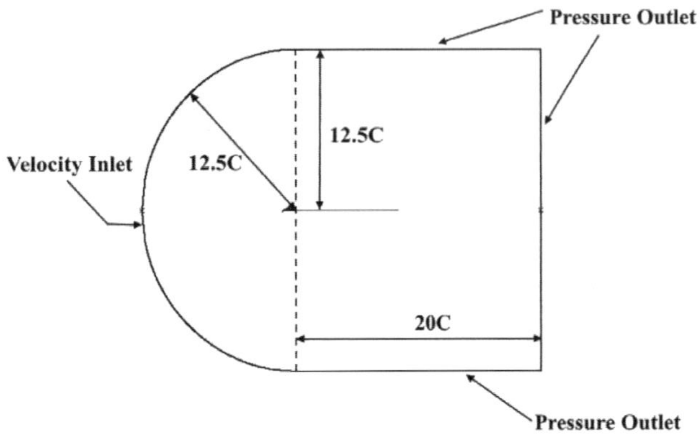

Figure 1 - geometry, computational domain size, and boundary conditions

Table 1 – Summary of grid parameters

Case	A	B
	Poor Res.	Medium Res.
total no. of cells	2.7×10^6	8×10^6
Max Δz^+	75	25
Max Δy^+	1.2	1

2 different grids are developed to investigate the effect of grid resolution on the aerodynamic and acoustic parameters as well. A poor resolution grid (Breuer, 2007) with $30<\Delta z^+<75$, $\Delta y^+<1.2$ and a medium resolution grid with $15<\Delta z^+<30$, $\Delta y^+<1$. The span-wise and stream-wise surface grid sizes are set to be equal ($\Delta x = \Delta z$), so that the acoustic sources will be consistently resolved across the surface of the airfoil (Wasala, et al., 2015).

2.2 Large Eddy Simulation

The idea of LES is to separate the scales of the flow to the small scales which are assumed to be universal, isotropic, and independent of flow boundary conditions, and large scales which are mainly dominated by the geometry and boundary conditions of the flow. Then for the small scales with universal behavior, a model can be used while the large scales can be resolved by solving Navier-Stokes equations (Saeedi, 2017). This will lead to a high computational saving while large energy-containing scales of the flow are directly resolved, and the net effect of small scales is modeled.

The separation of the flow scales accomplished by applying a low-pass spatial filter to the Navier-Stokes equations:

$$\frac{\partial \overline{u}_i}{\partial x_i} = 0 \qquad \text{Eqn. 1}$$

$$\frac{\partial \overline{u}_i}{\partial t} + \frac{\partial \overline{u}_i \overline{u}_j}{\partial x_j} = -\frac{1}{\rho}\frac{\partial \overline{p}}{\partial x_i} + \nu \frac{\partial^2 \overline{u}_i}{\partial x_j \partial x_j} - \frac{\partial \tau_{ij}}{\partial x_j} \qquad \text{Eqn. 2}$$

In which $\overline{u}_i$ and $\overline{p}$ represent the filtered velocity and pressure respectively, and τ_{ij} is the subgrid-scale stress (SGS) tensor which appears as the result of the filtering process and is defined as

$$\tau_{ij} = \overline{u_i u_j} - \overline{u}_i \overline{u}_j \qquad \text{Eqn. 3}$$

To close the above system of governing equations, modeling the SGS stress tensor is needed. Applying the Boussinesq hypothesis (Hinze, 1975) the trace-free form of SGS stress tensor is proportional to the mean resolved strain rate tensor through the following relation

$$\tau_{ij}^* = \tau_{ij} - \frac{1}{3}\tau_{kk}\delta_{ij} = -2\nu_{sgs}\overline{S}_{ij} \qquad \text{Eqn. 4}$$

in which δ_{ij} is the Kronecker delta, ν_{sgs} is the SGS eddy viscosity and $\overline{S}_{ij}$ is the filtered strain rate tensor defined as

$$\bar{S}_{ij} \equiv (\frac{\partial \bar{u}_i}{\partial x_j} + \frac{\partial \bar{u}_j}{\partial x_i})/2 \qquad \text{Eqn. 5}$$

Based on the Smagorinsky SGS stress model (Smagorinsky, 1963) SGS eddy viscosity can be evaluated as

$$\nu_{sgs} = C_S \bar{\Delta}^2 |\bar{S}|, \qquad \text{Eqn. 6}$$

where $\bar{\Delta}$ is the grid-level filter width, and $|\bar{S}| \equiv \sqrt{2\bar{S}_{ij}\bar{S}_{ij}}$ is the norm of $\bar{S}_{ij}$ and C_S is the model constant. Smagorinsky's model has a constant which is not universal and needs calibration.

Germano et al. (Germano, et al., 1989-1993) and subsequently Lilly (Lilly, 1992) convinced a procedure in which the Smagorinsky model constant, C_S, is dynamically computed based on the information provided by the resolved scales of motion. In this model, a test-grid-level filter is introduced with the size twice the size of the grid level filter such that $\tilde{\Delta} = 2\bar{\Delta}$.

In Dynamic Smagorinsky Model (DSM) the SGS stress tensor is calculated as

$$\tau_{ij}^* \overset{\text{def}}{=} \tau_{ij} - \frac{\tau_{kk}}{3}\delta_{ij} = -2C_s\bar{\Delta}^2 |\bar{S}|\bar{S}_{ij}, \qquad \text{Eqn. 7}$$

Where the dynamic model coefficient C_s can be obtained from

$$C_s = -\frac{M_{ij}\mathcal{L}_{ij}^*}{M_{mn}M_{mn}} \qquad \text{Eqn. 8}$$

Where $\mathcal{L}_{ij}$ is the resolved Leonard type stress defined as $\mathcal{L}_{ij} \overset{\text{def}}{=} \widetilde{\bar{u}_i\bar{u}_j} - \tilde{\bar{u}}_i\tilde{\bar{u}}_j$, and $M_{ij} \overset{\text{def}}{=} 2\tilde{\Delta}^2 \left|\tilde{\bar{S}}\right| \tilde{\bar{S}}_{ij} - 2\bar{\Delta}^2 |\bar{S}|\bar{S}_{ij}$ is the difference between test grid and grid level base tensors.

The C_s computed by DSM varies in time and space over a wide range. To avoid numerical instability in Fluent, C_s is clipped between 0.1 and 0.23 by default (ANSYS)

2.3 Acoustic Modelling

Lighthill's aeroacoustic analogy (Lighthill, 1952) (Lighthill, 1954) also known as acoustic analogy is derived by rearranging governing Navier-Stokes equations of fluid flow. The result is an inhomogeneous wave equation:

$$\frac{1}{a_0^2}\frac{\partial^2 p'}{\partial t^2} - \nabla^2 p' = \frac{\partial^2 T_{ij}}{\partial x_i \partial x_j} \qquad \text{Eqn. 9}$$

In which p' is the sound pressure in the far-field $(p' = p - p_0)$, a_0 is the speed of sound and $T_{ij} = \rho u_i u_j + P_{ij} - a_0^2(\rho - \rho_0)\delta_{ij}$ is defined as Lighthill stress tensor. The right-hand side of the equation represents the source term while the left-hand side terms show the spatial and temporal propagation of sound. The source term contains all types of sources as it is derived directly from the Navier-Stokes equations. The far-field sound pressure level can then be found by volume integrals over the domains containing sound sources. The main limitation of Lighthill's equation is that it is restricted to the unbounded fluid. Therefore, its application is limited to problems like jet noise where solid surfaces do not play a major role (Lighthill, 1963).

Ffowcs Williams and Hawkings (Ffowcs Williams, et al., 1969) included the influence of arbitrary moving surfaces. In this method, the fluid is partitioned into two different regions by a mathematical surface $(f = 0)$ which surrounded the sound sources. Same as Lighthill's analogy starting with continuity and Navier-Stokes equations results in an inhomogeneous wave equation as follows

$$\frac{1}{c_0^2}\frac{\partial^2 p'}{\partial t^2} - \nabla^2 p' = \frac{\partial^2}{\partial x_i\, \partial x_j}\{T_{ij}H(f)\} \qquad \text{Eqn. 10}$$

$$-\frac{\partial}{\partial x_i}\{[P_{ij}n_j + \rho u_i(u_n - v_n)]\delta(f)\}$$
$$+\frac{\partial}{\partial t}\{[\rho_0 v_n + \rho(u_n - v_n)]\delta(f)\}$$

Where u_i, v_i and n_i are fluid velocity, surface velocity, and the unit vector in $\mathrm{i}-$direction respectively, and u_n, v_n are fluid velocity and surface velocity component normal to the surface.

$$T_{ij} = \rho u_i u_j + P_{ij} - a_0^2(\rho - \rho_0)\delta_{ij} \qquad \text{Eqn. 11}$$

$$P_{ij} = p\delta_{ij} - \mu\left[\frac{\partial u_i}{\partial x_j} + \frac{\partial u_j}{\partial x_i} - \frac{2}{3}\frac{\partial u_k}{\partial x_k}\delta_{ij}\right] \qquad \text{Eqn. 12}$$

The wave equation can be integrated analytically under the assumptions of the free-space flow and the absence of obstacles between the sound sources and the receivers. The complete solution consists of surface integrals and volume integrals. The surface integrals represent the contributions from monopole and dipole acoustic sources and partially from quadrupole sources, whereas the volume integrals represent quadrupole (volume) sources in the region outside the source surface. The contribution of the volume integrals becomes small when the flow is low subsonic, and the source surface encloses the source region. In ANSYS Fluent, the volume integrals are dropped. Thus, we have

$$4\pi p_T'\left(\vec{x},t\right) = \int_{f=0}\left[\frac{\rho_0(U_n + U_r)}{r(1-M_r)^2}\right]dS + \int_{f=0}\left[\frac{\rho_0 U_n\{rM_r + a_r(M_r - M^2)\}}{r^r(1-M_r)^3}\right]dS \qquad \text{Eqn. 13}$$

$$4\pi p_L'\left(\vec{x},t\right) = \frac{1}{a_0}\int\left[\frac{\dot{L}_r}{r(1-M_r)^2}\right]dS + \int_{f=0}\left[\frac{L_r - L_M}{r^2(1-M_r)^2}\right]dS \qquad \text{Eqn. 14}$$
$$+\frac{1}{a_0}\int_{f=0}\left[\frac{L_r\{rM_r + a_d(M_r - M^2)\}}{r^2(1-M_r)^3}\right]dS$$

2.4 Solution Method

Instead of a uniform flow as initialization for LES, a steady-state RANS solution is used. It will reduce the computational costs of LES significantly. This will yield a quasi-stationary state for transient flow more rapidly. The SIMPLE algorithm is used for steady-state RANS and the k-ε model is applied as the viscous model. The steady-state solution is used as an initialization for LES. In LES, dynamic Smagorinsky-Lilly is selected as the SGS model, and the model's constant is defined dynamically in time and space. The non-iterative fractional step method is used for pressure-velocity coupling in transient LES with second-order central differencing for pressure and momentum and a Green-Gauss node-based method for the evaluation of pressure gradients. The time step size required in LES calculations is governed by the time scale of the smallest resolved eddies. That requires the local Courant-Friedrichs-Lewy (CFL) number to be of an order of 1. As it is generally difficult to know the proper time step size at the beginning of the simulation an adjustment is necessary after the flow is established. Also, for a given time step Δt, the highest frequency that the acoustic analysis can produce is f = 1/(2Δt). Hence the time step size of Δt = 5×10^6 *s* is selected here, and the maximum obtaining frequency is 100 kHz. The flow is solved for a physical time of 0.3 *s* where it has passed the domain one time. After that, the time marching is continued, and the acoustic data are obtained.

3. Results

The flow results are first analyzed by taking a closer look at the flow around the airfoil. The instantaneous velocity field around the airfoil is shown in Figure 2. As it can be seen laminar flow is accelerated on both sides of the airfoil, further towards the trailing edge, the flow becomes turbulent and transition occurs near the trailing edge. For a more accurate determination of the transition point, the instantaneous skin friction coefficient is plotted along the airfoil surface in Figure 3. As it can be seen transition occurs at x = 0.7C and it agrees with the reported location for the transition point in the literature (Kianoosh, et al., 2018).

Figure 2 – Velocity magnitude for NACA0012 at zero angle of attack, and Re=6.4×10^5

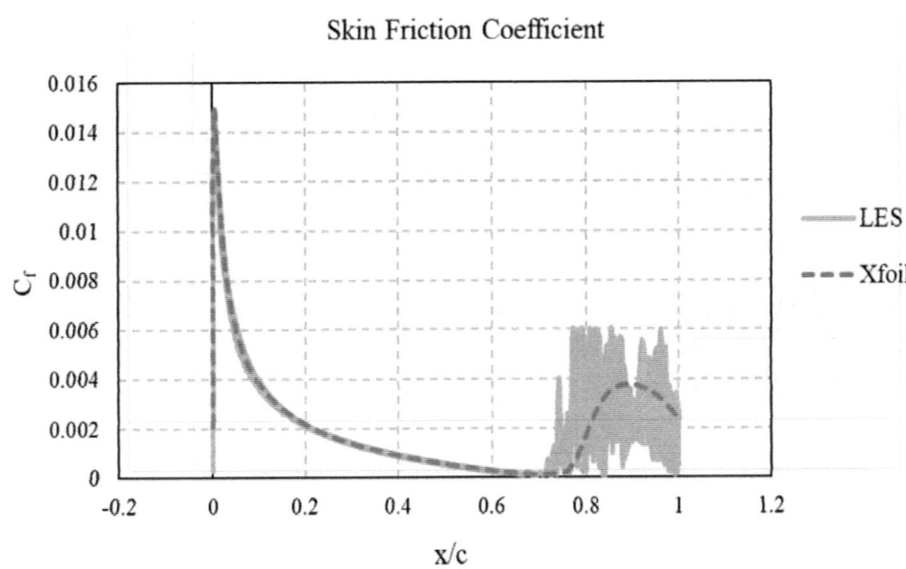

Figure 3 – Skin friction coefficient for NACA0012 at zero angle of attack, and Re=6.4×10^5

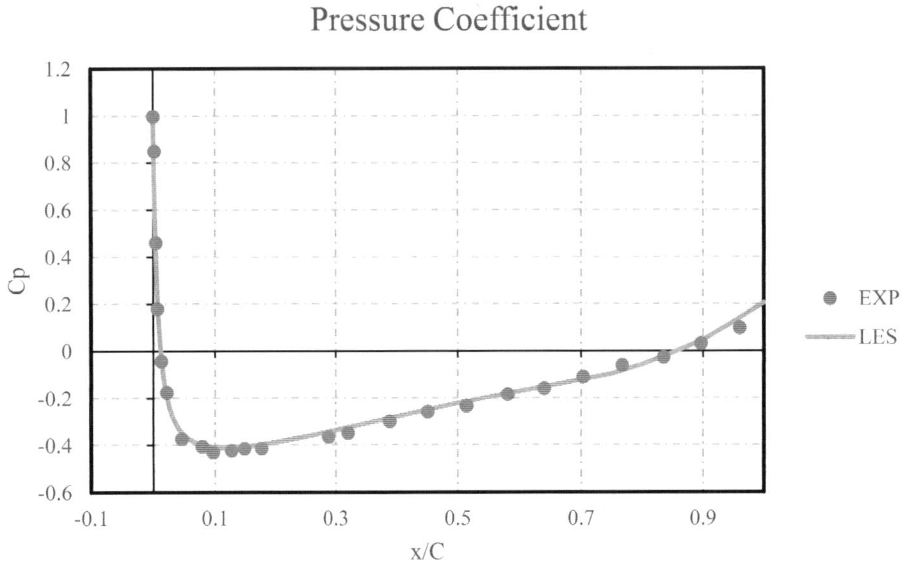

Figure 4 - Pressure Coefficient distribution of NACA0012 at zero angle of attack obtained from LES compared with experimental data from (Gregory, et al., 1970)

After achieving the quasi-stationary state, the aerodynamic properties of the airfoil are stored for computing the mean values. Figure 4 shows the time-averaged pressure coefficient of airfoil obtained from LES, compared with the experimental results of (Gregory, et al., 1970). The results show a good agreement between the pressure coefficient obtained from LES and the experimental data.

For far-field noise measurement, a receiver is positioned perpendicular to, and 1.22 m from the trailing edge at the model midspan to be comparable with the experimental data. Figure 5 shows the scaled 1/3 octave SPL spectra for 2 different grids compared to the experimental data of (Brooks, et al., 1989). Scaled 1/3 octave SPL and the Strouhal number are calculated as suggested by (Brooks, et al., 1989).

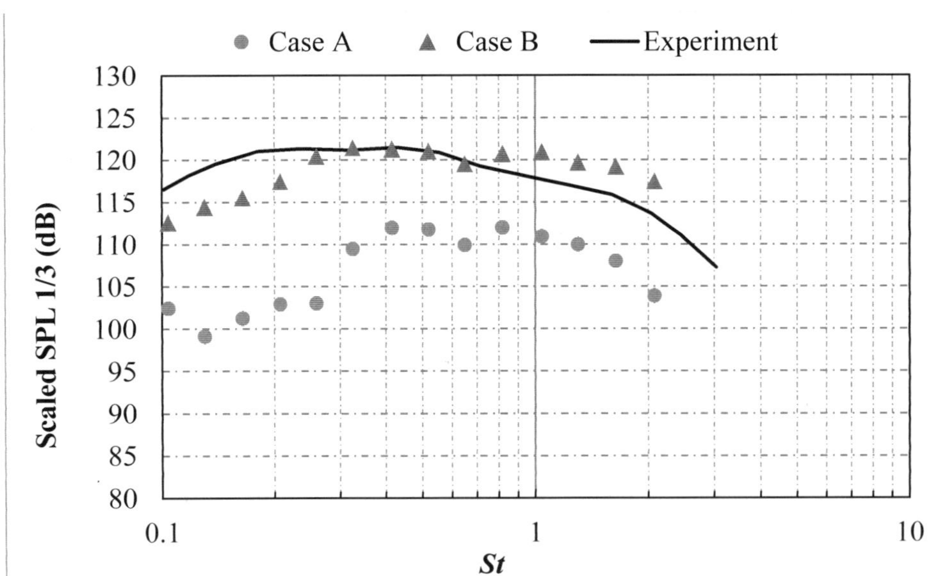

Figure 5 – Scaled 1/3 octave Sound Pressure Level spectra for NACA0012 airfoil at zero angle of attack and Re=6.4×10^5 compared with the experimental data of (Brooks, et al., 1989)

As it can be seen in Figure 5, In case A the frequency in which SPL has its local maximum is predicted correctly to be near 2 kHz. Case A underpredicts the sound pressure level as much as 20 and 15 dB in low and high frequencies respectively, and the difference with the experimental data is much larger in low frequencies. Whereas good agreement can be seen between results of Case B and experimental data both in low and high frequencies and predicting the frequency of maximum SPL. In Case B The maximum difference with the experimental data is less than 5 dB both in the low and high frequencies with the minimum difference in mid-range frequencies.

4. Conclusions

A large eddy simulation of fluid flow around a NACA0012 airfoil at zero angle of attack has been performed at $Re=6.4\times10^5$ with two different grid resolutions. Far-field noise has been estimated using FW-H acoustic analogy and compared with the experimental data extracted from the wind tunnel tests by (Brooks, et al., 1989). Results show a good agreement between aerodynamic properties of the airfoil and the flow behavior in comparison with the experimental data, but the acoustic results and the scaled SPL are significantly affected by grid resolution. The frequency at which maximum SPL occurs is predicted correctly by both grids. But in the grid with poor resolution ($30<\Delta z^+<75$), underprediction of 15-20 dB is observed in all frequencies, it might be due to the unresolved turbulence structures of the flow which are the sources of the TBL-TE noise. Acoustic results of medium resolution grid ($15<\Delta z^+<30$) show a good agreement in high and mid-range frequency with a difference of less than 3 dB, and an underprediction of less than 5 dB in high frequencies. It could be concluded that noise prediction using large eddy simulation is sensitive to the grid resolution and increasing the grid resolution leads to a better noise prediction.

References

ANSYS, I. (n.d.). *ANSYS Fluent Theory Guide.* Canonsburg.

Arakawa, C., Fleig, O., Iida, M., & Shimooka, M. (2005). Numerical approach for noise reduction of wind turbine blade tip with earth simulator. *J. Earth Simul., 2*(3), 11-33.

Breuer, M. (2007). Boundary Conditions for LES. In C. Wagner, T. Huttle, & P. Sagaut, *Large-Eddy Simulation for Acoustics* (pp. 201–215). Cambrifge University Press.

Brooks, T. F., Pope, D. S., & Marcolini, M. A. (1989). *Airfoil Self-Noise and Prediction.* NASA 1218.

Ffowcs Williams, & Hawkins, D. L. (1969). Sound Generation by Turbulence and Surfaces in Arbitrary Motion. *Philosophical Transactions of the Royal Society of London, 264*(No. A 1151), 321-342.

Germano, M., Piomelli, U., Moin, P., & Cabot, W. (1989-1993). A dynamic subgrid-scale eddy viscosity model. *Phys. Fluids A: Fluid Dyn.*, 1760-1765.

Gregory, N., & O'Reilly, C. L. (1970). *Low-Speed Aerodynamic Characteristics of NACA0012 Aerofoil Section, Including the Effects of Upper-Surface Roughness Simulating Hoar Frost.* Aeronautical Research Council Reports and Memoranda.

Hinze, J. O. (1975). *Turbulence.* New York: McGraw-Hill Publishing Co.

Kianoosh, Y., & Razeghi, A. (2018). Determination of the Critical Reynolds Number for Flow over Symmetric NACA Airfoils. *AIAA Aerospace Sciences Meeting.* Kissimmee, Florida.

Lighthill, M. (1952). On sound generated aerodynamically .I. General theory. *Proc. Roy. Soc. A, 211*, 564-587.

Lighthill, M. J. (1954). Sound generated aerodynamically. II. Turbulence as a source of sound. *Proc. Roy. Soc. A, 222*, 1-32.

Lighthill, M. J. (1963). Jet Noise. *AGARD Report*, 448.

Lilly, D. K. (1992). A Proposed Modification of the Germano Subgrid-Scale Closure Model. *Physics of Fluids, 4*, 633-635.

Lilly, G. (1974). On the Noise from Jets. *AGARD CP-131*, 13.1–13.10.

Saeedi, M. (2017). *A practice twords large-eddy simulation of incompressible turbulent flows.* Tehran: Amirkabir University Press.

Smagorinsky, J. (1963). General Circulation Experiments with the Primitive Equations. *Monthly Weather Review*, 99-164.

Wagner, C. A., Hüttl, T., & Sagaut, P. (2007). *Large-Eddy Simulation for Acoustics.* Cambridge: Cambridge University Press.

Wagner, S., BareiB, R., & Guidati, G. (1996). *Wind Turbine Noise.* Verlag, Berlin, Heidelberg: Springer.

Wang, M., & Moin, P. (2000). Computation of trailing-edge flow and noise using large-eddy simulation. *AIAA Journal, 38(12)*, 2201–2209.

Wasala, S. H., Storey, R. C., Norris, S. E., & Cater, J. E. (2015). Aeroacoustic noise prediction for wind turbines using Large Eddy Simulation. *J. Wind Eng. Ind. Aerodyn. 145*, 17-29.

Winkler, J., Moreau, S., & Carolus, T. (2009). Large-Eddy Simulation and Trailing-Edge Noise Prediction of an Airfoil with Boundary-Layer Tripping. *15th AIAA/CEAS Aeroacoustics Conference.*

Winkler, J., Moreau, S., & Carolus, T. (2010). Airfoil Trailing Edge Noise Prediction from Large-Eddy Simulation: Influence of Grid Resolution and Noise Model Formulation. *16th AIAA/CEAS Aeroacoustics Conference.*

INCEEUROPE

9th International Conference
on
Wind Turbine Noise
Remote from Europe – 18th to 21st May 2021

Turbulence Inflow Noise Prediction of Wind Turbine Rotors: The physically correct Representations of the Simplified Amiet and Lowson Model

C. Hornung, WRD GmbH: cordula.hornung@enercon.de
C. Scheit, WRD GmbH: christoph.scheit@enercon.de
N. Noffke, WRD GmbH: nils.noffke@enercon.de
M. Kamruzzaman, WRD GmbH: mohammad.kamruzzaman@enercon.de.

Summary

The sound in the lower frequency range of a wind turbines spectrum is increasingly gaining importance. The main aerodynamically induced contribution within the audible range of the lower frequencies is turbulence inflow noise. A number of papers is available addressing the prediction of inflow noise. Almost all of them are based on the theory of Amiet and Paterson, who provide a simplified and a complex version of their theory on the prediction of turbulent inflow noise at an airfoil. Many researchers employ the simplified approach in their models. However, a in depth literature research revealed that a number of equations are available, which differ considerably in applied constants as well as in the way the models are utilized. The goal of this paper is to sketch out potential points in the model prone to inconsistency and present a valid form of the simplified equation by deriving it from the original paper by Amiet and Paterson. The derived equation is then combined with an adjusted version of the Lowson model, consistent in its formulation to tackle the lower frequency deviations. The effect of airfoil thickness is taken into account via the thickness model proposed by Moriarty.

Subsequently, the corrected models are applied to predict the noise of a full size wind turbine. In order to verify the enhanced models – pure Amiet model and Amiet-Lowson model - a variation of length scale and turbulence intensity is performed, showing the expected physical behavior.

Finally, the results of the enhanced model are validated with measured spectra of a full-size multi megawatt turbine. In addition, they are compared to predictions determined with the complex model of Amiet. It can be shown that when employing the correct model equation, the simplified form can keep up with the complex form with respect to prediction accuracy.

1. Nomenclature

c_0	[m/s]	Speed of sound
$\overline{D_h}$	[-]	Directivity
d	[m]	Half-span
f	[Hz]	Frequency
I	[-]	Turbulent intensity
K	[1/m]	Wavenumber
$\widehat{K_x}$	[-]	Normalized wavenumber
r	[m]	Distance to an observer
$SPL_{\frac{1}{3}}$	[dB]	Sound Pressure Level (third octave band)
U_c	[m/s]	Convection velocity
z	[m]	Height above ground
α	[-]	Shear exponent
β	[-]	Prandtl - Glauert factor
λ	[m]	Acoustic wavelength
Λ	[m]	Length scale of eddy approaching the airfoil
ρ_0	[kg/m^3]	Density in the far field

2. Introduction

The attenuation of sound waves depends on their wavelength. Soundwaves with lower frequencies and thus higher wavelength are attenuated significantly less when propagating through the air. As an increasing number of authorities takes this into account and demand spectral predictions of emitted noise for tenders rather than accepting predictions based on the overall sound pressure result, the importance of a reliable prediction of the emitted noise in the lower frequency range increases.
The noise induced by the interaction of turbulent vortices in the atmosphere with an airfoil or blade section is the main aerodynamically induced contribution within the audible range of the lower frequencies. It is called in the following turbulence inflow noise (TIN).

In the generation process of TIN, an eddy of size Λ, which approaches the airfoil with a convection velocity U_c, leads to a disturbance at a frequency $f \approx U_c/\Lambda$. The frequency of the induced sound wave matches approximately this frequency with $f = c_0/\lambda$, where λ represents the acoustic wavelength. Hence, large eddies induce noise at lower frequencies, whereas the higher frequency noise stems from smaller eddies. Further, the response function of the airfoil or blade section to the respective disturbance depends on the size of the eddy. Large eddies with a length scale exceeding the chord lead changing loads and thus pressure along the entire chord. As a consequence, also the resulting acoustic wavelength is equal or greater than the chord. In this case, the airfoil is acoustically compact [1] and the induced sound waves radiate following a dipole characteristic. The resulting sound power level scales with Ma^6. If the size of the approaching eddy is smaller than the chord, only local pressure fluctuations in the region of the leading edge are imprinted. The global aerodynamic loads are not affected. As a result, the induced sound waves are smaller and the blade section is not acoustically compact. Thus, the directivity pattern changes to an edge noise source pattern. The intensity of these noise contribution scales with only the fifth power of the inflow Mach number. [1] [2]

In order to predict this type of sound different methods are available. They range from the computationally very expensive Computational Aero Accoustic (CAA) methods, which fully resolve the sources as well as the propagation, towards simple empirical models. Often, the models based on the theory of Amiet and Paterson are employed (commonly called Amiet model). Their models are based on analytical equations and thus provide a sensible compromise between computational effort and accuracy of the results.

Due to the different characteristics of the lower and higher frequency inflow noise, Patterson and Amiet [2] distinguish in between both. Although they provide an equation for both types, usually, only the higher frequency part is employed in wind turbine application. Due to relatively high chords, the low frequency part shifts towards frequencies of subordinate importance in legal regulations. In order to have a sensible and smooth transition in between both parts, then the higher frequency equation is instead corrected to account for deviations in the lower frequency range. Lowson [1] proposed one of the low frequency correction to the Amiet's high frequency model which is often utilized.

In addition, commonly correction functions are employed to account for the effects of finite thickness of the airfoils or influence of the angle of attack, since the theory of Paterson and Amiet [1] [2] is based on a flat plate assumption a 0° angle of attack.

In order to obtain the inflow noise model, Paterson and Amiet [2], [1] regard the turbulence as a random quantity and apply statistical measures such as a cross-power spectral density (PSD) to derive an equation of the far field noise induced by incoming eddies at a flat plate. The cross-PSD of the wall pressure fluctuations is thereby related to the far field noise by employing the theory developed by Kirchhoff and Curle [1]. The model considers the effects of compressibility and non-compactness, but ignores end effects [2]. In their papers, Paterson and Amiet derive a complex model based on this theory but also provide a simplified and thus easier to implement version of the model.

A number of papers is available addressing the prediction of TIN with their models, most of them focusing on the simplified version. However, a literature research revealed inconsistencies in between how different authors apply the simplified model resulting in dissimilar equations employed by the different researchers.

Hence, the first objective of the present paper is to provide an overview over the existing literature, i.e. over the different model variants the authors apply as well as the model utilized in the present work. Second, the behavior of the derived model is evaluated in the framework of a sensitivity study and the model is validated with measurements a multi-megawatt turbine as well as by comparison with the full, complex model by Amiet and Paterson.

The paper is structured as follows: First, the overview over the existing literature, i.e. over the model variants the different authors applied is given. Subsequently, the model employed in this paper based on the original Amiet equation is presented. Further, the model is also extended by the ideas of Lowson.
The derived model is then applied in the sensitivity study to show the effect of length scale as well as turbulence intensity on the inflow noise prediction. Lastly, the predictions by the simplified model are juxtaposed to those by the more complex model by Amiet and Paterson, as well as to sound measurements of a full-size rotor of a multi-megawatt turbine

3. Amiet and Lowson model

This section focuses on the variants of the Amiet model proposed in literature and presents the model employed in the current work. For this purpose, first the basic simplified Amiet model for high frequencies as well as Lowson model is sketched out.

3.1 The basic model

The basic equation for the higher frequency part of the TIN from the simplified Amiet model can be stated as follows for a flat plate in a turbulent stream [3]

$$SPL^H_{\frac{1}{3}} = 10\, log_{10}\, \overline{D_h} \frac{d \cdot \Lambda}{r^2}\, I^2\, \rho_0^2\, c_0^b\, Ma^a \frac{\widehat{K}_x^3}{\left(1+\widehat{K}_x^2\right)^{\frac{7}{3}}} + C. \tag{1}$$

In the equation Λ describes the length scale of the turbulent eddies, r represents the distance of the observer (normal to the source) and d the semi-span, ρ_0 and c_0 represent the density and speed of sound of the free flow, respectively. The turbulence Intensity I of the incoming flow is defined as

$$I = \sqrt{\left(\frac{<vv>}{U^2}\right)}, \tag{2}$$

where $< vv >$ is the Reynolds stress of fluctuations normal to the plate in the model and U is the inflow velocity at the flat plate. The wavenumbers of the Karman spectrum are described by $\widehat{K_x}$. The basic structure of the equation agrees in all literature sources. The constant C and the exponents a and b, however, vary according to the source.

As already mentioned, this equation is soften combined with the Lowson model to account for differences towards lower frequencies. The Lowson model is given as [1]:

$$SPL^L_{\frac{1}{3}} = SPL^H_{\frac{1}{3}} + 10\log_{10} LFC \tag{3}$$

The low frequency correction function, LFC, is is specified as:

$$LFC\ = 10\, S^2\, Ma\, K\, \beta^{-2} \tag{4}$$

In the equation S represents the Sears function (cf. eq.(5)), Ma the Mach number and K the wavenumber. The Prandtl-Glauert factor β accounts for the effects of compressibility at higher Mach number with $\beta^2 = 1 - Ma^2$.

$$S^2 = \left[2\,\pi \frac{K}{\beta^2} + \frac{1}{1+2.4\frac{K}{\beta^2}}\right]^{-1} \tag{5}$$

Then, with equation (3) to (5), the overall SPL for in terms of a third octave band spectrum results in

$$SPL_{\frac{1}{3}} = SPL^H_{\frac{1}{3}} + 10\log_{10} \frac{LFC}{1+LFC} \tag{6}$$

3.2 Amiet and Lowson model in literature

In the paper published originally by Paterson and Amiet [2] the constant C is set to 58.4, they however explicitly state the use of the density as well as the speed of sound to be in non-SI units. The normalized wavenumber $\widehat{K_x}$ is specified as $\widehat{K_x} = k/k_e$, where k_e is the wavenumber of the energy containing eddies in the van Karman spectrum. The exponents a and b are given as $a = 5$ and $b = 4$.

Lowson [1] then suggested the low frequency correction function for the Amiet model and therefore employed the model by Amiet and Paterson by utilizing a constant $C = 58.4$ and $\widehat{K_x} = K = \pi cf/U$ with $a = 5$ and $b = 4$. This matches also exactly the model variant given in Wagner et al. [4].

Fuglsang and Madsen [5] utilize the high frequency model with $C = 58.4$, $a = 5$, $b = 4$ and $\widehat{K_x} = \pi cf/U$. Moriarty et al. [6] employ a constant C of 58.4 together with the density and speed of sound in SI units in one of their earlier papers as well as $\widehat{K_x} = K$. However, they utilize a value of $C = 78.4$ in a later publication [7]. In the same paper they also give different definitions for the wavenumbers with $\widehat{K_x} = k/k_e$ and $K = \pi cf/U$. The exponents a and b have the same value as in the original model. With this, the model corresponds to the one from Buck et al. [8].

In the work by Zhu et al. [9], the value for C is not specified, they however utilize exponents of $a = 3$, $b = 2$.

In conclusion, in the different literature sources inconsistencies are found with respect to the constant C, the exponents of Mach number and speed of sound as well as with respect to the definition of the wavenumber in the Amiet model. The summary can be found in the following table:

Table 1: Overview on the different model variants in literature.

Source	a	b	C	$\widehat{K_x}$	Units
Paterson & Amiet [2]	5	4	58.4	k/k_e	Non-standard-SI units
Lowson [1]	5	4	58.4	$\pi cf/U$	Not specified (n.s.)
Wagner et al. [4]	5	4	58.4	$\pi cf/U$	Standard SI units
Fuglsang and Madsen [5]	5	4	58.4	$\pi cf/U$	Standard SI units
Moriarty et al. [6]	5	4	58.4	$\pi cf/U$	Standard SI units
Moriarty et al. [7]	5	4	78.4	k/k_e	Standard SI units
Buck et al. [8]	5	4	78.4	k/k_e	Standard SI units
Zhu et al. [9]	3	2	n.s.	n.s.	n.s.

3.3 Model used in present paper

In order to determine and understand the parameters employed in the present paper the derivation of the model equation by Amiet and Paterson is revoked. As suggested by Amiet and Patersion we also use the isotropic van Karman spectrum here to describe the turbulent fluctuations in the atmosphere. The reason for this is that, although large scale fluctuations in the atmosphere are far from being isotropic, for smaller scales the assumption of isotropy still holds (Kolmogorov hypothesis). In the frequency range of question for inflow noise in the present work, the corresponding scales are already located within this isotropic range of the atmospheric turbulence. Note however, that for lower frequencies an anisotropic approach might be valuable.
According to Amiet and Paterson, the approximate equation for the PSD of inflow noise fluctuations induced by a turbulent inflow characterized by a van Karman spectrum reads:

$$G_{pp}^{H}(z,\omega) = \frac{2d}{\pi c_0}\left(\frac{2\Lambda}{3\pi r}\right)^2 \frac{< vv >}{U^2} (\rho_0 U^2)^2 \left[\frac{\Gamma\left(\frac{1}{3}\right)}{\Gamma\left(\frac{5}{6}\right)}\right]^2 \frac{\widehat{K}_x^2}{(1+\widehat{K}_x^2)^{\frac{7}{3}}} \tag{7}$$

In the equation $\Gamma(x)$ is the Gamma Function and

$$\hat{K}_x^2 = \frac{\omega}{U k_e} \quad \text{with } k_e \text{ defined in the van Karman spectrum as } k_e = \frac{\frac{\sqrt{\pi}}{\Lambda}\Gamma\left(\frac{5}{6}\right)}{\Gamma\left(\frac{1}{3}\right)}. \tag{8}$$

In order to obtain third octave band levels, the equation is multiplied by the respective band width and normalized by the reference pressure, i.e.

$$SPL_{\frac{1}{3}} = 10\, log_{10} \frac{G_{pp} \Delta\omega}{p_{ref}^2} \qquad \text{with } \Delta\omega = 0.232\, \omega_{cen}. \tag{9}$$

Combining equation (7) and (9) then leads to an expression for the third octave band spectrum of the far field noise for the higher frequency TIN:

$$SPL_{\frac{1}{3}}^{H} = \tag{10}$$

$$10\, log_{10} \frac{d \cdot L}{r^2} \frac{\bar{u}^2}{U^2} \rho_0^2\, c_0^2\, Ma^5 \frac{\hat{K}_x^3}{(1+\hat{K}_x^2)^{\frac{7}{3}}} + \underbrace{10\, log_{10} \frac{2 \cdot 2^2 \cdot 0.232 \sqrt{\pi} \cdot \left[\frac{\Gamma\left(\frac{1}{3}\right)}{\Gamma\left(\frac{5}{6}\right)}\right]}{\pi \cdot (3\pi)^2 \cdot p_{ref}^2}}_{C}$$

Doing the math for the constant C results in $C = 78.4$ for a reference pressure of $p_{ref} = 2 \cdot 10^{-5}\, Pa$. Then, all other parameters have to be specified in SI units, i.e. the density should be given in [kg/m^3] and the speed of sound in [m/s]

The Amiet model in equation (10) is subsequently combined with the low frequency correction proposed by Lowson (cf. eq. (3) to (6)). In order to maintain consistency with respect to units the wavenumber K in the Lowson model is set to $= \pi c f / U$, while the wavenumber $\hat{K}_x$ in the Amiet model represents the normalized wavenumber of the van Karman spectrum and is determined following equation (8).

The Amiet model was derived for a flat plate with infinite thickness in a turbulent stream. In order to account for the effect of finite thickness on an airfoils, the model is combined with the thickness correction proposed by Moriarty et al. [10] in equation (11). Although the model proposed by Tian is a good alternative in case of a two dimensional case, it was shown, that for the three dimensional case of a wind turbine rotor, the model by Moriarty leads to better results [11].

$$\begin{aligned} \Delta_{SPL,\,Moriarty} = &-\Bigg(1.123\left(D_{rel,1\%} + D_{rel,10\%}\right) \\ &+ 5.317\left(D_{rel,1\%} + D_{rel,10\%}\right)^2 \left(\frac{2\pi f_c \cdot c}{U_{ref}} + 5\right)\Bigg) \end{aligned} \tag{11}$$

This combination of values and model equations also equals the model chosen by Buck et al. [8].

3.4 Application of 2D model to 3D Case

The set of equation presented in the previous section is able to describe the response of a 2.5 dimensional airfoil in a turbulent flow. However, in order to predict the inflow noise of a three-dimensional wind turbine rotor, the blade is cut into quasi-two-dimensional sections for each rotor position. The inflow conditions are determined for each section and have to be transformed from the fixed turbine coordinate system, in which usually wind speed and the atmospheric turbulence is defined, into the blade fixed, but moving coordinate system of each section. In case of isotropic turbulence, Reynolds stresses as well as length scales are invariant to a transformation of the reference coordinate system. However, this does not account for the turbulence intensity. The transformation from the inertial system towards the moving system affects the reference parameter of the turbulence intensity. As a results, rather than the wind speed, the reference parameter of the turbulence intensity is the local inflow velocity at each blade section, which is composed of the wind speed, the rotational speed and the influence of the induction at the blade. The equations of the Amiet model can then be evaluated in the moving system for each section separately. Subsequently the noise is propagated to a desired observer position. It is only at the observer position that the noise contributions stemming from the different blade sections and rotor positions are summed up.

4. Sensitivity study

This section is dedicated to discuss the influence of different length scales and turbulence intensities on the inflow noise emission of a wind turbine rotor. Therefore the model presented in the previous section is applied to predict the sound emission of a multi-megawatt wind turbine rotor at the IEC 61400-11 position, i.e. at an observer position downstream of the turbine with a distance to the towers nadir equal to the total turbine height.

First, the turbulence intensity is varied while the length scale is kept fixed and in the second part the length scale is constant while changing the turbulence intensity. Thereby the turbulence intensity refers to the turbulence intensity of the wind, with the reference velocity being equal to the wind speed. The resulting turbulence intensity at the blade is thus different for each blade sections for the reasons discussed above. The length scale describes the length scale of the energy containing eddies, i.e. the length scale corresponding to the wavenumber at the peak of the van Karman spectrum.

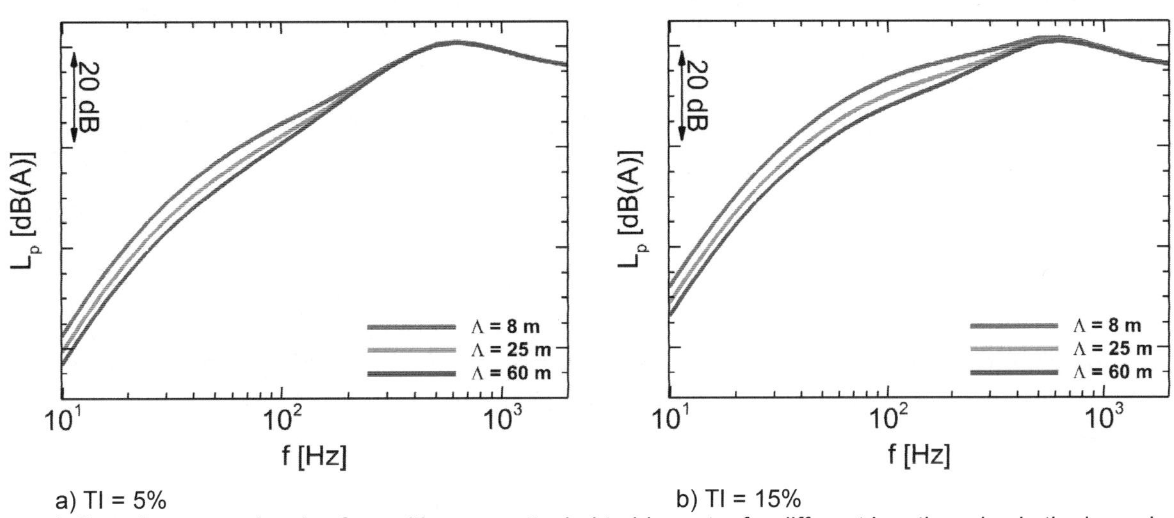

Figure 1: Sound pressure levels of a multi-megawatt wind turbine rotor for different length scales in the incoming flow at a rotational speed of 15.2 rpm.

Figure 1 shows the influence of the length scales on the predicted sound pressure spectra. The trailing edge noise for these spectra is predicted employing semi-empirical. One can see that the inflow noise contributes significantly to the total sound power levels approximately up to a frequency of 300Hz. For higher frequencies, the trailing edge noise dominates overall spectrum. Thus, changes in turbulence intensity or length scale have almost no effect. Figure 1 a) depicts the comparison for turbulence intensity in the wind of 5% and figure b) for 15%. It can be seen that the induced noise levels increase with smaller length scales in the atmosphere. The reason for this somewhat counterintuitive result is that for smaller length scales larger parts of the turbulent spectrum are in a length scale range relevant for inflow noise in the shown frequency range. Thereby, the influence is stronger for changes at smaller length scales. The delta between the smallest length scale of 8 m and 25 m is larger than the difference between 25 m and 60 m. This becomes even more apparent when taking into account the logarithmic scale of the ordinate. This behavior is also independent from the turbulence intensity.

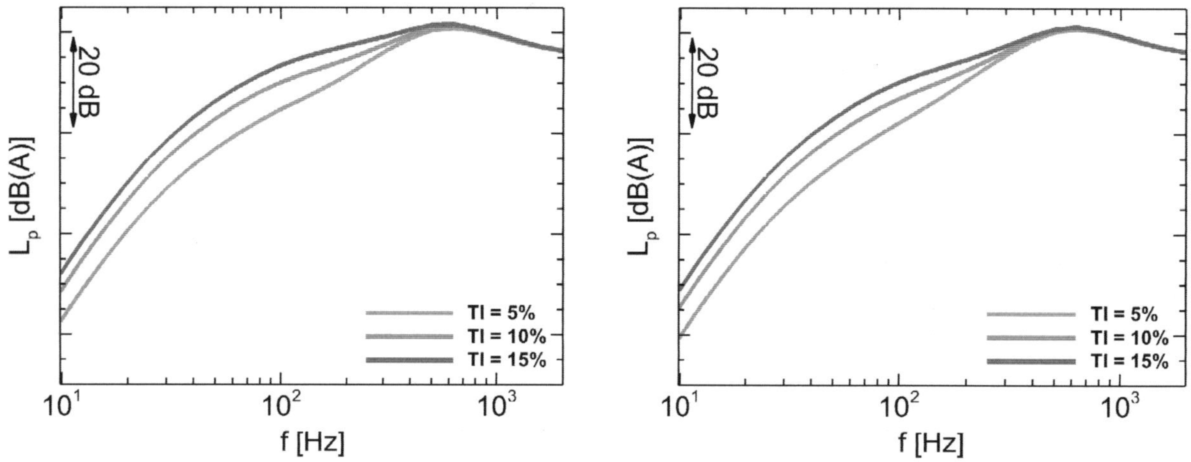

a) $\Lambda = 8$ m b) $\Lambda = 25$ m

Figure 2: Sound pressure levels of a multi-megawatt wind turbine rotor for different turbulent intensities in the incoming flow at a rotational speed of 15.2 rpm.

As expected, one can see that a higher turbulence intensity leads to an increase in noise levels and vice versa. This occurs for both depicted length scales in a similar manner. It can also be seen that due to the logarithmic scale the influence is non-linear.

Overall, it can be concluded that the set of equation employed for the model behaves as expected for a shift in turbulence intensity of length scale.

5. Comparison to experimental data and the complex Amiet model

This section now compares the simplified model to measurements at a turbine as well as to predictions with the complex Amiet and Paterson model. Therefore, upfront, the measurement setup is presented as well as the code utilized for the predictions with the complex Amiet model.

5.1 Measurements

In order to validate the model sound measurements from a multi-megawatt turbine are utilized, They were conducted in the framework of an extensive measurement campaign from September 2017 to March 2018. The measurement setup consisted of two microphones located downstream of the turbine at the same distance from the tower but different angular positions. With the microphones third octave band data from 20 Hz up to 6.3 kHz was recorded in 10s intervals. In order to be able to correct the noise levels for background noise, a cyclic shutdown concept was employed where 1h 40min operation alternated with 20min standstill. The minimum exceedance of measurements points taken into consideration was 3 dB above the background noise. All measurement points not fulfilling these requirements were excluded. The data is filtered for a location of the microphones of +/- 15° with respect to the direct wake of the turbine, for the present validation. The microphone was located on a sound reflecting measurement plate, thus the measured sound spectra are corrected for these effects, by subtracting 6 dB, assuming a sonically hard underground according to the IEC61400-11. This assumption is valid for high frequencies, however, there are some doubts about this assumption in the lower frequency range. The shown levels of the measured spectra are probably slightly higher in reality due to this assumption. [11]

Upstream of the turbine a wind met mast was installed to obtain the wind speed as well as the turbulence intensity at different heights above ground (37 m, 63 m, hub height, 99 m). The data utilized for the noise prediction in determined with a sliding window of 10 minutes (600 values at 1 Hz) placed symmetrically around the 10 s acoustic data points as far as possible. Subsequently, the resulting data sets were binned with respect to TI as well as according to the rpm of the turbine.

Table 2: Evaluated operation points of the multi-megawatt turbine

	15.2 rpm		
TI [%]	2.8	5.5	10.1
Λ [m]	6.23	6.05	39.08

In order to obtain the length scale, the methods proposed by Kelly et al. [12] were employed. Following their suggestions, the energy-containing eddies Λ can be determined from the measurements utilizing one of the two following equations:

$$\Lambda \approx \frac{\sigma_u}{dU/dz}$$

or by (12)

$$\Lambda \approx \frac{z\,\sigma_u}{U\,\alpha}.$$

In the equations σ_u represents the variance of the measured fluctuations, dU/dz is the gradient of the mean wind speed U over the height z and α the shear exponent.

In the present work Λ is evaluated at different heights with both approaches. Subsequently, the final value utilized for the prediction is derived based on a third order regression for all points. The results for the three operating points evaluated in the present paper are summarized in Table 2.

5.2 Simulations with IAGNoise+

IAGNoise+ [13] [14] is utilized for the simulation with the complex Amiet model. IAGNoise+ is a simulation tool to predict trailing edge noise, as well as inflow noise emission of a wind turbine.

An enhanced TNO model is employed for the first and the complex Amiet model for the latter. The code itself is developed and maintained at the Institute of aerodynamics and gas dynamics at the University of Stuttgart.

The complex Amiet model implemented in IAGNoise+ distinguishes between the lower and higher frequency regime and employs different model equations for both as proposed by Amiet. A solution for the airfoil response function based on the Sears function is used for the low frequency regime, i.e. $\mu < 0.4$. with

$$\mu = \frac{MK_x c}{2\beta^2} \; where \; K_x = \frac{\omega}{U}. \tag{13}$$

This approach is correct for $\mathcal{O}(\mu)$ but neglects all terms of $\mathcal{O}(\mu^2)$ and higher.

A series of corrections is made for the response function for the higher frequencies, i.e. $\mu > 0.4$. They take into account the finite length on the airfoil, by alternating in the corrections between leading edge and trailing edge. The expression for the effective lift can then be calculated as the sum of the corrections to obtain the final response function. The code utilizes the first two element of the series, which was suggested to be sufficient by Amiet [1].
Also within IAGNoise+ the inflow spectrum is modeled assuming an isotropic two-dimensional van Karman spectrum. To correct for the effect of airfoil thickness, which becomes important in particular for lower Mach numbers, similar to the present approach also IAGNoise+ employs the model by Moriarty et al. (cf. eq. (11)).

For further information on the inflow noise model employed within IAGNoise+ the reader is referred to [11].

5.3 Results

Figure 3 a) compares measurements, predictions by the simplified model from the present paper as well as the predictions by the complex model with IAGNoise+ with each other. Depicted are the sound pressure spectra matching the three inflow conditions specified in Table 2. The spectra are shifted by 10dB for the different inflow conditions to make the lines discernable.
Looking first at the measurements only, it stands out that the spread in measurement points is larger in the lower frequency range, i.e. the range dominated by the inflow noise, than in the higher frequency range which is ascribed to the trailing edge noise mainly. This shows that the inflow noise emission is very sensitive to changes in the incoming flow. Comparing then the predictions by the two models, it can be seen that the differences between simplified and complex model decrease with increasing turbulence intensity and related length scale. Overall, however, the delta between the two models is not too large taking into account the general uncertainty in inflow noise measurements. Still, the simplified model is able to capture the measurements very well, in particular when taking into account the potentially higher levels of the measured points due to the uncertainty in the correction for the sound reflecting plate (see section on the measurement setup).
Since both, simplified as well as complex model, include only the contributions of inflow noise they start to deviate from the measured spectrum already at around 60 Hz. The gap is closed when considering the trailing edge contribution, as depicted in Figure 3 b).

In Figure 3 b) the predictions of the total noise by the simplified Amiet model together with trailing edge noise contribution predicted with a semi-empirical model shown in comparison to the experimental measurement data. The dashed lines indicate the inflow noise. Thus, it can be concluded that the simplified Amiet model in combination with the low frequency correction of Lowson and the thickness correction by Moriarty is able to predict the inflow noise generated at a wind turbine with at least equal accuracy as the complex Amiet model. Further, the prediction with measurements shows that with this combination of models the prediction lays within the spread of measurement points.

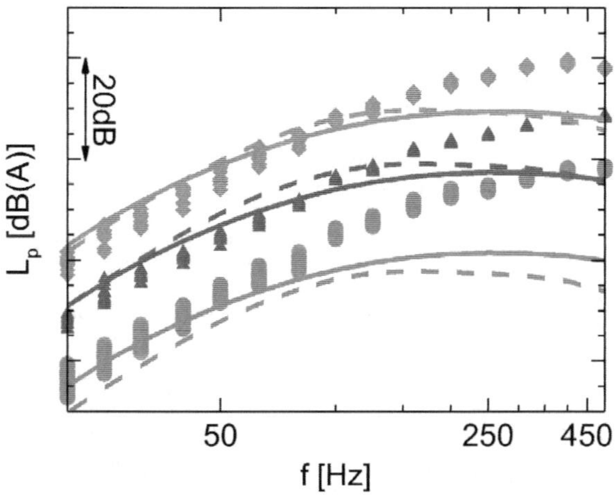

a) Inflow noise predicted with the simplified Amiet model (dashed lines) and complex model (solid lines)

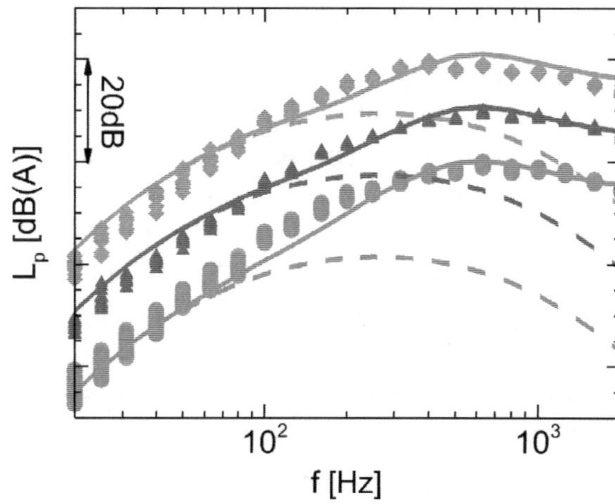

b) Inflow noise predicted with the simplified Amiet model (dashed lines) and total noise (solid lines) (trailing edge noise from semi-empirical model)

Figure 3: Sound pressure levels of a multi-megawatt wind turbine rotor for different turbulent intensities in the incoming flow according to Table 2 at a rotational speed of 15.2 rpm from measurements (symbols) and predictions with different models (lines). The spectra for the different TI are shifted by 10dB with TI = 2.8% in green, TI = 5.5% in pink and TI = 10.1 % in yellow.

6. Conclusions

Amiet and Paterson developed a theory for the prediction of inflow noise at a flat plate in a turbulent stream. They propose a simplified and a more complex version of their resulting model. A literature revealed that there are significant inconsistencies in between how different authors apply their simplified model. These inconsistencies were summarized in the present paper. Further, the model employed here was sketched out. For it, the simplified Amiet model was combined with the low frequency correction proposed by Lowson as well as with the thickness correction for airfoils proposed by Moriarty et al.. Subsequently a sensitivity study with respect to length scale as well as turbulence intensity was conducted to evaluate the behavior of the model for noise prediction at a full size wind turbine rotor and thereby verify the implantation and model combination. In a last step, the predictions of rotor noise by the simplified model was compared to predictions by the more complex model, employing the wind turbine noise prediction code IAGNoise+, developed at the University of Stuttgart. Further, the predictions by both models were compared to measurements. It could be shown, that in particular at higher turbulence intensities the two model lead to quite similar results, in particular when taking the natural spread of measurement in the frequency range for inflow noise into account. The comparison to measurements also showed that the set of model equations employed here is able to predict measurements with a good accuracy.

References

[1] R. K. Amiet, "Acoustic Radiation from an Airfoil in a Turbulent Stream," *Journal of Sound and Vibration,* vol. 4, no. 41, pp. 407-420, 1975.

[2] R. W. Paterson and R. K. Amiet, "Noise and Surface Pressure Response of an Airfoil to Incident Turbulence," *Journal of Aircraft,* vol. 8, no. 14, 1977.

[3] R. W. Paterson and R. K. Amiet, "Acoustic Radiation and Surface Pressure Characteristics of an Airfoil Due to Incident Turbulence," September 1976.

[4] S. Wagner, R. Bareiß and G. Guidati, Wind Turbine Noise, Berlin Heidelberg: Springer-Verlag , 1996.

[5] P. Fuglsang and H. Madsen, "Implementation and Verification of an Aeroacoustic Noise Prediction Model for Wind Turbine," Risoe National Lab. Denmark, 1996.

[6] P. Moriarty and P. Migliore, "Semi-Empirical Aeroacoustics Noise Prediction Code for Wind Turbines," 2003.

[7] S. Buck, S. Oerlemans and S. Palo, "Experimental Validation of a Wind Turbine Turbulent Inflow Noise Prediction Code," *AIAA Journal,* pp. 1495-1506, 2018.

[8] W. J. Zhu, N. Heilskov, W. Z. Shen and J. N. S. rensen, "Modelling of aerodynamically generated noise from wind turbines," *Journal of Solar Energy Engineering,* vol. 127, pp. 517-528, 2005.

[9] P. J. Moriarty, G. Guidati and P. Migliore, "Prediction of Turbulent Inflow and Trailing-edge Noise for Wind Turbines," in *11th AIAA/CEAS Aeroacoustics Conference (26th AIAA Aeroacoustics Conference)*, Monterey, 2005.

[10] C. Hornung, C. Scheit, N. Noffke, A. Altmikus, T. Lutz and E. Krämer, "Inflow Noise: Prediction and analysis of the relevance for a multi-megawatt turbine," in *DAGA*, Rostock, 2019.

[11] M. Kelly, "From standard wind measurements to spectral characterization: turbulence lenght scale and distribution," *Wind Energy Science,* 2018.

[12] C. Hornung, T. Lutz and E. Krämer, "A model to include turbulence-turbulence interaction in the prediction of trailing edge far field noise for high angles of attack or slightly separated flow," *Renewable Energy,* 2019.

[13] M. Kamruzzaman, D. Bekiropoulos, A. Wolf, T. Lutz and E. Krämer, "Rnoise: A RANS based airfoil trailing-edge noise prediction model," in *20th AIAA/CEAS Aeroacoustics Conference,*, Atlanta, GA, 2014.

[14] M. V. Lowson, "Assessment And Prediction Of Wind Turbine Noise," *Flow Solutions Report,* vol. 92, no. 19, 1992.

[15] P. J. Moriarty, G. Guidati and P. G. Migliore, "Recent Improvement of a Semi-Empirical Aeroacoustic Prediction Code for Wind Turbines," in *10th AIAA/CEAS Aeroacoustics Conference, AIAA 2004-3041*, 2004.

[16] R. K. Amiet, "Acoustic Radiation from an Airofil in a Turbulent Stream," *Journal of OSund and Vibration,* no. 41 (4), pp. 407-420, 1975.

[17] R. W. Paterson and R. K. Amiet, "Noise and Surface Pressure Response of an Airfoil to Incident Turbulence," *J. Aircraft,* vol. Vol 14, August 1977.

[18] Y. Tian, B. Cotte and A. Chaigne, "Wind Turbine Noise Modelling Based on Amiet's Theory," in *Theory 5th International Meeting on Wind Turbine Noise*, Denver, CO, United States, 2013.

[19] Y. Tian and B. Cotte, "Wind Turbine Noise Modeling Based on Amiet's Theory: Effects on Wind Shear and Athmospheric Turbulence," *Acta Acustica united with Acustica,* pp. 626-639, 2016.

INCEEUROPE

9th International Conference on Wind Turbine Noise Remote from Europe – 18th to 21st May 2021

Calculation of wind turbine noise uncertainty for downwind conditions

Bill Kayser, UMRAE, Cerema, Univ. Gustave Eiffel, Ifsttar, Strasbourg, 67035, France: bill.kayser@cerema.fr
Benoit Gauvreau, UMRAE, Univ. Gustave Eiffel, Ifsttar, Cerema, Bouguenais, 44344, France
David Ecotière, UMRAE, Cerema, Univ. Gustave Eiffel, Ifsttar, Strasbourg, 67035, France
Vivien Mallet, ANGE, INRIA, 2 rue Simone Iff, 75012 Paris, France

Summary

Wind turbine noise variability is significant for outdoor environments because of the influence of atmospheric conditions and ground properties that are both variable in time and space. Thus, realistic prediction of sound pressure levels involves to estimate the overall uncertainty induced by the most influential environmental phenomena on both acoustic emission and propagation. To do so, this study performs a method of propagation of uncertainty by using a quasi-Monte Carlo sampling of input data (*i.e.* environmental parameters) in order to feed an Amiet emission model coupled to a PE propagation model, and next to calculate the probability distribution of output data (*i.e.* sound pressure levels). As stochastic uncertainty propagation requires a high number of simulations, a metamodel of the global model (emission-propagation) for wind turbine noise was built using kriging technique in order to drastically reduce calculation duration. The results provide useful statistics and uncertainties information about sound pressure levels at neighborhood of a wind turbine: this information provide a better knowledge of sound pressure variability and will help to a better control of the quality of wind turbine noise prediction for inhomogeneous outdoor environment.

1. Introduction

Modelling wind turbine noise involves to take into account the aeroacoustic sources at blades (Bertagnolio et al., 2017; Buck et al., 2016; Oerlemans and Schepers, 2009), as well as meteorological effects (vertical wind and temperature gradients, atmospheric turbulence) and ground effects (acoustic impedance, roughness, topography). Such environmental effects fluctuate over both time and space, which can lead to high variability of Sound Pressure Level (SPL) at long range (Cheinet et al., 2018; Gauvreau, 2013; Kayser et al., 2018; Wilson, 2003; Zouboff et al., 1994), and thus to significant SPL uncertainty (Leroy et al., 2010; Pettit and Wilson, 2007; Wilson et al., 2014). For wind turbine noise context, several models are now capable of taking into account more and more effects (Barlas et al., 2017; Cotté, 2019; Heimann et al., 2018; Lee et al., 2016; McBride and Burdisso, 2017), but few studies has been done on quantifying environmental parameters sensitivities and uncertainties on the SPL estimation.

Thus, the objective of this paper is to conduct an uncertainty quantification of SPL estimation for wind turbine noise. The uncertainty considered in this paper deals with the very common situation of wind turbine noise assessment by SPL measurements near dwelling. Because these situations seldom imply other measurements than SPL or wind, the uncertainty on the SPL mainly comes from the possible variability of the environmental parameters that influence wind turbine noise. The uncertainty estimated here is then due a lack or a poor knowledge of these influential parameters during the assessment. This paper focus on downwind conditions because these conditions usually represent the most detrimental conditions for the neighbourhood, as the SPL is significantly lower in upwind and crosswind conditions at long range, and because these conditions do not require to take into account complex calculation involving the influence of turbulences. To do so, we are using stochastic technique based on quasi-Monte Carlo sampling in order to determine the distribution of the wind turbine noise model output (*i.e.* SPL) induced by the probability distribution of the uncertain inputs (*i.e.* environmental parameters).

In practice, thousands of simulations may be required to conduct an uncertainty propagation, which leads to prohibitive calculation cost. One solution is to replace the initial wind turbine noise model with a metamodel that reproduces the expected SPL with highly reduced computational costs and low error. For this purpose, we follow the same approach that (Mallet et al., 2018) used for urban air quality, or by (Lesieur et al., 2020) for urban noisemapping.

Section 2 of the paper reviews theories on the wind turbine noise emission model and acoustic propagation model, and explains the *coupling* technique (Cotté, 2019) for predicting wind turbine noise levels at long range. The construction and validation of the metamodel is presented in Section 3. The uncertainty analysis results are discussed in Section 4. Finally, Section 5 concludes.Wind turbine noise modelling.

2. Wind turbine noise modelling

2.1 The moving monopole technic

Following (Kayser et al., 2020), the moving monopoles approach (Cotté, 2019) is used to calculate wind turbine noise in outdoor environment. The SPL at a receiver location is calculated for a blade segment at an angular position Φ using the point source approximation (Salomons, 2001):

$$\mathrm{SPL}(\omega,\Phi) = \overbrace{\mathrm{SWL}(\omega,\Phi)}^{\text{emission}} - \underbrace{10\log(4\pi R_1^2)}_{\text{geometrical spreading}} + \overbrace{\Delta L(\omega,\Phi) - \alpha(\omega)R_1}^{\text{atmospheric and ground effects}}, \quad (1)$$

where ω is the angular frequency at the receiver, $\mathrm{SWL}(\omega,\Phi)$ is the angle-dependent source sound power level, $R_1 = \sqrt{x_R^2 + y_S^2 + (z_s - z_R)^2}$ is the distance between the source at $(0, y_s, z_s)$ and the receiver at $(x_R, 0, z_R)$, ΔL is the sound attenuation relative to the free field, and α is the atmospheric absorption coefficient in dB/m. The angle-dependent $\mathrm{SWL}(\omega,\Phi)$ can be obtained from the free-field SPL calculated using Amiet's theory (Amiet, 1975, 1976; Roger and Moreau, 2010) as done in (Cotté, 2017; Kayser et al., 2020; Tian and Cotté, 2016). The attenuation term $\Delta L(\omega,\Phi)$, can be calculated using any propagation model that consider meteorological effect and ground effect. A propagation model based on the Wide-Angle Parabolic Equation (WAPE)

(Collins, 1989; Gilbert and White, 1998) approximation of the wave equation is used here, as in (Kayser et al., 2020). The WAPE is solved using a Padé (1,1) approximants and solved with the method of (Collins, 1993).

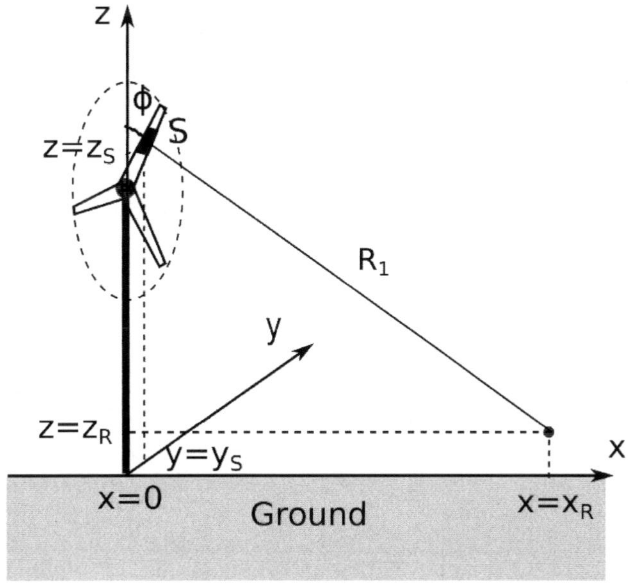

Fig.1 Schematics of the moving monopoles approach geometry with a receiver located at $(x_R, 0, z_R)$.

2.2 Ground effects

Acoustic properties of the ground are described by an effective admittance model defined as follows (Brelet, 2008):

$$\beta_{\text{eff}} = \beta + \beta_{\text{rough}} = \frac{1}{Z} + \beta_{\text{rough}}, \tag{2}$$

with Z the ground impedance calculated here by the Miki's impedance model (Miki, 1990):

$$Z = Z_0\left[1 + 5.50\left(\frac{f}{a_{\text{fr}}}\right)^{-0.632} + i8.43\left(\frac{f}{a_{\text{fr}}}\right)^{-0.632}\right], \tag{3}$$

$$k = k_0\left[1 + 7.81\left(\frac{f}{a_{\text{fr}}}\right)^{-0.618} + i11.41\left(\frac{f}{a_{\text{fr}}}\right)^{-0.618}\right], \tag{4}$$

where Z_0 is the specific impedance of air, k_0 is the acoustic wavenumber and a_{fr} is the ground airflow resistivity. The β_{rough} term gives the average effect of surface roughness on acoustic propagation. It is modelled by (Bourlier et al., 2013; Brelet and Bourlier, 2008):

$$\beta_{\text{rough}} = \int_{-\infty}^{+\infty}\left(\frac{d\kappa'}{(k_0 k_z(\kappa'))(k_0^2)} - \kappa\kappa'\right)W(\kappa\kappa'), \tag{5}$$

with $\kappa = k_0 \sin\theta_i$ where θ_i is the incidence angle of the acoustic wave with ground, $k_z(\kappa) = \sqrt{k_0^2 - \kappa^2}$ and W the roughness spectrum of ground. In this work, the probability density of the ground roughness heights is considered to follow a normal distribution, so that:

$$W(k) = \frac{\sigma_h^2 l_c e^{-\frac{k^2 l_c^2}{4}}}{2\sqrt{\pi}}, \tag{6}$$

with σ_h the standard deviation of the ground roughness heights and l_c the correlation length of the ground roughness heights.

2.3 Atmospheric effects

In the case of a ground with no obstacle, the atmosphere is considered homogeneous in the horizontal direction. It is then possible to neglect the dependence of the meteorological profiles on the distance. The atmospheric stratification effects on acoustic waves are then taken into account via the assumption of effective celerity, which means vertical profile is defined as follow:

$$\langle c_{\mathrm{eff}}(z)\rangle = \sqrt{\gamma R \langle T(z)\rangle} + \langle U(z)\rangle \cos\theta \,, \tag{7}$$

with γ the heat capacity ratio of dry air, R the specific gas constant, θ the angle between the wind direction and the direction of sound propagation. The mean wind and temperature vertical gradients (respectively $\langle U(z)\rangle$ and $\langle T(z)\rangle$) are assumed to have logarithmic shape such that:

$$\langle T(z)\rangle = T_0 + a_T \ln\left(\frac{z-d}{z_0}\right), \tag{8}$$

$$\langle U(z)\rangle = a_u \ln\left(\frac{z-d}{z_0}\right), \tag{9}$$

where T_0 is the atmospheric ground surface temperature, a_T and a_u are coefficients that determine the shape of the profiles (Gauvreau, 2013), $d = 0.66\ h_v$ is the displacement height accounting for the influence of vegetation height h_v on the entire vertical profile, and $z_0 = 0.13\ h_v$ is the roughness height of the flux profiles (Brutsaert, 1982).

The atmospheric absorption is taken into account in the model in accordance with the standard (ISO9613-1:1993, 1993) which depends on air temperature, atmospheric pressure and the relative humidity of air h_r.

2.4 Case study

The considered wind turbine has a nominal electrical power of 2.3 MW, a rotor diameter of 93 m, a hub height of 80 m and three blades of 45 m length. The speed of rotation increases linearly from 6 rpm at the cut-in wind speed of 4 m/s measured at the hub height, to 16 rpm at the wind speed of 12 m/s. The model output is the SPL time-averaged over one blade rotation (see Eq.1) expressed in dB, calculated for far field receiver positions in the 2D section: $x \in [500\,;\,1500]$ m and $z \in [0\,;\,10]$ m with a 0.5 m resolution.

The set of the inputs of the wind turbine noise model is composed of 9 environmental parameters which range intervals are chosen to be representative for temperate climate, and to ensure that the wind gradient coefficient interval cover the entire operating range of the turbine. The ground parameters a_{fr}, l_c and σ_h intervals are determined from values found in several previous *in situ* measurements (Blaes and Defourny, 2008; Borgeaud and Bellini, 1998; Davidson et al., 2000; Embleton et al., 1983; Gauvreau, 2013; Kayser et al., 2018; Nicolas and Berry, 1984). We focus on low vegetation heights so that the maximum of the h_v parameter is 1 m. We consider in this paper only downwind propagation condition ($\theta \in [0; 90]°$) because these conditions usually induce the most detrimental SPL to neighbourhoods, and because the influence of atmospheric turbulence on sound propagation can be neglected for these conditions. Nevertheless, in order to avoid unrealistic SPL in the shadow zone that can appear for some rare specific downwind conditions (theta close to 90° combined with negative temperature gradient, appearing less than 8 % for the (θ, a_T, a_u) combinations considered here), we follow the same approach as in (Kayser et al., 2020) which consists in introducing a limit of $\Delta L \geq -25$ dB (Heimann and Salomons, 2004).

We follow (Kayser et al., 2020) that concludes after a sensitivity analysis that the relative humidity of air h_r and the 2 ground roughness parameters l_c and σ_h have a negligible influence on the SPL for this specific configuration (source-receiver geometry, at the considered frequencies) related to wind turbine noise: < 1 dB of influence for each third octave band in the frequency range $[50; 1000]$ Hz of this study. Thus, the 3 parameters h_r, l_c and σ_h are set to a nominal average value which allow us to reduce the input set dimension of the model from 9 varying parameters to 6 varying parameters. The Table 1 summarises the wind turbine noise model input set.

Parameters	Description	Values
a_{fr} **(kN.s.m^{-4})**	Airflow resistivity of the ground	$\in [50; 5000]$
l_c **(m)**	Correlation length of the rough ground	0.5
σ_h **(m)**	Standard deviation of the roughness height	0.025
h_r **(%)**	Relative humidity of air	84%
T_0 **(°C)**	Atmospheric ground surface temperature	$\in [0; 30]$
a_u **(m/s)**	Wind profile coefficient	$\in [0.67; 1.67]$
a_T **(K/m)**	Temperature profile coefficient	$\in [-0.5; 0.25]$
h_v **(m)**	Vegetation height	$\in [0; 1]$
θ **(°)**	Wind and source-receiver angle	$\in [0; 90]$

Tab.1 Wind turbine noise model inputs set chosen to focus on downwind conditions for temperate climate.

3. Metamodeling

The computational cost of the model makes it unsuitable for a direct stochastic uncertainty quantification which requires thousands of simulations. The solution proposed here is to build a metamodel (or surrogate model) that accurately reproduce the behaviour of the model with a negligible computation time. The metamodel will emulate the SPL outputs in the target 2D section: $x \in [500 ; 1500]$ m and $z \in [0 ; 10]$ m with a 0.5 m resolution. In order to avoid a direct application of the emulation on the target area that would not drastically reduce the time computing because of a too big number of emulators required ($2001 \times 20 = 40020$ emulators), the output of the model is represented by a limited number of scalars (*i.e.* of the order of ten) that become the emulated quantities. To do so, the method consists in 3 steps (Lesieur et al., 2020; Mallet et al., 2018):

a. generation of a training sample to explore the inputs set space of the model,

b. reduction of dimension of the model outputs space through principal component analysis of the training sample noise maps, in order to defined any noise map as a linear combination of principal components onto a reduced subspace,

c. emulation of the relation between the projection coefficients and the inputs set by a fast statistical emulator based on kriging interpolation.

3.1 Training sample

The training sample is generated using a Latin Hypercube Sampling (LHS) (McKay et al., 1979) with a *maximin* criteria (Johnson et al., 1990) which distributes the sampling points much more uniformly than a standard Monte Carlo sampling. A preliminary convergence test showed that $N = 500$ training samples are enough to discretize correctly the 6D inputs space (a_{fr}, $T_0, a_u, a_T, h_v, \theta$) of this study. Concretely, $N = 500$ 2D noise maps are calculated at each central frequency of the third octave considered in the analysis ($f \in [50; 1000]$Hz). These noise maps are then used as a training sample for the output dimension reduction, as explained in the following section.

3.2 Output dimension reduction through principal component analysis

The wind turbine noise model can be seen as a function $\mathcal{M}$, whose inputs set $\mathbf{X} \in \mathbb{R}^6$ is a low-dimensional vector, and whose output $\mathbf{y} \in \mathbb{R}^{40020}$ is a large-dimensional vector. To express the model output $\mathbf{y}$ by a limited number of scalars, the approach is to project any output vector $\mathbf{y}$ onto a reduced basis subspace$(\mathbf{\Psi})_{j=1\cdots H}$, where $H \ll 40020$ is the number of dimensions of the reduced basis, and where $\mathbf{\Psi}_j \in \mathbb{R}^{40020}$. The vectors $\mathbf{\Psi}_j$ are determined by principal component analysis (PCA) (Jolliffe, 1986) of the centered training sample $\overline{\mathbf{Y}} = [\mathbf{y}_1 - \bar{\mathbf{y}} \cdots \mathbf{y}_N - \bar{\mathbf{y}}]$ where $\bar{\mathbf{y}} = \frac{1}{N}\sum_{l=1}^{N} \mathbf{y}_l$ is the mean of the training sample. It is then possible to express any output map $\mathbf{y}$ as a linear combination of principal component $\mathbf{\Psi}_j$: $\mathbf{y} \simeq \sum_{j=1}^{H} a_j \, \mathbf{\Psi}_j$ where $a_j = (\mathbf{y} - \bar{\mathbf{y}})^T \mathbf{\Psi}_j$ is the projection coefficient of the *j*-th principal component. The coefficients a_j are the emulated quantities by the metamodel as detailed below.

3.3 Kriging interpolation

Following the PCA, the coefficients $\boldsymbol{a}$ are known for each of the N noise maps of the training sample. However, for any new set of parameters, the $\boldsymbol{a}$ functions must be replaced by statistical emulators which allow to compute each a_j coefficient quickly for any combination of the $\mathbf{X}$ input parameters. An emulator $\widehat{a_j}$ is built using *ordinary* kriging (Matheron, 1965) for each $a_{j=1\cdots H}$ with the DiceKriging Package (Roustant et al., 2012) of the R software (R Core Team, 2013). Each emulator $\widehat{a_j}$ is expressed as the sum of a deterministic and stochastic part, such that:

$$\widehat{a_j}(\mathbf{X}) = m(\mathbf{X}) + \varepsilon(\mathbf{X}), \tag{10}$$

with $m(\mathbf{X})$ the unknown constant trend and $\varepsilon(\mathbf{X})$ the systematic error modelled by a centered stationary gaussian process. The *kernel* of the gaussian process (that aims to translate statistical dependencies between sample points) is the Matérn $\frac{5}{2}$ correlation function recommended in (Koehler and Owen, 1996).

3.4 Validation

This section evaluates the metamodel error by comparing outputs of the metamodel (built with $H = 15$ principal components) with outputs of the initial wind turbine noise model, for the same set of input using a testing sample composed of $n = 100$ new noise maps. The testing sample is generated using a *complementary* sample of the training LHS in the input set space. This strong constraint guarantees that the metamodel error is not underestimated. The error is evaluated using the different scores presented in table 2. The table 3 shows the results for all the frequencies of the analysis.

Score (dB)	Formula
Bias	$s = \frac{1}{n}\sum_{i=1}^{n} \Delta_i$
Standard deviation of the error	$\sigma = \sqrt{\frac{1}{n}\sum_{i=1}^{n} (\Delta_i - s)^2}$
Root mean square error	$\mathrm{RMSE} = \sqrt{\frac{1}{n}\sum_{i=1}^{n} {\Delta_i}^2}$

Tab.2 Scores for the performance evaluation of the metamodel. $n = 100$ is the total number of points in the testing sample and Δ_i is the error between the metamodel sequence and the corresponding model sequence.

	s (dB)	σ (dB)	**RMSE** (dB)
50 Hz	0.00	0.18	0.48
63 Hz	−0.03	0.20	0.56
80 Hz	−0.03	0.30	0.63
100 Hz	−0.03	0.33	0.65
125 Hz	−0.03	0.31	0.62
160 Hz	−0.01	0.32	0.65
200 Hz	0.00	0.32	0.69
250 Hz	−0.01	0.32	0.60
315 Hz	0.01	0.32	0.58
400 Hz	0.00	0.33	0.59
500 Hz	0.01	0.34	0.78
630 Hz	0.00	0.30	0.81
800 Hz	−0.02	0.35	0.98
1000 Hz	−0.02	0.36	1.2

Tab.3 Scores of the metamodel for $H = 15$ principal components. Restults are rounded to the hundredth of a dB.

The Table 3 shows that the bias s induced by the metamodel is negligible and can be considered as null. The error dispersion measured by σ is between 0.2 dB and 0.4 dB depending on the frequencies, which is very satisfactory for outdoor acoustics where the influence of environmental parameters on SPL are about several dB. The orders of magnitude of the **RMSE** below 800 Hz are about 0.6 dB which is also satisfactory. The slightly higher **RMSE** in high frequencies, *i.e.* about 1 dB at 800 Hz and 1000 Hz, are of little importance because the emission spectrum of a wind turbine is dominated by low frequencies. The intrinsic error committed at each frequencies by the metamodel is taken into account in the final uncertainty analysis that is presented in the section 4.

4. Uncertainty analysis

The objective of the uncertainty analysis is to determine the probability distribution of the output $\mathbf{y}$, denoted $\mathbb{P}(\mathbf{y})$, induced by the probability distribution of the uncertain parameters $\mathbf{X}$, denoted $\mathbb{P}(\mathbf{X})$. The distribution $\mathbb{P}(\mathbf{y})$ is numerically estimated by sampling the distribution $\mathbb{P}(\mathbf{X})$ in order to propagate the uncertainty of the inputs $\mathbf{X}$.

4.1 Probability laws of the input parameters

The minimum and the maximum values of the atmospheric condition parameters (T_0, a_T, a_u) are chosen to be representative of temperate climate conditions, and the distributions laws of these parameters are uniform within these intervals, in order to not favour any particular condition. Notice, however, that the choice of uniformity may lead to a possible overestimation of the uncertainties. The distribution of the propagation angle θ is uniform in the range $[0; 90]°$ in order to remain in the validity domain of the model. The values of the vegetation heights h_v are representative of low vegetation, and a uniform distribution in the range $[0; 1]$ m is used. The ground resistivity a_{fr} range is defined to cover the acoustic absorption properties of grounds usually encountered in the wind turbine context: the minimum and the maximum value are respectively 50 kN.s.m^{-4} and 5000 kN.s.m^{-4}. It should be noted that the ranges of characteristic values of the parameter a_{fr} for the different ground typologies all have an important dispersion (Attenborough et al., 2006; Carpinello et al., 2004; Embleton et al., 1983; Nicolas and Berry, 1984), with a distribution that is not uniform in a given interval, which reflects the spatial and temporal variability of this parameter as well as the experimental uncertainties (metrological and methodological) related to its measurement (Attal, 2016; Attal et al., 2019; Ecotière et al., 2015). A *lognormal* distribution is used (Ostashev and Wilson, 2015), whose parameters are empirically adjusted to obtain the predominance of $[200; 1000]$ kN.s.m^{-4} which are the most frequently encountered values around wind farms (Ecotière et al., 2018; Kayser et al., 2018).

The probability distributions $\mathbb{P}(\mathbf{X})$ are randomly sample by Quasi-Monte Carlo Sobol sequences (Sobol, 1967). The Sobol sequences are deterministic versions of the Monte Carlo method that provide a faster convergence, up to a factor of 10 compared to the classical Monte Carlo method with a lower discrepancy (Saltelli et al., 2008). A preliminary convergence test showed that using a Sobol sequences with 5000 points is an optimum to get an estimation of the statistical estimators with satisfactory precision.

4.2 Variability of sound pressure levels with distance

Following the uncertainty analysis, 5000 maps of SPL are calculated for the third octave bands $[50; 1000]$ Hz, from the calculation of SPL for each central frequency corrected by the width of the band. The figure 2 shows the 95% confidence interval of SPL with distance x to the source, for a receiver heigh of $z = 2$ m, at 2 third octave bands: $f_c = 50$ Hz and $f_c = 1000$ Hz. The confidence intervals are determined using the 0.025 and 0.975 quantiles of the SPL distribution. This statistic allows one to quantify the uncertainties on SPL estimation and that would be due to a lack of knowledge of the values of environmental parameters.

First of all, we notice that the confidence intervals at 95% are very wide, ranging from 23 dB at 1000 Hz, to 31 dB at 50 Hz, which reflects the great variability of SPL in wind turbine noise context in the absence of any knowledge of the environmental parameters. A large part of the observed dispersion is due to the variability of the acoustic emission (Kayser et al., 2020), which depends essentially on wind parameters. A higher variability is also observed at low frequency. The geometrical divergence influence is visible on the median levels (black line) which decrease with distance due to geometrical spreading. We notice an asymmetry of the distributions through the position of the median that is not centered in the confidence intervals, being closer to the upper bound than to the lower bound. This asymmetry can be related to the physical effects of environmental parameters that are not systematically linear on SPL, and by the fact that the acoustic emission presents a plateau when the nominal power of the wind turbine is reached. Furthermore, the variability of SPL with distance shows a behaviour that is different with frequency. For $f_c = 50$ Hz the width of the confidence interval does not vary significantly: there is 0.1 dB of variation with distance x which is negligible in environmental acoustics. On the other hand, the confidence interval at $f_c = 1000$ Hz increases with distance x: it is about 23 dB for $x = 500$ m and reaches 24 dB for $x = 1500$ m. This is related to some propagative effects that are more influential at higher frequency (*e.g.* ground absorption effect) as well as cumulative with distance.

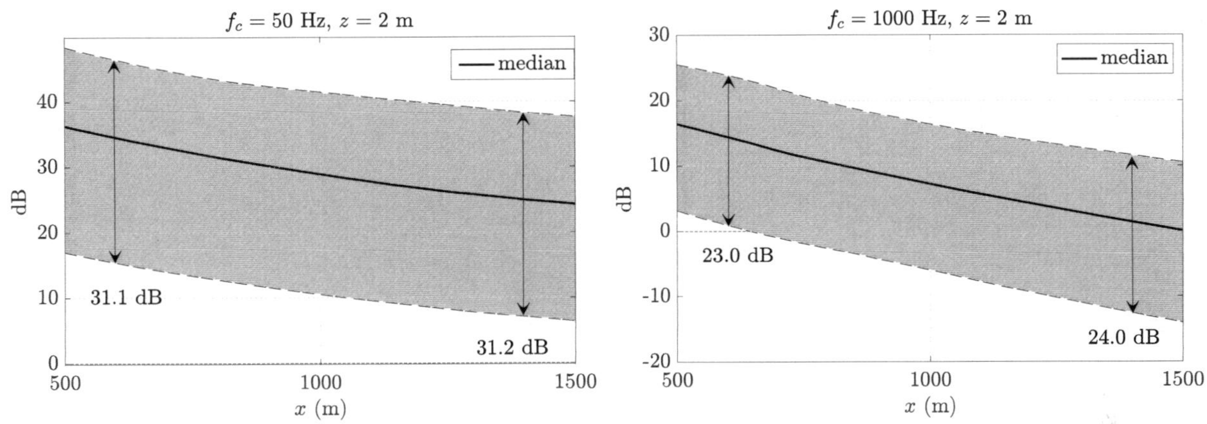

Fig.2 95% confidence intervals of SPL, for the third octave $f_c = 50$ Hz (left) and $f_c = 1000$ Hz (right). The median level is represented by the black line.

4.3 Global SPL dispersion for different wind conditions

Variabilities observed in the previous section is very large because of the hypothesis of no knowledge of environmental parameters values. However, wind properties (speed and direction), which explain the main part of the whole variability, are very often known thanks of measurements or because of some calculus hypothesis. This section investigates uncertainties on the estimation of SPL when only informations on wind is known, which is a representative situation of most measurements campaign of wind turbine noise.

In this application, wind speeds and wind directions are estimated at a reference height of 10 m. The SPL are classified according to wind speed classes of 1 m/s width and wind sectors of 30° width. The uncertainty propagation process is carried out by wind speed class and by wind direction sector, where the parameters a_{fr}, T_0, a_{T} and h_{v} are still considered as unknown and therefore vary in their initial range.

We consider 3 wind speed classes representative of several wind turbine operational modes: 6 m/s, 8 m/s and 12 m/s, accounting respectively to low, medium and maximum power. We consider 3 wind direction sectors: the wind sector $\theta \in [0; 30]°$ which corresponds to wind directions close to the source-receiver direction, the wind sector $\theta \in [30; 60]°$ which corresponds to an intermediate situation, and the wind sector $\theta \in [60; 90]°$ which corresponds to wind directions mostly transverse to the source-receiver direction.

The results are synthesized in figure 3, where the SPL are represented in *boxplot* format calculated in global level with A ponderation. The global levels are calculated from the third octave levels of the interval $f \in [50; 1000]$ Hz, and for 3 receiver locations: $(500{,}2), (1000{,}2), (1500{,}2)$ m.

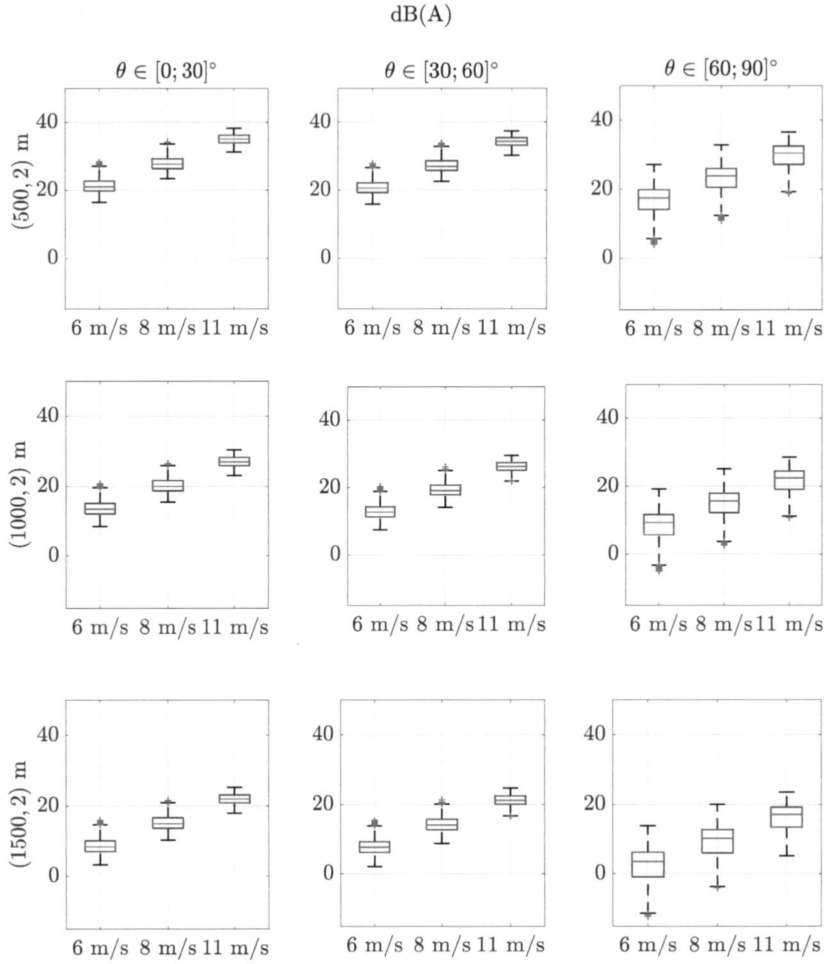

Fig.3 SPL Boxplots for each wind speed class ($6, 8$ and 11 m/s) and each wind sector ($\theta \in [0; 30], [30; 60], [60; 90]°$, for 3 receivers positions ($(500, 2), (1000, 2), (1500, 2)$ m). The median is represented by the red line. The bottom and top edges of the box indicate the 25th and 75th percentiles respectively. The whiskers extend to the most extreme data points not considered outliers. The outliers (greater than 99% of the data) are plotted individually using the red + symbol.

With no surprise, SPL decrease with distance x, and increase with wind speed (median sound levels increase by 10 to 15 dB(A) between wind speed class 6 m/s and wind speed class 11 m/s) for reasons related to both acoustic emission and acoustic propagation (refraction). The SPL dispersion increases slightly with distance x, especially between $x = 500$ m and $= 1000$ m: the interquartile range increases by about 0.3 dB(A) for $\theta \in [0; 30]°$, and increases by about 1 dB(A) for $\theta \in [60; 90]°$, whatever the wind speed class. This observation is essentially related to propagative effects which are cumulative with distance and will therefore influence the SPL variability at long range.

The SPL dispersion increases significantly for the wind sector $\theta \in [60; 90]°$: the interquartile ranges are of the order of 3 dB(A) for $\theta \in [0; 30]°$ and of the order of 7 dB(A) for $\theta \in [60; 90]°$. This can be explained by the directivity of wind turbines because the variations of the emission levels are more important for the cross wind sector $\theta \in [60; 90]°$ than for the wind sector $\theta \in [0; 30]°$ (Zhu et al., 2005). It should be noted that when $\theta \in [0; 30]°$, the refraction may tend to increase ground effects due to downwind conditions and thus increase SPL variability.

Nevertheless, it appears that the SPL dispersion is larger for $\theta \in [60; 90]°$ than for $\theta \in [0; 30]°$, which means that propagative effects seem to contribute relatively less to SPL dispersion than effects related to the directivity of the wind turbine.

Finally, the SPL dispersion is also influenced by the wind speed classes. Regardless of the wind sector or receiver position, the SPL dispersion is slightly lower for the 11 m/s wind speed class than for the other wind speed classes (6 m/s and 8 m/s). This is because the wind turbine operates at its rated power in the 11 m/s wind speed class, where the sound emission is relatively constant, thus reducing the SPL variability. This tendency is also related to propagative effects: the SPL dispersion is less important when the refraction is stronger. This observation is similar to previous work on long-range acoustic propagation for grazing acoustics sources (Zouboff et al., 1994). For instance, even if the effects of acoustic emission and acoustic propagation are not dissociated here, when observing the results at constant wind sector, one essentially distinguishes the influence of propagative effects on the SPL dispersion, and when observing the results at constant distance, one essentially distinguishes the influence of emission effects on the SPL dispersion.

5. Conclusion

The objective of this paper was to quantify the SPL uncertainties induced by the variability of environmental parameters, for the wind turbine noise context for downwind conditions. This required to accurately model the dominant aeroacoustic sources, as well as the acoustic propagation effects that fluctuate both with distance and time, and ultimately induces a significant SPL variability. In order to do so, the analysis carried out consisted in estimating the behavior of a wind turbine noise model output (*i.e.* sound pressure levels) when the uncertainties due to variability of the inputs (*i.e.* environmental parameters) are propagated in the model. The approach used in this work is stochastic and relies on Quasi-Monte Carlo input sampling technique in order to determine the output distribution. As this approach is very computationally intensive, a metamodel was built to be representative of the initial model in terms of approximation quality while presenting a negligible calculation cost.

Regarding the 95% SPL confidence interval by third octave band, a relatively constant SPL dispersion is observed with the distance to the source, higher in low frequencies: at 50 Hz the 95% confidence interval width is about 30 dB, than in high frequencies: at 1000 Hz the 95% confidence interval width is about 20 dB. The second kind of results is related to uncertainties on SPL estimation when only knowledge on wind properties (speed and direction) are available. The sound levels were expressed in global level (A-weighted). The results enable us to determine the environmental conditions and areas of the space where the issues are particularly high in terms of noise pollution and SPL uncertainty. As an example, this study showed that the SPL dispersion is greater for crosswind conditions (interquartile deviations of the order of 7 dBA) than for downwind conditions (interquartile deviations of the order of 3 dBA).

These results provide very useful information for the environmental acoustician, for example in the context of an impact study, because this research allows one to quantify the orders of magnitude of the SPL uncertainties associated with different wind conditions, and thus will help to a better control of the quality of wind turbine noise assessment

References

Amiet, R.K. (1975). Acoustic radiation from an airfoil in a turbulent stream. J. Sound Vib. *41*, 407–420.

Amiet, R.K. (1976). Noise due to turbulent flow past a trailing edge. J. Sound Vib. *47*, 387–393.

Attal, E. (2016). Caractérisation et optimisation d'assemblages d'éléments de murs végétalisés par méthodes acoustique et vibratoire. These de doctorat. Lille 1.

Attal, E., Côté, N., Shimizu, T., and Dubus, B. (2019). Sound Absorption by Green Walls at Normal Incidence: Physical Analysis and Optimization. Acta Acust. United Acust. *105*, 301–312.

Attenborough, K., Li, K.M., and Horoshenko, K. (2006). Predicting outdoor sound (CRC Press).

Barlas, E., Zhu, W.J., Shen, W.Z., Dag, K.O., and Moriarty, P. (2017). Consistent modelling of wind turbine noise propagation from source to receiver. J. Acoust. Soc. Am. *142*, 3297–3310.

Bertagnolio, F., Madsen, H.A., and Fischer, A. (2017). A combined aeroelastic-aeroacoustic model for wind turbine noise: verification and analysis of field measurements. Wind Energy *20*, 1331–1348.

Blaes, X., and Defourny, P. (2008). Characterizing Bidimensional Roughness of Agricultural Soil Surfaces for SAR Modeling. IEEE Trans. Geosci. Remote Sens. *46*, 4050–4061.

Borgeaud, M., and Bellini, A. (1998). A database for electromagnetic scattering studies of bare soil surfaces. In IGARSS '98. Sensing and Managing the Environment. 1998 IEEE International Geoscience and Remote Sensing. Symposium Proceedings. (Cat. No.98CH36174), pp. 1197–1199 vol.3.

Bourlier, C., Pinel, N., and Kubické, G. (2013). Method of Moments for 2D Scattering Problems: Basic Concepts and Applications (John Wiley & Sons).

Brelet, Y. (2008). Diffusion des ondes électromagnétiques par une surface rugueuse monodimensionnelle sous incidences modérées et rasantes : application au domaine maritime. thesis. Nantes.

Brelet, Y., and Bourlier, C. (2008). Bistatic Scattering from a Sea-Like One-Dimensional Rough Surface with the Perturbation Theory in HF-VHF Band. In IGARSS 2008 - 2008 IEEE International Geoscience and Remote Sensing Symposium, p. IV-1137-IV–1140.

Brutsaert, W. (1982). Evaporation into the Atmosphere: Theory, History and Applications (Springer Netherlands).

Buck, S., Oerlemans, S., and Palo, S. (2016). Experimental characterization of turbulent inflow noise on a full-scale wind turbine. J. Sound Vib. *385*, 219–238.

Carpinello, S., L'Hermite, P., Bérengier, M., and Licitra, G. (2004). A new method to measure the acoustic surface impedance outdoors. Radiat. Prot. Dosimetry *111*, 363–367.

Cheinet, S., Cosnefroy, M., Königstein, F., Rickert, W., Christoph, M., Collier, S.L., Dagallier, A., Ehrhardt, L., Ostashev, V.E., Stefanovic, A., et al. (2018). An experimental study of the atmospheric-driven variability of impulse sounds. J. Acoust. Soc. Am. *144*, 822–840.

Collins, M.D. (1989). Applications and time-domain solution of higher-order parabolic equations in underwater acoustics. J. Acoust. Soc. Am. *86*, 1097–1102.

Collins, M.D. (1993). A split-step Padé solution for the parabolic equation method. J. Acoust. Soc. Am. *93*, 1736–1742.

Cotté, B. (2017). Extended source models for long range wind turbine noise propagation. J. Acoust. Soc. Am. *141*, 3806–3806.

Cotté, B. (2019). Extended source models for wind turbine noise propagation. J. Acoust. Soc. Am. *145*, 1363–1371.

Davidson, M.W.J., Toan, T.L., Mattia, F., Satalino, G., Manninen, T., and Borgeaud, M. (2000). On the characterization of agricultural soil roughness for radar remote sensing studies. IEEE Trans. Geosci. Remote Sens. *38*, 630–640.

Ecotière, D., Glé, P., Gauvreau, B., Boittin, R., Lefèvre, H., and Lunain, D. (2015). Uncertainty of an in situ method for measuring ground acoustic impedance. In Proceedings of Internoise 2015, (San Francisco, California, USA: Internoise 2015), p.

Ecotière, D., Kayser, B., Gauvreau, B., Lebourdat, C., Bruneau, F., Guillaume, G., Lefèvre, H., Petit, A., Demizieux, P., Gary, V., et al. (2018). Emission et propagation du bruit des éoliennes : constitution d'une base de données expérimentale de référence. (Le Havre, France: 14ème Congrès Français d'Acoustique), p.

Embleton, T.F.W., Piercy, J.E., and Daigle, G.A. (1983). Effective flow resistivity of ground surfaces determined by acoustical measurements. J. Acoust. Soc. Am. *74*, 1239–1244.

Gauvreau, B. (2013). Long-term experimental database for environmental acoustics. Appl. Acoust. *74*, 958–967.
Gilbert, K.E., and White, M.J. (1998). Application of the parabolic equation to sound propagation in a refracting atmosphere. J. Acoust. Soc. Am. *85*, 630.
Heimann, D., and Salomons, E.M. (2004). Testing meteorological classifications for the prediction of long-term average sound levels. Appl. Acoust. *65*, 925–950.
Heimann, D., Englberger, A., and Schady, A. (2018). Sound propagation through the wake flow of a hilltop wind turbine—A numerical study. Wind Energy *21*, 650–662.
ISO9613-1:1993 (1993). Acoustics — Sound attenuation in free field — Part 1: atmospheric absorption calculation.
Johnson, M.E., Moore, L.M., and Ylvisaker, D. (1990). Minimax and maximin distance designs. J. Stat. Plan. Inference *26*, 131–148.
Jolliffe, I.T. (1986). Principal Components in Regression Analysis. In Principal Component Analysis, I.T. Jolliffe, ed. (New York, NY: Springer), pp. 129–155.
Kayser, B., Gauvreau, B., Ecotière, D., and Le Bourdat, C. (2018). A new experimental database for wind turbine noise propagation in an outdoor inhomogeneous medium. (Lyon, France: 17th International Symposium on Long Range Sound Propagation), p.
Kayser, B., Cotté, B., Ecotière, D., and Gauvreau, B. (2020). Environmental parameters sensitivity analysis for the modeling of wind turbine noise in downwind conditions. J. Acoust. Soc. Am. *148*, 3623–3632.
Koehler, J.R., and Owen, A. (1996). Computer experiments. In Design and Analysis of Experiments, (Elsevier), pp. 261–308.
Lee, S., Lee, D., and Honhoff, S. (2016). Prediction of far-field wind turbine noise propagation with parabolic equation. J. Acoust. Soc. Am. *140*, 767–778.
Leroy, O., Gauvreau, B., Junker, F., De Rocquigny, E., and Bérengier, M. (2010). Uncertainty assessment for outdoor sound propagation. (20th International Congress on Acoustics (ICA)), p.
Lesieur, A., Aumond, P., Mallet, V., and Can, A. (2020). Meta-modeling for urban noise mapping. J. Acoust. Soc. Am. *148*, 3671–3681.
Mallet, V., Tilloy, A., Poulet, D., Girard, S., and Brocheton, F. (2018). Meta-modeling of ADMS-Urban by dimension reduction and emulation. Atmos. Environ. *184*, 37–46.
Matheron, G. (1965). Les variables régionalisées et leur estimation: une application de la théorie des fonctions aléatoires aux sciences de la nature. (Masson et CIE).
McBride, S., and Burdisso, R. (2017). A comprehensive Hamiltonian ray tracing technique for wind turbine noise propagation under arbitrary weather conditions. In Seventh International Meeting on Wind Turbine Noise, (Seventh International Meeting on Wind Turbine Noise), pp. 1–12.
McKay, M.D., Beckman, R.J., and Conover, W.J. (1979). Comparison of Three Methods for Selecting Values of Input Variables in the Analysis of Output from a Computer Code. Technometrics *21*, 239–245.
Miki, Y. (1990). Acoustical properties of porous materials-Modifications of Delany-Bazley models-. J. Acoust. Soc. Jpn. E *11*, 19–24.
Nicolas, J., and Berry, J.L. (1984). Propagation du son et effet de sol. Rev. Acoust. 191–200.
Oerlemans, S., and Schepers, J.G. (2009). Prediction of Wind Turbine Noise and Validation against Experiment. Int. J. Aeroacoustics *8*, 555–584.
Ostashev, V.E., and Wilson, D.K. (2015). Acoustics in Moving Inhomogeneous Media (CRC Press).
Pettit, C.L., and Wilson, D.K. (2007). Proper orthogonal decomposition and cluster weighted modeling for sensitivity analysis of sound propagation in the atmospheric surface layer. J. Acoust. Soc. Am. *122*, 1374–1390.
R Core Team (2013). R: A Language and Environment for Statistical Computing (Vienna, Austria: R Foundation for Statistical Computing).
Roger, M., and Moreau, S. (2010). Extensions and limitations of analytical airfoil broadband noise models. Int. J. Aeroacoustics *9*, 273–305.
Roustant, O., Ginsbourger, D., and Deville, Y. (2012). DiceKriging, DiceOptim: Two R packages for the analysis of computer experiments by kriging-based metamodeling and optimization.
Salomons, E.M. (2001). Computational Atmospheric Acoustics (Netherlands: Kluwer Academic).

Saltelli, A., Ratto, M., Andres, T., Campolongo, F., Cariboni, J., Gatelli, D., Saisana, M., and Tarantola, S. (2008). Global Sensitivity Analysis: The Primer (John Wiley & Sons).
Sobol, I. (1967). On the distribution of points in a cube and the approximate evaluation of integrals. USSR Comput. Math. Math. Phys. *7*, 86–112.
Tian, Y., and Cotté, B. (2016). Wind turbine noise modeling based on Amiet's theory: Effects of wind shear and atmospheric turbulence. Acta Acust. United Acust. *102*, 626–639.
Wilson, D.K. (2003). The sound-speed gradient and refraction in the near-ground atmosphere. J. Acoust. Soc. Am. *113*, 750–757.
Wilson, D.K., Pettit, C.L., Ostashev, V.E., and Vecherin, S.N. (2014). Description and quantification of uncertainty in outdoor sound propagation calculations. J. Acoust. Soc. Am. *136*, 1013–1028.
Zhu, W.J., Heilskov, N., Shen, W.Z., and Sørensen, J.N. (2005). Modeling of Aerodynamically Generated Noise From Wind Turbines. J. Sol. Energy Eng. *127*, 517–528.
Zouboff, V., Brunet, Y., Bérengier, M., and Sechet, E. (1994). A qualitative approach of atmospheric effects on long range sound propagation. (Ottawa, Canada: 6th Inter. Symp. On Long Range Sound Prop.), p.

INCE EUROPE

9th International Conference

on

Wind Turbine Noise

Remote from Europe – 18th to 21st May 2021

If they are not being made ill by infrasound, then what is it?

Geoff Leventhall Acoustical Consultant geoff@activenoise.co.uk

Summary. Distress, assumed to follow exposure to wind turbine infrasound, has reduced, but not ceased and there remain groups of neighbouring residents who genuinely believe that their problems are due to wind turbine infrasound. Infrasound has been investigated both in the laboratory and in the field, but effects have not been confirmed, leading to the suggestion that alternative explanations for residents' problems should be explored. This paper offers one such explanation, based on what residents believe, how their beliefs may have developed and the influence of belief on health and wellbeing.

1.0 Introduction. Complaints of illness, said to be caused by wind turbines, continue from a small number of residents near windfarms, whose personal difficulties are real. Main causes of complaint are noise and visual disturbance, with emphasis often on infrasound. Some installations are visually oppressive for nearby residents or built to old noise criteria which are now recognised as too high. For example, 50dBA daytime.

Adverse publicity led residents to form protest groups to limit the progress of windfarms and rapid growth of these groups occurred through the developing internet. This growth was also assisted by local publicity and disturbing input from other districts which already had wind turbines. The outcome of these developments was to cause anxiety, and consequent stress, to susceptible individuals. A web page, "Quixote's last stand", which closed in January 2020, gave links to 2200 anti-wind groups internationally. Currently, coordinators of protest groups include the North American Platform Against Windpower (NAPAW), https://www.na-paw.org/ and the European Platform Against Windfarms (EPAW) https://www.epaw.org/, Also, Windwatch (www.wind-watch.org) is a centre for international information on wind turbine developments. The main purpose of the large number of web sites, which have been developed internationally, is to ensure that supposed problems of wind turbines, especially in relation to health, are kept high in public awareness. They have been successful at this, leading to high levels of anxiety in a small number of residents. These are the residents who are most likely to become complainants.

2.0 Early infrasound. In the early days of commercial wind turbine developments, around the beginning of the present century, residents who raised objections to turbines had little hard information on which to base their concerns about health and related effects and required a trusted leader to explain and interpret for them. At this point, Nina Pierpont, who led successful opposition to a wind farm in her own locality of Malone, New York State, came forward to fill the leader role. Pierpont promoted herself by taking space in her local newspaper, the Malone Telegram, in order to publish articles describing her beliefs and to stress her medical qualifications, For example an "advertisement" by her was titled "Health, hazard and quality of life near wind power installations. How close is too close" (Malone Telegram March 1st 2005). Another advertisement titled "Wind turbines and infrasound: What the latest research says" (Malone Telegram 26th March 2005), mainly consisted of quotations from a paper by van den Berg (van den Berg, 2004) in which Pierpont publicly confused the fluctuations of amplitude modulation with infrasound.

3.0 Development. Pierpont's views on harmful effects of infrasound, which became the basis for the Wind Turbine Syndrome . (Pierpont, 2009) spread rapidly through Pierpont's own web page (www.windturbinesyndrome.com) and other internet sources, helped by alarming press reports.(Deignan *et al.*, 2013) . Sarah Laurie (Australia) closely followed Pierpont's lead and developed the Waubra Foundation as an Australian anti-turbine organisation. https://waubrafoundation.org.au/

Both Pierpont and Laurie have medical qualifications and there remains the question of how medics, who are presumed to be knowledgeable on safe and unsafe doses of medication, could not grasp the concept of safe and unsafe doses of infrasound. The presence of infrasound was seen as a danger, even at large distances from a turbine and at low levels of infrasound.

Pierpont's views on infrasound became a main "illusory truth" [1] in residents' complaints and continue to be quoted (Leventhall, 2017). For example, Sherri Lange, who is CEO of NAPAW writes in a piece headed , "Infrasound: A growing liability for wind power" that:

> *"Wind turbines appear to be at the apex of producing human discomfort, annoyance, and harm. In particular, infrasound and low-frequency noise (ILFN) harm because of impacts unique to this concoction of noise."*
> *May 29th 2019* https://www.masterresource.org/wind-power-health-effects/infrasound-growing-liability-windpower/

Many residents near to proposed wind turbines have been influenced by the "infrasound is harmful" theme of objector publicity. A recent press report from Ireland, a country which has had a rapidly developing wind industry, with associated adverse publicity, includes:

---

[1] An Illusory Truth arises when repetition of a statement leads to belief in that statement, even though the statement may be false.

"We cannot stress enough how much we are opposed to this development which will have detrimental effects regarding the health and wellbeing of our family and our neighbours," Linda said. "We know from listening to horrifying stories of the profoundly negative impact that infrasound has, knowing that it has already cost families tens of thousands of euro to replace their windows in their home to try to reduce the impact of infrasound on their bodies and, more notably, when trying to sleep at night.

"Worse is the effects of infrasound on the human body which we have discovered from the renowned biomedical engineer, Dr Mariana Alves-Pereira, who has done extensive research and studies on this type of noise exposure.

"We urge everyone who reads this to look at her research. It is a very real, low frequency noise that can have debilitating health effects due to its exposure. You never know, someday a turbine farm could be built beside you" Linda said." February 11th 2021
https://avondhupress.ie/arrival-of-twins-fuels-parents-fight-against-wind-farm/

This resident has absorbed the illusory truth surrounding infrasound and is expressing considerable anxiety at the prospect of the windfarm, an anxiety which will only heighten over the next few years. She has been exposed to "horrifying stories" on the effects of infrasound and is transmitting her anxiety to others. She may be in the group which becomes ill due to the future wind farm.

The internet blog, Stop These Things, https://stopthesethings.com, has a daily anti-wind output and includes warnings of infrasound in its posts. For example:

Home Wreckers: Finnish Study Finds Wind Turbine Infrasound Unsafe For Residents Living Within 15 Km. February 1 2019

Sick Again: Wind Turbine Infrasound Delivers Daily Torment to Wind Farm Neighbours. August 2 2019

There are well-organised groups spreading misinformation and anxiety on health effects of infrasound and on wind turbines generally. The public develops its views under the influence of this misleading background, especially so in regions where new windfarms are proposed, but not yet built.

The present paper deals only with the infrasound from windfarms. Audible noise and other effects are not considered here.

4.0 Brief review of infrasound from wind turbines. It is helpful to review the basics of infrasound from wind turbines. Infrasound was associated with wind turbines through the early work of NASA and SERI. The NASA work is mainly linked to Hubbard, Shepherd and collaborators and is summarised in (Hubbard and Shepherd, 1991). The SERI/NREL (Solar Energy Research Institute/National Renewable Energy Laboratory) work is mainly linked to Kelley It is summarised in

https://www.academia.edu/6213488/Overview_of_NREL_Wind_Turbine_Acoustics_Research?email_work_card=title

Kelley and collaborators also produced detailed reports on the MOD-1 and MOD-2 turbines. (Kelley *et al.*, 1985; Kelley *et al.*, 1988)

Frequency analysis of the microphone signal from wind turbine noise showed spectra rich in harmonics with a fundamental at the blade passing frequency of 1~2Hz. NASA showed that infrasound, related to the interaction of the downwind blades and the tower, was radiated from all turbines studied. The infrasound spectrum had a large number of harmonics and originated in the pressure pulses radiated from blade-tower interaction.

The MOD-1 turbine, which Kelley investigated, was also downwind, mounted on a square base, four splayed legs, lattice tower - like a small Eiffel Tower. When the wind blew across a diagonal of the base, the tip of the blade came near to a splayed leg and there was maximum pulse amplitude.

A small number of residents were disturbed by these maximum pulses, which imparted transient energy to building components. The energy was then dissipated by resonant decay of the vibration of the building components, sometimes leading to low frequency noise, which might be audible. Thus, there could be transfer of acoustic energy from infrasonic to audible frequencies, with the potential for disturbance. To quote Kelley:

> *"Complaints of noise emanating from the operating MOD-I were confined to about a dozen families living within a 3-km radius of the turbine, about half of whom were annoyed frequently. These families represented a very small fraction of the total households within this radial distance, a number exceeding 1000 homes, including most of the town of Boone itself."*

Thus, about 1% of neighbours complained. Kelly concentrated attention on two homes, referred to as #7 and #8. House #7 was a double-width mobile home. House #8 was a 1-1/2 storey conventional timber building. Both these are lightweight structures, which respond to impulsive excitation.

Kelley next investigated the MOD-2 upwind turbine mounted on a cylindrical tower and concluded that the developments of

- moving the blades upwind of the tower, so that they passed the tower in a region of lower air disturbance and
- using a cylindrical tower to avoid the effects of splayed legs of a lattice tower

showed that

> *"annoyance to the community from the 1983 configuration of the MOD-2 turbine can be considered very unlikely at distances greater than 1 km (0.6 mile) from the rotor plane."*

Modern turbines incorporate the design changes of MOD-2 . Developments have further reduced the sound power output of turbines by improvements in design.

There have been a number of more recent measuremtns of infrasound from wind turbines, including (Zajamšek *et al.*, 2016) (Hansen *et al.*, 2013; Zajamsek *et al.*, 2014), (Hansen *et al.*, 2015),(Keith *et al.*, 2018).

5.0 Impulsive infrasound. Over recent years, those who claim adverse effects from infrasound have described wind turbine infrasound as "impulsive infrasound". An obvious interpretation is that "impulsive infrasound" describes pulses of infrasound. For example, if a continuous 10Hz infrasound is switched on for a short time and then off, repeating, it becomes impulsive infrasound.

The pulses from wind turbines are rapid variations in the atmospheric pressure. In the time domain they are often around 0.1 pascals peak pressure at residences, with a duration of about one-tenth of the repetition period. (Vanderkooy and Mann, 2015) It appears that the term "impulsive infrasound" is being applied incorrectly to these pulses since, if they are considered in the frequency domain, they are formed from summation of **continuous** infrasound components.[2]

The preferred description of the pulses is in terms of their time variation, and that they are air pressure pulses occurring at the blade repetition frequency, with typical peak about 0.1 pascal and duration about 0.1 seconds. In terms of frequency components, there is a harmonic spectrum based on the blade repetition frequency. "Pulsed" is time domain terminology whilst "infrasound" is frequency domain terminology. When combined, "pulsed infrasound" is pulses of infrasound due to infrasound being switched on and off, which is not what is generated by wind turbines.

6.0 Misinformation. Opponents of wind turbines have developed a successful misinformation campaign on infrasound. They have influenced not only the public, but administrators and politicians, particularly in Australia, (Australia, 2015)whilst some acousticians who have commenced a study of wind turbine infrasound, possibly did so influenced by a persuasive, but likely incorrect, background of supposed infrasound problems. A mythology has developed around wind turbine infrasound (Leventhall, 2005)

7.0 Citizen sponsored research and Government sponsored research. An example of differing conclusions reached by citizen research and government research is illustrated in Finland, where a citizen organisation, the Finnish Environmental Health Association, supports work on effects of changes to the environment. Recent emphasis has been on infrasound from wind turbines. Government sponsored research has also been completed recently.

8.0 Research from the Finnish Environmental Health Association. In this citizen sponsored research, which was based on a questionnaire study investigating the effects of infrasound on those living near turbines, plus infrasound measurement, included 46 families, comprising 193 people, in regions where wind turbines had been installed from 0.5 to 3 years before the survey. (Anon, 2019) The most common health change in the preceding year was sleep disturbance, but "very few" residents blamed this on windfarms. There were about three times more "severe symptoms" up to 15km from wind farms than further away and it is claimed that 15-20km is a typical distance at which "pulsating infrasound" from turbines can be measured. Some of the publications on the work, which are in Blogs on the internet, are translated into English, but these do not give clear decibel levels for the

---

[2] There are also aerodynamic sources of infrasound in a wind turbine, but these are not included in an "impulsive infrasound" description

infrasound, although the determining factor adopted by the investigators appears to be that turbine infrasound could still be measured.

The Finnish citizen work has a basic assumption that infrasound from wind turbines causes health problems and investigates how far away from turbines the infrasound can be detected by sensitive measuring instruments. It also assumes that any effects are due to the infrasound component of turbine noise, not from the audible noise.

9.0 Government sponsored and University Research in Finland Turunen and colleagues used an unmasked questionnaire on how residents felt about the effect of wind turbines on their health (Turunen *et al.*, 2021b). The title of the resulting paper was "Symptoms intuitively associated with wind turbine infrasound". Intuition implies the ability to reach a conclusion without conscious reasoning or acquiring additional knowledge, but there has been so much public exposure to claims on the effects of infrasound from wind turbines, that it is likely that residents have developed a wind turbine belief chain of "wind turbines >> infrasound >> health effects". Turunen et al conclusions included that:

> *It is not possible to explain why some people intuitively associated their symptoms with wind turbine infrasound. However, it can be speculated that interpretations of symptoms are affected by many other factors in addition to actual exposure.*

Consideration of widespread illusory truth on effects of infrasound from wind turbines, indicates that previous exposure to this may have been a main determinant of the residents' beliefs.

A second paper by Turunen et al (Turunen *et al.*, 2021a) investigated self-reported health effects of the total exposure i.e broadband sound and inaudible infrasound. Known effects of infrasound occur at high levels, such as in excess of 110dB for vestibular disfunction - levels which would be perceived by other body sensors. Turunen also raises the question of biological plausibility. That is, whether there is a causal relationship between the known physical exposures and resulting severe symptoms. Turunen et al conclude that because of the limited number of epidemiological studies and insufficient research on the subject

> "*It is not possible to conclusively rule out the possibility that long-term exposure to infrasound produced by wind turbines could have negative health effects. It should also be noted, that audible broadband noise from wind turbines can deteriorate health and wellbeing if the sound pressure level is high enough to cause annoyance and sleep disturbance.*"

An investigation by Stickler, leading to a Masters Thesis, (Stickler, 2020) compared the responses of two groups who lived near wind turbines - symptomatic and asymptomatic individuals – and their abilities to discriminate between wind turbine noise with and without its normal infrasound content. A total of 24 participants took part, comprising 11 who were symptomatic and 13 asymptomatic. Symptomatic participants found the experiment to be more stressful than did asymptomatic. Final conclusions included:

"The current study investigated responses to realistic wind turbine infrasound in individuals who feel that wind turbines cause them discomfort or to feel ill. These individuals were not found to be perceptually sensitive to WTIS compared with controls who did not attribute symptoms to wind turbines, and thus no evidence was found for a mechanism through which WTIS could induce symptoms. Physiological reactions to WTIS exposure remain a doubtful cause of symptoms attributed to WTIS considering the results of this study, previous literature on the perceptibility and health effects of wind turbine noise and the plausibility of mechanisms by which WTIS has been proposed to cause symptoms. Symptom attributions appear more likely to be explained by cognitive processes such as symptom misattribution or by disturbance caused by clearly perceptible consequences of wind turbine operation such as audible low-frequency noise".
[WTIS = wind turbine infrasound]

The Finnish Government supported a major investigation on infrasound from wind turbines (Maijala *et al.*, 2020). The aims included investigation of:

1. the full spectrum characteristics of wind turbine noise
2. symptoms intuitively associated with infrasound
3. human response to wind turbine sounds
4. actual sensitivity to infrasound of those who attribute health problems to it.

The main outcomes of the research included:

1. Installation of wind turbines in a rural environment increased the background infrasound to levels similar to those in an urban environment.
2. The peak values of the lowest infrasound frequencies came close to human detection threshold.
3. The symptoms intuitively associated with infrasound were relatively common, as was annoyance from audible wind turbine noise.
4. General health problems were more common amongst those who were already concerned about health effects of wind turbines. However, detection experiments did not show sensitivity to the infrasound in wind turbine noise.
5. Annoyance experiments did not show infrasound as a factor.
6. Physiological measurements showed no effect of wind turbine noise on heart rate, heart rate variability or skin conductance.

The conclusions of the project are summarised as

"Taken together, the behavioural findings of the current study suggest that wind turbine infrasound cannot be reliably perceived and it does not result in increased annoyance. Participants that showed health effects did not show signs of increased infrasound sensitivity and did not rate wind turbine sounds more annoying. These findings do not support the hypothesis that infrasound is the element of wind turbine sound that caused annoyance."
Maijala et al page 73.

And

"No support was found for the proposition that wind turbine infrasound could increase arousal and elicit physiological stress responses even in situations where the infrasound is not perceived" Maijala et al page 74

The work from Finland illustrates how scientific evidence is moving towards rejecting infrasound as a cause of health problems from wind turbines. This does not mean that turbines do not cause problems, as there is a complex entanglement of induced beliefs, anxiety, stress and potential illness, which causes harm to vulnerable residents

10.0 Vulnerable residents. The press report given earlier in this paper (Section 3) is one recent illustration from the examples of stressed responses of residents who anticipate installation of wind turbines in their locality. There are clear reasons why residents might object to wind turbines, but health effects and fears of infrasound are often a main feature of complaints.

In late 2015, the Australian Government appointed a National Wind Farm Commissioner to deal with problems from wind farms. The Commissioner's Annual Report for 2019 includes totals for all complaints received, as follows:

> *"From the Office's inception in November 2015 through to 31 December 2019, the Office has received a total of 361 complaints, comprising:*
> - *70 matters relating to 14 operating wind farms*
> - *234 matters relating to 58 proposed wind farms*
> - *six matters relating to five proposed solar farms, and*
> - *51 matters that did not specify a particular project or development.*
>
> *Of the total of 361 complaints received by the Office as at 31 December 2019, 349 of those complaints had been closed. The remaining 12 complaint matters are at various stages of the complaint handling process."*

https://www.nwfc.gov.au/sites/default/files/nwfc-annual-report-2019.pdf?v=1602650948

Complaints have largely been directed at **proposed** windfarms, whilst a reduction in year by year complaints has also been noted. The focus of complaints on *proposed* wind farms illustrates the anxieties which prospective residents feel. The anxieties reflect genuine beliefs and feelings, so we do not question these, but we can legitimately question the paths by which the beliefs arose.

11.0 Illusory Truth and Nocebo [3] The Illusory Truth Effect, also known as the Truth Effect, describes the influence of repetition on development of beliefs. (Dechene *et al.*, 2010; Henkel and Mattson, 2011; Polage, 2012; Fazio *et al.*, 2015; Brashier *et al.*, 2020) The strength of the Illusory Truth Effect depends on repetition of a false statement, gradually leading to its acceptance.

When there is a choice, we tend to believe what we prefer to believe, so stabilising our personal equilibrium, which is supported by our beliefs. We feel comfortable when our existing beliefs are confirmed and, if we feel antagonistic towards wind turbines, we are receptive to negative statements about them.

The basis of the Illusory Truth Effect is:

- Repetition of false statements leads to development of belief in them.

[3] Parts of this section are updated from the author's papers at WTN2017 and WTN2019

- Repetition of similar statements from different sources strengthens this belief.
- The development of a belief is made easier by each repetition.

Misinformation, which is a path to illusory truth, has recently become widespread on the internet.

The Nocebo Effect is related to the Illusory Truth Effect. The development of a nocebo response is the step which follows acceptance of an illusory truth. Nocebo was originally described in the medical context, where symptoms and treatments are influenced by expectations and conditioning. Recognition of the effect goes back to at least the early 1960s (Kennedy, 1961) . The outcome of communication to patients, either directly or by implication, illustrates the crucial importance of information transfer in creating expectations(Benedetti *et al.*, 2007) (Bensing and Verheul, 2010; van Laarhoven *et al.*, 2011; Reicherts *et al.*, 2016; Chavarria *et al.*, 2017).

These papers make clear that the nocebo responses are well established in general clinical work and are powerful in their operation, an operation which is largely based on expectations and conditioning. It is a short step to consider nocebo as an element in health related responses to wind turbines, within the widespread "background noise" of assertions that wind turbine infrasound is harmful to health.

The first direct application of the Nocebo Effect to wind turbines was by Chapman et al and has been supported by follow-up work (Chapman *et al.*, 2014) (Crichton *et al.*, 2014a; Crichton *et al.*, 2014b; Crichton and Petrie, 2015; Tonin *et al.*, 2016; Chapman and Crichton, 2017) However, the importance of expectations had been investigated earlier (Crichton *et al.*, 2013) with the following results:

> *During exposure to audible windfarm sound and infrasound, symptoms and mood were strongly influenced by the type of expectations. Negative expectation participants experienced a significant increase in symptoms and a significant deterioration in mood, while positive expectation participants reported a significant decrease in symptoms and a significant improvement in mood.*

An existing Nocebo Effect can be minimised by developing positive expectations through verbal suggestion e.g. talking therapies. In certain circumstances these newly developed positive expectations have been sufficiently strong to overturn the Nocebo Effect (Bartels *et al.*, 2017).The personality traits of the subject are an important factor (Kern *et al.*, 2020)

Of course, objectors to wind turbines are not comfortable with the Nocebo Effect as it points responsibility back to them. For example, Pierpont implies that Nocebo Effects are not valid when, referring to sufferers, she writes:

> *"They are not fabricating these symptoms. Their symptoms are not Simon Chapman's silly "nocebo effect." The symptoms are — real! Really and truly caused by IWT infrasound."* (Pierpont, 2017)

Statements like this indicate Pierpont's misunderstanding of the Nocebo Effect, implying that symptoms induced by the effect are, in some way, unreal. This is not what the literature says.

When we do not have arguments to counter a concept, we may resort to making fun of the idea or using false logic against it. This appears to be what Pierpont is doing here. It is remarkable that Pierpont, a medic, does not understand the Nocebo Effect.

12.0 Supersensitive persons? There are two groups of people who may be supersensitive to noise:

1. They have a lower than average hearing threshold. The measured hearing threshold has a standard deviation of 5~6dB amongst subjects over all frequencies, including infrasound. (Kurakata and Mizunami, 2008). Assuming a normal distribution, about 2% of people might have a threshold which is 10~12dB lower than the standard threshold and about 0.15% could have their threshold 15-18dB lower than the standard. Of course, we do not have sufficient information to be sure of how the hearing threshold varies in the extremes of the distribution, but there is the possibility that the introduction of a noise source close to a person with highly sensitive hearing, could cause problems to that person, whilst not affecting others.

2. They have normal hearing, but a low psychological tolerance to noise. This is partly a question of personality. Although personality may be fixed, reactions to a noise can be modified in order to help a sufferer desensitise from an annoying noise. (Leventhall *et al.*, 2012)

Both sensitive groups present problems for criteria, since criteria are usually set to protect 90-95%, not 100%, of a population from annoyance.

13.0. A way forward? This paper has shown how negative publicity on the effects of infrasound from wind turbines, spread by persons who are strongly motivated in their opposition to wind turbines, may have a distressing effect on psychologically and emotionally vulnerable residents in the vicinity of windfarms. Infrasound is used as a tool to unite opposition to windfarms through raising anxiety on health effects.

The title of this paper asks "If they are not being made ill by infrasound, then what is it?". Objector use of the Illusory Truth and Nocebo Effects induces stress and anxiety about infrasound, sufficient to cause distress in vulnerable people. Prolonged stress my lead to illness.(Benton and Leventhall, 1994) People are not known to be made ill by exposure to infrasound, but may succumb to fear and anxiety about the presence of infrasound, heightened by objector's use of illusory truth. Section 3.0 contains an example of this.

It has been shown that the Nocebo Effect can be reversed (Bartels *et al.*, 2017), whilst talking therapies, such as Cognitive Behavioural Therapy (CBT), have been successful in helping sufferers to cope with their noise problems. (Leventhall *et al.*, 2012). These may be a way forward.

14.0. References

Anon, 2019. The pilot study shows that infrared nuisance will only be significantly reduced more than 15 kilometers from wind farms. Finnish Environmental Health SYTe ry, https://syte.fi/2019/2001/2010/pilottitutkimus-osoittaa-infraaaanihaitan-vahenevan-merkittavasti-vasta-yli-2015-kilometrin-paassa-tuulivoimaloista/.

Australia, S., 2015. Report Select Committee on Wind Farms https://www.aph.gov.au/parliamentary_business/committees/senate/wind_turbines/wind_turbines/final_report.

Bartels, D., van Learhoven, A., Stroo, M., Hijne, K., et al, 2017. Minimizing nocebo effects by conditioning with verbal suggestion: A randomized clinical trial in healthy humans. PLOS|ONE, https://doi.org/10.1371/journal.pone.0182959.

Benedetti, F., Lanotte, M., Lopiano, L., Colloca, L., 2007. When words are painful: unravelling the mechanism of the Nocebo effect. Neuroscience 147, 260-271.

Bensing, J.M., Verheul, W., 2010. The silent healer: The role of communication in placebo effects. Patient Education and Counseling 80, 293-299.

Benton, S., Leventhall, H.G., 1994. The role of "background stressors" in the formation of annoyance and stress responses. Jnl Low Freq Noise Vibn 13, 95-102.

Brashier, N.M., Eliseev, E.D., Marsh, E.J., 2020. An Initial Accuracy Focus Prevents Illusory Truth Cognition 194.

Chapman, S., Crichton, F., 2017. Wind turbine syndrome A communicated disease. Sydney University Press.

Chapman, S., Joshi, K., Fry, L., 2014. Fomenting sickness: nocebo priming of residents about expected wind turbine health harms. Frontiers in Public Health, doi: 10.3389/fpubh.2014.00279.

Chavarria, V., Vian, J., Pereira, C., Data-Franco, J., Fernandes, B.S., 2017. The Placebo and Nocebo Phenomena: Their Clinical Management and Impact on Treatment Outcomes. Clinical Therapeutic 39, 2017.

Crichton, F., Chapman, S., Cundy, T., Petrie, K.J., 2014a. The link between health complaints and wind turbines: support for the nocebo expectations hypothesis. Frontiers in Public Health 2, doi: 10.3389/fpubh.2014.00220.

Crichton, F., Dodd, G., G, S., Gamble, G., Petrie, K.J., 2014b. Can Expectations Produce Symptoms From Infrasound Associated With Wind Turbines? Health Psychology 33, 360-364.

Crichton, F., Dodd, G., Schmid, G., Cundy, T., Petrie, K.J., 2013. The Power of Positive and Negative Expectations to Influence Reported Symptoms and Mood During Exposure to Wind Farm Sound. Health Psychology . doi: 10.1037/hea0000037.

Crichton, F., Petrie, K.J., 2015. Health complaints and wind turbines: The efficacy of explaining the nocebo response to reduce symptom reporting. Environmental Research 140, 449-455.

Dechene, A., Stahl, C., Hansen, J., Wanke, M., 2010. The Truth About the Truth: A Meta-Analytic Review of the Truth Effect. Pers Soc Psychol Rev 14, 238-258.

Deignan, B., Harvey, E., Hoffman-Goetza, L., 2013. Fright factors about wind turbines and health in Ontario newspapers before and after the Green Energy Act. Health, Risk and Society, 234-250

Fazio, L.K., Brashier, N.M., Payne, B.K., Marsh, E.J., 2015. Knowledge Does Not Protect Against Illusory Truth. Journal of Experimental Psychology: General 144, 993-1002.

Hansen, K., Walker, B., Zajamsek, B., Hansen, C., 2015. Perception and annoyance of low frequency noise versus infrasound in the context of wind turbine noise Proceedings Wind Turbine Noise 2015, Glasgow UK.
Hansen, K., Zajamsek, B., Hansen, C., 2013. Analysis of Unweighted Low Frequency Noise and frasound Measured at a Residence in the Vicinity of a Wind Farm Proceedings of Acoustics 2013 – Victor Harbor 17-20 November 2013, Australian Acoustical Society, .
Henkel, L.A., Mattson, M.E., 2011. Reading is believing: The truth effect and source credibility. Consciousness and Cognition 20, 1705-1721.
Hubbard, H.H., Shepherd, K.P., 1991. Aeroacoustics of large wind turbines. J. Acoust. Soc. Am. 89, 2495-2508.
Keith, S.E., Daigle, G., Stinson, M., 2018. Wind turbine low frequency and infrasound propagation and sound pressure level calculations at dwellings. Jnl Ac Soc America 144, 981-996.
Kelley, N.D., McKenna, H.E., Hemphill, R.R., Etter, C.I., Garrelts, R.I., 1985. Acoustic Noise Associated with the MOD 1 Wind Turbine: Its Source, Impact, and Control. Report No. SERI/TR-635-1166 Solar Energy Research Institute.
Kelley, N.D., McKenna, H.E., Jacobs, E.W., Hemphill, R.R., BirkenHeuer, N.J., 1988. The MOD-2 Wind Turbine - Aeroacoustical Noise Sources, Emissions, and Potential Impact Solar Energy Research Institute. Task No. WE721201, .
Kennedy, R.S., 1961. The nocebo reaction. . Med. World (Lond.) 95, 203-205.
Kern, K., Kramm, C., Witt, C.M., Barth, J., 2020. The influence of personality traits on the placebo/nocebo response. A systematic review. Journal of Psychosomatic Reseach.
Kurakata, K., Mizunami, T., 2008. The statistical distribution of normal hearing thresholds for low frequency tones. Jnl Low Freq Noise Vibn and Active Control 27, 97 - 104.
Leventhall, G., 2017. Why do some people believe that they are "made ill" by wind turbine noise. Proc 7th International Confernce Wind Turbine Noise.
Leventhall , G., Robertson, D., Benton, S., Leventhall, L., 2012. Helping sufferers to cope with noise using distance learning cognitive behaviour therapy. J. Low frequency Noise, Vibration and Active Control 31, 193-204.
Leventhall, H.G., 2005. How the "mythology" of infrasound and low frequency noise related to wind turbines might have developed. Proc Wind Turbine Noise 2005 INCE/Europe, Berlin September 2005.
Maijala, P., Turunen, A., Kurki, I., Pakarinen, S., Kaukinen, C., al, e., 2020. Infrasound Does Not Explain Symptoms Related to Wind Turbines. Government Publication Finland.
Pierpont, N., 2009. Wind Turbine Syndrome. K-Selected Books.
Pierpont, N., 2017. ISO acknowledges motion sickness from low-frequency oscillatory motion below 1 Hz. North American Platform Against Wind Power, http://www.na-paw.org/Pierpont-ISO-9996.php.
Polage, D.C., 2012. Making up History: False Memories of Fake News Stories. Europe's Journal of Psychology 8, 245–250.
Reicherts, P., Gerdes, A., Pauli, P., Wieser, M., 2016. Psychological Placebo and Nocebo Effects on Pain Rely on Expectation and Previous Experience. Journal of Pain 17 (2), 203-214.
Stickler, E., 2020. Perceptual Sensitivity to Infrasound is Not Associated With Wind Turbine Related Symptoms MSc Thesis University of Helsinki.

Tonin, R., Brett, J., Colagioiri, B., 2016. The effect of infrasound and negative expectations to adverse pathological symptoms from wind farms. Jnl Low Freq Noise Vibn and Active Control 35, 77-90.
Turunen, A., Tiittanen, P., Yli-Tuomi, T., Taimisto, P., Lanki, T., 2021a. Self-reported health in the vicinity of five wind power production areas in Finland Environment International 106419 151, hrtps://doi.otg/10.11 OI 16/j.envint.2021.106419.
Turunen, A., Tiittanen, P., Yli-Tuomi, T., Taimisto, P., Lanki, T., 2021b. Symptoms intuitively associated with wind turbine infrasound. Environmental Research, hrtps://doi.otg/10.11 OI 16/j.envint.2021.106419.
van den Berg, G.P., 2004. Effects of wind profile at night on wind turbine sound. Jnl Sound Vibration 277, 955 - 970.
van Laarhoven, A.I.M., Voglaar, M., Wilder-Smith, O.H., van Riel, P.L.C.M., 2011. Induction of nocebo and placebo effects on itch and pain by verbal suggestions. PAIN 152, 1486-1494.
Vanderkooy, J., Mann, R., 2015. Measuring Wind Turbine Coherent Infrasound Proceedings Wind Turbine Noise 2015, Glasgow.
Zajamsek, B., Hansen, K., Hansen, C., 2014. Investigation of the time dependent nature of infrasound measured near a wind farm. Proceedings Internoise 2014.
Zajamšek, B., Hansen, K.L., Doolan, C.J., Hansen, C.H., 2016. Characterisation of wind farm infrasound and low-frequency noise. Jnl Sound Vibration 370, 176–190.

INCEEUROPE

9th International Conference on
Wind Turbine Noise
Remote from Europe – 18th to 21st May 2021

A meta-analysis on the impact of wind turbine noise on sleep using validated objective sleep assessments

Tessa Liebich[1,2], Leon Lack[1], Kristy Hansen[3], Branko Zajamsek[1], Nicole Lovato[1], Peter Catcheside[1], Gorica Micic[1]

[1]Adelaide Institute for Sleep Health: A Flinders Centre of Research Excellence, College of Medicine and Public Health, Flinders University, Bedford Bark, SA, Australia
[2]College of Education, Psychology and Social Work, Flinders University, Adelaide, SA, Australia
[3]College of Science and Engineering, Flinders University, Adelaide, SA, Australia

Corresponding author's email address: tessa.liebich@flinders.edu.au

Summary

Very little is known surrounding the impact of wind turbine noise on sleep. Previous research is limited to cross-sectional study designs reporting on self-reported and anecdotal impacts on sleep rather than commonly used and validated objective and subjective sleep assessments. More recently, experimental studies using the gold standard measure of sleep, polysomnography as well as actigraphy and validated sleep questionnaires have been conducted. This paper provides a quantitative summary of data on the meta-analytical findings of wind turbine noise effects on sleep using validated objective sleep assessments. Search terms involved “wind farm noise”, “wind turbine noise”, “wind turbine sound” “wind turbine noise exposure” and “sleep”. Eligible studies for inclusion needed to be published in English after the year 2000 and had to use either polysomnography or actigraphy measures to assess sleep in the presence of wind turbine noise and uniform outcomes were meta-analysed. Five studies were eligible for a meta-analysis. Meta-analyses (Hedges g; 95% confidence interval [CI]) revealed no significant differences in objective sleep onset latency (0.03, 95% CI -0.34 to 0.41), total sleep time (-0.05, 95% CI -0.77 to 0.67), sleep efficiency (-0.25, 95% CI -0.71 to 0.22) or wake after sleep onset (1.25, 95% CI -2.00 to 4.50) in the presence versus absence of wind turbine noise (all $p > .05$). This suggests that wind turbine noise does not significantly impact some of the key indicators of objective sleep parameters. Given the variable measurement methodologies, limited sample sizes and mixed wind turbine noise interventions, cautious interpretation remains warranted. Well controlled experimental studies with reasonable sample sizes and ecological wind turbine noise exposures are needed to give more conclusive evidence surrounding the impact of wind turbine noise on objective sleep measurement.

This conference proceeding is a small extract of the full systematic review and meta-analysis investigating the effect of wind turbine noise on sleep using validated objective and subjective sleep assessments. Please see the following reference for the fully published paper:

Liebich, T., Lack, L., Hansen, K., Zajamšek, B., Lovato, N., Catcheside, P., & Micic, G. (2020). A systematic review and meta-analysis of wind turbine noise effects on sleep using validated objective and subjective sleep assessments. *Journal of Sleep Research*, e13228.

Introduction

It is no surprise that good sleep is vital for health and quality of life and that conversely, insufficient sleep can result in daytime functioning impairments (Janssen, Basner, Griefahn, Miedema, & Kim, 2011; Micic et al., 2018). Whilst there are several environmental advantages associated with wind farms, some residents living near them have reported experiencing sleep disruption (Basner et al., 2014; Crichton et al., 2014; Janssen et al., 2011; Krogh, Gillis, Kouwen, & Aramini, 2011; Muzet, 2007; World Health Organization, 2011). Pre-existing psychosocial stress and exposure to potentially annoying noise can lead to maladaptive coping strategies and further conditioned responses such as increased alertness, potentially resulting in insomnia. Wind turbine noise can propagate longer distances and penetrate through buildings more readily compared to other common environmental noise sources such as traffic noise and could thus be a potential psychosocial stressor for residents who live near wind turbines.

Previous studies have primarily used cross-sectional study designs (i.e., observational studies at a specific point in time). There is only one other meta-analysis that has investigated the impact of wind turbine noise on sleep but at the time this meta-analysis was published, no studies had included objective sleep measurements such as the gold standard polysomnography or actigraphy (Onakpoya, O'Sullivan, Thompson, & Heneghan, 2015). Since then, objective studies have emerged, thus calling for an updated analysis. This paper is a succinct summary of the already published systematic review and meta-analysis investigating the impact of wind turbine noise on sleep using validated objective and subjective sleep assessments. This conference paper is reporting only on the meta-analysis, please see the above reference for the fully published paper.

1. Methods

1.1 Design

The meta-analysis was written in accordance with the Preferred Reporting Items for Systematic Reviews and Meta-analyses (PRISMA) Statement (Moher, Liberati, Tetzlaff, & Altman, 2009).

1.2 Data sources and search strategies

Electronic searches were performed from January until April 2020 and conducted in PubMed, Scopus, Science Direct, CINAHL, PsycARTICLES, Web of Science and MEDLINE. Title and abstract search terms included “wind farm noise”, “wind turbine noise”, “wind turbine sound”, “wind turbine noise exposure” and “sleep”.

1.3 Study selection criteria

Following duplicate removal, the primary author (TL) screened all retrieved abstracts according to the selection criteria seen in Table 1. Given previous reviews and meta-analyses revealing a lack of publications prior to 2000, the search strategy was limited to publications >2000 that used objective measures of sleep (polysomnography or actigraphy) to assess the effect of wind turbine noise on sleep. The fully published paper also includes psychometrically validated sleep questionnaires, but this information is beyond the scope of this conference paper.

Table 1. Study selection criteria.

Article Criteria
1. Original, full text, peer-reviewed article 2. Contains terms "wind farm noise, wind turbine noise, wind turbine noise exposure, or wind turbine sound" and "sleep" in the title/abstract 3. Written in English 4. Published after 2000
Sample Characteristics Criteria
• Adults, >17 years of age • Reportedly living/working within 15km from a wind farm or exposed to WFN as part of the study procedure
Primary Outcome Criteria
Evaluated the impact of WFN on any of the following objective and/or psychometrically validated subjective sleep parameters: • Sleep onset latency (SOL), total sleep time (TST), wake after sleep onset (WASO), sleep efficiency. • Global scores of PSQI, ISI and/or ESS.
Meta-analysis Criteria
Examined the presence versus absence of WFN on any of the aforementioned objective and/or psychometrically validated subjective sleep parameters (i.e., included a control group/condition and WFN exposure condition).

Note. SOL = Sleep Onset Latency; TST = Total Sleep Time; WASO= Wake After Sleep Onset; ESS = Epworth Sleepiness Scale; ISI = Insomnia Severity Index; PSQI = Pittsburgh Sleep Quality Index.

1.4 Objective sleep measurement

1.4.1 Polysomnography

Polysomnography uses direct electroencephalography (EEG) measurements and a widely accepted scoring criteria to describe sleep-wake timing, sleep stages, sleep onset latency (the time taken to fall asleep), wake after sleep onset (the total time spent awake across the sleep period), total sleep time (the total amount of sleep time), sleep efficiency (the total time spent asleep as a percentage of time available for sleep between lights-off and wake time), and brief arousals from sleep (Martin & Hakim, 2011). Studies were deemed eligible if the protocol involved an experimental night (a night with wind turbine noise exposure) and a control night (a quiet, wind turbine noise free night) and also reported the aforementioned sleep parameters.

1.4.2 Actigraphy

Actigraphy algorithmically infers sleep and wake periods from gross body movements and the measurement device is often worn on the wrist (Smith et al., 2018). This form of objective sleep measurement provides estimates of the timing and duration of sleep and awakenings from which sleep onset latency, total sleep time, wake after sleep onset and sleep efficiency can be inferred. Eligible actigraphy studies needed to report these sleep parameters and also needed to have a wind turbine noise exposure condition and a control condition to be considered in the meta-analysis.

1.5 Statistical analyses

1.5.1 Data extraction

Relevant data fields for extraction were identified by the primary author (TL) and are shown in Table 2. The statistics reported in the retrieved articles included mean (SD) or mean (SE) from which pooled variance were determined where possible.

1.5.2 Meta-analyses, heterogeneity and risk of biases

All analyses were conducted using Meta-Essentials: Workbooks for meta-analysis, version 1.5 (Hak, van Rhee, & Suurmond, 2016) for estimation of pooled mean effects and 95% CI using random effects models. Given the small sample size of eligible studies, Hedges' g was deemed the appropriate effect size to analyse group differences (Borenstein, Hedges, Higgins, & Rothstein, 2011). Prediction intervals (PI) were also reported which describe the range in which 95% of future studies are predicted to fall. Meta-analyses were conducted on all eligible studies that used polysomnography or actigraphy sleep assessments and reported on uniform objective sleep parameters. The *Q*-statistic (Cochrane's Q) was also reported to indicate the average variability of the effect size of each sleep parameter. Corresponding with the *Q*-statistic, the I^2 statistic was also reported which describes the proportion of variance of real differences in effect sizes (Hak et al., 2016). Lastly, publication bias was assessed and displayed via funnel plots of the effect size in comparison to the standard error of each sleep parameter.

2. Results

2.1 Selection of studies

Figure 1 demonstrates the PRISMA flow diagram, outlining the study selection process at each stage of screening. The search strategy revealed 451 records and after consulting with other authors and screening pertinent reference lists, another 7 records were included. 324 records remained after duplicates were removed and 49 remained after abstract screening. Full-text screening excluded 41 studies, leaving eight studies that met the inclusion criteria. One of these studies described two separate pilot studies that were thus treated as two separate records, making nine eligible studies for qualitative synthesis after abstract screening. For quantitative synthesis, four studies were excluded, leaving five studies eligible for meta-analysis.

2.2 Study demographics

Table 2 describes the sample and testing characteristics, outcomes, and main findings of the meta-analysed studies. Three of the five studies were conducted experimentally in a sleep laboratory in Sweden, with the remaining two studies conducted in the field in Canada. The mean (SD, range) age of the five meta-analysed studies was 40.8 (16.3, 22.2-55.9) years. Studies included not only those who lived near wind turbines but also unexposed individuals. Three studies used synthesised WTN recordings (Ageborg Morsing et al., 2018a, 2018b; Smith et al., 2020) and two studies used 8-10 hour records of actual WTN measured inside participant's homes (Jalali et al., 2016a; Lane, Bigelow, Majowicz, & McColl, 2016). All five studies reported sleep onset latency, total sleep time, wake after sleep onset and sleep efficiency, four of which included polysomnography measurements and one that used actigraphy measurements.

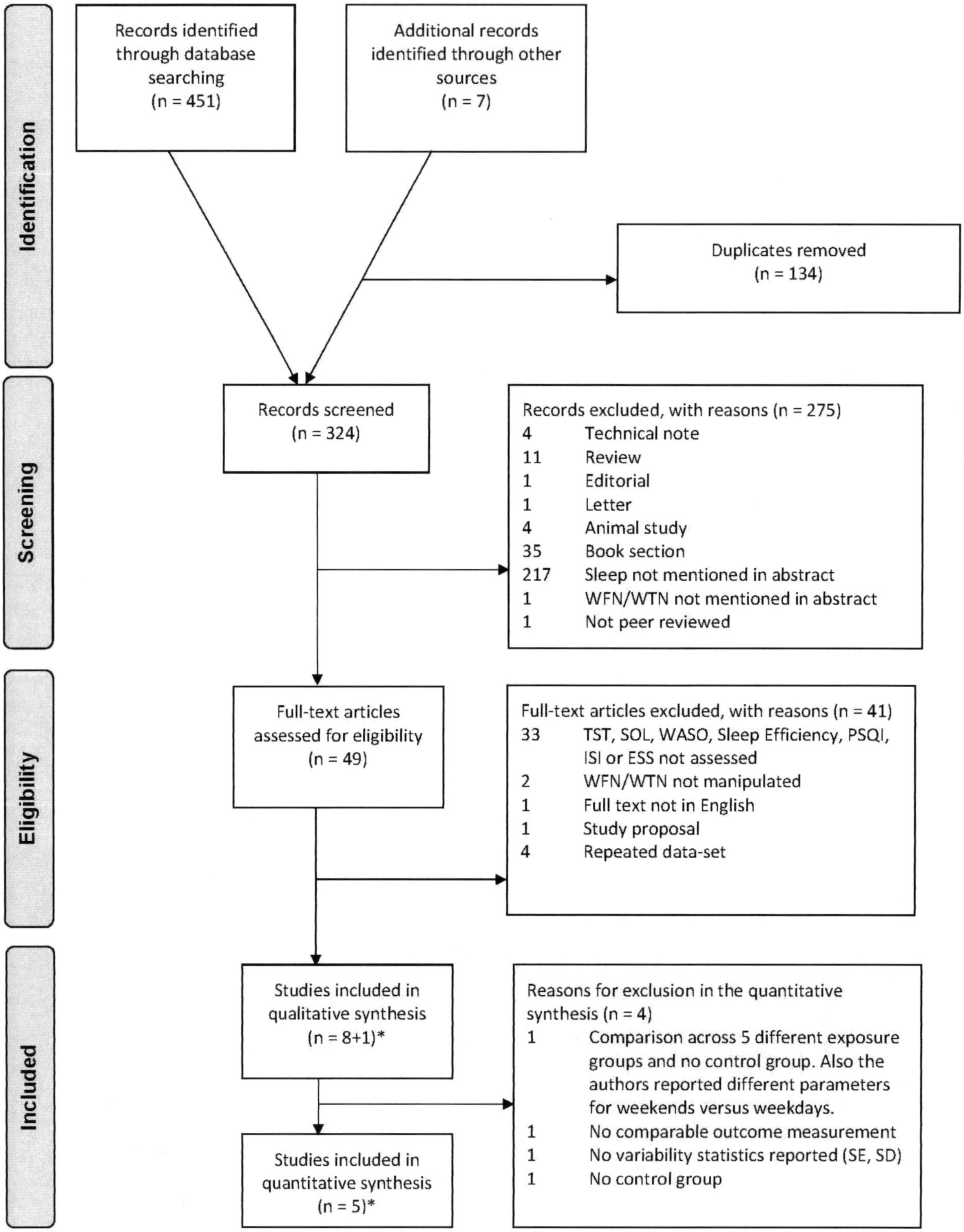

Figure 1. PRISMA flow diagram showing the process for inclusion. *Note.* *1 additional study was included as one record conducted and analysed two separate studies.

Table 2. Study characteristics, outcomes and findings of the meta-analysed studies.

Study	Location, environment	Sample size	Mean age, years	Design	SPL of WTN exposure	Method of SPL measurement	Distance measurements	Outcomes	Tools	Study findings
Ageborg Morsing et al. (2018a)	Sweden, laboratory	6	22.2	5-night experimental laboratory study	29.5 dBL_{Aeq}, 34.1 dBL_{Aeq}, 33.7 dBL_{Aeq} indoor WTN (varying frequencies and AM characteristics)	Three 8-hour night-time synthesised WTN exposures with varying filtering, frequency bands and AM beats	N/A	Sleep onset latency, sleep efficiency, total sleep time. *	PSG	No significant effects of sleep onset latency, sleep efficiency or total sleep time between control night and any of the WTN exposure nights.
Ageborg Morsing et al. (2018b)	Sweden, laboratory	6	24	5-night experimental laboratory study	32.8 dBL_{Aeq} 32.8 dBL_{Aeq}, 30.4 dBL_{Aeq}, indoor WTN (varying frequencies and AM characteristics)	Three 8-hour night-time synthesised WTN exposures with varying filtering, frequency bands and AM beats	N/A	Sleep onset latency, sleep efficiency, total sleep time. *	PSG	No significant effects of sleep onset latency, sleep efficiency or total sleep time between control night and any of the WTN exposure nights.
Jalali et al. (2016a)	Canada, open flat agricultural fields	16	55.9	Pre-post field study	Time 1: 36.55 dB(A) Time 2: 36.50 dB(A)	10-hour noise measurements at two participant's residences for 16 nights before and after wind turbine operation.	10 individuals <1,000m from a turbine and 6 individuals >1,000m from a turbine.	Sleep onset latency, sleep efficiency, total sleep time, wake after sleep onset.	PSG	No significant difference between sleep onset latency, sleep efficiency, total sleep time or wake after sleep onset at Time 1 compared to Time 2.
Lane et al. (2016)	Canada, rural matched areas	32	Exposed group: 60.4; Unexposed group: 41.4 (adjusted mean age: 50.9)	Cross-sectional field study	N/A	8-hour equivalent A-weighted sound level.	Exposed group: mean (SD) distance of 794.6m (264.1) from a turbine. Unexposed group: mean (SD) distance of 2,931m (1,015.6) from a turbine.	Sleep onset latency, sleep efficiency, total sleep time, wake after sleep onset.	Actigraphy	No significant differences in sleep onset latency, sleep efficiency, total sleep time or wake after sleep onset for WTN-exposed individuals compared to unexposed individuals.
Smith et al. (2020)	Sweden, laboratory	50	51.2	Experimental laboratory study	32 dBL_{Aeq} indoor WTN including AM	Continuous synthesised WTN based on short- and long-term recordings including AM.	Exposed group = resided <1km from a turbine of reported sleep disturbance or annoyance from wind turbines in the past month, unexposed group.	Sleep onset latency, sleep efficiency, total sleep time, wake after sleep onset.	PSG	No significant differences in sleep onset latency, sleep efficiency, total sleep time or wake after sleep onset between the control night and the WTN exposure night.

AM, amplitude modulation; dBL_{Aeq}, equivalent continuous sound pressure level; dB(A), A-weighted decibel; PSG, polysomnography, SD; standard deviation; SPL, sound pressure level; WTN, wind turbine noise. *denotes no WASO data were analysed in the study. TL contacted the authors to obtain mean (SD) values to include in the meta-analysis.

2.3 Meta-analyses of objective sleep parameters

Table 3 shows the range in effect size between studies, the combined effect size of all studies reporting on each objective sleep parameter and the 95% confidence intervals, the significance of the effect, the 95% prediction interval and the heterogeneity between studies. As shown in Table 3, when all included studies were combined, no statistically significant effects of wind turbine noise exposure on sleep onset latency, total sleep time, sleep efficiency or wake after sleep onset were found when compared to no wind turbine noise exposure. The heterogeneity between studies was low and not statistically significant for sleep onset latency, total sleep time and sleep efficiency but was high for wake after sleep onset. This suggests that the effect of wake after sleep onset is not generalizable across the studies. Given the small number of included studies, a meaningful subgroup analysis was not possible. However, once the actigraphy study was removed from the analysis, the heterogeneity decreased from 89.77% (p <.001) to 12.82% (p = .328) and was no longer significant. This finding was not surprising given this was the only study that used actigraphy measurement in comparison to the four polysomnography studies.

Table 3. Hedges' *g* [95% CI] and associated meta-analytic statistics.

Objective Sleep Parameter	Range of Hedges' *g* between studies	Combined Hedges' *g*[95% CI]*	*p*-value	95% Prediction Interval (PI)**	Heterogeneity (Q, I^2, *p*-value)
SOL (mins)	0.44-0.62	0.03[-0.34 to 0.41]	p =0.403	-0.34 to 0.41	Q = 3.29, I^2 = 0%, p =0.510
TST (mins)	-0.02-7.19	-0.05[-0.77 to 0.67]	p =0.849	-1.30 to 1.20	Q = 7.81, I^2 = 48.8%, p =0.099
Sleep efficiency (%)	-1.21-0.43	-0.25 [-0.71 to 0.22]	p =0.139	-0.82 to 0.32	Q = 4.4, I^2 = 9.16%, p =0.354
WASO (mins)	-0.02-7.19	1.86[-7.54 to 11.27]	p =0.394	-9.77 to 13.50	Q =36.62, I^2 = 94.54%, p = <.001

Note. *95% CI = 95% confidence interval; **PI = 95% prediction interval; 95% of future studies effects are predicted to fall within this range.

3. Discussion

A meta-analysis was conducted that investigated the impact of wind turbine noise exposure on sleep using validated objective sleep assessments. Only one systematic review and meta-analysis exists to date that has investigated the impact of wind turbine noise on sleep but was limited to self-report and cross-sectional study outcomes only (Onakpoya et al., 2015). More recently, experimental studies that have used objective measures of sleep have been conducted, calling for an updated meta-analysis. Five studies uniformly reported key objective sleep outcomes including sleep onset latency, total sleep time, wake after sleep onset and sleep efficiency. The meta-analysis showed no evidence to support that sleep onset latency, sleep efficiency, total time spent asleep and awake during the night are significantly impacted by the presence of wind turbine noise exposure in comparison to no wind turbine noise exposure.

However, Jalali et al. (2016a) reported no significant difference in sleep parameters between pre- versus post-operational wind farm stages, but also reported no significant differences in A-weighted wind turbine noise exposure between the pre- and post-stage and thus it is perhaps not surprising that sleep outcomes were not affected. Furthermore, whilst field studies are the most representative of real-world wind turbine noise conditions, they lack control over other external variables including wind speed, wind direction, study blinding, and visual impacts that also have the potential to impact sleep disruption. For example, wind speed and wind direction can influence the airflow, turbulence, and propagation of wind turbine noise, leading to variability in amplitude modulation, infrasound, tonality, and swish components of wind turbine noise and could contribute to reported sleep disruption. Therefore, inconsistent study designs have the potential to influence different findings across studies (Micic et al., 2018).

Whilst there are various types of sleep measurement available that are widely used (actigraphy, sleep diaries, devices), polysomnography is considered the gold standard measure of sleep as it allows for the direct measurement of brain activity. In comparison, actigraphy relies on sleep-wake inferences via gross body movements and is based on pre-defined activity thresholds. As a result, whilst actigraphy is an objective measure of sleep, it has demonstrated poor specificity for discriminating wake from sleep when activity is low and is thus not the gold standard of sleep measurement. Contrastingly, whilst polysomnography is technical, intrusive, and expensive, studies using polysomnography allow for the superior control of most external variables that could confound sleep outcomes (Aziz, 2017). Polysomnography also allows for the measurement of microstructural changes and sleep stage shifts that extend beyond basic sleep macrostructure parameters. The studies by Ageborg Morsing et al. (2018a, 2018b) have shown that by using a repeated measures design and PSG under controlled experimental conditions, there are some significant impacts of wind turbine noise exposure on the timing of the first awakening, frequency of awakenings per hour, reductions in deep sleep and less continuous time spent in stage 2 sleep). Whilst these studies did not show significant impacts on the more standard sleep metrics (sleep onset latency, total sleep time, wake after sleep onset, sleep efficiency), they do suggest that there may be some more detailed changes in cortical activity, sleep stage and physiological changes. Further investigation of the effect of wind turbine noise exposure on these finer grained features is warranted in the future, using larger sample sizes.

This was the first meta-analysis that has investigated the impact of wind turbine noise on objective measures of sleep. Limitations involved retrieving a small number of studies with varied methodologies and outcome variables which prevented a more comprehensive meta-analysis. Studies that were retrieved also used mixed methods and variable measurement and model-based estimates of wind turbine noise exposure levels (worst-case wind turbine noise including amplitude modulation, infrasound, and tonality through to more typical wind turbine noise exposure levels). As a result, one of the main findings of this study is that carefully controlled experimental studies are needed to provide a more definitive answer regarding the impact of wind turbine noise on sleep.

4. Conclusions

In summary, five studies were included in the meta-analysis, which showed no significant impacts of wind turbine noise exposure versus non-exposure on some key objective sleep parameters, including sleep onset latency, total sleep time, wake after sleep onset and sleep efficiency. Due to the mixed methodologies, study designs and overall limited studies in this area, future experimental studies are needed.

Acknowledgements

Special thanks to Hannah Scott and Claire Dunbar for their ongoing support.

References

Ageborg Morsing, J., Smith, M., Ögren, M., Thorsson, P., Pedersen, E., Forssén, J., & Persson Waye, K. (2018a). Wind turbine noise and sleep: Pilot studies on the influence of noise characteristics. *International Journal of Environmental Research and Public Health, 15*(2573), 1-14. doi:10.3390/ijerph15112573

Ageborg Morsing, J., Smith, M., Ögren, M., Thorsson, P., Pedersen, E., Forssén, J., & Persson Waye, K. (2018b). Wind Turbine Noise and Sleep: Pilot Studies on the Influence of Noise Characteristics. *International Journal of Environmental Research and Public Health, 15*(2573), 1-14. doi:10.3390/ijerph15112573

Aziz, H. A. (2017). Comparison between field research and controlled laboratory research. *Archives of Clinical and Biomedical Research, 1*, 101-104. doi:https://doi.org/10.26502/acbr.50170011

Basner, M., Babisch, W., Davis, A., Brink, M., Clark, C., Janssen, S., & Stansfeld, S. (2014). Auditory and non-auditory effects of noise on health. *The Lancet, 383*(9925), 1325-1332. doi:10.1016/s0140-6736(13)61613-x

Borenstein, M., Hedges, L. V., Higgins, J. P., & Rothstein, H. R. (2011). *Introduction to meta-analysis*: John Wiley & Sons Ltd.

Crichton, F., Dodd, G., Schmid, G., Gamble, G., Cundy, T., & Petrie, K. J. (2014). The power of positive and negative expectations to influence reported symptoms and mood during exposure to wind farm sound. *Health Psychol, 33*(12), 1588-1592. doi:10.1037/hea0000037

Hak, T., van Rhee, H., & Suurmond, R. (2016). How to interpret results of meta-analysis. *Available at SSRN 3241367*.

Jalali, L., Bigelow, P., Nezhad-Ahmadi, M. R., Gohari, M., Williams, D., & McColl, S. (2016a). Before-after field study of effects of wind turbine noise on polysomnographic sleep parameters. *Noise Health, 18*(83), 194-205. doi:10.4103/1463-1741.189242

Janssen, S., Basner, M., Griefahn, B., Miedema, H., & Kim, R. (2011). Environmental Noise and Sleep Disturbance. In F. Theakston (Ed.), *Burden of disease from environmental noise* (pp. 55-70). Europe: World Health Organization.

Krogh, C. M. E., Gillis, L., Kouwen, N., & Aramini, J. (2011). WindVOiCe, a Self-Reporting Survey: Adverse Health Effects, Industrial Wind Turbines, and the Need for Vigilance Monitoring. *Bulletin of Science, Technology & Society, 31*(4), 334-345. doi:10.1177/0270467611412551

Lane, J. D., Bigelow, P. L., Majowicz, S. E., & McColl, R. S. (2016). INTERNATIONAL PERSPECTIVES: Impacts of industrial wind turbine noise on sleep quality: Results from a field study of rural residents in Ontario, Canada. *Journal of Environmental Health, 79*, 8-13.

Martin, J. L., & Hakim, A. D. (2011). Wrist actigraphy. *Chest, 139*(6), 1514-1527. doi:10.1378/chest.10-1872

Micic, G., Zajamsek, B., Lack, L., Hansen, K., Doolan, C., Hansen, C., . . . Catcheside, P. (2018). A Review of the Potential Impacts of Wind Farm Noise on Sleep. *Acoustics Australia, 46*(1), 87-97. doi:10.1007/s40857-017-0120-9

Moher, D., Liberati, A., Tetzlaff, J., & Altman, D. G. (2009). Preferred reporting items for systematic reviews and meta-analyses: the PRISMA statement. *PLoS Med, 6*(7), e1000097. doi:10.1371/journal.pmed.1000097

Muzet, A. (2007). Environmental noise, sleep and health. *Sleep Med Rev, 11*(2), 135-142. doi:10.1016/j.smrv.2006.09.001

Onakpoya, I. J., O'Sullivan, J., Thompson, M. J., & Heneghan, C. J. (2015). The effect of wind turbine noise on sleep and quality of life: A systematic review and meta-analysis of observational studies. *Environ Int, 82*, 1-9. doi:10.1016/j.envint.2015.04.014

Smith, M. G., Ogren, M., Thorsson, P., Hussain-Alkhateeb, L., Pedersen, E., Forssen, J., . . . Persson Waye, K. (2020). A laboratory study on the effects of wind turbine noise on sleep: results of the polysomnographic WiTNES study. *Sleep, 43*(9). doi:10.1093/sleep/zsaa046

Smith, M. T., McCrae, C. S., Cheung, J., Martin, J. L., Harrod, C. G., Heald, J. L., & Carden, K. A. (2018). Use of Actigraphy for the Evaluation of Sleep Disorders and Circadian Rhythm Sleep-Wake Disorders: An American Academy of Sleep Medicine Systematic

Review, Meta-Analysis, and GRADE Assessment. *J Clin Sleep Med, 14*(7), 1209-1230. doi:10.5664/jcsm.7228
World Health Organization. (2011). *Burden of disease from environmental noise: Quantification of healthy life years lost in Europe*. Regional Office for Europe: World Health Organization.

INCEEUROPE

9th International Conference
on
Wind Turbine Noise
Remote from Europe – 18th to 21st May 2021

Subjective responses to wind turbine noise amplitude modulation: pooled analysis of laboratory listening studies and synthesis of an AM character rating penalty

Michael J B Lotinga, WSP: michael.lotinga@wsp.com
Toby Lewis, WSP: toby.lewis@wsp.com

Summary

Laboratory listening studies investigating human exposure-response relationships with amplitude-modulated wind turbine sound are re-examined using a common measure for modulation. Cross-study similarities and differences are observed and discussed within the context of establishing an evidence-based AM rating method for application in noise assessments.

1. Introduction

The periodic amplitude modulation (AM) of wind turbine noise (WTN) has long been identified as a factor in wind farm noise disturbance [1, 2, 3]. A previous investigation led by WSP included a UK government-funded systematic review of relevant research on the response to AM in WTN, which concluded that AM contributes to expected annoyance responses to WTN [4, 5]. One output from the WSP study was a proposed AM character 'rating penalty adjustment' intended to represent the expected subjective equivalence of WTN featuring perceptible AM, compared to responses to WTN with imperceptible AM. The magnitude of the penalty was based on an AM rating scale first proposed by RenewableUK [6] (which also reviewed results from the same core research base of listening tests), the distinction in [4] being the AM metric used to quantify the depth of modulation, which is described below. One of the aims of the present study is to reconsider the findings of the previous work, in view of the results from more recent research. The principal contributions of this study include: a) reanalysis of the exposure stimuli used in the studies to produce results scaled to a common AM metric (described below), in order to enable pooling of comparable results, and consideration of the acoustic factors affecting the response to AM; and b) the development of a practical, evidence-based model for the equivalent (acute, or short-term) annoyance of perceptible AM in WTN, which could be useful in quantifying the impact of AM for the purpose of wind farm planning controls and compliance determination.
This study is based on three fundamental assumptions: i) that it is valid to characterise the subjective effect of sound characteristics (such as AM) using a dB-based rating penalty approach (which is supported by existing standardised methods, eg, ISO 1996-2 [7] and BS

4142 [8]); ii) that it is justifiable to combine the data from separate listening studies with the aim of developing a unified approach to predict responses to WTN (ie, that any variation between the pooled studies is not too dissimilar to the natural variation that occurs between different sites and respondents); and iii) that it is reasonable to characterise expected responses using the mean values from the study participant samples.

2. Amplitude modulation quantification

There is no standardised method for quantifying AM in WTN, and a variety of approaches are currently in use internationally (eg [9, 10, 11, 12]). As shown below, these differences in the approach used lead to varying results. For example, some procedures are based on the variation in the total A-weighted sound level, while others use filtering to focus the analysis on spectral regions expected to correspond with WTN AM characteristics. While some studies have indicated that subjective responses to AM may be relatively insensitive to the precise value of the modulation depth [12, 13], in order to directly compare studies it is important to minimise the uncertainty associated with the use of different quantification methods.

In this study, the common metric adopted is based on the UK Institute of Acoustics (IOA) 'AMWG Reference Method' [14]. This method was developed following a detailed review of alternative approaches to measuring and evaluating AM, which identified the benefits and shortcomings of other known methods [15]. It has been selected as it is designed for to reliably detect and quantify modulation depth (MD) and modulation frequency (MF) in the presence of extraneous noise, and to enable efficient automated analysis of large datasets without requiring continuous recording of audio data. These features make it a strong candidate for use in field applications, and consequently within regulatory controls for AM WTN impact.

2.1 IOA AM metric implementation

The open-source Python kernel for the IOA method [16] has been used in this study, with modifications to accommodate AM signals with higher modulation frequencies than the original method allows. The published method is based on the calculation of a 'modulation spectrum' from an A-weighted time-averaged L_{eq} evaluated over 100 ms intervals. The modulation spectrum is used to identify the fundamental MF and its corresponding 2nd and 3rd harmonics, which are isolated and inverse-Fourier transformed to generate a time-domain representation of the underlying modulation within the signal; this 'reconstructed' signal is then analysed to quantify the MD. To avoid signal aliasing, the original method is limited by the 100 ms averaging to a MF of ≤1.6 Hz. To go beyond this limitation requires the input signal to be time-averaged over shorter intervals. In this study of synthesised WTN-like sounds, the highest MF analysed is 4 Hz, for which a →40 ms interval would theoretically be appropriate to avoid aliasing. However, it is well known from signal processing theory that to closely reconstruct a time-domain signal, the discrete-time form must in general be oversampled [17]. Therefore, since the IOA metric relies on the time-domain representation, this means that with increasing MF (even while staying within the 1.6 Hz Nyquist limit), the method as originally defined could return smaller MDs than would be the case if the signal was suitably oversampled to obtain a more accurate discrete time approximation. During the pooled stimuli analysis detailed below, this issue was investigated using L_{eq} averaging intervals between 100 ms and 20 ms. The expected effect was observed to a lesser or greater extent depending on the particular stimuli set; in some cases the differences between 100 ms averaging and shorter intervals were negligible. However, as exemplified in Figure 1, in cases with higher MFs and MDs, observed differences in MD could be more than 0.5 dB. On the basis of this analysis, an L_{eq} averaging interval of 25 ms was selected – this both avoids aliasing for the signals analysed, and yields good convergence in MD. For field work involving MFs up to 2 Hz, an averaging time of 50 ms could be a more practical approach that also addresses the potential underestimation of MD. The IOA metric has been employed in other recent theoretical and field-based AM WTN studies [11, 18, 19]. In [11], modifications to the method have also been applied, intended to quantify modulated tonal components, which are not the focus of our study. However, Hansen et al [11] also propose a reduced AM prominence criterion for field data analysis using the IOA method, to optimise the balance between 'false positives' and 'false negatives'. In our study the original IOA-recommended prominence criterion is retained, as false detection rates are not a concern with the synthesised stimuli investigated here.

In this study, the descriptor $\Delta L'_{Aeq,25msBP(max)}$ is used to indicate the IOA MD metric; here, the prime indicates that the level difference ΔL is calculated from a reconstructed signal (and not directly from the absolute L_{Aeq}), while the subscript 'BP(max)' indicates that the reconstruction is derived from the bandpass filtered $L_{Aeq,25ms}$, and the bandpass frequency range yielding the maximum MD value is selected from three different partially overlapping spectral regions – see [14] for details.

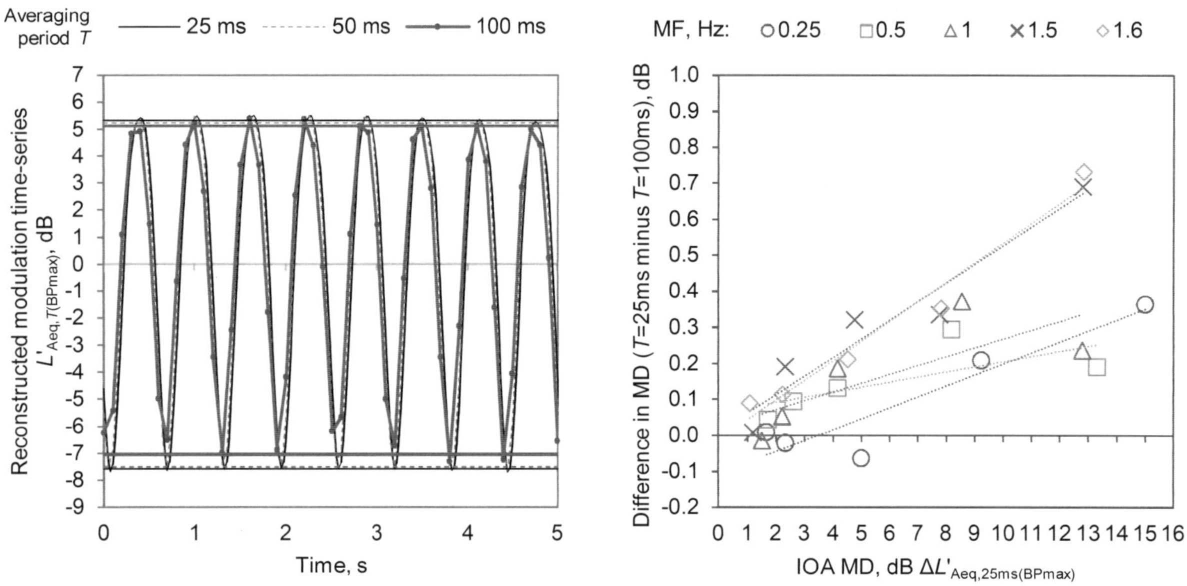

Figure 1: Example of comparative analysis of IOA MD with varying L_{eq} averaging time T for synthesised WTN-like signal; (a, left) modulation time series for MF=1.6 Hz, MD≃13 dB, horizontal lines indicate corresponding 10-second L_5 and L_{95} values used to evaluate IOA MD; (b, right) comparison of MD results obtained using T=25 ms vs T=100 ms for a range of MFs and MDs (linear regression trend lines shown for visualisation)

3. Study selection and features

Table 1: Summary of Studies Selected for Pooled Analysis

Key publication	Study group (ref)	Participant sample				Participant response measurements
		Size	Age			
		N	$\bar{x}$	s	[a, b]	
von Hünerbein et al (2013) [20]	Salford (SL)	20	~32	~8.7	~20-54	i) magnitude scaling estimation: annoyance ii) paired comparison adjustment: equivalent annoyance
Yokoyama et al (2015) [21]	Tokyo (TY)	17	N/A	N/A	21-26	i) onset of fluctuation sensation ii) paired comparison adjustment: equivalent perception of 'noisiness'
Hafke-Dys et al (2016) [22]	Poznan (PZ)	19	N/A	N/A	19-24	magnitude scaling estimation: annoyance
Virjonen et al (2019) [23]	Turku (TK)	35	27	5	20-39	magnitude scaling estimation: annoyance
a) Schäffer et al (2016) [24] b) Schäffer et al (2018) [25]	EMPA (EM)	a) 60 b) 52	37 42	11.7 11.9	18-60 18-62	magnitude scaling estimation: annoyance

The studies addressed in this paper are listed in Table 1, for which the relevant data (stimuli audio files with corresponding aggregated response data) are available. All the studies took place under controlled laboratory conditions and identified AM WTN exposure-response as a subject of the investigation.

The original stimuli files generated for the Turku study [23] could not be accessed (due to lack of archiving), so have been carefully resynthesised using the same approach and computational tools as the original study, undertaken in consultation with the study authors to minimise any associated uncertainty.

3.1 Exposure Stimuli

The study groups took different approaches to the synthesis of the AM WTN, with varying levels of complexity. In particular, for the Salford and EMPA studies, the researchers went to considerable lengths in attempting to simulate features of wind turbine sound (details in [26] and [27] respectively). Less-complex approaches were taken in the Tokyo, Poznan and Turku studies, which were based on modulating random noise signals and spectral shaping to broadly represent a simplified frequency spectrum of far-field AM WTN. Listening tests of AM WTN synthesis quality have indicated that the complexity of the approach may not necessarily equal increased authenticity, although for perceptual congruity it was found to be important to ensure similarity of the synthesised spectrum with real WTN [28].

Three aspects of frequency spectrum can be discerned in AM WTN:

i) the overall spectral content in the audio range (which in general may contain both modulated and steady components);
ii) the 'modulation spectrum', here defined as the Fourier transform of a sound pressure level time-series, which exhibits spectral peaks at the MF and its harmonics (ie in the infrasound region); and
iii) the spectral modulation, ie the regions of the audio spectrum in which the modulated content of the signal resides – the IOA metric assumes that this is concentrated in one of three defined ranges: 50-200 Hz, 100-400 Hz or 200-800 Hz. Frequency spectra analysed for steady stimuli ('imperceptible AM') for each study are shown in Figure 2, and discussed further below.

The stimuli for each study were analysed using a Python script (based on the SciPy library), to calibrate the audio, apply third-octave band filters (compliant with IEC 61260-1:2014 [29]), and AM processing. Where necessary, stimuli files were truncated to remove fade-in and fade-out portions, which otherwise would have affected the signal parameters and AM calculations. Example spectrograms for each study are shown in Figure 3. The files were also auditioned aurally – these steps enabled objective and subjective evaluation of specific signal features that are discussed further below.

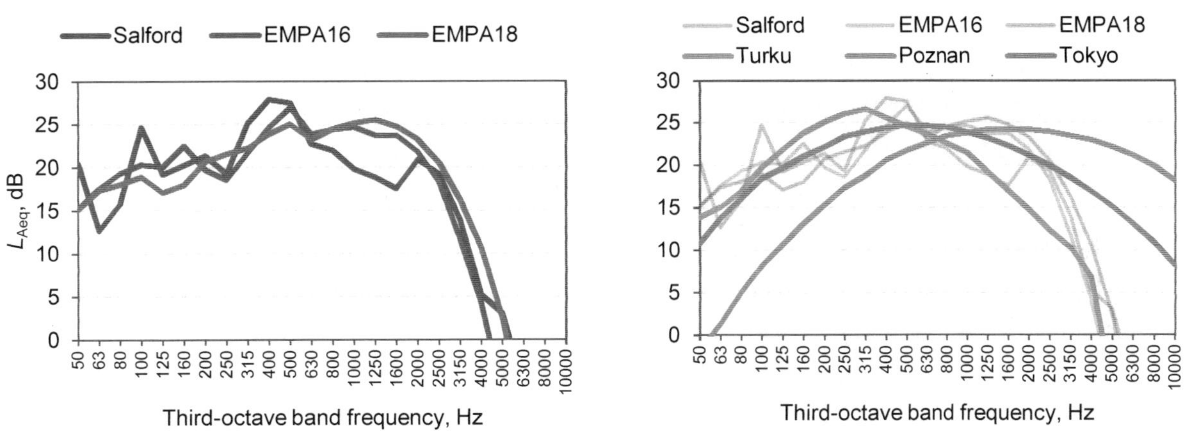

Figure 2: Comparison of A-weighted spectra: (a, left) Salford, EMPA 2016 and 2018 studies; (b, right) Turku, Poznan and Tokyo (with other studies indicated for comparison)

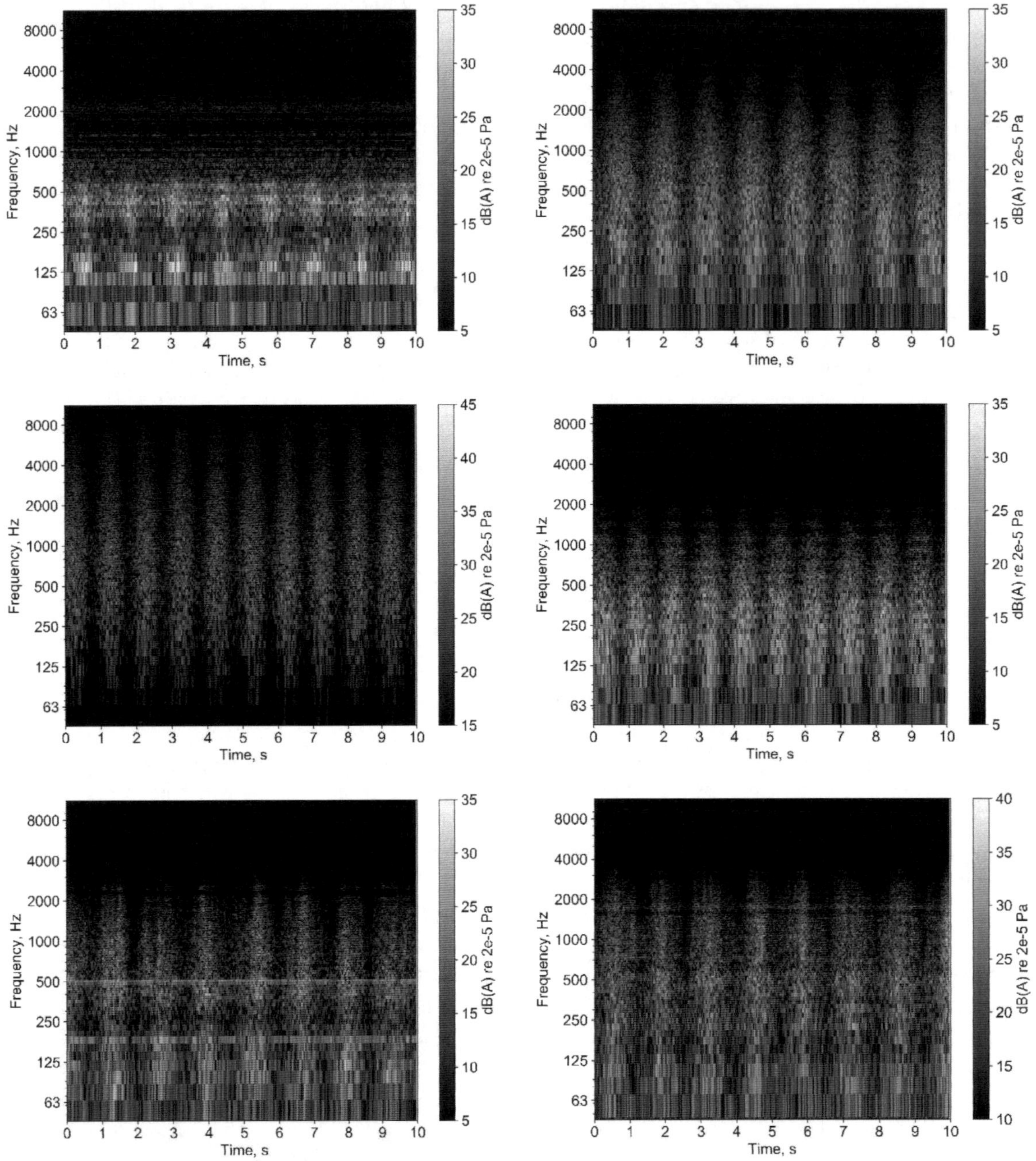

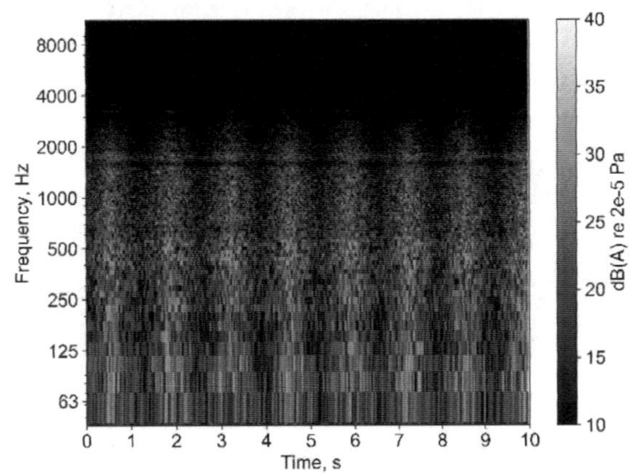

Figure 3: Spectrograms of example exposure stimuli from: (a, top left) Salford [35 dB L_{Aeq}, 8.0 dB $\Delta L'_{Aeq,25ms(50\text{-}200Hz)}$]; (b, top right) Tokyo [35 dB L_{Aeq}, 8.1 dB $\Delta L'_{Aeq,25ms(50\text{-}200Hz)}$]; (c, row 2 left) Poznan [45 dB L_{Aeq}, 8.6 dB $\Delta L'_{Aeq,25ms(200\text{-}800Hz)}$]; (d, row 2 right) Turku [35 dB L_{Aeq}, 8.6 dB $\Delta L'_{Aeq,25ms(50\text{-}200Hz)}$]; (e, row 3 left) EMPA16 [35 dB L_{Aeq}, 8.2 dB $\Delta L'_{Aeq,25ms(50\text{-}200Hz)}$]; (f, row 3 right) EMPA18 [40 dB L_{Aeq}, 9.2 dB $\Delta L'_{Aeq,25ms(50\text{-}200Hz)}$, with turbulence]; (g, row 4 left) EMPA18 [40 dB L_{Aeq}, 9.1 dB $\Delta L'_{Aeq,25ms(50\text{-}200Hz)}$, without turbulence]

3.2 Response Measurement

Most of the studies asked participants to rate 'annoyance' on an 11 point numerical (0-10) scale, typically in accordance with ISO 15666 (the 'ICBEN scale') , using a magnitude scaling estimation approach. It has been previously noted [4] that the kind of annoyance rated in a laboratory listening test is unlikely to closely represent that experienced in real exposure settings, which leads [24] to a distinction between acute (or short term) annoyance measured in listening tests (addressed here), and chronic (or long term) annoyance, usually measured in epidemiological field surveys.

The Turku study [23] used the absolute annoyance ratings to indirectly derive a relative measure of 'equivalent annoyance' (EqA) on a decibel scale – the method employed to do this is described in more detail below. As a response outcome, EqA is of primary interest to our study, as it enables the practical approach of 'rating' (or 'penalising') the AM characteristic in WTN, making it compatible with regulatory limits based on aggregated level.

The Salford study also investigated EqA (in addition to absolute annoyance), but directly, by using a paired comparison method, in which participants adjusted the level of a steady stimulus to match their judged annoyance for an AM stimulus.

Similarly, the Tokyo study directly examined subjective equivalence using a paired comparison method, by asking subjects to judge the 'perceived noisiness' of an AM stimulus and adjust its level against an unmodulated equivalent.

3.3 Salford Study

The Salford study investigated a wide range of possible features within AM WTN with potential links to annoyance. These included spectral content, spectral modulation, AM envelope shape, modulation frequency, masking sound, overall average level (L_{Aeq}), and MD. The range of variable parameters was narrowed via sensitivity studies to final testing focussed on the effect of changes in L_{Aeq} and MD using an AM WTN spectrum weighted towards lower frequencies – Figure 3a indicates that the corresponding stimuli have the strongest low-frequency skew of the studies examined.

The study concluded that L_{Aeq} had a significant association with increased annoyance, while MD (ie the 'design MD' defined as the level difference between averaged extrema in the $L_{Aeq,100ms}$) showed an observable trend that was not found to be significant [20].

The results of reanalysing the stimuli in terms of the IOA metric are illustrated in Figure 4a, in which a systematic increase in the MD value (compared with the 'design MD', defined using the broadband $L_{Aeq,100ms}$) of 2-3 dB is observed for the 50-200 Hz bandpass range. Reference to Figure 3a indicates that this is because the (A-weighted) sound energy with greatest apparent modulation in the signal resides below 200 Hz – this is effectively 'diluted' when the modulation is analysed over the wider frequency range used in the original study.

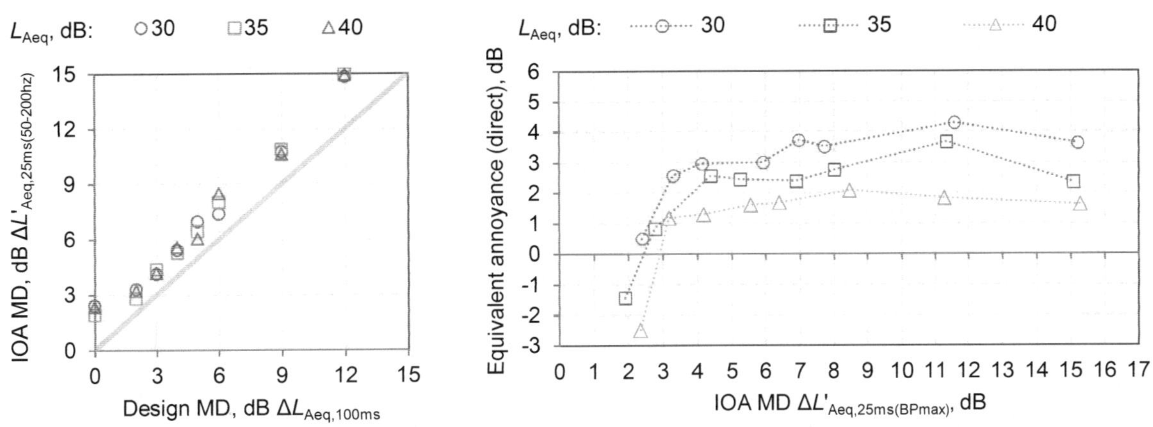

Figure 4: Results from Salford study reanalysis; (a, left) comparison of IOA MD with study design MD; (b, right) equivalent annoyance

The EqA dB values (derived directly from paired comparisons) plotted against MD are shown in Figure 4b. An important observation is that there appears to be an interaction between MD and L_{Aeq}, with lower levels eliciting greater adjustments to make the sounds subjectively equivalent. Another feature is the apparent flattening and even reduction of the values at higher MDs, suggesting either difficulty in discernibility between stimuli with high MD, or (as noted in the original study), difficulty in direct comparisons between increasingly dissimilar stimuli pairs.

3.4 Tokyo Study

The Tokyo study focussed on the perception of AM. A simplified approach to the derivation of the stimuli was applied by using a smooth spectrum with a -4 dB/octave slope, which was based on an approximate mean from a database of field measurements [21]. As shown in Figure 2b, this differs from some of the other studies examined – it contains a greater proportion of high-frequency energy and subjectively sounds less realistic as a synthetic representation of WTN.

The study examined an onset threshold for fluctuation sensation corresponding with varying MD. The reanalysis of these results with the IOA MD metric is shown in Figure 5a, which for reference also shows the original logistic regression curves (corresponding with the original study MD metric). Figure 5a shows that the fluctuation sensation onset determined using the IOA metric is ~0.6 dB higher than the 'AM index' used in the study to quantify MD (defined using an expression from [30]), resulting in a 50% probability of fluctuation sensation at around 2.3 dB $\Delta L'_{Aeq,25ms(BPmax)}$, and 95% probability at MDs around 3 dB (45 dB L_{Aeq}) and 3.6 dB (35 dB L_{Aeq}). The explanation for this is that the IOA metric tends to yield slightly higher values than the AM index used, especially around 1-4 dB $\Delta L'_{Aeq,25ms(BPmax)}$, which corresponds with the transition region in Figure 5a. The apparent onset threshold of perception in the region of 2 to 3 dB (IOA MD) has also been indicated in field measurements [19]. It should be noted once again that the IOA method selects the bandpass range with greatest MD, which for these stimuli was 50-200 Hz. The Tokyo study also showed that fluctuation sensation in WTN depends on the spectral modulation, and that lower spectral modulation regions contribute less to the sensation than higher regions [21]. This suggests that lower AM perception thresholds could be expected for WTN with higher frequency ranges of spectral modulation.

The reanalysis for the equivalent 'perceived noisiness' dB values are shown in Figure 5b. In contrast with the Salford study (which concerned affective responses), these results do not show a similar effect of L_{Aeq} on the results, although only two levels were tested. Inspection of associated confidence intervals (CIs) indicates the differences between the responses for the 35 and 45 dB L_{Aeq} stimuli were typically not statistically significant (as shown in Figure 5b, estimated 84% CIs were used to examine this aspect, as 84% CI overlap inspection can be useful as a rapid and convenient approximation for difference of means hypothesis tests at a 95% significance level [31]).

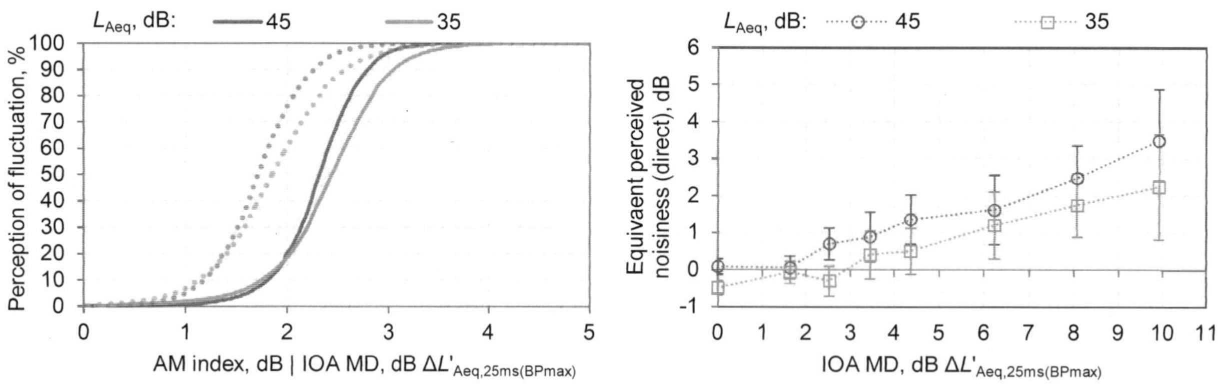

Figure 5: Results from Tokyo study reanalysis; (a, top left) fluctuation sensation onset probability logistic regression curves (solid lines: IOA MD; dotted lines: AM index MD from original study); (b, top right) equivalent perceived noisiness – errorbars indicate estimated 84% confidence intervals (see text regarding CI overlap and [31])

This apparent disparity between the Tokyo and Salford results may be due to the differences in the response types obtained, or how participants interpreted the tests. Alternatively, it could be related to cultural differences in the judgements of sound characteristics, as has been previously noted in psychoacoustical research [32, 33].
The Tokyo study did not investigate annoyance ratings, and so cannot be included in the pooled analysis of annoyance effects.

3.5 Poznan Study

The Poznan study investigated a range of different types of modulated sound, including narrowband and broadband noise with varying MF and MD. Responses comprised absolute ratings of annoyance, and the study indicated that, for broadband noise, annoyance was significantly affected by MF, MD and the interaction between them [22].
During the aural audition undertaken for our reanalysis, it was noted that the stimuli sounded different to those from the other studies. Figure 3c and Figure 2b show that there is a much greater proportion of mid- and high-frequency energy in the sounds (and lesser low-frequency energy), which, subjectively, results in a pink noise-like sound that is not representative of WTN. As such, the results of the experiment are of limited relevance to this study, as the focus of this investigation is AM in WTN. Also, the study did not cover steady (imperceptible AM) sounds at more than one L_{Aeq}, which would be necessary to estimate a reference curve from which to evaluate EqA dB values. Therefore, the results from the Poznan study have not been included in the pooled analysis.

3.6 Turku Study

The Turku study fixed the L_{Aeq} of the AM stimuli to focus on the effects of MF and MD, using a wide range of values for each, including MFs up to 16 Hz. Since this is outside the relevant range of values for WTN, only the stimuli for MFs up to 2 Hz have been included in the pooled analysis. However the study notes that annoyance responses did not tend to increase for MFs ≥2 Hz [23]; the responses for 4 Hz MF are shown for comparison in Figure 6a. The study found that both MF and MD had a significant effect on annoyance, with MD exhibiting a logarithmic association with annoyance; this can be seen in Figure 6b, and a similar logarithmic relationship with MF is also observed in Figure 6a.

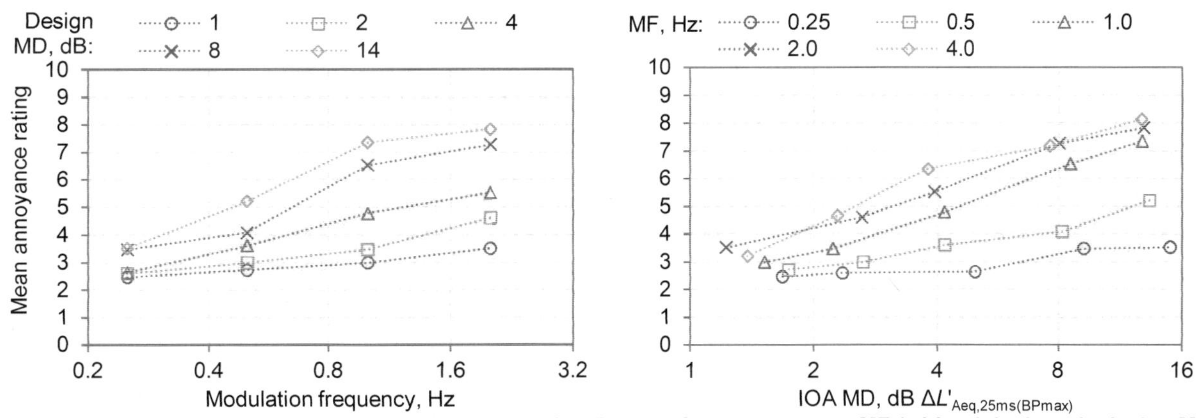

Figure 6: Results from Turku study reanalysis (L_{Aeq}: 35 dB); (a, left) annoyance vs MF (with original study design MD as parameter); (b, right) annoyance vs IOA MD

3.7 EMPA 2016 Study

The EMPA 2016 study investigated the effects of overall level, modulation presence (with/without), and modulation type (periodic or non-periodic). The MF was fixed at 0.75 Hz. Our stimuli reanalysis using the IOA metric yielded MDs of ~1.5 dB and 7.4-8.2 dB $\Delta L'_{Aeq,25msBP(max)}$ for the with/without AM conditions respectively.

During our aural audition of the stimuli it was noted that the sounds contain tones, which is confirmed in the synthesis description reported in [27], and is also clear from the spectrogram in Figure 3e. To investigate this further, tone audibility and corresponding tonality character rating penalty analysis has been undertaken using three standardised quantification methods:

i) Joint Nordic Method version 2 (JNM2 [34], which is also used in BS 4142:2014 [8]);
ii) ISO/PAS 20065:2016 [35], as applied in ISO 1996-2:2017 [7]; and
iii) ETSU-R-97 [36] (which employs a method based on the original Joint Nordic Method).

The analysis results are shown in Figure 7, plotted against both L_{Aeq} and MD.

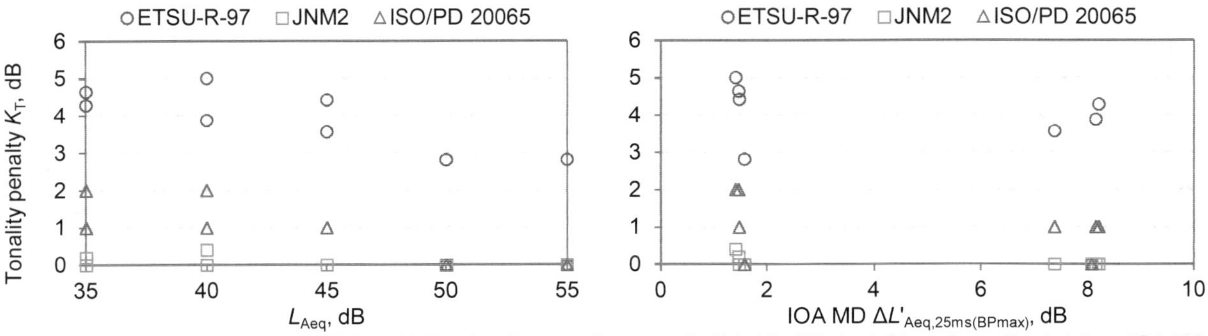

Figure 7: Tonality analysis of EMPA 2016 stimuli according to refs [34, 35, 36]; (a, left) vs L_{Aeq}; (b, right) vs IOA MD

Although all three methods identified the tones (at 172 Hz and 487 Hz), the analysis shows differences of up to 5 dB between the tone penalty adjustments derived using each method (which, critically with modulated signals, take varying approaches to time-averaging of narrowband spectra). In this case, the ISO/PAS 20065 method (which employs 3-second averaging) corresponded most closely with subjective assessment of the tonal prominence, while the ETSU-R-97 method was found to be oversensitive and JNM2 relatively insensitive, in this case.

The results in Figure 7 also exhibit interactions between the tonality, L_{Aeq} and MD (also confirmed aurally). Since tonality is a well-known sound characteristic that affects annoyance, these interactions could have influenced the judgements of annoyance due to AM; accordingly, the results of this study have not been included in the pooled analysis.

3.8 EMPA 2018 Study

The EMPA 2018 study examined the effect of MD, balance of spectral content (quantified as the L_{Aeq} minus L_{Ceq}) and the influence of a synthesised representation of the effect of atmospheric turbulence on the sound propagation, which introduced random level fluctuation components to the periodic modulations. Subjectively, the stimuli with turbulence sound more realistic (compared with field recordings of WTN), and the aural effect can be inferred visually by comparison of Figure 3f and Figure 3g – a slight 'fluttering' can be observed in the upper frequencies ≥2 kHz.

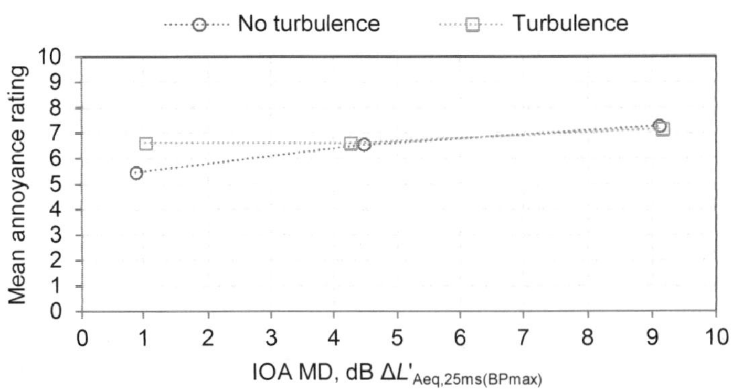

Figure 8: Results from EMPA 2018 reanalysis (L_{Aeq}: 40 dB)

The study used the same synthesis method as the 2016 study, although no tonal components were generated for the 2018 experiment. Stimuli with strong low frequency content (considerably stronger than would generally be expected for WTN) were included alongside pink noise sounds, and spectral balance was found to have a significant influence on annoyance [25]. Since these LF and pink noise stimuli are not representative of WTN, they have not been included in the analysis presented here.
The study also reconfirmed the influence of L_{Aeq}, and MD, although the observed effect of MD was relatively slight – possible reasons for this are discussed below. The MF used was fixed (0.75 Hz). The reanalysis of the 40 dB L_{Aeq} WTN stimuli is shown in Figure 8, which also indicates that any effect of turbulence on annoyance was only apparent at low MD.

4. Pooled analysis

4.1 Methodology

Conventional meta-analytic methods for pooled data generally comprise deriving a form of weighted mean effect value from the included studies [37, 38]. The type of weighting applied depends on the underlying assumption made about the relationship between the pooled studies. In the fixed-effect paradigm, the assumption is that a single, true effect underlies the various studies, which are inherently closely similar but represent varying degrees of (within-study) sampling error in identifying the common effect. The fixed-effect weighting applied is based on the study inverse-variance, which tends to give greater weight to larger studies (with smaller variance). Under the alternative random-effect model, the assumption is that each study may have its own 'true' effect (eg, this could occur if the participant sample is drawn from significantly different populations), but there is value in synthesising the studies to derive a combined effect as a mean of the multiple true effects. Effectively, this means that a study with smaller variance that would be given greater weight in a fixed-effect meta-analysis may have its weight reduced under the random-effect approach if it happens to differ from the overall (unweighted) mean by more than other studies (this is the between-study variance).
Application of such methods (which have their roots in clinical intervention studies) to acoustical laboratory exposure-response listening tests poses challenges. Pooled meta-analysis (or meta-regression) for these datasets are complicated by the multiple exposure variables expected to contribute to the response effect (some of which are expected to interact), which are also not at consistent magnitudes across the studies – this multidimensionality is not easily addressed using such techniques.
We have applied a semi-heuristic method, which is considered to represent a preliminary approach, and accordingly leads to tentative conclusions. In this analysis, we have opted not to apply weights to the different studies. This is because:

i) the participant samples for the included studies are of not dissimilar (and relatively small) sizes, and the within-studies variances observed are of similar magnitudes;
ii) the study features indicate that there might be expected to be differences in the underlying effects (due to differences in the stimuli, and differences in the participant samples), suggesting a random effects paradigm as the appropriate assumption, which implies the relationship being sought lies closer to the between-studies mean than in a fixed-effect model; and
iii) the need for a between-study mean in a random-effects model would lead to extensive interpolation work at the outset of the analysis, to estimate response values at a consistent magnitude for all relevant variables in each study, which could introduce substantial uncertainties that would be difficult to evaluate or control.

Statistical analysis has been performed in Python, using the Statsmodels and SciPy libraries, alongside exploratory data analysis and further modelling in Microsoft Excel. Following initial exploratory data analysis, the influence of several variables on the mean (absolute) annoyance ratings for the data from the three pooled studies (Salford, EMPA 2018 and Turku) has been tested in an iterative multiple linear regression exercise. Numerous candidate models were fitted to the data, and an optimal fit evaluated by comparing values for the *F*-statistic, Adjusted R^2, standard error of the estimate (S_E) and Bayesian Information Criterion (BIC). Since the aim of this part of the analysis was to confirm combinations of significant variables, potential multicollinearity has also been investigated via variance inflation factors, with candidate models exhibiting high multicollinearity (VIFs>5) disregarded as unreliable.

The tested variables included: study (categorical), mean participant age (per study), overall level (L_{Aeq}), spectral balance (determined as the difference between C-weighted and A-weighted statistical levels L_{eqC-A}, L_{10C-A}, L_{50C-A}, L_{75C-A}, or L_{90C-A}), MF and log-transformed MF, MD and log-transformed MD, MD 'strength category', and multiple variable interactions. The MD strength category was tested as an alternative variable to numerical MD, on the basis of the aforementioned observation that absolute MD (over ranges relevant to WTN) has a relatively weak proportionate influence on annoyance. The MD strength categories 'low', 'mid' and 'high' were therefore defined as <3 dB, 3-7.5 dB and >7.5 dB $\Delta L'_{Aeq,25ms(BPmax)}$. These intervals were simple approximations based on 84% CI inspection as described above.

The identified significant variables have then been used to develop a model for the pooled EqA values derived from the measured ratings.

The EqA values for each of the included studies have been derived based on the concept applied in [23]. The method reported here differs from [23] insofar as in that study, a linear regression model was used with the steady (imperceptible AM) stimuli to derive the 'apparent L_{Aeq}' for the AM stimuli. However, in [39], it is highlighted that for otherwise identical sounds that vary only in overall level, annoyance ratings on a finite scale are expected to follow a sigmoidal relationship, with extreme values asymptotically approaching the upper and lower limits of the scale. Examination of the regression residuals for the resulting linear model in [23] indicates the presence of endogeneity, in this case caused by fitting a linear model to data exhibiting curvature. Accordingly, we have applied a sigmoidal expression to predict the expected mean annoyance *A* as shown in Equation 1, which (as is shown later) provides a better fit for the observed values (ie improved residuals exogeneity and smaller S_E):

$$A = \frac{10}{1 + \exp\left[-k \ln(L/L_0)\right]}$$

Equation 1

where *L* is the level (L_{Aeq}), and *k* and L_0 are parameters affecting the slope and the centralisation of the curve respectively. The parametric values for Equation 1 corresponding with each study stimuli set have been optimised using a Levenberg-Marquardt least squares algorithm implemented in SciPy.

Models for EqA have been developed based on the insights from the annoyance rating regression analysis by testing possible expressions with least squares optimisation, performed using an evolutionary algorithm in Excel Solver.

Finally, the AM character rating penalties based on the EqA analysis are verified by application to the study exposure stimuli and examining the correlation of the 'AM rated' levels with the mean annoyance ratings.

4.2 Results

Variables affecting mean annoyance ratings

The variables and performance of the strongest model identified from the specification process are summarised in Table 2. The relatively high (Adjusted) R^2 value of 0.937 should be interpreted with caution, as inflated R^2 values are naturally expected when working with aggregated dependent variables [40]. However, this statistic remains useful as a relative measure of fit amongst candidate models.

The model summarised in Table 2 indicates that mean annoyance ratings in the pooled dataset are significantly affected by L_{Aeq}, $L_{50C\text{-}A}$, MD and the interaction of MF with MD. The observation of a significant influence of mean age should also be interpreted with some caution, as with these data it is naturally a form of proxy categorical dummy variable for 'study' (which was itself also shown to be a significant variable in some weaker candidate models, when controlling for L_{Aeq} and MD). However, there are good reasons to expect that age could have had some effect on responses, as transportation noise-related annoyance is known to exhibit an inverted-u relationship with age (peaking around 45 years) [41], and there were different age distributions in each pooled study, with mean values ranging from 27 to 42 years (see Table 1). The positive coefficient for the mean age predictor variable is therefore consistent with expected behaviour. Nonetheless, since the age of receptors is neither a realistic nor practical factor to include for WTN regulatory purposes, the model for EqA described below omits this variable.

Table 2: Annoyance regression model summary

Regression model predictor variable	Coefficient $\hat{\beta}$ (95% CI)		p-value	
Mean age	0.139 (0.105 – 0.174)		<0.001	
L_{Aeq}	0.207 (0.159 – 0.255)		<0.001	
$L_{50C\text{-}A}$	-1.092 (-1.216 – -0.968)		<0.001	
ln(MD)	1.371 (1.163 – 1.579)		<0.001	
ln(MF)	0.383 (-0.203 – 0.970)		0.195	
ln(MD) × ln(MF)	0.651 (0.311 – 0.992)		<0.001	
Overall model performance	Adjusted R^2	F-statistic	S_E, dB	BIC
A ~ Age + L_{Aeq} + $L_{50C\text{-}A}$ + ln(MD) + ln(MF) + ln(MD)×ln(MF)	0.937	124.3 (p <0.001)	0.231	89.91

Analysis of equivalent annoyance

The results of fitting the nonlinear model described by Equation 1 to the 'imperceptible AM' reference stimuli in the studies is shown in Figure 9. Linear regression fits are also indicated in Figure 9 for comparison.

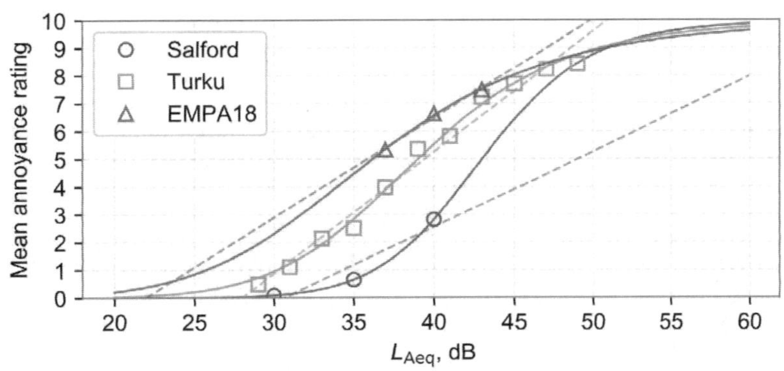

Figure 9: 'Imperceptible AM' stimuli reference curves for indirect derivation of equivalent annoyance (markers: observed values; solid lines: applied sigmoidal model; dashed lines: linear model, for comparison)

The reference curves have been used to derive the corresponding 'apparent L_{Aeq}' and associated EqA for the AM stimuli, which are plotted against MD in Figure 10.

The values in Figure 10 exhibit some consistency between studies where exposure stimuli have similar AM properties. For example, the EqA values for the Salford 40 dB L_{Aeq} (0.78 Hz MF) stimuli are within 1.5 dB of the EMPA 2018 values (for 40 dB L_{Aeq}, 0.75 Hz MF), although, as discussed further below, the latter are consistently lower.

The behaviour for the Salford 35 dB L_{Aeq} values is also comparable with those for the Turku 35 dB L_{Aeq}, 1 Hz MF stimuli, for MDs of up to ~8-9 dB $\Delta L'_{Aeq,25ms(BPmax)}$.

There is an observable trend with increasing EqA for decreasing overall level, which is explained by the observation that, at higher overall levels, the annoyance associated with the 'imperceptible AM' reference is higher; it seems that the *additional* annoyance caused by the AM component diminishes somewhat as the overall levels increase. A similar influence of L_{Aeq} on (indirect) EqA can also be seen in the results of [24], albeit with a smaller apparent effect. There are no EqA values in Figure 10 greater than ~1 dB for MDs below ~2.5 dB $\Delta L'_{Aeq,25ms(BPmax)}$, which is consistent with the perceptual onset threshold curves identified from the Tokyo results (see Figure 5a) – it is expected that some of the participants would not have perceived AM at such low MDs.

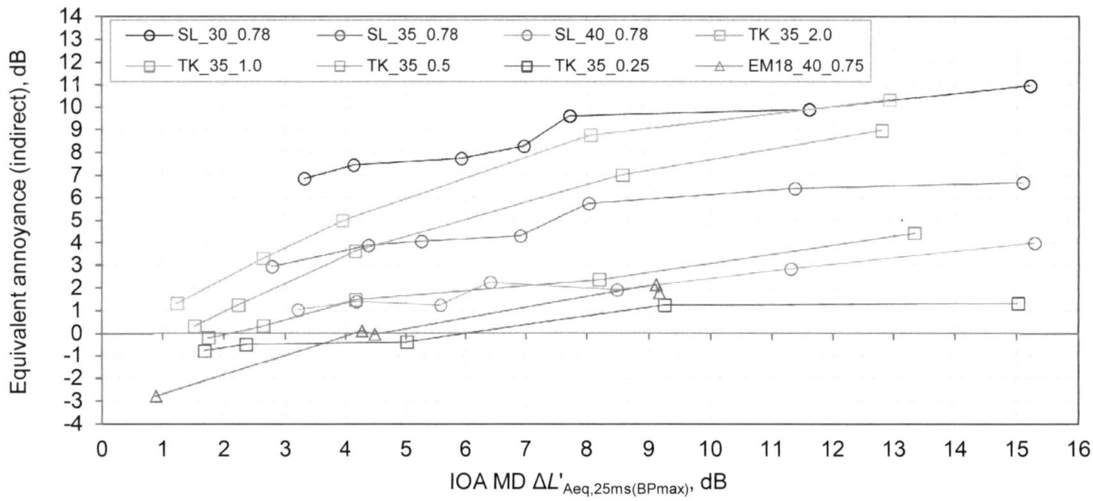

Figure 10: Equivalent annoyance values (indirect derivation) with modulation depth (legend coding: Study_L_{Aeq}_MF)

The relatively large negative value shown for the EMPA 2018 study at ~1 dB $\Delta L'_{Aeq,25ms(BPmax)}$ is explained by the inclusion of the turbulence feature within the 'imperceptible AM' reference stimuli – this particular data point corresponds with a low-AM stimulus that was not used in the reference curve derivation and did not feature synthesised turbulence, leading to a lower annoyance rating compared with the reference (and negative EqA) – this point has therefore been excluded from the models for an AM rating penalty developed below.

Model for AM rating penalty based on equivalent annoyance

The optimised model for an AM rating penalty, K_{AM} based on EqA is shown in Equation 2. The rating is truncated at 0 and →11 dB (to avoid negative penalties and extrapolation beyond the range of the supporting data).

$$\begin{aligned} K_{AM} &= \ln(\text{MD})[3.48 + 1.92\ln(\text{MF})] - 3.46\tanh\left(-1.07L_{Aeq} + 37.4\right) + 0.11L_{50C\text{-}A} - 2.23, && 0 < K_{AM} < 10.6,\ \text{MD} \geq 2.2 \\ K_{AM} &= 0, && K_{AM} \leq 0,\ \text{MD} < 2.2 \\ K_{AM} &= 10.6, && K_{AM} > 10.6 \end{aligned}$$

Equation 2

The validation of Equation 2 is shown in Figure 11a, which demonstrates a good agreement with the EqA dB values derived from measured annoyance ratings.

Although it is apparent that spectral balance has some effect on responses, in the cases of the included stimuli, the effect appears to be relatively small. For practical field applications, it would be desirable to eliminate the $L_{50C\text{-}A}$ term, as C-weighted measurements in the outdoors and in windy environments could be prone to extraneous low frequency noise, potentially affecting the reliability of the rating method. Accordingly, a simplified expression that omits this term is shown in Equation 3, with corresponding validation shown in Figure 11b – it can be seen that the omission of the $L_{50C\text{-}A}$ term leads to a negligible reduction in prediction accuracy.

$$K_{AM} = \ln(MD)[3.4 + 1.8\ln(MF)] - 3.6\tanh(-0.6L_{Aeq} + 21) - 0.9, \quad 0 < K_{AM} < 10.6, MD \geq 2.2$$
$$K_{AM} = 0, \quad K_{AM} \leq 0, MD < 2.2$$
$$K_{AM} = 10.6, \quad K_{AM} > 10.6$$

Equation 3

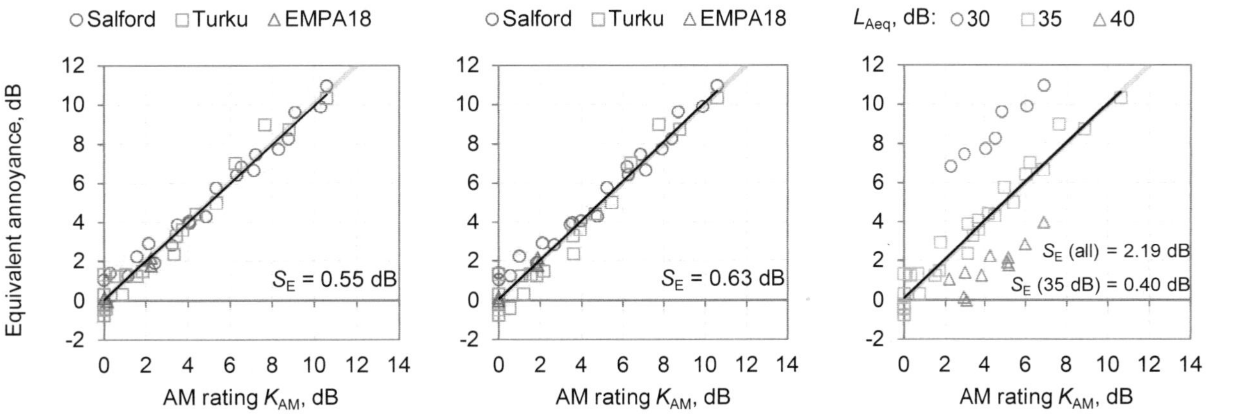

Figure 11: Validation of AM equivalent annoyance rating models; (a, left) optimised Equation 2; (b, centre) simplified Equation 3; (c, right) simplified Equation 4

In the pursuit of field practicability, it is also recognised that it is often not possible to accurately determine WTN L_{Aeq} at long range over short periods (eg, 10 minutes is used as a reference evaluation interval in [36]) using conventional measurement methods and technology. This is typically due to relatively small differences between the levels of WTN and background sound (such as that generated by wind interaction with vegetation and structures). As a result, it is common practice to evaluate the WTN level by surveying over long periods and deriving aggregated values using statistical methods. This type of approach makes the inclusion of L_{Aeq} as a variable within an AM rating penalty impractical. A further EqA model has therefore been developed that omits this term (despite its apparent importance in the subjective judgements of annoyance related to AM in the laboratory tests), as shown in Equation 4.

$$K_{AM} = \ln(MD)[3.5 + 2\ln(MF)] - 1.3, \quad 0 < K_{AM} < 10.6, MD \geq 2.2$$
$$K_{AM} = 0, \quad K_{AM} \leq 0, MD < 2.2$$
$$K_{AM} = 10.6, \quad K_{AM} > 10.6$$

Equation 4

As is evident in the corresponding validation shown in Figure 11c, the model of Equation 4 represents a substantial reduction in overall prediction accuracy compared with the other models. However, this model is based entirely on the main features of AM, ie the MF, the MD, and (implicitly by the use of the IOA AM metric) the spectral modulation. Figure 11c indicates that the reduction in agreement with measurements is due to the divergence of the Salford study data into its three L_{Aeq} groups, and the 'overrating' of the (40 dB L_{Aeq}) EMPA 2018 stimuli compared with the values derived from the measurements. The observed agreement along the 1:1 reference line is best for stimuli with an L_{Aeq} of 35 dB (for which the S_E is <0.5 dB), which fall in the middle of the levels examined here. Figure 11c also indicates that the differences between the L_{Aeq} groups are limited to the intercept; the slope and scatter of the data points for each level group are otherwise well-modelled by Equation 4 – this shows the difference is effectively a shift in the total EqA value (of ~3 dB) due to the apparent L_{Aeq}-modifier effect.

Application of equivalent annoyance AM rating model to absolute annoyance ratings

The models for EqA presented in Equation 3 and Equation 4 have been applied as AM rating penalty adjustments added to the L_{Aeq} for the stimuli for each included study and used to test correlation with the mean (absolute) annoyance ratings, as plotted in Figure 12. For comparison, the AM rated level according to the penalty adjustment established in [4], (which is dependent only on MD) is shown in Figure 12c.

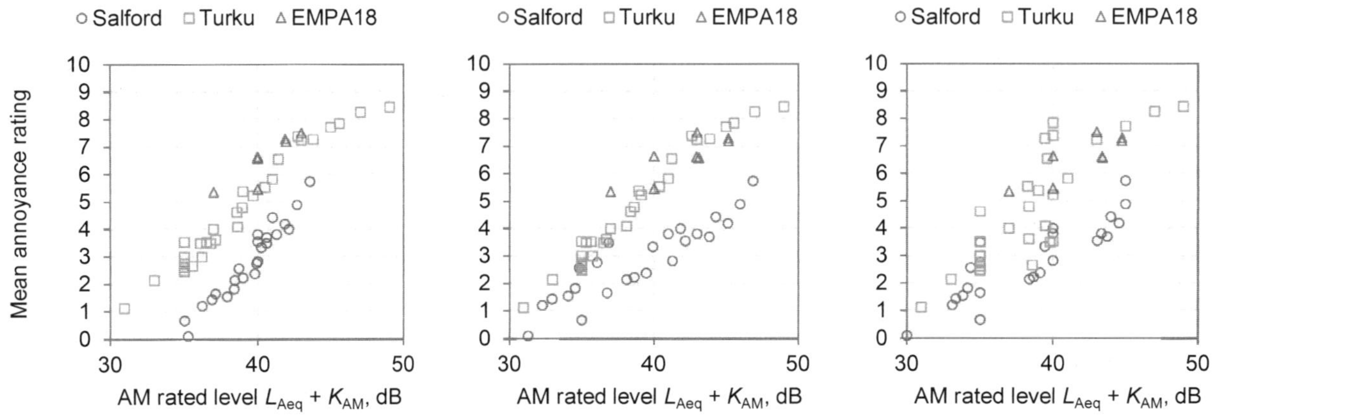

Figure 12: AM rated levels vs mean annoyance ratings; (a, left) $L_{Aeq} + K_{AM}$ Equation 3; (b, centre) $L_{Aeq} + K_{AM}$ Equation 4; (c, right) $L_{Aeq} + K_{AM}$ ref [4]

The correlation coefficient results for the AM rating penalty adjustment schemes shown in Table 3 indicate that the EqA models developed here represent an improvement in the correlation between the AM-rated levels ($L_{Aeq} + K_{AM}$) and the mean absolute annoyance ratings for the test stimuli, when compared with the equivalent rating from [4].

Table 3: AM rated L_{Aeq} with penalty adjustment K_{AM} vs annoyance correlation comparison (*$p<0.05$)

Study	AM rated level $L_{Aeq} + K_{AM}$ Equation 3 vs mean annoyance		AM rated level $L_{Aeq} + K_{AM}$ Equation 4 vs mean annoyance		AM rated level $L_{Aeq} + K_{AM}$ [4] vs mean annoyance	
	Spearman's ρ	Pearson's r	Spearman's ρ	Pearson's r	Spearman's ρ	Pearson's r
Salford	0.975*	0.965*	0.920*	0.912*	0.907*	0.875*
Turku	0.939*	0.980*	0.945*	0.981*	0.631*	0.633*
EMPA18	0.906*	0.862*	0.631	0.829*	0.523	0.813*

5. Discussion

There are observable differences between EqA derived using either the direct (paired comparison) method, or the indirect approach (derivation from magnitude scaling estimation). Although these differences are in part explained by variation in the magnitude of the AM parameters within the exposure stimuli, the Salford study is unique among those examined here in that it included both types of response test, which enables a fair comparison of results from both the direct and indirect approaches. Examining the Salford results in Figure 10 alongside the associated values in Figure 4b, the same broad features are evident, ie the relative influences of L_{Aeq} and MD. However, the direct approach yields values that are considerably smaller than the indirect approach; values for the latter extend up to 2-3× the directly-derived equivalents. An explanation for this might lie in the difficulty posed to participants in directly comparing two increasingly different stimuli and making quite complex judgements about which makes them feel more annoyed – this was also an observation made in the Salford study [20], and has been noted in previous annoyance listening test studies using paired comparisons [42].

The EqA values for the EMPA 2018 study shown in Figure 10, while comparable with the 40 dB L_{Aeq} Salford study stimuli, do tend to be lower. As shown in Figure 9, the numerical reason for this is the higher mean annoyance ratings given to the reference (negligible AM) stimuli. The underlying explanation may be related to the differences between the stimuli, but another possible causal factor could be the higher age range of the EMPA 2018 study, which is focussed more closely around the peak of the aforementioned expected noise (in)tolerance-age curve. The older participant sample may have tended to rate the 'imperceptible AM' reference stimuli as somewhat more annoying than the younger participant samples in the other studies for comparable L_{Aeq}, leading to a smaller observed difference between the reference stimuli and the AM stimuli. Indeed, a significant correlation between the individual participant annoyance ratings and age was noted in the EMPA 2018 study [25].

The EqA models developed indicate that, as shown in the results of [20] and [24], there appears to be some modification of affective responses to AM by the exposure level, which is also observed in direct comparisons (Figure 4b). The observed effect is somewhat counterintuitive, insofar as lower levels appear to result in heightened adverse response to the AM characteristic. The psychoacoustic model for fluctuation strength [30] indicates that sound pressure level is proportional to the perception of fluctuation caused by AM, which suggests AM-induced additional annoyance would also be expected to increase with level. Reasons for the apparently opposite observed behaviour are not immediately clear, and further investigation into this result could be valuable. Incorporating the this modifier effect into an AM character rating penalty (as per Equation 2 and Equation 3) results in a system in which quieter AM WTN is penalised more than louder WTN with the same AM characteristics, which consequently could result in similar 'AM rated levels' for both scenarios. It could be difficult to justify a rating system with this feature in regulatory applications, unless further evidence (including field experiments), supports such an approach. The AM rating penalty derived using Equation 4 (which omits the L_{Aeq}-modifier effect, depending on AM parameters only at the expense of prediction accuracy), is compared with the equivalent scheme from [4], which is dependent only on MD, in Figure 13. This shows that Equation 4 results in larger penalty values at MFs higher than 0.5 Hz, and smaller values at lower MFs. It also extends the penalisation system down towards ~2 dB MD. The rating system of [4] is most similar to the 0.75 Hz MF curve from Equation 4 shown in Figure 13.

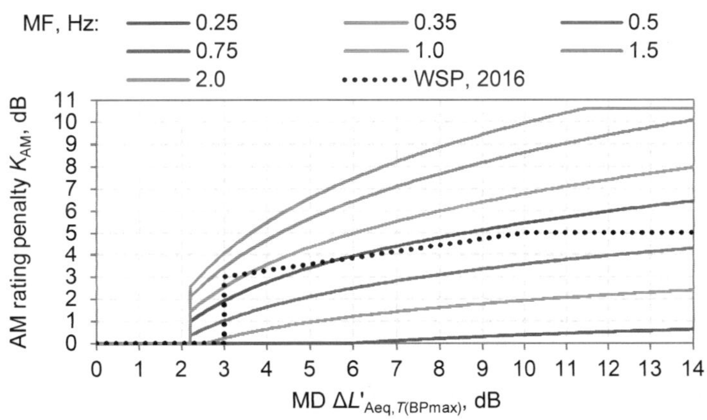

Figure 13: AM rating penalty K_{AM} (Equation 4) compared with reference [4] (NB: $T \leq 50$ ms is recommended for MF ≤ 2 Hz)

6. Conclusions

Previous AM studies employ different quantification procedures, which leads to uncertainties in comparing results, and may not reliably represent the AM present in analysed signals. The IOA AM metric has been found to be an effective and reliable common metric for MD, with widespread applications for laboratory and field experiments, as well as in WTN regulation. Our study indicates that its scope of applicability and its robustness can both be increased by reducing the L_{eq} averaging period used from 100 ms to 50 ms, which ensures that:

i) the metric can accurately measure AM at MFs up to 2 Hz (eg for smaller turbines with higher rotation speeds); and
ii) that relatively large MD values are not underestimated, due to the time domain resolution of the reconstructed AM signal.

Notwithstanding the above, our analysis does indicate that 100 ms averaging is expected to be acceptably accurate while MDs and MFs remain relatively low.
Implementing a 50 ms averaging would be more onerous in field surveys, but is in line with other quantification methods used for sound characteristics with a rate of level change >10 dB/s [8] (a rate which can be realised for AM WTN at high MDs or MFs).
The investigation in [11] also indicates that further work may be necessary to confirm whether refinement of the IOA-advised prominence criterion is appropriate to optimise the metric for field applications. Based on experience with AM field data, it is suggested that such optimisation should include reference to a benchmark that combines aural audition of survey recordings alongside visual signal inspections, to ensure reliability.
Using the IOA metric as a common AM quantification method, the reanalysis of the laboratory studies has enabled a pooled regression of the annoyance responses. The analysis confirms significant associations of annoyance with L_{Aeq}, MD, MF and spectral balance, while highlighting the interactions between these factors that require consideration. Logarithmic associations with annoyance are observed for MD and MF. It is also inferred from the results that respondent age may have an influence on annoyance ratings of AM WTN.
Models for EqA (in terms of dB) have been developed from the pooled data using an indirect method, in which annoyance responses to AM stimuli are referred to response curves generated from reference stimuli with imperceptible AM. It is apparent that the indirect approach results in higher EqA values than the direct method of paired stimuli comparison, and this observation merits further research – it remains unclear which approach carries greater ecological validity, although it is noted that direct comparison of two increasingly different sounds in the context of an annoyance judgement may be difficult for participants.
Features of the EqA models have been considered within the context of practicality for regulatory application. On this basis, an AM rating penalty K_{AM} that depends only on the MF and MD (quantified using the IOA AM metric, which implicitly also accounts for variation in the spectral modulation) is presented in Equation 4, and compared with an earlier approach [4] (see Figure 13). The approach taken here results in larger penalties with increasing MF (and vice versa), which was not previously included as a parameter in [4]. The new rating yields improved correlation with annoyance responses.
It is recognised that the annoyance ratings analysed here represent acute (short-term) responses measured in artificial contexts. The relevance for chronic (long-term) responses in realistic situations remains unclear, and controlled, field-based experiments in responses to AM would be highly valuable in addressing this lack of clarity. Nonetheless, the results from these lab-based tests will be useful in informing the design of future studies.

Acknowledgements

The authors are immensely grateful for the assistance of Dr Petra Virjonen, Dr Honorata Hafke-Dys, Professor Anna Preis, Dr Beat Schäffer, Dr Sabine von Hünerbein, Dr Sakae Yokoyama and Christina Ioannidou.

Declarations

This research has not received any external funding.
CRediT: MJBL – conceptualisation, methodology, software, analysis, visualisation, writing (original draft); TL – writing (review and editing).

References

[1] E. Pedersen and K. Persson Waye, "Perception and annoyance due to wind turbine noise–a dose-response relationship," *Journal of the Acoustical Society of America,* vol. 116, no. 6, pp. 3460-3470, 2004.

[2] G. P. van den Berg, "The beat is getting stronger: the effect of atmospheric stability on low frequency modulated sound of wind turbines," *Journal of Low Frequency Noise, Vibration and Active Control,* vol. 24, no. 1, pp. 1-24, 2005.

[3] F. van den Berg and D. Bowdler, "Amplitude modulation," in *Wind turbine noise*, D. Bowdler and L. G, Eds., Brentwood, Multi-Science Publishing, 2011, pp. 121-151.

[4] WSP | Parsons Brinckerhoff, "Wind turbine AM review: Phase 2 report," Department for Business, Energy and Industrial Strategy, 2016.

[5] M. J. B. Lotinga, R. A. Perkins, B. Berry, C. J. Grimwood and S. A. Stansfeld, "A review of the human exposure-response to amplitude-modulated wind turbine noise: health effects, influences on community annoyance, methods of control and mitigation," in *12th ICBEN Congress on Noise as a Public Health Problem*, Zurich, 18-22 June, 2017.

[6] RenewableUK, "The development of a penalty scheme for amplitude modulated wind farm noise," RenewableUK, 2013.

[7] ISO, "ISO 1996-2:2017 Acoustics — Description, measurement and assessment of environmental noise — Part 2: Determination of sound pressure levels," International Organization for Standardization, 2017.

[8] BSI, "BS 4142:2014+A1:2019 Methods for rating and assessing industrial and commercial sound," British Standards Institution, 2019.

[9] H. Tachibana, H. Yano, A. Fukushima and S. Sueoka, "Nationwide field measurements of wind turbine noise in Japan," *Noise control engineering journal,* vol. 62, no. 2, pp. 90-101, 2014.

[10] J. Pohl, J. Gabriel and G. Hübner, "Understanding stress effects of wind turbine noise - the integrated approach," *Energy Policy,* vol. 112, pp. 119-128, 2018.

[11] K. L. Hansen, P. Nguyen, B. Zajamšek, P. Catcheside and C. H. Hansen, "Prevalence of wind farm anplitude modulation at long-range residential locations," *Journal of Sound and Vibration,* vol. 455, pp. 136-149, 2019.

[12] C. Ioannidou, S. Santurette and C.-H. Jeong, "Effect of modulation depth, frequency, and intermittence on wind turbine noise annoyance," *Journal of the Acoustical Society of America,* vol. 139, no. 3, pp. 1241-1251, 2016.

[13] S. von Hünerbein and B. Piper, "Affective response to amplitude modulated wind turbine noise," in *6th International Meeting on Wind Turbine Noise*, Glasgow, 20-23 April, 2015.

[14] IOA Noise Working Group (Wind Turbine Noise), "Amplitude Modulation Working Group Final Report: A method for rating amplitude modulation in wind turbine noise," Institute of Acoustics, 2016.

[15] IOA Noise Working Group (Wind Turbine Noise), "Amplitude Modulation Working Group Discussion Document: Methods for rating amplitude modulation in wind turbine noise," Institute of Acoustics, 2015.

[16] IOA AMWG, "IOA AM code," 22 03 2017. [Online]. Available: https://sourceforge.net/projects/ioa-am-code/.

[17] A. Brandt, Noise and vibration analysis, John Wiley & Sons: Chichester, 2011.

[18] B. Barlas, W. J. Zhu, W. Z. Shen, K. O. Dag and P. Moriarty, "Consistent modelling of wind turbine noise propagation from source to receiver," *Journal of the Acoustical Society of America,* vol. 142, no. 5, pp. 3297-3310, 2017.

[19] S. Jennings and J. Kennedy, "Detection and evaluation of amplitude modulation from a wind energy development, Ireland," *Acta Acustica united with Acustica,* vol. 105, no. 6, pp. 1042-1052, 2019.

[20] S. von Hünerbein, A. King, B. Piper and M. Cand, "Wind Turbine Amplitude Modulation: Research to Improve Understanding as to its Cause & Effect. Work Package B(2): Development of an AM Dose-Response Relationship," RenewableUK, 2013.

[21] S. Yokoyama, T. Kobayashi, S. Sakamoto and H. Tachibana, "Subjective experiments on the auditory impression of the amplitude modulation sound contained in wind turbine noise," in *6th International Meeting on Wind Turbine Noise*, Glasgow, 20-23 April, 2015.

[22] H. Hafke-Dys, A. Preis, T. Kaczmarek, A. Biniakowski and P. Kleka, "Noise annoyance caused by amplitude modulated sounds resembling the main characteristics of temporal wind turbine noise," *Archives of Acoustics,* vol. 41, no. 2, pp. 221-232, 2016.

[23] P. Virjonen, V. Hongisto and J. Radun, "Annoyance penalty of periodically amplitude-modulated wide-band sound," *Journal of the Acoustical Society of America,* vol. 146, no. 6, pp. 4159-4170, 2019.

[24] B. Schäffer, S. J. Schlittmeier, R. Pieren, K. Heutschi, M. Brink, R. Graf and J. Hellbrück, "Short-term annoyance reactions to stationary and time-varying wind turbine and road traffic noise: a laboratory study," *Journal of the Acoustical Society of America,* vol. 139, no. 5, pp. 2949-2963, 2016.

[25] B. Schäffer, R. Pieren, S. J. Schlittmeier and M. Brink, "Effects of different spectral shapes and amplitude modulation of broadband noise on annoyance reactions in a controlled listening experiment," *International Journal of Environmental Research and Public Health,* vol. 15, no. 5, p. 1029, 2018.

[26] P. White, "Wind Turbine Amplitude Modulation: Research to Improve Understanding as to its Cause & Effect. Work Package B(1): Development of an objective AM measurement methodology," RenewableUK, 2013.

[27] R. Pieren, K. Heustchi, M. Müller, M. Manyoky and K. Eggenschwiler, "Auralization of wind turbine noise: emission," *Acta Acustica united with Acustica,* vol. 100, no. 1, pp. 25-33, 2014.

[28] D. P. Nguyen, K. Hansen, B. Zajamšek and P. Catcheside, "Evaluation of wind farm noise amplitude modulation synthesis quality," *Applied Acoustics,* vol. 166, p. 107349, 2020.

[29] IEC, "IEC 61260-1:2014 Electroacoustics - Octave-band and fractional-octave-band filters - Part 1: Specifications," International Electrotechnical Commission, 2014.

[30] H. Fastl and E. Zwicker, Psychoacoustics: facts and models, Berlin: Springer-Verlag, 2007.

[31] M. E. Payton, M. H. Greenstone and N. Schenker, "Overlapping confidence intervals or standard error intervals: what do they mean in terms of statistical significance?," *Journal of Insect Science,* vol. 3, no. 1, p. 34, 2003.

[32] S. Kuwano, S. Namba and H. Fastl, "On the judgment of loudness, noisiness and annoyance with actual and artificial noises," *Journal of Sound and Vibration,* vol. 127, no. 3, pp. 457-465, 1988.

[33] S. Kuwano, H. Fastl and S. Namba, "Loudness, annoyance and unpleasantness of amplitude modulated sounds," in *Inter-noise 99*, Fort Lauderdale, Florida, 06-08 December, 1999.

[34] T. H. Pedersen, M. Søndergaard and B. Andersen, "AV 1952/99 Objective method for assessing the audibility of tones in noise: Joint Nordic Method – version 2," DELTA, 1999.

[35] ISO, "ISO/PAS 20065:2016 Acoustics — Objective method for assessing the audibility of tones in noise — Engineering method," International Organization for Standardization, 2016.

[36] Energy Technology Support Unit Noise Working Group on Wind Turbines, "ETSU-R-97 The assessment and rating of noise from wind farms," Department of Trade and Industry, 1996.

[37] M. Borenstein, L. V. Hedges, J. P. T. Higgins and H. R. Rothstein, Introduction to meta-analysis, Chichester: John Wiley & Sons, 2009.

[38] J. P. T. Higgins, J. Thomas, J. Chandler, M. Cumpston, T. Li, M. J. Page & V. A. Welch (editors), "Cochrane Handbook for Systematic Reviews of Interventions version 6.2 (updated February 2021)," [Online]. Available: https://training.cochrane.org/handbook/current.

[39] DELTA, "AV 1102/07 The "Genlyd" noise annoyance model: dose-response relationships modelled by logistic functions," DELTA, 2007.

[40] C. Camacho Martínez-Vara de Rey, M. P. Galindo Galindo and M. A. Arias Velarde, "Effects of using mean scores in regression models: an example from environmental psychology," *Quality & Quantity,* vol. 35, no. 2, pp. 191-202, 2001.

[41] P. W. M. van Gerven, H. Vos, M. P. J. van Boxtel, S. A. Janssen and H. M. E. Miedema, "Annoyance from environmental noise across the lifespan," *Journal of the Acoustical Society of America,* vol. 126, no. 1, pp. 187-194, 2009.

[42] A. Niedzwiecki and H. S. Ribner, "Subjective loudness and annoyance of filtered N-wave sonic booms," *Journal of the Acoustical Society of America,* vol. 65, no. 3, pp. 705-707, 1979.

INCE EUROPE

9th International Conference
on
Wind Turbine Noise
Remote from Europe – 18th to 21st May 2021

Physics-based auralization of wind turbine noise

David Mascarenhas, Benjamin Cotté, Olivier Doaré
Institute of Mechanical Sciences and Industrial Applications (IMSIA), ENSTA Paris, CNRS, CEA, EDF, Institut Polytechnique de Paris, France
david.mascarenhas@ensta-paris.fr

Summary

Amplitude modulation of wind turbine noise is known to be a potential source of annoyance for people living in the vicinity of wind farms. To better understand this auditory annoyance, we propose to auralize the sound that is generated by the wind turbines, rather than to observe a visual representation of the sound levels. It is desirable for the developed auralization tool to be physically-based rather than sample-based. This allows control over the prevailing physical parameters. In our work, the auralization tool is based on Amiet's theory in the frequency domain, and considers the main broadband aerodynamic noise sources, namely trailing edge noise and turbulent inflow noise. For the auralization of the full wind turbine noise, the power spectral density for each blade segment and each position is considered along with the appropriate time shift due to the propagation between the moving blades and the fixed observer. In this study, an efficient method is discussed for the conversion of the frequency-domain power spectral density into a time domain signal. The appropriate time delay due to propagation is accounted for. Finally, a proper implementation of energy conserving cross-fading between consecutive signal grains is proposed. The complete auralized signal for the wind turbine noise in free field is then computed with different receiver orientations and meteorological conditions and compared with the original results in the frequency domain. This auralization tool combined with Virtual Reality/ Augmented Reality can help in building the wind farms while also accounting for auditory annoyance factor in the design phase.

1. Introduction

Noise from modern wind turbines is dominated by broadband aeroacoustic mechanisms generated by the wind interacting with the rotating blades. The amplitude modulation of this broadband noise is known to be a potential source of annoyance for people living in the vicinity of wind farms. The main sources of the broadband noise are to turbulent inflow noise (TIN), trailing edge noise (TEN) and stall noise (Bowdler and Leventhall (2011); Oerlemans and Schepers (2009); Bertagnolio et al. (2017); Buck et al. (2018)). Among these noise sources, stall noise is the most intense and is caused only when the angle of attack of the blade is large, whereas TIN and TEN occur for all angles of attack. Thus, in the simplest case scenario it can be said that TIN and TEN are observed at all times for an operating wind turbine and are the dominant sources of wind turbine noise. The synthesis of these noise sources allows to understand the auditory perception and psychoacoustics that relates to the annoyance caused by the wind turbine noise. Sample-based auralization previously done for wind turbine noise covers a few prevailing parameters and cannot be extended further to compensate for other settings (Pieren et al. (2014)). This restriction occurs because the sample-based synthesis is directly dependent on the noise recordings. There is hence a need for a physics-based auralization, as it allows for a greater control over the desired physical parameters that contribute to the noise generating acoustics. Such an auralization tool can be useful to study the noise generation in the design phase itself and can also helpful to predict the generation of the wind turbine noise in the vicinity.

The frequency-domain modelling of the wind turbine noise sources is studied by Tian and Cotté (2016) based on Amiet's theory (1975; 1976) for TIN and TEN. This noise prediction model is used as the base physical model for the auralization. The main objective of the article is to discuss an auralization tool which converts the frequency-domain model into a time-domain signal, while also accounting for appropriate physical parameters such as time delay due to propagation and geometrical spreading. The two noise mechanisms are presumed to be uncorrelated and are synthesized separately. The resulting synthesized signal is thus a summation of the TIN and TEN, generated by the wind turbine blades in rotation that is experienced by an observer in the free field.

The paper first explains briefly the frequency-domain prediction models for TIN and TEN with few important parameters. The conversion of the frequency spectra to the time-domain signal is explained followed by the description of how the propagation delay is accounted for. Some crucial parameters which are subjective but essential for the auralization are studied in detail. The final auralization tool is presented with a few test cases and the auralized signal are available here: `http://sites.google.com/view/david-mascarenhas/wtnc-2021`.

2. Frequency-domain modelling of broadband airfoil noise

For a fixed airfoil of span L and chord c, the original models proposed by Amiet (1975; 1976) predict the noise generated by the leading edge and trailing edge of the airfoil assimilated to a thin plate interacting with the turbulent gusts of uniform velocity. The noise observed in the far field (x_R, y_R, z_R) is predicted in the form of a Power Spectral Density (PSD).

2.1 Turbulent inflow noise (TIN)

The atmospheric turbulence convected at the inflow velocity U, interacting with the leading edge of the airfoil produces turbulent inflow noise. The PSD of the turbulent inflow noise observed in the far-field for an airfoil of large aspect ratio ($L > 3c$), is given by (Amiet (1975); Roger and Moreau (2010)):

$$S'^{\,\mathsf{TIN}}_{pp}(x_R, y_R, z_R, \omega) = \left(\frac{\rho_0 k c z_R}{2S_0^2}\right)^2 \pi U \frac{L}{2} \Phi_{ww}\left(\frac{\omega}{U}, \frac{k y_R}{S_0}\right) \left|\mathcal{L}_{TI}\left(x_R, \frac{\omega}{U}, \frac{k y_R}{S_0}\right)\right|^2, \tag{1}$$

where ω is the angular frequency, $k = \omega/c_0$ is the acoustic wavenumber, ρ_0 is the air density, c_0 is the speed of sound, S_0 is a modified distance between the source and the observer, and $\mathcal{L}_{TI}$ is the turbulent inflow noise transfer function, connecting the airfoil surface pressure fluctuation to the acoustic pressure at the far-field point. Φ_{ww} is the two-dimensional energy spectrum, modeled by a von Kármán spectrum for homogeneous and isotropic turbulence (Amiet (1975); Tian and Cotté (2016)).

2.2 Trailing edge noise (TEN)

The turbulent boundary layer fluctuations convected at the velocity U_c interact with the trailing edge of the airfoil and generate trailing edge noise. The PSD of trailing edge noise observed in the far-field for an airfoil of large aspect ratio ($L > 3c$), is given by (Amiet (1976); Roger and Moreau (2010)):

$$S'^{\text{TEN}}_{pp}(x_R, y_R, z_R, \omega) = \left(\frac{kcz_R}{4\pi S_0^2}\right)^2 \frac{L}{2}\Phi_{pp}(\omega) l_y\left(\omega, \frac{ky_R}{S_0}\right)\left|\mathcal{L}_{TE}\left(x_R, \frac{\omega}{U_c}, \frac{ky_R}{S_0}\right)\right|^2, \tag{2}$$

where Φ_{pp} is the wall pressure fluctuation spectrum, l_y is the spanwise correlation length estimated by the Corcos model, and $\mathcal{L}_{TE}$ is the transfer function for trailing edge noise. The wall pressure fluctuation spectrum Φ_{pp} is calculated using Goody's model for the pressure side and Rozenberg's model for the suction side of the airfoil (Tian and Cotté (2016)).

2.3 Extension to a full size modern wind turbine

By dividing a wind turbine blade into segments with the appropriate aspect ratio and twist, Tian and Cotté (2016) extended these models to a full size wind turbine as can be seen in Fig.1a. As a rotating blade experiences non-uniform flow along the span, with the incoming velocity strongest at the blade tip, the segmentation of the blade allows for the implementation of different inflow velocities for each segment. The segmentation is done ensuring the segment span is greater than the spanwise turbulence correlation length.

To account for the rotational motion of the blade, the model approximates the complete rotation of the blade as a series of translations between discrete angular positions. The convective amplification and Doppler effect caused by the rotating blades is also accounted for following Sinayoko et al. (2013). The instantaneous PSD, $S_{pp}(x_0, \omega, \beta)$ at the observer for an azimuthal blade position β is given by:

$$S_{pp}(\mathbf{x_0}, \omega, \beta) = \left(\frac{\omega_e}{\omega}\right) S'_{pp}(\mathbf{x}, \omega_e, \beta), \tag{3}$$

where $S'_{pp}(\mathbf{x}, \omega_e, \beta)$ given by Eq. (1) and (2) is the PSD for a fixed blade, ω_e and ω are the emitted and observed frequencies, $\mathbf{x_0}$ and $\mathbf{x}$ are the observer coordinates in the hub and blade coordinate systems.

The wind profile power law relation $U(z) = U_{ref}(z/z_{ref})^\alpha$ is implemented to consider the influence of different wind inflow velocities at different heights of the atmosphere. The reference inflow velocity U_{ref} = 8 m/s is taken at the hub height H_0 = 80 m, chosen as the reference height z_{ref}. The implemented source model of Tian and Cotté (2016) for a full size wind turbine thus gives the frequency-domain response of each segment of each blade at each discrete angular position β (Fig.1a). This response is obtained for an observer at a position defined by the angle τ with respect to the wind direction and by the horizontal distance R_0 from the base of the wind turbine tower (Fig.1b).

3. Auralization method

As each of the segments of the wind turbine blades contribute individually to the total noise, they can be synthesized separately and then summed together at each time step. The synthesis of one

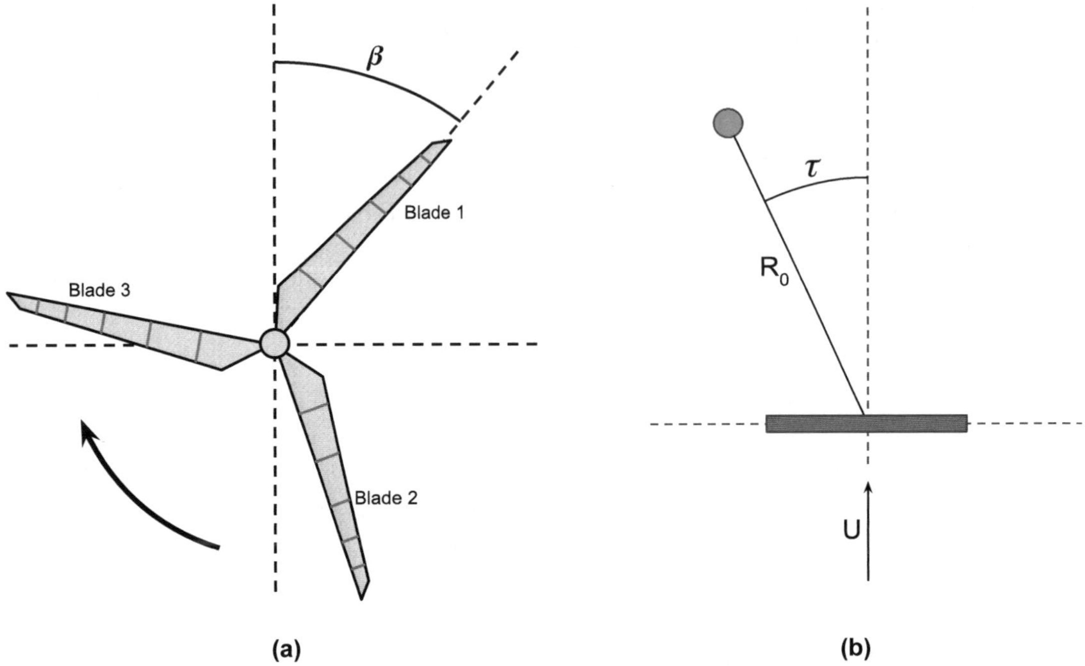

Figure 1: (a) Approximated segmentation of the blades and discrete angular position β (b) Observer position with respect to the wind turbine rotational plane (top view). R_0 is the horizontal distance between the wind turbine hub and the observer, τ is the angle made by R_0 and the wind direction.

segment at a particular blade angular position is referred here as a grain. The noise observed at the receiver is the summation of all the uncorrelated grains at the corresponding time step. We first discuss in Section 3.1 an efficient method to auralize a single grain, which consists in converting the PSD to a time-domain signal. The propagation delay is accounted for in Section 3.2, and the choice of the cross-fading window function is discussed in Section 3.3.

3.1 Conversion from frequency spectra to time-domain signal

The PSD in the model of Tian and Cotté is a numerical approximation of the airfoil noise for a set of frequencies. The pressure amplitude corresponding to a particular frequency can be directly calculated as:

$$p(f) = \sqrt{1\ \mathrm{Hz} \cdot S_{pp}(f)} \qquad (Pa). \tag{4}$$

This conversion for the all of the frequencies, gives the pressure amplitude-spectra correspondingly. The inverse Discrete Fourier transform (IDFT) converts a frequency-domain spectra into a time domain signal, while conserving the same number of points. The number of points of the concerned pressure amplitude spectrum is insufficient to obtain a time signal of sufficient length. To obtain the desired duration of the signal, the amplitude spectrum is interpolated between the two frequency limits, while the pressure amplitudes for the other frequencies are taken as zero. The total number of frequency points that include the interpolated amplitude spectrum and the frequencies with zero amplitude corresponds to the number of points of the one-sided frequency spectra. As the noise is assumed to be stochastic, a random phase between 0 and 2π is assigned to each of the complex amplitudes in this one-sided frequency spectrum. To obtain a real-valued signal, the one-sided frequency spectrum is converted to a symmetric double-sided frequency spectrum and the IDFT is taken. The schematic approach of the method is shown in Fig.2a.

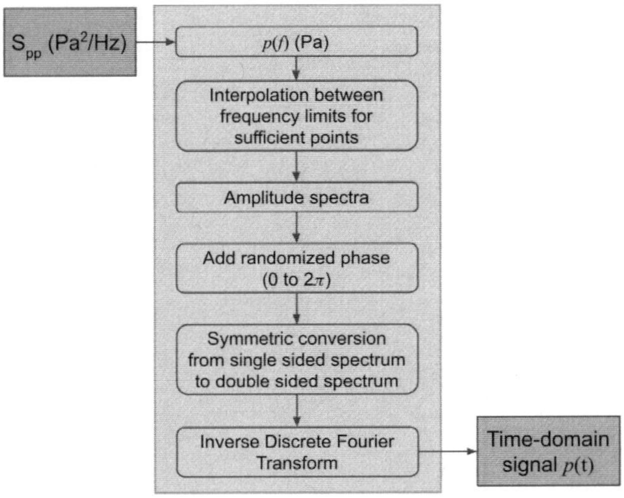

(a)

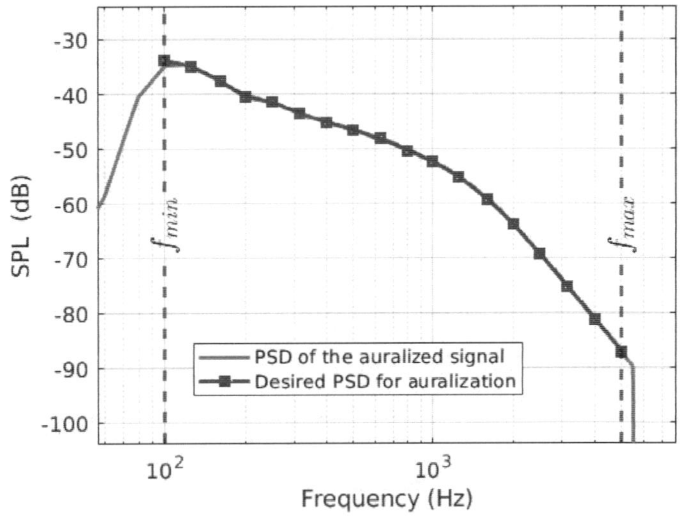

(b)

Figure 2: (a) Schematic approach of the method for the conversion from the frequency-domain PSD to the time-domain signal, (b) PSD of the resultant auralized grain along with the desired PSD

To avoid edge effects which are observed due to the IDFT, the signal is synthesized to an extraneous length and the truncated to the desired length. This synthesized signal which is obtained from the PSD corresponding to one segment is the grain under consideration. A grain auralized between f_{min}= 100 Hz and f_{max}= 5000 Hz shows a good replication of the desired PSD as seen in Fig.2b. Beyond the limits of the frequency range set by f_{min} and f_{max} there is a smooth decay of the amplitude of the auralized signal.

3.2 Propagation time and grain length

As the rotation of the blades is discretized into a number of angular positions of equal angular steps $\Delta\beta$, the time duration of a single angular step is given by:

$$\Delta t_s = \frac{\Delta\beta}{\Omega}, \tag{5}$$

where Ω is the rotational speed of the wind turbine blade. Depending on the position of the receiver, the propagation distance is different for each segment of the blade at each angular position β. If the distance between the source and the receiver is $R(\beta)$, then the propagation time is $R(\beta)/c_0$. The time duration of the emitted noise at the receiver is thus given by:

$$T_{\Delta\beta} = \Delta t_s + \frac{\Delta R}{c_0} = \frac{\Delta\beta}{\Omega} + \frac{\Delta R}{c_0}, \tag{6}$$

where, $\Delta R = R(\beta + \Delta\beta) - R(\beta)$ is the difference between the propagation distances related to the corresponding successive angular positions that can be positive or negative.

3.3 Cross-fading window function

Successive allocation of the auralized grains is observed to produce artifacts in the form of clicks during the transition of one grain to another (Fig: 3a). To avoid this form of artifact, the transition between grains has to be done with a certain amount of overlap between each grain, while still

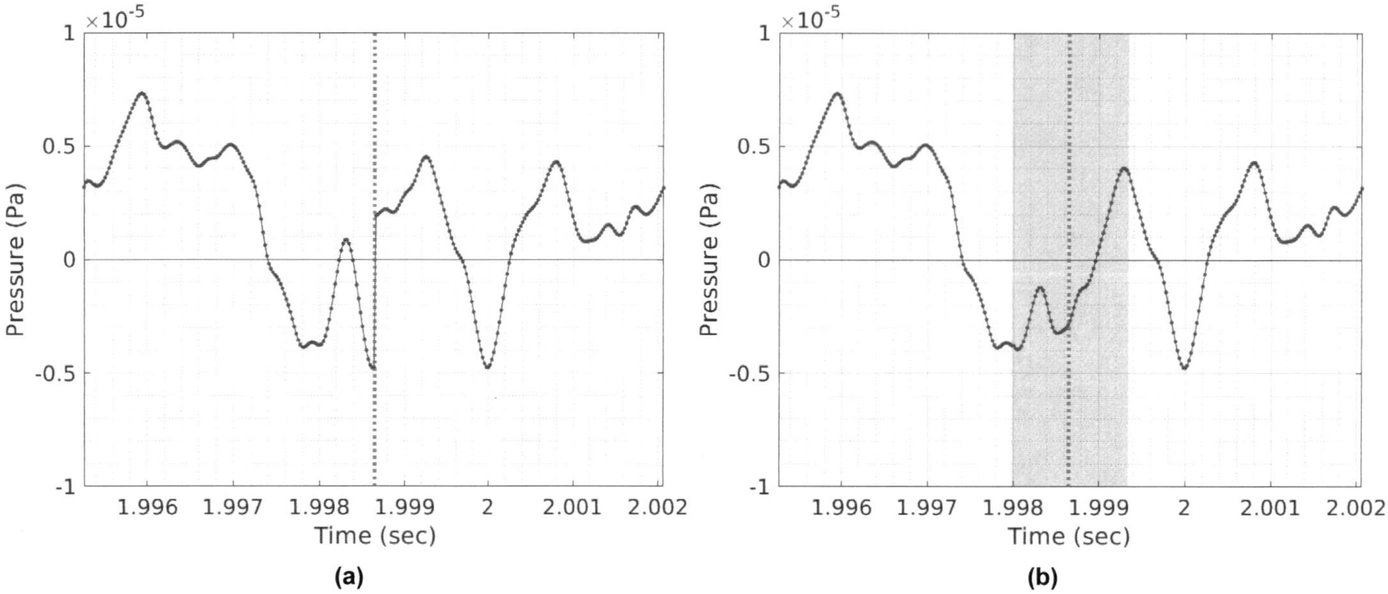

Figure 3: (a) Signal without an overlapping window function. Artifacts in the form of clicks are observed at the transitions. (b) Signal with an overlapping window function applied to the shaded region.

conserving the absolute size of each grain. Such transitions of audio signals are known as cross-fading. To facilitate the cross-fading between two grains, a window function is designed and applied to each grain. The window function $W[k]$ of N samples is composed of the overlapping functions $f[k]$ and $g[k]$ with a unit response between them and can be defined as:

$$W[k] = \begin{cases} f[k] & \text{for} \quad 1 \leq k < w_l \\ 1 & \text{for} \quad w_l \leq k \leq N - w_l \\ g[k] & \text{for} \quad N - w_l < k \leq N \end{cases} \tag{7}$$

where w_l is the desired length of the overlap function that is to be set. The overlapping functions $f[k]$ for the fade-in and $g[k]$ for the fade-out, are required to serve for the purpose of overlapping between two grains, such that the original power is conserved during the overlap.

It is necessary to set the desired length of the overlapping function as a constant for all grains, while also noting that the size of each grain may differ. This is done by setting a constant length w_l for the functions $f[k]$ and $g[k]$ for all the grains, while the variability of the grain lengths are achieved by the length of the unit response between the overlap functions: $N_{unit} = N - 2w_l$.

To be consistent with the length of the overlap functions and the size of the grains, the length w_l is divided equally between the two successive grains. The total length of the window N thus consists of the length of the grain $N(T_{\Delta\beta})$, the length for the preceding overlap $w_l/2$ and the length for the successive overlap $w_l/2$ (Fig. 4). The relation between the length of the window N and the desired length of the overlap function, w_l is thus defined as:

$$w_l/2 + N(T_{\Delta\beta}) + w_l/2 = N \Rightarrow N(T_{\Delta\beta}) + w_l = N_{unit} + 2w_l \Rightarrow w_l = N(T_{\Delta\beta}) - N_{unit}. \tag{8}$$

This gives us the maximum length of the overlap function w_l, that corresponds to the smallest grain in the system and to $N_{unit} = 0$: $w_l = \min(N(T_{\Delta\beta}))$. The length of the overlap function w_l is thus restricted to $0 \leq w_l \leq \min(N(T_{\Delta\beta}))$. The amount of overlap is defined as:

$$\Psi = \frac{w_l}{\min(N(T_{\Delta\beta}))}, \tag{9}$$

with $0 \leq \Psi \leq 1$. The defined window function for $\Psi = 20\%$ is plotted in Fig. 4a.

For two signals to cross-fade while maintaining the required power level during the transition, the cross-fade functions g and f, must satisfy the equation given by (Fink et al. (2016)):

$$f^2 + 2 \cdot f \cdot g \cdot r_{(p_1,p_2)} + g^2 = 1, \tag{10}$$

where $r_{(p_1,p_2)}$ is the correlation coefficient of the two overlapping signals p_1 and p_2, which is zero for uncorrelated signals and one for completely correlated signals. As we assume that successive grains are uncorrelated, $r_{(p_1,p_2)} = 0$ and Eq. (10) also satisfies the Princen-Bradley criterion (Bäckström (2019)). Following Fink et. al. (2016), a number of functions satisfying these conditions are available for the selection of the cross-fading overlap function. The simplest and efficient functions for the windows are given by $f(\chi) = \sin(\frac{\pi\chi}{2})$ and $g(\chi) = \cos(\frac{\pi\chi}{2})$, where $\chi \in [0, 1]$ is the normalized time index. These functions used in Eq. (7), with the desired length of the overlap function w_l determined by Eq. (9) define the window function used to facilitate the cross-fading between two successive grains.

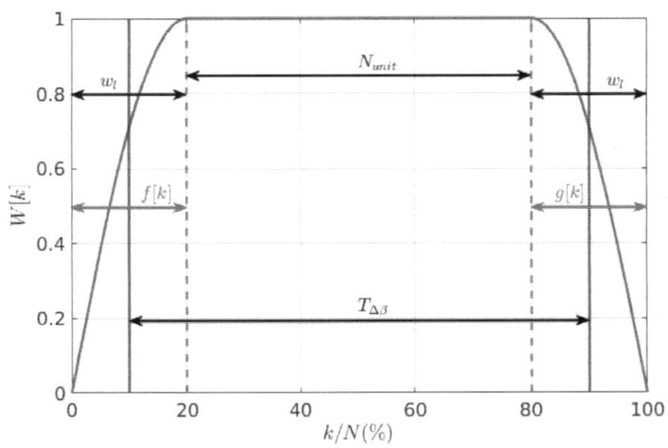

(a)

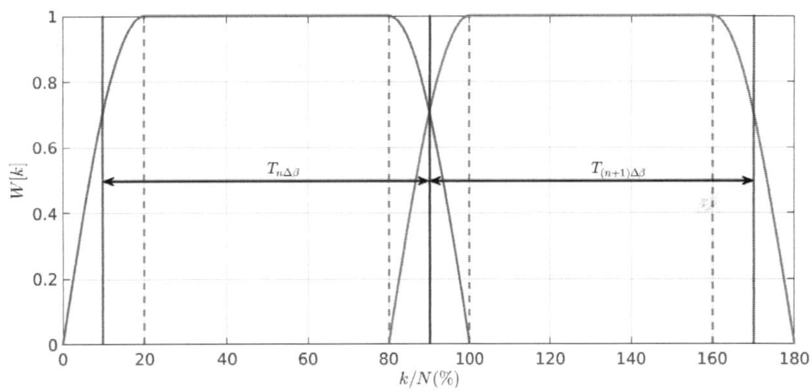

(b)

Figure 4: (a) The total window function $W[k]$ with the overlapping window functions $f[k]$ and $g[k]$ with Ψ=20% of overlap (b) Overlap of the window functions.

4. Results

4.1 Influence of overlap amount in the cross-fading between grains

The principal purpose for the cross-fading window is to facilitate a smoother transition from one grain to the next while maintaining the correct power level. However, the overlap amount determined by Ψ, influences the qualitative realism of the auralized signal. To understand the influence of the parameter Ψ, a single segment is auralized for the simplest case.

The trailing edge noise emitted by the tip segment of one wind turbine blade at a radial distance of 45 m, rotational speed Ω= 1.47 rad/sec and hub height of H_0 = 80 m is auralized with N_β=12 discrete angular positions. The wind inflow velocity for all position is taken to be 8 m/s (no wind shear). The maximum change in amplitude of the auralized noise will be observed for a receiver at the crosswind position ($\tau = 90°$). Taking the distance of the receiver as R_0 = 100 m from the base of the hub for this position, the system is auralized for different values of Ψ, between 1% and 100%.

An RMS with the moving window of duration 50 ms is used to calculate the SPL which is used as an envelope function to detect the structural changes in the auralized signal related to the individual grains. This time duration is well adapted to detect the structural differences in the

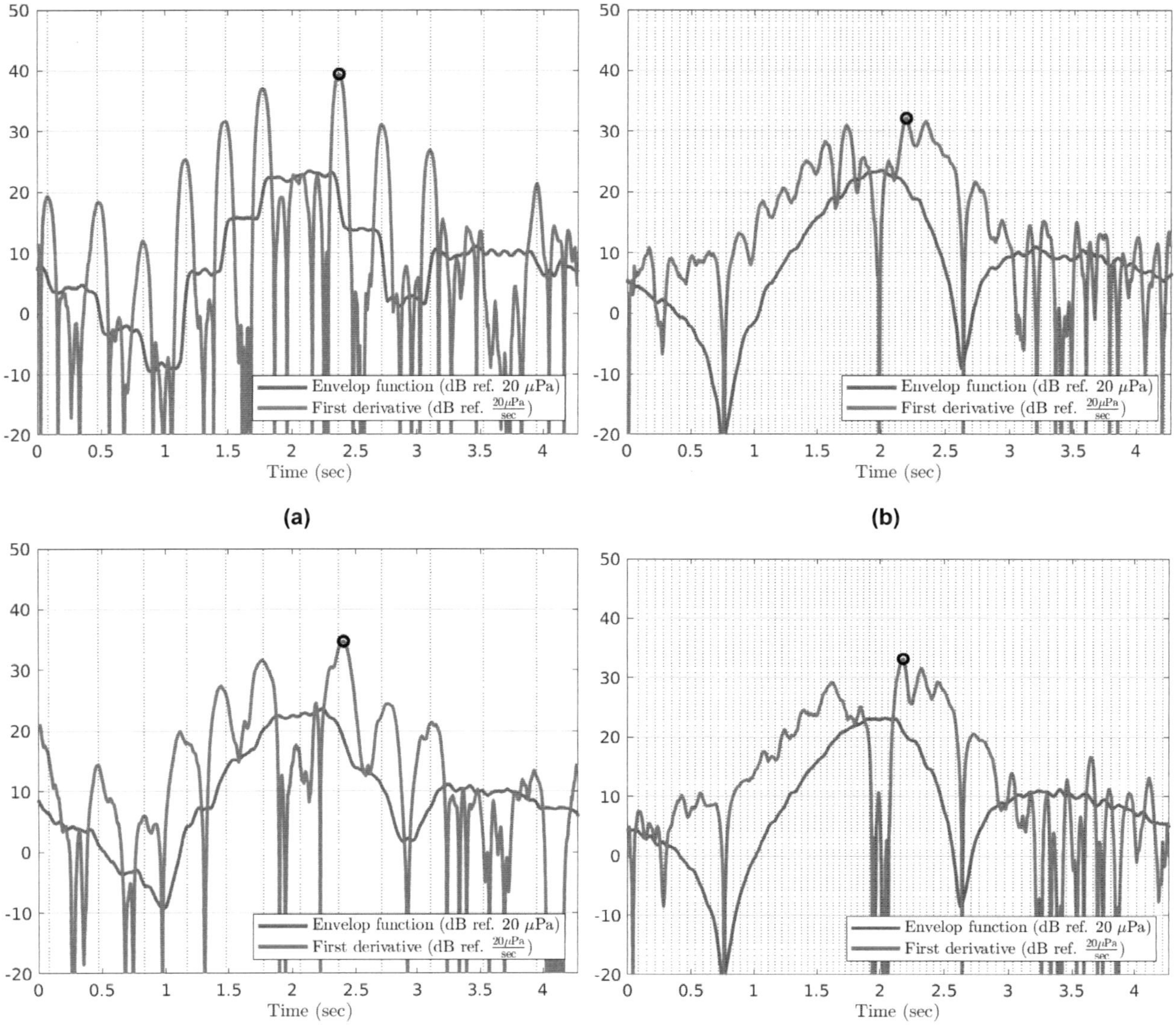

(a) (b) (c) (d)

Figure 5: The envelope function $L_1(t)$ with the SPL of the time derivative $L_2(t)$. The vertical lines indicate the transitions in time between the grains and the black circle showing the peak of the derivative. (a) $\Psi = 1\%$ and N_β= 12, (b) $\Psi = 1\%$ and N_β= 72, (c) $\Psi = 100\%$ and N_β= 12, (d) $\Psi = 100\%$ and N_β= 72.

signal because the minimum grain duration in the system is greater than 50 ms. The SPL of the envelop function used is given by:

$$L_1(t) = 10\ \log_{10}\left(\frac{p^2_{rms,50\text{ms}}(t)}{p^2_{ref}}\right) \quad (\text{dB re. } 20\ \mu\text{Pa}), \tag{11}$$

where $p_{rms,50\text{ms}}$ is the moving RMS over 50 ms. The change in the amplitude of the grains is understood by taking the SPL of the time derivative of the moving RMS which is given by:

$$L_2(t) = 10\ \log_{10}\left(\frac{(dp_{rms,50\text{ms}}(t)/dt)^2}{(p_{ref}/1\ s)^2}\right) \quad (\text{dB re. } 20\ \mu\text{Pa/s}). \tag{12}$$

The variations in the amplitude between successive grains is captured as peaks as seen in Fig. 5. The maximum rate of change in the amplitude in the auralized signal is used to quantify the quality of the transitions for different amounts of the overlap. For different values of Ψ, the maximum rate of amplitude change between grains is seen in Fig. 6a.

The larger the amount of the overlap between two grains, the smoother is the transition, with Ψ= 100% facilitating the smoothest transition between grains. A difference of $\sim 6\ dB$ of the rate of amplitude change is seen between the maximum and minimum values of the overlap. This difference between the change in the amplitudes of each grain for different values of Ψ is clearly audible: [link]. The average SPL calculated by Eq. (11) for the entire signal is the same for every value of Ψ with the standard deviation of less than 1 dB, which indicates that the overall power level is conserved for all values of Ψ. As the computational cost is the same for any value of Ψ, choosing the optimal amount of Ψ = 100% is beneficial for the synthesis of the signal.

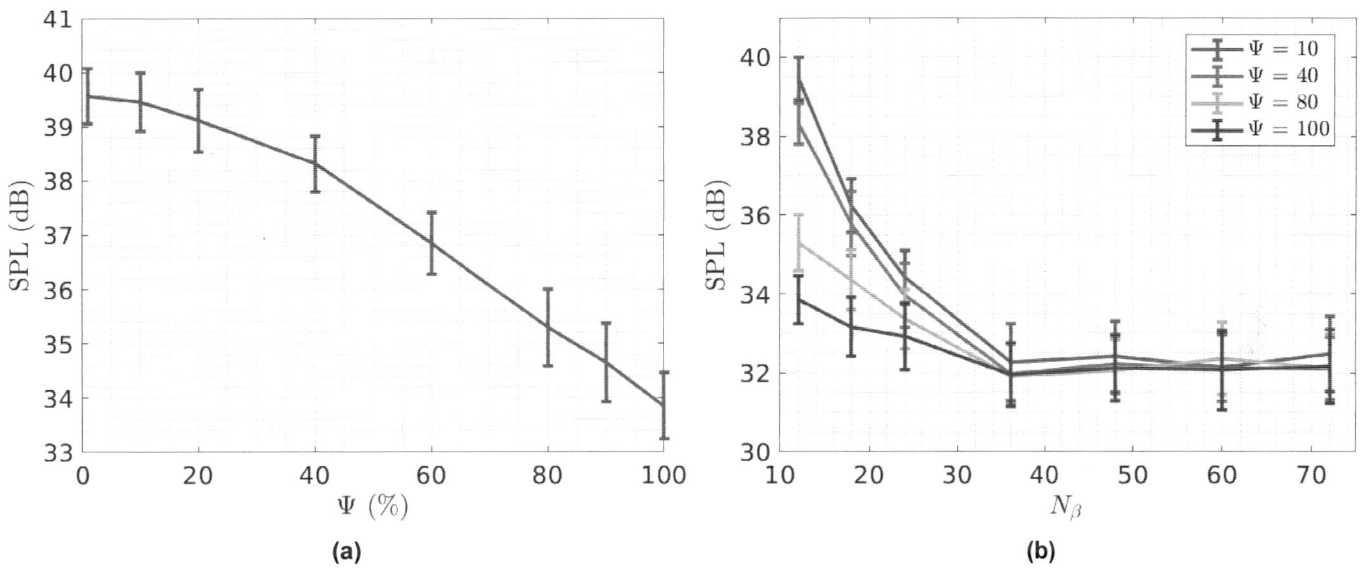

Figure 6: Maximum of the calculated SPL L_2 for (a) different values of Ψ with $N_\beta = 12$, (b) different values of N_β. The error bars show the standard deviation calculated over 50 realizations.

4.2 Influence of the number of grains

The number of grains in the auralized signal is equal to the number of discrete angular positions N_β set for the rotation of the blade. It is apparent that the larger the number of angular positions, the closer the system approaches the continuous rotational motion of the blade. The difference in the amplitude between two adjacent grains changes with the number of discretized angular positions in a single rotation. For a larger number of discretized angular positions, the amplitude change between adjacent grains is less, resulting in a smoother transition between grains in the auralized signal. The influence of N_β on the quality of the auralized signal can thus be related to the rate of the amplitude change between grains. To understand how the quality of the auralized signal is influenced by N_β, the tip segment is considered as in the previous section (Section 4.1) using different values of N_β.

The SPL of the envelop function as defined by Eq. (11) is taken again over 50 ms which is still smaller than the smallest grain size used for the analysis. The rate of change in the amplitude is captured by taking the SPL of the moving RMS which is defined by Eq. (12). The maximum rate of change in the amplitude of the auralized signal is used to quantify the influence of N_β on the quality of the signal. For different values of N_β, the maximum rate of amplitude change between grains is seen in Fig. 6b. It is clear from Fig. 5 and 6b that the quality of the transitions in the auralized signal is influenced by the number of discrete angular positions, N_β. The value of the curves of the maximum rate of change converges for N_β greater than 36. This difference of the quality of the transitions is also audibly distinct [link], with the largest value of N_β approaching the smoothest signal.

Increasing N_β comes with an increase of the computational cost. To resolve this particular trade-off between the realism and computational cost, a lower value of N_β can be used with the largest possible value of Ψ. Using $N_\beta = 36$ and $\Psi = 100\%$ the signal can be auralized approaching the quality that is attained using $N_\beta = 72$ and $\Psi = 10\%$ [link]. Note that the auralization done for this analysis concerns a single blade segment. Accounting all the segments of the wind turbine blades will induce less audible artifacts, as the transitions of each grain occur at a different time for each segment, thus the transitions between grains are not very noticeable when a complete wind turbine is auralized.

4.3 Test cases

Using the optimal values of the parameters $\Psi = 100\%$ with $N_\beta = 36$, we auralize the wind turbine noise for different receiver positions between the frequency limits f_{min}= 100 Hz and f_{max}= 6000 Hz. As the trailing edge noise and turbulent inflow noise have independent contribution to the noise of the wind turbine, they are auralized separately and then summed together. Similarly the contribution of each of the 8 segments of the blade is considered to be independent and thus also auralized separately and finally summed. The resultant auralized signal from one blade is shifted correspondingly in time to add the response of the second and the third blade of the wind turbine, thus achieving the desired auralization for the wind turbine noise.

For a wind turbine with rotational speed Ω = 1.47 rad/sec, blade span of 45 m and hub height of H_0 = 80 m, the test cases are done for no wind shear and different wind shear exponents, α and various receiver positions on the ground. The turbulence parameters that are considered correspond to a neutral atmosphere, as given in Tian and Cotté (2016). The spectrograms of the auralized signals for the receivers at the downwind and crosswind positions are seen in Fig. 7. Strong amplitude modulation is seen in the crosswind position, while there is less variation downwind. The auralized signals for the test cases of different parameters are available here: `http://sites.google.com/view/david-mascarenhas/wtnc-2021`.

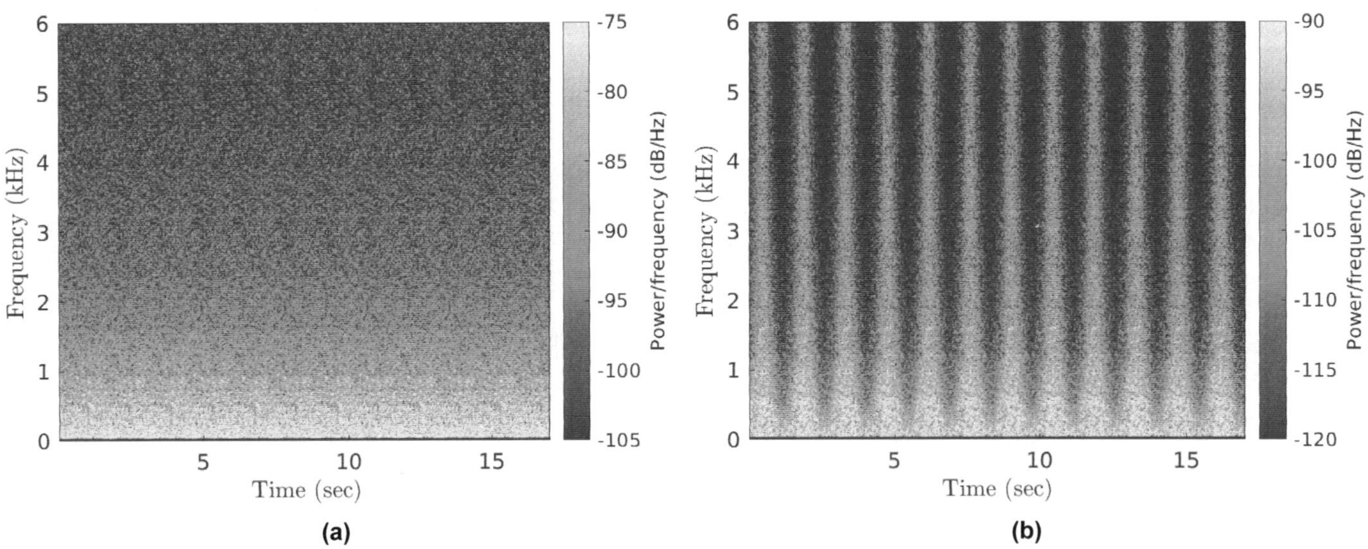

Figure 7: The spectrograms of the auralized wind turbine noise inclusive of turbulent inflow noise and trailing edge noise for no wind shear. Receiver position at R_0 = 100 m and (a) Downwind (τ = $0°$), (b) Crosswind (τ = $90°$).

5. Conclusions

The auralization tool discussed in this article converts the frequency-domain model of the wind turbine noise to a time-domain signal. The auralization is based on the physical models of the airfoil noise generated by the leading edge and the trailing edge of the wind turbine blades. As this model is physics-based, modification in the parameters such as the receiver position, rotational speed, wind speed etc. are easily achieved to alter the requirements for the desired auralized signal. The important parameters that alter the quality of the auralization such as the overlap amount Ψ and the number of grains N_β are also described. From the analysis, we conclude that the optimum value for Ψ is 100% and induce no additional computational cost. We also conclude that the maximum rate of change in the amplitude for different values of N_β converges above N_β = 36. Considering the computational cost and the audible difference which is noticeable, N_β = 36 and Ψ = 100% provides the optimum values for the auralization.

The auralization is done based on the source emitting the airfoil noise in free field. This does not include the atmospheric and propagation effects such as absorption, refraction, ground reflection. Other sources of noise such as vegetation and background noise also need to be added to simulate a realistic environment of the required scenario. The estimated values of the parameters that are used are on the basis of the results obtained by signal processing. In the future, we can envisage to optimize the synthesis parameters on the basis of psycho-acoustic analysis in order to minimize computational costs while achieving the most realistic wind turbine auralization.

Acknowledgement

This project has received funding from the European Union's Horizon 2020 research and innovation programme under the Marie Sklodowska-Curie grant agreement No 812719.

References

Amiet, R. K. (1975). Acoustic radiation from an airfoil in a turbulent stream. *Journal of Sound and vibration*, 41(4):407–420.

Amiet, R. K. (1976). Noise due to turbulent flow past a trailing edge. *Journal of sound and vibration*, 47(3):387–393.

Bäckström, T. (2019). Overlap-add windows with maximum energy concentration for speech and audio processing. In *ICASSP 2019-2019 IEEE International Conference on Acoustics, Speech and Signal Processing (ICASSP)*, pages 491–495. IEEE.

Bertagnolio, F., Madsen, H. A., and Fischer, A. (2017). A combined aeroelastic-aeroacoustic model for wind turbine noise: verification and analysis of field measurements. *Wind Energy*, 20(8):1331–1348.

Bowdler, D. and Leventhall, G. (2011). *Wind Turbine Noise: how it is produced, propagated measured and received*. Multi-Science Publishing, Brentwood, UK.

Buck, S., Oerlemans, S., and Palo, S. (2018). Experimental validation of a wind turbine turbulent inflow noise prediction code. *AIAA Journal*, 56(4):1495–1506.

Fink, M., Holters, M., and Zölzer, U. (2016). Signal-matched power-complementary cross-fading and dry-wet mixing. *Proceedings of the 19th International Conference on Digital Audio Effects (DAFx-16)*.

Oerlemans, S. and Schepers, J. G. (2009). Prediction of wind turbine noise and validation against experiment. *International journal of aeroacoustics*, 8(6):555–584.

Pieren, R., Heutschi, K., Müller, M., Manyoky, M., and Eggenschwiler, K. (2014). Auralization of wind turbine noise: emission synthesis. *Acta Acustica United with Acustica*, 100(1):25–33.

Roger, M. and Moreau, S. (2010). Extensions and limitations of analytical airfoil broadband noise models. *International Journal of Aeroacoustics*, 9(3):273–305.

Sinayoko, S., Kingan, M., and Agarwal, A. (2013). Trailing edge noise theory for rotating blades in uniform flow. *Proceedings of the Royal Society A: Mathematical, Physical and Engineering Sciences*, 469(2157):20130065.

Tian, Y. and Cotté, B. (2016). Wind turbine noise modeling based on amiet's theory: Effects of wind shear and atmospheric turbulence. *Acta Acustica united with Acustica*, 102(4):626–639.

INCEEUROPE

9th International Conference
on
Wind Turbine Noise
Remote from Europe – 18th to 21st May 2021

Comparison of tonality analysis methods for wind turbine receptor based long-term monitoring data sets

Allan Munro, Aercoustics : AllanM@Aercoustics.com
Adam Suban-Loewen, Aercoustics : AdamS@Aercoustics.com
Payam Ashtiani, Aercoustics : PayamA@Aercoustics.com

Summary

The purpose of this study is to assess whether the IEC 61400-11 (IEC) and ISO/PAS 20065 (ISO) standards with different averaging time intervals result in significant variations in tonal audibility results, by comparing an analysis of long-term data sets using both tonality methods for time intervals of 3, 10 and 60 seconds.

For steady tones, the analysis revealed similar Mean Audibility for different time intervals using a given methodology, and different Mean Audibility results between ISO and IEC, when compared. The study also revealed different Tonal Prevalence Trends for ISO and IEC. For non-stationary tones, the analysis revealed higher Mean Audibility for shorter time intervals for IEC, and variable Mean Audibility for different time intervals for ISO.

To better understand the variation in ISO tonality results for non-stationary tones, the authors examine case studies showing: (1) “tone blur” - where the tones frequency variation causes a tone to blur; (2) “diluted tone” - where high audibilities are diluted from low audibilities at other frequencies; and (3) “crossing tones” - two tones crossing in frequency.

These case studies demonstrate the complex nature of evaluating tonal audibility measured at the receptor location near to wind turbines and the factors that could influence the mean tonal audibilities for different averaging time intervals.

1. Introduction

The results of tonal assessments are of concern to both people who reside near wind turbines, and to wind turbine operators who must demonstrate compliance with local regulatory noise limits. Although methods exist for assessing tonality for wind turbines and other industrial noise, there is some uncertainty about how to assess wind turbine tonality at receptor locations specifically.

The measured tonal audibility at the receptor location may be influenced by factors at the source (e.g., varying tonal frequency and amplitude from the turbine) or at the receptor (e.g., variation in masking noise or acoustic interactions between multiple turbines).

Due to the potential for variation of tonal frequency and other factors that affect tonal audibility, the averaging time is of specific import. For example, it is not clear if averaging all 1-minute periods at a given wind speed can underestimate the tonal audibility at a residence and if a shorter averaging time is more appropriate.

Previous studies adapted either the IEC 61400-11 or the ISO/PAS 20065 standard for receptor-based wind turbine measurements by modifying the narrowband time interval (3, 10, and 60 seconds) to apply to long-term monitoring data sets that can last for weeks or months.

The purpose of this study is to assess whether the IEC 61400-11 and ISO/PAS 20065 standards with different averaging time intervals result in significant variations in tonal audibility results, by comparing an analysis of long-term data sets using both tonality methods for time intervals of 3, 10 and 60 seconds.

1.1 Overview of Standards

There are multiple standards for the assessment of tones in sound pressure recordings. Narrowband methods are frequently used in common standards such as ISO/PAS 20065 (2016) [1] (ISO), which targets general purpose detection of tones from any source in broadband noise; and IEC 61400-11 Edition 3.1 Section 9.5 (2012) [2] (IEC), which specifically targets measurement of tones from a nearby wind turbine source. While IEC addresses sound emission from a location close to the turbine, investigation has been done to its application to immission measurements at receptors [3].

A common approach underlying both IEC and ISO is to assess a critical band around each possible tone, where lines are classified as “tone”, “masking”, or “neither”. The audibility of the tone is calculated as the energy ratio of summed tone energy and an effective sum of masking energy, with a frequency-dependent adjustment applied. This “audibility criterion” (IEC) or “masking index” (ISO) curve has been determined experimentally from listening tests and reflects the subjective response of a typical listener to time-invariant tones of different frequencies. ISO also includes additional checks such as tone distinctness, and cases of exactly two tones in a critical band.

Some differences between IEC and ISO are summarized in Table 1.

Table 1: Comparison of IEC 61400-11 Edition 3.1 Section 9.5 (2012) [2] and ISO/PAS 20065 (2016) [1] tonality standards

	IEC	ISO
Time interval over which each narrowband spectrum is calculated	10 seconds	3 seconds
Line spacing or frequency resolution	[1.0, 2.0] Hz	[1.9, 4.0] Hz
Masking Level	A single energy average from the lines below the 70th percentile	Iterative procedure, which eliminates possible masking lines, conservatively reducing the level as low as possible
Tone Grouping	Tones of the same "origin" within ±25% of the critical band are energy averaged together	Audibility is recalculated using all lines classified as tone in the critical band
Combining audibilities from multiple spectra over time	Energy average	Energy average, after replacement of audibility <= 0 dB, or no detected tone, with -10 dB

Some regional jurisdictions use 1-minute measurement intervals, which facilitate matching the measured Sound Pressure Level and corresponding turbine operational parameters. This introduces the question of how the prescribed methods of combining multiple spectra to 60-second intervals compares with longer 60-second time intervals for narrowband calculation. The effect on non-stationary or frequency modulated tones is of particular interest.

1.2 Considerations for Measurement of Tones in Wind Turbine Noise

The measurement of far-field noise from wind farms is typically complicated by a few factors of note. The perceived prominence of a tone is driven by the difference between tonal energy and masking energy within a critical frequency band. Because the audibility is a relative measure of these differences, changes in each aspect (changes in the tone level, and changes in the masking energy) have the potential to change the perceived audibility simultaneously and can make for tonal perception that is highly variable in time.

Tone Level

Factors that affect the tone level have to do with the mechanism of tone generation, and its characteristics. Tones caused by rotating machinery such as gearboxes will have an inherent level and directivity depending on the mechanism of sound emissions. The main mechanism for gearbox noise is usually affected by a combination of forced and resonant response of the nacelle/gearbox/tower interactions and can have a complex directivity pattern. Other examples have been pumps, fans or blowers in the nacelle used for cooling, and emanating sound from openings in the nacelle. These would have directivity patterns related to the opening locations.

Masking Noise

Wind turbines require the presence of wind to operate, resulting in audible noise being generated by both the turbine and the wind at the same time. As a result, the background sound levels due to wind interaction with objects, such as trees, can be significant sources of masking noise. Such levels of wind induced background noise are routinely comparable and often exceed the sound from wind turbines alone at higher wind speeds (i.e., 5-7 m/s) and can often mask the tone. At lower wind speeds (i.e., 0-4m/s), the data collected is less dependent on wind speed as the wind interaction with trees etc. is intuitively less prevalent when compared to higher wind speeds. Vegetation noise can also be correlated to wind direction depending on the type, orientation and distance of the vegetation to the microphone.

2. Investigation into Tonality Analysis for Long-Term Data

2.1 General Background on Long-Term Data Sets

Two long-term data sets were chosen for the analysis because of the presence of prominent tones, and high-quality data as demonstrated by a high signal to noise ratio.

Tones identified in the emission-test of the turbine type closest to the measurement location or prominent tones identified in the long-term data sets were used to determine "frequencies of interest" which were used as a basis for a search for tones at the receptor which were likely to have been generated by the closest turbine rather than an external source. These emission tests conformed to the methodology prescribed by IEC 61400-11 (edition 3.0), "wind turbine generator systems – Part 11: Acoustic noise measurement techniques".

Specifically, for long-term data set #1, a "438 Hz tone" was prominent. The monitor was placed in an open field with multiple turbines of the same type, where the closest turbine was 629 metres away from the monitoring location, in the downwind direction of the prevailing wind condition. A secondary turbine was located 750 metres away in the same general direction, and other turbines were more than 1 kilometre away. In this data set, wind bins 4 – 7 are presented and have a high signal to noise ratio ranging from 12.3 – 4.6 dB, respectively.

For long-term data set #2, a "110 Hz" tone was prominent from one turbine that was 428 metres away from the monitoring location, in the downwind direction of the prevailing wind condition. The monitor was approximately 10 metres from some nearby trees, and more than 1 kilometre away from other turbines in the area. High shear conditions were prevalent for this data set with high hub height wind conditions present concurrently with low 10m height wind speeds at the monitoring location. In this data set, wind bins 1 – 4 are presented and have a high signal to noise ratio ranging from 13.4 – 6.2 dB, respectively.

2.2 Overview of Data Collection and Filtering

For the duration of the measurement campaign, acoustic and anemometer data was logged simultaneously in one-minute intervals. The acoustic data included A-weighted overall equivalent sound levels (LAeq) and 1/3 octave band levels between 20 Hz and 10,000 Hz. Additionally, audio recordings of each interval were saved for the purposes of listening/analysis and source verification. The audio recordings were recorded with a sample rate of 25600 Hz to facilitate the tonality analysis.

The microphone was placed at a measurement height of 1.5m for long-term data set #1 and 4.5m above grade for long-term data set #2. In addition, the microphone was placed at least 5 metres away from any large reflecting surfaces, in direct line of sight to the nearest turbines, and as far as practically possible from trees or other foliage.

The recorded weather data included average wind direction and wind speed. To account for the effect of wind speed on the measured sound level, measurement intervals are sorted into integer wind bins based on the measured 10m windspeed, into full integer wind bins. Each wind bin ranges from 0.5 m/s below to 0.5 m/s above each respective wind bin (i.e., 5m/s wind bin represents all intervals with average wind speeds between 4.5m/s and 5.5m/s).

The data sets were filtered to exclude measurement intervals where the data reliability is reduced due to transient noise intrusions (such as vehicle pass-bys), environmental conditions, or equipment operating outside of its specifications.

Intervals that pass the filtering criteria are sorted into "Turbine ON" periods based on if the nearby turbines were rotating and generating high electrical output (i.e., >85% rated power).

As required by regional jurisdictions, the data was filtered by wind direction for downwind conditions. The downwind condition is considered a worst-case assessment for overall sound pressure level. It should be noted that the downwind condition may not necessarily be the worst case for tonal audibility, and it is possible that crosswind conditions and cases with lower electrical power output may also have significant tonal audibility.

2.3 Methodology for Narrowband Calculation

The sound pressure waveform was recorded with 25600 Hz sample rate and 32-bit floating point precision and calibrated to units of Pa. Recordings were divided into 3, 10, and 60-second intervals. The waveform for each interval was A-weighted with a digital filter, then sliced into 0.5-second segments, with 50% overlap. Each segment was multiplied by the Hann window, then its one-sided DFT magnitude was calculated with the FFT, resulting in lines of 2 Hz frequency resolution. Normalization was done for the number of samples in the segment, the coherent gain of the Hann window, and the one-sided nature of the DFT. Note that no normalization was done for the effective frequency resolution, so values were not represented as Power Spectral Density. The narrowband lines were energy averaged across all segments for a given interval. The values were converted to dB with a reference of 20μPa.

2.4 Methodology for Tonality Assessment

The Decisive Audibility as per ISO and the Mean Audibility as per IEC and ISO was calculated for every 60 second measurement interval for long-term data set #1 and #2. This was repeated for the 10 and 3 second measurement intervals. The tonal prevalence was also calculated for each wind bin for each respective time interval. Tones of the same origin are identified using the guidance of IEC, which stipulates that tones are considered tones of the same origin if they are within an interval of ±25% of the critical band centred at the frequency of interest.

- **Decisive Audibility (ISO)** is the maximum audibility for that interval within a frequency range of interest corresponding to tones of the same origin, to eliminate contamination and unrelated tones from other sources.
- **Mean Audibility (ISO)** is the energy average of the decisive audibilities for each interval. To ensure sufficient distance from the positive audibilities for all spectra in which no tone is found, or the audibility is <= 0 dB, a -10 dB placeholder is applied. The mean audibility is aggregated over 60-second intervals.
- **The overall tonal audibility or "Mean Audibility" (IEC)** is the energy average of all data points with an identified tone of the same origin for each wind speed bin. Only intervals with identified tones are included.
- **The tonal prevalence** is the percentage of the data points within a wind bin with a tone detected for a respective time interval.

2.5 Results for Long-Term Data Sets for Downwind and >85% Power

The results of the tonality analysis for long-term data set #1 are summarized in Tables 2 (ISO) and 3 (IEC), and the results for long-term data set #2 are summarized in Tables 4 (ISO) and 5 (IEC).

Table 2: Long-term data set #1- ISO "438 Hz tone" - 60s, 10s and 3s Interval - Mean Audibility and Tonal Prevalence

	Wind Bin (m/s)	Mean Audibility [dB]	Tonal Prevalence [%]
60 seconds	4	-4.9	25
	5	2.9	84
	6	1.5	73
	7	-1.9	37
10 seconds	4	-3.4	29
	5	2.7	80
	6	1.4	67
	7	-1.3	40
3 seconds	4	-2.4	36
	5	2.6	75
	6	1.4	63
	7	-1.2	37

Table 3: Long-term data set #1 – IEC "438 Hz tone" - 60s, 10s and 3s interval - Mean Audibility and Tonal Prevalence

	Wind Bin (m/s)	Mean Audibility [dB]	Tonal Prevalence [%]
60 seconds	4	-2.7	50
	5	1.9	76
	6	0.7	58
	7	0.2	28
10 seconds	4	-1.6	88
	5	1.9	91
	6	0.9	83
	7	-0.8	70
3 seconds	4	-1.1	86
	5	2.1	95
	6	1.3	91
	7	-0.3	80

Table 4: Long-term data set #2 – ISO "110 Hz tone" - 60s, 10s and 3s - Mean Audibility and Tonal Prevalence

	Wind Bin (m/s)	Mean Audibility [dB]	Tonal Prevalence [%]
60 seconds	1	-1.5	67
	2	1.3	83
	3	1.2	76
	4	-1.1	54
10 seconds	1	-1.7	56
	2	1.3	79
	3	1.3	75
	4	-1.0	51
3 seconds	1	-0.1	75
	2	1.3	73
	3	1.4	71
	4	-0.9	49

Table 5: Long-term data set #2 – IEC "110Hz tone" - 60s, 10s and 3s - Mean Audibility and Tonal Prevalence

	Wind Bin (m/s)	Mean Audibility [dB]	Tonal Prevalence [%]
60 seconds	1	0.8	100
	2	2.2	98
	3	2.4	92
	4	0.8	88
10 seconds	1	0.8	100
	2	2.2	100
	3	2.2	99
	4	0.5	99
3 seconds	1	1.2	100
	2	2.3	100
	3	2.4	98
	4	0.7	98

Similar Mean Audibility for different time intervals for a particular methodology
For respective wind bins with positive tonal audibilities, the mean audibility was comparable for time intervals of 60, 10 and 3 seconds, only differing by 0.1 dB for ISO and 0.1-0.3 dB for IEC for long-term data set #1. For example, the 5m/s wind bin mean audibility as per ISO for 60, 10 and 3 seconds was 2.9 dB, 2.7 dB and 2.6 dB.

These results indicate that analysis using a time interval of 60, 10 or 3 seconds produces very similar mean audibility results for a particular methodology (either IEC or ISO). This is likely due to the fact that the data is filtered for worst-case overall sound pressure level which corresponds to steady turbine operational parameters (i.e., steady RPM and >85% power output) when the tone level is relatively stable, and data is grouped into wind bins with similar masking noise.

Different Mean Audibility between ISO and IEC, when compared.
While similar mean audibility for different time intervals for a particular methodology was observed. There is no clear answer of which standard would produce higher mean audibility results. For data set #1, ISO produced higher mean audibility results compared to IEC across all time intervals. For example, the 5m/s wind bin mean audibility for data set #1 as per ISO for 60, 10 and 3 seconds was 2.9 dB, 2.7 and 2.6 dB and the mean audibility as per IEC for 60, 10 and 3 seconds was 1.9 dB, 1.9 and 2.1 dB.

Conversely, for data set #2, IEC produced higher mean audibility results than ISO across all time intervals. For example, the 2m/s wind bin mean audibility for data set #2 as per ISO for 60, 10 and 3 seconds was 1.3 dB, 1.3 and 1.3 dB and the mean audibility as per IEC for 60, 10 and 3 seconds was 2.2 dB, 2.2 and 2.3 dB. This is likely due to differences in the calculation methods for ISO and IEC.

Different Tonal Prevalence Trends for ISO and IEC.
This analysis indicates that the ISO methodology has lower tonal prevalence for shorter time intervals. For example, for data set #1 (Table 2), Wind Bin 5 reduced in tonal prevalence from 84 to 80 to 75, over time intervals of 60, 10 and 3 seconds, respectively. Likewise, for data set #2 (Table 4), Wind Bin 2 reduced in tonal prevalence from 83 to 79 to 73, over time intervals of 60, 10 and 3 seconds, respectively.

Conversely, the IEC analysis displayed increasing prevalence for data set #1 for shorter time intervals and high prevalence for all time intervals for data set #2. For example, for data set #1 (Table 3), Wind Bin 5 increased in tonal prevalence from 76 to 91 to 95, over time intervals of 60, 10 and 3 seconds, respectively. For data set #2 (Table 5), there was generally high tonal prevalence for all time intervals.

3. Investigation into Non-Stationary Tones

3.1 Methodology for Non-Stationary Tones

To determine the impact of non-stationary tones, which have a changing frequency over time, the following steps were used to identify them in the long-term monitoring data sets. For the investigation of non-stationary tones, the filters for downwind and high electrical power output were removed.

The 3-second tonality results were filtered to ensure continuous periods of tone detection for at least 12 seconds (4 intervals) and used to calculate the standard deviation of tone frequency within each 60-second interval. The 1-minute data points were filtered within each wind bin for a standard deviation of tone frequency between 7 and 11 Hz. The lower bound was used to ensure sufficient variation, and the upper bound was used to avoid including separate tones in most cases. The remaining data points were investigated to confirm that the detected tones were non-stationary tones.

3.2 Results for Long Term Data Sets for Non-Stationary Tones

The results of the tonality analysis after filtering for non-stationary tones for long-term data set #1 are summarized in Table 6 (ISO) and Table 7 (IEC), and for long-term data set #2 are summarized in Table 8 (ISO) and Table 9 (IEC).

Table 6: Long-term data set #1 - ISO - 60s, 10s and 3s Interval - Mean Audibility and Tonal Prevalence

	Wind Bin (m/s)	Mean Audibility [dB]	Tonal Prevalence [%]
60 seconds	4	2.1	100
	5	3.4	100
	6	2.9	100
	7	2.0	100
10 seconds	4	1.5	65
	5	3.0	85
	6	2.1	79
	7	1.5	73
3 seconds	4	1.4	61
	5	2.8	79
	6	2.0	71
	7	1.3	62

Table 7: Long-term data set #1 - IEC - 60s, 10s and 3s Interval - Mean Audibility and Tonal Prevalence

	Wind Bin (m/s)	Mean Audibility [dB]	Tonal Prevalence [%]
60 seconds	4	2.2	63
	5	1.8	86
	6	0.8	69
	7	0.2	65
10 seconds	4	1.5	90
	5	2.1	94
	6	1.4	89
	7	1.0	85
3 seconds	4	1.6	93
	5	2.4	97
	6	1.7	95
	7	1.3	93

Table 8: Long-term data set #2 - ISO - 60s, 10s and 3s Interval - Mean Audibility and Tonal Prevalence

	Wind Bin (m/s)	Mean Audibility [dB]	Tonal Prevalence [%]
60 seconds	1	3.0	100
	2	4.6	100
	3	4.3	100
	4	4.8	100
10 seconds	1	2.9	78
	2	4.8	86
	3	4.4	84
	4	4.9	84
3 seconds	1	3.3	83
	2	5.1	88
	3	4.7	86
	4	5.3	86

Table 9: Long-term data set #2- IEC - 60s, 10s and 3s Interval - Mean Audibility and Tonal Prevalence

	Wind Bin (m/s)	Mean Audibility [dB]	Tonal Prevalence [%]
60 seconds	1	4.1	100
	2	5.4	91
	3	4.8	100
	4	4.9	100
10 seconds	1	4.3	94
	2	6.0	91
	3	5.5	96
	4	6.2	95
3 seconds	1	4.4	90
	2	6.6	90
	3	5.8	95
	4	6.5	95

Higher Mean Audibility for shorter time intervals for IEC

For respective wind bins, the mean audibility was higher for decreasing time intervals of 60, 10 and 3 seconds, differing by 0.6-1.1 dB for IEC for data set #1. For example, the 5m/s wind bin mean audibility as per IEC for 60, 10 and 3 seconds was 1.8 dB, 2.1 dB and 2.4 dB. These results indicate that analysis using a shorter time interval of 60, 10 or 3 seconds produces higher mean audibility results for IEC.

Variable Mean Audibility for different time intervals for ISO

Using the ISO analysis, data set #1 had higher mean audibility results for the 60-second time interval, as compared to the shorter 10 and 3 second time intervals, across all wind bins. However, using the ISO analysis on data set #2 produced the opposite result, i.e., lower mean audibility for the 60-second time interval, as compared to the shorter 10 and 3 second time intervals, across all wind bins. For example, for data set #1, the mean audibility results for wind bin 5 were 3.4, 3.0 and 2.8 dB for time intervals of 60, 10 and 3 seconds, respectively. Conversely, for long-term data set #2, the mean audibility results for wind bin 2 were 4.6, 4.8 and 5.1 dB for time intervals of 60, 10 and 3 seconds, respectively.

The main contributor to the variability found in the results of the ISO method was the application of a -10 dB placeholder for intervals during which a tonal audibility was negative. This can have the effect of reducing the mean audibility of a data set with lower tonal prevalence.

4. Case Studies of Non-Stationary Tones

To better understand the possible underlying causes for the tonality results with non-stationary tones, case studies were selected for ISO tonality analysis, representing divergent samples:

- **Tone Blur:** The first case shows a non-stationary tone with high 3-second decisive audibilities and mean audibility of 3-second intervals, and low 60-second decisive audibility, from the tone being blurred in the 60-second narrowband spectrum.
- **Diluted Tone:** The second case shows a non-stationary tone with some very high 3-second decisive audibilities but moderate mean audibility and 60-second decisive audibility, as a result of the high audibilities being diluted from low audibilities at other frequencies.
- **Crossing Tones:** The third case shows tones crossing in frequency, which results in a higher 60-second decisive audibility than mean audibility.

Table 10: Summary of tonality results for case studies of non-stationary tones

	Tone Frequency (Hz)	3-second Decisive Audibility (dB)	60-second Decisive Audibility (dB)	Mean Audibility of 3-second Intervals (dB)	Prevalence of 3-second Tones with Positive Audibility (%)
Tone Blur	[86.0, 108]	[2.6, 9.0]	0.5	5.8	100
Diluted Tone	[418, 452]	[0.5, 10.3]	5.1	5.2	90
Crossing Tones	[420, 446]	[0.2, 5.4]	3.5	1.3	75

4.1 Tone Blur

The spectrogram for a non-stationary tone with frequency varying from 86 to 108 Hz can be seen in Figure *1*. A summary of tonality results, including decisive, mean audibility and prevalence is shown in the first row of Table 10. The worst case 3-second audibilities on the lower and upper range of frequencies are shown in Figure 2 and Figure 3 respectively. Figure 4 shows the 60-second audibility calculated from a single narrowband spectrum, in which the tone is blurred across frequency as a result of the energy average in each narrowband line. As a result, the lower frequency portion of the tone is classified as "neither" or "masking"; the high masking level combined with the low levels of the lines classified as "tone" results in a low 60-second decisive audibility of 0.5 dB, especially compared to the mean audibility of 3-second intervals of 5.8 dB.

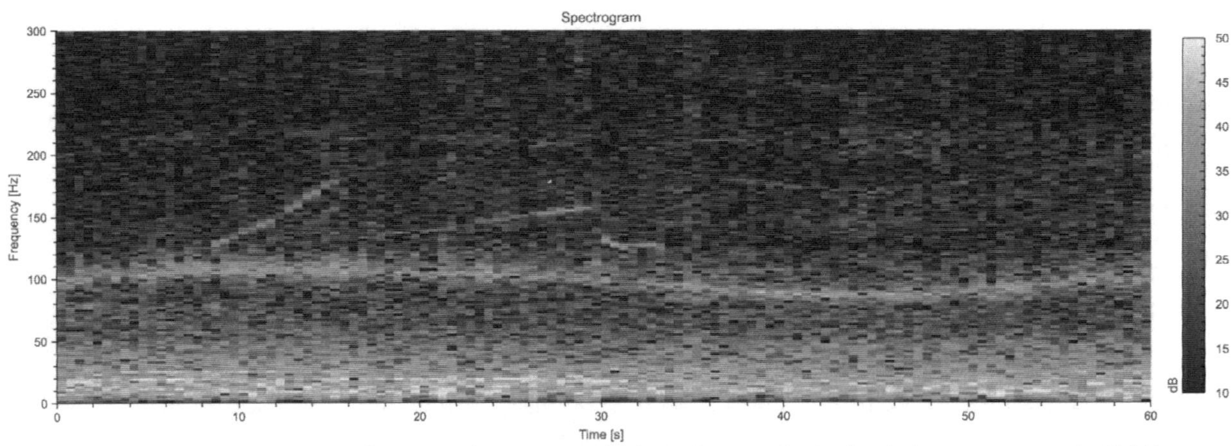

Figure 1: Narrowband spectrogram of a non-stationary tone, which can be seen fluctuating in frequency around ~100 Hz

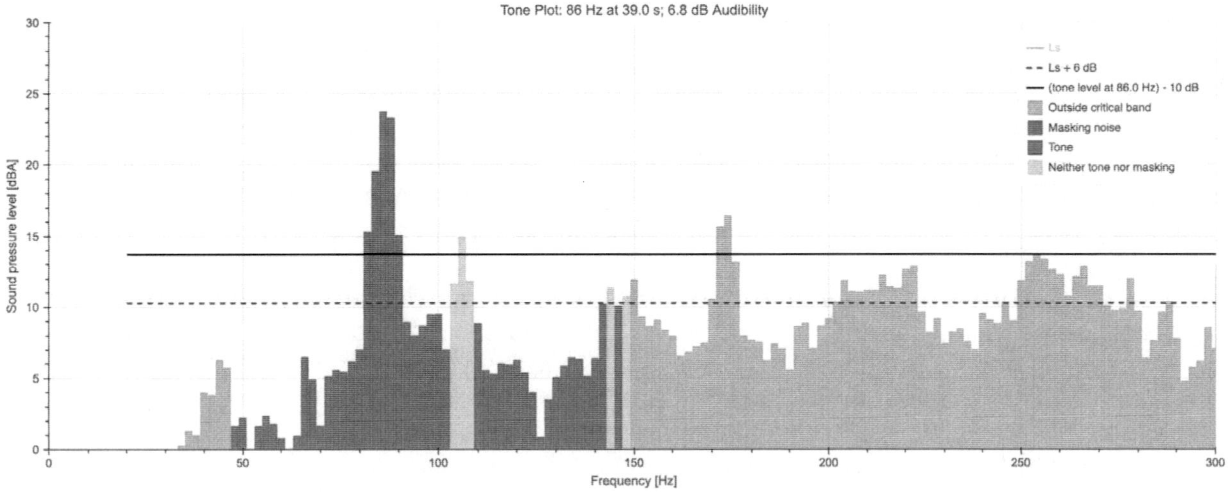

Figure 2: Narrowband line classification for worst case 3-second decisive audibility at 86 Hz

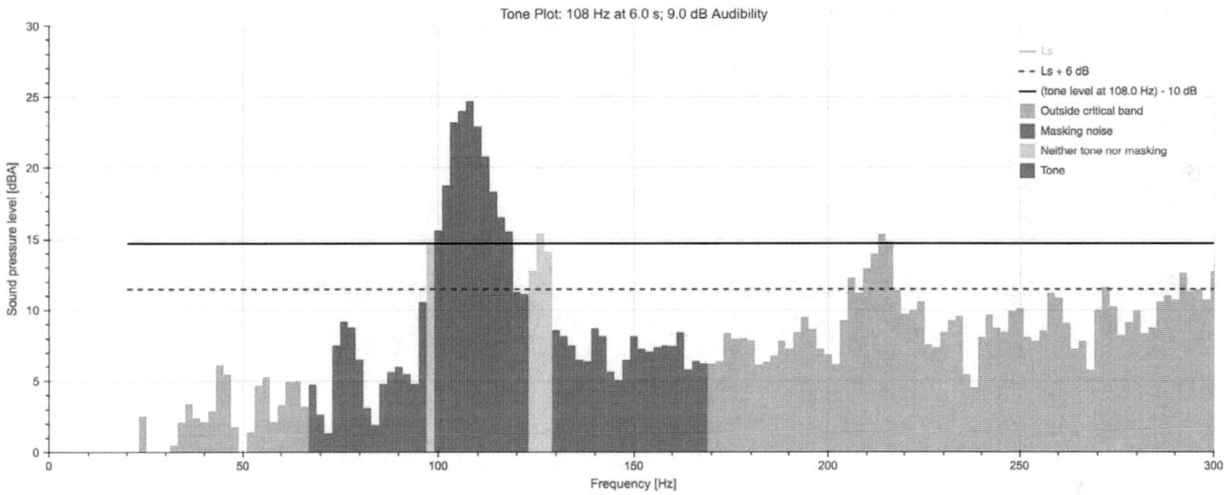

Figure 3: Narrowband line classification for worst case 3-second decisive audibility at 108 Hz

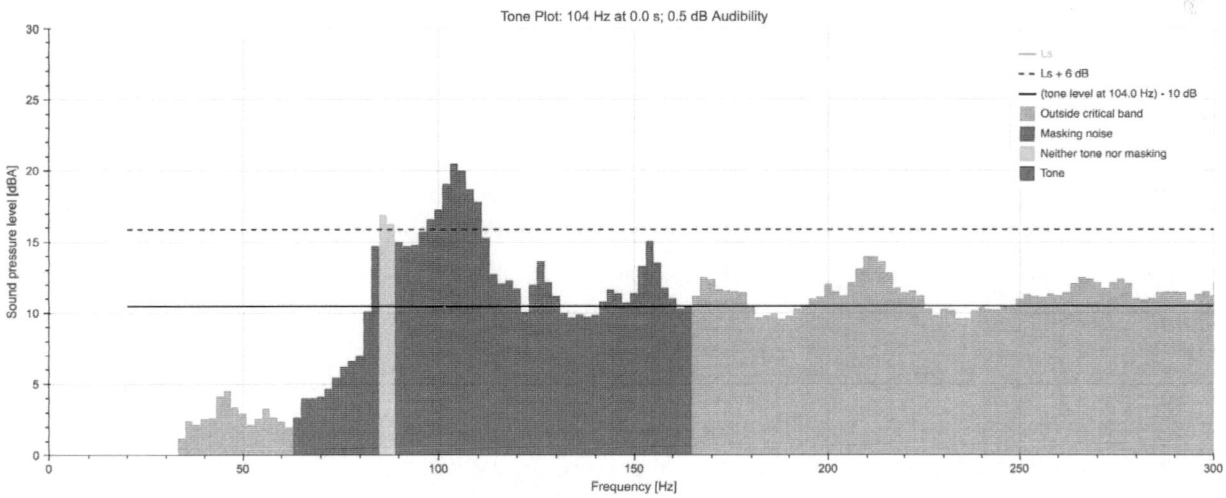

Figure 4: Narrowband line classification for 60-second decisive audibility

4.2 Diluted Tone

The spectrogram for a non-stationary tone with frequency varying from 418 to 452 Hz can be seen in Figure 5, and a summary of tonality results is provided in the second row of Table 10. The worst case 3-second audibilities on the lower and upper range of frequencies are shown in Figure 6 and Figure 7 respectively. Figure 8 shows the 60-second audibility calculated from a single narrowband spectrum, in which only the higher frequency lines are prominent enough to be classified as tone, and the lower frequency lines are classified as neither.

The consistently high levels of the upper frequency lines, along with the lower frequency lines being classified as "neither" rather than "masking", still results in a relatively high 60-second decisive audibility of 5.1 dB. Interestingly, this is comparable to the mean audibility of 3-second intervals of 5.2 dB as the very large 3-second audibilities become diluted in the energy average, both from smaller audibilities at different frequencies and from the -10 dB replacement for negative audibilities.

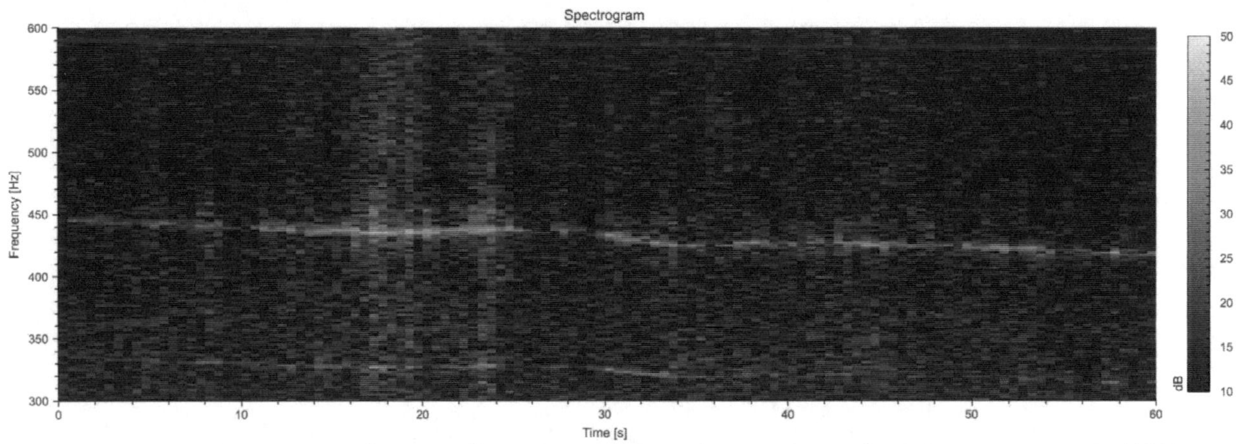

Figure 5: Narrowband spectrogram of a non-stationary tone, which can be seen descending in frequency around ~440 Hz

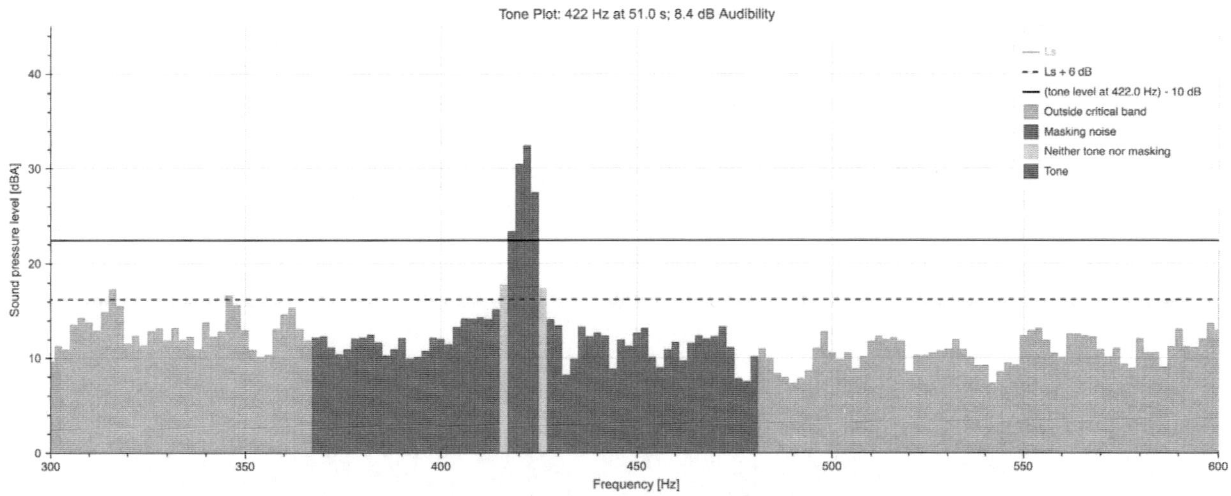

Figure 6: Narrowband line classification for worst case 3-second decisive audibility at 422 Hz

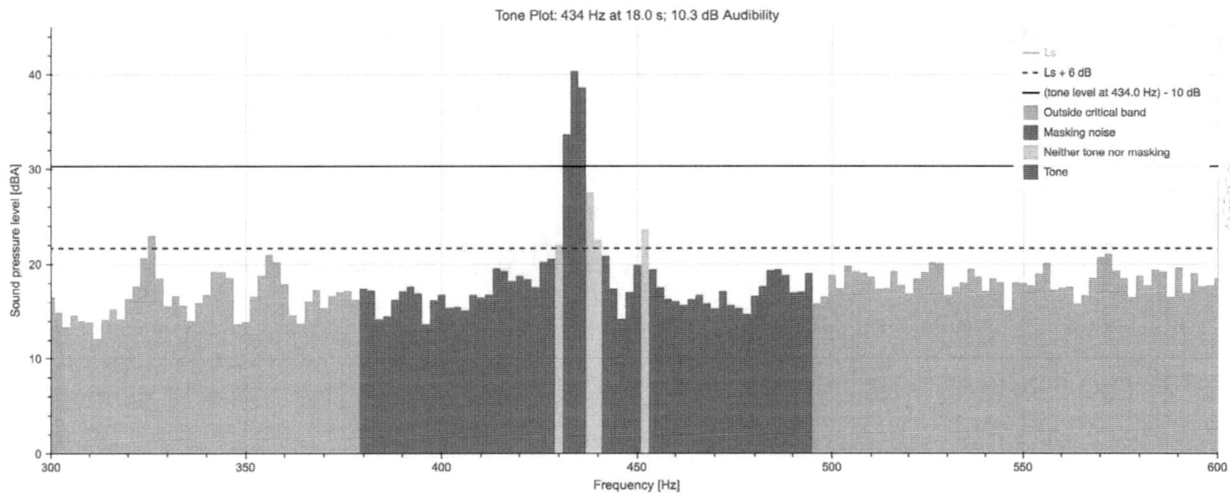

Figure 7: Narrowband line classification for worst case 3-second decisive audibility at 434 Hz

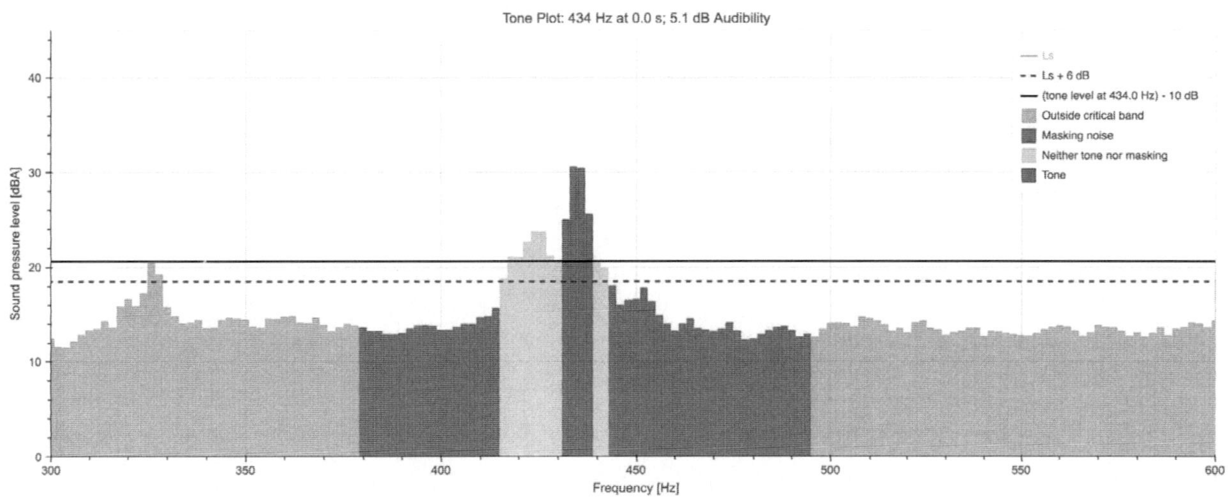

Figure 8: Narrowband line classification for 60-second decisive audibility

4.3 Crossing Tones

Figure 9 shows the spectrogram for two non-stationary tones crossing in frequency, varying from 420 to 446 Hz. The third row in Table 10 provides a summary of tonality results. The worst-case 3-second decisive audibilities during the crossover at 430 Hz and immediately after can be seen in Figure 10 and Figure 11 respectively. Figure 12 shows the 60-second audibility calculated from a single narrowband spectrum, in which both tones appear as a single very high bandwidth tone, consisting of 14 adjacent lines.

This results in a relatively high 60-second audibility of 3.5 dB, compared to the mean audibility of 3-second intervals of 1.3 dB, caused by the lower prevalence of positive 3-second audibilities.

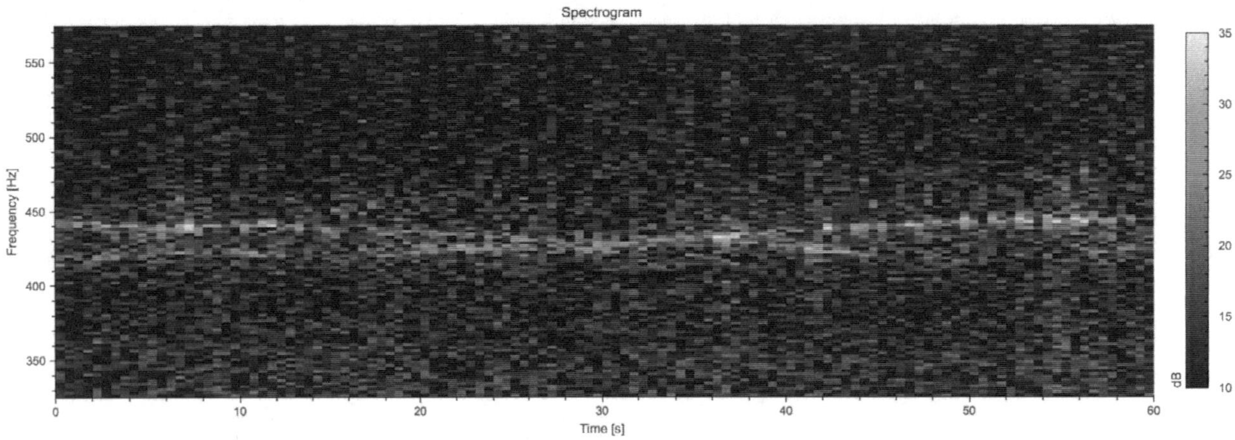

Figure 9: Narrowband spectrogram of a non-stationary tone, which can be seen crossing frequency around ~430 Hz

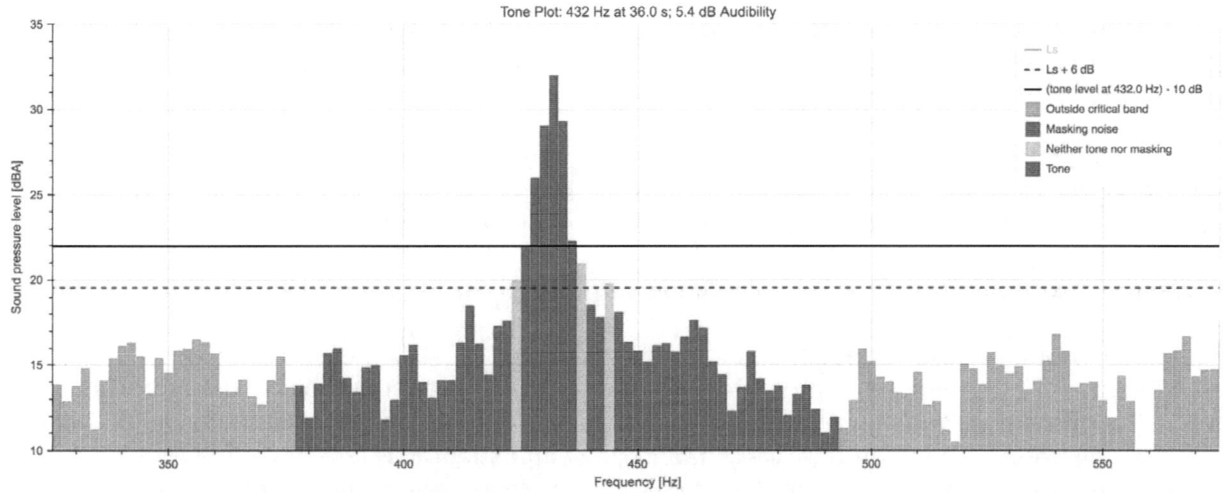

Figure 10: Narrowband line classification for worst case 3-second decisive audibility at 432 Hz, during cross-over

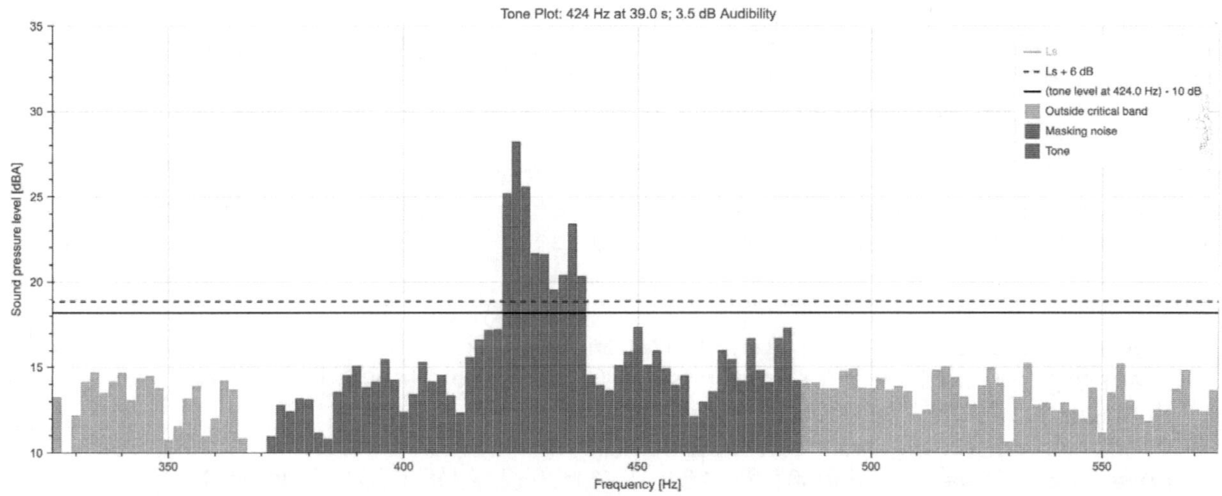

Figure 11: Narrowband line classification for worst case 3-second decisive audibility at 424 Hz (and 436 Hz), immediately following cross-over

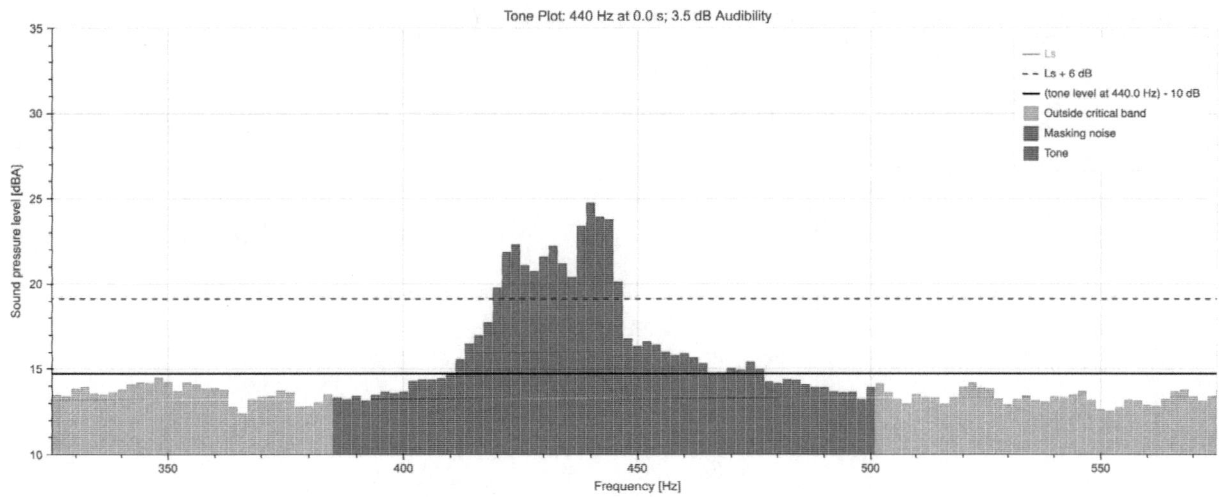

Figure 12: Narrowband line classification for 60-second decisive audibility

5. Conclusions

The purpose of this study was to assess whether the IEC 61400-11 (IEC) and ISO/PAS 20065 (ISO) standards with different averaging time intervals result in significant variations in tonal audibility results, by comparing an analysis of long-term data sets using both tonality methods for time intervals of 3, 10 and 60 seconds.

For steady tones, the analysis revealed similar Mean Audibility for different time intervals using a given methodology, and different Mean Audibility results between ISO and IEC, when compared. The study also revealed different Tonal Prevalence trends for ISO and IEC. For non-stationary tones, the analysis revealed higher Mean Audibility for shorter time intervals for IEC, and variable Mean Audibility for different time intervals for ISO.

The main contributor to the variability found in the results of the ISO method was the application of a -10 dB placeholder for intervals during which a tonal audibility was negative. This has the effect of changing the decisive audibility of a data set in unpredictable ways. While the philosophy behind this approach likely has to do with discounting times when a tone is inaudible, the application of the method is at risk of poor repeatability over a long-term measurement campaign.

These case studies demonstrate the complex nature of evaluating tonal audibility measured at the receptor location near to wind turbines and the factors that could influence the mean tonal audibilities for different averaging time intervals.

This study focused on the results of applying these methodologies to long-term data measured at wind farms. Further studies may consider psychoacoustic evaluation of the methods regarding how people would rate their perception of steady or non-stationary intermittent tones.

References

[1] International Organization for Standardization (2016) *Acoustics — Objective method for assessing the audibility of tones in noise — Engineering method*. Publicly Available Specification ISO/PAS 20065:2016(E).

[2] International Electrotechnical Commission (2012) *Wind turbines — Part 11: Acoustic noise measurement techniques*. IEC 61400-11:2012+AMD1:2018 Edition 3.1 Section 9.5

[3] T. Kobayashi, S. Yokoyama, A. Fukushima, T. Ohshima, S. Sakamoto, H. Tachibana (2015) *Assessment of tonal components contained in wind turbine noise in immission areas*. Proc. 6th International Wind Turbine Noise Conference.

INCE EUROPE

9th International Conference
on
Wind Turbine Noise
Remote from Europe – 18th to 21st May 2021

Wind Turbine Sound Quality Rating

Isaac Old, RSG: isaac.old@rsginc.com

Summary

The current state of practice for wind turbine sound emission rating is specification of the A-weighted sound power and tonal audibility pursuant to IEC 61400-11. While this allows for an accurate characterization of the overall sound emissions of a given wind turbine, there are other metrics that can help inform how a turbine sound. This presentation explores ways that sound quality metrics could be implemented to provide a more comprehensive acoustical rating for wind turbines.

1. Introduction

Wind turbine sound emissions are currently characterized using the A-weighted and/or 1/3 octave sound power and tonal audibility, measured and calculated according to IEC 61400-11. This sound power is then used, in part, in sound propagation modelling for noise assessments in the preconstruction phase of a project and is typically part of the turbine's warranty. It is also used for research on wind turbines, where modelled sound levels from wind turbines are compared with subjective response.

Most wind turbines sound relatively similar, so comparing modelled A-weighted sound levels at receivers can be used as a proxy for sounds with not only the same magnitude, but also similar sound quality. An issue can occur if a particular turbine has an atypical sound characteristic, such as a creak, squeak, grind, or knock, which I term a "sound defect". Similar sound defects are briefly discussed in Annex A of IEC 61400-11, but there is no measure of method given to quantify them, limiting its usefulness for rating or specification. While sound defects can be an indicator of a functional problem with the turbine, that is not always the case and, a noticeable defect does not always result in a sufficient increase in overall sound emissions to exceed the warranted range. These sound defects can nevertheless cause sound from a particular turbine to be more noticeable, and/or intrusive than a "normal" sounding turbine.

The purpose of the paper is to explore ways that wind turbine sound emission ratings could consider abnormal sound emissions in a way that could be included in a warranty. The idea of this paper is to start the conversation surrounding this topic.

While there are few standardized methods for rating how a given product sounds in addition to how much sound it produces. Product manufacturers have been using techniques to rate the "sound quality" of products . For example, automobile manufacturers have been using sound quality rating methods to quantify sound that is considered desirable to consumers. A variety of metrics have been derived over the years to quantify sound quality as are discussed in the next section.

2. Sound Quality Terminology

Quantifying how a source sounds instead of just how much sound it produces is most often discussed in the field of psychoacoustics. Several metrics have been developed to quantify sound quality. The four most common are Loudness, Sharpness, Roughness, and Fluctuation Strength. These are described in more detail below. Some less-common metrics that have been described in the literature are also mentioned.

Loudness

The sound quality metric of Loudness corresponds with a sensation of intensity. Loudness has the unit of sones. Loudness can also be converted into the loudness level. The unit of loudness level is the phon. Both are a function of level and frequency in a way that is described by equal loudness curves [1]. These were derived by comparing the perceived intensity of a sound to a pure tone of the same magnitude at 1 kHz. Equal loudness-curves went into the derivation of the A-weighting network. As a result, Loudness is already partially considered in IEC 61400-11 through the use of A-weighted sound levels. The difference is that equal loudness spectral curves change shape based on the overall magnitude of the sound. The A-weighting does not change shape based on the magnitude of the sound being measured. Changing the shape of the A-weighting for a *sound power* measurement would not make sense due to *sound power* not actually being a perceived measure. A-weighting also does not consider masking effects, while more recent loudness calculation methods do. A loudness calculation is standardized in ISO 532-1.

Sharpness

Sharpness is often considered to be related to how dense a sound is. Sharpness is most dependent upon the bandwidth of a sound and the center frequency of the band. Sharpness generally increases with increasing band center frequency and increasing critical band rate [1] [2]. The unit for Sharpness is the Acum. Unlike Loudness, Sharpness is largely uninfluenced by level. Increasing the upper cut-off frequency of a sound will tend to increase sharpness, as will increasing high frequency content. Decreasing the lower cut-off frequency will tend to reduce Sharpness.

Roughness

Roughness is the quality of sound that is related to rapid fluctuation of the sound and is dependent upon the extent that a sound modulates, as well as the frequency at which it modulates [1] [2]. At low modulation frequencies the fluctuation is perceived as a change in sound level, which peaks at about 4 Hz. By about 15 Hz a sound is considered "rough," a sensation that increases until about 70 Hz and vanishes above about 300 Hz. Roughness is also often related to increasing bandwidth of a sound. Frequency modulation, in addition to amplitude modulation can provide the sensation of roughness. The unit of roughness is the asper.

Fluctuation Strength

Fluctuation strength is related to the sensation of sound fluctuating in level. It is dependent on fluctuation depth, fluctuation rate, and the frequency of the sound that is fluctuating. At fluctuation rates above 20 Hz, the sensation transitions to roughness and the maximum fluctuation strength is at a rate of approximately 4 Hz [1] [2]. Fluctuation strength also increases with increasing level and increasing modulation depth. Fluctuation strength does not increase until a fluctuation depth of about 3 dB and reaches its maximum at a depth of about 30 dB. Fluctuation depth also reaches a maximum in the mid frequency range, decreasing with frequencies above and below 1 kHz. The unit of fluctuation strength is the bark.

Other Metrics

Several other metrics have been derived that do not appear to have gained widespread acceptance [1] [2]. These include:

- Impulse occurrence rate,
- Spectrum rotation,
- Sensory pleasantness,
- Spectrum rotation,
- High frequency modulation,
- Mid frequency modulation,
- Low frequency modulation,
- Kurtosis, and
- Pitch value.

Although they do not appear to be widely used, it does not mean they will be irrelevant for describing wind turbine sound quality.

3. Rating Sound Quality

Sound emissions ratings that include sound quality indicators is not a new idea, with a history that dates back at least 25 years in some form [3], though the purpose is often different. Assessing product sound quality has been part of the automotive industry for a while. Similar techniques have been used for brush cutters and vacuum cleaners, among others [2] [3]. In these cases, the intent is to make a product more desirable to a consumer for its particular use. This is different from trying to limit intrusive sound in a standardized manner. As a result, some of these methods may not work for rating wind turbines. Rating methods are described below, along with strengths and weaknesses of each method.

3.1 Jury Rating

The conventional way of determining the quality of a sound source is by presenting a series of sounds to a jury of listeners and to record their subjective reactions or have them compare the sound to a series of pre-selected descriptors. This allows actual feedback on the acceptability of a sound from listeners. There are a number of issues that would make it clumsy for rating product sound quality. The first is that that convening a series of listeners for each time a sound emissions test is performed would be costly and time consuming [2]. There is also the risk that different juries may have different responses to some characteristics.

3.2 Unacceptable Sound Specification

This method is currently used in simplistic noise ordinances and consists of banning certain sound characteristics such as “clicks,” “creaks,” “banging,” etc. Annex A of IEC 61400-11 specifies that similar descriptions should be included in reporting. The issue with this is that it is purely subjective. None of those terms are well defined and it is not clear the level of prominence needed for a defect to be deemed unacceptable.

3.3 Metric Range Specification

The range of "normal" wind turbine noise could be determined for a variety of psychoacoustic metrics. Wind turbine sound measurements could then be compared to these ranges to assess whether a turbine has a normal or abnormal sound quality. The benefit of this method, relative to the first two is that it is less subjective and would be relatively easy to implement as a specification. The downside is that it does not take into account potential interactions between the metrics or the relative importance. For example, if a turbine sound has low sharpness, and low roughness, but relatively high fluctuation strength, is it more or less intrusive than a turbine sound with slightly higher than average values for all three? This difficulty could be mitigated through careful study of the most relevant metrics for wind turbine noise.

3.4 Acceptability Function

Functions have been previously devised for products using sound quality metrics as independent variables. In some cases the resulting dependent variable is an overall weighting of the acceptability of the sound [5] [6] and in others it is a rating of the overall annoyance of the sound [2] [3] [4]. The benefits of this method are that it takes into account relative importance of different metrics and the interactions of metrics. The downside is that there will always be some variability in the relative importance of different sound characteristics for different people, so the function will more or less representative for some.

4. Discussion

There are a number of benefits in using sound quality metrics to rate sound emissions in a standardized manner as they are a better reflection of human response, and can serve to characterize noises such as creak and squeal. In this section, I discuss some of the methodological details that must be considered.

Sound quality metrics are intended to mimic human response, so intuitively they should be measured at a place where humans are likely to hear the sound, such as at a residence and not at the IEC standardized position. This is particularly the case for a metric such as sharpness, which has a frequency dependency. This means that sharpness will have a distance dependency, although this can be compensated for.

If measurements are conducted at the residence, then whether or not a turbine complies with specification will be dependent upon the distance from the turbine to the residence, the particular soundscape of the area, and even current weather conditions. In short it would adversely affect repeatability of any test. With either location, any guarantee or specification would need to explicitly state the distance measurements were made.

Each of the rating methods will require a certain amount of subjectivity, through use of jury trials. This will be intrinsic to any attempt to rate a sound source based on its sound quality. The way this can be mitigated to some extent is to make sure that jury trials have sufficient participants to cover a number of people with different sensitivities.

It should be mentioned that IEC 61400-11 and other standards already make some attempt to address sound quality through including a tonal audibility calculation, A-weighting, and description of unusual sounds. The methods discussed above should be seen as complementing the standard in its current form.

5. Next Steps

Of the methods discussed in Section 3, the metric range specification and acceptability function methods, provide the more objective and repeatable options and are the most promising. Both will require a better understanding of the most intrusive wind turbine sound characteristics and how they relate to various sound quality metrics. This can be accomplished through jury ratings of both "normal" and "abnormal" wind turbine sound [4] [9]. After the jury ratings, values could be calculated for various sound quality metrics and an analysis should be performed to determine metrics that are the most relevant for wind turbine sound subjective response.

6. Conclusions

This paper discusses ways that wind turbine sound could be rated based on sound quality in addition to the current ratings of sound power and tonality pursuant to IEC 61400-11. The purpose is to take into account abnormal wind turbine sound that could increase its relative annoyance and intrusiveness. This could be used to better inform developer purchasing decisions and could be used in the turbine warranty, in addition to current sound power guarantees.

This paper finds that, although there are currently few standardized methods to rate overall sound quality, several authors have used methods to quantify sound quality, that could be adapted for wind turbines.

References

[1] H. Fastl and E. Zwicker, Psychoacoustics: Facts and Models, 3rd ed., Berlin: Springer Verlag, 2007.

[2] U. Widmann, "Aurally Adequate Evaluation of Sounds," in *Euronoise 98*, Munich, German, 1998.

[3] D. Bowen, "Use of Principal Components Analysis in Establishing Correlations Between Sound Qality Metrics and Jury Ratings of Product Sound," in *SQS 2008*, Dearborn, Michigan, 2008.

[4] W. Blazier, "Sound Quality Considerations in Rating Noise From Heating, Ventilating, and Air-conditioning (HVAC) Systems in Buildings," *Noise Control Eng. J.,* vol. 43, no. 3, pp. 53-63, 1995.

[5] J.-G. Ih, D.-H. Lim, S. Sung-Hwan and Y. Park, "Experimental Design and Assessment of Product Sound Quality: Application to a Vacuum Cleaner," *Noise Control Eng. J.,* vol. 51, no. 4, pp. 244-252, 2003.

[6] M. Takada, K. Iida, S. Tsunekawa and S.-i. Iwamiya, "Sound Quality Evaluation of Noise Emitted from Brush Cutters," in *Inter-Noise 2018*, Chicago, Illinois, 2018.

[7] S.-i. Sato, J. You and J. Y. Jeon, "Sound Quality Characteristics of Refrigerator Noise in Real Living Environments With Relation to Psychoacoustical and Autocorrelation Function Parameters," *J. Acousti. Soc. Am.,* vol. 122, no. 1, pp. 314-325, 2007.

[8] W. Sung, P. Davies and J. S. Bolton, "Sound Quality Evaluation of Residential HVAC&R Equipment," in *Inter-Noise 2018*, Chicago, Illinois, 20108.

[9] S. Krishnamurthy, A. Christian and S. Rizzi, "Psychoacoustic Test to Determine Sound Quality Metric Indicators of Rotorcraft Annoyance," in *Inter-noise 2018*, Chicago, Illinois, 20108.

[10] G. Cerrator-Jay, D. Collings and D. Lowery, "Implementation of Sound Quality Measurements in Component Rating Tests," in *SQS 2002*, Dearborn, Michigan, 2002.

[11] *IEC 64100 Wind Turbines - Part 11: Acoustic Noise Measurement Techniques,* Geneva, Switzerland: International Electrotechnical Commission, 2012.

INCEEUROPE

9th International Conference
on
Wind Turbine Noise
Remote from Europe – 18th to 21st May 2021

Establishing Sound Limits for Wind Energy: What is the Role of Annoyance?

Christopher Ollson, Ollson Environmental Health Management (OEHM), Canada, christopher.ollson@gmail.com
Mark Bastasch, Jacobs, United States, Mark.Bastasch@jacobs.com

Summary

Internationally sound standards for wind facilities were largely based on existing regulations of industrial noise. We now have a significant body of wind turbine specific annoyance and health research. Globally, researchers have asked survey participants to report their level of annoyance to a number of facets living with wind energy projects (e.g., sound, shadow flicker, aviation lighting, and aesthetics). Despite difference in study methodologies, only 9-13% of the annoyance is correlated to the sound level or distance setback to the closest turbine. People's attitude towards wind energy, concerns with aesthetic change to landscape and perception of fairness of the permitting process are far stronger predictors of annoyance.

Some argue 10% high annoyance (HA) is the appropriate threshold for setting wind turbine sound standards. This target, endorsed by the WHO, was purportedly intended to limit the casual pathway for sound-annoyance to lead to cardiovascular disease. However, numerous high-quality cross-sectional studies of wind energy facilities have found no association between sound levels and stress, sleep quality, objective or self-reported health impacts (including cardiovascular disease), or quality of life indicators at sound levels up to the 45-50 dBA L_{eq} range. Therefore, annoyance should not be considered in isolation when establishing sound limits. Given that non-acoustical factors have been found to play a more significant role in self-reported annoyance, policy makers find themselves in the typical position of deciding how best to balance the benefits afforded by wind development and the concerns raised by such developments.

1. Introduction

Globally, we continue to see exponential growth in the installation of wind turbines for the production of renewable energy. Over the years onshore wind turbines have grown from 1.5 MW machines (~120 m tall) to the current models typically ranging from 2 to 6 MW (150 m to >260 m). In 2019, there were over 60,000 operating wind turbines in the United States, generating 105 GW of energy, enough to power 32 million homes and seven percent of the nation's electricity [1]. It is anticipated that demand for wind energy will continue to increase.

There are no overarching federal guidelines that govern wind turbine installations and their interaction with local residents. Although, several States have established siting guidelines or regulations, many leave decisions on establishing siting criteria to local Counties. This has resulted in a diverse range of siting criteria implemented and being discussed across the country.

Community concerns discussed during the regulatory process include, but are not limited to, establishing minimum distances between homes and/or property lines, the resulting change to the landscape, the levels of sound or shadow flicker as well as physical safety concerns (ice, structural failures, etc.). Internet sources are often cited by those opposing developments and locally elected officials must evaluate all evidence before them in establishing a regulatory balance for their community.

Over the past fifteen years there has been extensive research evaluating public health and welfare concerns of those living in proximity to wind turbines. This body of independent research conducted by university professors, consultants and government agencies has taken place in multiple countries and studied a range in turbine and project sizes. The potential (or lack thereof) for direct and indirect health impacts has been robustly investigated and it is acknowledged that issues surrounding level of community annoyance are far more subjective. Consistently the concept of 'annoyance' has been demonstrated to increase with an apparent increase in sound level at the exterior of homes. However, more detailed review by these researchers has identified that sound levels themselves do not explain the reported increase in annoyance. Thus, what role should reported levels of annoyance play in establishing sound criteria for wind energy projects? Or phrased differently, all else being equal, will lower sound level limits necessarily result in lower levels of reported annoyance?

2. Methods

Information for this paper was obtained through a search of the PUBMED health sciences database for the scientific peer-reviewed literature. Government publications, siting criteria and statutes were retrieved through an Internet search. Peer-reviewed publication of primary studies on wind turbine noise, annoyance, sound and potential health effects for those living in proximity to wind turbines form the basis for information synthesis and results for this paper. However, valuable information on community attitude and the authors' personal experience of working on siting projects in local communities across the United States serve to inform the context around public opinion and issues.

3. Public Concern and Involvement in Developing Sound Standards in the United States

The public generally favours expanding the development of wind energy across the US. In a recent national wide pole 84% percent of respondents believed that more or the same emphasis should be placed on producing domestic energy from wind [2]. At the same time, community acceptance challenges are increasingly recognized as important obstacles to developing new renewable energy projects. This emerging field of research and study has shown that there is increased need for developers to engage individuals and communities early and often in an attempt to educate and alleviate concerns from landowners in the project area [3,4,5].

Somewhat unique to the United States is that decisions on appropriate siting criteria for wind turbines, including sound levels, often rest with the local elected government. It is common that local counties will enact, or update, ordinances that prescribe a wind turbine sound limit at homes. This may involve months and in some cases over a year of public meetings and hearings to establish permissible sound levels for wind turbines. These public meetings may draw some people expressing very strong opinions advocating for sound limits so low that projects are no longer viable.

Those opposing developments are often armed with what they believe to be credible scientific information to support requiring wind turbine setbacks greater than 2 km and sound levels less than 35 dBA at a non-participating dwellings. Most recently there are those that have latched on to the concept of annoyance as a potential health effect. They rely in part on the WHO Environmental Noise Guidelines for the European Region conditional sound guideline [6].

Although elected officials often find that direct and indirect health effects are not a concern, the question remains framed as what sound limit appropriately addresses amenity, quality of life and annoyance concerns? Review of the scientific literature indicates that over reliance on sound limits as the primary or sole means to addressing these concerns may not result in the desired outcome, lower levels of annoyance and increased community acceptance.

4. Potential Health Impacts of Living in Proximity to Wind Turbines

Over the past two decades there has been a considerable amount of attention focused on the investigation of potential health impacts for people living in proximity to wind turbines. This research first emerged in Europe in the 2000s [7] and then expanded in earnest across the globe. Multiple governments have funded multimillion dollar, multiyear, multidisciplinary research programs that have focused on annoyance, sleep, subjective self-reported health effects (e.g., migraines, dizziness, chronic pain), objective measures of health (e.g., hair cortisol, blood pressure), stress and quality of life surveys.

At the time Knopper and Ollson (2011) published the first literature review on wind turbines and health effects there were 15 papers published in the field [8]. Since that time there have been over 80 peer-reviewed articles published investigating the relationship between wind turbine and community health. A number of subsequent literature reviews that have collated the published research in the field are available [9,10,11,12].

The focus of this paper is not on a review of the published health literature in the field. Rather the authors submit that in general there has not been an established causal pathway of wind turbine noise and impact on either direct or indirect health impacts as the weight of scientific evidence and public health researchers conclude:

- There is no association between wind turbine sound levels of up to 46 dBA at the exterior of non-participating homes and impact on sleep [12,13,14,15].
- The level of low frequency noise or infrasound from wind turbines at non-participating homes does not cause sleep disturbance or other health effects. The levels are typically within background levels at homes and are well below levels that could induce health impacts [16,17].
- Studies have not shown any statistically significant increase in the self-reported prevalence of chronic pain, asthma, arthritis, high blood pressure, bronchitis, emphysema, chronic obstructive pulmonary disease (COPD), diabetes, heart disease, migraines/headaches, dizziness, or tinnitus in relation to WTN exposure up to 46 dB. In other words, individuals with these conditions were equally distributed among people living at all sound levels in the study [18,11].

- In 2019, the most recent literature review of the published research on wind turbine and health found no association between noise and stress effects and biophysical variables of sleep. The high-quality studies found wind turbine noise was not associated with restricted quality of life, anxiety and/or depression [11].
- The 2019 published study of the U.S. population living around wind turbines by Lawrence Berkeley National Laboratory found that wind turbine annoyance and related stress effects are not a widespread problem [4].

However, all studies do point to an increase level of self-reported annoyance and living in proximity to wind turbines. If one accepts the premise that living in proximity to wind turbines does not result in direct or indirect health impacts then how does one evaluate the importance of community annoyance?

5. Annoyance and Living in Proximity to Wind Turbines

Perhaps the most contentious issue involving living in proximity to wind turbines and potential effects on human health revolves around the concept of annoyance. The term annoyance has been used for decades in the research of sound sources. This is in no way a new or novel term being used to measure those living in proximity to wind turbines and their feelings towards the sounds they emit. Excessive annoyance is used in the study of sound as a primary indicator to ascertain whether further health studies are warranted [19].

The International Standards Organization (ISO) has a Technical Specification - *ISO/TS 15666:2003(en) Acoustics — Assessment of noise annoyance by means of social and socio-acoustic surveys* [20]. This Technical Specification provides details and questionnaires for conducting socio-acoustic surveys to ascertain annoyance levels from a variety of sound sources.

> ***Noise-induced annoyance:*** *one person's individual adverse reaction to noise*
>
> *Note 1 to entry: The reaction may be referred to in various ways including, for example, dissatisfaction, bother, annoyance and disturbance due to noise.*
>
> *Note 2 to entry: Community noise annoyance is the prevalence rate of this individual reaction in a community, as measured by the responses to questions specified in Clause 5, and expressed in appropriate statistical terms.*

The investigation of self-reported noise annoyance is a relatively straightforward and cost-effective means to conduct initial investigations. In the absence of further health studies, annoyance surveys are an important primary indicator as it is postulated that annoyance may be part of the causal pathway of adverse health outcomes [6]:

> Annoyance » Increased stress » Sleep disturbance » Noise-induced cardiovascular and metabolic diseases

However, as discussed by Lazlo et al. (2012) the mere reporting of noise-induced annoyance should be evaluated with caution as non-acoustical factors are noted to play an important role [21]:

> *"Annoyance as a reaction indicator should be evaluated with caution as non-acoustical factors play an important role in annoyance ratings. Technical interventions reducing noise levels may therefore not have impacts on annoyance proportionate to their impacts on sound levels. Further studies, investigating impacts on health endpoints (e.g. blood pressure) in changed noise situations are needed.[21]"*

In fact, as described in Section 4, it is exactly these type of investigations on direct and indirect health impacts that have been completed for living in proximity to wind turbines.

5.1 Levels of Self-Reported Annoyance for Populations Living in Proximity to Wind Turbines

Considerable research has been conducted in an attempt to understand the role of sound in reported annoyance from those living around wind turbines. This work began in Europe in the early 2000s [7,22,23]. It has since been expanded to large scale studies such as those in Canada [18,24,25] and the United States [4,26]. As part of these larger health surveys, researchers included questions intended to assess the level of annoyance.

One of the challenges of evaluating the data from the various studies is that the questions regarding annoyance are not uniform across the studies. In general, each of the research studies asked participants to recall their annoyance to wind turbine sound and other factors (e.g., aesthetic feeling to turbines, shadow flicker, navigation lights, etc.) over the course of the past year.

Table 1 provides the scales and descriptors of annoyance used in the various studies. Although the descriptors between Health Canada and European studies differ, they both group the Likert scale 4 and 5 results together to quantify the "highly annoyed" population. The worked conducted by the Lawrence Berkeley National Laboratory (LBNL) for projects in the United States provides a descriptor of "very annoyed" or only the Likert scale 5 in their analysis.

Table 1. Likert Scale Responses to Describe Self-Reported Annoyance to Living in Proximity to Wind Turbines. Responses in Bold Indicate the Values Used to Represent % Highly Annoyed (%HA) in Reporting of Results.

Likert Scale	1	2	3	4	5
European Studies (2000s)	Do not notice	Notice, but not annoyed	Slightly annoyed	**Rather annoyed**	**Very annoyed**
Health Canada (2014)	Not at all	Slightly	Moderately	**Very**	**Extremely**
LBNL (2015)	Not at all	Slightly	Somewhat	Moderately	**Very**

Each of these researcher undertook a detailed statistical analysis to determine if the reported annoyance is strongly linked to the sound levels or if other subjective or objective variables better describe the relationship between reported annoyance and wind turbines. This is a critical linkage, as Lazlo et al. (2012) [21] described that establishment of sound limits as the primary means to limit annoyance may not reduce annoyance as intended.

5.2 What are the Strongest Factors that Predict Wind Turbine Annoyance?

European Work

In 2004, Swedish researchers Pedersen and Persson Waye first reported on wind turbine annoyance [7]. This was the first study that attempted to tease out the relationship between self-reported wind turbine noise annoyance and subjective attitude factors towards the turbines themselves.

The authors report:

> *"The three subjective factors of attitude to visual impact, attitude to wind turbines in general, and sensitivity to noise were forced into the model one-by-one, two-by-two, and finally all together. When adding the subjective factor of attitude to visual impact as an independent variable, the influence of the noise exposure decreased, but was still statistically significant. The pseudo-R2 increased from 0.13 to 0.46, indicating that the new model explained 46% of the variance in annoyance. Adding the two remaining subjective factors did not improve the model as the coefficients did not reach statistical significance. [7]"*

This suggests that only 13% of the annoyance could be attributed to the wind turbine sound itself. That far more dominant a factor in reporting of annoyance was the 'attitude to visual impact' that the wind turbines had on the landscape.

Overall, they concluded:

> *"The unexpected high proportion of annoyance could be due to visual interference, influencing noise annoyance, as well as the presence of intrusive sound characteristics. The respondents' attitude to the visual impact of wind turbines on the landscape scenery was found to influence noise annoyance.[7]"*

This ground-breaking research suggests that the wind turbine noise itself is only a small contributor to the self-reported annoyance of living around wind energy projects. It helped shape all future research in this field.

Another important finding of the European work was that self-reported annoyance was dependent on whether one economically benefits from the wind energy project [27]. Bakker et al. (2012) clearly indicate that the percent of people reporting being rather/very annoyed by outdoor wind turbine noise (up to 54 dBA) that did not economically benefit was 12%, while it was only 3% for those who did economically benefit [27]. In addition, no one who economically benefited from the wind project was rather/very annoyed with the resulting indoor noise levels. This further supports the notion that it is not the wind turbine noise itself that drives the annoyance state; rather, subjective factors such as visual cue and attitude play an important role.

Over the next two decades a number of researchers continued to administer self-reported questionnaire surveys to a number of European populations living in proximity to wind turbines [22,23,27,28,29,30]. This additional body of literature was consistent with, and continued to support, the findings of the original work.

Health Canada Finding on Annoyance

Between 2012 and 2014 Health Canada undertook the Wind Turbine Noise and Health Study, which is the most comprehensive epidemiological study conducted to date [31]. The Health Canada studies provide the most detailed statistical interpretation of wind turbine noise (WTN) and annoyance and resulted in multiple peer-reviewed publications on wind turbine and annoyance [18,24,25,32,33,34].

Health Canada studied two separate populations living near wind projects, those in Ontario and Prince Edward Island (PEI). Across the entire study group an average of 10% of those living with WTN levels between 35 and 40 dBA reported being highly annoyed (HA), which increased to 13.7% HA between 40 to 46 dBA. There was a clear difference in prevalence of annoyance from wind turbines between the two provinces. The level of highly annoyed between 40-46 dBA were <10% in PEI group and considerably higher in the Ontario cohort. Those living in Ontario were 3.29 times greater to report being highly annoyed than those in PEI, especially above 35 dBA [18].

Through their multiple regression models Health Canada was able to explain 58% of the reported %HA through a number of factors, with only approximately 9% of the reported annoyance attributed to the wind turbine noise level itself. The subjective variables (e.g., personal benefit, physical safety concerns, blinking lights, and province) accounted for a much greater proportion of the annoyance reporting in the model [24]. The authors go on to opine that efforts aimed at mitigating the community response to wind turbine noise would profit from considering these non-acoustical factors associated with annoyance.

In 2018, Health Canada also reported on their findings with respect to self-reported annoyance and economic benefit [32]. They found that annoyance was effectively zero among those that receive economic benefit from the wind project. This finding is similar to that of the previously reported European work where those who receive direct compensation did not report annoyance with wind turbine sound [27].

In response to work published by Barry, et. al. (2018) [35], Health Canada evaluated how a home's distance to nearest turbine influenced annoyance levels [33]. They found that the outdoor wind turbine sound levels was highly correlated with distance to the nearest turbine. Therefore, no additional information would be gained by using distance as a proxy for sound. Hence, the conclusions from their earlier papers on WTN, annoyance, health outcomes and quality of life would equally apply to distance to nearest turbine.

It is important to understand whether stress related responses are causally related to high levels of self-reported annoyance or wind turbine sound levels. Of the 1238 study participants 1077 (87%) agreed to have blood pressure measurements, 917 of 1043 (87.9%) participants with hair consented to sampling for cortisol analysis (a stress hormone) [25]. In the Concluding Remarks the authors report:

> *"The results provide no evidence that self-reported or objectively measured stress reactions are significantly influenced by exposure to increasing levels of WTN up to 46 dB."*[1]

Health Canada's comprehensive studies identify that wind turbine annoyance is not solely related to sound levels. Rather, there are numerous factors that contribute to the lived experience and reported level of annoyance. Thus, if one wishes to reduce annoyance, focusing solely on sound limits is unlikely to result in the desired outcome.

United States National Survey of Attitudes of Wind Power Project Neighbours – Lawrence Berkeley National Laboratory (LBNL)

In 2015, the U.S. Department of Energy (DOE) funded LBNL to conduct a 4-year investigation to collect attitude data on a nationally representative sample of individuals living near U.S. wind energy projects [36]. The objective was to better understand how U.S. communities are reacting to living in proximity to wind turbines. This work has resulted in three key publications with respect to attitude, annoyance and stress-mediated responses to wind energy projects [3,4,26].

In their first publication *Attitudes of U.S. Wind Turbine Neighbors: Analysis of a Nationwide Survey [3]* LBNL details the variables influencing neighbours attitudes toward their local wind project. Similar to European and Canadian research they found that perceptions about visual aspect of turbines on the landscape and community, and perceptions of potential impact on their property values influenced their attitude.

Perhaps the most striking finding of this initial paper was that only 8% of overall respondents report having a very negative or negative attitude towards the local wind project after construction [3]. In fact, almost counter intuitively once the authors controlled for project attributes, whether a wind turbine can be seen or heard, and demographics they found that those living closer to wind turbines have significantly more positive attitudes than those living further away [3]. This appears to contradict a higher level of annoyance being reported at greater sound levels, and by default those living in closer proximity to wind turbines [26].

LBNL in cooperation with sound experts at RSG and health experts from Germany reported on annoyance and its driving variables in *Wind turbine audibility and noise annoyance in a national U.S. survey: Individual perception and influencing factors* [26]. Haac et al. (2019) provides the overall annoyance levels in 5 dBA sound increments for both the entire response sample (including participating and non-participating residents) and for only non-participating residents. These unadjusted annoyance levels are slightly different that the European and Canadian metric in that they report the % very annoyed population and a % not highly annoyed (Table 1). There was only a slight difference of % very annoyed in the entire response sample vs. only the non-participating residents at the 35-40 dBA level (13% vs

[1] Note the reference to 46 dBA was the maximum modeled level at the exterior of homes in the Health Canada study. It is not intended to suggest that this be an appropriate upper bound limit on wind turbine sound.

14%) and the 40-45 dBA level (19% vs 21%). It is not until the 45-50 dBA level (21% vs 32%) that a divergence is seen whereby non-participating residents are clearly reporting a greater annoyance with the wind project [26]. However, the U.S. respondents showed the exact same trend of increasing annoyance with exterior sound level as those found by the Europeans and Canadians.

The researchers attempted various analytical approaches to determine what variables most influence people's reporting of annoyance. In the U.S. sample population only 12% ($R^2 = 0.12$) of the Very Annoyed responses could be predicted in the Basic Noise Annoyance model (i.e., correlated solely with wind turbine sound level) [26]. However, the Subjective model that included a number of attitudinal variables resulted in a considerable increase in the models ability to predict annoyance ($R^2 = 0.56$) . Similar to the European and Canadian results, noise annoyance was best explained by visual disapproval of wind turbines on the landscape [26].

The LBNL research led to a companion paper on *Monitoring annoyance and stress effects of wind turbines on nearby residents: A comparison of U.S. and European samples* [4]. This research was an attempt to ascertain the significance of the reported annoyance on stress effects for those in the U.S. and compare the results to a set of European results from Germany and Switzerland. The researchers developed a novel Annoyance Stress-Scale (ASScale) to characterize stress-impacted individuals living near wind turbines.

The U.S. sample included 1441 residents living near 231 wind farms across 24 states. People living between 0.08 to 4.8 km (average 1.3 km) from a turbine were included in the research. Sound levels in the study ranged from <30 dBA to >50 dBA. They found that there was a low prevalence of annoyance, stress symptoms and need for coping strategies. In terms of the noise ASScale (NAAScale), annoyance was correlated with increasing concerns about the perception of lack of fairness in the permitting process amongst other subjective variables. Similar to the European and Canadian work, distance from nearest turbine and exterior sound level at homes were not found to be correlated to reported noise annoyance [4]. The researchers conclude:

> *"Our findings provide evidence that WT annoyance and related stress effects are not a widespread problem. Average annoyance levels of residents near wind farms in Europe and the U.S. were low, with the levels for noise similar across both samples, with European levels slightly higher for shadow-flicker, lighting and landscape change. In all cases the annoyance levels were comparable to the levels associated with traffic noise.[4]"*

The findings of LBNL US research are similar, albeit using a different methodology, to Health Canada. That although residents may report an increase in annoyance with increasing sound levels that this annoyance is not dependent on the sound level itself but more importantly to subjective factors. Importantly that wind turbine annoyance and any related stress effects are not a widespread problem for those living in proximity to wind energy projects.

5.3 Conclusions on the Weight of Scientific Evidence for Health and Annoyance

Wind turbine annoyance is a well studied phenomenon. There have been almost twenty years of international research that time and again result in similar findings. What can be seen from these peer-reviewed articles (and many others) is that the relationship between wind turbines and human responses to them (as measured by annoyance) is extremely complex and influenced by numerous variables. Key points that have come out of these studies are:

- Annoyance is not a medical condition. It is not a recognized medical disease and it is not classified in the World Health Organization's International Statistical Classification of Disease and Related Health Problems 11th revision (ICD 11). This does not mean that efforts to reduce annoyance are not laudable and it is acknowledged that lack of community acceptance presents substantial barriers to future developments.

- A proportion of people that notice sound from wind turbines self-report finding it annoying. Approximately 10% of people with sound levels between 35 dBA and 40 dBA Leq report high annoyance levels with the wind project.

- There is at best only a weak association between wind turbine sound at the exterior of people's homes and the reported level of annoyance. The results of European, Canadian and U.S. studies suggest that only 9-13% of the annoyance can be correlated to the sound level itself.

- Self-reported annoyance is much more strongly related to the subjective factors of attitude of the visual impact of turbines on the landscape and fairness in the siting process.

- People who economically benefit from wind turbines often experience higher sound levels outside their homes than non-participants and have significantly decreased levels of annoyance (virtually non-existent) compared to individuals that received no economic benefit.

- The level of annoyance is not correlated to stress mediated health outcomes.

- Based on the results of the LBNL study U.S. wind turbine annoyance and related stress effects are not a widespread problem.

- Wind turbine sound levels at the exterior of homes does not impact quality of life outcomes [37].

The weight of scientific evidence suggests that there will be a non-trivial level of annoyance in local populations living in proximity to wind turbines. Given the scientific evidence, one could not set a sound level from wind turbines that would completely eliminate annoyance from living in proximity to wind turbines. This is because only a fraction of reported annoyance is associated with sound. Thus over emphasizing the importance of sound limits (be it through establishing low sound level requirements or draconian compliance measures) is unlikely to result in the desired outcome.

6. World Health Organization Conditional Sound Annoyance-based Guideline

The WHO (2018) relied heavily on self-reported annoyance its Environmental Noise Guidelines for the European Region [6] when it specifically addressed wind turbines for the first time:

> *"For average noise exposure, the GDG [Guideline Development Group]* ***conditionally*** *recommends reducing noise levels produced by wind turbines below* ***45 dB*** $\boldsymbol{L_{den}}$*, as wind turbine noise above this level is associated with adverse health effects." [emphasis in original document]*

This conditional guideline has been a source of confusion as well as anxiety in some communities. It is for example unclear how an L_{den} type approach would be practically applied to wind turbines, does this mean 35 dBA at night and 45 dBA during the day or perhaps a fixed limit of 38 dBA?

It is important to note that while published in 2018, the WHO's systematic literature review of wind turbine noise and health had established a publication cut-off date of 2014. This review focused on scientific studies of wind turbine sound and the potential health outcomes of:

- Ischaemic heart disease (IHD) (including angina pectoris and/or myocardial infarction), incidence of hypertension, prevalence of highly annoyed population, permanent hearing impairment, reading skills and oral comprehension in children, and sleep disturbance

To set annoyance as a potential adverse health effect, the WHO relied on a brief review by Eriksson et al. (2018)[38] and proposed self-reported annoyance may be part of the causal pathway between annoyance, stress, sleep disturbance, which could lead to noise-induced cardiovascular and metabolic diseases.

The WHO determined that although annoyance is considered a less serious health effect than sleep disturbance, that an absolute risk associated with the sound guideline value should be closest to, but not above 10% HA (highly annoyed), to be health protective. However, nowhere in the Eriksson et al. (2018) [38] or the WHO (2018) [6] is there a scientific justification for the use of a 10% HA threshold value to protect against direct adverse health effects.

Additionally, the WHO assigned each of the 'critical health outcomes' an impact of disease and disability weight (DW) to prioritizing amongst them when deriving guideline exposure levels. The DWs are ratings between 0 and 1, in which 0 indicates no disability and 1 indicates the maximum amount of disability and are commonly used in calculating burden of disease. Annoyance received a very low DW rating of 0.02, while for example ischaemic heart disease was given a DW of 0.405.

WHO reviewed only two studies on self-reported annoyance and wind turbine noise, including Janssen et al. (2011) [28; that included 7,22,23] and Kuwano et al. (2014) [39]. They indicate the evidence was given a 'low quality' GRADE based on inconsistency and lack of precision in the papers. In addition, a quantitative risk assessment for adverse health effects for annoyance could not be provided given that the annoyance results could not be pooled because of heterogeneity between the studies.

Through plotting the Janssen et al. (2011) [28] and Kuwano et al. (2014) [39] annoyance curves a 10% HA was seen at approximately 45 dB L_{den}. The WHO briefly acknowledged that subjective factors of attitude and visual cue play a role in self-reported annoyance. Furthermore, they recognized that effects related to attitudes towards wind turbines are hard to discern from those related to noise and maybe partly responsible for the associations between annoyance and sound [8].

The WHO stated that a low quality GRADE score indicates that: *"further research is very likely to have an important impact on the certainty of the effect estimate and is likely to change the estimate"*. However, they make no mention that at the time of publication in 2018 subjective factors were already known to play a much larger role in wind turbine annoyance than wind turbine sound.

They did acknowledge that public engagement, consultation and communication are key components for future wind developments. However, the GDG could not assess feasibility of, or discern whether any beneficial effects of noise reduction, would outweigh the cost of intervention or setting sound regulations.

The WHO 2018 guideline was based on evidence from four papers dealing with annoyance and did not account for the role, or lack thereof, noise played in the annoyance evaluation. Research available prior to publication of the WHO guideline in 2018 but prior to their 2014 literature review cut-off has identified that sound alone explains very little of the reported annoyance. Thus, caution is advised in relying too strongly on the WHO guideline, in keeping with the WHO's explanation a conditional recommendation:

- A **conditional** recommendation requires a policy-making process with substantial debated and involvement of various stakeholders. There is less certainty of its efficacy owing to lower quality of evidence of a net benefit, opposing values and preferences of individuals and populations affected or the high resource implications of the recommendation, meaning there may be circumstances or settings in which it will not apply.

Unfortunately, the release of the WHO guideline in 2018 has resulted in more confusion and tension in communities looking to establish siting regulations or guidelines. The concept of 10% HA as a level associated with adverse health effects was not robustly discussed nor clearly established for wind turbines, particularly with respect to sound levels. In addition, the delay between the literature cut-off date (2014) and publication (2018), a period of robust growth in the scientific literature, represents a missed opportunity.

7. Discussion and Conclusions

It is true that we often regulate objectionable behaviours based on what different societies deems to be acceptable/tolerable, even if there is no direct health risk. Setting noise standards requires the balancing of development objectives with community annoyance. This is certainly true of aircraft noise, where "*The costs of aircraft noise regulation, on the other hand, are seen as great enough, and to affect enough people, that they outweigh the costs of merely annoying a relatively small number of people.* [40]"

All of this is not to say that acoustics can be ignored. Acoustics is certainly a very important considerations in siting of wind projects. There is always the potential for legitimate sound concerns and complaints should be addressed promptly – be they a result of a maintenance issue, a turbine malfunction or some other factor. As said by Bowdler (2012) "*Bad noise management doesn't just affect your project, but future ones.[41]*" However, available evidence does not identify acoustics as the key determinate in ensuring community acceptance, which is helpful to acknowledge that when establishing regulatory and compliance requirements.

It is also not to say that research on wind turbine annoyance should be ignored. There is no question that establishing siting requirements or guidelines is a complicated undertaking. Community concerns need to be taken seriously, are complex and take time and effort to address. Not all communities will appreciate the benefits that wind projects may provide. Understanding community concerns is invaluable and appropriately responding to them will facilitate future development. It will also help temper expectations of the community during the permitting process.

Initially, as wind turbines increased in nameplate capacity and height there was also an increase in sound power levels. However, in recent years there has been a decrease in sound power level with increased turbine size. Thus, we pose the question:

> All else being equal, if a project was quieted by 3 or 5 dBA, would self-reported community annoyance be substantially reduced?

We do not believe that this would be the case. There will always be a percentage of people that self-report annoyance regardless of the sound limit or setback distance required. People's attitude towards wind energy, concerns with aesthetic change to landscape and perception of fairness of the permitting process are far stronger predictors of annoyance and would likely outweigh a perception in noise reduction.

Additionally, given the lack of clear or substantial linkage between sound level an annoyance, what is the necessary level of precision or accuracy required for wind turbine predictions or compliance assessments? Given models of transportation sources, which are substantially louder and more prevalent, are considered to be validated in the USA by the Federal Highway Administration when measured vs modelled are within 3 dBA, should similar consideration be afforded to wind turbines?

Wind energy projects bring clear socio-economic health benefits to host communities. These are in the form of taxes (which may support schools, parks, etc.), landowner payments (which may drought proof or otherwise financially stabilize the local farmer), provide jobs in typically rural communities, and an offset for the need for fossil fuel derived energy. These all have indirect health benefits at the individual and community level.

Regulators must balance the societal desire for wind energy, economic benefits and potential for community annoyance. This is not unlike other policy decisions that require balancing of important objectives. It should be recognized, that focusing solely on reducing sound levels or draconian enforcement of sound limits may not address the underlying complexity of the source of people's self-reported annoyance.

References

[1] AWEA (American Wind Energy Association). 2019. Wind Powers America Annual Report

[2] Gallup, 2021. Energy Poll. https://news.gallup.com/poll/2167/energy.aspx

[3] Hoen, B., Firestone, J., Rand, J., Elliot, D., Hübner, G., Pohl, J., Wiser, R., Lantz, E., Haac, R., Kaliski, K. 2019. Attitudes of U.S. Wind Turbine Neighbors: Analysis of a Nationwide Survey. Energy Policy 134 (2019) 110981

[4] Hübner G, Pohl J, Hoen B, Firestone J, Rand J, Elliott D, Haac R. 2019. Monitoring annoyance and stress effects of wind turbines on nearby residents: A comparison of U.S. and European Samples Environ Int. 2019 Nov;132:105090. doi: 10.1016/j.envint.2019.105090. Epub 2019 Aug 19.

[5] Besesette, D.L. & Mills, S.B. 2021. Farmers vs. Lakers: Agriculture, amenity, and community in predicting opposition to United States wind energy development. Energy Research & Social Science. Vol 72.

[6] World Health Organization (WHO) Regional Office for Europe. 2018. Environmental Noise Guidelines for the European Region.

[7] Pedersen E, Persson Waye K. 2004. Perception and annoyance due to wind turbine noise – a dose–response relationship. *J Acoust Soc Am* (2004) **116**:3460-70.

[8] Knopper, L.D. and **Ollson, C.A.** 2011. *Health Effects and Wind Turbines: A Review of the Literature.* Environmental Health. 10:78. Open Access. Highly Accessed.

[9] McCunney, R.J., Mundt, K.A., Colby, D., Dobie, R., Kaliski, K., Blais, M. 2014. Wind Turbines and Health A Critical Review of the Scientific Literature. JOEM Volume 56, Number 11

[10] Knopper, L.D., **Ollson, C.A.,** McCallum, L.C., Aslund, M.L., Berger, R.G, Souweine, K., and McDaniel, M. 2014. *Wind turbines and Human Health.* Front. Public Health, Vol. 2, Art. 63

[11] Freiberg et al. 2019 Health effects of wind turbines on humans in residential settings: Results of a scoping review. Environmental Research 169 (2019) 446–463

[12] Liebich et al. 2020. A systematic review and meta-analysis of wind turbine noise effects on sleep using validated objective and subjective sleep assessments. Journal of Sleep Research.

[13] Michaud DS, Feder K, Keith SE, Voicescu SA, Marro L, Than J, Guay M, Denning A, Murray BJ, Weiss SK, Villeneuve PJ, van den Berg F, Bower T. 2016a. Effects of Wind Turbine Noise on Self-Reported and Objective Measures of Sleep. Sleep. 2016 Jan 1;39(1):97-109

[14] Jalali, L.,Bigelow P., Nezhad-Ahmadi M.R., Gohari M., Williams D., McColl S. 2016. Before–after field study of effects of wind turbine noise on polysomnographic sleep parameters. Noise Health; 18:194-205.

[15] Michaud D.S., Keith S.E., Guay M., Voicescu S., Denning A., McNamee J.P. 2021. Sleep actigraphy time-synchronized with wind turbine output. Sleep. March 2021.

[16] Berger R.G., Ashtiani P., **Ollson C.A.**, Whitfield Aslund M., McCallum L.C., Leventhall G., Knopper L.D. 2015. *Health-based audible noise guidelines account for infrasound and low frequency noise produced by Wind Turbines.* Front Public Health. Vol 3, Art. 31

[17] Ministry for the Environment, Climate and Energy of the Federal State of Bade Wuerttemberg Germany. 2016. Low-frequency noise including infrasound from wind turbines and other sources

[18] Michaud DS, Feder K, Keith SE, Voicescu SA, Marro L, Than J, Guay M, Denning A, McGuire D, Bower T, Lavigne E, Murray BJ, Weiss SK, van den Berg F. 2016b. Exposure to wind turbine

noise: Perceptual responses and reported health effects. J Acoust Soc Am. 2016 Mar;139(3):1443-54.
[19] Berglund B, Lindvall T. (Eds.) *Community Noise*. Center for Sensory Research, Stockholm University and Karolinska Institute (1995).
[20] International Standards Organization (ISO). 2003. Technical Specification. ISO/TS 15666:2003(en) Acoustics – Assessment of Noise annoyance by means of social and social acoustics survey.
[21] Laszlo HE, McRobie ES, Stansfeld SA, Hansell AL. Annoyance and other reaction measures to changes in noise exposure - A review. *Sci Total Environ* (2012) **435**:551-562.
[22] Pedersen E. Health aspects associated with wind turbine noise-Results from three field studies. *Noise Control Eng J* (2011) **59**:47-53.
[23] Pedersen E, Persson Waye K. Wind turbine noise, annoyance and self-reported health and well-being in different living environments. *Occup Environ Med* (2007) **64**:480-6.
[24] Michaud DS, Keith SE, Feder K, Voicescu SA, Marro L, Than J, Guay M, Bower T, Denning A, Lavigne E, Whelan C, Janssen SA, Leroux T, van den Berg F. 2016c. Personal and situational variables associated with wind turbine noise annoyance. J Acoust Soc Am. 2016 Mar;139(3):1455-66.
[25] Michaud DS, Feder K, Keith SE, Voicescu SA, Marro L, Than J, Guay M, Denning A, Bower T, Villeneuve PJ, Russell E, Koren G, van den Berg F. 2016d. Self-reported and measured stress related responses associated with exposure to wind turbine noise. J Acoust Soc Am. 2016 Mar;139(3):1467-79.
[26] Haac, R.T., Kaliski, K., Landis, M., Hoen, B., Rand, J., Firestone, J., Elliott, D., Hübner, G., Pohl, J. 2019 Wind turbine audibility and noise annoyance in a national U.S. survey: Individual perception and influencing factors. Journal of the Acoustical Society of America 146.
[27] Bakker RH, Pedersen E, van den Berg GP, Stewart RE, Lok W, Bouma J. Impact of wind turbine sound on annoyance, self-reported sleep disturbance and psychological distress. *Sci Total Environ* (2012) **425**:42–51.
[28] Janssen SA, Vos H, Pedersen E. A comparison between exposure-response relationships for wind turbine annoyance and annoyance due to other noise sources. *J Acoust Soc Am* (2011) **130**:3746-53.
[29] Klaeboe & Sundfor (2016) Windmill Noise Annoyance, Visual Aesthetics, and Attitudes towards Renewable Energy Sources Int. J. Environ. Res. Public Health 2016, 13, 746
[30] Pohl et al. 2018. Understanding stress effects of wind turbine noise – The integrated approach Energy Policy 112 (2018) 119–128
[31] Health Canada. 2014. Wind Turbine Noise and Health Study: Summary of Results. https://www.canada.ca/en/health-canada/services/health-risks-safety/radiation/everyday-things-emit-radiation/wind-turbine-noise/wind-turbine-noise-health-study-summary-results.html
[32] Michaud DS, Marro L, McNamee J. 2018. The association between self-reported and objective measures of health and aggregate annoyance scores toward wind turbine installations. Can J Public Health. 2018 Apr;109(2):252-260.
[33] Michaud, DS., Guay, M., Marro, L., Than, J. 2018 Response to: "Using residential proximity to wind turbines as an alternative exposure measure to investigate the association between wind turbines and human health," by Barry, Sulsky, Kreiger (2018) The Journal of the Acoustical Society of America. 144, 330
[34] Michaud DS, Marro L, McNamee J. 2018. Derivation and application of a composite annoyance reaction construct based on multiple wind turbine features Can J Public Health. 2018 Apr;109(2):242-251.
[35] Barry, R., Sulsky, S., Kreiger. N. 2018. Using residential proximity to wind turbines as an alternative exposure measure to investigate the association between wind turbines and human health. The Journal of the Acoustical Society of America. 143, 3278
[36] Lawrence Berkley National Laboratory. National Survey of Attitudes of Wind Power Project Neighbors. https://emp.lbl.gov/projects/wind-neighbor-survey

[37] Feder K, Michaud DS, Keith SE, Voicescu SA, Marro L, Than J, Guay M, Denning A, Bower TJ, Lavigne E, Whelan C, van den Berg F. 2015. An assessment of quality of life using the WHOQOL-BREF among participants living in the vicinity of wind turbines. Environ Res. 2015 Oct;142:227-38. doi: 10.1016/j.envres.2015.06.043. Epub 2015 Jul 11
[38] Eriksson, C. and Pershagen, G. 2018 for the WHO. Biological Mechanisms Related to Cardiovascular and Metabolic Effects by Environmental Noise
[39] Kuwano, S., Yano, T., Kageyama, T., Sueoka, S., and Tachibana, H. (2014). "Social survey on wind turbine noise in Japan," Noise Control Eng. J.62(6), 503–520.
[40] Fidell, S and Mestre, V. 2020. A Guide to U.S. Aircraft Noise and Regulatory Policy. Springer. ISBN 978-3-030-39908-5
[41] Bowdler, d. 2012. Wind Turbine Noise Why Does it Annoy? Not just the decibels. Presentation EWEA Oxford.

Annex A

Recently Established Sound Siting Criteria in the United States

A number of U.S. jurisdictions have recently undertaken extensive review of their permissible sound guidelines for wind turbine (e.g., North Dakota, South Dakota and New York). State siting commissions were presented with expert testimony on potential health impacts, annoyance and quality of life concerns. Considerable attention was focused on the body of wind turbine noise annoyance research, the WHO 2018 conditional sound guideline and what role annoyance should play in establishing sound standards,

State officials evaluated the weight of scientific evidence to establish sound level standards that they deemed to be protective of public health and would result in tolerable annoyance within their communities. As a result, several states have lowered their permitted sound standards from 50 to 45 dBA L_{eq} at the exterior of non-participating homes, uniform across day and nighttime hours (Table 3).

For example, since 2018 the New York State Siting Board has adjudicated over 10 wind energy permit hearings. It has consistently ruled that 45 dBA Leq (8-hour nighttime) is sufficient to protect health and welfare, while balancing the issue of annoyance. For example, in a recent decision the Examiners found [Deer River Wind (Case 16-F-0267, June 2020)]:

> *"The Examiners do not find that the Berkeley Studies warrant a change in the Siting Board's established WTN limits [45 dBA Leq (8-hour)]. The Berkeley Studies do indeed show a correlation between higher WTN levels and higher annoyance levels, but they do not establish any specific regulatory targets, and more importantly they do not establish a quantified link between WTN levels and health impacts."*

Table 3. State level wind turbine sound standards at non-participating homes.

State	MW and Turbines Installed (2019)	Allowable Sound Limit	Basis for Standard
North Dakota	3,640 MW 1,875 turbines	A wind energy conversion facility site must not include a geographic area where, due to operation of the facility, the sound levels within one hundred feet of an inhabited residence or a community building will exceed **forty-five (45) dBA**. The sound level avoidance area criteria may be waived in writing by the owner of the occupied residence or the community building	Updated 2020 – The North Dakota Public Service Commission adopted a statewide standard of 45 dBA through a formal Rule Making and public comment period. The standard was set to balance between annoyance / complaints and responsible wind energy development.

South Dakota	1,848 MW 899 turbines	**45 dBA** Leq at the exterior of non-participating homes	Updated 2018 - South Dakota county ordinances for sound typically range from 45 to 50 dBA at the exterior of homes. However, the SD Public Utilities Commission set a precedent for a 45 dBA sound standard in permitting projects all new development in the state since 2018.
New York	1,987 MW 1,128 turbines	A maximum noise limit of forty-five **(45) dBA** Leq (8-hour), at the outside of any existing non-participating residence, and **fifty-five (55) dBA** Leq (8-hour) at the outside of any existing participating residence	Since 2018 the sound standard in NY was lowered from 50 dBA to 45 dBA through state permitting of wind projects. "The Examiners do not find that the Berkeley Studies warrant a change in the Siting Board's established WTN limits [45 dBA Leq (8-hour)]. The Berkeley Studies do indeed show a correlation between higher WTN levels and higher annoyance levels, but they do not establish any specific regulatory targets, and more importantly they do not establish a quantified link between WTN levels and health impacts."

It is important to remember that 50 dBA Leq was the permitted sound level for over a decade in North Dakota, South Dakota, and New York when the majority of the currently operating wind projects were installed. In all cases the states rejected that sound standards needed to be lowered to the WHO 2018 conditional sound guideline to continue the ongoing protection of their citizens.

INCEEUROPE

9th International Conference
on
Wind Turbine Noise
Remote from Europe – 18th to 21st May 2021

Stymied by Standards? Arguments for wind turbine noise standards that actually measure irritant drivers

William K.G. Palmer P. Eng. **TRI-LEA-EM** **trileaem@bmts.com**

Summary

Reviews of wind turbine noise standards in many jurisdictions have been previously reported at Wind Turbine Noise Conferences by Fowler, [1] Cooper, [2] and others. The reviews show that the criteria for acceptability is usually based on a dBA sound level rating at the home of a resident, sometimes conditioned by time of day, or by tonality. However, soundscapes assessed as acceptable by current standards are often identified by residents as very irritating and disruptive to sleep. This paper presents the analysis of different soundscapes assessed as acceptable in the vicinity of different types of wind turbines and in the natural environment. Examination of differences that are neither readily perceived by A-weighting nor by tonal assessment per IEC 61400-11 [3] are presented to demonstrate that the currently accepted criteria may miss factors that are contributors to irritation reported by residents. Recommendations for revisions to noise standards to address adverse human impacts are made to assist legislators and regulators. The advantage of adopting the recommended revisions to noise standards will be that residents will be less likely to report disturbance from wind turbines and as a result there will be less impediment to proponents planning to install wind turbines as dissent will be reduced.

1. Introduction

The driving force behind this paper were presentations by citizens living in the footprints of wind power developments in Ontario, Canada, to the Multi Municipal Wind Turbine Working Group, comprised of municipally elected and appointed representatives. One particular case stood out, when the citizens identified that their sleep, and overall enjoyment of their homes had been adversely impacted by tonal emissions from wind turbines erected in their municipality. On visiting their homes, a tonal signature from the turbines could be heard. A narrow band analysis of recordings taken at the site showed a clear spike in the sound pressure levels at the residences at about 445 Hz, corresponding to the sound that could be distinguished coming from the wind turbines. The residents characterized the sound as a clear tonal "woo, woo, woo" rising and falling cyclically at the blade pass frequency. Field staff from the provincial Ministry responsible for the environment visited the homes of several residents, and agreed there was indeed an audible tonal emission from the turbines. The catalyst to this presentation was that after the collection of compliance audit data for the ministry by acoustical consultants, the audit report concluded the wind turbines were neither tonal nor audible. How could an evaluation consistent with international standards conclude that a clearly audible, tonal, irritating sound was neither tonal nor audible? The corollary to this statement is how could responsible professional acoustical consultants prepare and regulators issue the audit report without question?

The impasse between reality and the reported conclusion led to this investigation into the standards. It may be said that there are two mutually exclusive purposes of wind turbine noise standards. These are generally considered to be environmental standards. One point of view might be that the purpose of the environmental noise standards for wind turbines is to protect citizens from harm caused by noise emissions into the environment where the citizen lives, or values for its natural amenities and soundscape. A second point of view is that the purpose of the wind turbine environmental noise standards is to enable the erection of wind turbines considered by the proponent as a benign (and profitable) source of energy without constraint from those who do not want intrusion of wind turbines "into their backyard." The purpose of this paper is to try to resolve the impasse between the two viewpoints by presenting factual evidence about the noise emitted by wind turbines that may be used to assess the harm to humans. Thus, the intent is to provide a guide to legislators who try to listen to both sides of the argument, and to regulators, who put the wishes of the legislators into manageable guidelines.

Reviews of wind turbine noise standards in many jurisdictions show that the criteria for acceptability is usually based on a dBA sound rating at the home of a resident, sometimes conditioned by the time of day. In some jurisdictions, the dBA limits are adjusted by tonality, applying a penalty to the A weighted limit.

We can read that the dBA weighting criteria for sound was established considering voice communications and gives a lower weighting factor to higher or lower frequencies than for the mid-range frequencies used for most human communications. Just as "lumens" is a measure of what we consider is the most visible lighting spectrum, "A-weighting" gives us a measure of what we consider the most audible portion of the audio spectrum. But, just as what are considered invisible parts of the light spectrum with a higher (ultraviolet) or lower (infrared) frequency can impact a human, for example giving a sunburn quickly to sensitive individuals, so too some present evidence that the inaudible parts of the sound pressure frequency above or below the normally considered A-weighted range might impact humans differently.

This paper will present results of different soundscapes, both near different types of wind turbines, and in the natural environment, to compare the properties of the spectrum, so as to identify where very similar A-weighted sound levels can have very different audible properties and sound pressure levels. From this, the paper will discuss the concept of irritation. Irritation and anger (a synonym to annoyance) is a recognized cause of disease by the International Classification of Diseases, Tenth Revision (ICD-10 R45). [4]

Finally, the paper will present arguments for revision to noise standards to address adverse human impacts.

2. Why 40 dBA is not a sufficient criteria

As described in the introduction, a common criterion for acceptability of wind turbine sound Is a 40 dBA threshold. In some cases, such as Ontario [5] the criterion is based on a 40 dBA limit any time of the day or night, although the limit is permitted to increase as high as 51 dBA if the ambient wind speed at 10 metres is measured at 10 metres per second. While there are higher limits in some states in the United States, the World Health Organization "Environmental Noise Guidelines" for the European Region [6] recommends the use of a 45 Lden (day, evening, night) limits for wind turbines. The Ontario constant 40 dBA limit would result in an Lden rating for 46.2 dB Lden. A comprehensive listing of recent advances in wind turbine noise research is given by Hansen and Hansen [7] in a paper issued in 2020.

Four simple scoping examples are presented in Figures 1 - 4 showing analysis of recordings made using a calibrated MOVO MA 2000 electret condenser microphone with an omni-directional response flat from 35 Hz to 18kHz recorded by making a video record using an iPhone SE or 7 Plus. The examples are all within ±10 dB of 40 dBA. Two were recorded in an environment > 5 km distant from wind turbines with an output > 500 kW, and two at residences in different wind power developments in order to demonstrate how there are considerable differences in the quality of the sound even though the dBA values may appear similar.

For each case, the analysis presented in the figure is from the Faber Acoustical Electroacoustics Toolbox V3.9.10 analysis of the sound recording, showing:

- the A weighted FFT output from 100 to 10,000 Hz (dealing only with audible sound, not infrasound due to microphone limitations)
- the sound level meter reading for flat, C-Weighted, and A-Weighted data
- the one-third octave band analysis in flat weighting and A-Weighting

2.1 Figure 1 - Wind in Trees – Nearest Vestas V82 Wind Turbine > 6 km

The one-third octave analyser and A-Weighted FFT output shows a "smooth bump" between about 500 Hz and 1500 Hz, indicative of the broadband wind "swish" in the needles of the nearby pine trees. The presence of bird songs can be seen in the small peaks between about 3 kHz and 5 kHz. The A-weighted one-third octave band analyser shows the one-third octave at 100 Hz is 15 dB below the one-third octave at 1000 Hz, and the A-weighted FFT analysis also shows a drop of about 5 dB below the level at 1000 Hz.

2.2 Figure 2 – Nearest Vestas V82 Wind Turbine ~ 500m

The A-weighted one-third octave analyser and FFT do not show the same drop in magnitude below 1000 Hz that is seen in the Figure 1- Wind in Trees example. The FFT analysis shows that in the wind turbine influenced environment the values below 500 Hz have not fallen below the 1000 Hz case, but have instead risen to be up to 5 dB higher than the level at 1000 Hz.

If one looks at the time variance of sound pressure level (Lp) for the dBA or unweighted dB levels (as will be demonstrated in the Conference presentation) the presence of the cyclical variation of the sound pressure level at the blade passage frequency is clear. For this particular example, the dBA weighted sound pressure level cycles up and down about 2 dB at the blade pass frequency. The unweighted dB level variation is higher, typically about 4 dB in this example confirming the dominance of lower frequency components. The issue of the irritation from the cyclical variation of the sound was described in a previous paper. [8]

2.3 Figure 3 – Nearest Siemens SWT-101 Wind Turbine ~ 700m

This example was recorded during the period of the audit performed referred to in the introduction that concluded that tonality was neither present nor audible at this particular home. While the FFT display indicates a rise at 442 Hz to be some 13 dB above the average values at frequencies above or below the indicated rise. A second smaller rise is noted at 108 Hz. What the one-third octave analysis does demonstrate, is the challenge in using the method that Cooper and Evans [9] described common in many Australian state standards, or in French standards [10], of determining tonality if a one-third octave is higher than the adjacent one-third octaves (or two adjacent octaves in the case of the French NF S 31-010) by a value of (typically) 5 dB. This demonstrates that using one-third octave analysis can be a problem if the tonality is near to the division point between two one-third octaves, raising the value of both.

In this case, examining the time variance of sound pressure level (Lp) for the dBA or unweighted dB levels the presence of the cyclical variation of the sound pressure level at the blade passage frequency is present as described in Section 2.2 but in addition to there being a broadband sound variation, the tonal presence makes the cyclical variation even more clear.

2.4 Figure 4 – Lake Waves – Nearest Vestas V82 Wind Turbine > 5 km

This example demonstrates the challenge of simply using dBA as a determinant of concern. The lake waves example demonstrates the highest dBA rating of the 4 examples shown, and yet, both the one-third octave band analysis and the fft analysis demonstrate the largest drop in sound pressure level between 100 Hz and 100 Hz, with both the one-third octave and the fft analysis being about 30 dB less at 100 Hz than 1000 Hz. A listening test also shows that the fluctuation in sound pressure level caused by the lake waves is not mechanically cyclical as for the turbine cases, and has a longer period, making it appear more restful for many.

3. Considering the evaluation of tonality by international standards

Following the initial publication of the research article by W. Palmer, "Confirming Tonality at Residences Influenced by Wind Turbines" in May 2020, [11] Colin H. Hansen PhD identified an error in the original paper, which was appreciated. To help verify the corrections made to the article, before they were reissued, Kristy L. Hansen PhD suggested that the tonality evaluation and calculation performed in the paper be checked by running the set of examples provided for the Round Robin test of the IEC 61400-11 and ISO/PAS 20065 [12] methods for analysing tonality content reported at the 8th Wind Turbine Noise Conference in 2019. Kristy Hansen provided a copy of the sound files used for the 10 mandatory and 21 optional evaluations, as well as the A-weighted spectrum evaluation in an excel spreadsheet as supplied during the original round robin test. All 31 files were evaluated to verify the IEC-61400-11 evaluation method and 20 files were evaluated to verify the ISO/PAS 20065 evaluation methods used in this paper.

A summary of the tonal audibility evaluations is included in Table 1. Initially, the "Spectra-A" data provided in the Round Robin Tests were run through the analysis software developed (excel spreadsheets) to verify they were calculating the Tonal Audibility consistently with the average values identified by Søndergaard. The evaluation models showed good consistency when evaluating the provided Spectra-A for both the IEC-61400-11 case and the ISO-PAS 20065 case.

When the provided sound files were evaluated using the Electroacoustics Toolbox, although over half of the tests showed good correlation (tonal audibility within 1 dBA of the average reported by Søndergaard) others showed a greater variation, although within the standard deviation identified by Søndergaard. As a test, all samples were re-run using the open source Audacity audio editor, which showed a closer correlation to the average results identified in the WTN 2019 paper.

As identified by Søndergaard, complex waveforms typified by the M04 example in the Round Robin Test resulted in the least consistency. Examples of the fft outputs for that case are shown in Figures 5 to 7, for the case of evaluating the provided Spectra, the audacity evaluated sound file and the electroacoustics toolbox evaluated sound file. It does suggest some caution in the evaluation of complex waveforms as the fft software are indeed only calculational models. This initial evaluation did provide confidence though that the audio processing equipment and modelling being used would present consistent results for assessing tonal audibility.

3.1 Further challenges to the evaluation of tonality and tonal audibility

The flag raised when evaluating complex waveforms has been recognized by other international acousticians. The comment was received from a reviewer of the tonality paper issued in 2020 that rather than evaluating tonality using complex calculational models such as the IEC 61400-11 or ISO/PAS 20065 standards, their preference was to use a much simpler one-third octave method. A similar argument is presented in the ECMA-74 Standard [13] for evaluation of tonality, although the ECMA standard breaks the sound spectrum in "barks" rather than "octaves." Effectively both methods recognize that an average listener groups sounds together over a fairly consistent ratio of frequencies. The cases presented in the 2017 WTN presentation, "A Rigorous Method of Evaluating Wind Turbine Noise", [14] demonstrated that tonality in the 450 Hz environs can be a problem with some turbines. The challenge in using one-third octave analysis, as noted previously in Figure 3, is that the tonality at about 450 Hz becomes split into the 400 and 500 Hz one-third octave bands, which transition at 447 Hz, and it can hide the presence of tonality. While the frequency band transitions identified in the ECMA-74 standard based on the bark scale have transitions at 300, 400, and 510 Hz and thus keep all of the tonality in a 450 Hz case in one specific bark range, that too might be problematic for a tonality nearer the transition point.

Other than perhaps those individuals with "perfect pitch" who can recognize specific frequencies such as the piano tuner who can tune to 440 Hz "by ear", most humans do not have specific transition "bins" in their cochlea, they simply group sounds together. Rather than using artificial boundaries as in the one-third octaves, or the bark scale, a better approach may simply to pick a centre frequency of a predominant sound, and then to calculate "pseudo" one-third octaves or barks about that predominant sound. Then, the assessment of tonality would be simpler, as the ear would recognise the group of sounds together, and if one central range is higher than either of the two surrounding ranges, then the identification of tonality would be simpler than using either the one-third octave or bark scale.

3.2 Evaluation of specific example of irritating tonality

The problem of both the IEC-61400-11 Standard and the ISO/PAS 20065 Standard is the way they treat the average sound level in the critical band. Three simple examples shown in Figures 8, 9, and 10 will help to explain the problem. Figure 8 shows a tone evaluated by both IEC 61400-11 and ISO PAS 20065 as strongly audible. It is a fairly simple, narrow bandwidth tone. Evaluation of the energy in the critical (central) bark as per ECMA-74 shows it is 7.6 dB higher than the lower bark, while only 4.2 dB higher than the upper bark.

Figures 9 and 10 show two comparable tones, which are broader in width than the one shown in Figure 8. They differ though in that the energy is not in one smooth curve, but in a more jagged one. Yet, all the energy in the central 450 Hz bark segment will be perceived by the listener as related. The energy in the central bark is over 5 dB higher than either the upper or lower bark in each case. Yet, the IEC 61400-11 and ISO/PAS 20065 codes dismiss both of these cases as inaudible. Curiously, when one listens to the recordings of all three, there is little perceptible difference. In fact, all are clearly audible.

Both the IEC 61400-11 and the ISO/PAS 20065 codes reduce the impact of tones in the critical bandwidth by focusing on the critical tone, and then lumping the rest of the energy in the central bandwidth into masking. This is inappropriate as that energy contributes to the sound perceived by the listener. The US Standard [15] cautions care is needed in selecting the bandwidth to delineate the discrete tone, as if the band is too narrow, the mean-square sound pressure of the tone (or the tone sound pressure) will be underestimated and the mean-square sound pressure of the noise will be overestimated. A more appropriate assessment technique is to compare the energy in the critical bandwidth – either by one-third octave analysis, or by bark analysis, or even more appropriately by calculating a specific central band similar in width to the bark or one-third octave but centred on the tonal peak.

4. Conclusions

Address the issue of cyclical noise which is a significant irritant by assessing the sound variation not only in Leq, but also in Lp. Where Lp is varying by more than 2 to 3 dB in a cyclical manner related to blade pass frequency, apply a penalty.

Calculation of tonality for wind turbines could be greatly simplified, and be far more effective by adopting a simplified approach similar to what is used in ECMA-74 to determine tonality. Compare the energy in the discrete energy spectrum in the bark (or better, a calculated bark centred on the tone) to the energy in the upper and lower adjacent frequency bark units. Where the central bark exceeds adjacent barks by 5 dB, apply a penalty.

Simple examples have been used to show that narrow band noise situations can be dismissed as inaudible by both the IEC 61400-11 and ISO/PAS 20065 codes, even while they are very similar sounding to a listener to a noise condition that is calculated to be very tonal and audible.

Recognize the limitations of the current codes in IEC 61400-11 and ISO/PAS 20065 in evaluation of complex waveforms. While the work fine for a simple narrow tone, they are unable to cope with the more typical narrow range noise produced by wind turbines that generates the irritation to those living in the environment surrounded by wind turbines.

References

[1] Fowler, K., Koppen, E., Matthias, K., *International Legislation and Regulations for Wind Turbine Noise*, 5th International Conference on Wind Turbine Noise, Denver 28-30 August, 2013.

[2] Cooper, J., Evans, T., *Accuracy of noise predictions for wind farms.* 5th International Conference on Wind Turbine Noise. Denver 28-30 August, 2013

[3] International Electrotechnical Commission (IEC). (2018). *IEC 61400-11 Edition 3.1 Wind Turbines - Part 11: Acoustic noise measurement techniques.* Geneva: International Electrotechnical Commission.

[4] World Health Organization, ICD-10, *International Classification of Diseases*

[5] Ontario Ministry of the Environment. (2018). *Noise Guidelines for Wind Farms.* Retrieved from https://www.ontario.ca/page/noise-guidelines-wind-farms

[6] World Health Organization Regional Office for Europe 2018, *Environmental Noise Guidelines for the European Region,* https://www.euro.who.int/en/publications/abstracts/environmental-noise-guidelines-for-the-european-region-2018

[7] Hansen, C., Hansen, K., *Recent Advances in Wind Turbine Noise Research*, *Acoustics* **2020**, *2*, 171–206; https://doi.org/10.3390/acoustics2010013

[8] Palmer, W. K. (2017, Nov. 07). *Why wind turbine sounds are annoying and why it matters. Global Environment, Health & Safety, 1*(No. 2:12), 1-17.

[9] Evans, T. C., Cooper, J., *A Comparison of Tonal Noise Regulations in Australia.* Acoustics 2015 Hunter Valley. AAS2015. Australian Acoustical Society.

[10] (AFNOR) Association Française de Normalization, NF S 31-010, *Acoustique - Caracterisation et Mesurage des bruit dans l'environment.* AFNOR, Paris.

[11] Palmer, W.K.G., *Confirming Tonality at Residences Influenced by Wind Turbines*, Journal of Energy Conservation, Vol. 1 Issue 3 DOI: 10.14302/issn.2642-3146,jec-20-3359

[12] Søndergaard, L.S., Thomsen, C., Pedersen, T.H., Prominent tones in wind turbine noise – Round-robin test of the IEC 61400-11 and ISO/PAS 20065 methods for analysing tonality content, 8th International Conference on Wind Turbine Noise, Lisbon, 12-14 June, 2019.

[13] ECMA International. (2019). *Measurement of airborne noise emitted vy information technology and telecommunications equipment.* ECMA International. Geneva: ECMA International.

[14] Palmer, W.K.G., *A Rigorous Method of Addressing Wind Turbine Noise*, 7th International Conference on Wind Turbine Noise, Rotterdam, May 2017.

[15] Acoustical Society of America. (2020). ANSI/ASA-1-13 *Measurement of Sound Pressure Levels in Air (American National Standard).* Melville, NY: Acoustical Society of America.

Figure 1 – Wind in Trees – Nearest Vestas V82 Wind Turbine > 6 km

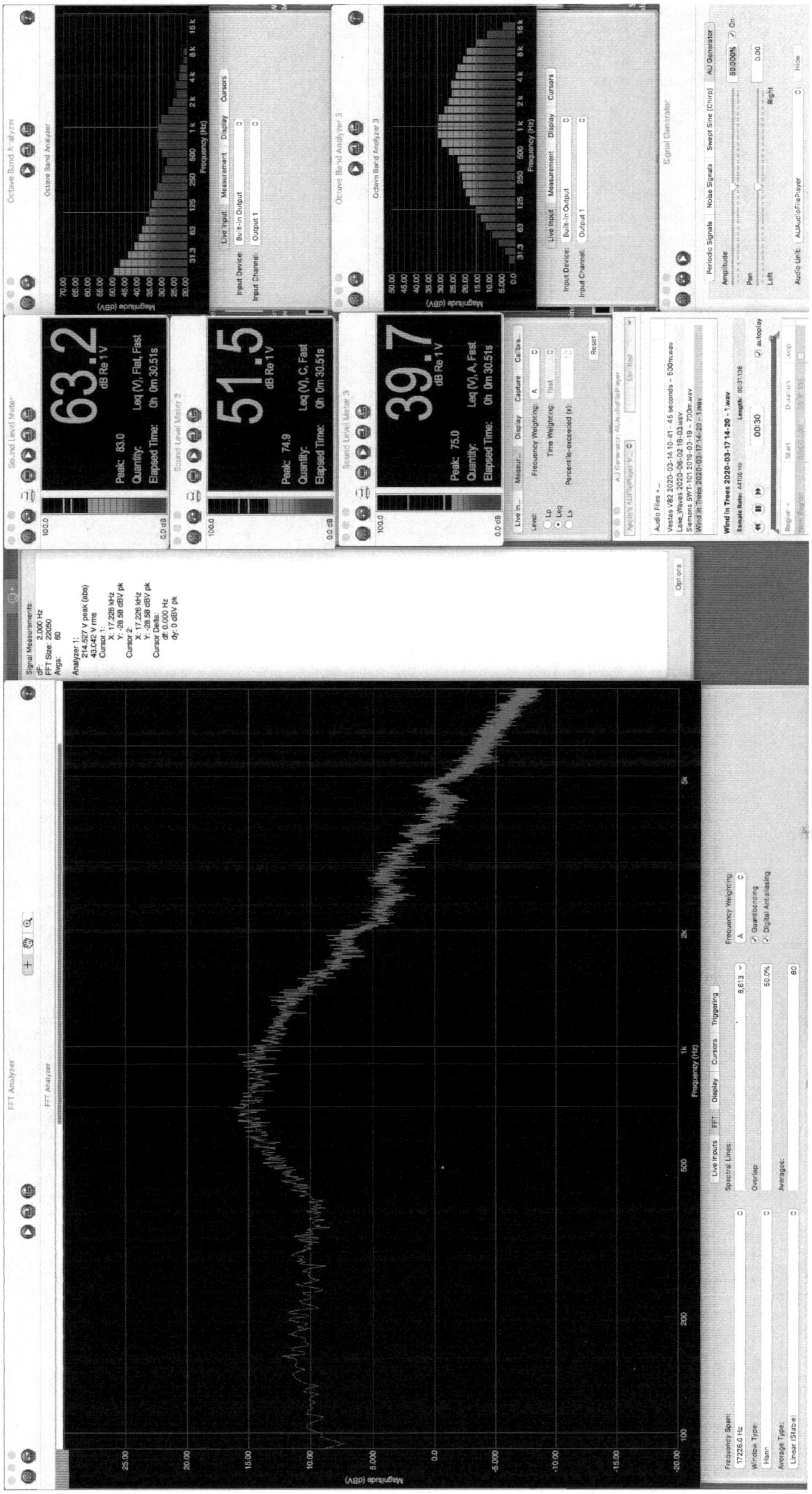

Figure 2 – Nearest Vestas V82 turbine ~ 500m

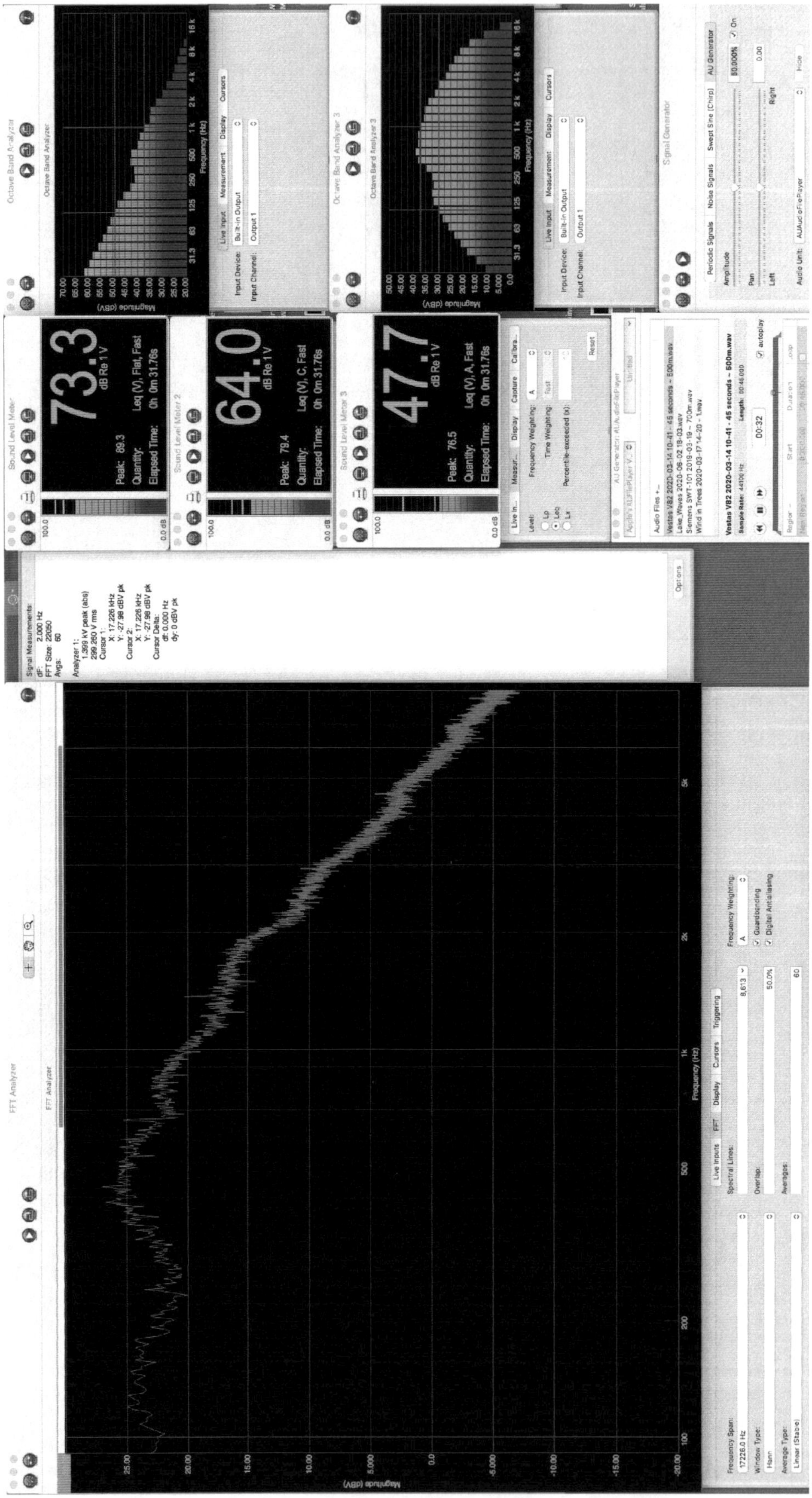

Figure 3 – Nearest Siemens SWT-101 Turbine ~ 700 m

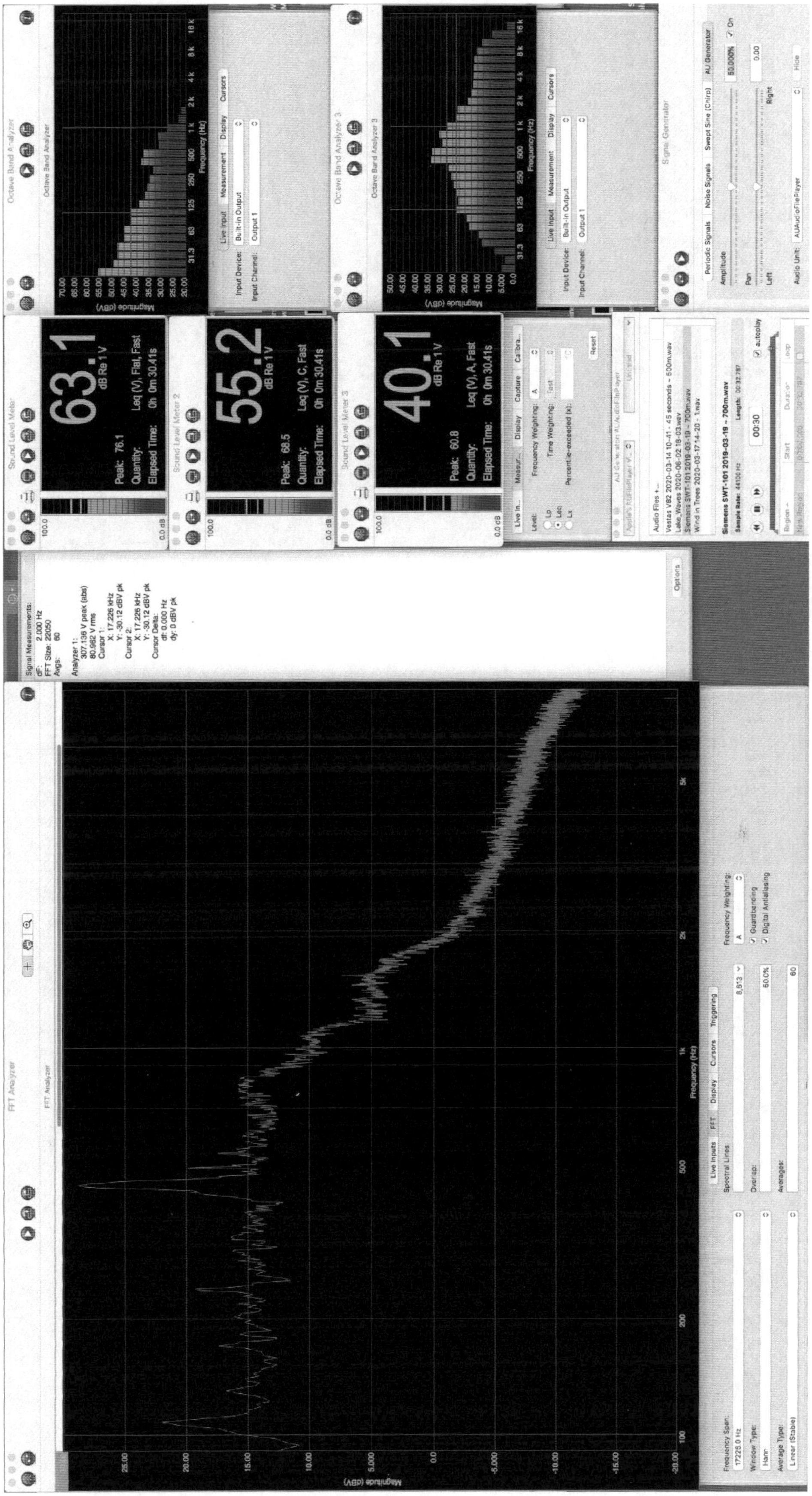

Figure 4 – Lake Waves – Nearest Vestas V82 turbine > 5 km

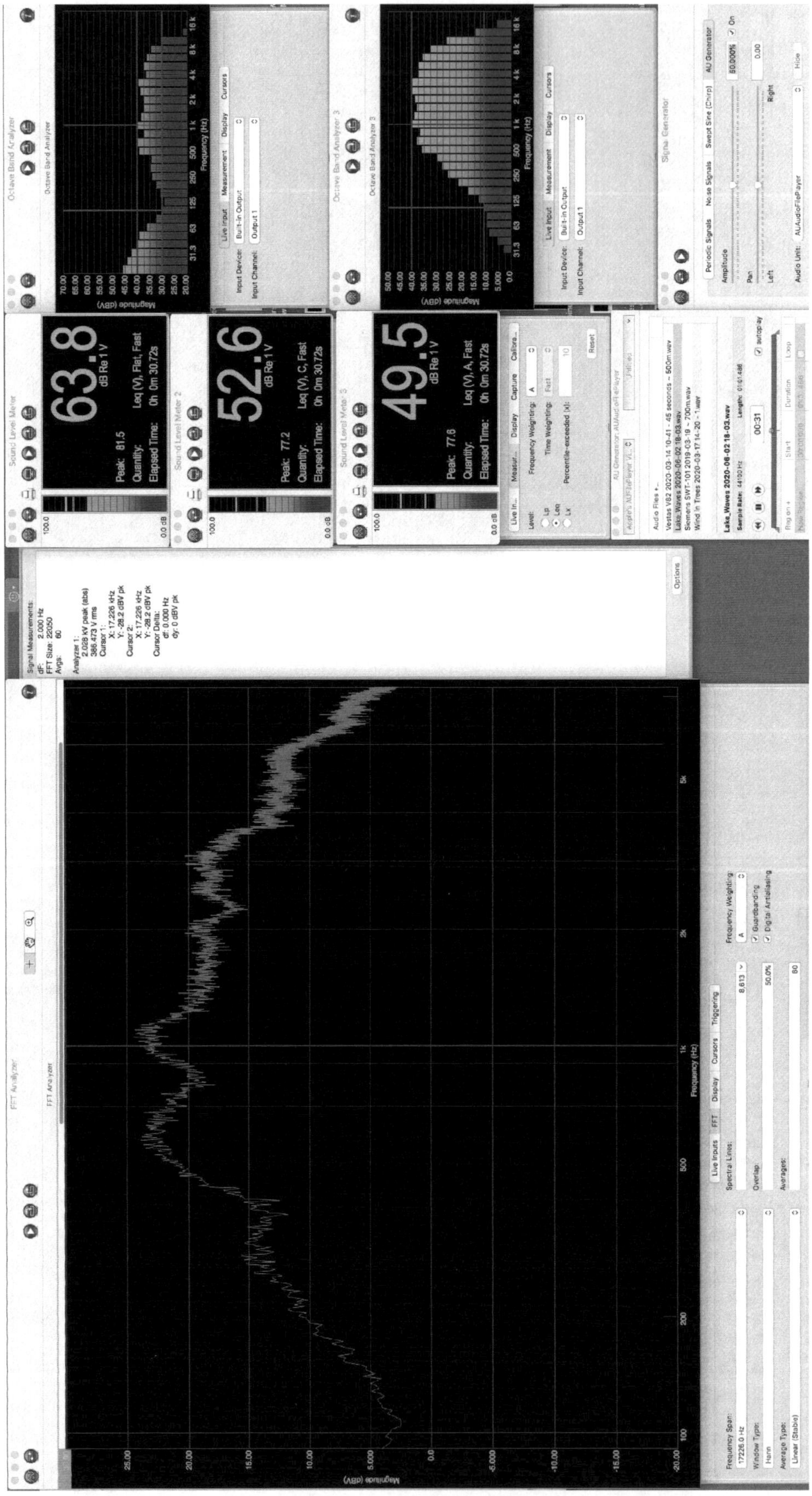

Table 1 – Verification of Tonal Audibility Models and FFTs using Round Robin Test Files

	IEC – 61400-11				ISO/PAS 20065			
RRT Test	WTN 2019	Model Spectra	Toolbox Eval	Audacity Eval	WTN 2019	Model Spectra	Toolbox Eval	Audacity Eval
M01	11.1	11.4	7.2	11.4	11.2	11.2	7.2	10.9
M02	-5.4	-5.4	-4.6	-5.4	-5.0	-4.9	-4.9	-5.4
M03	13.5	13.5	14.5	13.5	12.7	12.3	12.3	12.5
M04 *	6.5	6.5	3.9	7.1	7.8	1.6	4.1	9.4
M05	9.1	9.1	8.1	9.1	9.1	8.9	8.0	8.9
M06	12.6	12.6	13.5	13.2	12.7	12.6	13.1	12.7
M07	5.2	5.2	5.1	5.5	4.9	4.7	4.7	4.9
M08	12.1	12.1	8.9	12.3	12.3	12.3	8.7	12.1
M09	1.7	1.7	0.3	2.0	2.0	2.0	0.8	2.2
M10	14.6	14.6	15.6	14.5	14.5	14.6	15.7	14.4
O01	14.5	15.1	13.9	15.1	14.7	14.4	13.3	14.7
O02	8.0	8.0	7.8	8.2	8.2	8.2	7.8	8.1
O03	-	-	-	-	-9.3	-8.7	-	-
O04	5.1	5.6	6.1	5.6	5.3	5.2	5.7	5.2
O05	9.4	9.6	8.0	9.7	9.4	9.5	8.2	9.2
O06	4.5	4.5	5.0	4.3	4.4	4.4	6.5	3.9
O07	-0.2	-0.2	0.7	-0.8	-0.1	-0.1	0.7	-0.3
O08	12.3	12.6	11.5	12.4	12.1	12.3	13.0	11.7
O09	15.0	15.8	11.8	15.8	14.6	15.2	11.2	15.1
O10	9.5	9.5	10.6	9.6	9.6	9.6	10.7	9.4
O11	7.5	7.5	8.1	7.5	7.5			
O12	8.2	8.8	4.2	8.7	7.5			
O13	9.7	9.9	11.0	10.0	9.7			
O14	5.5	5.7	5.6	5.6	5.6			
O15	4.8	4.9	4.5	5.1	4.9			
O16	4.6	4.6	5.6	4.5	4.7			
O17	8.1	8.1	3.0	8.3	6.4			
O18	4.4	4.5	-0.7	4.8	3.6			
O19	10.0	10.1	8.3	10.1	9.7			
O20	10.8	10.8	11.7	10.7	10.7			
O21	-	-	-	-	-9.9			

Round Robin Test 4 was noted as unique in the paper by Søndergaard for WTN 2019.
See next pages for FFT charts, all nominally calibrated the same, yet with varying result.

Figures 5 through 7

Output from Calibrated FFT Display

Figure 5 – Round Robin Test M04 - Spectra A (as provided)

- **Spans 15 to 40 dBA at frequency below critical frequency**
- **Spans 10 to 40 dBA at frequency above critical frequency**

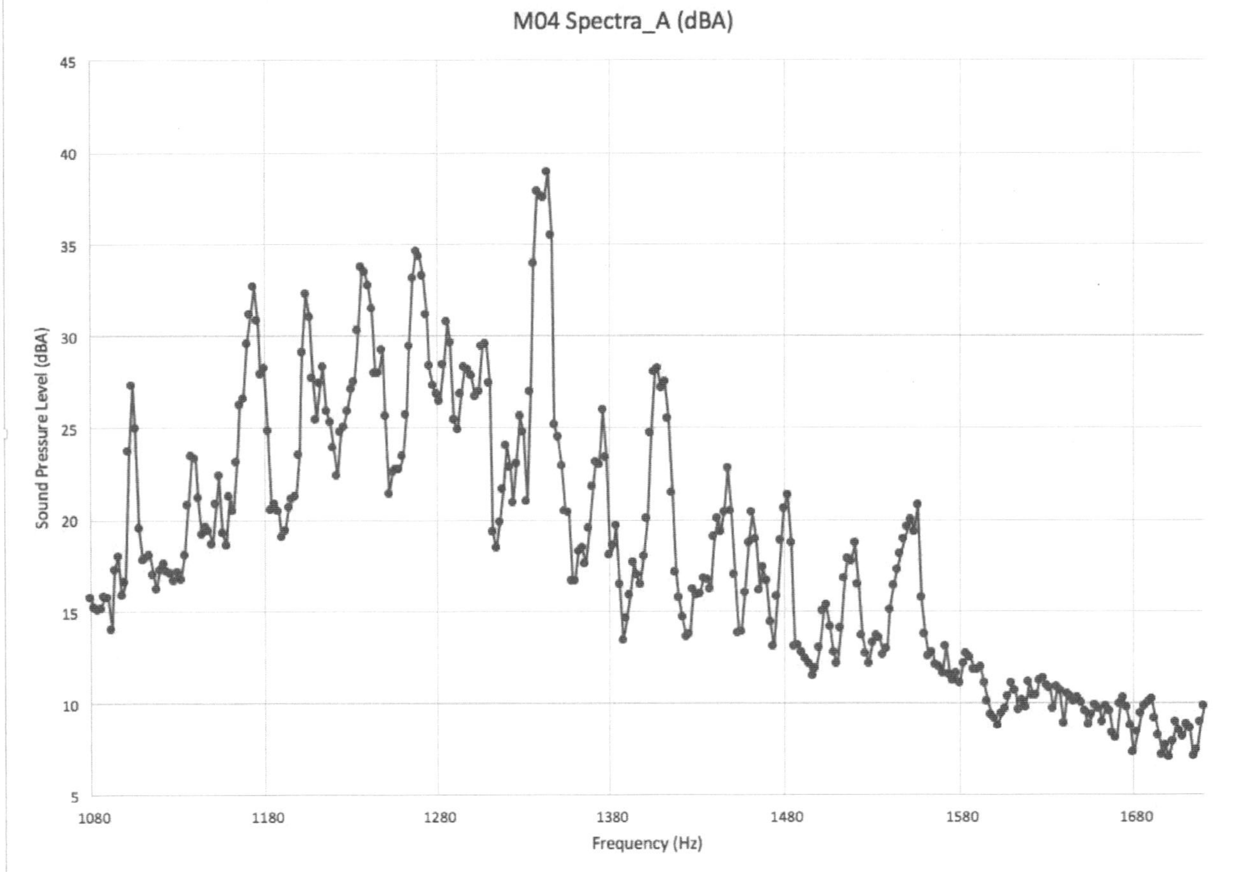

Figure 6 – Round Robin Test M04 – Audacity evaluation (next page)

- **Spans 20 to 45 dBA at frequency below critical frequency**
- **Spans 15 to 45 dBA at frequency above critical frequency**

Figure 7 – Round Robin Test M04 – Electroacoustics Toolbox evaluation (next page)

- **Spans 17 to 35 dBA at frequency below critical frequency**
- **Spans 10 to 35 dBA at frequency above critical frequency**

All evaluation tools calibrated using the 84 dB calibration signal provided with Round Robin files.

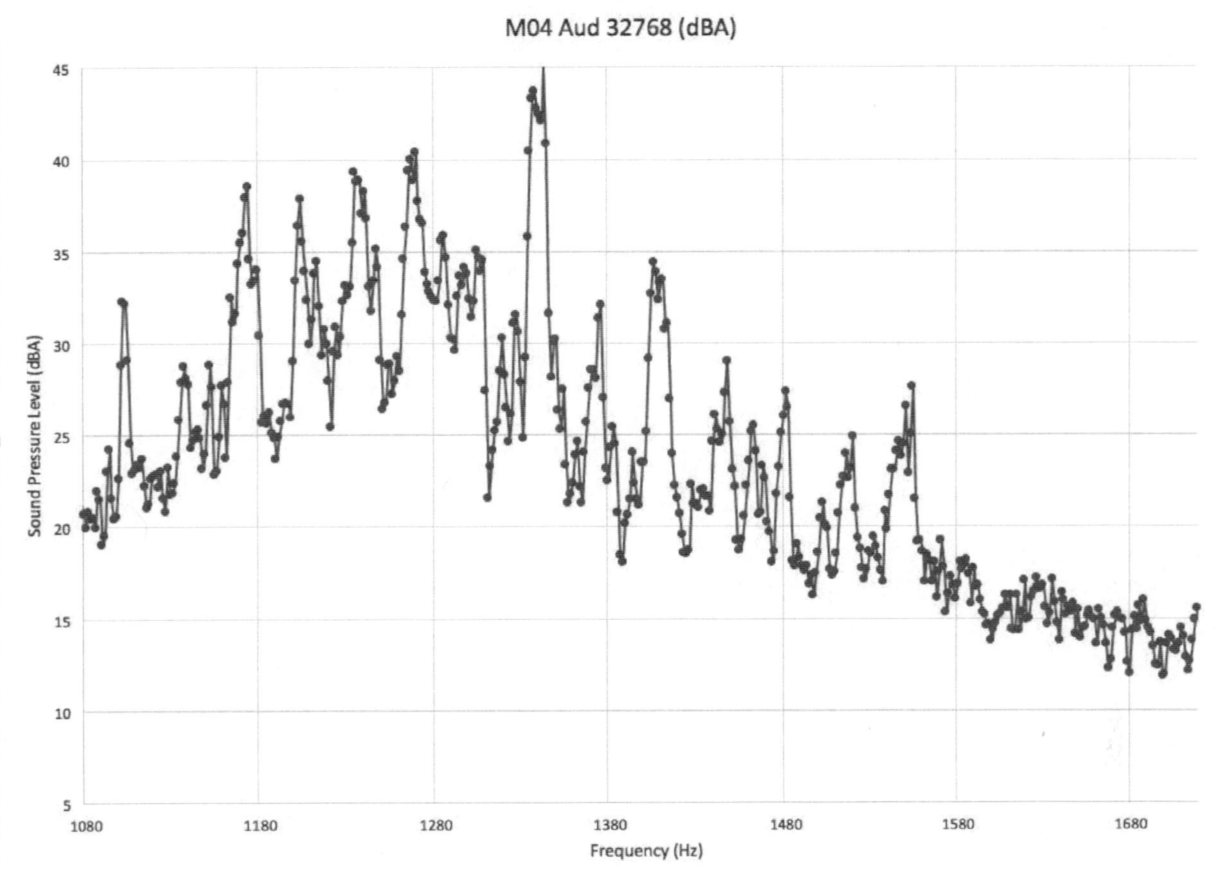
M04 Aud 32768 (dBA)
Sound Pressure Level (dBA)
Frequency (Hz)
45
40
35
30
25
20
15
10
5
1080
1180
1280
1380
1480
1580
1680

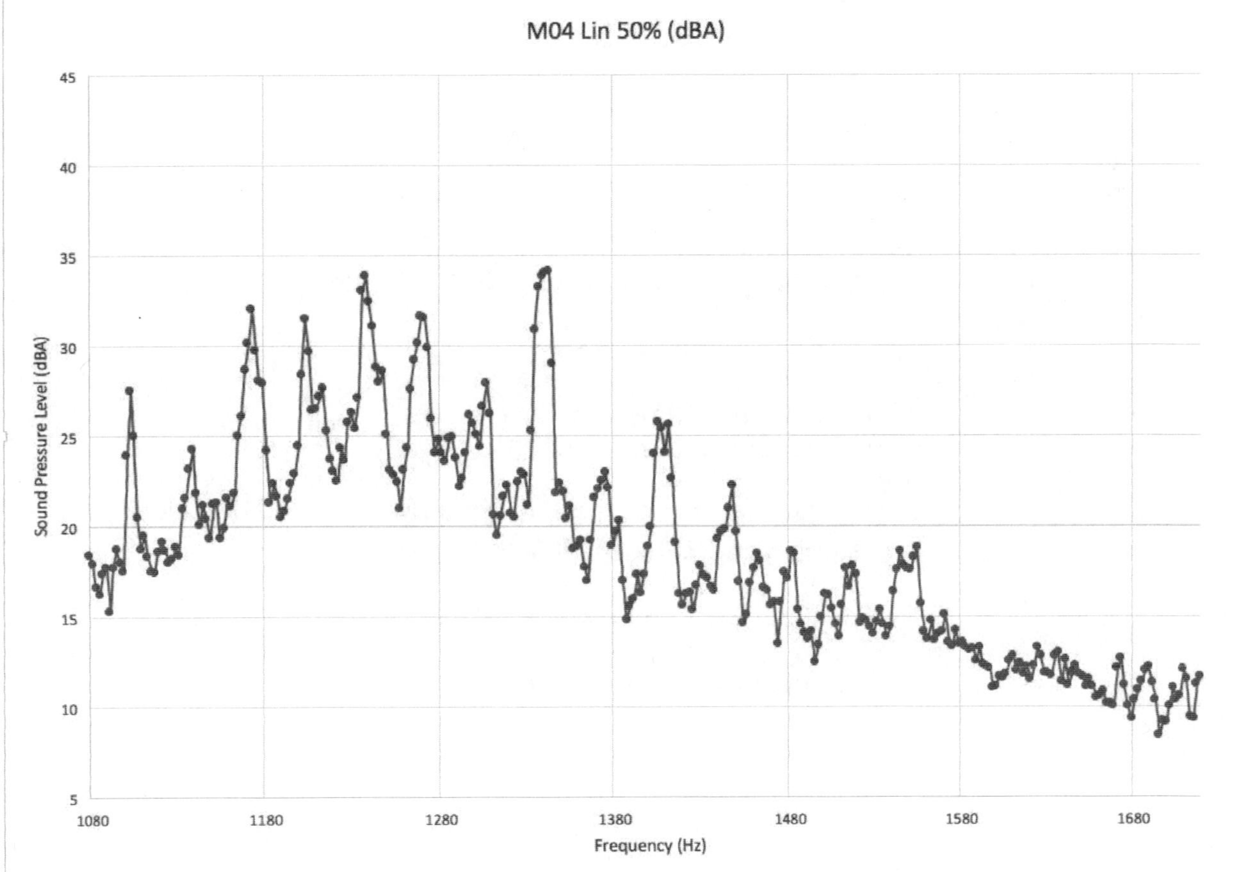
M04 Lin 50% (dBA)
Sound Pressure Level (dBA)
Frequency (Hz)
45
40
35
30
25
20
15
10
5
1080
1180
1280
1380
1480
1580
1680

Figures 8
An Audible Tone

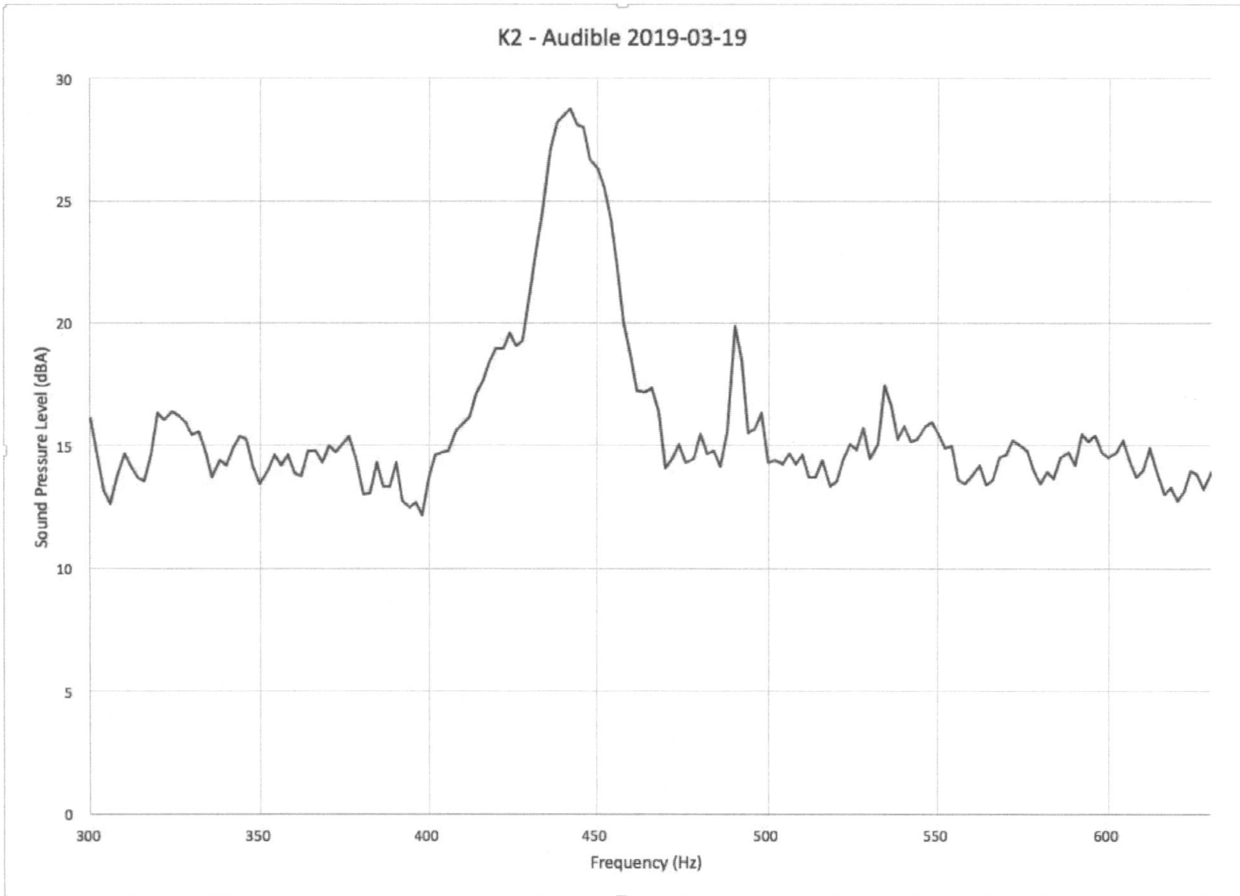

K2 – 2019-03-19

Tonal Audibility per IEC 61400-11 + 5.8 dB
Tonal Audibility per ISO / PAS 20065 + 6.2 dB

Tone 442 Hz

Tonality per IEC 61400-11 + 3.6 dB
Tonality per ISO/PAS 20065 + 4.0 dB

Masking per IEC 61400-11 + 32.4 dB
Masking per ISO/PAS 20065 + 32.2 dB

Critical Bark to Lower + 7.6 dB

Critical Bark to Upper + 4.2 dB

Figure 9
An Inaudible Tone (??)

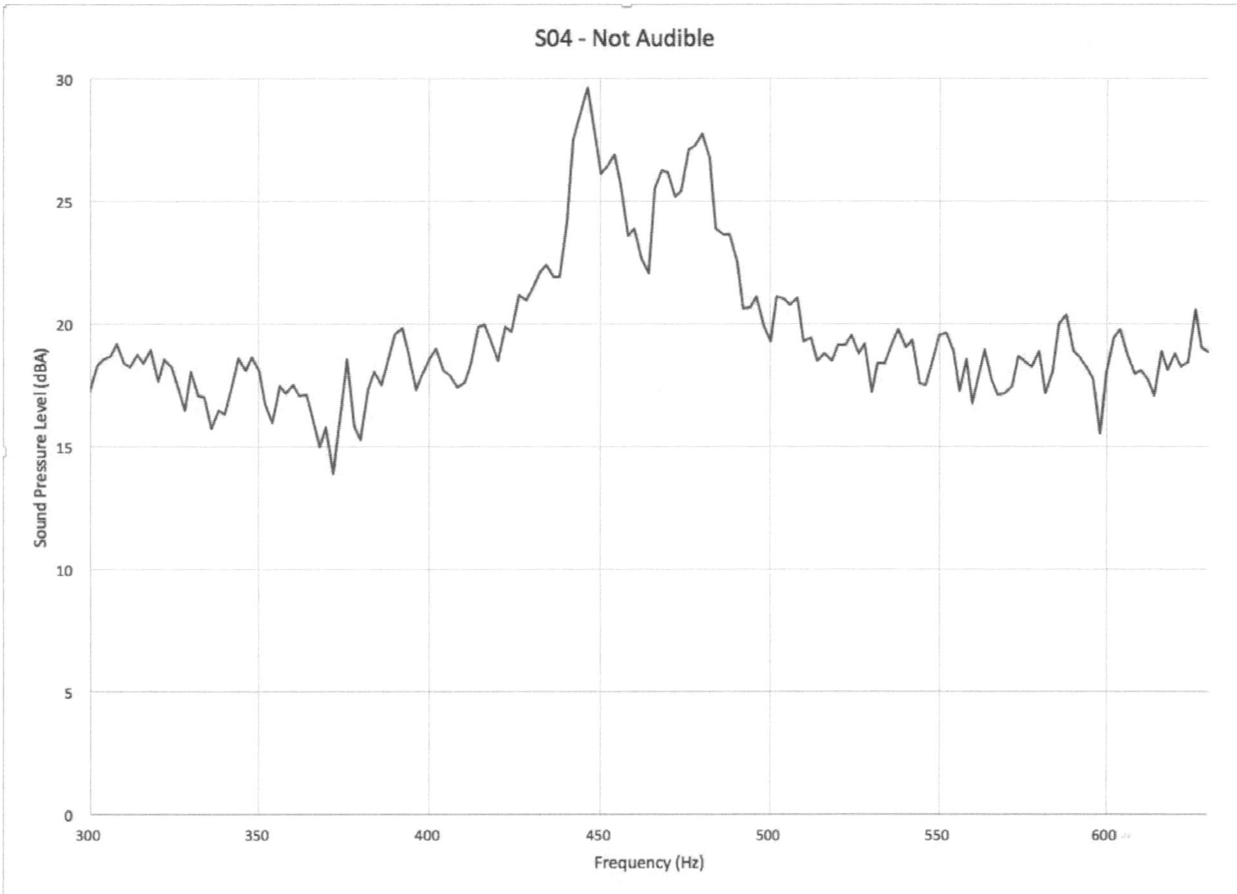

K2 – 2019-11-17

Tonal Audibility per IEC 61400-11 - 5.5 dB
Tonal Audibility per ISO / PAS 20065 - 9.2 dB

Tone 446 Hz

Tonality per IEC 61400-11 -7.8 dB
Tonality per ISO/PAS 20065 -11.5 dB

Masking per IEC 61400-11 + 38.2 dB
Masking per ISO/PAS 20065 + 39.3 dB

Critical Bark to Lower + 6.2 dB

Critical Bark to Upper + 5.3 dB

Figure 10
An Inaudible Tone (??)

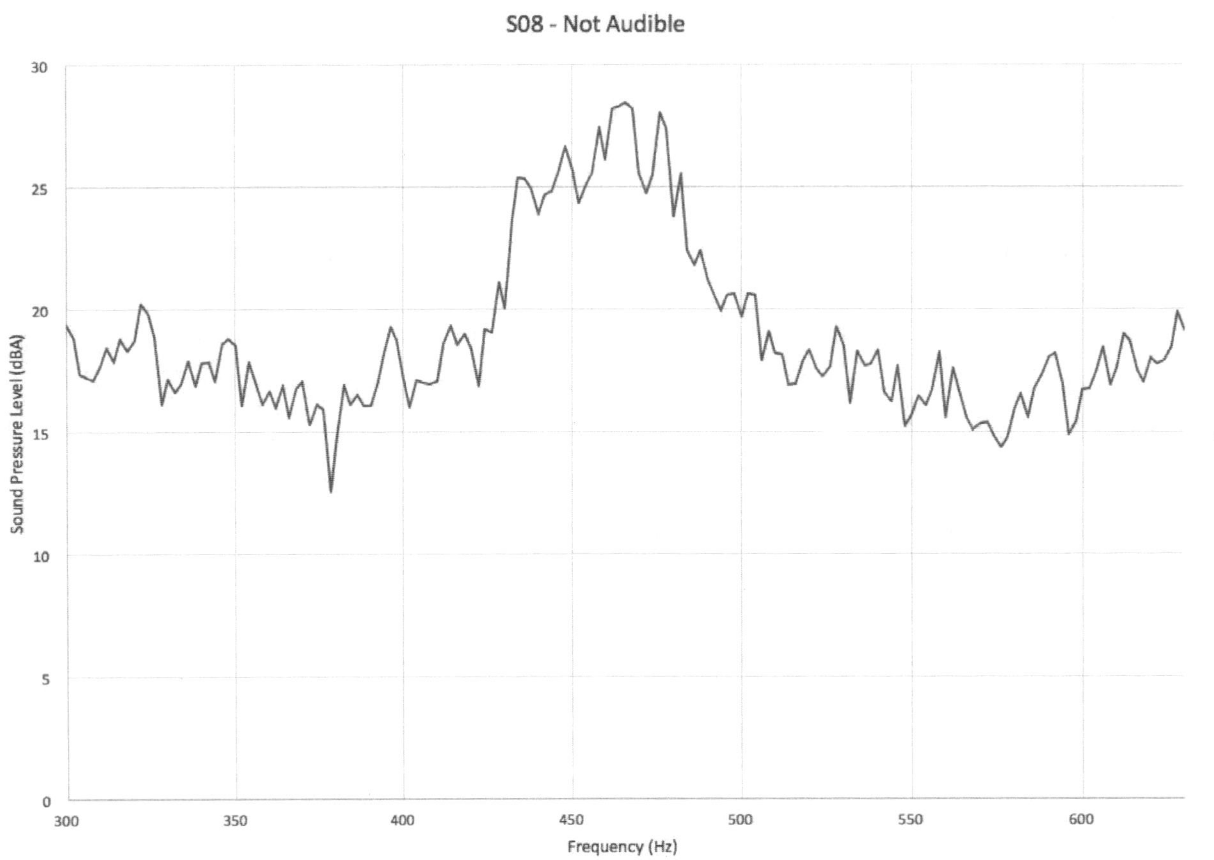

K2 – 2019-11-27

Tonal Audibility per IEC 61400-11 – Not detectable
Tonal Audibility per ISO / PAS 20065 – 2.9 dB

Tone 466 Hz

Tonality per IEC 61400-11 – Not detectable
Tonality per ISO/PAS 20065 – 5.2 dB

Masking per IEC 61400-11 + 38.3 dB
Masking per ISO/PAS 20065 + 38.6 dB

Critical Bark to Lower + 6.3 dB

Critical Bark to Upper + 6.5 dB

INCEEUROPE

9th International Conference
on
Wind Turbine Noise
Remote from Europe – 18th to 21st May 2021

Assessment and rating of Wind turbine noise immission at dwellings – the influence of amplitude modulation, aerodynamic noise sources and the Doppler effect

Kai Pies, Schalltechnisches Ing.-Büro Pies GbR, kpies@schallschutz-pies.de
Sergio Martinez, TÜV Rheinland Energy GmbH, sergio.martinez@web.de

Summary

The study on the effects of amplitude modulation (AM) is based on multiple long-term measurements of wind turbines (WTG) in Germany. Additional measurements in short distances to wind turbines were performed. In addition to the noise level, the characteristics of aerodynamic noise sources and the Doppler effect are responsible for the annoyance of residents. This effect is called amplitude modulation (AM) and is dependent on the rotation speed and the synchronisation of several wind turbines. The effect can be simulated with calculation models. In addition to the sound pressure level, AM is another noise effect that annoys people in the periphery of WTG. The study shows that the noise effect can be related to the L_{Aeq} and the modulation depth p_{AM5}. p_{AM5} can be derived e.g. using the wavelet analysis. The annoyance can be described in terms of a dose-effect relationship. The results of listening tests show a good agreement to the found relationship. It can also be shown that a surcharge can be determined on the basis of the relationship.

1. Introduction

The assessment of noise in the neighbourhood of wind turbines and wind farms is based on a combination of measurement, calculations and a defined assessment procedure. For manufacturers and planners, the rating level is an essential parameter for limiting the number of possible wind turbines that can be installed. The focus is on the night time because of the higher sensitivity when sleeping. The focus here is not on protecting the interior of a building, but on compliance with a guide value outside the building. An external value is independent of the building stock and the location of the living and bed rooms and is therefore a suitable value for planning.
Both the prognosis and the measurements are carried out by accredited test laboratories. A high level of experience in dealing with measuring systems or calculation software is required.
The results presented in this publication are based on scientific studies on 3 different wind parks and 3 individual wind turbines. The measurements at the wind farms were made due to complaints from residents. The study is based on a total of 27 data sets of one hour each. The measurements at the wind farms were carried out over a period of more than 6 weeks. The data

include different meteorological boundary conditions at night, mostly between midnight and 05:00 a.m. The distances between the wind turbines and the relevant measuring points (dwellings) are 600 m, 800 m and 1000 m.
In addition to these measurements, further measurements were carried out on 3 individual wind turbines. The measurements were done for several hours. The measurement setup corresponds to DIN EN 61400-11 with a reverberant plate and secondary wind screen. The sampling rate for the sound pressure level curve was 8 Hz or 10 Hz. In addition, sound files (sampling rate up to 48 kHz) were saved for which narrowband spectra were calculated.
The results of this paper help to assign the causes of noise effects. In addition, the impact of noise from wind turbines on people is investigated. These results should be included in the assessment of the noise in immission control.

2. Analysis of the noise of single WTG in short distances (cf. [1])

The noise from wind turbines (WTG) is measured according to the procedures and boundary conditions of DIN EN 61400-11 [1]. For example, the reference measuring points and the equipment are specified. One aspect is relevant here: the sound source is modelled as a point source in the centre of the rotor. However, the azimuth and the radiation angle to the measuring microphone are defined within a narrow range. The noise levels are measured according [1] as average values over a duration of at least 10 seconds. The statistics of the measurement data should cover the entire operating range of the WTG. This means that the multi-dimensional relationship between wind inflow, meteorological conditions, electrical power and parameters of the rotor movements and their aerodynamics must be taken into account. This is only partially possible with [1].
The following Fig.1 shows an evaluation - based on 30-second mean values - of the measured sound pressure level (Original recordings with a sampling rate of 1 Hz.) close to a single WTG. The measurements were done in the lee wind direction at a reference measuring point according to [1]. The operating data of the wind turbine are listed below:

- max. electrical power P_{el} = 2.4 MW
- rotor diameter RD = 117 m
- hub height NH = 91 m
- WTG equipped with a gearbox

The typical curve of the sound pressure level is shown in **Fig.1** as a function of the electrical power (10-second data points) of the wind turbine (cf. Martinez et al. 2016, 2001). The different percentile level ranges are represented by different shapes / colours. Above levels of ~ 52 dB (A), the data points lie in a band with a width of approx. 2 dB. Acoustically, this corresponds to a quasi-stationary state of the wind turbine with a noise level that only fluctuates by ± 1 dB.

Fig.2 shows the relationship between third octave levels $L_{Aeq,\text{third octave}}$ and the total level.L_{Aeq}. The curves correlate well in the range above 500 Hz. In the frequency range between 63 Hz and 200 Hz, however, the WTG develops a peculiar, uncorrelated, even anti-correlated frequency behaviour. This fact is responsible for higher emissions in some level ranges of the L_{Aeq}. This may be attributed to the gearbox and the speed-dependent aerodynamic behaviour of the wind turbine. At greater distances the higher frequencies components quickly lose their importance because of the air damping. With regard to this low-frequency effect, however, it can be questioned whether the overall level alone is sufficient as an assessment variable. This behaviour of WTG is typical in practice and not limited to the WTG examined.

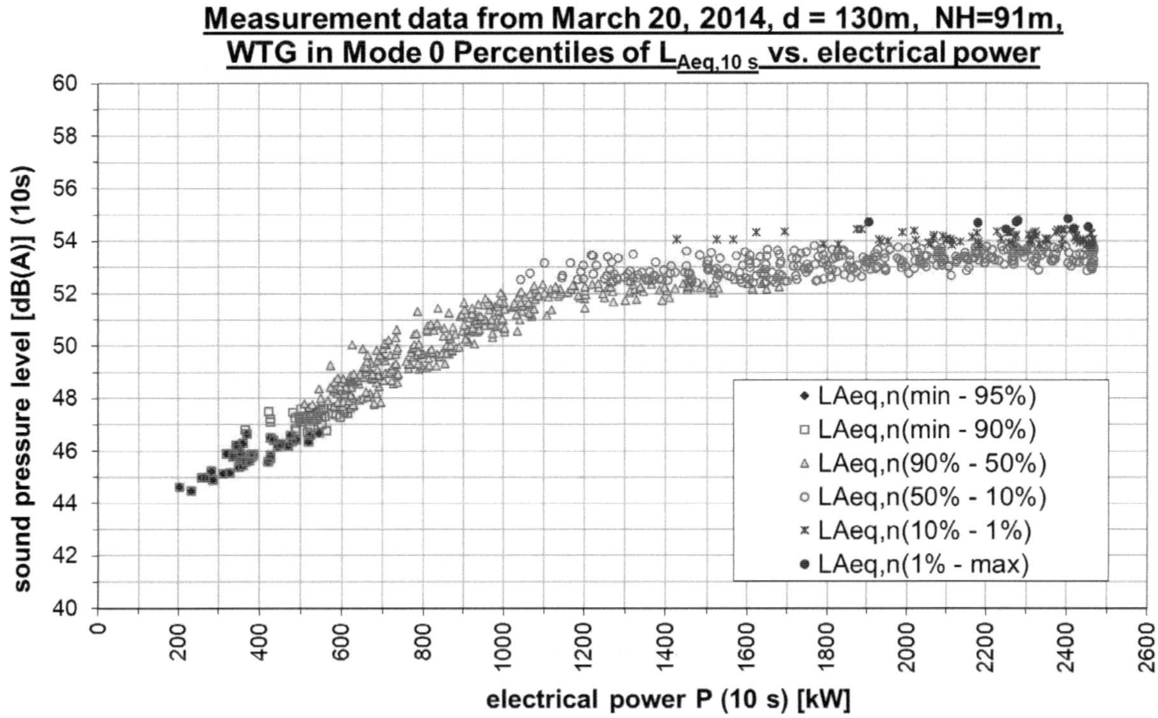

Fig 1: Sound pressure level $L_{Aeq,10s,n}$ (N= 1 to 810) at the reference measuring point (RMP), (color) percentile statistics $L_{Aeq,10s,n}$ vs. Power output (10s)

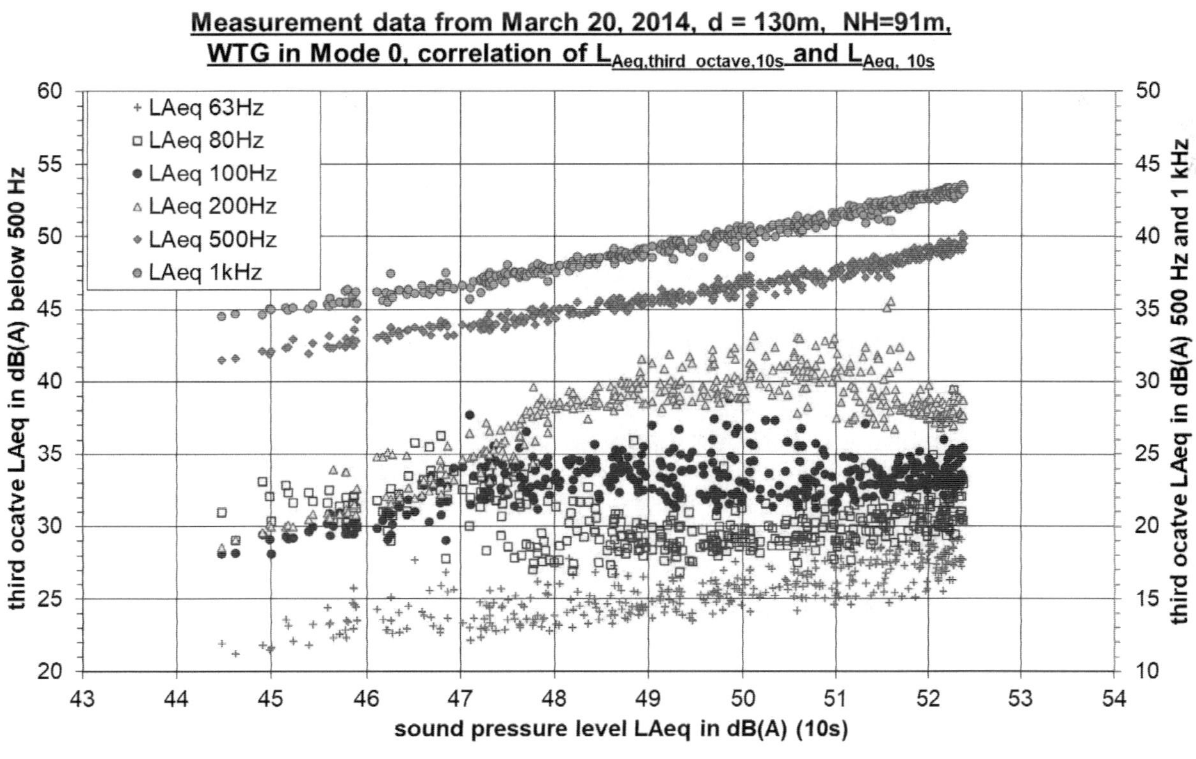

Fig 2: Correlation of the third-octave band levels (10s) with the total level $L_{Aeq,10s}$ (under $L_{Aeq,30s,50}$)
A total of approx. 400 measurement data

Another investigation on a single wind turbine deals with the acoustic properties that are not recorded by the measurement method according to [1]. A typical property is the amplitude modulation (AM) of the WTG. This effect is present and measurable in small and larger distances to wind turbines if the sampling frequency for the L_{Aeq} is set to 10 Hz or higher. The investigation is based on measurement data from individual wind turbines as well as from wind farms. **Fig.3** and **Fig.4** show the results of measurements at a single WTG simultaneously at two measuring points and in a distance of 170 m. The wind turbine has a maximum electrical power of P = 3 MW, a rotor diameter of RD = 113 m and a hub height of NH=142.5 m. The WTG is not equipped with a gearbox.

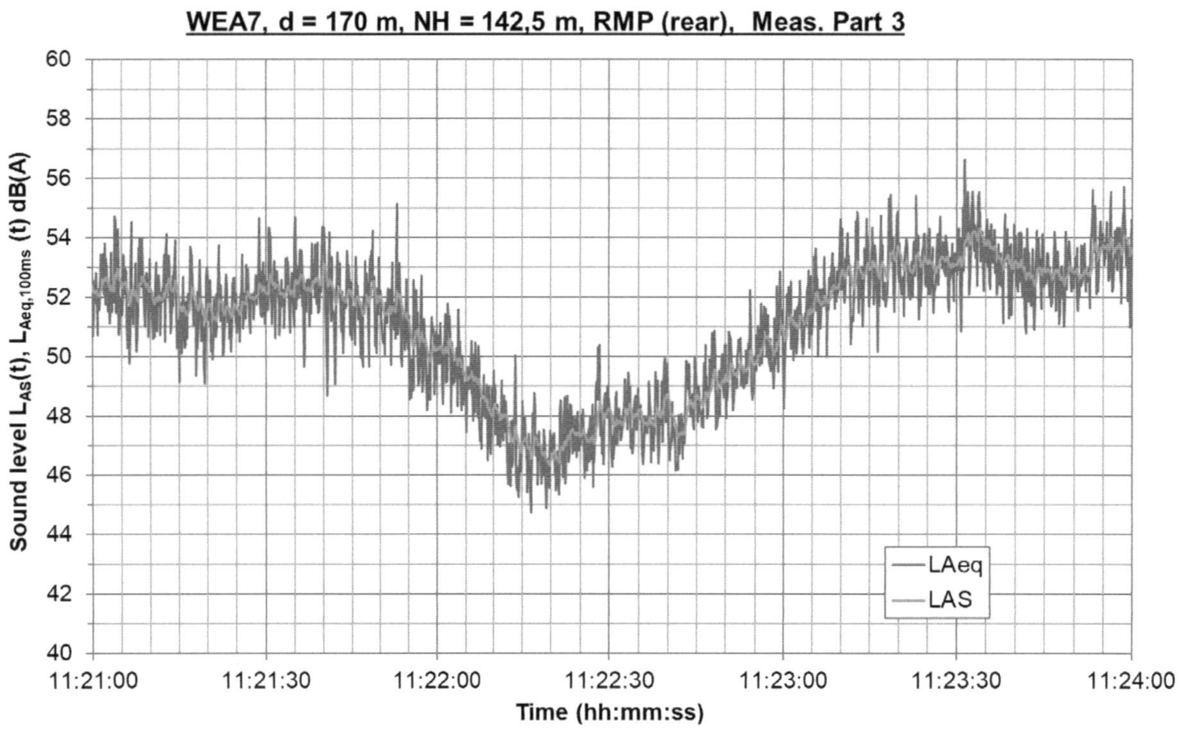

Fig 3: Time history $L_{Aeq,100ms}(t)$ and $L_{AS}(t)$ at the ref. measuring point RMP according to DIN EN 61400-11, boundary microphone (!). Evaluations: AM: $p_{AM5}(11{:}21)$ = 3.64 dB; $p_{AM5}(11{:}22)$= 3.16 dB; $p_{AM5}(11{:}23)$ = 3.26 dB, average $p_{AM5,m}$ = 3.4 dB. $L_{Aeq,m}$ = 51.5 dB(A), power output variable: 1.6 MW

Fig.3 shows the results from a measuring point that is located in the lee of the WTG. The measurements were done in compliance with DIN EN 61400-11 with the microphone located on a plate. The measuring point in **Fig. 4** is located beside the WTG, slightly in front of the rotor and at a height of 4 m above the ground. The measurements show that the sound pressure level, according to the rotational movement of the blades, varies relatively strongly in time. The AM fluctuation width of this signal modulated on the basis of the rotation (in this case the overall A level) is direction-dependent. The effect is stronger in the lateral direction than in the slipstream of a wind turbine, even if the average overall level L_{Aeq} is approx. 1,1 dB laterally lower. The AM fluctuation range is quantified by the parameter p_{AM5}. (cf. Martinez, Pies et al (2018) and (2019)). In this case, the trend correction is not carried out with the wavelet method - as explained more in detail in Martinez et al (2019), but according to the Japanese method in Fukushima et al (2017). E.g. the fluctuation width p_{AM5} of the AM is determined from the statistics of the level differences $\Delta L_{Aeq,100ms}$ (t) = $L_{Aeq,100ms}$ (t) - L_{AS}(t). In this case, the AM modulation frequency results from three times the rotational frequency of the WTG.

Fig.5 and **Fig.6** show the amplitude modulation of the wind turbine without the long-term fluctuations. The data come from the same measurements as shown in **Fig.3** and **Fig.4**.

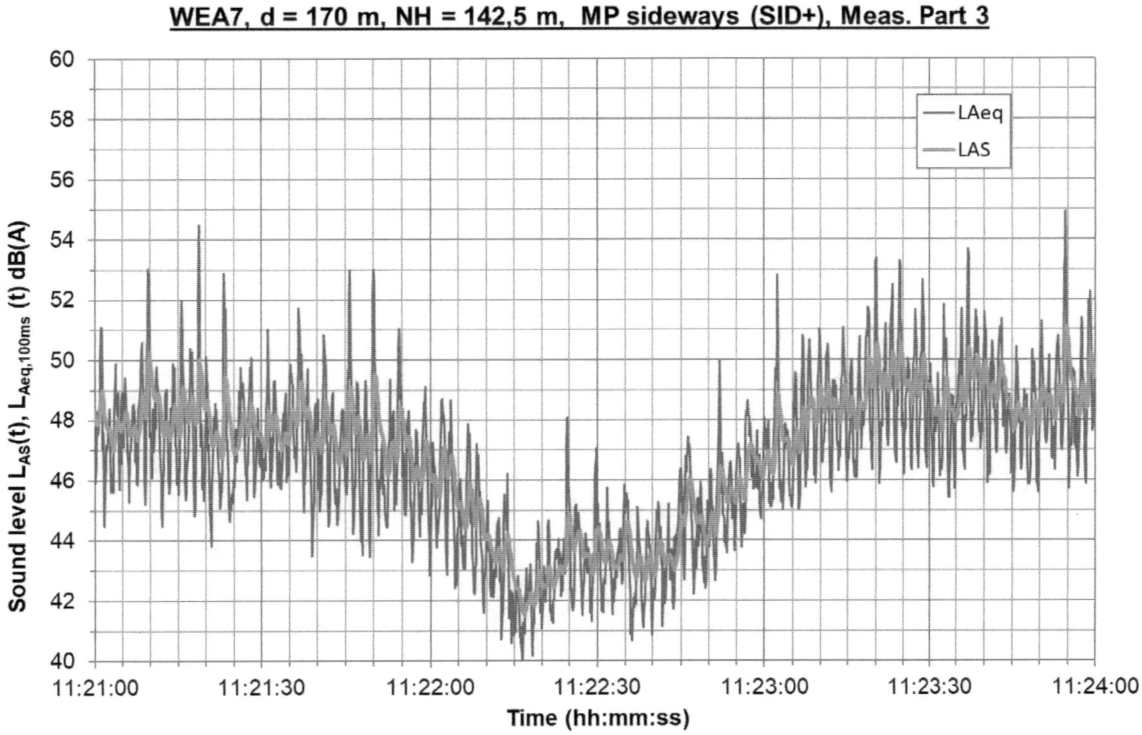

Fig. 4: (Simultaneous to Fig.3) Time history $L_{Aeq,100ms}(t)$ and $L_{AS}(t)$ at the measuring point sideways (SID, aprox. 90°+) at a height of 4 m. Evaluation: AM: $p_{AM5}(11:21) = 6.24$ dB; $p_{AM5}(11:22) = 4.68$ dB; $p_{AM5}(11:23) = 5.47$ dB; average $p_{AM5,m} = 5.5$ dB. $L_{Aeq,m} = 47.4$ dB(A)

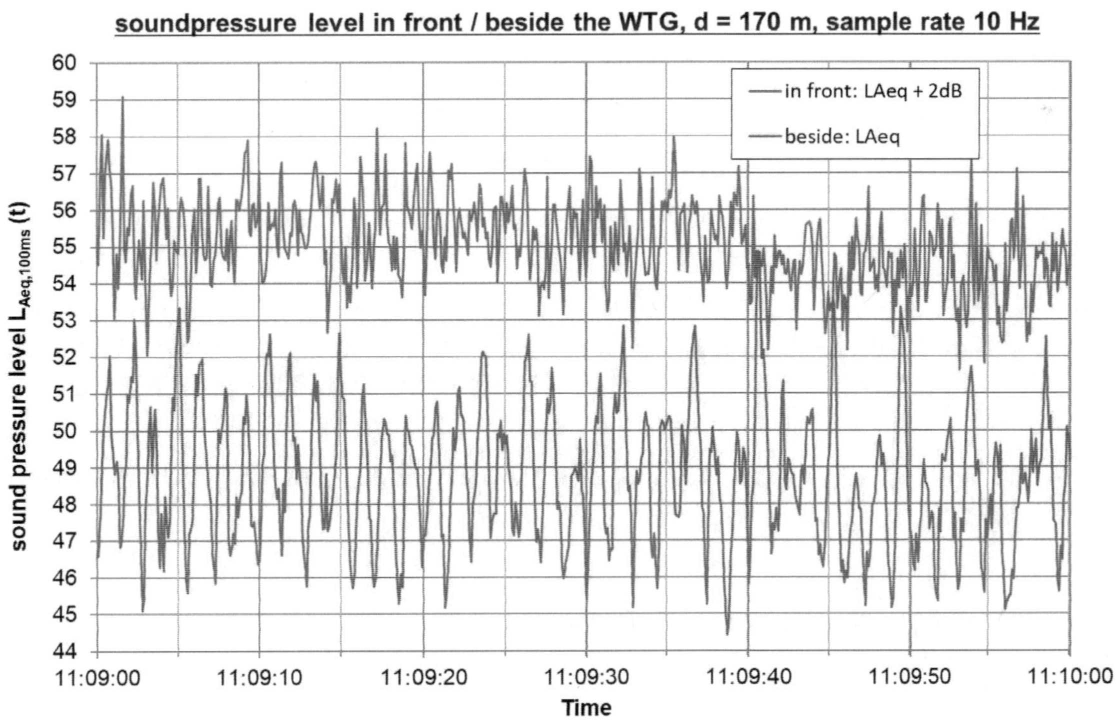

Fig. 5: Time history sound pressure level $L_{Aeq,100ms}(t)$ at the RMP and simultaneously at the measuring point sideways (SID+), approx. 90°+. Evaluations: $L_{Aeq,m}$ (RMP) = 53.2 dB(A); $L_{Aeq,m}$ (SID+) = 49.1 dB(A); $\Delta L/\Delta t$ (10/90) at RMP: 13.6/-13.5 dB/s ; $\Delta L/\Delta t$ (10/90) at SID+: 14.2/-13.6 dB/s; Power output ≈ 2.2 MW

The measurements at both measurement points were done simultaneously. The extent of the AM can be clearly seen. The short-term fluctuations can lead to variable noise level change speeds. They can be determined using the derivative $dL_{Aeq,\ 100ms}$ / dt of the noise level. From the 10% or 90% quantile of this quantity it can be estimated that level change speeds above ± 11 dB/s can occur.

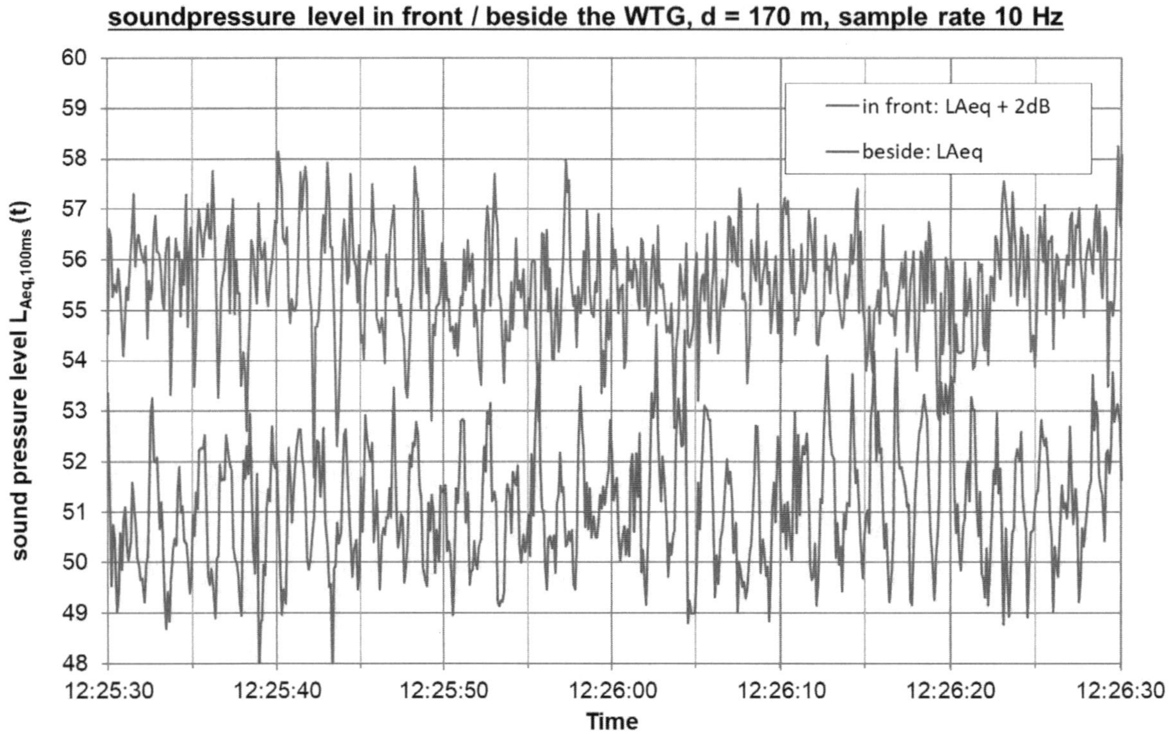

Fig. 6: Time history sound pressure level $L_{Aeq,100ms}(t)$ at RMP and simultaneously at measuring point Front 45°(VN45). Evaluations: $L_{Aeq,m}$ (RMP) = 53.7 dB(A); $L_{Aeq,m}$ (VN45) 51.3 dB(A); ΔL/Δt (10/90) at RMP: 13.1/-12.8 dB/s ; ΔL/Δt (10/90) at the VN45: 11.6/--11.5 dB/s. Power output ≈ 2.4 MW

This is caused by the Doppler effect and also by speed- and direction-dependent effects in the aerodynamic noise. After all, blades with a length of over 55 m glide past the fixed receiver in the air at different radiation angles.

The "signatures" of each individual rotor blade can be guessed from the curves shown in **Fig.5** and **Fig.6**. If you calculate mean values for e.g. 10 second or 1 minute intervals, these effects become invisible. However, these noise effects propagate even at greater distances.

3. Noise of wind farms - determination of amplitude modulation - wind farm effect

The occurrence of AM in wind farms is known from studies by many authors (see, among others, [8], [9], [10]). To determine the AM of wind farms, we developed our own measurement and evaluation method in [4] as an alternative to [6] and [11].

The procedure is as follows:

- Before starting the actual measurements at wind farms, samples have to be taken which indicate the presence of AM. Afterwards, suitable measuring point(s) are selected. The measuring stations are planned in such a way that monitored continuous measurements implying subsequent evaluation are possible.

- The sound pressure level is recorded at a sampling rate of 8 Hz to 10 Hz with a time weighing of FAST. Spectral information (e.g. third octave band level) are recorded simultaneously. Interfering extraneous noises are to be faded out so that only the relevant contributions of WTG are taken into account.
- The measurement duration is one hour. This time frame is divided into time blocks of approx. 1 minute to 2 minutes duration (in our measurements data blocks with n=1024 points, approx. 2 minutes were used) and the average level p_{EM} is determined.
- The individual time series are subjected to a wavelet analysis (9 levels at 1024 points). In this process, the measurement signal is trend-corrected by restricting it to the higher levels of the analysis (i.e. a high-pass) by means of back-transformation.
- The trend-corrected measurement signal is then statistically evaluated. The resulting level distribution is usually restricted, often similar to a Gaussian distribution. The values are often between +/- 2 dB to +/- 4 dB.
- From this distribution, the level values of the 95% quantile (both sides) are determined and this AM fluctuation range is referred to as p_{AM5}.
- The results of the data blocks of one hour are summarized to resolve the mean values: p_{EM-m} (mean value of p_{EM}), p_{AM5-m} (mean value of p_{AM5}).
- For each data block, a spectral analysis of the reconstructed signal is also possible to determine the dominant AM frequency. In many data blocks, the AM frequency is clearly visible, in other blocks one observes several, neighbouring AM frequencies (wind farm effect) or a kind of noise when no strong "AM" frequency is present.

Fig.7a-7c shows a schematic summary of such evaluations (cf. Martinez et al (2017)). This concerns a wind farm with 6 WTG. Levels are determined by 3 WTG with a rated electrical power of Pel = 3.3 MW, RD = 112 m and NH = 140 m at a distance of approx. 565-635 m from the measuring point at a height of 4 m.

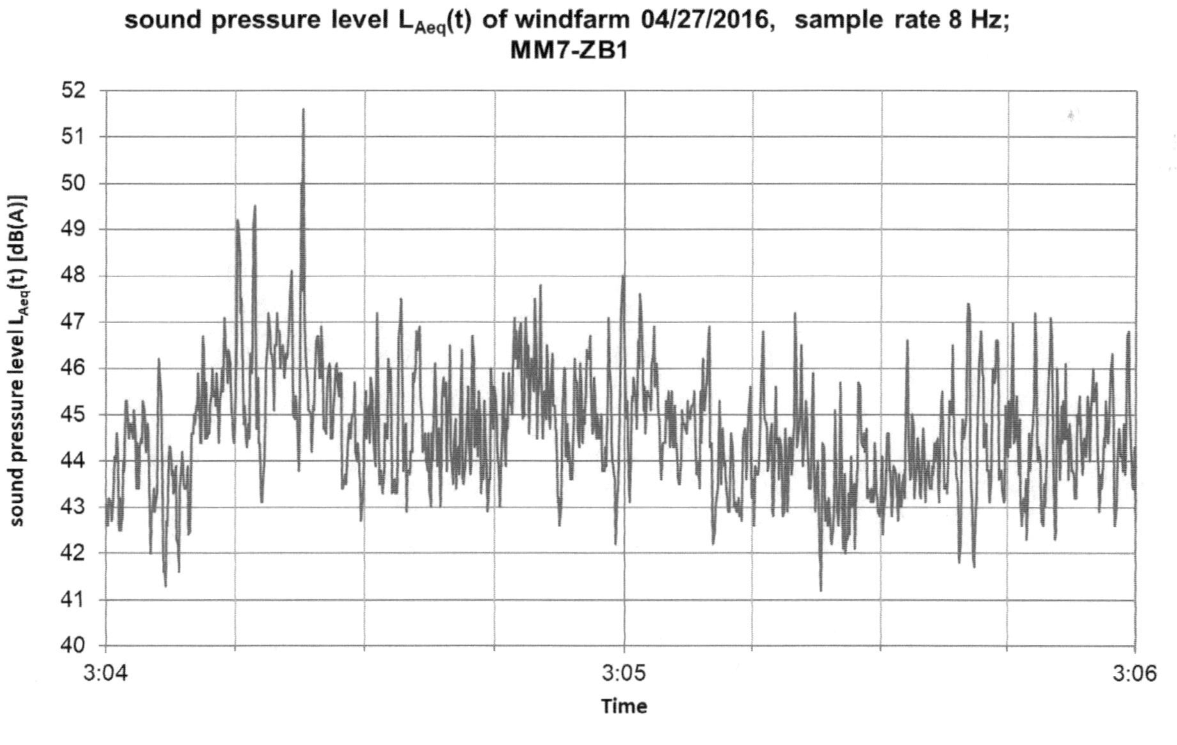

Fig. 7a: Level time history, wind farm, measurements at MM7-ZB1

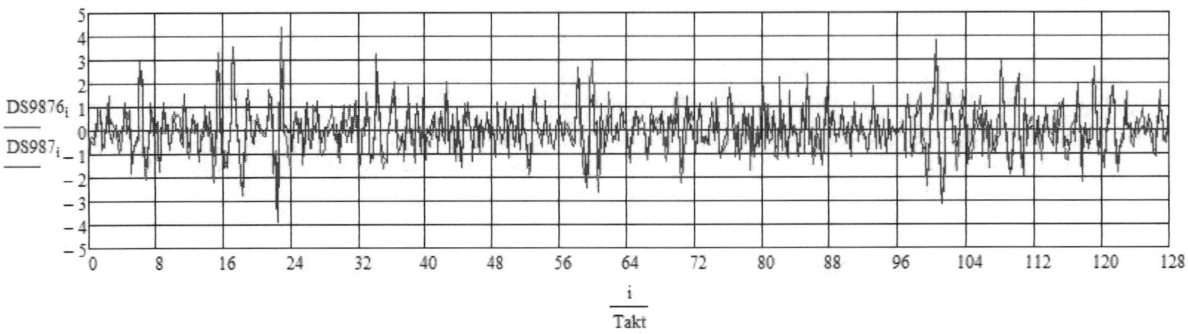

Fig. 7b: Noise level time history, trend-corrected, after wavelet-analysis. (Color: red, high pass best)

The original measurement data can be reconstructed well by means of wavelet analysis (Pearson correlation coefficient is 0.94) and the trend-corrected signal is extracted by summing only the higher levels. The extracted signal leads to p_{AM5} = 3.7 dB. The analysis of the AM frequencies results in at least three components between 0.45 Hz and 0.65 Hz (Fig. 7c, FDS87). This results from the wind park effect, visible in the "beat" character of the signal in **Fig.7a**. This indicates a synchronization of the noises of identical WTG. These results are similar to the results of van den Berg (2006).

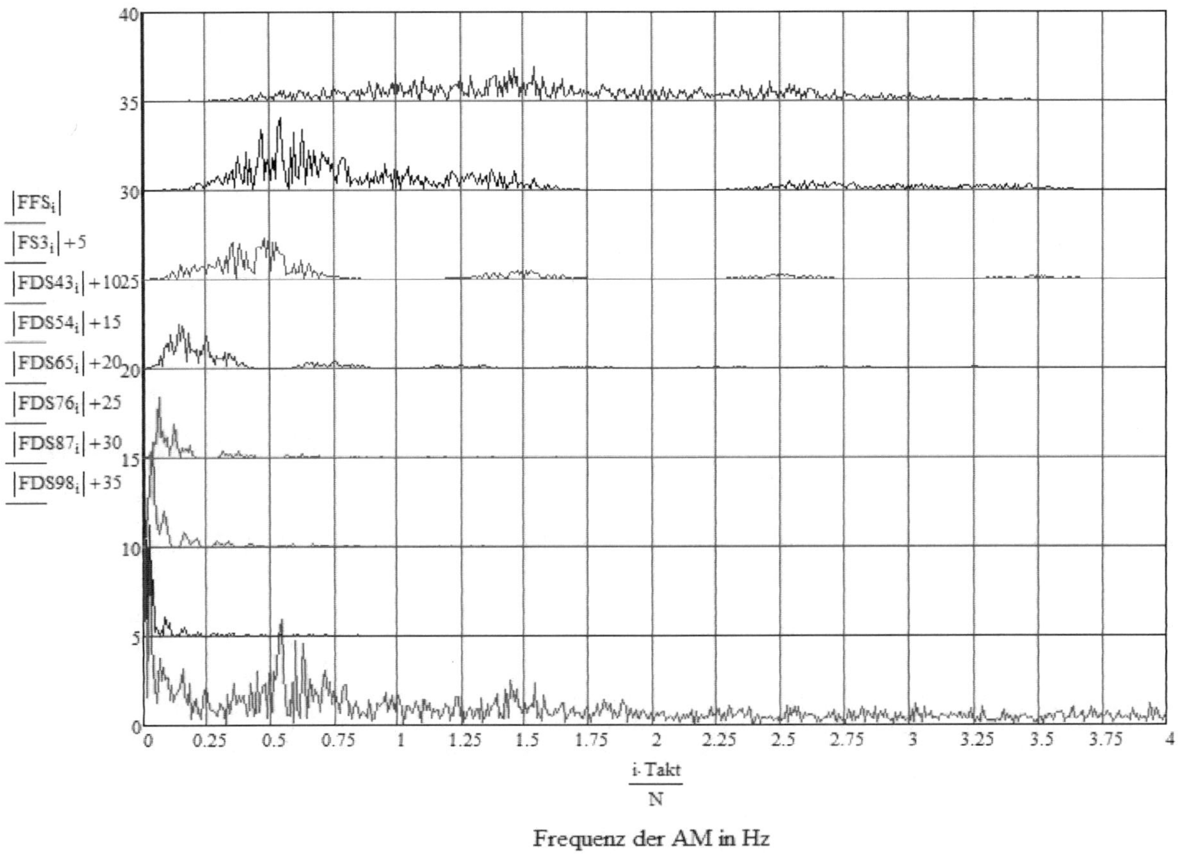

Fig.7c: FFT analysis of the noise level time history after the wavelet transformation. FFS (red, bottom) is the usual FFT analysis of the orig. signal, without trend correction. The FDS87 plane (grey, second up) shows the relevant modulation frequencies (see text), similar to FFS.(cf. Text)

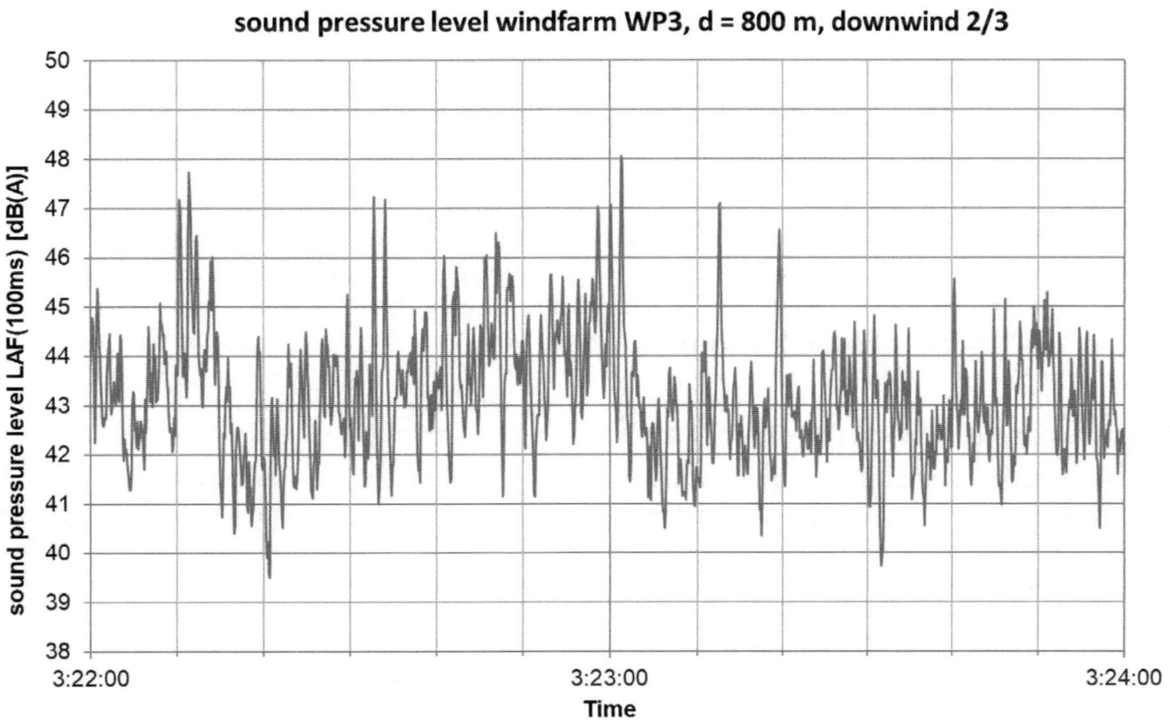

Fig. 8: Sound level time history from a wind farm at a distance of 800m. $L_{Aeq,m}$ = 43.4 dB(A), p_{AM5} = 4 dB, with *"wind farm effect"*, downwind

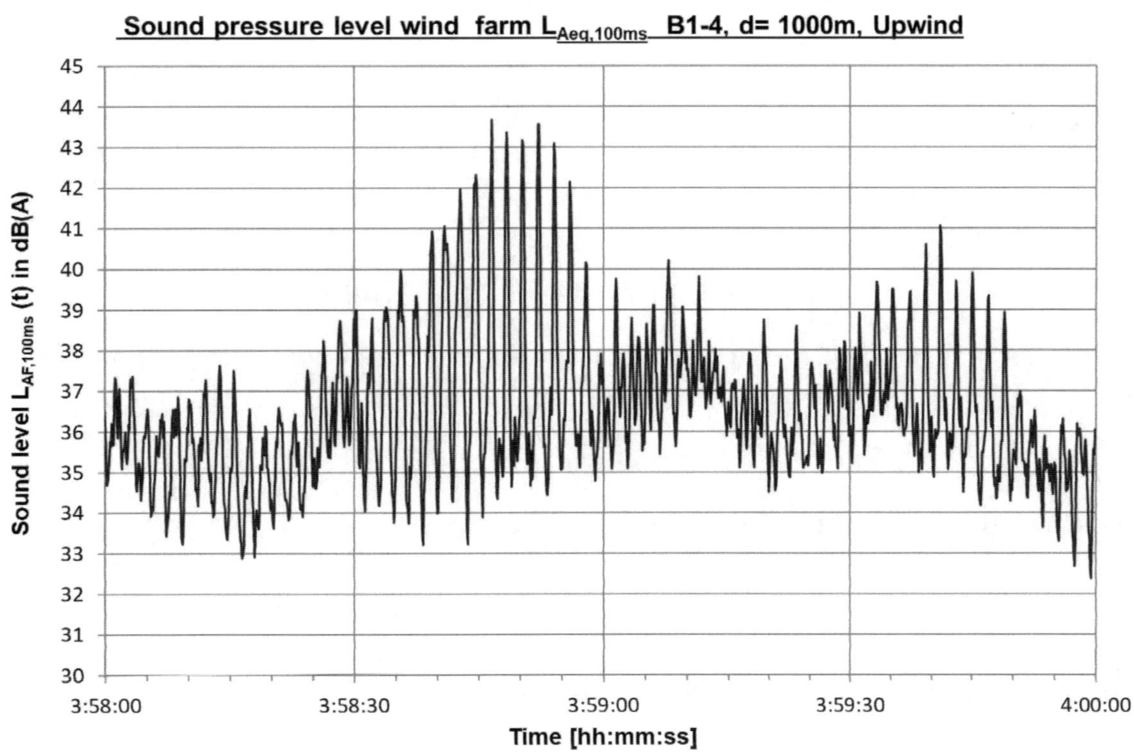

Fig. 9: Sound level time history from a wind farm at 1000m distance and updwind. $L_{Aeq,m}$ = 37.1 dB(A), p_{AM5} = 6.5 dB, with *"wind farm effect"*

Fig.8 and **Fig.9** show further typical measurement data from two different wind farms, each with 5 WTGs. For wind farm WP3 (**Fig. 8**), the receiver is located at a distance of 800 m and in the "downwind". In the case of measurement B1-4 (**Fig. 9**), the distance is approx. 1000 m to the receiver and the "meteo" condition is "upwind". The result is that AM and wind farm effect occur frequently. The wind farm effect amplifies the characteristic of AM and Doppler effect.

For the auralization of the wind turbine noise, it is useful to have original sound files with AM from wind turbines or wind farms, especially with regard to the Doppler effect. From the measurement data up to a distance of 1000 m it can be seen undoubtedly that the acoustic signature of the rotor blades can propagate over greater distances.

4. Annoyance and dose-effect relationship of the WTG

A basic dose-effect relationship to quantify the annoyance from WTG noise was developed by Janssen S.A. et al [13]. In this publication, the effect of amplitude modulation was only considered indirectly. For a detailed description of AM, we refer to the work of Schäffer B. et. Al [14]. The authors developed an approach with logistic functions instead of the "polynomial" dose-effect relationship as a function of L_{den} (from the EU Environmental Noise Directive) [13]. Bockstael A. et al. [15] proceed in a similar way to achieve noise reduction using the dose-effect relationship. A different approach was followed in Switzerland based on the studies [14]. Here the amplitude modulation is considered to be impulsive, which leads to an impulse adjustment. Another study by Lotinga M. et al [16] suggests an impulsive adjustment due to the AM of up to 5 dB. The studies by Virionen P. et al [17] also show comparable results.
Based on these results, an "alternative" logistic annoyance function was developed. For this purpose, practical parameters from noise measurements are used. The relevance of these parameters for the effect on humans is suggested in [14], among others.
These are the average sound pressure level for a defined period of time L_{Aeq} and the mean AM level associated with this period, expressed by the value p_{AM5}.

It can be assumed that the AM modulation frequencies (WTG with 3 rotor blades) are usually between 0.5 Hz and 1 Hz (note: this isn't infrasound). The development of the function is documented in [4] and [5] and practically presented in [18]. The approach is the extended logistic function from the publication [15]:

$$P(A\%; X) = a\,(1+b\,\exp(-\,c\,X))\text{-}1 \quad (1a)$$

X is defined as a superposition of a linear function containing the variables L_{Aeq} and p_{AM5} as follows:

$$X = f_1(L_{Aeq}) + f_2(p_{AM5}) \quad (1b)$$

The functions f_1 (L_{Aeq}) and f_2 (p_{AM5}) are defined as follows:

$$f_1(L_{Aeq}) = -2.5 + 0.09\,L_{Aeq} \quad (1c)$$
$$f_2(p_{AM5}) = 0.25\,p_{AM5} \quad (1d)$$

Adjusting the other constants leads to the following values:

a = 0.97 $\quad$ b = 0.466 10^2 $\quad$ c = 2.631

The constants were adjusted on the basis of a large number of measurement results from wind farms to fit the equation 1a. The procedure was based on the approach of Schäfer, B. et al. [14]. The following conditions were specified according to [14] and [15]:

- The function in Eq. (1a) should be limited at about L_{Aeq} = 50 dB(A). This is the upper limit in populated areas.
- the function $f_1(L_{Aeq})$ is only positive from L_{Aeq} = 28 dB.
- the function $f_2(p_{AM5})$ takes into account the amount of p_{AM5} according to the results in [14]
- The function P(A%; X) can be scaled.

The functions (1a) - (1d) thus formed were first presented and verified by Pies, K. et al [22].
To test the annoyance function, listening tests were carried out on 10 test persons, who were/are not annoyed by the sounds of WTG.

For the scope of the listening tests conducted here, it is ensured that the results are not influenced by a preconception of the test persons (cf. [19, 20, 21]). In a preliminary test with 3 test persons, the duration of the listening samples was set at 25 seconds.
The sound files were selected on the basis of the available real measurements at 5 immission locations. A total of 15 sound samples with different average levels L_{Aeq} (= p_{EM}) and width of the amplitude modulations p_{AM5} were selected. Table 1 shows the corresponding L_{Aeq} and p_{AM5} values for the 15 sound samples, as well as the predicted values of the sensitivity function P(A%;X) (see equation (1)).

Table 1: Characteristics of the 15 sound files / audio samples with 25s duration

Audio sample	Parameters			
	L_{Aeq} [dB(A)]	p_{AM5} [dB]	P(A%;X)	P(A%;X) scaled
1	44	4,3	0,91	4,5
2	44	4,1	0,91	4,6
3	43	3,9	0,89	4,5
4	44	3,6	0,89	4,5
5	37	6,6	0,90	4,5
6	37	10,5	0,97	4,8
7	38	8,4	0,96	4,8
8	37	4,5	0,76	3,8
9	38	5,6	0,87	4,4
10	34	3,1	0,38	1,9
11	34	2,5	0,32	1,6
12	35	3,0	0,43	2,1
13	34	3,1	0,43	2,2
14	35	3,5	0,55	2,7
15	35	3,4	0,53	2,6

Preliminary tests with audio samples showed that a scale from 0 to 5 is useful for classifying the level of annoyance. On this basis, the calculated values of the annoyance function P(A%;X) according to Eq. (1) were scaled from 0 to 5. These scaled values are listed in the last column of Table 1.
The sound samples were presented in an anechoic half-room via loudspeakers that were positioned at a distance of 1.5 m from the test persons. The audio files were leveled exactly to the L_{Aeq} in the second column in Table 1 at the position of the test person.

At the beginning of the hearing test, the test persons were asked the following question: "If this noise were present in front of your house, how much would it annoy you? The rating was to be made on a scale from 0 (not annoyed) to 5 (strongly annoyed). Each respondent was given 5 fixed audio samples. In addition, 3 out of 10 possible further audio tests were added at random. The samples could be played as often as desired without any order being specified. At the beginning of the audio test, each subject was presented with a audio sample that had a value of the scaled annoyance function of P(A%;X) ~ 0. This was an outdoor background noise without operation of wind turbines with an average level of L_{Aeq} = 26 dB(A). A total of 90 listening tests were evaluated.

Fig.10 shows the result of the entire listening test (see also [22]). The ratings for each of the 15 sound files were averaged and plotted. It was found that the ratings of the individual sound samples deviate from the calculated function by a maximum of one rating level. The result shows that the annoyance function found (see eq. (1a)-(1d)) can be represented as a function of the two acoustic parameters L_{Aeq} and p_{AM5}. The proposed function is thus in principle suitable for predicting the annoyance of test persons.

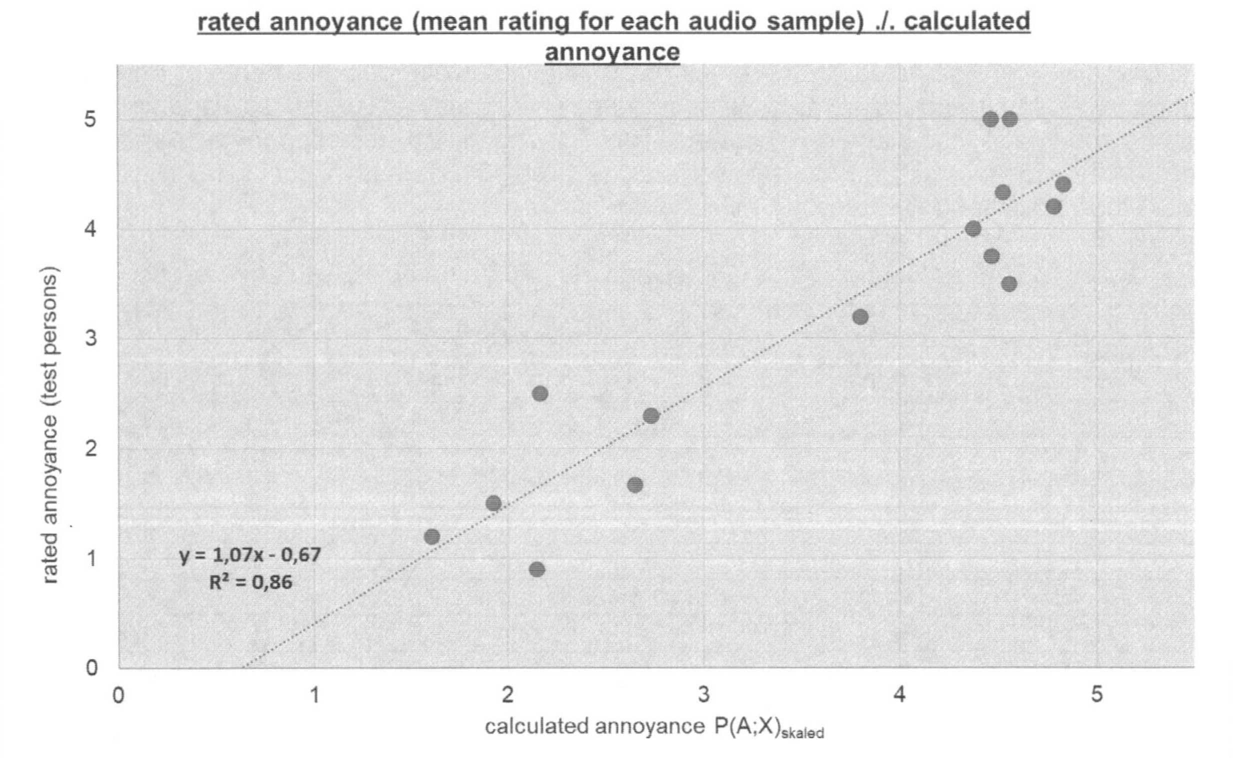

Fig. 10: Comparison of the calculated annoyance (s. Eq. [(1a)-(1d)] scaled and Table 1, (cf. Pies et al (2019)) and the mean values of the subjective evaluation in the listening tests.

In the study by P. Virjonen et al [17], sounds with AM were investigated in a laboratory. Comparative annoyance studies were carried out on noises without AM (in the sound pressure level range of 29 dB (A) and 49 dB (A)) and with AM. The relationship between the noises with and without AM was developed. The difference can then be interpreted as the "surcharge" k that is caused by AM. The procedure was applied for noise at sound pressure levels L_{Aeq} = 35 dB(A). The AM was synthetically superimposed with a variable amplitude width Dm corresponds to approximately p_{AM5} in [4,5] and [22]) in steps of 1 dB, 2 dB, 4 dB, 8 dB and 14 dB as well as different modulation frequency f_m (0.25 Hz, 0.5 Hz, 1 Hz, 2 Hz, 4 Hz, 8 Hz and 16 Hz). The procedure and the results of the authors are comprehensible and documented in detail (see e.g. Table III in [17]).

The measurement data according to [17] can also be evaluated using our method for different amplitude widths Dm = 1 dB, 2 dB, 4 dB and 8 dB. The results for wind farms or modern WTG allow a limitation to modulation frequencies fm of 0.5 Hz up to approx. 1 Hz. This allows a limitation to modulation frequencies fm of 0.5 Hz up to approx. 1 Hz and thus an averaging of the surcharges k for the parameter fm. The calculation of P(A%;X) according to Eq. (1) only requires the L_{Aeq} at 35 dB(A) and p_{AM5}, that can be considered equivalent to Dm.
The result of the comparison for different values Dm (1 dB, 2 dB, 4 dB, 8 dB) and fm = 0.5 Hz and 1 Hz is presented in Fig.11. It shows good significance which underlines the linear relationship between surcharge k and the annoyance function according to Eq. (1a)-(1d) [22]. It can be seen that no surcharge k is necessary for level 1. At level 2 ist is k = 1 dB. For level 3, the surcharge is at least k = 2 dB. For the higher levels, results at least k = 3 dB evident.

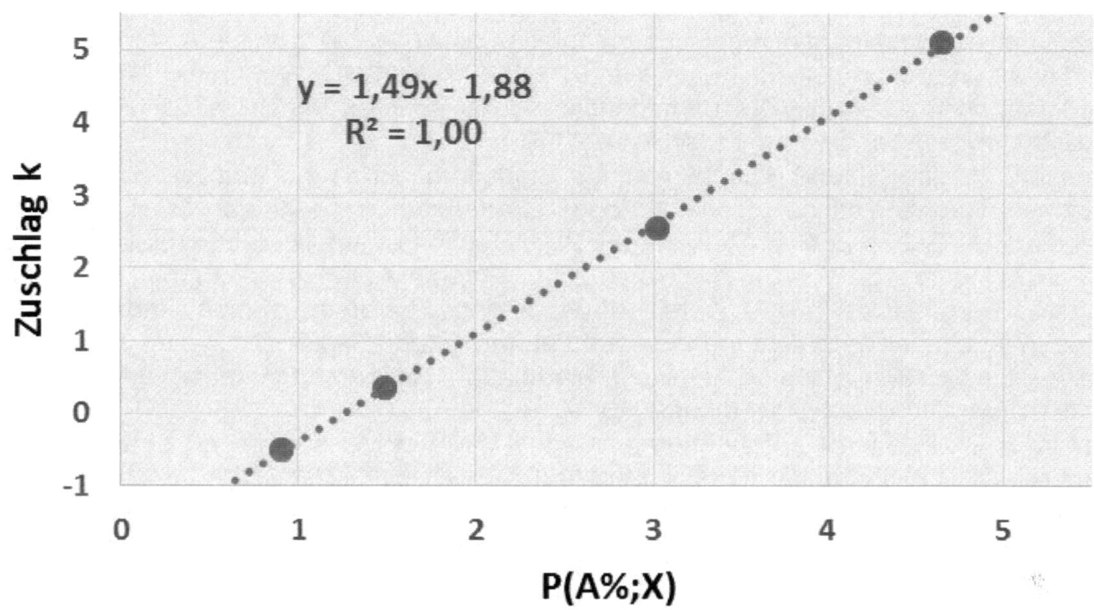

Fig. 11: Comparison of surcharge **k** according to P.Virjonen et al (2019) with the annoyance function according to Eq. [(1a-1d)], Pies (2019) scaled.

5. Conclusion

The present study shows that the amplitude modulation (AM) depends on many factors. The directional dependence is an important parameter for the sound radiation from wind turbines. The Doppler effect is also characteristic of the noises caused by the moving rotors. The level curve of the amplitude modulation has a short-term characteristic that only becomes visible at a sampling frequency of >10 Hz. Longer averaging times are therefore unsuitable for describing amplitude modulation and the equivalent sound pressure level is not sufficient to describe the annoyance. The amplitude modulation is particularly noticeable in rural areas with low background noise. If several wind turbines act together, the wind park effect occurs and low-frequency noise emissions can occur particularly at part load. Measurements show that these components can radiate over greater distances.
Based on this knowledge, reductions in the noise source are conceivable. The approach according to [22] already shows corresponding options.

References

[1] DIN EN 61400-11 (VDE 0127-11): 2013-09 Windenergieanlagen – Teil 11 : Schallmessverfahren (IEC 61400-11:2002); Deutsche Fassung EN61400-11:2013 (Edition 3) und Vorversionen der DIN EN 61400 (2007-03 oder 2003-11) (Edition 2)

[2] Martinez, S., Fennel, F., Perzentil-Spektren: Merkmalbezogene Kenngrößen bei Geräuschmessungen, Fortschritte der Akustik, DAGA 2016 Aachen, S. 699-702

[3] Martinez, S., Statistische Methoden bei Geräuschmessungen im Freien, Zeitschrift für Lärmbekämpfung, 48(2001) S. 43-55.

[4] Martinez, S,, Fennel, F., Pies, K., Amplitudenmodulation als besondere Geräuschcharakteristik der Windenergieanlagen, Lärmbekämpfung 13 (2018), S. 171-182.

[5] Pies, K., Martinez, S., Pies, D., Fennel, F., Ferreira, A.-K., Maleklo, A., Amplitudenmodulation- Wirkung auf den Menschen, Richtcharakteristik und Dopplereffekt, Lärmbekämpfung 14 (2019), S. 159-173.

[6] Fukushima A., Tachibana H.; Comparison of the IOA method and Japanese F-S method for quantitative assessment of amplitude modulation of wind turbine noise- A study based on the field measurement results in Japan, Wind Turbine Noise 2017, Rotterdam 2-5. May 2017

[7] Oerlemanns S.; Detection of aeroacoustic sound sources on aircraft and wind turbines, Thesis University of Twente, Entschede, 2009

[8] Larsson, C. und Öhlund, Olof, Amplitude modulation of sound from wind turbines under various meteorological conditions, J. Acoust. Soc. Am. 135(1), January 2014, S. 67-73.

[9] Halstead D., Suban-Loewen A., Ashtiani P.; Detection of amplitude modulation in Southern Ontario wind farms, Wind Turbine Noise 2015, Glasgow 20-23. April 2015

[10] van den Berg G.P. (2006), University of Groningen, The sound of high winds: The effect of atmospheric stability on wird turbine sound an microphone noise.

[11] Renevable UK, Wind Turbine Amplitude Modulation: Research to improve understanding as to its Cause and Effect, December 2013

[12] Martinez, S., Fennel, F.; Windenergieanlagen: AM bei Messungen an Einzelanlagen und Windparks, Fortschritte der Akustik, DAGA 2017, Kiel, S. 1230-1233

[13] Janssen S. A., Vos H., Eisses A. R. und Pedersen E., A comparison between exposure-response relationships for wind turbine annoyance and annoyance due to other noise sources, J. Acoust. Soc. Am. Vol 130, Nr. 6, 2011, S. 3746-3753

[14] Schäffer B., Schlittmeier S.J., Pieren R., Heutschi K., Brink M., Graf R. und Hellbrück J., Short-term annoyance reactions to stationary and time-varying wind turbine noise and road traffic noise: A laboratory study, J. Acoust.Soc. Am. 139(5) 2016 S. 2949-2963

[15] Bockstael A., Dekonink L., Can A., Oldoni D., De Coensel B. und Boteldooren D.; Reduction of wind turbine noise annoyance: An operational approach, Acta Acustica united with Acustica, Vol 98(2012) 392-401

[16] Lotinga M., Perkins R., Singleton R., Wind Turbine AM Review, Phase 2 Report, Departement of Energy & Climate Change, WSP Parsons Brinkerhoff, August 2016

[17] Virjonen P., Hongisto V., Radun J.; Annoyance penalty of periodically amplitude-modulated wide-band sound, J. Acoust. Soc. Am. 146(6), Dec. 2019, S. 4159-4170.

[18] Martinez, S., Pies,K., Fennel,F., Ferreira, A.-K., Richtungs- und Entfernungsabhängigkeit der Amplitudenmodulation(AM) bei Windenergieanlagen, Fortschritte der Akustik, DAGA 2020, Hannover, S. 484-487

[19] Kastka J., Bursch-Galetke E., Buchta E., Krauth J., Mau U., Muth T., Schümer R., Siegmann S., Untersuchung zum Lärmkontingenzkonzept: Akzeptanz, Belästigungsverhalten und Meinungsbild der Bevölkerung am Beispiel des Düsseldorfer Flughafens 1987-1995, Fortschritte der Akustik, DAGA1996, Bonn

[20] Pedersen, E., Health aspects associated with wind turbine noise - Results from three field studies, Noise Control Eng. J. 59, 2011, 47–53.

[21] Pedersen, E., and Persson Waye, K. (2004). Perception and annoyance due to wind turbine noise—a dose-response relationship, J. Acoust. Soc. Am. 116, 2004, 3460–347

[22] Pies K., Martinez S., Pies D.; Wirkung der WTG-Geräusche auf den Menschen, DAGA 2019 Rostock, S. 724-727.

INCE EUROPE

9th International Conference
on
Wind Turbine Noise
Remote from Europe – 18th to 21st May 2021

Assessing Wind Turbine Noise Perception by means of Contextual Laboratory and Online Studies

Stephan Preihs, Jakob Bergner, Daphne Schössow, and Jürgen Peissig
Institute of Communications Technology (IKT), Leibniz University Hannover, Germany
Correspondence: {preihs, bergner, schoessow, peissig}@ikt.uni-hannover.de

Summary

In the course of an interdisciplinary research project at Leibniz University Hannover, perceptual properties of wind turbine noise were investigated. In this scope we conducted a series of listening experiments for assessing wind turbine noise perception under controlled and reproducible conditions in an immersive and context-sensitive laboratory setup. Due to the corona pandemic situation we were not able to perform continuative large scale laboratory studies and therefore were forced to switch to online surveys.

In this paper we report about our methodology and the general layout of the studies, including the collection of empirical data and the assessment of the participants' attitude towards wind energy as well as the calibration of the respective playback systems. We show how our stimuli were collected, preprocessed and reproduced. Our work focuses especially on the maintenance of ecological validity of the stimulus presentation including correct levels, non-acoustic context and participant attention. We address specific advantages and drawbacks of both laboratory and self-reporting online studies. Finally, we show some exemplary results and draw some final conclusions from our study designs.

1. Introduction: Wind Turbine Noise Perception

Human perception of environmental sound sources is an important topic and receives more and more attention nowadays due to an increased sensitivity to noise pollution. In the case of wind turbines, there have been additional efforts in recent years to include perceptual aspects in the assessment and prediction of noise immissions in supplement to standardized sound measurements. For gaining

further insights on different aspects of wind turbine noise (WTN) annoyance, perception-based studies are an accepted methodology in current research.
When doing such studies usually a tradeoff regarding parameters influencing realism and controllability has to be found. This also includes thoughts on the common quality criteria objectivity, validity and reliability. Whilst the validity of the answers in noise perception questionnaires is highest in field studies as persons concerned are asked directly, the reproducibility and generalizability is very limited. Laboratory studies on the other hand have a good reproducibility and a larger and more diverse group can participate. However validity of context is harder to reach. With online surveys, which are rather uncommon in this field, even more participants can be reached in short time, but there is little control over whether they answer carefully or follow predetermined guidelines and the context validity even harder to reach. For auditory scenes, the term *context* encompasses acoustic, environmental, and personal aspects that potentially influence perception, interpretation, and response of and to the respective scene. In this work, we refer to context in terms of background noise, visual cues, as well as imagined activity during exposure (walking vs. leisure at home).
Most studies on WTN are based on presenting auditory stimuli only. Presenting visual stimuli, especially dynamic, is a more recent approach [13], [9]. In most of the past studies headphones or stereo loudspeakers were used for reproduction. Spatial reproduction for example by using Ambisonics is not common in WTN research (see [11], [5]) while it is more widely used in more general noise annoyance research (e.g. [15]). Lab studies can be conducted unfocussed or focussed. In unfocussed studies the participants are not told what kind of stimuli they will hear and often get a different task to concentrate on. In focussed studies on the other hand the question of annoyance is assessed directly by letting the participants rate a number of short stimuli. For the evaluation for the perceived annoyance the ISO 15666 [7] is widely used. Often subject-related parameters such as sensitivity to noise (e.g. [20] [17]), demographic data or the attitude towards the noise source (e.g. [13]) are surveyed as well to gain further insights into the reasoning behind the annoyance ratings.
Direct comparisons between residents and non-residents are seldom made, but a general higher tendency of residents to detect WTN was found [4], [10]. Whilst significant effects between age and annoyance rating were not found in WTN lab studies ([10], [11], [13]), van Gerven found an inverted u-shaped relationship between age and the percentage of highly annoyed persons looking at the data of several large-scale transportation noise annoyance studies [16]. The personal attitude towards wind farms seems to have an influence on the annoyance rating. Schäffer's results show that a more positive attitude goes along with a lower annoyance rating [13].
The studies described in this paper were not conducted solely to gain new findings in the field of WTN annoyance; rather we intended to evaluate different methodologies that could be used to perform studies which include immersive presentation of audiovisual stimuli in a reproducible way. The general longterm motivation for our investigations is to find relevant factors and signal parameters of WTN soundfields which allow to build a data driven annoyance model.

2. Immersive Wind Turbine Recordings

An immersive presentation of stimuli is not very common in the research on WTN such that a possible validation gap exists between the presented sound and reality. Thus our studies aim to present the stimuli as realistically as possible in the respective surroundings (lab, online).
For the collection of stimuli we developed a dedicated tool chain which includes audio and video recordings as well as meta data acquisition. The core of our recording system consists of a Neu-

mann KU100 dummyhead and Ambisonics microphone arrays (either Sennheiser Ambeo Mic or mhacoustics Eigenmike). The microphones are mounted on tripods at a reasonable height of head and are equipped with proprietary primary, secondary an possibly tertiary windscreens. In addition to immersive soundfield recordings we log reference sound pressure levels for calibration of the stimuli with a measuring plate and a Norsonic Nor145 levelmeter in accordance with the IEC 61400-11 standard. The video recordings are either 360° videos shot with a Insta360 One X camera or regular 16:9 videos using a Sony Alpha 7R III with a Sigma 14-24 mm wide angle lens. The metadata collection includes wind speed and direction measurements at the immission site using a Calypso ultrasonic anenometer as well as GPS position data gathered by the NOR145.

The postprocessing of stimuli includes level adjustment on the basis of the reference sound pressure at the immission site. For Ambisonics-based loudspeaker reproduction we apply a highpass with $f_{lim} = 160\,\mathrm{Hz}$ for Ambisonics orders $n > 0$ to reduce low-frequency pseudo noise due to wind turbulences at the microphone array membranes [19]. Additionally, for our lab studies, the loudspeaker setup is equalized for a flat frequency response at the listening position by the usage of a self developed measurement and equalization procedure [1].

3. Conducted Studies

The design of our studies aims to present the acoustical immission of WTN not isolated but instead in a plausible context regarding background noise, visual cues and general scene setting. The acoustical properties of the presented audio stimuli were therefore also tested for plausibility [3]. The wind turbine recordings are rated by the participants in terms of annoyance using questions based on the ISO 15666 [7]. This results in an all negative rating scale, where the rating is carried out on an 11-point Likert scale where 0 stands for "not at all annoying" and 10 for "extremely annoying". Further explored parameters are the noise sensitivity by [20], attitude towards wind farms by [13], questions about hearing [13], as well as demographic data.

3.1. Laboratory Studies: Methodology

In 2019 a lab study was conducted in the *Immersive Media Lab* (IML) of the institute. For more details about the room see [2] and [6]. In the IML higher order Ambisonics reproduction with a 36.4 loudspeaker system was utilized. Additionally, three projectors and screens were used such that the field of view was completely covered when sitting in the sweetspot of the room (see Figure 1). The questionnaire was presented as a Python application on a tablet and could be answered through the touchscreen.

The goal of the lab study was to evaluate and validate the reproduction method, the developed questionnaire and the creation of context as well as its effect on the participants in general. Another examined point was the comparison between Ambisonics and binaural presentation using the same base stimuli. The relations found in the results were also compared to other studies like the ones described in the introduction. The experimental setup as well as the questionnaire was tested with 10 students and staff members before the actual study with 35 participants was conducted. Most of the stimuli used were real recordings. The audio was captured with the Eigenmike (sound field microphone) and the KU-100 (binaural dummyhead). This study used 360° video recordings which were converted, cut and got a fade-in/-out applied. Afterwards, to present them on the three screens, the equilinear videos were rectified.

Fig. 1 The experimental setup for the lab study in the Immersive Media Lab of the IKT.

As visual stimuli 7 scenes with a similar setting (wind turbines in a field) and one interior scene were used. The latter and one of the outdoor scenes were created and animated with Blender and used synthetic audio while the others were real recordings. An exemplary selection of scenes can be seen in Figure 2. The six real recordings differ in how many wind turbines are visible and in the distance to the nearest turbine. Most of those scenes used real audio recordings, but one used the same synthetic audio as the 3D-modelled outdoor scene. One scene was used as reference as no wind turbines were audible or visible. All scenes had a length of one minute and the participants were asked to watch the full video, but could stop after at least 5 seconds or watch the video multiple times. This set of scenes was chosen to compare different real settings and distances to wind turbines but also to compare real and synthetic stimuli. The indoor scene in addition was chosen to compare the effect of different contexts.

Distances to the nearest turbine ranged from 150 to 553 m respectively 36.8 to 47 dB(A) and the number of wind turbines from 1 to 11 visibly respectively 1 to 4 audibly. Equally to the annoyance another similar 11-point Likert scale was used to rate the perceived realism of the scenes. As question "How realistically did this scene seem to you?" with the scale between "extremely realistic" (0) and "not at all realistic" (10) was used. There was also a question about subjective influencing factors. Each scene was presented twice: once in Ambisonics format using loudspeakers and once binaurally using headphones. The order of the scenes was randomized for each participant by the method of latin squares to counterbalance habituation effects and fatigue. The procedure of the questionnaire of the lab study was: an introduction, questions about hearing, sight and well-being, the noise sensitivity questionnaire, a training example, the randomized listening examples, feedback on the listening examples, the attitude questionnaire, demographic data, and feedback on the experiment itself.

3.2. Laboratory Studies: Results

There were 35 participants in the lab study, but four data sets had to be excluded from further analysis due to missing data points. Six residents took part in the study. Participants were aged between 23

Fig. 2 Examples of visual stimuli for the lab study. From top to bottom: real recording, virtual outdoor scene, virtual indoor scene.

and 73 (μ = 44.0). 15 were male and 16 female. Two participants used a hearing aid. All but two persons rated their sight at least average. Eight participants lived in a rural surrounding, 14 in a town and 9 in a city. The noise sensitivity according to [20] was between 25 and 85.2 (μ = 56.6; scaled between 0 as not noise sensitive and 100 as extremely noise sensitive) and the attitude towards wind farms according to [13] between 32.5 and 90.0 (μ = 66.1, which equals a rather positive attitude - with 0 being an extremely negative and 100 an extremely positive attitude). Four participants declared they had not heard a wind turbine ever before.
Many participants reported that the presented audio-visual recordings made it possible to plausibly imagine the scenes. Particularly the visual presentation helped with the context as 29 participants stated. The experimental procedure was rated positive by 28 participants and the length of the stimuli was deemed adequate by 27. This means that the length of 1 minute per stimuli should be considered for future focused lab studies while most other studies use only 20-30 seconds.
Besides acoustical measurable factors, cues of correlations between annoyance ratings and personal factors could be found. To find significant differences between groups hypothesis tests were used. As the data was not normally distributed Kruskal-Wallis' test (with Bonferroni as post-hoc p-value correction) was used instead of a one-way ANOVA when comparing more than two groups, and Mann-Whitney's test instead of a t-test. The level of significance for all cases is $p \leq 0.05$. In the following a few selected parameters are mentioned:
A significant difference in the rating between residents ($\mu = 4.875, SD = 2.848, median = 4$) and non-residents ($\mu = 3.421$, $SD = 2.804$, $median = 3$) showed that residents tend to give higher annoyance ratings: Mann-Whitney $U = 15344.5$, $n_1 = 64$, $n_2 = 368$, $p < 0.001$, see Figure 3a.
Regarding the age there were also significant differences in the ratings: Kruskal $H(4) = 30.733601$,

$p < 0.001$, see Figure 3b. Participants aged 18-24 showed a lower annoyance than the other groups with the exception of participants aged 35-54, which were underrepresented in this study. The other groups showed no significant differences.
Another trend was found between the attitude towards wind farms and the annoyance. A negative attitude correlates with a higher annoyance ($\rho = -0.178$ if all ratings are considered or $\rho = -0.466$ when only the average rating per participant is considered). The noise sensitivity showed no linear correlation with the average annoyance rating ($\rho = 0.027$).
The influence of the reproduction method and the type of stimuli was also analyzed. For example significantly higher but so far not explainable annoyance ratings were found using Ambisonics. More details can be found in [14] as they are not the focus of this paper.
General feedback on the study showed that participants could easily immerse themselves in the scene due to the high level of context given. Whilst minor improvements in the wordings can be made, the overall questionnaire seems suitable. Known relationships with the annoyance rating like the influence of the personal attitude or being a resident could be reproduced with the results gathered.

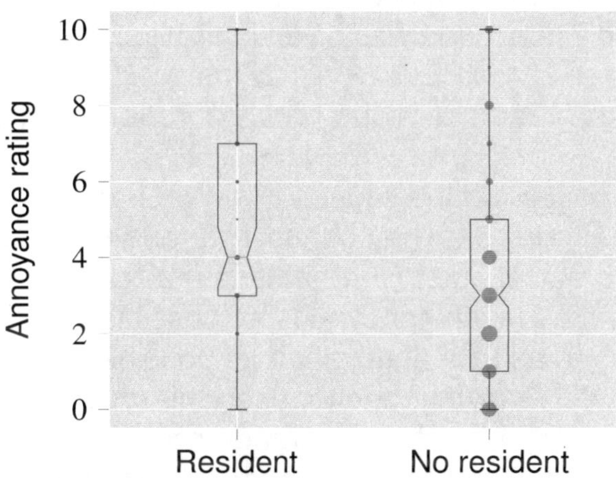

(a) Annoyance rating as a function of being a resident.

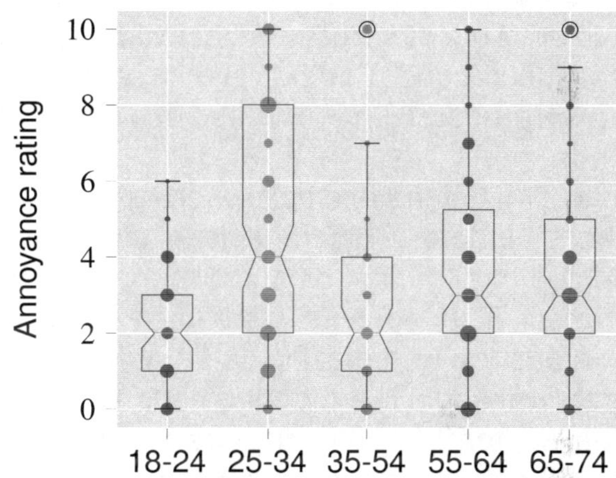

(b) Annoyance rating as a function of grouped age.

Fig. 3 Selected subject-related parameters of the lab study (Box Plots with 25 and 75% quartiles and 95% CI of the median as well as the number of participants for each rating illustrated).

3.3. Online Studies: Methodology

The online study was based on the feedback and results gathered in the laboratory study described in the last section and refined with respect to stimuli selection and question wording. Basic questions like the annoyance rating scale, the statements to determine the noise sensitivity and attitude towards wind farms, and the demographic questions were kept. For keeping the immersion on the highest possible level while reaching a broad audience, videos with binaural audio were used.
The online survey was split into two parts which were executed at different dates with approximately 20 minutes editing time each. The first part included questions about hearing ability, noise sensitivity, attitude towards wind energy and turbines, demographic data and two training examples for the listening examples. The second part included another training example and the actual listening

examples as well as some questions on concentration and perceived realism afterwards. In both parts a volume calibration routine preceded the listening examples.

As the playback setup of the participants could not be controlled directly, they were asked to give information about their used headphones and other playback related parameters. At this point the operating system was also queried since e.g. Windows, Android and iOS use system settings to change the audio reproduction which can result in unwanted influences on the playback level. For each affected operating system a step-by-step guide to check and modify the settings was given during the survey. Third party software was not considered directly. The participants were asked to execute the survey in quiet surroundings and, if applicable, to disable noise cancellation of their headphones. Participants using smartphones were excluded as the webpage of the survey would not show correctly on a small screen. Users of loudspeakers were also excluded as the auditory spaciousness is less than for headphones. Wireless headphones were excluded as well as audio coding artifacts could influence the perception of stimuli and the playback of lower frequencies might not work correctly.

The survey was created with LimeSurvey [8]. An exemplary screenshot can be seen in Figure 4. To link the answers of participants between the two parts, an individual code was given in the first part to access the second. For the listening examples mp4 videos (H264 MPEG4 AVC, 25fps, 1920x1080 pixel) were used with FLAC audio (48kHz, stereo, lossless). Each stimulus had a duration of 30 seconds.

In total 17 listening examples were given, which were composed of nine scenes with WTN and two “anchors”. In total 5 different visual stimuli were used (see Figure 5). The scenes 1, 3, and 9 were presented three times: binaural, mono (same signal on both ears), and in the end again as binaural for a retest. The “anchors” were used to generate as little and as much annoyance as possible. Both show a pasture with grazing cattle. While the scene generating as little annoyance as possible used natural rural audio like tweeting birds and wind, the “extremely annoying” anchor used the recording of a helicopter flying over the scene and had to be preprocessed with dynamic range compression to keep the audio level in line with the other stimuli. The recordings of the WTN scenes were taken from three distances to the nearest turbine: 150 m, 390 m, and 700 m. Some of the recordings had some background noise or visually distinctive features (see Table 1). The order of the stimuli was randomized for each participant.

To get an estimate of the volume the participants used, a row of questions was asked to approximate their auditory threshold. First of all they were asked to set a pleasant volume with a supplied music excerpt. It was supposed that a mean LAeq level of 68 dB was used (see [18]). After that, with the request not to change the volume, noise stimuli that were filtered according to the mean spectrum of the music excerpt and attenuated by -40 to -70 dB below the RMS value of the music excerpt were played in 3 dB steps (see Figure 6). For each stimulus the participants should state whether they could hear faint noise or not. If at -40 dB the answer of no audible noise was given, the participant was asked to turn the volume up slightly until they could hear the noise. If at -70 dB the answer of audible noise was given, the participant was requested to lower the volume until the noise was not perceivable any more. Participants who could not hear the noise stimulus of -30 dB in comparison to the music excerpt were excluded as they supposedly would not be able to hear the WTN as well. Another check for “happy clickers” (participants who just fill in the survey randomly to get their payment) was to ask for each scene how many wind turbines were visible in the foreground. However in all groups of participants the results showed that the understanding of the word *foreground* varied as some participants even counted small turbines at the horizon.

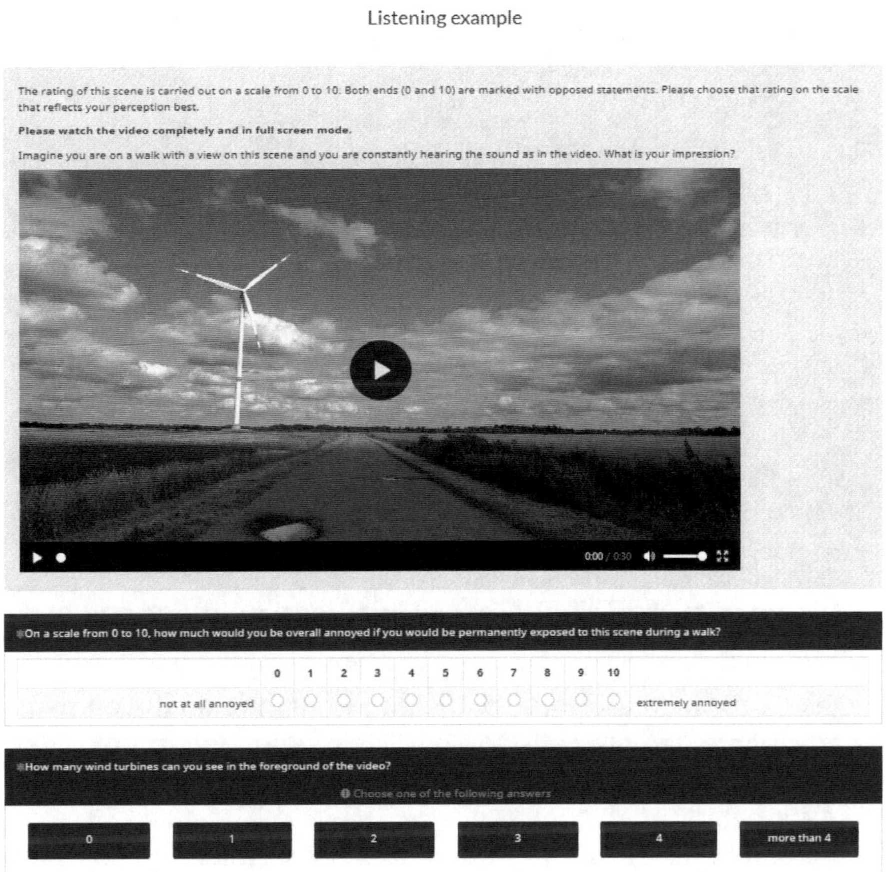

Fig. 4 Screenshot of a stimulus rating page of the online survey.

After a test-run of the survey, three independent groups of participants were recruited. The first group was recruited from Mo'Web [12] which is a provider of subject panels of desired composition regarding demographic details. For the first part of the study 150 participants (75 each living in rural or urban surroundings) were acquired. Of those 100 qualified for part two and in the end 60 complete datasets were gained. Exclusion criteria during the survey were the compliance to the guidelines (usage of wired headphones and a PC/tablet/laptop) and a sufficient hearing ability. For the latter the ability to distinguish stereo channels was also tested. For this two test signals were played with the question whether the signal could be heard first on the right ear and then on the left or in reverse. There was also the answer possibility "there is no difference" to find participants with hardware mono playback. While the first stimulus consisted of two speech fragments the second consisted of two tonal sounds.

The second group of participants took part under controlled conditions at the institute. A PC with Windows 10, a Fireface UC audio interface and Sennheiser HD25 on-ear headphones were used for this. The group consisted of 19 staff members as it was not possible to invite participants due to the corona pandemic. The goal for this experiment was to check whether the results were influenced by the setup. Unlike the other two groups these participants attended both parts of the survey with only a short break in between.

(a) Scene 1: 150 m

(b) Scenes 2-4: 390 m

(c) Scenes 5, 6, 9: 700 m

(d) Scenes 7, 8: 700 m

(e) Anchors

Fig. 5 Visual stimuli of the WTN scenes used in the online survey.

The third group consisted of 15 voluntary participants and went through the same process as the first group. Some of the participants had also already participated in the laboratory study in 2019. The reason for this group was to gain more answers since the resonance for the second part in the first group was smaller than expected.

In comparison to the laboratory study the wording of the instructions and the scenario context was changed based on the results and feedback. In the lab study the description of the imaginative situation was "Imagine you are sitting in your garden or you are on a walk with a view of the following scene...". The participants stated it would make a difference for them to imagine being on their own property or being on a walk as in the first case it is impossible "to walk away from the noise". Thus it seemed important to use only one fixed imaginative situation. As the scenes of this study were all filmed from dirt roads and relatively short distances to the turbines, the situation was changed to "It's Sunday afternoon and you have the day off. To enjoy the nice weather you'd like to go on a walk. You decide to have a walk around the nearby fields.". In the statements 1, 5, and 9 about the attitude towards wind farms of [13] the phrase "wind turbine" was exchanged with "wind energy".

3.4. Online Studies: Results

The group of participants from Mo'Web resulted in 60 complete answers. However 12 of those had to be discarded as either the variability of the annoyance ratings was very low (standard deviation of the ratings from all scenes < 1) or it could be assumed that the task on hand was not understood correctly. That was the case when the anchor with the helicopter had an annoyance rating lower than 2 and at the same time lower than the pleasant anchor. Nine participants stated they live or lived near a wind farm. The age of the participants ranged from appr. 18 to 74. 30 men and 18 women took part. 26 participants lived in rural surroundings, 11 in a town and city each (those two groups are combined into "urban" in the following). The noise sensitivity according to [20] ranged from 29.2 to 96.3 ($\mu = 52.7$) and the attitude towards wind farms according to [13] from 32.5 to 95.0 ($\mu = 63.1$). 21 participants stated that they had never heard a wind turbine before.

Table 1 Overview of the WTN scenes used in the online survey.

Scene	Distance	Special features	Retest
1	150 m		yes
2	390 m	tractor + car passing	
3	390 m		yes
4	390 m	more vegetation	
5	700 m	farming	
6	700 m	car passing	
7	700 m	view towards street (no WT), car passing	
8	700 m	view towards street (no WT)	
9	700 m	audio: no.8, video: no.5	yes

Stimulus-related parameters:
Known relations are reproducible as the results show. For example the annoyance rating decreases with increased distance to the nearest wind turbine (150 m: $\mu = 5.840$, 390 m: $\mu = 3.900$, 700 m: $\mu = 3.110$): Kruskal $H(2) = 91.170$, $p < 0.001$, see Figure 7a.
As the scenes 8 ($\mu = 2.958$, $SD = 2.843$, $median = 2.5$) and 9 ($\mu = 3.313$, $SD = 2.784$, $median = 3$) use the same auditory stimulus it is possible to investigate whether the number of visible turbines has an influence on the annoyance ratings while keeping the distance the same. Hypothesis testing did not show significant differences between the two visual presentations in the annoyance rating: Mann-Whitney $U = 1039.0$, $n_1 = n_2 = 48$, $p = 0.403$.
To check the intra-rater reliability the scenes 1, 3, and 9 were rated twice with binaural audio by each participant. The second rating took part in a separated randomized block after the other stimuli. The correlations between the associated ratings are high (scene 1 $\rho = 0.853$, scene 3 $\rho = 0.825$, scene 9 $\rho = 0.883$). The intra-rater reliability using Krippendorff's alpha is high too (1: $\alpha = 0.851$, 3: $\alpha = 0.809$, 9: $\alpha = 0.816$). However in the control group there were some outliers in the correlation (1: $\rho = 0.356$, 3: $\rho = 0.704$, 9: $\rho = 0.843$) as well as in the intra-rater reliability (1: $\alpha = 0.388$, 3: $\alpha = 0.663$, 9: $\alpha = 0.731$) which could not be explained so far. For the third group the values were similar to the first group (all $\rho > 0.82$ and all $\alpha > 0.73$). The high correlations and intra-rater reliability values indicate that the participants filled in the questionnaire carefully and the effect of training and fatigue was rather weak.

Subject-related parameters:
In contrast to the lab study the online survey did not show any significant differences between annoyance ratings of residents ($\mu = 3.844$, $SD = 2.833$, $median = 3$) and non-residents ($\mu = 3.937$, $SD = 2.827$, $median = 3$): Mann-Whitney $U = 38421.0$, $n_1 = 135$, $n_2 = 585$, $p = 0.622$.
The trend between the attitude towards wind farms and the annoyance rating is slightly higher with $\rho = -0.255$ than in the lab study; the correlation using the average rating per participant is slightly lower ($\rho = -0.325$). Further it shows that participants with a *rather negative* ($\mu = 6.36$, $SD = 2.346$, $median = 6$) attitude show a significantly higher annoyance rating than those participants with a *neutral* ($\mu = 3.900$, $SD = 2.600$, $median = 3$), *rather positive* ($\mu = 3.824$, $SD = 2.509$, $median = 3$) or *very positive* ($\mu = 2.907$, $SD = 3.207$, $median = 1$) attitude (Kruskal $H(3) = 81.402$, $p < 0.001$; pairwise

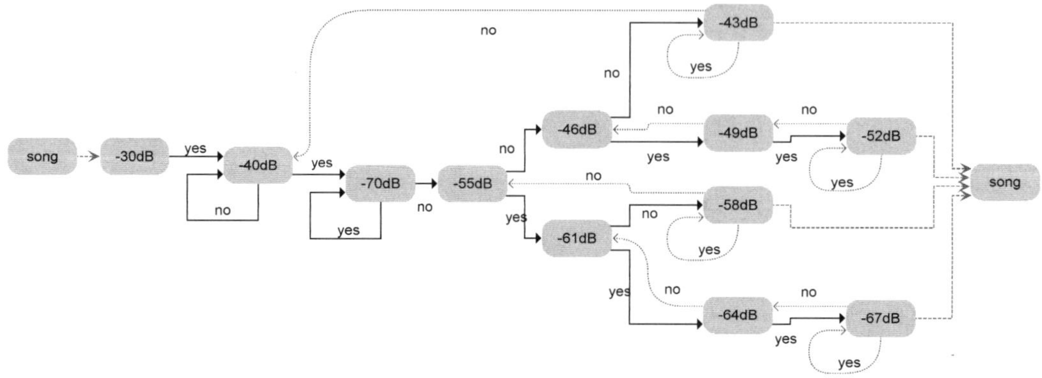

Fig. 6 Procedure to approximate the auditory threshold. Black arrows show the path through the survey based on the given answer; grey dashed arrows are not influenced by any given answer; blue arrows show the evaluation of the answers to infer the maximum heard audible threshold.

Games-Howell: $p = 0.006$ for pairs including *rather negative*, $p > 0.05$ else).
The noise sensitivity showed only a very weak linear correlation with the average rating: $\rho = 0.108$.
Concerning the age there are significant differences like in the lab study: Kruskal $H(3) = 16.910$, $p < 0.001$. Younger participants (18-34) showed less annoyance than older (45-74). The significance according to Games-Howell between the groups 18-34 and 45-54 with Bonferroni correction was $p = 0.006$ and between 18-34 and 55-74 $p = 0.047$.
There were significant differences between participants living in rural ($\mu = 4.682$, $SD = 2.828$, $median = 4.5$) and urban ($\mu = 3.018$, $SD = 2.549$, $median = 3$) surroundings: Mann-Whitney $U = 85882.0$, $n_1 = 390$, $n_2 = 330$, $p < 0.001$.

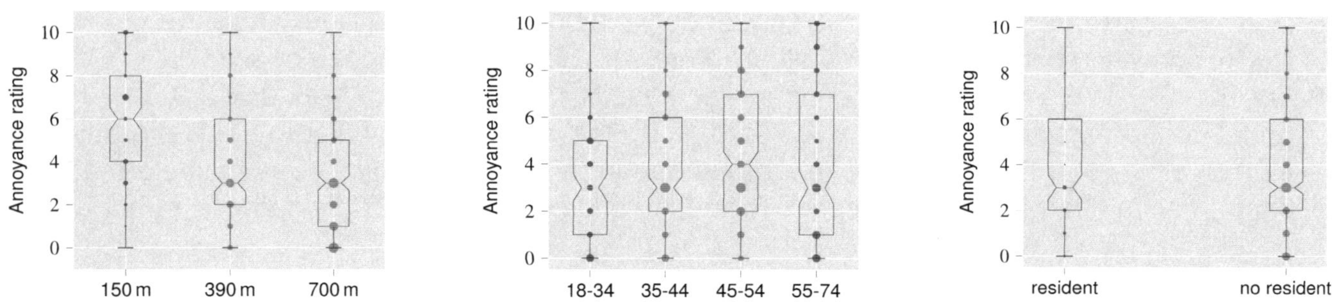

(a) Annoyance rating as a function of WT distance.

(b) Annoyance rating as a function of grouped age.

(c) Annoyance rating as a function of being a resident.

Fig. 7 Selected stimulus- and subject-related parameters of the online study (Box Plots with 25 and 75% quartiles and 95% CI of the median as well as the number of participants for each rating illustrated).

Playback-related parameters:
A weak linear correlation could be found between the approximated auditory threshold and the annoyance rating ($\rho = -0.227$). In this context the questions arises if the auditory threshold can be identified reliably at all with the given methods.
If all Mo'Web participants are considered, the deviation of the thresholds between part one and two has a mean of 1.375 dB and a standard deviation of 9.586 dB, the highest deviation is ± 27 dB. But many participants used different playback systems or headphones for the two parts although they were asked not to do so. In other cases a difference in the initial volume was assumed as in at least one part the participant was requested to change their volume. When only those participants with a (potentially) unchanged setup and no prompted change in volume are considered, the mean is 0.6 dB and the standard deviation 9.156 dB with the highest deviation between the two parts at ± 18 dB.
In the control group 13 participants had a deviation of 0 dB between the two parts. Another five showed a deviation of ± 3 dB. One participant had a deviation of ± 9 dB, but this person also showed large difference in the initial volume.
In the third group the standard deviation of those using a (potentially) unchanged setup and without prompted changes in volume is rather small with $\sigma = 4.192$ dB.
The distributions can be seen in Figure 8. The assumption can be made that, if the setup is not changed, the auditory threshold can be approximated with this method.
When comparing the same scenes with mono and binaural reproduction the general trend shows a strong correlation between the annoyance ratings: scene 1 $\rho = 0.707$, scene 3 $\rho = 0.828$, scene 9 $\rho = 0.884$. The intra-rater reliability using Krippendorff's alpha is high too (1: $\alpha = 0.681$, 3: $\alpha = 0.833$, 9: $\alpha = 0.855$). In the control group the correlation was partially weaker (1: $\rho = 0.208$, 3: $\rho = 0.705$, 9: $\rho = 0.719$) as well as the intra-rater reliability (1: $\alpha = 0.370$, 3: $\alpha = 0.720$, 9: $\alpha = 0.515$). In the third group the values were similar to the first group (all $\rho > 0.68$ and all $\alpha > 0.72$). The correlation values indicate that, as expected, with increasing distance the influence of binaural features decreases, however this was only deducted from three examples and thus is only a provisory result (which should be studied further). The binaural cues might not have a strong influence on the annoyance rating, but from the feedback of the participants the conclusion can be drawn that they help to get a better feeling fo the context of a scene.

4. Conclusion and Discussion

In this paper we report about methodologies for assessing wind turbine noise perception in immersive and context-sensitive listening experiments. We describe the general layout of the studies including the assessment of empirical data and the calibration of the respective playback systems. Our work focuses especially on the maintenance of ecological validity of the stimulus presentation including correct levels, non-acoustic context and participant attention.
Our lab study used an Ambisonics-based reproduction of WTN-scenes and included 360° video presentation. It resulted in insightful feedback and the participants reported that they could easily put themselves in the scenes. In addition known relations regarding subject-, stimuli- and playback-related parameters could be reproduced. However, a laboratory study can only be held with a small number of participants, the gathering of answers is quite time consuming and it is also more difficult to find participants with certain criteria.
With our online survey a lot of answers can be collected in short time, but the methods for presenting stimuli are a lot more limited and thus lose immersion. Furthermore the setup for an online auditory

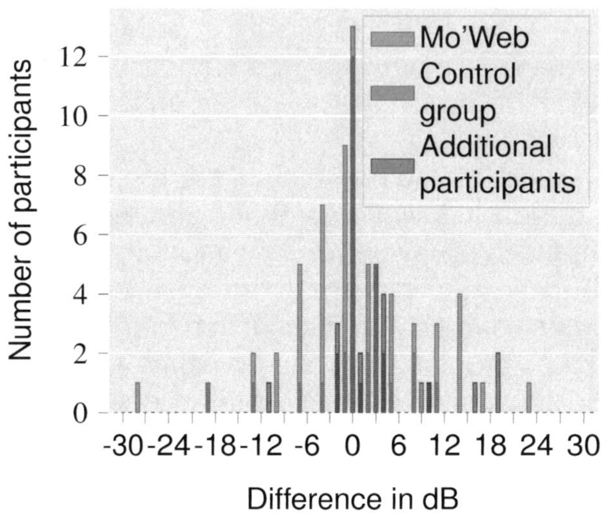

Fig. 8 Distribution of the differences in the auditory threshold in dB between the two parts of the online study for all three participant groups. The darker coloured areas correspond to the fraction of participants who (potentially) used the same setup in both parts.

test is limited in traceability. Therefore the results had to be checked for validity by including pre- and postscreening of subjects with the use of additional listening tasks and side questions.
Nevertheless our methodology for an online survey can be used to get fast feedback on scenes, especially under specific conditions since it is much easier to recruit targeted participant groups. Thus our suggestion would be to gather general trends regarding certain parameters of interest with an online survey before conducting a laboratory study. For example an online survey could be used to select a number of evaluation criteria which should be further investigated in a lab study.
Summarizing we can say that the four independent groups in our studies showed generally similar trends for the different factors that were investigated, such that we think that the methodology of immersive and context-sensitive listening experiments has passed a first step of validation. As significant factors during the evaluation proofed especially the attitude towards wind farms, the residence location and the age. Noise sensitivity or the approximated auditory threshold did not show a high correlation with the annoyance ratings in the available data. In terms of signal characteristics the influence of the distance to the nearest wind turbine was also confirmed.

Acknowledgement

The work described in this publication was part of the project WEA-Akzeptanz (FKZ 0324134A) which was funded by the German Federal Ministry for Economic Affairs and Energy (BMWi). We thank the BMWi for the funding we received. We would also like to thank all subjects and colleagues who participated in the studies.

References

[1] Bergner, J., Preihs, S., Hupke, R., and Peissig, J. (2019a). A system for room response equalization of listening areas using parametric peak filters. In *Proc. of AES Int. Conf. on Immersive and Interactive Audio.*

[2] Bergner, J., Preihs, S., and Peissig, J. (2019b). Investigation on ecological validity within higher order ambisonics reproductions of wind turbine noisescapes. In *Proc. of Internoise.*

[3] Bergner, J., Preihs, S., and Peissig, J. (2019c). On wind turbine noisescape reproduction for perceptual evaluation. In *45. Jahrestagung für Akustik DAGA.*

[4] Bolin, K., Bluhm, G., and Nilsson, M. E. (2014). Listening test comparing a-weighted and c-weighted sound pressure level as indicator of wind turbine noise annoyance. *Acta Acustica United With Acustica*, 100(5):842–847.

[5] Butera, F., Burgemeister, K., Fisher, K., and Mounter, D. (2014). Using wind farm noise auralisations for effective community consultation. In *INTER-NOISE and NOISE-CON Congress and Conference Proceedings*, volume 249-5, pages 2993–3000. Institute of Noise Control Engineering.

[6] Hupke, R., Ordner, J., Bergner, J., Nophut, M., Preihs, S., and Peissig, J. (2019). Towards a virtual audiovisual environment for interactive 3D audio productions. In *Proc. of AES Int. Conf. on Immersive and Interactive Audio.*

[7] ISO (2003). ISO/TS 15666: Assessment of noise annoyance by means of social and socio-acoustic surveys.

[8] LimeSurvey GmbH (2020). URL: `https://www.limesurvey.org/de/`. Online; last access 26-11-2020.

[9] Maffei, L., Iachini, T., Masullo, M., Aletta, F., Sorrentino, F., Senese, V. P., and Ruotolo, F. (2013). The effects of vision-related aspects on noise perception of wind turbines in quiet areas. *International Journal of Environmental Research and Public Health*, 10(5):1681–1697.

[10] Maffei, L., Masullo, M., Gabriele, M. D., Votsi, N.-E. P., Pantis, J. D., and Senese, V. P. (2015). Auditory recognition of familiar and unfamiliar subjects with wind turbine noise. *International journal of environmental research and public health*, 12(4):4306–4320.

[11] Manyoky, M., Hayek, U. W., Pieren, R., Heutschi, K., and Grêt-Regamey, A. (2016). Evaluating a visual-acoustic simulation for wind park assessment. *Landscape and Urban Planning*, 153:180–197.

[12] mo'web research (2020). URL: `https://www.mowebresearch.com/de/`. Online; last access 26-11-2020.

[13] Schäffer, B., Pieren, R., Wissen Hayek, U., Biver, N., and Grêt-Regamey, A. (2019). Influence of visibility of wind farms on noise annoyance – A laboratory experiment with audio-visual simulations. *Landscape and Urban Planning*, 186:67 – 78.

[14] Schössow, D., Bergner, J., Preihs, S., and Peissig, J. (2020). Audiovisuelle Laborstudie zur Lästigkeit von WEA-Schall. In *46. Jahrestagung für Akustik DAGA.*

[15] Taghipour, A. and Pelizzari, E. (2019). Effects of background sounds on annoyance reaction to foreground sounds in psychoacoustic experiments in the laboratory: limits and consequences. *Applied Sciences*, 9(9):1872–1890.

[16] Van Gerven, P. W., Vos, H., Van Boxtel, M. P., Janssen, S. A., and Miedema, H. M. (2009). Annoyance from environmental noise across the lifespan. *The Journal of the Acoustical Society of America*, 126(1):187–194.

[17] Weinstein, N. D. (1978). Individual differences in reactions to noise: a longitudinal study in a college dormitory. *Journal of Applied Psychology*, 63:458–466.

[18] Wycisk, Y. (2018). Bestimmung der Wiedergabeeigenschaften von Kopfhörern und Lautsprechern in musikpsychologischen Internetexperimenten. Master's thesis, Hanover Univ. of Music, Drama and Media.

[19] Zhao, S. (2018). *An investigation of the wind noise reduction mechanism of porous microphone windscreens.* PhD thesis, RMIT University Melbourne.

[20] Zimmer, K. and Ellermeier, W. (1998). Ein Kurzfragebogen zur Erfassung der Lärmempfindlichkeit. *Umweltpsychologie*, 2(2):54–63.

INCE EUROPE

9th International Conference
on
Wind Turbine Noise
Remote from Europe – 18th to 21st May 2021

The Quasi-3D Blade and Rotor Noise Prediction Methodology for the PNoise Code and Results

Sara Rodriguez
Sao Paulo State University (Poli-USP):sara.rodriguez@alumni.usp.br

Joseph Youssif Saab Jr.
Maua Institute of Technology (IMT):saab@maua.br

Alexandre Martuscelli Faria
Sao Paulo State University (Poli-USP):martuscellifaria@usp.br

João Paulo Barreto
Sao Paulo State University (Poli-USP): jopa84@gmail.com

Marcos de Mattos Pimenta
Sao Paulo State University (Poli-USP):marcos.pimenta@poli.usp.br

Summary

Owing to the large and steady number of downloads of the *Qblade* open source wind turbine design environment, some improvements are being implemented in order to broaden the capacity of the embedded aeroacoustic module, *PNoise*, which originally consisted of a 2D airfoil trailing-edge (TE) noise prediction tool. Other self-noise sources other than leading-edge are being added to the tool, plus inflow noise modeling and a quasi-3D tool for full rotor noise estimation. The aim of the upgrade is to provide the user with the approximate sound pressure levels and related spectra for all self-noise sources plus inflow noise, for 2D and 3D prediction in steady flow. The current models available for the calculations are the modified-BPM, with boundary layer data provided by the XFLR5 integrated code for the self-noise sources and the Von Kármán and the modified-Lowson model by the inclusion of the Rapid Distortion Theory model. Due to the recursive nature of the quasi 3D method, it relies on the same underlying boundary layer thickness hybrid calculation of the 2D calculation, but with local flow conditions adjusted to the local Reynolds and Mach number of each spanwise section of the blade as calculated by the Blade Element Momentum method. The method has a resulting accuracy limited by the combined errors inherent to each method, plus the error deriving from the use of a finite number of blade sections and a discrete number of angular positions of the blade rotating along the azimuthal plane. This paper describes the methodology for both blade and rotor noise prediction, along with the preliminary results for the predicted blade and rotor noises calculated for a 31 m diameter WT rotor.

LIST OF ABBREVIATIONS AND ACRONYMS

AOA angle of attack

BEM Blade Element Momentum

BPM The initial letter of the surname of Thomas F. Brooks, D. Stuart Pope and Michael A. Marcolini

BRS Blade Reference System

CCS Chord Coordinate System

DMS Double Multiple Streamtube

HFCS Hub, Fixed Coordinate System

HR hub radius

LBL-VS laminar-boundary-layer-vortex shedding

TBL-TE turbulent-boundary-layer-trailing-edge

LE leading-edge

LTRS Lower Tower Reference System

OASPL Overall Sound Pressure Level

PNoise Poli-USP noise prediction code inside the QBlade software

POLI-USP Polytechnic School - São Paulo State University - Brasil

RDT Rapid Distortion Theory

QBlade TU Berlin-developed, public domain, Wind Turbine Performance and Structural analysis software

TBL Turbulent Boundary Layer

TE trailing-edge

TU-Berlin Technische Universität Berlin

RRCS Rotor, Rotating Coordinate System

RSRS Rotated Section Reference System

RTERS Rotated Trailing Edge Reference System

SRS Section Reference System

TE-Blunt-VS trailing-edge-bluntness-vortex-shedding

TERS Trailing Edge Reference System

TTH Tower To Hub

UTRS Upper Tower Reference System

XFLR5 An analysis tool for airfoils, wings and planes operating at low Reynolds Numbers

XFOIL An interactive program for the design and analysis of subsonic isolated airfoils

WT wind turbine

YBRS Yaw-Bearing Reference System

1. Introduction

The need for a quick noise prediction tool in the early stages of the wind turbine (WT) design, due to the cost, is interesting to both research and the wind industry. The QBlade software was developed at Technical University Berlin following the principles of the General Public License, it was made to design different types of airfoils with reliable and robust forecasts simulating the flow conditions around the blade and with the partnership of the POLI-USP, by the Poli Wind Group, it was possible to implement an aeroacoustic noise prediction tool called PNoise integrated in the QBlade software.

The QBlade is an user-friendly software that integrates the XFOIL/XFLR5 functionality and allows the user to rapidly design custom airfoils and compute their performance polars and directly integrate them into a wind turbine rotor design and simulation, and it also includes the Blade Element Momentum (BEM) and Double Multiple Streamtube (DMS) correction algorithms.

The *quasi-3D* method it was also implemented in the *PNoise* module The quasi-3D methodology was also implemented in the PNoise module, based on [3, 11, 12, 24, 25, 29, 30], using the user input data and calculated by the QBlade software, it was possible to analyze the noise emitted by the WT rotor and not just analyzing an airfoil segment of the WT blade, according to the well-known semi-empirical or simplified theoretical models and bringing to a next level the prediction of the noise emission by the WT.

The development of the quasi-3D method in the *PNoise* code, previously envisioned in the publications [16–20, 22] includes the following features:

- Overall and spectral sound pressure level quick estimate;
- User selection of flow conditions specifications and calculation according the noise models validity range;
- Application for any airfoil, blade and rotor geometry;
- User input of the source of the vertical turbulent scale to predict some of the noise sources;
- User selection of airfoils self-noise sources to be modeled, according the BPM method;
- Inflow noise source selection;
- An auto-adjustment for the BEM and XFOIL polar calculations with a warning for the user to verify errors considering the validity limitations in a table view;
- Arbitrary, user-defined observer position;
- Open-source, free software download, under GPL.

The methodology applied for the *quasi-3D*, the implementation on the *PNoise* module and the validation tests are presented in the following sections.

2. Quasi-3D Methodology

The quasi-3D methodology was first devised for the calculation of torque and later adapted for noise prediction. The *PNoise* employs the quasi-3D method for the blade and rotor, that is based on the 2D method for noise prediction from discretized-blade segments, with the blade divided by different spanwise segments, with quantitative-rotation positions or angles in the azimuthal plane.

The axial and radial induction factors are calculated iteratively using the BEM [13] to calculate the values of the relative speed, angle of attack (AOA), Mach and Reynolds numbers. According to the self-noise employed it will be made some correction methods.

For precision in the highly induced rotor flow, the contribution from each blade segment must be calculated with the local flow, i.e., the flow corrected with the locally induced parameters. For this reason, the quasi-3D method needs additional flow information that may be provided by the BEM method, for instance. Some methods may have corrections for tip loss, 3D rotational augmentation, etc., which may be considered by each blade segment position.

For each noise source selected by the user it will be calculated the spectrum in $1/3$ octave band for each radial blade segment, considering a fixed observer position related to the corresponding segment, i.e. the observer rotates with the blade and is positioned at a fixed distance from the blade segment according to the Figure 1.

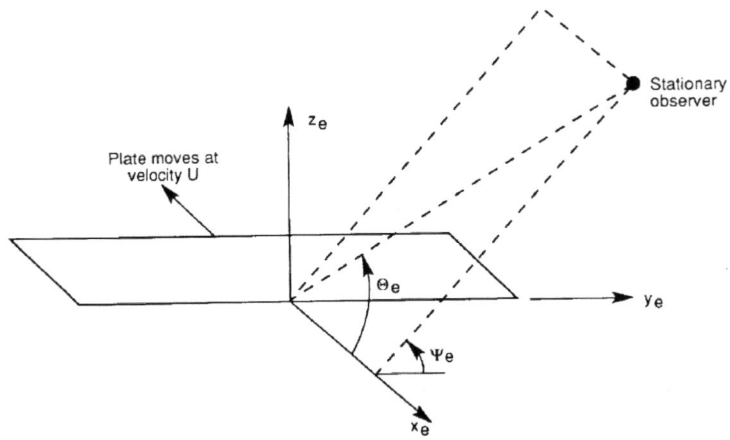

Fig. 1 The 3D coordinate system postponed related to the TE considering a moving plate of a velocity U in the opposite direction of the abscissa x_θ and angles position $\Theta \rightarrow \theta$ and $\Psi \rightarrow \phi$. Source: [3].

For the quasi-3D blade is necessary to make a coordinate system transformation for the observer fixed in relation to the ground using a directivity function. For the quasi-3D rotor it will be considered the movement of all the blades according to the input user data.

The total Overall Sound Pressure Level (OASPL) spectrum for the quasi-3D method is computed according to the effects of the directivity function. For the quasi-3D blade it will be summed all of the noise spectrum contributions of each blade segment and for the quasi-3D rotor it will be applied for all of the moving blades, according equation 1. If the observer position is distant, in the far-field, a propagation model shall be employed.

$$SPL_{total} = 10 \cdot \log\left(\frac{1}{n}\left(\sum_{i=1}^{n} 10^{SPL_i/10}\right)\right) \quad (1)$$

2.1. Definition of the Reference System

The quasi-3D methodology used in the *PNoise* script uses the transformation matrices proposed by Saab [22] and based on the German Guideline for Certification of Wind Turbines [1], for considering the reference system most simple and the best to start comparing to the existent reference systems adopted by Broux, Vargas, and Zhu [5, 29, 30].

For the selected base-system, the WT geometry representation is simplified by the absence of rotor axis tilt and cone angles, with the main advantage of that the blade and chord coordinate systems have one axis aligned with the pitch axis of the blade, while for the other systems studied some unusual spanwise axis were adopted, passing through the blade camber line or the blade centerline, for instance.

Considering to take the pitch axis as part of the reference system we get a simplifying assumption, i.e., the pitch axis is assumed to intersect the chord line of all sections at 25% of its length from the TE, so it is possible to make the pitch axis intersect the thickest section of each airfoil, (or close to it), the region where the center of pressure and structural spar should be preferably located and often the axis through which different airfoil sections are stacked in order to assemble a blade. A few decimeters tolerance in the TE position due to the assumption described will not have any significant impact at the Sound Pressure Levels (SPLs) for an observer position located at the far field.

For the quasi-3D calculation in the *PNoise* module, five reference systems were adopted. The number of coordinate reference systems is large due to the decision to employ only one transformation operation (translation or rotation) between two adjacent systems, to keep the overall transformation traceable and didactic as follows.

2.2. Quasi-3D Blade and Rotor Noise Prediction Transformation Matrices

2.2.1. Blade Reference System

Denominated (X_B, Y_B, Z_B), its origin is at the blade root, at hub radius, and it rotates with the rotor. Its orientation to the hub is fixed. X_B in direction of rotor axis, Z_B radially (spanwise axis), and Y_B (chordwise axis) is set so that the system rotates clockwise, according Figure 2.

The observer position should be first determined in relation to the BRS. The default observer position is 10 m downstream of the rotor plane, 10 m downstream of the blade in the rotation plane and in the mid-span of the blade.

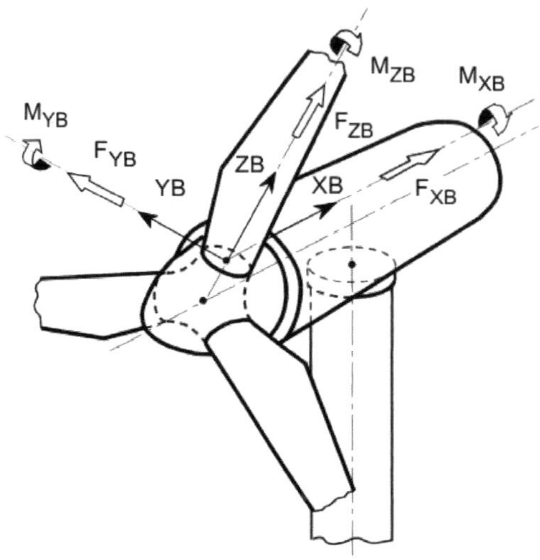

XB in direction of the rotor axis
ZB radially
YB so that XB, YB, ZB rotate clockwise

Fig. 2 Blade Reference System (BRS). Source: [1], p. 4-31, fig 4.A.1.

2.2.2. A Section Reference System
Denominated (X_S, Y_S, Z_S), is obtained by the translation of the BRS to the mid-section of the considered blade segment, along the pitch axis. This system is based on the Chord Coordinate System (CCS) [1], as presented in Figure 3.

The translation shall be done up to the blade section under consideration.

$$\begin{aligned} X_S &= X_B \\ Y_S &= Y_B \\ Z_S &= Z_B - (r_{i+1} + r_i)/2 \end{aligned} \tag{2}$$

, where r_{i+1} is the outboard radius and r_i is the inboard radius of the blade segment considered.

2.2.3. A Rotated Section Reference System
Denominated (X_{RS}, Y_{RS}, Z_{RS}), this system is obtained by rotating the SRS along the 'Z' (pitch) axis, by an angle a, in order to make its 'Y' (chordwise) axis aligned with the chord of the mid-section of the blade section considered.

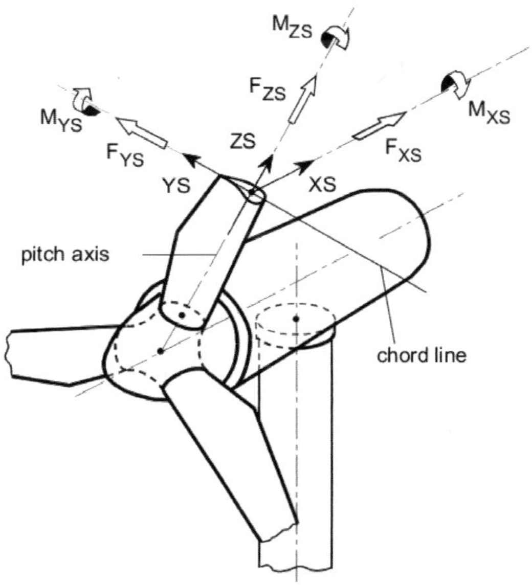

YS in direction of the chord, orientated to blade trailing edge
ZS in direction of the blade pitch axis
XS perpendicular to the chord, so that XS, YS, ZS rotate clockwise

Fig. 3 Section Reference System (SRS). Source: [1], p. 4-31, fig 4.A.2.

$$\begin{bmatrix} X_{RS} \\ Y_{RS} \\ Z_{RS} \end{bmatrix} = \begin{bmatrix} \cos a & \sin a & 0 \\ -\sin a & \cos a & 0 \\ 0 & 0 & 1 \end{bmatrix} \cdot \begin{bmatrix} X_S \\ Y_S \\ Z_S \end{bmatrix} \tag{3}$$

The angle a is the total angle between the $Y_B Z_B$ blade reference system plane and the local midsection chord line. If the $Y_B Z_B$ plane is set equal to the rotation plane, then the angle a is the equivalent to the θ angle:

$$\theta = \theta_p + \beta \tag{4}$$

, where θ_p is the pitch angle and β the local twist angle.

2.2.4. A Trailing Edge Reference System

Denominated (X_T, Y_T, Z_T), it is obtained by translating the Rotated Section Reference System (RSRS) chordwise (along the 'Y' axis) up to the TE. That is accomplished by sliding it 75% of the chord length of the mid-section, towards the TE.

$$X_T = X_{RS} \tag{5}$$
$$Y_T = Y_{RS} - 0.75(C_{r_{i+1}} + C_{r_i})/2$$

, where $C_{r_{i+1}}$ is the chord of the outboard section and C_{r_i} is the chord of the inboard section of the blade segment considered.

$$Z_T = X_{RS} \tag{6}$$

2.2.5. A Rotated Trailing Edge Reference System

Denominated (X_{RT}, Y_{RT}, Z_{RT}), is obtained by rotating the Trailing Edge Reference System (TERS) along the 'X' axis so that the 'Z' axis is aligned with the local TE line and the 'Y' axis is perpendicular to the TE line in the chord-pitch plane.

$$\begin{bmatrix} X_{RT} \\ Y_{RT} \\ Z_{RT} \end{bmatrix} = \begin{bmatrix} 1 & 0 & 0 \\ 0 & \cos b & \sin b \\ 0 & -\sin b & \cos b \end{bmatrix} \cdot \begin{bmatrix} X_T \\ Y_T \\ Z_T \end{bmatrix} \tag{7}$$

, where

$$b = \arctan\left(\frac{C_{r_i+1} - C_{r_i}}{r_{i+1} - r_i}\right) \tag{8}$$

This last reference system coincides (except per axis notation), with the BPM coordinate system for the directivity function, relating the observer position to the TE of the section currently considered in the calculation loop.

The BPM TE noise and directivity model requires the following coordinates for calculation: (Θ_e, ψ_e, r_e).

After all the reference system transformation, these coordinates may be found with the aid of following expressions:

$$r_e \equiv r_{RT} = \sqrt{X_{RT}^2 + Y_{RT}^2 + Z_{RT}^2}$$
$$\Theta_e = \pm \arctan\left(\frac{Z_{RT}}{Y_{RT}}\right) \tag{9}$$
$$\psi_e = \pm \arctan\left(\frac{X_{RT}}{Z_{RT}}\right)$$

2.3. Quasi-3D Rotor Noise Prediction Transformation Matrices

2.3.1. A Lower Tower Reference System

Denominated (X_{LT}, Y_{LT}, Z_{LT}), for the quasi-3D simulations, both steady State and transient regimes, the observer position should be first determined in relation to the Lower Tower Reference System (LTRS) [1]. Its origin is at the tower base centerline and it is completely fixed, not rotating with the nacelle.

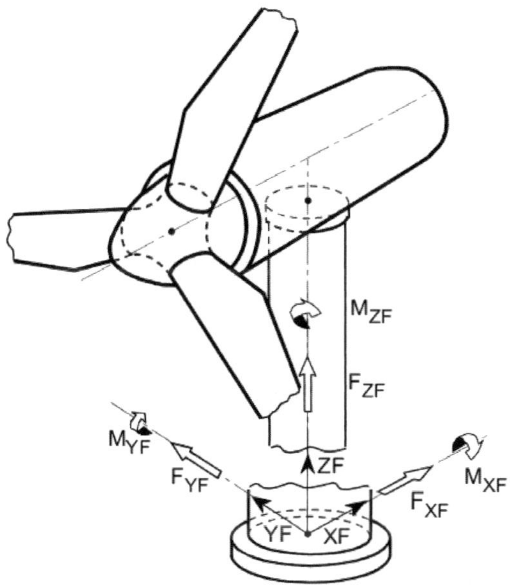

XF horizontal
ZF vertically upwards in direction of the tower axis
YF horizontally sideways, so that XF, YF, ZF rotate clockwise

Fig. 4 LTRS. Source: [1], p. 4-33, fig 4.A.6.

The default values for the observer position is for a straight downstream observer position, distant one and a half rotor diameter from the tower base. It is supposed to have also a default rotor yaw angle of 0°.

2.3.2. An Upper Tower Reference System

Denominated (X_{UT}, Y_{UT}, Z_{UT}), this reference is obtained by purely translating vertically up (along 'Z' axis) the LTRS by the height of the tower.

$$
\begin{aligned}
X_{UT} &= X_{LT} \\
Y_{UT} &= Y_{LT} \\
Z_{UT} &= Z_{LT} - H
\end{aligned}
\tag{10}
$$

, where H is the height of the WT tower.

2.3.3. A Yaw-Bearing Reference System

Denominated (X_{YB}, Y_{YB}, Z_{YB}), the yaw bearing coordinate system has its origin at the intersection of the tower axis and the upper edge of the tower top.

Although the origin coincides with that of the Upper Tower Reference System (UTRS), the Yaw-Bearing Reference System (YBRS) system is not fixed, it rotates with the nacelle.

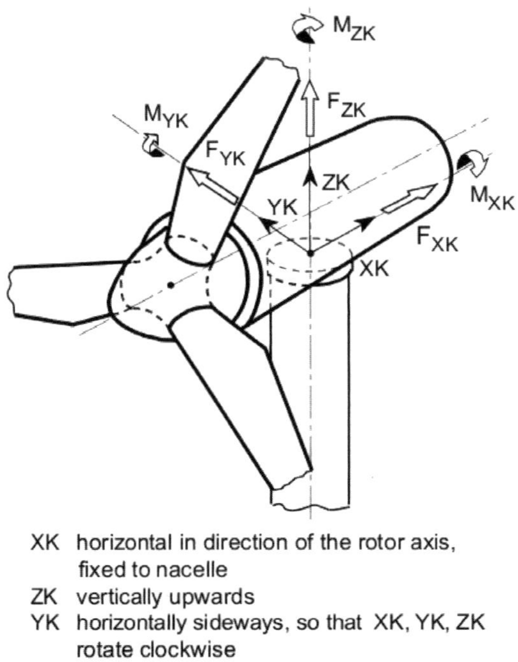

Fig. 5 YBRS. Source: [1], p. 4-33, fig 4.A.5.

$$\begin{bmatrix} X_{YB} \\ Y_{YB} \\ Y_{YB} \end{bmatrix} = \begin{bmatrix} \cos\gamma & \sin\gamma & 0 \\ -\sin\gamma & \cos\gamma & 0 \\ 0 & 0 & 1 \end{bmatrix} \cdot \begin{bmatrix} X_{UT} \\ Y_{UT} \\ Z_{UT} \end{bmatrix} \tag{11}$$

, where γ is the yaw angle, measured in the horizontal 'XY' plane, from the 'X' axis, positive when the nacelle rotation is from 'X' to 'Y' direction. By default is equal 0°.

2.3.4. A Hub, Fixed Coordinate System

Denominated (X_{HF}, Y_{HF}, Z_{HF}), according the Figure 6. As mentioned before, in this simplified WT model we are dropping any tilt and cone angles. Also the rotor plane is defined by the co-planar pitch axis of the blades.

$$X_{HF} = X_{YB} + TTH_x$$
$$Y_{HF} = Y_{YB} \tag{12}$$
$$Z_{HF} = Z_{YB} - TTH_y$$

, where TTH_x is the horizontal and TTH_z is the vertical distance between the tower center axis and the rotor plane (Tower To Hub (TTH))

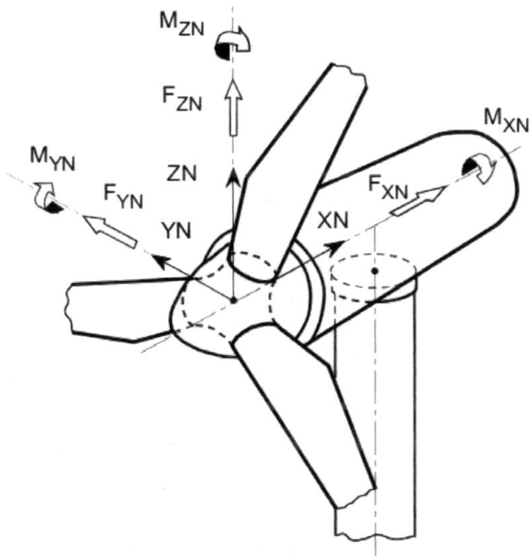

XN in direction of the rotor axis
ZN upwards perpendicular to XN
YN horizontally sideways, so that XN, YN, ZN rotate clockwise

Fig. 6 Hub, Fixed Coordinate System (HFCS). Source: [1], p. 4-32, fig 4.A.3.

2.3.5. A Rotor, Rotating Coordinate System

Denominated (X_{RR}, Y_{RR}, Z_{RR}), has its origin at the rotor plane center, and rotates with the rotor, along the 'X' axis.

$$\begin{bmatrix} X_{RR} \\ Y_{RR} \\ Z_{RR} \end{bmatrix} = \begin{bmatrix} 1 & 0 & 0 \\ 0 & \cos\epsilon & \sin\epsilon \\ 0 & -\sin\epsilon & \cos\epsilon \end{bmatrix} \cdot \begin{bmatrix} X_{HF} \\ Y_{HF} \\ Z_{HF} \end{bmatrix} \tag{13}$$

, where ϵ is the rotation angle, measured in the vertical 'ZY' plane. The angle is measured from a blade in the upper plane departing the vertical reference line and is positive when the rotation is from

'Y' to 'Z' direction, i.e. for a rotor spinning clockwise as viewed from upstream.

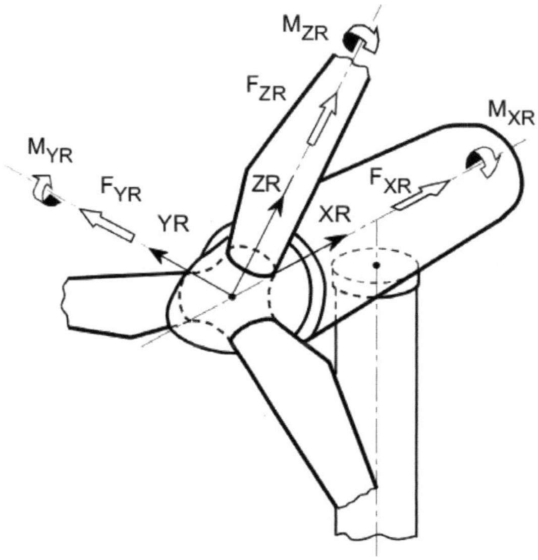

XR in direction of the rotor axis
ZR radially, orientated to rotor blade 1 and perpendicular to XR
YR perpendicular to XR, so that XR, YR, ZR rotate clockwise

Fig. 7 Rotor, Rotating Coordinate System (RRCS). Source: [1], p. 4-32, fig 4.A.4.

Plus all the local reference systems described for the quasi-3D calculation described above (BRS, SRS, RSRS, TERS and Rotated Trailing Edge Reference System (RTERS)).

The transformation from the RRCS to the BRS is obtained by purely translating spanwise (along 'Z' axis) the RRCS by the hub radius (HR).

$$\begin{aligned} X_B &= X_{RR} \\ Y_B &= Y_{RR} \\ Z_B &= Z_{RR} - HR \end{aligned} \tag{14}$$

2.4. Code Flow Diagrams for the Quasi-3D in PNoise

Two flow diagrams were made to guide the realization of the PNoise code, they are presented in the Appendices **??** and B.

2.4.1. Quasi-3D Blade Flow Diagram

The Figure A.1 presents the flow diagram developed by Saab [22] for the estimation of the quasi-3D blade noise, considering the observer rotating with the blade in a constant distance postponed the

TE.

2.4.2. Quasi-3D Rotor Flow Diagram

The Figures B.1 and B.2 presents the flow diagram developed by Saab for the estimation of the quasi-3D rotor noise, considering the observer related to the ground level.

3. PNoise Simulations

The following sections presents the simulation conditions and the simulations results in order to validate the PNoise code, using the data presented by the literature [11, 15, 30] of the Bonus Combi B31/300 wind turbine, with a rated power of 300 kW. The validation conditions are presented in the Appendix C.

3.1. Selected Wind Turbine Datasheet

The WT tested has the following descriptions:

- Rotor radius: 15.5 m;
- Tower height: 31 m;
- Blade number: 3;
- Rotation speed: 35.2 rpm;
- Airfoil: Profile series NACA 63_1-212, according [15], Appendix 1, p. 67-75;
- Tip pitch angle: -1.0 degree; TE-Bluntness: 0.5% $\times$ chord;
- Cut-in wind speed: 3.5 $m\,s^{-1}$;
- Rated wind speed: 8 $m\,s^{-1}$;
- Cut-off wind speed: 25 $m\,s^{-1}$;
- Wind direction: Upwind;
- Receiver position: 40 mm in the downwind direction at the ground level;
- Ground surface roughness: 10 mm.

3.2. Directivity

The sound pressure level changes according to the position of the observer because of the effect of the directivity of the sound, the effect can be observed considering an observer walking around the WT in a given radius and getting the OASPL in discretized angles around the WT.

Figure 8 presents a comparison between the simulations performed in the PNoise software with those presented in the research by Zhu. The directivity was simulated for the Bonus Combi B31/300 considering an observer at the ground level at distances of 200 m, 100 m, and 80 m from the WT.

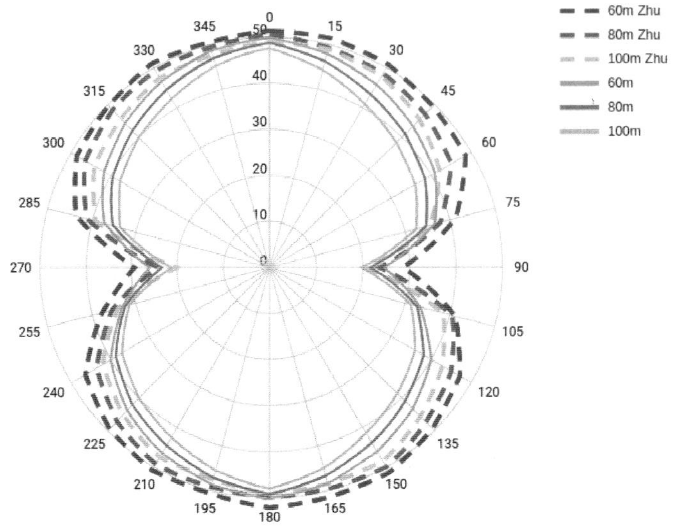

Fig. 8 Comparison of the directivity simulation of PNoise and the Zhu research. Source: [30],p. 57, fig. 5.8

The shape of the directivity curves define sound radiation as a dipole, there are not symmetrical due to twist along the blades. The simulated curves in the PNoise are similar to those presented in Zhu's work, so it is possible to validate the directivity factors used in the self-noise simulations in the quasi-3D blade and rotor code of PNoise.

3.3. Quasi-3D Rotor Simulations

It was implemented to the *PNoise* code the following BPM models of aerodynamic self-noises:

- Turbulent-boundary-layer-trailing-edge for pressure side;
- Turbulent-boundary-layer-trailing-edge for suction side;
- Separation stall;
- Laminar-boundary-layer-vortex shedding;
- Trailing-edge-bluntness-vortex-shedding;
- Tip vortex formation noise;
- The total sum of the noise source models simulated.

Initially the PNoise was implemented with the prediction of a 2D airfoil TE noise prediction from the aerodynamic airfoil self-noise according the BPM model [3]. The model includes the TBL-TE pressure and suction sides and separation-stall. This method was implemented by Joseph Saab [20, 26, 27] and it is integrated into the current version of the software QBlade. After, it was implemented in the PNoise module the airfoil turbulent inflow noise according the prediction method based on Amiet's theory by Alexandre Martuscelli [6, 8–10]. The software provides that the calculation may be given by Von Kármán or by the Rapid Distortion Theory (RDT). The remaining noise sources mentioned above was implemented in 2D and the quasi-3D was coding for all them for blade and rotor.

Considering the work of Stefan Oerlemans and Peter Fuglsang [28] it was simulated in the same conditions for the WT Bonus Combi 300 kW, considering the quasi-3D rotor implementation, and presented in individual graphics to comparison according Figures 9 to 16 and from [28] p. 23. figure 4-2.

Observing the graphs, it is possible to verify that the results obtained by Stefan Oerlemans and Peter Fuglsang [28] and the PNoise module are compatible and consistent with the expected for the models implemented.

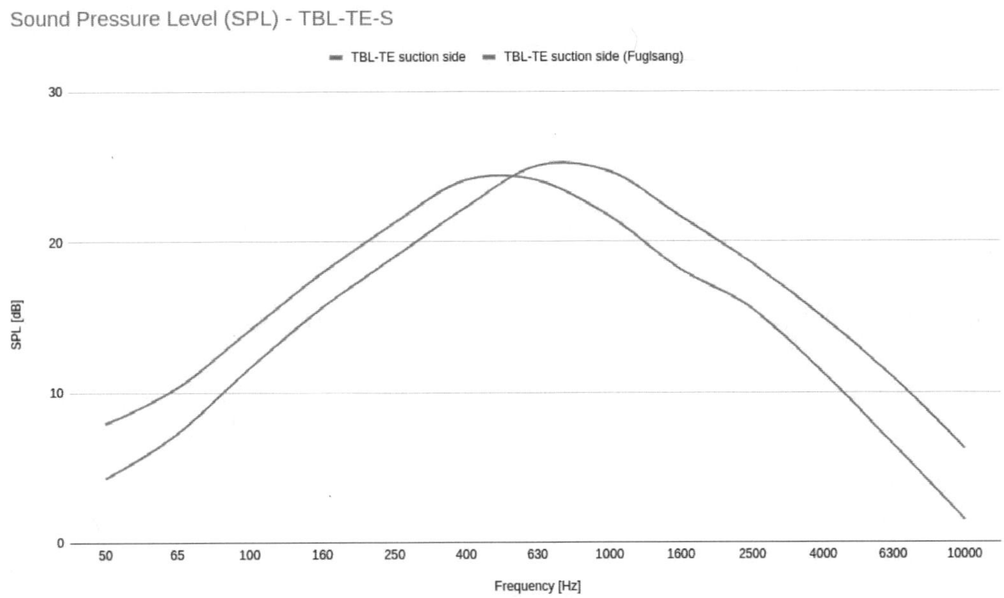

Fig. 9 Comparison between Fuglsang [11] and PNoise quasi-3D rotor noise for turbulent-boundary-layer-trailing-edge (TBL-TE) suction.

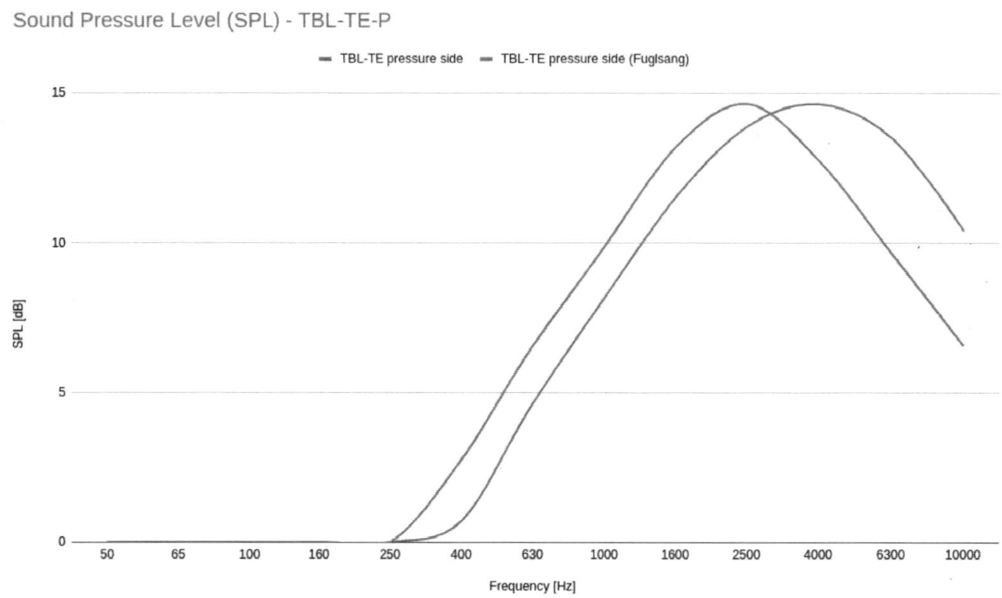

Fig. 10 Comparison between Fuglsang [11] and PNoise quasi-3D rotor noise for TBL-TE pressure.

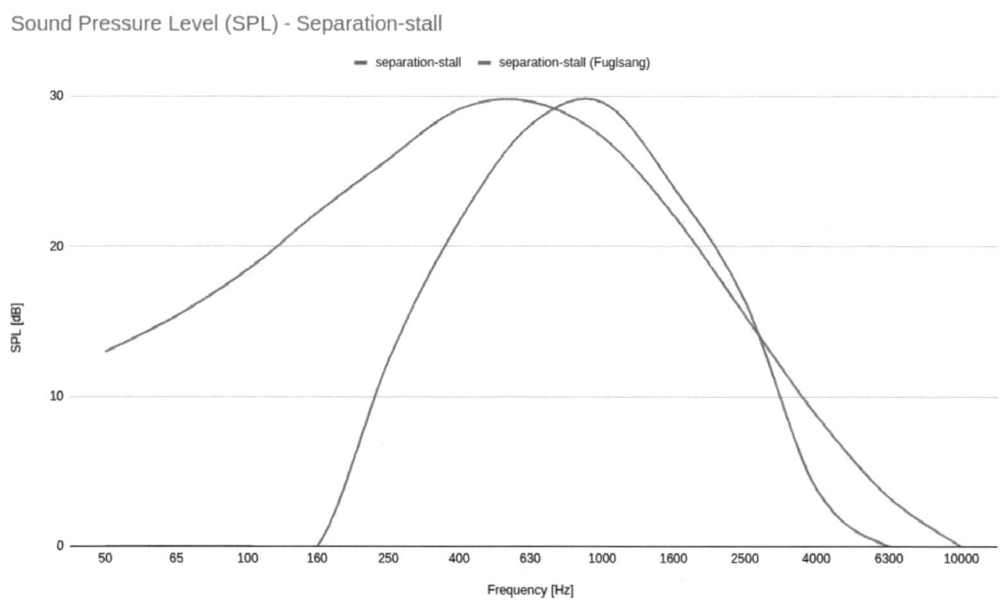

Fig. 11 Comparison between Fuglsang [11] and PNoise quasi-3D rotor noise for separation-stall.

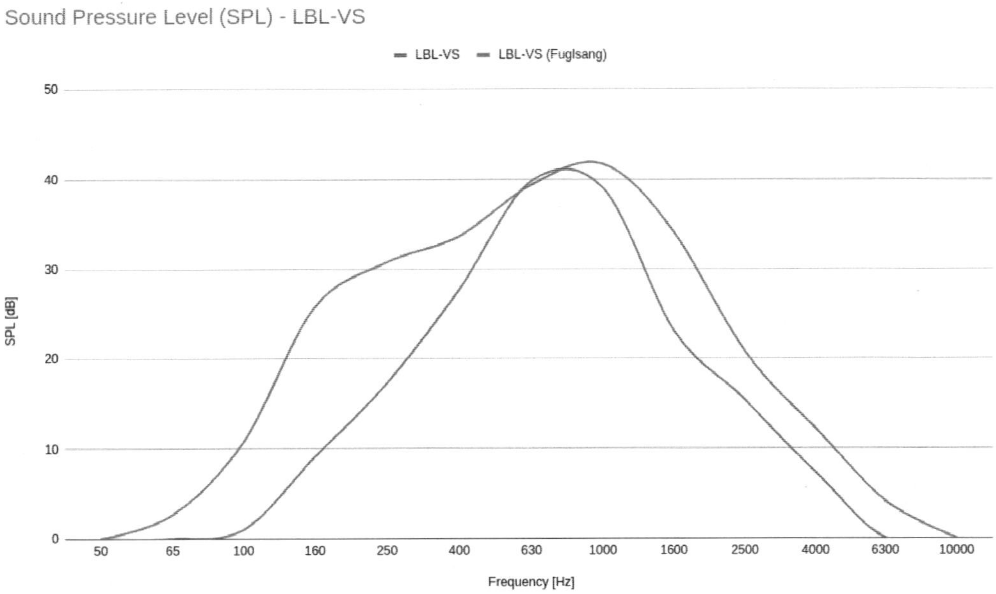

Fig. 12 Comparison between Fuglsang [11] and PNoise quasi-3D rotor noise for laminar-boundary-layer-vortex shedding (LBL-VS).

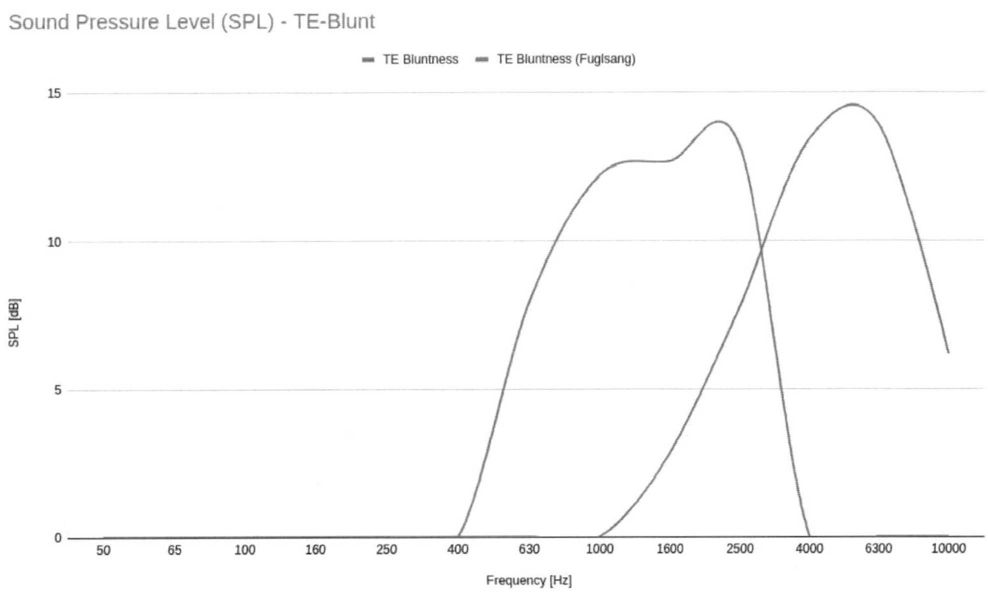

Fig. 13 Comparison between Fuglsang [11] and PNoise quasi-3D rotor noise for Bluntness.

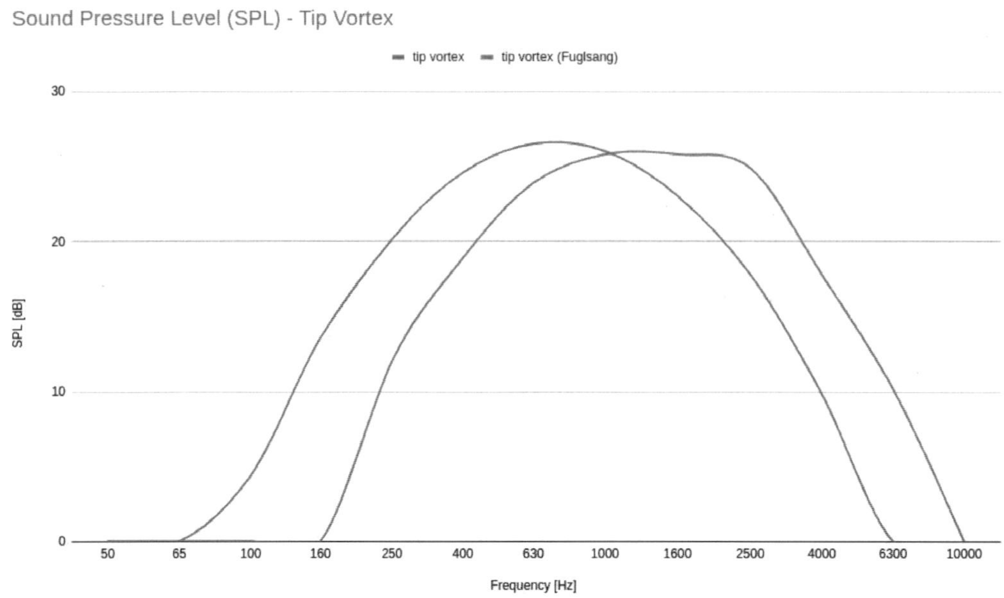

Fig. 14 Comparison between Fuglsang [11] and PNoise quasi-3D rotor noise for Bluntness.

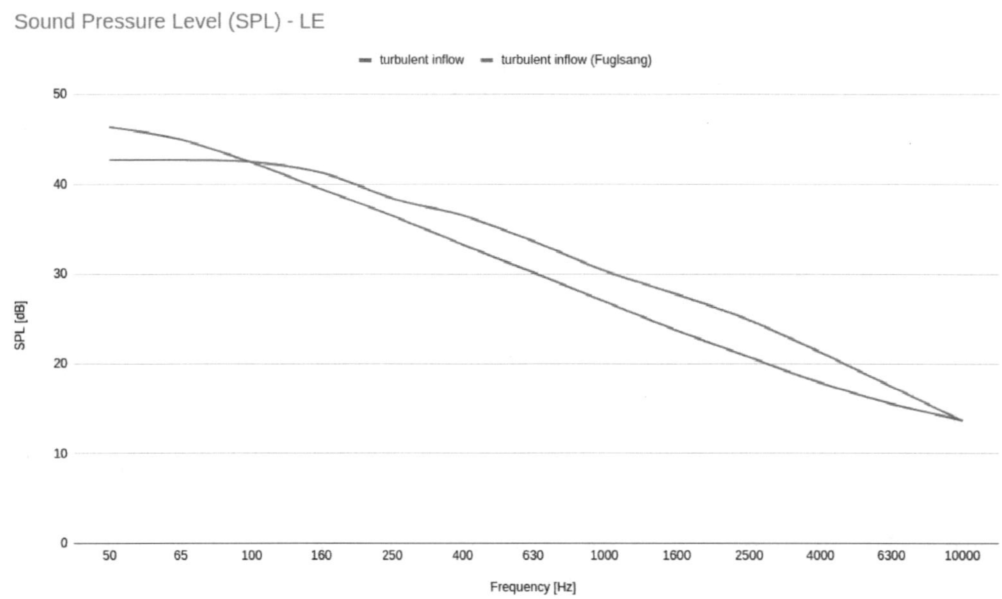

Fig. 15 Comparison between Fuglsang [11] and PNoise quasi-3D rotor noise for Leading-edge.

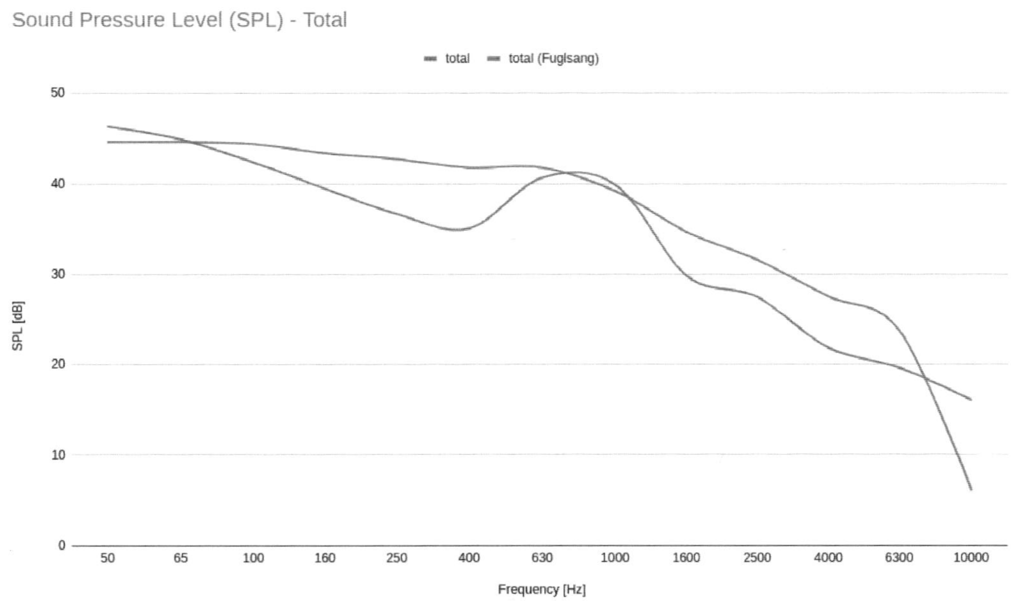

Fig. 16 Comparison between Fuglsang [11] and PNoise quasi-3D rotor noise for the total noise.

4. Conclusion

The work consisted in the implementation of the quasi-3d noise model for the existent QBlade software developed by the TU-Berlin and after it was implemented the PNoise module with new tools tested and implemented by the POLI-Wind group*.

The quasi-3d method was initially sketched by Joseph and in this job there was presented the first results of the implementation, with the addiction of the remaining aerodynamic noise sources from the BPM method using as reference the work of Zhu [30] for testing the directivity factors and Stefan Oerlemans and Peter Fuglsang [28] for the quasi-3D rotor spectra. These works were used because of the lack of experimental data for newer and larger WTs, important for the validation of the PNoise module.

The results presented were consistent with those in the literature and the quasi-3D method in addition to obtain a good accuracy using low computational time, i.e. the acs PNoise takes minus than one minute to simulate, proved to be a strong choice to make a conceptual design and also to optimize the acs WT blade shape.

The PNoise is open source which gives flexibility to anyone to customize it according to their needs and is possible to implement other methods and tools to improve the researches and contribute to the wind industry in order to mitigate noise from WTs.

*The group formed by the authors and are also presenting other works to review [7] and develop new WT blades and [21]

References

[1] (2010). Guideline for the certification of wind turbines. *Germanischer Lloyd, Hamburg*. www.gl-group.com/GLRenewables.

[2] Bampanis, G., Roger, M., Ragni, D., Avallone, F., and Teruna, C. (2019). Airfoil-turbulence interaction noise source identification and its reduction by means of leading edge serrations. In *25th AIAA/CEAS Aeroacoustics Conference*. American Institute of Aeronautics and Astronautics.

[3] Brooks, T., Pope, D., and Marcolini, M. (1989). Airfoil self-noise and prediction. 1218.

[4] Brooks, T. F. and Marcolini, M. A. (1986). Airfoil tip vortex formation noise. *AIAA Journal*, 24(2):246–252.

[5] Broux, F. (2008). 3d extension of the wind turbine noise prediction tool iagnoise. Master's thesis, University of Stuttgart.

[6] Faria, A., Pimenta, M., Jr., J. S., and Rodriguez, S. (2018). Selection of a leading edge noise prediction method for PNoise. *The Academic Society Journal*, pages 214–223.

[7] Faria, A. M., Jr., J. Y. S., Rodriguez, S., Barreto, J. P. S., and Pimenta, M. (2021). A review on the development of airfoils for wind turbine blades. *INCE Europe - Wind Turbine Noise*.

[8] Faria, A. M., Jr., J. Y. S., Rodriguez, S., and de Mattos Pimenta, M. (2020). A rapid distortion theory-based airfoil turbulent inflow noise prediction method. *Journal of the Brazilian Society of Mechanical Sciences and Engineering*, 42(8).

[9] Faria, A. M., Jr., J. Y. S., Rodriguez, S., and Pimenta, M. (2019a). Airfoil le noise prediction supplement for pnoise code. *INCE Europe - Wind Turbine Noise*.

[10] Faria, A. M., Jr., J. Y. S., Rodriguez, S., and Pimenta, M. (2019b). Discussion on turbulence inflow noise prediction. *Brazil Windpower*.

[11] Fuglsang, P. and Bak, C. (2004). Development of the risø wind turbine airfoils. *Wind Energy*, 7(2):145–162.

[12] Glegg, S. A. L. (1987). Significance of unsteady thickness noise sources. *AIAA Journal*, 25(6):839–844.

[13] Hansen, M. (2008). *Aerodynamics of Wind Turbines: second edition*. Earthscan, 2 edition.

[14] Herrig, A., Wurz, W., Krämer, E., and Wagner, S. (2008). New cpv-results of naca 0012 trailing-edge noise.

[15] Jakobsen, J. and Andersen, B. (1993). Aerodynamical noise from wind turbine generators - experiments with modification of full scale rotors.

[16] Jr., J. Y. S. (2016). *Trailing-Edge Noise – Development and Application of a Noise Prediction Tool for the Assessment and design of Wind Turbine Airfoils*. PhD thesis, University of Sao Paulo - EPUSP. http://www.teses.usp.br/teses/disponiveis/3/3150/tde-14032017-140101/pt-br.php.

[17] Jr., J. Y. S. and de Mattos Pimenta, M. (2015). Displacement thickness evaluation for bpm-type airfoil-te noise prediction model. *INCE Europe 6th International Conference on Wind Turbine Noise, Glasgow*.

[18] Jr., J. Y. S. and de Mattos Pimenta, M. (2016). Displacement thickness evaluation for semi-empirical airfol trailing-edge noise prediction model. *Journal of the Brazilian Society of Mechanical Sciences and Engineering*.

[19] Jr., J. Y. S. and de Mattos Pimenta, M. (2017). Verification and validation of the "pnoise" airfoil trailing-edge noise prediction module inside "qblade". *INCE Europe 7th International Conference on Wind Turbine Noise, Rotterdam.*

[20] Jr., J. Y. S., de Mattos Pimenta, M., Faria, A. M., and Rodriguez, S. (2018). Determination of local flow conditions and preliminary investigation on the validity of the PNoise code for large-size wind turbine. *Journal of the Brazilian Society of Mechanical Sciences and Engineering*, 40(10).

[21] Jr., J. Y. S., Pimenta, M., Faria, A. M., and Rodriguez, S. (2021). Developing new airfoils for larger wind turbine blades. *INCE Europe - Wind Turbine Noise.*

[22] Jr., J. Y. S., Rodriguez, S., Faria, A. M., and de Mattos Pimenta, M. (2019). The quasi-3d te rotor noise prediction tool of the pnoise code. *INCE Europe 8th International Conference on Wind Turbine Noise, Lisbon.*

[23] Juknevicius, A. and Chong, T. P. (2018). On the leading edge noise and aerodynamics of thin aerofoil subjected to the straight and curved serrations. *Journal of Sound and Vibration*, 425:324–343.

[24] Kamruzzaman, M., Lutz, T., Nuebler, K., and Krämer, E. (2011). Implementation and verification of an aeroacoustic wind turbine blade analysis tool. *Fourth International Meeting on Wind Turbine Noise, INCE Europe, Rome.*

[25] Moriarty, P. and Migliore, P. (2003). Semi-empirical aeroacoustic noise prediction code for wind turbines. Technical report.

[26] Saab Jr, J. (2015). Displacement thickness evaluation for semi-empirical airfoil trailing-edge noise prediction model. *Journal of the Brazilian Society of Mechanical Sciences and Engineering*, 38.

[27] Saab Jr, J. (2016). Trailing-edge noise: development and application of a noise prediction tool for the assessment and design of wind turbine airfoils.

[28] Stefan Oerlemans, P. F. (2012). Low-noise wind turbine design.

[29] Vargas, L. d. (2008). Wind turbine noise prediction. *Instituto Superior Técnico, Lisboa.*

[30] Zhu, W. J. (2004). *Modelling of noise from wind turbines.* PhD thesis, Technical University of Denmark. Lyngby, Denmark.

Appendices

A. Quasi-3D Blade Flow Diagram

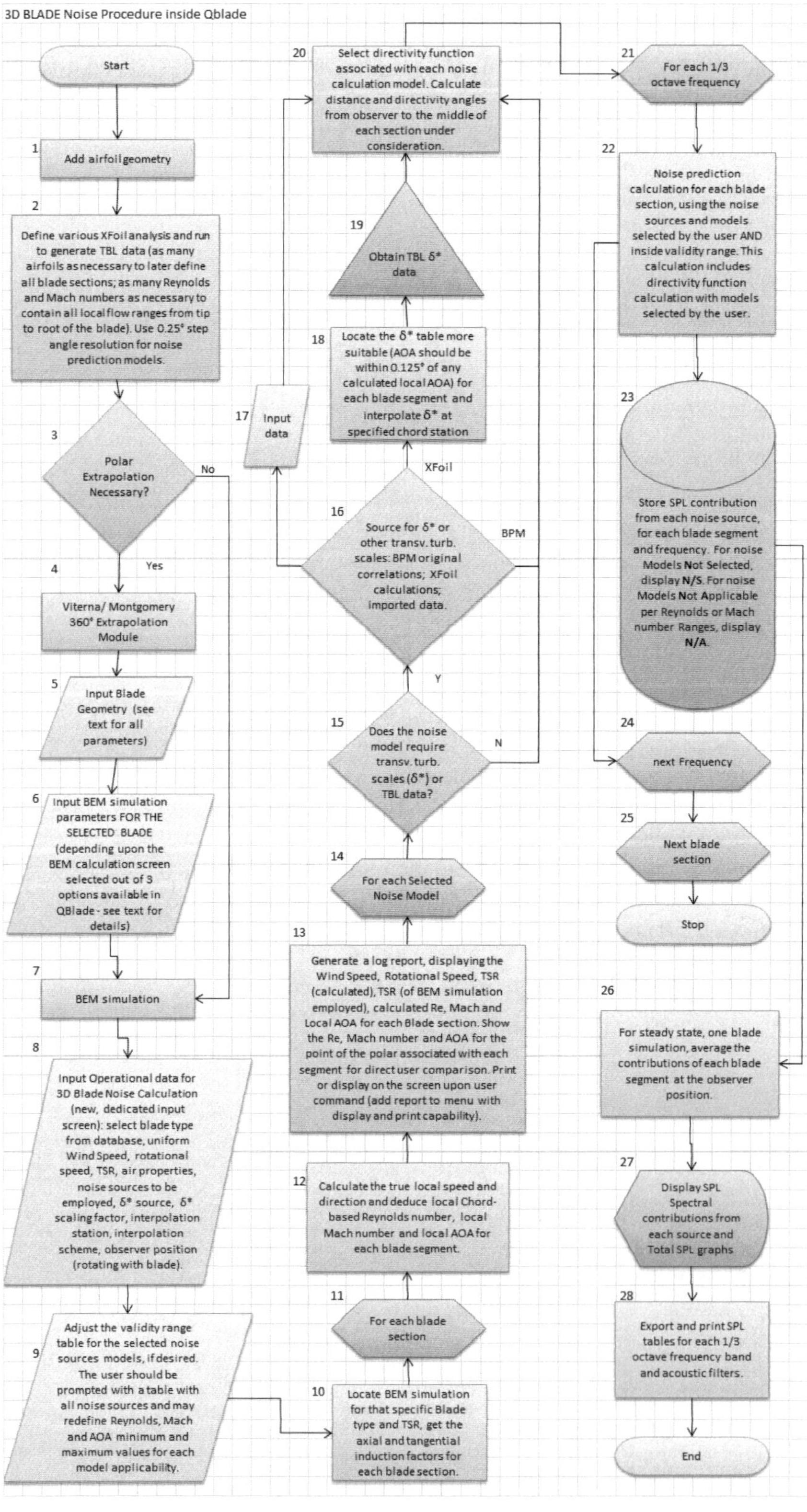

Fig. A.1 Quasi-3D blade noise procedure for the PNoise module. Source: [22], p. 8, fig 2.

B. Quasi-3D Rotor Flow Diagrams

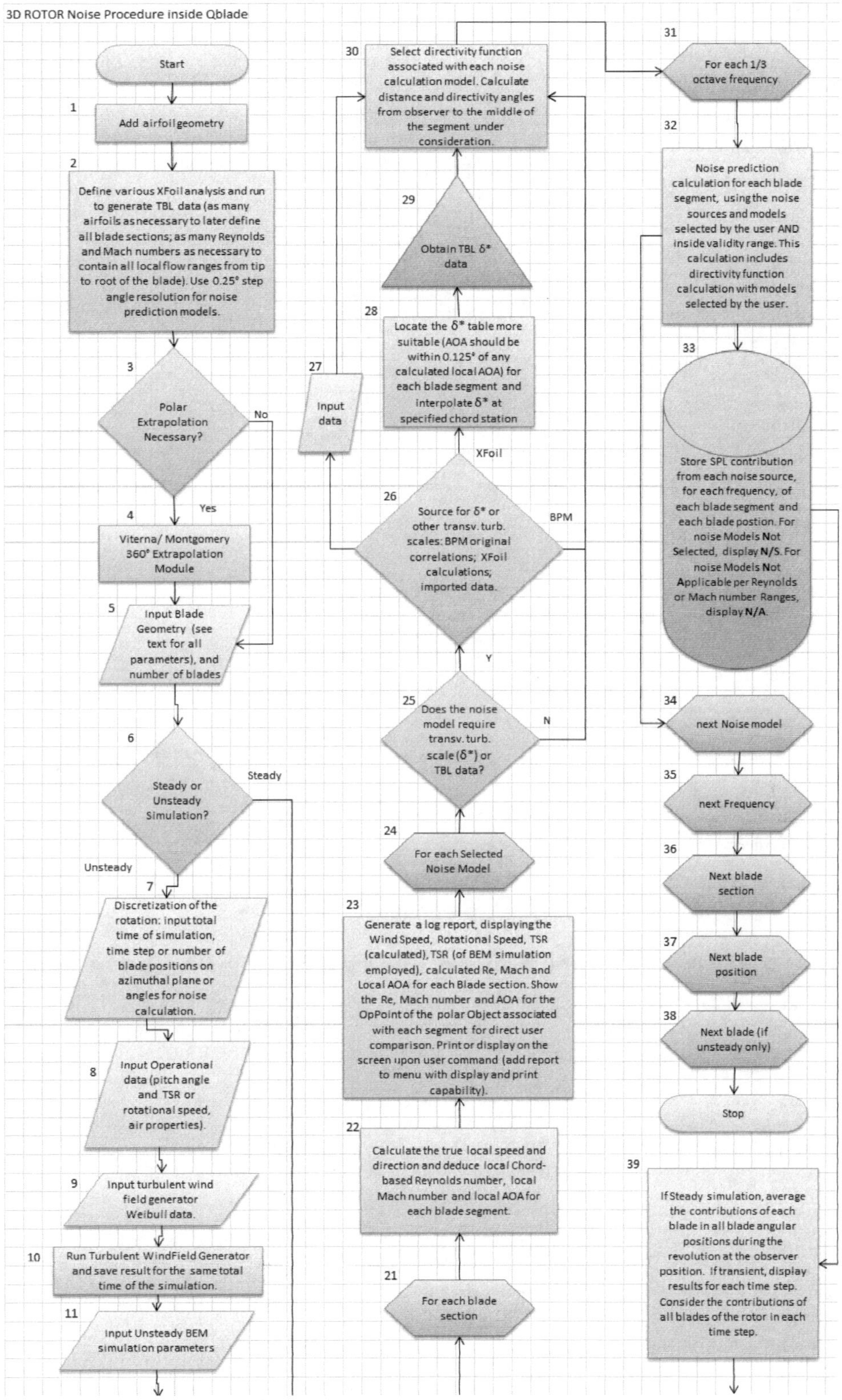

Fig. B.1 Quasi-3D rotor noise procedure for the PNoise module.

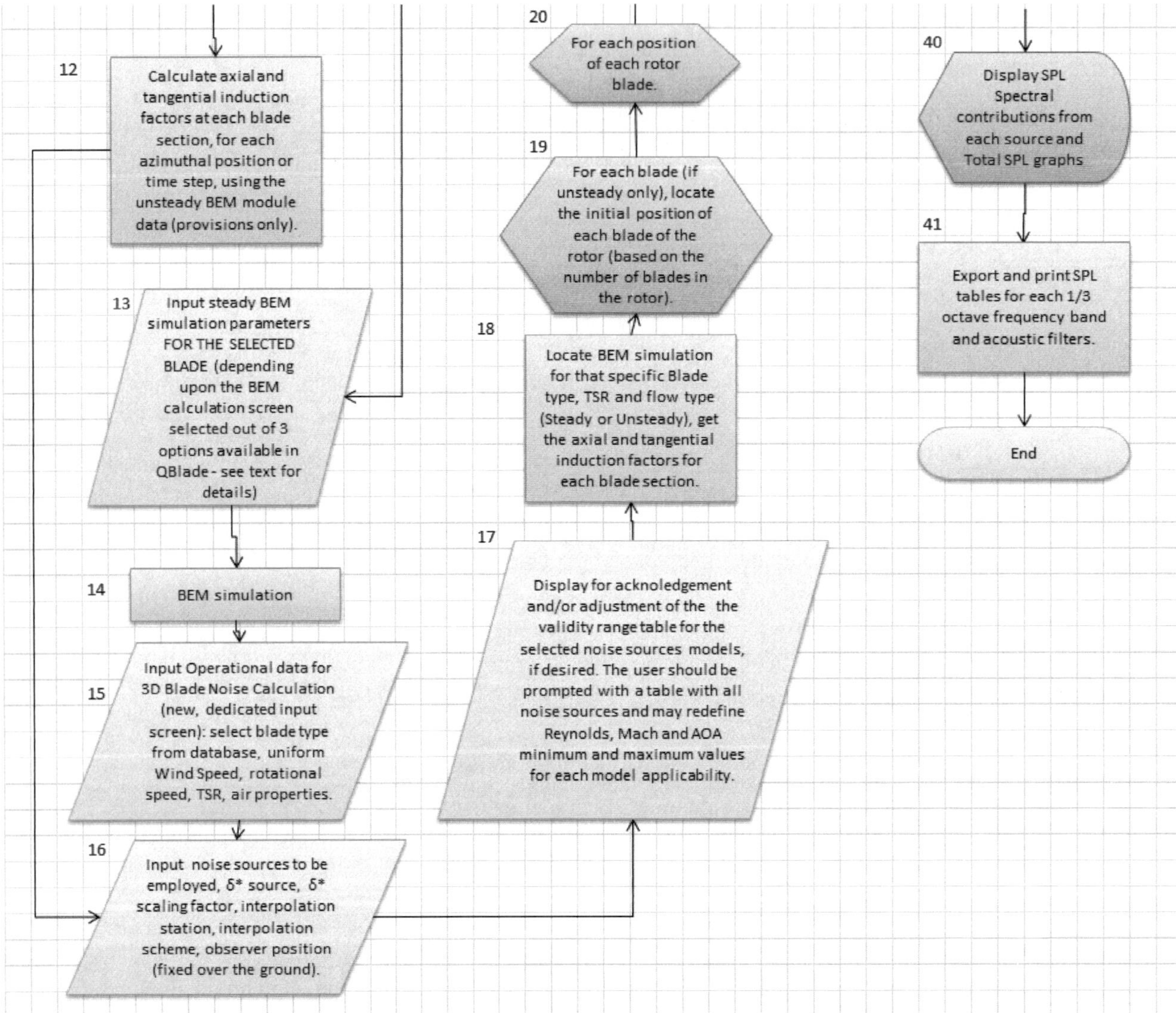

Fig. B.2 Quasi-3D rotor noise procedure for the PNoise module.

C. Model Validity

To guarantee the physical fidelity of the simulation, it is necessary to stay in accordance with a validity-range for all the active noise sources models. The prediction must be in accordance with the following values of AOA, Mach, and Reynolds numbers according to the Table C.1.

Noise model	Reference	Reynolds Upper	Reynolds Lower	Mach Lower	Mach Upper	AOA (abs) Upper
BPM TE	[3]	2.400,000	600,000	-	0.21	19.8º
RDT-Modified LE	[2, 23]	1,300,000	600,000	0.06	0.18	-
	[14]	2,400,000	-	-	0.21	-
BPM LBL-VS	[3]	1,600,000	45,000	0.09	0.21	15.4º
BPM Tip Vortex	[3, 4]	1,300,000	120,000	0.12	0.21	14.4º
BPM TE-Blunt-VS	[3]	2,600,000	1,100,000	0.12	0.21	6.1º

Table C.1 Validity range according the mentioned references for each noise source.

INCEEUROPE

9th International Conference
on
Wind Turbine Noise
Remote from Europe – 18th to 21st May 2021

Developing New Airfoils for Larger Wind Turbine Blades

Joseph Youssif Saab Jr. - Mauá Institute of Technology (IMT-Brazil) saab@maua.br [1]
Marcos de Mattos Pimenta - São Paulo State University (Poli-USP) mpimenta@usp.br

Summary

A review on performance data published for selected airfoils of horizontal-axis wind turbines (HAWT) suggests that they were designed for equipment smaller than the current standard and future trends in HAWT sizes. The resulting gap in airfoils readily available for HAWT research and development is further stretching as new equipment is designed for diameters of 200 m and above, which imply in much higher local[2] Reynolds number flows. This text briefly illustrates the development process and preliminary results of new, dedicated HAWT airfoils designed for the 7 to 9 million Reynolds number and 0.14 to 0.21 Mach number flows, typical of 100-m diameter HAWTs, achieved with the use of the open-source *QBlade* platform and its integrated *PNoise*, TE noise code. The use of the combined codes enabled the design of a new family of airfoils based on the Somers/NREL S830 main airfoil and the first estimates predict preserved aerodynamic characteristics and comparable or lower Trailing-Edge noise emission performance under higher Reynolds and Mach numbers. It is implied that the use of these tools in a combined form might lend themselves to the design of new families of larger airfoils, suitable for 200+ m diameter rotors, with both aerodynamic and aeroacoustic design requirements.

1. Introduction

A review on the major airfoils developed for large-size horizontal axis wind turbines over the last twenty years [1] seem to point out that airfoils in public domain were designed for local Reynolds number flows up to 6 million and with major design restrictions based on aerodynamic requirements, such as high lift coefficients, low sensitivity of the transition point with regards to the coefficient of lift, high aerodynamic efficiency, and prescribed pitching moment and thickness as further restrictions. However, none of them were apparently designed with requirements based on noise emissions.

While many other airfoils certainly include noise emission as a design requirement, they are not in public domain, like the Somers/NREL "S" family of airfoils [2], [3], [4], which were designed with Trailing Edge (TE) noise restrictions in mind. Also, the Somers/NREL "S" family of airfoils

---

[1] Corresponding author.
[2] Chord-based Reynolds number.

were designed for blades with a recommended maximum span of 25 m and local Reynolds numbers flows around 4 million.

By designing some conceptual HAWT equipment with rotor diameters of 100, 180 and 220 m [5] with the aid of a purpose-build BEM code based on [6], and by analysing the induced flow field, it was possible to ascertain that local flow conditions for large HAWT rotors are more likely of the order displayed in Table 1.

Table 1 – Estimation of more realistic local flow conditions for HAWT main airfoils, at 85% blade span, for large diameter HAWT. Source: [5]. For assumptions, airfoils, chord distribution, twist and angular speed, please see reference.*

HAWT	Radius @ 85% blade span [m]	Chord-Based Reynolds number	Mach number
Poli-100	42.5	7,700,000	0.21
Poli-180	76.5	13,000,000	0.19*
Poli-220	93.5	15,500,000	0.19*

With a better idea of the real operating flow conditions around large-size equipment, it was decided to investigate the feasibility of deploying the integrated *QBlade* [7], [8] and PNoise [9] codes in order to (i) make a sensitivity study on the shape of the airfoils with regards to TE noise emission, and (ii) try and propose new airfoils designed under both aerodynamic and TE noise requirements for a 100 m diameter HAWT, under more realistic flow conditions.

2. Methodology

This section describes the models embedded in the codes and their limitations, the design strategy, the workflow, and the design requirements laid out for the new airfoil family.

2.1 Codes, Models and restrictions.

The *PNoise* TE noise code is based on a modified BPM TE noise model [10] with the *XFLR5* [11] providing the turbulent boundary layer (BL) data, both integrated inside the unique wind-turbine-design, graphical interface and user-friendly environment provided by the *QBlade* software [7], [8], [12], [13].

In the original BPM research paper [10], the TE noise model was introduced and validated for turbulent (tripped) flow up to $Re_C \leq 1.5 \times 10^6$, $M < 0.21$ and 19.8^0 AOA (angle-of-attack). All experiments were made for the NACA 0012 airfoil, based on the acoustic spectra measured in this range.

The *PNoise* code implementation verifications [9] were accomplished within the original limitations of the BPM model. However, since in this new form it has the flexibility for assessing the TE noise of any airfoil geometry[3] and at any prescribed Reynolds and Mach number flows, it should be stressed that the use of a model beyond the original validation scope is a *prediction* [14], and implies no guarantees on the predicted results prior to experimental validations.

[3] The PNoise is not restricted to the NACA0012 airfoil geometry.

A NREL validation study for the BPM model [15] showed good agreement of the BPM prediction when compared with data taken from a series of wind tunnel tests performed at the NLR [16]. The comparison was made at $M = 0.21$ and AOA ranging from 0^0 to 13.1^0. The agreement was good for frequencies near 3 kHz but for lower frequencies (~800 Hz) the differences found were up to 6 dB. The study did not expand the validation range of the model.

Doolan and Moreau [17] have plotted SPL spectra as a function of Strouhal number for some experiments against BPM predictions. For the case of the IAG Wind tunnel data [18] at chord-based Reynolds number $Re_C{\sim}2.9 \times 10^6$, it has shown good agreement with BPM prediction at $M = 0.20$, for peak Strouhal number and higher frequencies. However, by verifying the IAG Wind Tunnel Data made available by [18], it seems that the higher chord-based Reynolds number of the experiment was limited to $Re_C{\sim}2.4 \times 10^6$.

2.2 Design Philosophy and Strategy.

Two important questions were addressed at the beginning of the design process:

- Did the semi-empirical modified-BPM TE noise model represent the aeroacoustic phenomena enough, from the physical point of view, so as to provide guidance for designing quieter and efficient airfoil geometries?
- How should the new airfoil geometry be designed for reduced TE noise emission, based on the sources modeled by the code and their relative acoustic strength, obtained from preliminary and validation simulations, without disregard to other known noise sources not modeled into the code?

Based on the BPM TE model validity previous discussed, it was assumed, as a hypothesis, that the mathematical model represented the underlying physical phenomena[4] in sufficient detail to allow comparison of yielded results from different geometries in similar flows. Thus, a two-way, "causality" relation between geometry and the TE noise emission model is assumed.

Although the *PNoise* model was, at the time of this airfoil development, equipped solely with the TE noise model, good design practices were also adopted in the procedure for prevention or attenuation of mechanisms responsible for other self-noise sources plus inflow noise. On the one hand, the adoption of a more extensive list of restrictions than those controlled by the code prevented the use of an automated optimization algorithm for the development of the new airfoil geometry, forcefully restricting the design of experiment (DOE) to fewer cases. On the other hand, the careful profile designing with special concerns to smoothness and pressure gradient control to avoid other noise sources would increase the chances of producing an airfoil with practical application qualities. Also, the manual iteration process would be supported and made viable by the detailed sensitivity analysis performed.

The initial application of the aerodynamic and aeroacoustic analysis tools at hand to the S830 baseline airfoil for both the original NREL design point (Case #100 - Reynolds number of 4 million) and the new operating point (Case # 103, Reynolds number of 7.7 million), resulted in the benchmark values shown in Table 2.

---

[4] After all, the BPM TE noise model is made up of three independent sub-source contributions, carefully modelled to predict peak frequency and spectral roll-off noise.

Table 2 – Initial results from the numerical aerodynamic and T.E. noise analysis of the S830 baseline airfoil at the design and new operating conditions.

Airfoil	S830, original	S830, original
Design point	NREL design point	Poli-100 WT @ 85% span
Reynolds number:	4,000,000	7,700,000
Mach number:	0.21	0.21
Airfoil / case number	**#100**	**#103**
cl max:	1.65	1.72
Angle, cl)max	18.0	18.4
Max cl/cd value:	77.1	91.7
Angle, cl/cd)max	5.7	6.1
cl @ cl/cd)max:	1.16	1.20
cd @ cl/cd)max:	0.0150	0.0131
cm @ cl/cd)max:	-0.146	-0.144
Tripping (%chord), upper/lower surface:	5%/10%	5%/10%
AOA test range:	-3 to +20	-3 to +20
Convergence range	-3 to +20	-3 to +20
OASPL (dB), @ (cl/cd)max.	94.02	94.30
Peak Frequency:	80 Hz (87.7 dB)	100 Hz (87.7 dB)
OASPL (dB(A)):	80.30	81.00
OASPL (dB(B)):	88.20	89.30
OASPL (dB(C)):	93.40	93.60
SPL_α (dB)	92.10	92.80
SPL_s (dB)	89.20	88.80
SPL_p (dB)	77.60	77.00

The figures shown for case #100 in Table 2 represent the estimated aerodynamic and noise goals for the S830 airfoil at the original NREL/Somers design conditions. When compared to case #103, which better represents the higher speed flow condition for the larger WT of the Poli-100 WT, it is possible to see the aerodynamic performance improvement to be expected from the increase in Reynolds number, for both $(cl/cd)_{max}$ and cl_{max}. Also, the unweighted OASPL level is expected to increase very little, by 0.28 dB under the higher speed flow, while the A-weighted SPL is expected to increase more sensibly, by 0.70 dB.

By analyzing the three independent noise sources of case #103 that make up the TE noise in the BPM model, the contributions are estimated to be 92.8 dB, 88.8 dB and 77.0 dB from the angle-of-attack, suction side and pressure side sources, respectively.

For this and consistently over the many simulations accomplished, the angle-of-attack dependent TE noise source has appeared in this model to be the dominant TE noise source. The suction side contribution is 4.0 dB less than the largest source, adding only 1.4 dB to the total SPL. The pressure side contribution is 15.8 dB less than the largest source, adding approximately 0.2 dB (virtually no contribution) to the total SPL. The log sum of the three sources resulted in the displayed total SPL of 94.3 dB.

It is also possible to see that the C-weighted overall SPL (SPL(C)) minus the A-weighted overall SPL (SPL(A)) is significant (13.1 dB for case #100 and 12.6 dB for case #103), which implies that a significant amount of the noise energy is emitted in the low frequency range, which is supportive of Bowdler and Leventhall [19] findings.

Also, since SPL(A) increment is larger than the overall SPL increase, it could be possible to infer that, for AOAs close to 6°, some acoustic energy migrates from infrasound to audible frequencies with the increased Reynolds number, provided the math model has physical adhesion.

Although audible frequencies are the most targeted for noise reduction due to direct human annoyance concerns, the impact of infrasound in human health may be also significant, with noise propagation occurring over large distances and the potential for high annoyance effects, which may include the induction of window and door panes vibration in audible frequencies or, more significant, physiological effects, such as those discussed by Schomer et al. [20].

This first benchmark analysis also raised some additional questions which helped refining the strategy for the direct design of quieter airfoils while preserving the aerodynamic performance:

- Is it possible to reduce the intensity of the largest TE noise source, the one associated with the angle-of-attack (AOA)?
- Is it possible to attenuate the suction side noise while transferring some of the airfoil loading (with opposite signal) to the pressure side, so that noise is reduced but overall aerodynamic efficiency is preserved?
- Is it possible to reduce at the same time the unweighted sound pressure level (SPL) and the A-weighted SPL (SPL(A)), for the Poli-100, 85% span flow conditions?

2.3 Workflow

Below it is shown the workflow sequence for the direct[5] airfoil design method selected for the development.

2.3.1 Baseline calculations:

- Select airfoil baseline geometry for 85% chord, with fixed 21% relative thickness due to structural considerations for main HAWT airfoils.
- Set up a WT blade and rotor geometry, plus operating conditions, as a reference.
- Using the BEM method, calculate the operating conditions at 85% length.
- Test and save aerodynamic performance data of baseline airfoil geometry at required Reynolds and Mach numbers, using the XFoil, for future reference. Since the blades operate in dirt (also noisier) condition most of the time, the primary focus is on performance under tripped flow.
- Test and save the selected aerodynamic key performance indicator (the Power Coefficient), using the QBlade code.
- Test and save aeroacoustic performance data of the baseline airfoil geometry at the required flow conditions, using the modified-BPM method implemented (PNoise).

[5] As opposed to the *inverse design technique*, accomplished by obtaining the desired geometry from the previous specified Euler number distribution over the upper and lower surfaces of the airfoil.

2.3.2 2D Design Loop:

- Modify the selected baseline airfoil geometry, working on variations of the most relevant parameters (while keeping the specified maximum thickness prescribed), using the sensitivity study performed for guidance.
- Check 2D aero and acoustic performance. Do they meet the requirements? Is performance improved over last iteration?
- Next geometry modification.

2.3.3 3D Verification:

- Design a variation of the baseline WT blade, with the new airfoil in selected blade sections replacing the baseline airfoil. Verify the rotor Power Coefficient and energy conversion performance and compare to baseline data.

2.4 Airfoil design requirements

Based on extensive literature review, the design requirements for the new airfoils were laid out as shown in Table 3.

Table 3 – Airfoil design requirements selected for a main airfoil with 21% thickness.

#	Requirement	Reason	Reference
1	Airfoil suitable for a large size HAWT, outboard section @85% radius Reynolds and Mach numbers.	WT diameter increase trend displayed over time. Noise sources are concentrated on outer radius, but not at the very tip.	[21]
2	Suitable for pitch-controlled WT.	Deep stall noise may be a dominant noise source and there is currently no mature noise prediction model for this regime. It must be avoided by a pitch control blade, which, except for the transient regime, will drive the blade to operate far from stall conditions. This is the current trend for machines with higher efficiency and power coefficients, C_p.	Current trends for quiet operation control options.
3	The LE thickness (radius) should be significant.	To allow for smooth stall characteristics, to avoid deep stall in low angles and to reduce inflow noise.	[22]. [3]. [23]
4	Relative insensitive to roughness.	According to [2], this is necessary in order to keep efficiency in the presence of LE soiling and debris accumulation.	[2]
5	Early laminar to turbulent transition in a controlled point.	To reduce overall SPL; to eliminate tonal noise from the pressure side; to avoid laminar bubbles. This will be imposed by early tripping of the BL.	[24], p. 84
6	To have low profile drag, but to preserve airfoil thickness determined by structural requirements.	Compatible with the structural stiffness required at the radial position in which the airfoil will be implemented. The maximum thickness was set to 21%.	[25].
7	The pitching moment coefficient about ¼	In order to limit aerodynamic torsion loads, which could drive the weight of the	[3].

	chord should be close to -0.145.	blade up during the subsequent structural sizing procedure.	

3. Sensitivity analysis

A "splined S830" airfoil - a S830 airfoil reproduced with the aid of spline curves, suitable for later parametrical modification - was validated against the original S830 performance parameters and then systematically subjected to independent, limited-amplitude parametrical changes incorporated into the baseline configuration, with subsequent observation of the aerodynamic and aeroacoustic response to each alteration:

- LE radius variation.
- Upper surface forward side convexity variation "forward convexity-FC".
- Camber variation.
- Flap angle variation.
- Maximum thickness position variation.
- Upper surface aft side convexity variation "aft convexity-AC".
- Trailing Edge gap.

The results were systematically compared with the calculated performance of the original S830 airfoil at original design conditions (case #100) and to the splined S830 baseline configuration, at the larger Reynolds number of 7.7 million (case #103).

Due to the extensive nature of the analysis performed only some of the outputs of the sensitivity analysis are discussed here. For the good practices adopted throughout the XFLR5 and PNoise simulation, detailed input for each step and the complete results, please refer to the original reference [5].

The case numbers from 105 through 121, with the associated parametrical change for each case are shown in Table 4 and the analysis results are shown in Figure 1.

Table 4 - Case # identification for the parametric sensitivity study

Case #	Description of parametric change imparted to the baseline S830 geometry.
105	LE radius doubled.
108	LE radius halved.
110	Less upper forward side convexity
111	More upper forward side convexity
112	Decreased camber
113	Increased camber
114	Less flap angle
115	More flap angle
116	Thickness shifted towards LE
117	Thickness shifted towards TE
118	Less upper aft side convexity
119	More upper aft side convexity
120	Smaller TE gap
121	Larger TE gap

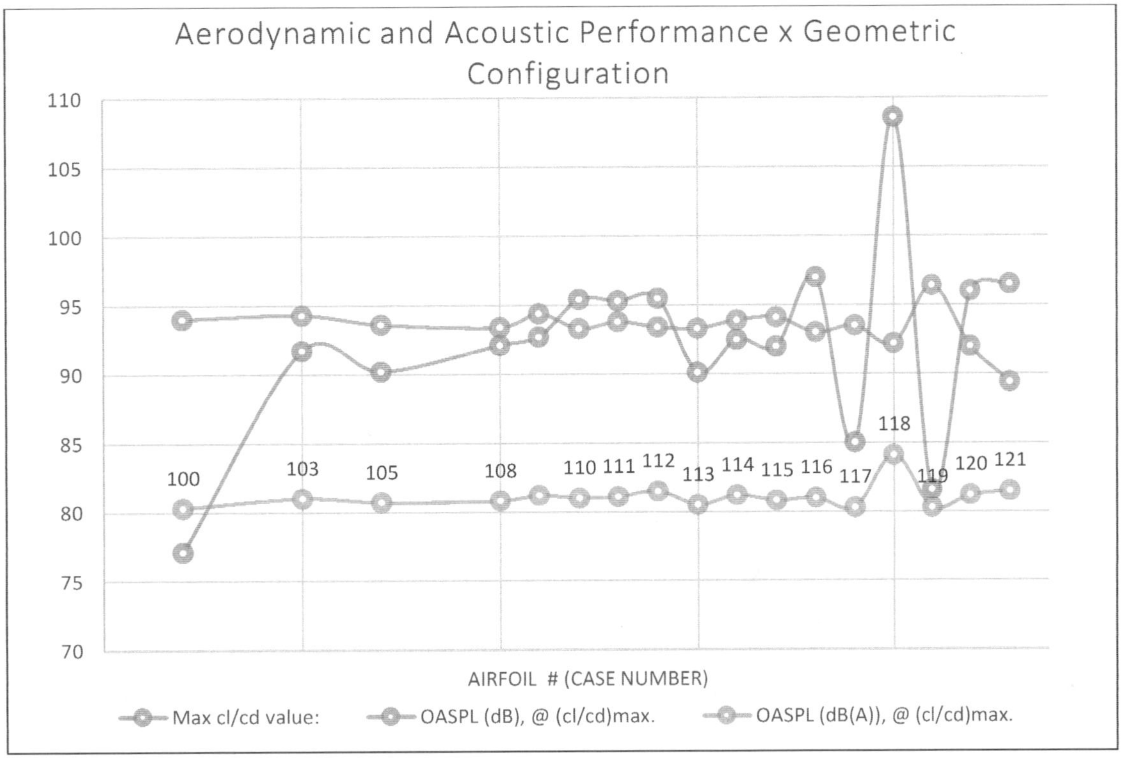

Figure 1 – Aerodynamic Efficiency (blue circles) and Overall Sound Pressure Levels (orange and grey circles) at the maximum efficiency points for the parametric study cases. Cases 100 and 103 are the reference cases, at Reynolds Numbers of 4 million and 7.7 million, respectively.

Figure 1 shows that, throughout the range of cases simulated, no reduction in the SPL(A) was noticed during the sensitivity analysis. Although larger reductions in SPL(A) would be most desirable, it turned out to be physically unlikely to happen according to these findings: the geometry size (chord, wetted edge size) and flow conditions (AOA, velocity, turbulent boundary layer displacement thickness) of the cases presented have led to peak frequencies close to the 100 Hz 1/3 octave band, i. e., close to the lower end of the audible range. Since the SPL(A) scale is attenuated up to the 800 Hz band and it is particularly buffered at the 100 Hz band (-19.1 dB), even significant SPL reductions near this peak frequency would not translate into significant SPL(A) reductions.

As a confirmation of this hypothesis, the calculation of the peak frequency for an AOA below the switching angle was done explicitly. In the BPM model, the peak-frequency is determined by the peak-Strouhal number, St_1, as a function of the Mach number:

$$St_1 = 0.02 \cdot M^{-0.6} \qquad (3-1)$$

For a Mach number of 0.21, a local flow velocity of $72.32\, m/s$ and a displacement thickness of $3.16 \times 10^{-2}\, m$, evaluated with the aid of XFLR5 at the suction side of the reference case #109, at 6°, the resulting peak-frequency was:

$$St_1 = 0.02 \cdot (0.21)^{-0.6} = 0.0510 = {f \cdot \delta^*}/{U} \qquad (3-2)$$

$$f \sim 117\ Hz \qquad (3-3)$$

This fact confirms the argument that, for such airfoil and flow scales, a “quiet profile”, as far as TE noise is concerned, is probably an airfoil with reduced sound pressure level emissions in the lower frequencies of the audible range and possibly in the infrasound range, only.

Further, as the AOA is increased towards the deep stall region, the turbulence scale is also increased, with more noise energy being emitted in the lower frequencies and infra-sound portion of the spectrum. This is compatible with another conclusion of the BPM experiments [10], p. 3, that for mildly separated flows, the dominant noise was emitted from the TE area, while for deep stall flow, the noise was radiated from the whole chord. This finding was also consistent with the findings of Fink and Bailey [26], who introduced the concept of *universal noise spectrum*.

Finally, the peak SPL(A) level was observed for case #118, with an increase of 4 dB (A) above the average level. This may only be achieved through a significant BL thickness reduction ($\sim 50\%$), which indicates a decrease in the turbulence vertical scale with the associated increase in the eddy frequencies. This would also explain the spectrum peak frequency shift up, towards the 200 Hz band in this particular case. This BL layer thinning effect was achieved due to a significant decrease in camber derived from the smoothed convexity on the upper aft side of the #118 airfoil.

These preliminary results suggested that overall TE noise reduction was possible through geometry manipulation; however, TE noise reduction in the audible range for this scale of chord and flow (i.e., for this vertical turbulence scale) would be very limited at the best, and would probably have to be traded for unweighted SPL.

With all the information derived from the sensitivity analysis performed, it was possible to try the design some airfoils deriving from a step-by-step combination of the parametrical variations, and associated amplitudes, that proved successful at producing:

- Aerodynamic efficiency comparable to case #109 but with reduced unweighted overall SPL.
- Aerodynamic efficiency comparable to case #109 but with reduced audible SPL.
- Aerodynamic efficiency comparable to case #100 (original aerodynamic requirements) but emitting less noise than case #109.

Many combinations were tried in this case, and the result proved sensitive to the order of application of the modifications, which would make an automated optimization process quite challenging, prone to finding many local minimums.

4. Results

4.1 The SP4621 HP, a “Silent Profile” aimed at high aerodynamic performance.

This analysis has led to the development of the SP4621HP airfoil, with high aerodynamic efficiency and reduced unweighted, overall SPL in relation to the S830 operating in its design condition and also in the new condition.

The airfoil coding is intended to represent:

- 4 – The first digit stands for the nominal airfoil camber, as a % of the chord.
- 6 – The second digit stands for the nominal maximum camber position divided by 10, as a % of the chord.

- 21 – The maximum nominal airfoil thickness, expressed as a % of the chord.
- HP – High Performance WT airfoil.

From the twenty or so acceptable airfoil combinations produced, the SP4621HP stood out for revealing a high aerodynamic performance standard, with $(cl/cd)_{max}$~110; cl_{max}~2.0 and almost 3 dB less overall SPL over the baseline case.

Other configurations were obtained with even lower overall SPL, but the coefficient of moment and/or the TE gap exceeded the specifications by a far margin, or the aerodynamic efficiency or lift coefficient were poor. The SP4621HP airfoil profile is shown in Figure 2, along with the S830 profile. It is possible to see the preserved thickness of the original airfoil, considered a fixed requirement. The developed airfoil showed larger $(cl/cd)_{max}$, cl_{max} and less unweighted TE noise than the S830 in any of the reference flow conditions.

In spite of the fact that the SP4621HP also has an operating point $cl(\sim 1.43)$ higher than the regular S830 $cl(\sim 1.2 - 1.3)$ range, it emits lower unweighted overall TE noise than the original S830 in both flow conditions.

Also, the new airfoil is expected to display superior performance concerning inflow-noise, although no advantages could be quantified with the model, which was restricted to TE noise capabilities at the time.

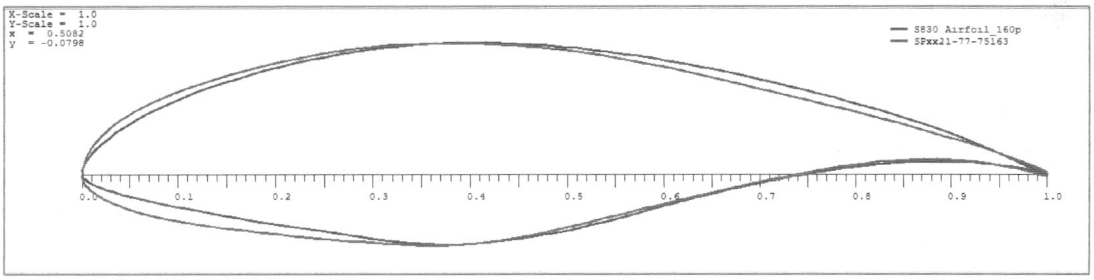

Figure 2 – The SP4621HP profile (red), with preserved original thickness in relation to the S830 airfoil (blue).

One of the limitations of the SP4621HP airfoil is that it is designed with a TE gap of 0.235% of the chord length and this implies, at the design position of 85% span, in thickness of ~3.5 mm only, which is a quite tight production thickness to specify for a 1.55 m chord airfoil. However, in order to avoid a tonal noise emission from the TE, also known as *airfoil singing* [27], the TE thickness/boundary layer thickness ratio must be less than 0.25 for all spanwise stations where the airfoil is to be employed and this might impose higher manufacturing costs to the design.

The clear advantages of the new design however, are in the reduction of SPL_α and SPL_s levels from the reductions in the correspondent sources. However, as discussed in other cases, the reduced camber decreased the BL thickness, which increased the TE noise frequency, as can be confirmed by the higher, 200 Hz peak in the spectrum. This resulted in increased A-weighted SPL which runs from 0.8 to 2.4 dB(A) larger than the baseline case. This fact once more confirmed the main tradeoff revealed in the course of this research.

Due to structural, production cost and noise concerns, the SP4621HP airfoil may be operated below optimum conditions, hereby called the *"Quiet Mode of Operation"*, or QMO, defined as the point of $(cl/cd) < (cl/cd)_{max}$ that corresponds to cl~1.2. This situation would not impose aerodynamic[6] loads larger than expected at the original design condition of the S830 airfoil, provided the WT is operating in a steady state. At QMO, the SP4621HP airfoil operates under partial loading with an AOA of 4.0°, as opposed to 6.2° in the maximum efficiency situation.

4.2 The SP4721LA.

The number coding for the airfoil designed for attenuated TE noise emission in the audible range is similar to the previous one, except for the suffix *"LA"*, suggesting "A-Level" attenuation.

During this particular development, the objective was to investigate the feasibility of designing an airfoil geometry with flatter ${cl}/_{cd}$ $versus\ AOA$ curve at the top. This would allow the airfoil to operate closer to the $(cl/cd)_{max}$ for a broader range of angles. Also, the plateau should start for a low AOA, which would allow the airfoil to operate at the QMO setting with little penalty in relation to the maximum efficiency AOA.

After some trials, it was found that the quick air acceleration over the suction side, with consequent formation of laminar bubble close to the LE, was one of the mechanisms capable of increasing the slope of the $\partial(cl/cd)/\partial\alpha$ curve, also known as $cl\alpha$. However, a "horn" appeared systematically at the polar diagram signaling the angles of formation and bursting of the bubble, with corresponding flow reattachment. Unfortunately, while this bubble may help reduce one of the sources associated with the TE noise, it may also originate another, unmodeled source, which is the laminar boundary layer vortex-shedding (LBL-VS), noise. However, despite the limitations of the low order numerical simulation, it was noticed that most of the flat-top $cl\alpha$ curve could be preserved by advancing the tripping position towards the LE, to 1% of the chord, at the suction side. The resulting effect was a dedicated airfoil with a somewhat less-smooth polar behavior, but with good TE noise attenuation obtained for a $cl{\sim}1.2$ at a low 4.5° AOA, which also corresponds to the $(cl/cd)_{max}$ angle. This is, so to describe, an airfoil for which the HP and the QMO modes of operations have merged together.

The main feature of the SP4721LA airfoil, whose outline is shown in Figure 3, is that it radiates less SPL(A) TE noise than the original S830 airfoil does at a 7.7 million Reynolds number flow.

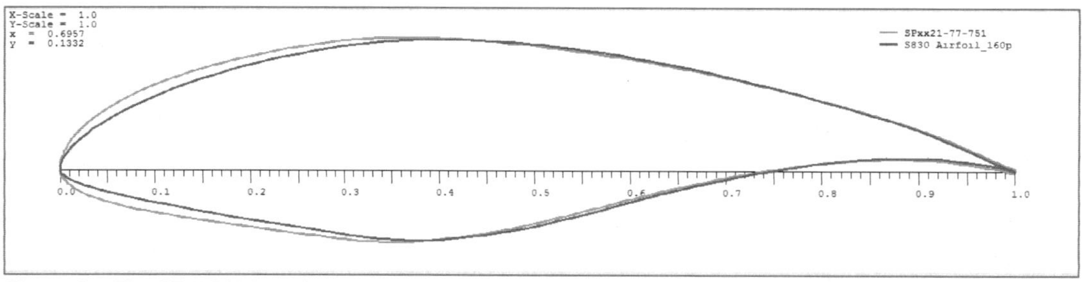

Figure 3 – The SP4721LA profile (green), with preserved original thickness in relation to the S830 airfoil (blue).

The SP4721LA has associated values of $(cl/cd)_{max}$ and cl slightly below those of the reference case #103, however, this happens in exchange for a significant 4.74 dB unweighted SPL reduction and 0.5 dB SPL(A) noise reduction.

[6] For larger blades, the main structural concern may be the loading derived from gravitational and centrifugal stresses, not necessarily aerodynamic [25].

4.3 Predicted SPL and spectra for the proposed airfoils.

Figure 4 shows that both airfoils proposed, the SP4621HP and the SP4721LA, have lower predicted overall noise emissions than the S830. Also, most of the noise supposed to be emitted in the lower frequencies, with noise in the audible range displaying much smaller levels for the specific chord (1.55 m), Mach (0.21) and Reynolds number (7.7 million) flow. All airfoils were simulated at $(cl/cd)_{max}$.

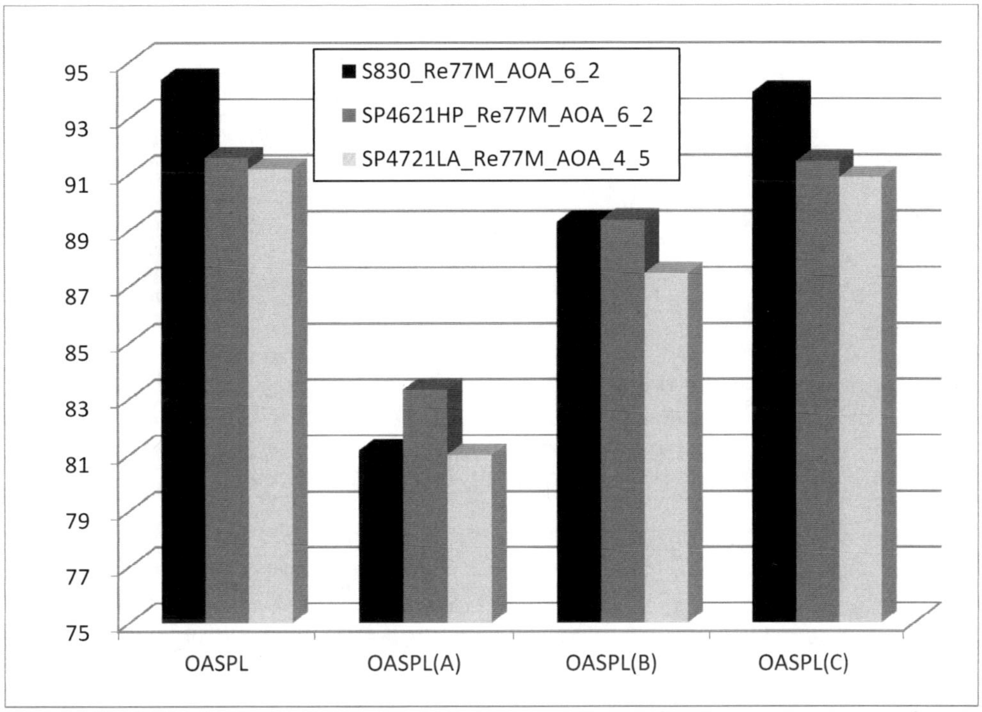

Figure 4 – Weighted and unweighted predicted overall SPL for the proposed airfoils, compared with the reference S830 airfoil.

Figure 5 shows that the baseline airfoil emitted SPL is dominated by the new geometries up to the frequency of 150 Hz. The SP4621HP has better aerodynamic characteristics than the S830 airfoil but higher noise in the 150 Hz to 1,000 Hz frequency range, while the SP4721LA has comparable noise emission in this frequency range. All airfoils have a similar TE noise emission behavior above 1,000 Hz, when analyzed by this model.

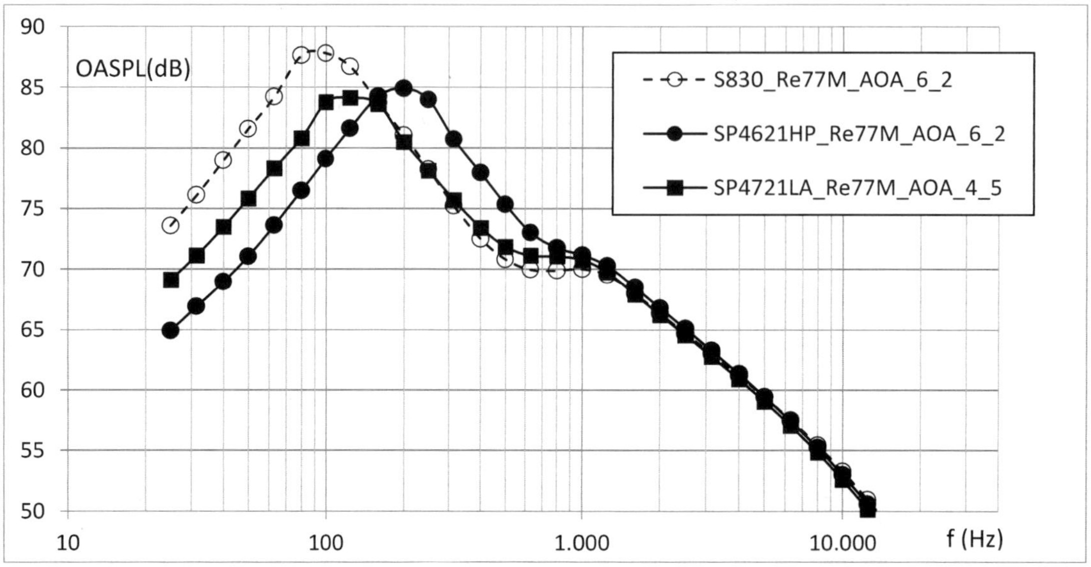

Figure 5 – OASPL numerical spectra predicted for the for the proposed airfoils, compared with the reference S830 airfoil

Figure 6 reveals higher SPL(A) levels predicted for the SP4621HP airfoil from 160 Hz through 800 Hz, while the SPL(A) for the SP4721LA should be close to the S830 for mid and high frequencies, with lower emission at lower frequencies.

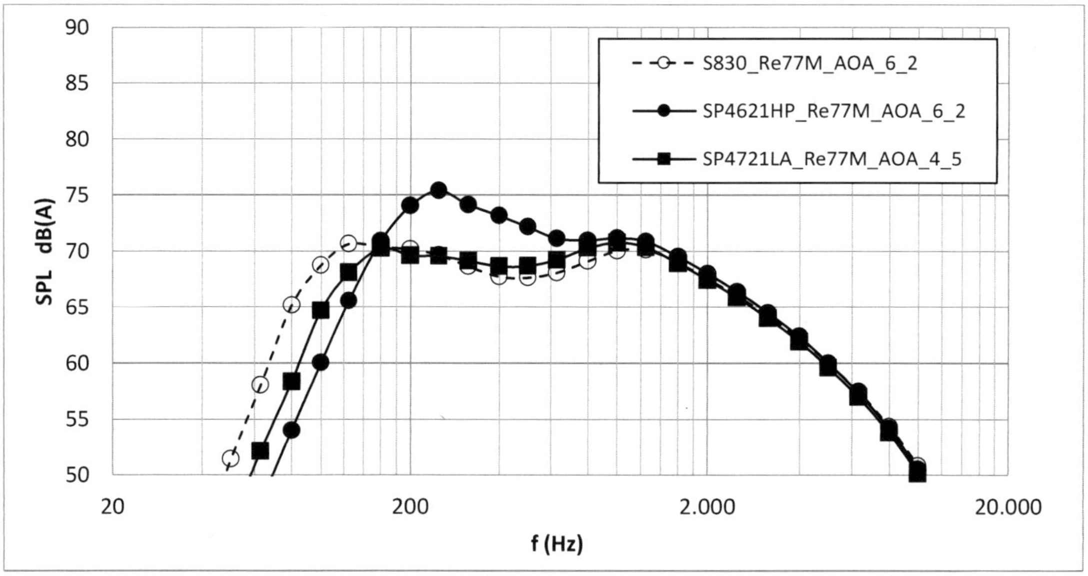

Figure 6 – Predicted A-weighted SPL spectra for the new and reference airfoils.

Figure 7 illustrates the impact of the contribution of the AOA-dependent source in the model (SPL_α) for two different noise abatement strategies: geometric redesign (SP4621HP) which results in a more pronounced horizontal shifts of the spectrum and off-peak operation (SP4621-QMO), which results in both significant horizontal and vertical shift of the reference spectrum.

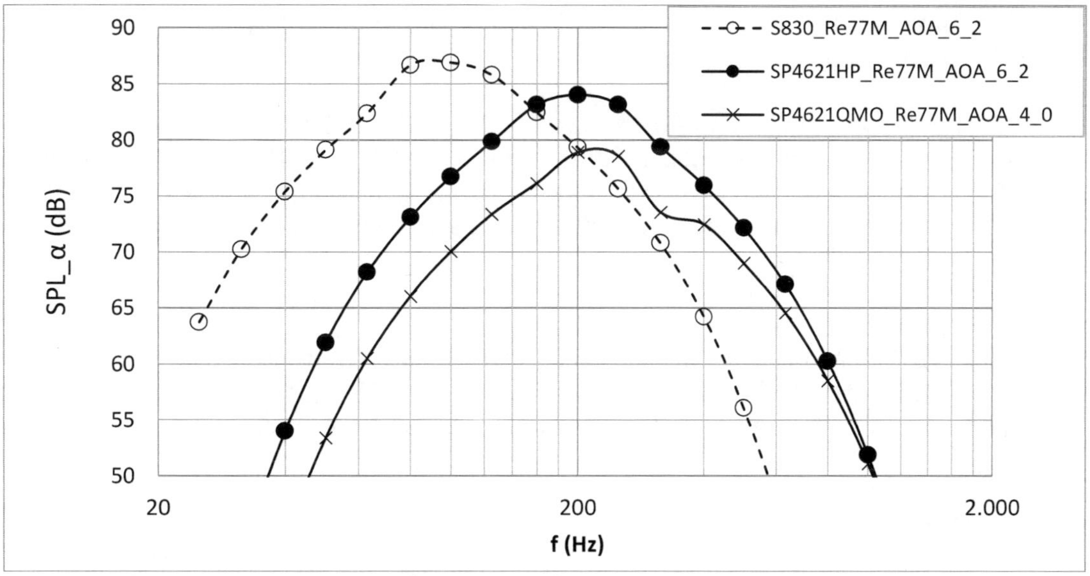

Figure 7 – Predicted spectra for the angle-of attack dependent TE noise source (SPL_α).

4.4 Airfoil profile coordinates.

The proposed SP4621HP and SP4721LA airfoil profiles may be downloaded from the worldwide web sites listed in the following references and freely used for research purposes:

- The Poli-Wind Group home page: [28]
- The UIUC airfoil database: [29], "updates" page.
- The Airfoil Tools database: [30]

In case of commercial applications of the airfoils, please contact either transtec@usp.br or the corresponding author.

5. Conclusions

It is mathematically possible to generate new airfoils from the combination of boundary layer/far field solvers and noise emission codes in order to abide to both aerodynamics and aeroacoustic requirements simultaneously.

This development process may be automated with the aid of multi-disciplinary optimization, genetic, neural and other optimization tools, provided all relevant noise sources for the flow situation are modeled. In the present case, the development was manually performed and guided by an extensive sensitivity analysis.

However, as for any new airfoil developed from theory, the need remains for validation of the aerodynamic and aeroacoustic performance against experimental data, which is very difficult to achieve due to the high speeds required in an aeroacoustic wind tunnel. One possible alternative and future development opportunity is the verification of the proposed airfoils by comparing the available results to those predicted by higher order methods, like computational fluid dynamics (CFD-RANS), detached-eddy simulation (DES), large-eddy simulation (LES) and eventually direct numerical simulation (DNS), the last of which would arguably replace the need for experimental validation altogether.

Considering the two new airfoils proposed, namely the SP4621HP and the SP4721LA, both provide a predicted overall SPL reduction close to 3 dB. The SP4621HP may be operated in a shallower AOA, with an associated predicted overall SPL reduction in excess of 6 dB in relation to the S830 airfoil, and with comparable aerodynamic performance under original NREL design requirements ("QMO mode").

Once the new PNoise code beta version with all airfoil self-noise sources plus inflow noise sources modelled [31] has been satisfactorily verified, a more automated and broader-scope airfoil design method is intended to be developed by the Poli-Wind team, which would welcome any collaborations both in this matter and on the validation of the existing airfoils performance.

Acknowledgments

The authors would like to thank our colleagues Sara Rodriguez and Alexandre Martuscelli Faria, who have embraced the challenge to expand the PNoise capabilities to the full domain of airfoil self-noise, plus inflow noise and have also recently integrated the upgraded code along with the planed quasi-3D module into the QBlade BEM model.

To Dr. David Marten, at T.U. Berlin, our advanced thanks for the upcoming release of the Qblade with the embedded upgraded PNoise module, that will allow it to be readily downloaded and used by researchers and manufactures, at no cost, under the General Public Licence policy.

To Dr. Michael M. Selig (UIUC Airfoil database) and to the curator of the Airfoil Tools data site, our gratitude for storing and indexing the airfoil coordinates.

References

[1] A. M. Faria, J. Y. Saab Jr, S. Rodriguez, J. Barreto and M. M. Pimenta, "A Review on the Development of Airfoils for Wind Turbine Blades," in *Wind Turbine Noise, 2021*, Online, 2021.

[2] D. Somers and J. Tangler, "The S822 and S823 Airfoils," Golden, Co., 2005a.

[3] D. Somers and J. Tangler, "The Airfoils S830, S831 and S832," Golden, Co., 2005b.

[4] D. Somers and J. Tangler, "The Airfoils S833, S834 and S835," Golden, Co., 2005c.

[5] J. Y. Saab Jr, "Trailing-Edge Noise – Development and Application of a Noise Prediction Tool for the Assessment and design of Wind Turbine Airfoils," 2016. [Online]. Available: http://www.teses.usp.br/teses/disponiveis/3/3150/tde-14032017-140101/pt-br.php. [Accessed 19 07 2017].

[6] M. Hansen, Aerodynamics of Wind Turbines, Second Edition ed., London: Earthscan, 2008.

[7] G. Pechlivanoglou, D. Marten, G. Weinzierl, N. Moesus and J. Wendler, "QBlade," Berlin, 2009.

[8] D. Marten, "Extension of an Aerodynamic Simulator for Wind Turbine Blade Design and Performance Analysis," TU Berlin, Berlin, 2010.

[9] J. Y. Saab Jr, M. M. Pimenta, J. R. C. Piqueira, D. Marten, G. Pechlivanoglou, C. N. Nayeri, C. O. Paschereit and A. M. Faria, "Verification and Validation of the "PNoise" Airfoil Trailing-Edge Noise Prediction Module inside "QBlade".," Rotterdam, The Netherlands, 2017.

[10] T. F. Brooks, S. Pope and M. A. Marcolini, "Airfoil Self-Noise and Prediction," NASA, Langley, 1989.

[11] M. Drela, H. Youngren and A. Deperrois, "XFOIL/XFLR5," Cambridge, Ma., 2009.

[12] D. Marten and J. Wendler, "Qblade Guidelines v0.6," TU Berlin, Berlin, 2013.

[13] D. Marten, "Qblade Short Manual V0.8," TU Berlin, Berlin, 2014.

[14] W. Oberkampf and C. Roy, Verification and Validation in Scientific Computing, Cambridge: Cambridge University Press, 2012.

[15] P. Moriarty and P. Migliore, "Semi-Empirical Aeroacoustic Noise Prediction Code for Wind Turbines," Golden, CO, 2003.

[16] S. Oerlemans, "Wind Tunnel Aeroacoustic Tests of Six Airfoils for Use on Small Wind Turbines.," NREL SR-500-34470, Golden, Co., 2003.

[17] C. Doolan and D. Moreau, "Review of NACA 0012 Turbulence Trailing Edge Noise Data at Zero Angle of Attack," in *5th International Meeting on Wind Turbine Noise*, Denver, Co., 2013.

[18] A. Herrig and W. K. E. W. S. Würz, "New CPV-Results of Naca 0012 Trailing-Edge Noise," in *International Conference on methods of Aerophysical Research, ICMAR 2008*, 2008.

[19] D. Bowdler and G. Leventhall, Wind Turbine Noise, Brentwood: Multi-Science Publishing Co Ltd., 2011.

[20] P. Schomer, J. Erdreich, J. Boyle and P. Pamidighatam, "A Proposed Theory to Explain Some Adverse Physiological Effects of the Infrasonic Emissions at Some Wind Farm Sites," in *5th Int. Conf. on Wind Turbine Noise*, Denver, 2013.

[21] S. Oerlemans, "Wind Turbine Noise: Primary Noise Sources," Amsterdam, 2011.

[22] W. Timmer and R. van Rooji, "Summanry of the Delft University Wind Turbine Dedicated Airfoils," *Transactions of the ASME, Vol. 125,* pp. 488-496, Nov. 2003.

[23] P. Migliore and S. Oerlemans, "Wind Tunnel Aeoracoustic Tests of Six Airfoils for Use in Small Wind Turbines.," NREL/CP-500-35090, Golden, CO., 2003.

[24] S. Wagner, R. Bareiß and G. Guidati, Wind Turbine Noise, Berlin: Springer, 1996.

[25] D. Griffith, T. Ashwill and B. Resor, "Large Offshore Rotor Development: Design and Analysis of the SANDIA 100 m Wind Turbine Blade," in *53rd AIAA/ASME/ASCE/AHS/ASC Structures, Structural Dynamics and Materials Conference*, Honolulu, 2012.

[26] M. Fink and D. Bailey, "Airframe Noise Reduction Studies and Clean-Airframe Noise Investigation.," NASA Report CR-159311, 1980.

[27] W. K. Blake, Mechanics of Flow-Induced Sound and Vibration, vol. I, Orlando, FL.: Academic Press, 1986 I.

[28] Universidade de São Paulo - USP, "Poli-WInd (USP)," [Online]. Available: https://sites.usp.br/poli-wind/update-new-poli-wind-website/. [Accessed 10th march 2021].

[29] M. Selig, "UIUC Applied Aerodynamics Group," [Online]. Available: https://m-selig.ae.illinois.edu/ads/coord_updates/. [Accessed 10th 03 2021].

[30] Airfoil Tools, [Online]. Available: oiltools.com/airfoil/details?airfoil=sp4621hp-po. [Accessed 10th 03 2021].

[31] S. Rodriguez, J. Y. Saab Jr., A. M. Faria and M. M. B. J. P. Pimenta, "The Quasi-3D Blade and Rotor Noise Prediction Methodology for the PNoise Code and Preliminary Results," in *8th International Conference on Wind Turbine Noise*, Europe (Remote), 2021, to appear.

[32] J. Y. Saab Jr and M. M. Pimenta, "Displacement Thickness Evaluation for Semi-Empirical Airfol Trailing-Edge Noise Prediction Model," *Journal of the Brazilian Society of Mechanical Sciences and Engineering, V.38, N.2,* pp. 385-394, 2016.

9th International Conference
on
Wind Turbine Noise
Remote from Europe – 18th to 21st May 2021

Numerical Study of the Impact of Vortex Generators on Trailing Edge Noise

F. Seel, T. Lutz, E. Krämer
University of Stuttgart, Institute of Aerodynamics and Gas Dynamics, Germany

Corresponding Author: ferdinand.seel@iag.uni-stuttgart.de

Summary

The continuous increase in rotor size of horizontal axis wind turbines leads to a chain of structurally-driven aerodynamic adaptations. Especially the increase of airfoil thickness necessitates a strong and steep pressure recovery in the rear part of the airfoil. Thus, modern wind turbines are increasingly equipped with Vortex Generators (VGs) to reduce the risk of separation due to the strong adverse pressure gradients. Those vane-like passive aerodynamic devices fixed on the blade's surface induce streamwise vortices to reenergize the lower part of the boundary layer. In terms of aeroacoustics, VGs reduce airfoil noise for high angles of attack by delaying separation and consequently stall noise. In attached flow conditions the VGs, however, may lead to an increase in trailing edge noise due to their influence on boundary layer development. In this paper the influence of the VGs on trailing-edge noise is investigated with numerical methods and compared with published experimental results. For this purpose, a NACA0018 airfoil with counter rotating common down vortex generators is simulated using RANS-methods. The output is utilized to compute the trailing-edge noise with the inhouse acoustic code IAGNoise+ based on an enhanced TNO-Method. The numerical results show good agreement with the experiment. Especially the increase in sound pressure level in the frequency range up to 5 kHz due to the modified boundary layer shape agrees well for moderate angles of attack. Furthermore, it was found that vortex correction and grid refinement in the propagation area of the VG induced vortices has a significant impact on the computed noise spectrum.

1. Introduction

In order to get an acceptable revenue of medium and low wind sites, the rotor radius of wind turbines has to be increased which leads to a chain of structure driven aerodynamic adaptations. The larger span of the blades aggravates the structural loading. To overcome this new structural challenge, the airfoil thickness of multimegawatt turbines is increased up to more than 30% of chord length for inner and mid blade areas [1]. Since high relative thickness results into strong adverse pressure gradients (APG), these blade regions are prone to separation which reduces power outcome and enhances noise emission. This is why these blade sections are more and more equipped with VGs. These vane-like passive aerodynamic devices, fixed perpendicular to

the blade surface, induce streamwise vortices to reenergize the lower area of the boundary layer (BL). From an aerodynamic point of view, the micro- and macroscale phenomena produced by VGs and VG arrays have been investigated through experiments and simulations by many authors since Taylor in 1947 [2, 3, 4, 5].
Till today VGs have mainly been used in inner blade regions where the inflow velocity is low. Therefore, the noise emission is not relevant. Due to the trend towards higher radii the aerodynamically VG relevant areas are shifted towards outer positions where the inflow velocities are higher. In this area the VGs may change the noise footprint of wind turbines. However, till now only few publications are related to this subject [6, 7].
In terms of aeroacoustics, the VGs could be beneficial at high angle of attack (AoA) due to the delayed rise of broadband stall noise which emits much higher sound levels than the dominant trailing edge (TE) noise in non-stalled conditions. According to Fink and Bailey [8] more than $10\ \mathrm{dB}$ difference in sound pressure level (SPL) were observed when comparing a stalled airfoil with a non-stalled one. Nevertheless, for low AoA the topology of the boundary is still massively changed by the streamwise vortices and affects the dominant TE noise. To the best knowledge of the authors, no specific published research on the influence of VG induced streamwise vortices on TE noise has been undertaken so far. Therefore, this paper addresses the evaluation of a numerical setup, which computes the TE noise of an airfoil section equipped with VGs. To do so, the hereafter presented results are compared qualitatively to the experimental studies from Kolkmann et al. [6], who investigated the impact of VGs on the sound pressure spectra from $1\ \mathrm{kHz}$ up to $10\ \mathrm{kHz}$ by means of a beamforming method. Kolkmann identified two major noise mechanisms: For $f\ <\ 4\ \mathrm{kHz}$ TE noise is dominant whereas for frequencies above $f\ <\ 6.35\ \mathrm{kHz}$ the dominant noise is emitted from the area around the VG-array itself.

2. Numerical Methods

In this paper the TE noise of a NACA0018 airfoil with and without VGs at a chord-based Reynolds number of $Re_c = 660{,}000$ is investigated numerically. For this purpose, steady RANS computations were performed with the CFD solver FLOWer and evaluated with the inhouse noise prediction code IAGNoise+.

2.1 CFD Computations

The computations were performed as fully turbulent steady RANS simulations with the CFD solver FLOWer developed by the German Aerospace Center DLR [9]. For closure the two equation linear eddy viscosity model from Menter known as Menter-SST turbulence model [10] was used. As it is well established that linear eddy viscosity models overestimate the eddy viscosity within vortices, a vortex correction for the SST Model after Brandsma et al. [11] was investigated. Roughly explained, this correction is able to detect vortex cores by comparing the magnitudes of the rate-of-strain tensor and the vorticity tensor. In case of much higher magnitude of the vorticity tensor, the model increases the production term in the dissipation equation of the turbulence model to reduce the eddy viscosity. This reduction leads to a lower dissipation of vorticity in the vortex core. By doing so, the physical effect of reduced turbulence production in vortex cores [12] is modelled.

2.2 CFD Setup

The numerical setup is a NACA0018 airfoil extruded in spanwise direction. The extrusion length corresponds to one VG interspacing, i.e. $5.5\ h_{VG}$ $(= 0.11\ c)$, where h_{VG} is the VG height and c the airfoil chord length. At the spanwise boundaries periodic boundary conditions are used. These choices aim to reproduce the experimental test section of Kolkmann et al. [6], who used a tripped NACA0018 airfoil with $0.2\ \mathrm{m}$ chord length and $0.7\ \mathrm{m}$ span. To reduce extraneous noise the focus region had a reduced span of $0.3\ \mathrm{m}$ which was equipped with counterrotating VG pairs. As shown in Figure 1, the setup consists of several structured meshes brought together with the Chimera

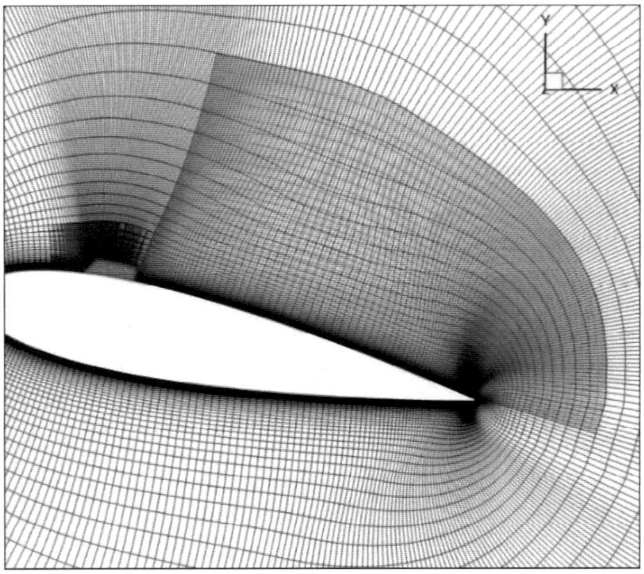

Figure 1: Side view of the chimera setup. Airfoil mesh (black), VG mesh (red), propagation mesh (green).

overlapping mesh technique. The mesh of the fully resolved VG pair (red) was designed to have high resolution in the area of vortex formation, which is important in order to tackle the viscous effects and obtain accurate vortex topology [13]. The low dissipative vortex propagation is crucial in order to evaluate the influence of the vortices on TE noise. Therefore, several so called "propagation meshes" (green) with increasing refinement were built and evaluated (see Table 1). To keep the computational costs reasonable, the refinement in spanwise direction is constant for all propagation meshes. The BL of the airfoil is resolved in a very fine manner ($y^+ < 0.1$) in order to resolve the steep near-wall gradients, which appear due to the redistribution process of the streamwise vortices across the BL.

Table 1: Overview of the refinement parameters for the different propagation meshes

Propagation mesh	**Number of cells overall setup (Mio.)**	**Number of cells propagation mesh (Mio.)**	**Number of cells chordwise**	**Number of cells spanwise**	**Number of cells wall normal**
Coarse	9.2	1.44	128	128	88
Medium	12.4	3.01	184	128	128
Fine	15.6	5.77	256	128	176
Ultra Fine	22.1	11.49	362	128	248

2.3 Noise Prediction Tool

The prediction of the TE noise has been performed with the inhouse noise code IAGNoise+ [14, 15], which is an enhanced TNO-based model by Parchen [16]. The code extracts field and surface quantities from a RANS solution. To do so, extrusion lines are positioned in wall-normal direction in the vicinity of the TE (in the following: $x/c = 0.990$) at predefined spanwise positions on suction and pressure side. Along these lines 100 points are defined, following a typical BL growth rate. For each point a trilinear interpolation in space with the 27 surrounding cells is computed. In a next step, the wall pressure fluctuations (WPFs) are computed by means of statistical methods. The corresponding equation for the power spectral density of the WPFs according to Hornung et al. [15] is:

$$P(k_1, k_3, \omega) = 4\rho^2 \left(\frac{1}{k_1^2 + k_3^2}\right) \cdot \int_0^\infty \left(k_1^2 \left(\frac{\partial \overline{U_1}}{\partial x_2}\right)^2 + \frac{2(k_1^2 + k_3^2)}{15\upsilon}\varepsilon\right) \cdot \Lambda_2 \cdot \Phi_{22} \cdot \overline{\upsilon\upsilon} \cdot \Phi_m \cdot e^{-2|k|x_2}\, dx_2 \qquad (1)$$

where k_1 and k_3 are corresponding to the wave number in mean flow and spanwise direction respectively, ω represents the angular frequency, ρ the density, dU_1/dx_2 the velocity gradient in wall-normal direction, υ the kinematic viscosity, ε the dissipation, Λ_2 (i.e. $\Lambda_{22,2}$) the turbulent integral length scale in vertical direction, Φ_{22} the normalized spectrum of the vertical velocity

fluctuations, $\overline{vv}$ the vertical velocity fluctuations (i.e. Reynolds stresses), Φ_m the moving axis spectrum and k the wavenumber vector.
The original TNO model considers only the interaction of the mean shear with turbulence (MTI), whereas the enhanced model also accounts for the turbulence-turbulence interaction (TTI) [15]. Further implementation details can be found in [14, 15].
As a last step of the noise computation, the modelled WPFs are propagated my means of the following far field equations based on Chase [17] and Brooks & Hodgson [18]:

$$G_{ff}(f) = \frac{L\bar{D}}{2\pi R^2} \int_0^\infty \frac{\omega}{c_0 k_1} \frac{P(k_1,0,\omega)}{1-\frac{\omega}{c_0 k_1}} dk_1 \quad \text{and} \quad SPL(f) = 10 \cdot \log\left[\frac{G_{ff}(f) \cdot df}{(2 \cdot 10^{-5})^2}\right] \tag{2}$$

which finally provides the overall SPL and SPL spectra at a given observer point in space. In the following the observer is located at a distance of $1.5\ \text{m}$ to be in accordance with experimental data from Kolkmann [6].

3. Results and Discussion

3.1 Boundary Layer Analyses at Trailing Edge

The noise prediction code IAGNoise+ uses several BL quantities for the computation of TE noise. Hence, these quantities have to be assessed. The main drivers according to the model are the velocity gradient in wall-normal direction dU_1/dx_2 relevant for the MTI, the dissipation ε driving the TTI, the turbulent integral vertical length scale Λ_2 and the normal Reynolds stresses in wall-normal direction $\overline{vv}$. These quantities are evaluated for different refinements of the propagation mesh at 5 spanwise positions equally distributed across one VG-induced main vortex. As the VGs are placed on the suction side, the pressure side will not be considered in the following.

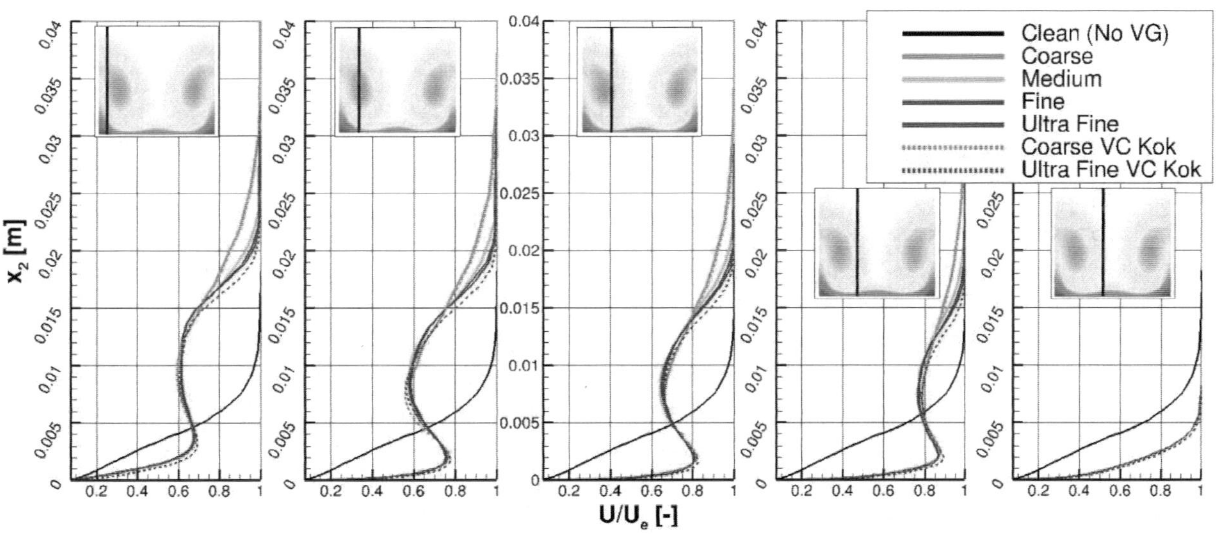

Figure 2: NACA0018 suction side streamwise velocity profiles at $x/c = 0.99$, $\alpha = 6°$, $Re_c = 660{,}000$, $Ma = 0.146$, for different equidistant spanwise positions of one main vortex.

Normalized streamwise velocity

First of all, the normalized streamwise velocity U/U_e in Figure 2 shows very good agreement for all meshes in the lower part of the BL. The typical S-shaped topology as experimentally observed in [12, 19, 20] is matched by all grid refinements and the considerable redistribution of kinetic energy towards the lower parts of the BL compared to the clean case is obtained. In the upper BL ($x_2 > 0.015\ \text{m}$) coarse and medium meshes show only weak gradients while the finer grids exhibit convergent behaviour. As one can see from equation (1), the influence of the BL quantities on the emitted noise decreases with increasing wall distance x_2. This is why the deviations in the outer BL can be expected to have only minor influence on the computed noise levels according to the acoustic model in use. Concerning the vortex correction (VC) only minor changes for the considered refinements are obtained.

Dissipation ε

The dissipation level ε (Figure 3) increases compared to the clean case in the outer BL due to the higher BL thickness. The different computations show very good agreement in the lower BL with small deviations for the coarse mesh. At the outer part of the BL ($x_2 > 0.02$ m) a converging behaviour towards lower values for increasing mesh refinement is observed. This complies with previous studies, where BL thickness was overestimated in case of insufficient mesh refinement. The added VC does not lead to major deviations for the considered cases.

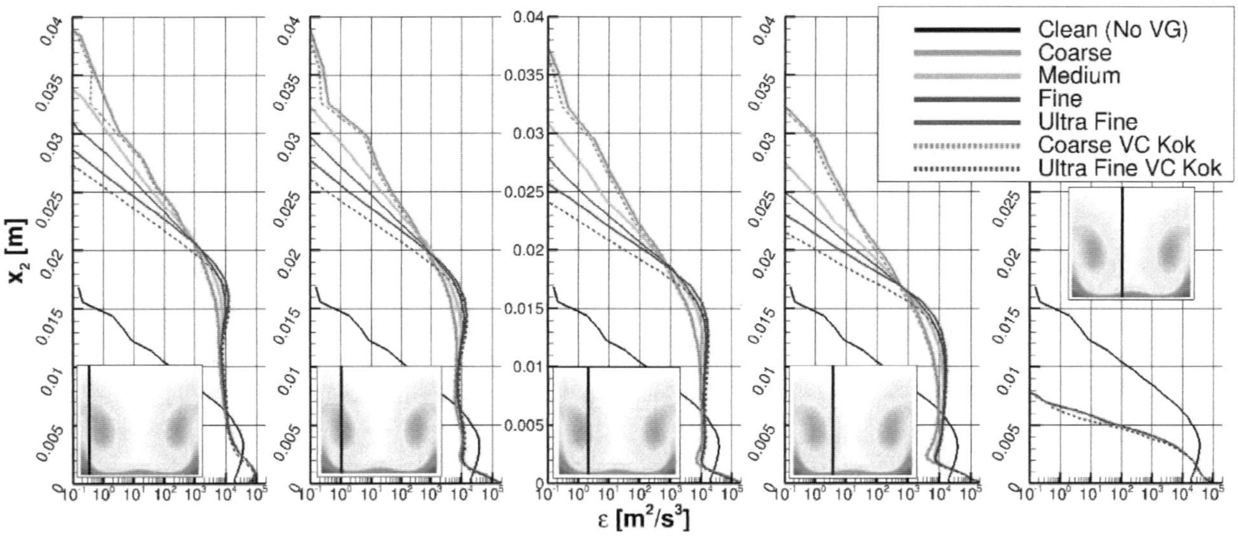

Figure 3: NACA0018 suction side dissipation profiles at $x/c = 0.99$, $\alpha = 6°$, $Re_c = 660{,}000$, $Ma = 0.146$, for different equidistant spanwise positions of one main vortex.

Vertical Reynolds stresses $\overline{vv}$

The presented study is using a linear eddy viscosity model (Menter-SST model), hence only turbulence kinetic energy k_t gives information about the Reynolds stresses. To counteract the consequent modelling error, due to isotropic turbulence assumption, an anisotropy factor is used in this work. This factor was developed by Kamruzzaman et al. [14] and improved by Hornung et al. [15] towards use in slightly separated flows.

As shown in Figure 4, the anisotropic normal Reynolds stresses $\overline{vv}$ differ considerably from the clean case. Instead of only one peak, several extrema emerge around the main vortex. This is in accordance with results from Angele et al. [12] who conducted PIV experiments of a turbulent BL subjected to an APG. Angele observed a $\overline{vv}$-peak in the outer part of the BL at spanwise positions of outflow (i.e. where the vortex induces a flow away from the surface). The outflow area corresponds to the two plots on the left of Figure 4.

Considering the different mesh refinements, the result from the coarse mesh is not matching the other results except for the very low BL area. In contrast to the observations from above (streamwise velocity and dissipation), VC has a considerable influence on $\overline{vv}$ also in the near-wall region. Especially at the outflow region and in the center of the main vortex (two left plots respectively) $\overline{vv}$ is reduced. This is in accordance with published experimental observations [11, 21] about vanished turbulent production in the outflow area and in the vortex core which reduces $\overline{vv}$ as well as the turbulent length scale which will be discussed hereafter.

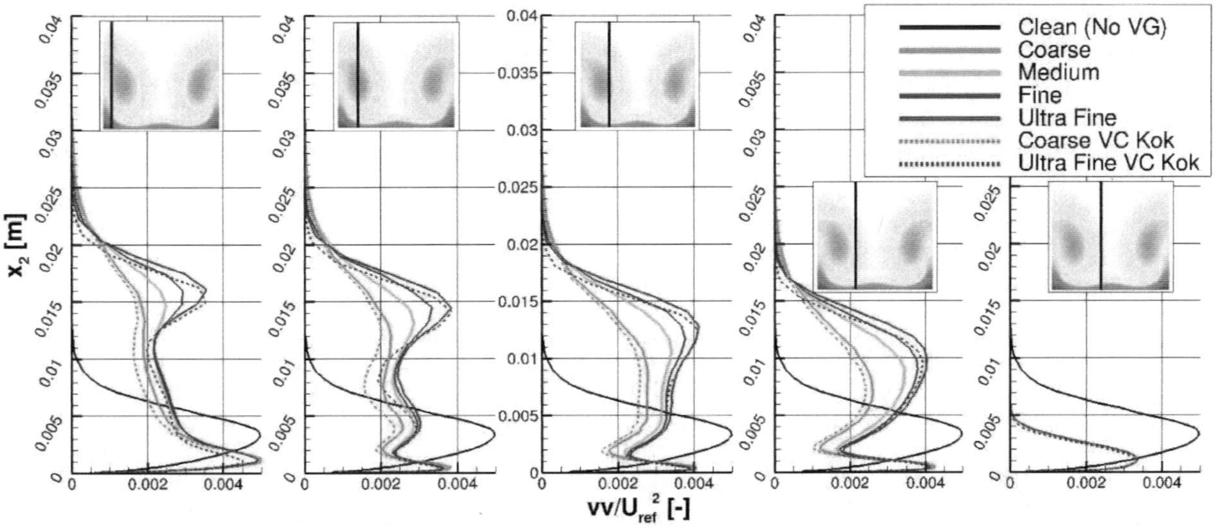

Figure 4 : NACA0018 suction side wall normal Reynolds stresses at $x/c = 0.99$, $\alpha = 6°$, $Re_c = 660{,}000$, $Ma = 0.146$, for different equidistant spanwise positions of one main vortex.

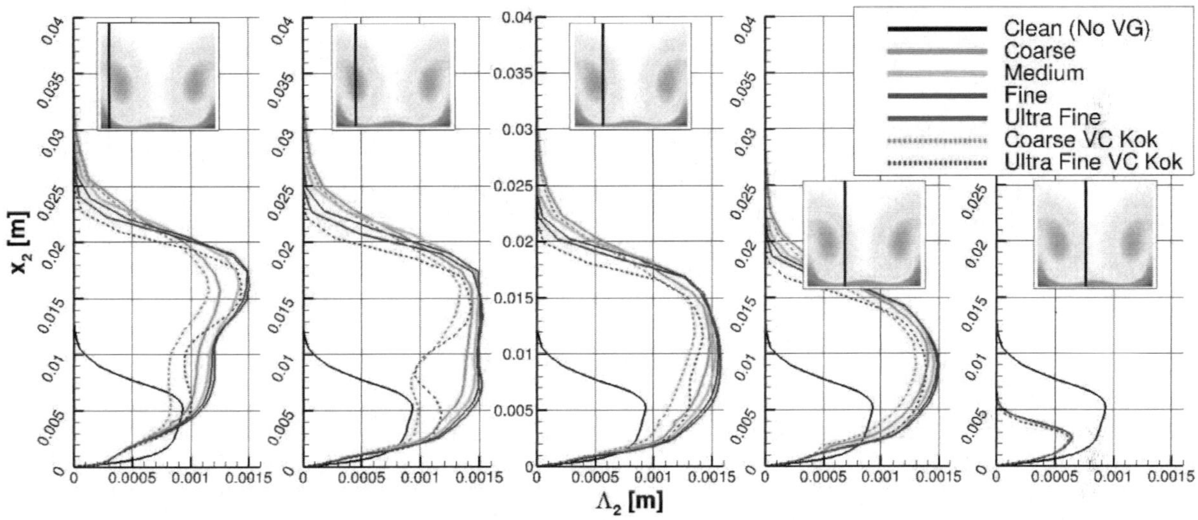

Figure 5: NACA0018 suction side turbulent vertical integral length scale profiles at $x/c = 0.99$, $\alpha = 6°$, fo different equidistant spanwise positions of one main vortex.

<u>Turbulent vertical integral length scale Λ_2</u>

The VC deviates considerably from the other results in the center of the main vortex for both refinements. This behaviour is explained, once again, by the reduction of turbulence in the vortex core. Since Λ_2 is calculated out of $\overline{vv}$, the behaviour is visible for both. Comparison of the herein presented VC results with a further study of VC on medium grid showed an overlap between two dissipation mechanisms concerning the main vortex: As highlighted in Figure 7 by blue circles the first dissipation mechanism is an intentional reduction of turbulence by the VC in the vortex sub core and the second mechanism (black circles) is the dissipation of turbulence depending on mesh refinement in the outer BL area. The latter is much smaller for the ultra fine grid. This is why noise output from ultra fine mesh with VC will be considered as most trustful. Nevertheless, even within coarser grids, VC might improve reliability of the noise computation at least for moderate AoA. Regarding higher AoA, where the BL is encountering stronger APG resulting in higher Reynolds stresses (Figure 6), the impact of VC has almost vanished. As Westphal et al. [22] concluded from experimental investigations of one isolated VG-induced vortex exposed to APG, the increased distortion of the Reynolds stresses of vortices encountering high APG in turbulent BL could lead to modelling difficulties. Furthermore, Westphal showed that APG leads to flatter vortices (stretched in spanwise direction). This is not the case in the present study. One reason for this disagreement could be that the interaction between the different streamwise vortices dominates over the APG-induced flattening, which finally creates vortices stretched in wall-normal direction. Recent publications (e.g. Spalart et al. [23]) agree on the fact that such multiple viscous aerodynamic interactions might be too complex for RANS. For this reason the RANS-based noise results for high AoA might contain inaccuracies. In order to overcome these limits, further studies with higher fidelity methods (e.g. DES) are required.

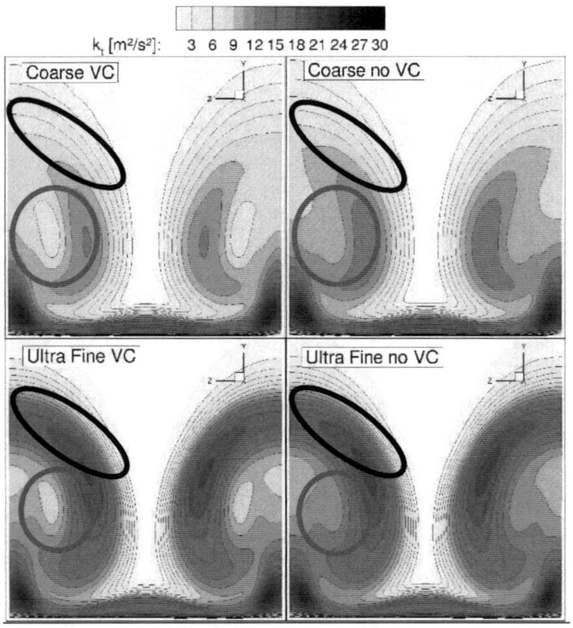

Figure 7: Turbulence kinetic energy k_t at $x/c = 0.99$, $\alpha = 6°$

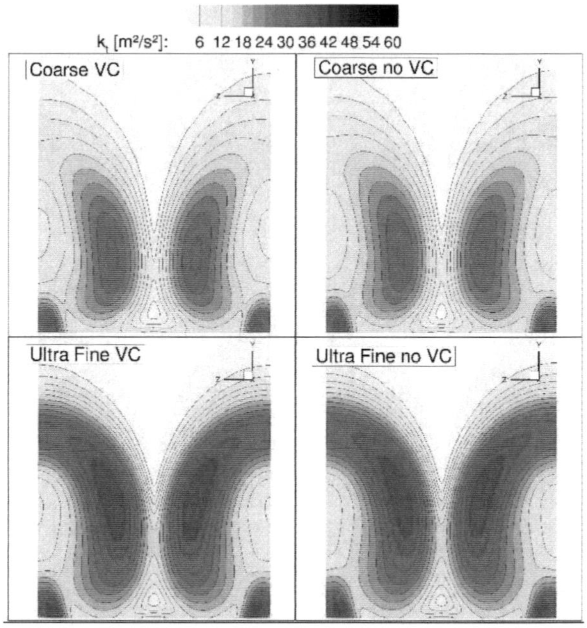

Figure 6: Turbulence kinetic energy k_t at $x/c = 0.99$, $\alpha = 12°$

3.2 Trailing Edge Noise

TE noise is investigated for two AoA. First $\alpha = 6°$ is considered to examine the impact of VGs on SPL while flow is attached. This choice is motivated by the fact that VGs are passive devices, which cannot be retracted while not required. Furthermore, for this moderate AoA the noise model has been deeply validated [14, 15]. However, as VGs are devices to delay separation, a second AoA of $\alpha = 12°$ is investigated hereafter.

<u>Influence of slice selection</u>

The noise code IAGNoise+ is only considering the 2D turbulence state and was improved in order to capture spanwise emission by considering multiple slices and performing trilinear interpolations for each data point as explained in section 2.3. Due to the streamwise vortices the structure of the BL is highly three-dimensional. This is why an assessment of the minimum number of slices required per VG is done hereafter. To do so the Sound Pressure Level (SPL) of the suction side for medium refinement was computed for different numbers of slices placed within an equidistant spanwise spacing. As shown in Figure 8 for $\alpha = 6°$, convergent behaviour for increasing number of slices over the entire frequency range is obtained. In case of 5 slices per VG the peak SPL at 1250 Hz is underestimated and the frequency range from 4 to 10 kHz is overestimated. For 15 slices or more per VG, the SPL delivers only minor deviations over the entire spectrum. As a consequence, 25 slices per VG are used for all noise computations in the following.

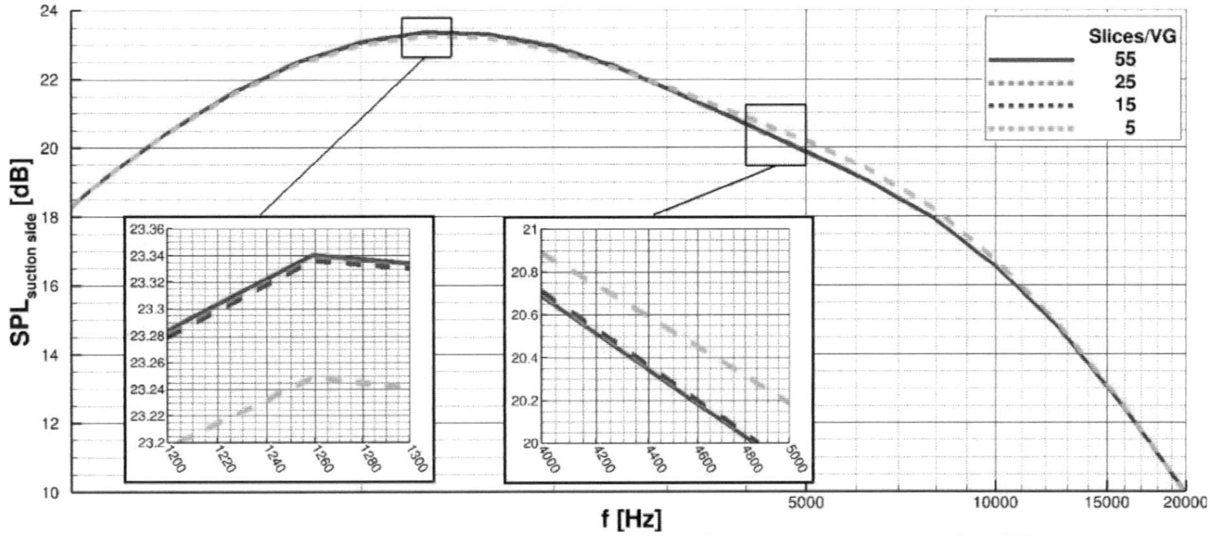

Figure 8: 1/3 octave band spectra of NACA0018 with VGs of suction side, $\alpha = 6°$ for different number of span wise slices.

Influence of grid refinement

In Figure 9 and Figure 10 the differences in SPL spectra $\mathrm{dSPL}(f) = \mathrm{SPL}_{\mathrm{VG}}(f) - \mathrm{SPL}_{\mathrm{NoVG}}(f)$ of computed TE noise for $\alpha = 6°$ and $\alpha = 12°$ respectively are compared qualitatively to the beamforming results of the overall noise (all noise sources including TE noise) from Kolkmann et al. [6]. This comparison can be justified by the fact that, according to Kolkmanns observations, TE noise is the main noise source for frequencies below 4 kHz. For this reason no accordance to the experiments for the frequency range above 5 kHz can be expected.

Regarding $\alpha = 6°$, convergent behaviour of dSPL for increasing grid refinement is obtained with dSPL reduction for VC cases. As mentioned earlier, this is because of reduced turbulence in the vortex sub core and the resulting lower k_t in this area. Good qualitative agreement is obtained between computed and experimental results: The increasing slope of dSPL from 1 to 4 kHz in the experiment is very well represented for the highest resolved computation (i.e. ultra fine with VC). The results underline the importance of the refinement area and show clearly that state of the art mesh requirements in terms of minimum grid resolution are not sufficient for TE noise computation of airfoils equipped with VGs.

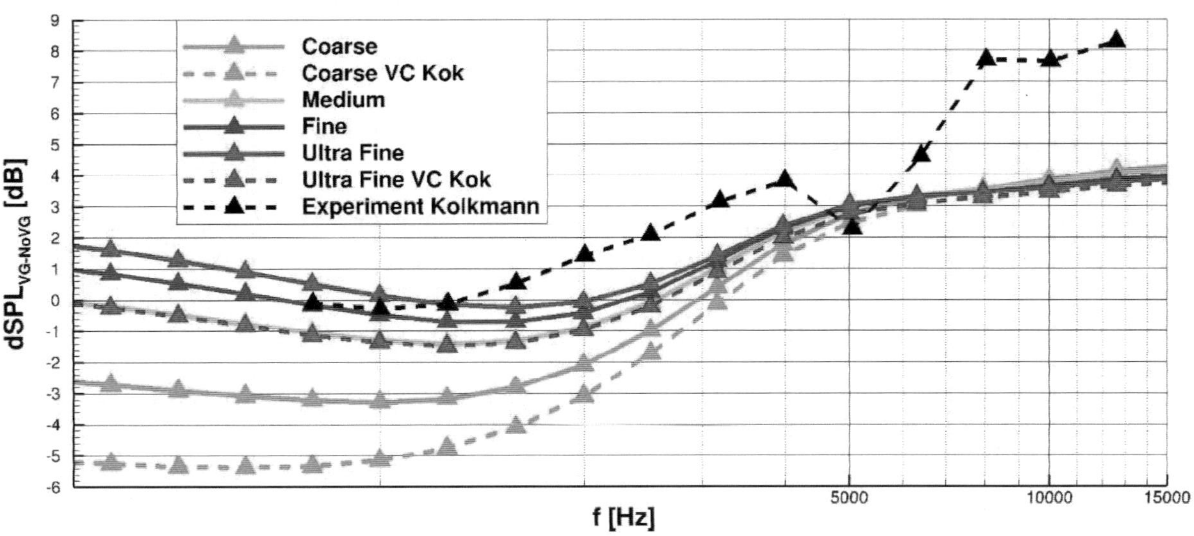

Figure 9: Difference of 1/3 octave band spectra between NACA0018 with and without VGs, $\alpha = 6°$, 25 Slices per VG.

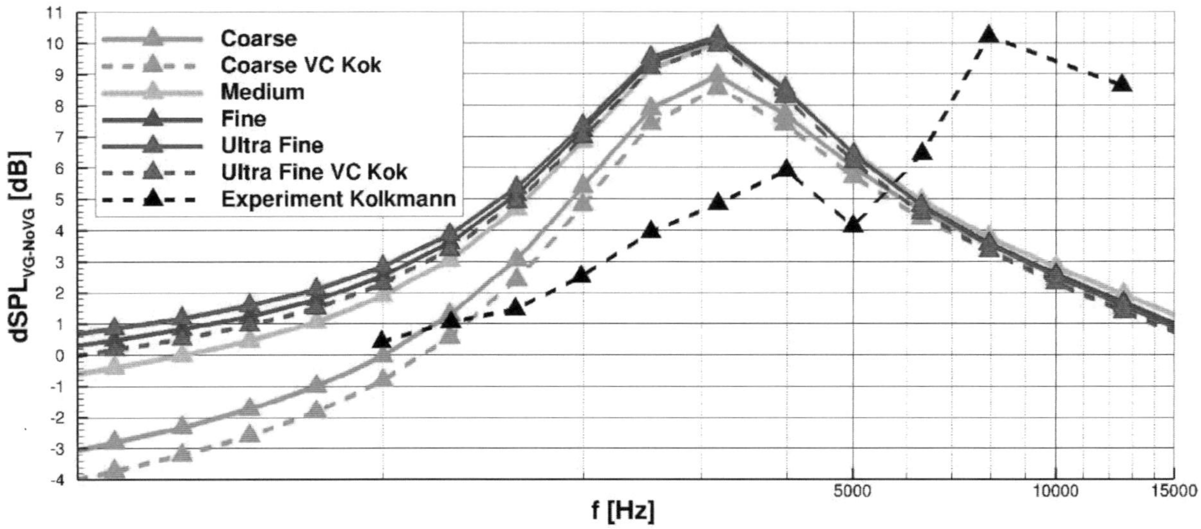

Figure 10: Difference of 1/3 octave band spectra between NACA0018 with and without VGs, $\alpha = 12°$, 25 Slices per VG.

Regarding $\alpha = 12°$ (Figure 10), the computed dSPL slope for $f < 3{,}5\,\mathrm{kHz}$ is represented satisfactorily for all meshes except coarse refinement. However, an offset up to $\mathrm{dSPL} \approx 5\,\mathrm{dB}$ is visible. Due to the dSPL representation, the reason for this offset could result from two sources:

- Underestimation of clean airfoil TE noise: Firstly, at $\alpha = 12°$ a separated region is apparent in the numerical results. Consequently the steady RANS simulations have only limited reliability. Nevertheless, good numerical convergence was reached and the computed separation line is in good agreement with Kolkmann's observations (RANS: $x_{sep.}/c = 0.82$; Kolkmann: $x_{sep}/c = 0.85$). Secondly, the validation of IAGNoise+ from Hornung et al. [15] was only done for weak separation around $x_{sep}/c > 0.97$.

- Overestimation of the VG airfoil TE noise: The considered AoA leads to a strong APG with which the embedded VG vortices are interacting. This could lead to an unphysical increase of modelled turbulence and consequently to noise increase. Furthermore, linear eddy viscosity modelling of streamwise vortices embedded in strong APG boundary layers could lead to poor results [22] and has to be validated with higher fidelity methods as DES (more precisely wall modelled LES) in order to entirely resolve the VG-vortices.

4. Conclusions

Trailing edge noise of a NACA0018 airfoil has been investigated numerically by means of an enhanced TNO-based method for two AoA (6° and 12°) with and without counter rotating common down flow triangular vortex generators. First the relevant BL quantities for noise emission were analysed, then the SPL resulting from TE noise was qualitatively compared to the published experiment from Kolkmann et al. [6]. A BL assessment for $\alpha = 6°$ has been undertaken with special interest on vortex dissipation and the improved computation of vortices through a VC method embedded in the Menter-SST turbulence model. It was found that mesh refinement of the BL (in wall-normal and streamwise direction) over the entire vortex propagation area till the trailing edge improves considerably the conservation of turbulence ($\overline{vv}$ and Λ_2) as well as its distribution in space. This conservation is essential in order to capture TE noise over the entire frequency spectrum. The noise computation showed that mesh refinement increases the SPL whereas the VC reduces it. This behaviour emerges from two different mechanisms: Firstly, the refinement reduces the overestimated decay of turbulence in the upper BL particularly in the

spanwise locations of outflow. Secondly, VC appears to be useful to model the reduction of turbulence in the vortex core.
For high AoA (herein $\alpha = 12°$) which contain areas of flow separation without VGs but fully-attached flow with VGs, the VC doesn't significantly reduce the modelled turbulence in the vortex center. Further work needs to be done to analyse if this behaviour emerges from physics or from the VC modelling weakness in high APG. For this reason, further investigations with higher fidelity methods (e.g. WMLES) have to be made.
In terms of TE noise, the computational setup is able to capture VG-induced SPL increase compared to the clean airfoil. The slope of this increase fits qualitatively well in the TE noise dominated frequency band ($f < 4\,\mathrm{kHz}$) for moderate and high AoA ($\alpha = 6°$ and $12°$). For $\alpha = 12°$ the dSPL-spectrum overshoots the experiment. The reason for this behaviour could be the underestimation of the clean airfoil TE noise, the overestimation of the airfoil equipped with VGs, or a combination of both. Nevertheless, the presented method is able to give qualitative insights without requiring huge computational resources and could help to examine the numerous geometrical VG-parameters towards an aeroacoustically optimized design.

Acknowledgments

This work could be realized thanks to the project Schall_KoGe (ref: 0324337C) founded by the German Federal Ministry of Economic Affairs and Energy (BMWi). The authors gratefully acknowledge the High-Performance Computing Center Stuttgart (HLRS) for providing computational resources within the project Schall_KoGe.

References

[1] G. Sieros, "Redesign of the AVATAR Reference Wind Turbine rotor," Technical Report, 2017.

[2] H. D. Taylor, "The elimination of diffuser separation by vortex generators," *United Aircraft Corporation,* Vols. Report No. R-4012-3, 1947.

[3] H. H. Pearcey, "Shock induced separation and its prevention by design and boundary layer control," in *Boundary layer and flow control*, In Lachmann, 1961.

[4] C. Yao, J. Lin and B. Allen, "Flowfield Measurement of Device-Induced Embedded Streamwise Vortex on a Flat Plate," *1st AIAA Flow Control Conference,* 2002.

[5] C. Velte, "Characterization of Vortex Generator Induced Flow," DCAMM Special Repport No. S118, Dissertation, Technical University of Denmark, 2009.

[6] D. Kolkman, L. Santana and M. Sanders, "Experimental characterization of vortex generators induced noise of wind turbines," in *AIAA/CEAS Aeroacoustics Conference AIAA AVIATION Forum*, 2018.

[7] Q. Ye, F. Avallone, W. Van der Velden and D. Casalino, "Effect of Vortex Generators on NREL Wind Turbine: Aerodynamic Performance and Far-Field Noise," *Journal of Physics: Conference Series,* vol. 1618 052077, 2020.

[8] M. R. Fink and D. A. Bailey, "Airframe Noise Reduction Sutdies and Clean-Airframe Noise Investigation," NASA CR159311, 1980.

[9] N. Kroll, C. Rossow, D. Schwamborn, K. Becker and G. Heller, "MEGAFLOW – a Numerical Flow Simulation Tool for Transport Aircraft Design," *Proceedings of the 23rd International Congress of Aeronautical Sciences, Toronto,* 2002.

[10] F. R. Menter, "Two-equation eddy-viscosity turbulence models for engineering," *AIAA Journal,* vol. Vol. 32 (8), 1994.

[11] H. S. Brandsma, J. C. Kok, H. S. Dol and A. Elsenaar, "Leading Edge Vortex Flow Computations and Comparison with DNW-HST Wind Tunnel Data," in *Advanced Flow*

Management: Part A – Vortex Flows and High Angle of Attack for Military Vehicles, Loen, Norway, 2001.

[12] K. P. Angele and B. Muhammad-Klingmann, "The effect of streamwise vortices on the turbulence structure of a separating boundary layer," *European Journla of Mechanics B/Fluids,* vol. 24, pp. 539-554, 2005.

[13] E. Wik and S. T. Shaw, "Numerical Simulation of Micro Vortex Generators Erik," in *2nd AIAA Flow Control Conference* , Portland, Oregon, 2004.

[14] M. Kamruzzaman, D. Bekiropoulos, A. Wolf, T. Lutz and E. Krämer, "Rnoise: A RANS based airfoil trailing-edge noise prediction model," in *20th AIAA/CEAS Aeroacoustics Conference*, Atlanta, GA, 2014.

[15] C. Hornung, T. Lutz and E. Krämer, "A model to include turbulence-turbulence interaction in the prediction of trailing edge far field noise for high angles of attack or slightly separated flow," *Renewable Energy,* 2019.

[16] R. Parchen, "Progress report DRAW, a prediction scheme for trailing-edge noise based on detailed boundary-layer characteristics," TNO Institute of Applied Physics, The Netherlands, 1998.

[17] D. M. Chase, "Noise Radiated from an Edge in Turbulent Flow," *AIAA Journal,* vol. 12 (8), pp. 1041-1047, 1975.

[18] T. F. Brooks and T. H. Hodgson, "Trailing edge noise prediction from measured surface pressures," *Sound Vib.,* vol. 78 (1), pp. 69-117, 1981.

[19] K. P. Angele and F. Grewe, "Instantaneous Behavior of Streamwise Vortices for Turbulent Boundary Layer Separation Control.," *ASME J.Fluids Eng.,* vol. 129, pp. 226-235, 2007.

[20] G. Godard and M. Stanislas, "Control of a decelerating boundary layer. Part 1: Optimization of passive vortex generators," no. Aerospace Science and Technology 10, p. 181–191, 2006.

[21] Z. Leitzau, M. Ramasamy, R. Jain, J. Leishman and J. Ekaterinaris, "Comparison of Advanced RANS Modeling with Dual-Plane PIV Measurements for a Hovering Rotor," *Journal Of The American Helicopter Society,* vol. 63, no. 022003, 2018.

[22] R. V. Westphal, J. K. Eaton and W. R. Paulex, "Interaction Between a Vortex and a Turbulent Boundary Layer in a Streamwise Pressure Gradient," in *Turbulent Shear Flows 5*, Heidelberg, 1987.

[23] P. R. Spalart, M. L. Shur, M. K. Strelets and A. K. Travin, "Direct Simulation and RANS Modelling of a Vortex Generator Flow," *Flow Turbulence Combust,* vol. 95, pp. 335-350, 2015.

[24] C. Hornung, T. Lutz and E. Krämer, "Development of Design Guidelines for Low Noise But High Yield Wind Turbines," in *Wind turbine Noise Conference Proceedings*, Lisbon, 2019.

INCEEUROPE

9th International Conference
on
Wind Turbine Noise
Remote from Europe – 18th to 21st May 2021

On the need for improved prediction models and updated noise regulations to utilize the advanced controls strategies that are available for modern wind turbines

Bo Søndergaard SWECO Denmark A/S: Bo.Sondergaard@Sweco.dk

Summary

Modern wind turbines have a capacity for operating flexible according to specific requirements. Simple operating conditions like running in a discrete low noise mode during nighttime with low noise demands have been possible for many years. Wind turbines of today can change operation continuously and hence the sound output similarly if the conditions change. It is possible based on feedback from the environment (wind speed, wind direction, wind shear, temperature, temperature gradient etc.) to operate a wind farm to an optimized power production with constraints on the noise level in the surroundings whether the noise constraints are a constant level or a wind speed dependent noise level. What is required is a reliable prediction model, that can handle the meteorological variations, a detailed set of noise output data for the wind turbines and noise regulations that allows for using these principles. The principles behind and corresponding requirements will be discussed.

1. Introduction

The development of wind energy around the world is subject to local regulations. That includes regulations on noise exposure in the environment. Most of these regulations are based on experiences from industry noise and does not reflect the fact that wind energy production and noise generated in the process is strongly related to meteorology like wind speed, wind shear, turbulence etc. Where regulations include noise prediction models it is based on simple empirical models with limited or no options for handling long range propagation and elevated sources, not to mention wind direction, wind shear, temperature gradients and complex terrain. Since regulations are simple, development and operation of wind farms stays equally simple.

2. Development of wind turbines

Wind turbines have changed from simple wind turbine generators to energy production facilities. Instead of passively reacting to the wind speed the wind turbines of today interacts with the situation through algorithms developed by the manufacturers along with the development of blade design. This leads to a lot of possibilities in park control strategy. The energy output of a wind farm can be changed according to demand either by shutting individual wind turbines down or changing the mode of operation for a group of wind turbines if that is considered more feasible.

3. Development of prediction models

The most used prediction model for wind turbine noise assessment is ISO 9613-2. This method is developed from a good knowledge on sound propagation from theory and measurements. It is designed for low wind speeds below 5 m/s at 10 m height, downwind propagation and a moderate temperature inversion. It works well for the design range, but for elevated sources, higher wind speeds and complex terrain it becomes more unreliable. Some quick fixes in the modelling procedures have made it possible to achieve reasonable results in the order of what is measured in the vicinity of wind turbines/wind farms. However, this does not change the fact that ISO 9613-2 is not adequate for detailed evaluation of noise from wind farms.
Already in the late 1990's work was initiated in the Nordic countries to develop a new prediction model named Nord2000. Not specifically for wind turbines but to create a more general model that would work for varying meteorology, complex terrain and longer distances. This would give a better estimate for the annual average for e.g. road traffic noise. Nord2000 is a semi-analytical ray-tracing model (ie. part engineering and part scientific). Harmonoise is a similar model developed in the EU framework, but work on that model is stopped.

There are simplifications in the Nord2000 model, but it has been validated through measurements with results in good agreement. The results are not very different from what is achieved by the ISO 9613-2 model with the quick fixes and for conditions where they are both expected to work well. But Nord2000 works well for conditions outside the design range for ISO 9613-2. It is possible to include varying wind directions, varying wind speeds and complex terrain in the modelling. This also increase the requirements on the user of the model and probably also the user of the results. The Nord2000 model is commercially available in different software version and could be included in noise regulations.

Nord2000 have made some shortcuts which still leaves room for improvement on precision and reliability. Especially for long range and upwind predictions or when there are some untypical meteorological conditions on a site. Work on this is typically research projects at universities and we have seen Universities present models that can move us to the next level. However, these models are not finalized to a level where they can be brought to commercial use and be integrated in noise regulations.

Improving prediction models is not (only) a question of predicting the noise with more decimals but more on extending the range of validity and reliability to more situations by including more details in the prediction models. Validation of the models based on measurements have an inherent limit on the accuracy due to small scale variations in meteorology, background noise from other sources, wind induced noise in the measurement equipment, instrumental uncertainty etc. Within this limit it is not possible to say which prediction model is better. This makes it difficult to validate models precisely at large distances. However, it is possible to validate the model's dependency on different environmental parameters if they are measured in the validation process. This is the benefit of advanced prediction models to models like ISO 9613-2.

Nord2000 was a governmental initiative, but no similar initiatives are made these days so where will the next advanced prediction model come from. From time to time universities presents advances in prediction models including different environmental effects and apparently the potential is there. However, these advances seem more to be part of the educational process at the universities more than an ambition to finalize, validate and commercialize the prediction model to a level where it is generally accepted.

4. The smart principle

Years ago, we were considering the principle "Intelligent Noise Monitoring System for wind farms". We had just spent time validating the Nord2000 model and saw the perspective in operating wind farms according to more flexible constraints based on valid reliable noise predictions. Unfortunately, others saw the same and took a general patent on these principles and it was not possible to get funding for the idea. However, the principle we put up at the time was as described in Figure 1.

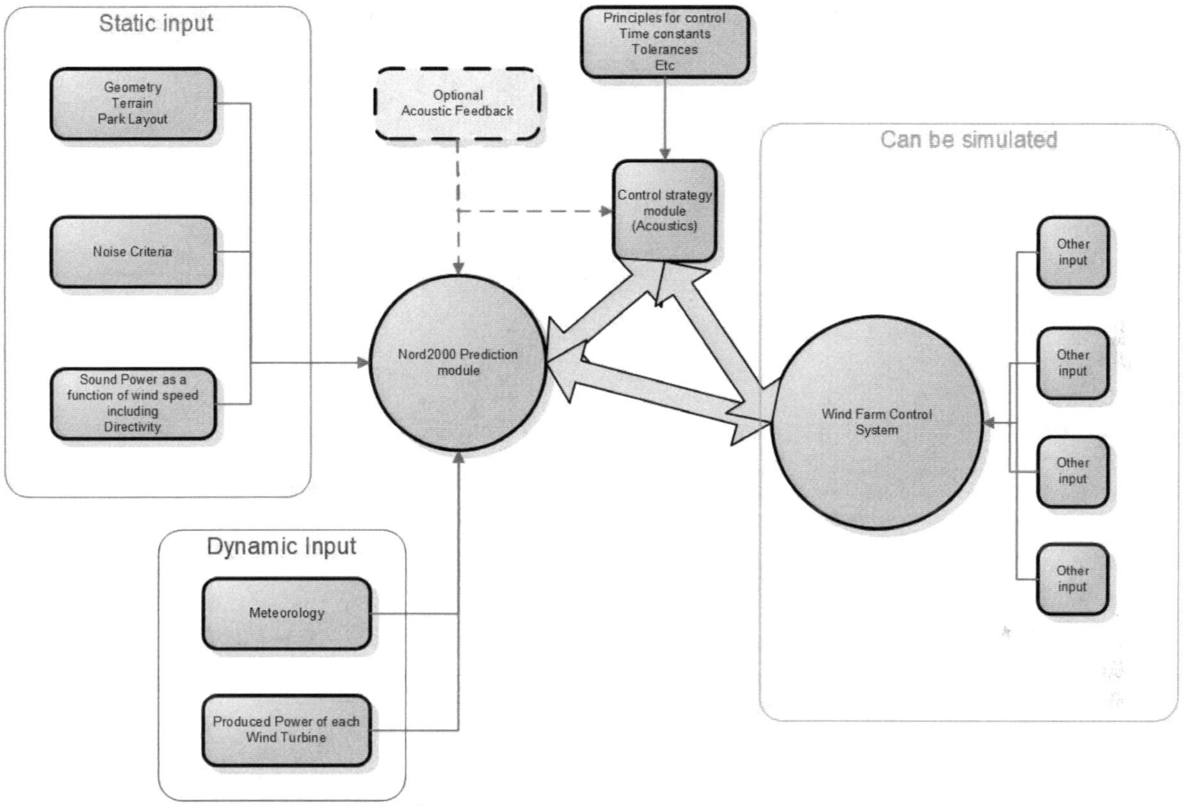

Figure 1. A principle for smart regulation of a wind farm

The basic principle is that the operation of the wind farm will be adjusted dependent on the predicted noise in the environment and the corresponding constraints from the smart noise criteria. The control strategy defines the basic principles for the wind farm in question. This is a first draft and other inputs and constraints are likely to be included in a real-life situation.

5. Smart regulations

Many regulations are simplified. This is a tradition, you could say, because it makes it easier to communicate and easier to verify compliance/no compliance. If the regulations stay simple It will not be possible to utilize the full potential of modern wind turbines. If regulations were allowing for including the varying meteorology in the assessment of the noise it would be possible to increase the overall capacity of a wind farm (e.g. more wind turbines) by changing the operational configuration of the wind farm continuously so the noise criteria are never violated but with different wind turbines operating at different power and noise output depending on the actual conditions.

Smart regulations will have to include strategies for testing for compliance. This will obviously not be simple but logging of the wind farms operational output could be part of the strategy.

6. Some consequences

Obviously, one of the consequences of smart regulation could be that the noise in the environment could be increased compared to a situation with simple regulations and will often be at or close to the noise limit. This is a major change in the level of protection which is secured with the simple regulations where in some situations the noise level is low, and the noise limit is the maximum noise level experienced.
The smart regulations should therefore be based on an ambition to improve the level of protection of the environment.

7. Conclusions

With modern wind turbines the possibilities for operating wind farms smart is present. This can result in improved performance of the wind farms and possibly to a lesser impact of the environment.

However, to utilize this feature four things are required.

1. Smart wind turbines
2. Smart control systems/strategies
3. Advanced prediction models. Validated and available.
4. Smart noise regulations.

Bullet 1 is already fulfilled and most likely bullet 2 is too. But for bullet 3 and 4 a lot of work is still to be done.

INCE EUROPE

9th International Conference
on
Wind Turbine Noise
Remote from Europe – 18th to 21st May 2021

Tonality content analysed with both 1/3 octave band and narrowband methods with comparison to listening test

Lars S. Søndergaard, FORCE Technology: lss@force.dk
Mark Bastasch, Jacobs, Mark.Bastasch@jacobs.com

Summary

When analysing signals for tonality content typically either a narrowband method or a 1/3 octave band method is used. This paper presents the results of the analysis of both methods and compares the results with listening test. The stimuli are 31 files, 29 of them which have tones. Twenty-four of stimuli are based on wind turbine noise. The tested 1/3 octave band methods are ISO 1996-2:2017, ANSI S12.9-2005/Part 4 and ANSI S12.9-2013/Part 3. The tested narrowband method is ISO PAS 20065.

When compared to prominence in a subjective listening test the percentage of explained variance of the linear regression (83 %) was good for the narrowband method. For the 1/3 octave band methods it is seen that 10-13 of the 29 stimuli have a tone frequency which falls in the boundary area between two neighbouring 1/3 octave bands. When tones are proximate to the boundary of the band the ability of 1/3 octave band methods on identifying a tone is limited by the implementation of the 1/3 octave band filters. This was an anticipated limitation of the 1/3 octave band methods. With the tested implementation of the 1/3 octave band filters a reasonable correlation is seen with the prominence of the listening test when tones do not fall near the boundary of the band.

While narrowband methods are preferred by many, this finding indicates that for the type of stimuli evaluated in this study the 1/3 octave band methods may also be effective provided that the tone frequency is not near the boundary of the band. If one can ensure that the tone frequency will not be in the boundary region, the 1/3 octave band methods are simple to implement, particularly when evaluating large long-term datasets.

1. Introduction

It is generally believed that audible tones in noise increase the annoyance relative to a situation where there are no audible tones in the noise [1]. In many countries a penalty is added if the noise contains audible tones, for example Denmark [2].

While subjective observations may be relied on in some jurisdictions, many evaluate whether the tones are audible with an objective method. Presently, IEC 61400-11[3] is the widely used method for analysing the tonality content in wind turbine noise for sound power level measurements which are made close to the turbine and on a reflective ground board. In some regulatory proceedings, if tones are not identified in these source sound power level measurements, tones are presumed not to occur at more distant receiver locations. Other methods are used worldwide for analysing the tonality content in wind turbine noise at the receptor position. One of the most used methods at the receptor position is the ISO 1996-2 Annex C method [4]. However, in 2016 the new ISO/PAS 20065 method [5] was published, replacing the ISO 1996-2 Annex C method.

In 2018 a new standardization working group was formed, PT-61400-11-2, with the scope "Measurement of wind turbine noise characteristics in receptor position", see section 2. Within the framework of the standardization work a round robin was conducted in 2018-2019 comparing the IEC 61400-11:2012 and ISO/PAS 20065:2016 for analysing tones in wind turbine noise [6]. The analysis showed comparable results by comparing the two narrowband methods. When compared to prominence in a subjective listening test the percentage of explained variance of the linear regression (83 %) was good for these two narrowband methods.

However, 1/3 octave band analysis are still relied on in some jurisdictions to determine the presence of audible tones for a variety of source, not just wind turbines. As such, a re-analysis was conducted as part of the PT 61400-11-2 effort to compare the effectiveness of 1/3 octave band methods in identifying prominent tones in the 31 listening test stimuli. While potentially not as robust as narrowband methods, 1/3 octave band methods are the basis for evaluating tones for non-wind turbine sources in some jurisdictions, particularly in the USA [7] [8]. Alberta, Canada also relies on a 1/3 octave band method for the evaluation of low frequency tones [9]. As 1/3 octave band spectra are readily data logged by most modern sound level instrumentation, the challenges associated with audio recordings and narrowband data processing methods may be avoidable in some circumstances.

2. Working group PT 61400-11-2

In 2018 the working group PT 61400-11-2 was established "Wind energy generation systems - Part 11-2: Measurement of wind turbine noise characteristics in receptor position" [10], The working group currently has 35 members from Europe, North America and Asia. The working group is a part of IEC TC 88 "Wind energy generation systems".

2.1 Scope

The scope of the working group is as follows:
"To establish standardized techniques and methods applicable for measurement of noise immission. The technical specification focuses on compiling methods for quantification of wind turbine noise in one or a set of documents in order to ensure a common reference taking the characteristics of wind turbine operational parameters into account.

Quantification of the acoustics phenomena at the receiver position includes the following topics:
- *amplitude modulation;*
- *low frequency noise;*
- *impulsivity;*
- *tonality;*
- *sound pressure levels;*
- *rating level, which includes adjustments for the above-mentioned phenomena, should be suggested if possible."*

3. Methods for analysing tonality

The following section describes the 1/3 octave band and narrow band analysis methods used in this analysis.

3.1 1/3 Octave Band Methods

The 1/3 octave band methods evaluated in this study include ISO 1996-2: 2017[11], ANSI S12.9-2005/Part 4[12] and ANSI S12.9-2013/Part 3[13] each which are briefly described below.

ISO 1996-2:2017

The informative Annex K of ISO 1996-2 (2017) provides guidance for identifying a prominent discrete tone, by comparing the time-average sound pressure level in some one-third-octave band with the time-average sound pressure level in the adjacent two one-third-octave bands. For a prominent discrete tone to be identified as present, the time-average sound pressure in the one-third-octave band of interest is required to exceed the time-average sound pressure levels of both adjacent one-third-octave bands by some constant level difference. The constant level difference may vary with frequency. Possible choices for the level difference are:

- 15 dB (25 Hz to 125 Hz)
- 8 dB (160 Hz to 400 Hz)
- 5 dB (500 Hz to 10 kHz)

A-weighting is not mentioned in the informative Annex K. Therefore, it is presumed this analysis is conducted on the linear spectra and the A-weighted version of the analysis included in this paper for completeness considerations.

ANSI S12.9-2005/Part 4

The informative Annex C of ANSI S12.9 Part 4 (2005) provides guidance for identifying a prominent discrete tone consistent with ISO 1996-2 and uses the same criteria.

ANSI S12.9-2013/Part 3

The informative Annex B of ANSI S12.9 Part 3 (2013) uses the same criteria to that of ISO 1996-2 (2017) and S12.9 Part 4. It differs in that it requires that that "the equivalent-continuous sound pressure level in the one-third octave band of interest is required to **exceed the arithmetic average** of the equivalent-continuous sound pressure level for the two adjacent one-third octave bands by [the criteria]..." [emphasis added].

While not explicitly stated, it is expected that one first confirms that the 1/3 octave band level of the linear spectra exceeds the two adjacent bands and then compares average level difference to prominence criteria. The lack of a requirement identifying that the octave band of interest must also exceed each of the adjacent bands can result in a steeply declining spectrum being inadvertently identified as containing prominent discrete tones. For example, in a 1/3 octave band sequence of 50 dB, 45 dB and 15 dB, the 45 dB band would be identified as a prominent tone. This is likely an oversight as it was expected that one would only be utilizing this method in response to a tonal complaint, thus some level of a tone would be present. The situation where one is automating the analysis of large datasets for tonal analysis was likely not envisioned when this was drafted.

The States of Illinois and Oregon utilize a 1/3 octave band approach in their regulations for determining pure tones for all sources, not just wind turbines.[1] The overall sound limits in Illinois are based on linear octave bands (e.g., there are no overall A-weighted limits, solely octave band limits). Oregon also has optional linear octave band criteria in Oregon [7]. Therefore, the one-third octave band tonal limits are evaluated based on linear spectra given there is no mention of A-weighting in pure tone part of the regulations. For reference, the Illinois regulations are stated below and clearly identify that the 1/3 octave band of interest must exceed each of the adjacent bands[2]:

> *"...sound, having a one-third octave band sound pressure level which, when measured in a one-third octave band at the preferred frequencies, exceeds the arithmetic average of the sound pressure levels of the two adjacent one-third octave bands on either side of such one-third octave band by:*
>
> 1. *5 dB for such one-third octave band with a center frequency from 500 Hertz to 10,000 Hertz, inclusive. Provided: such one-third octave band sound pressure level exceeds the sound pressure level of each adjacent one-third octave band, or;*
>
> 2. *8 dB for such one-third octave band with a center frequency from 160 Hertz to 400 Hertz, inclusive. Provided: such one-third octave band sound pressure level exceeds the sound pressure level of each adjacent one-third octave band, or;*
>
> 3. *15 dB for such one-third octave band with a center frequency from 25 Hertz to 125 Hertz, inclusive. Provided: such one-third octave band sound pressure level exceeds the sound pressure level of each adjacent one-third octave band."*

Summary of 1/3 octave band approaches evaluated

All of the evaluated 1/3 octave band options in this study utilize the same criteria (e.g., 5, 8, 15 dB). Potential differences include if the spectra evaluated is A-weighted or linear and both cases have been evaluated in this study. It is presumed that lack of specificity with regards to satisfying the "*Provided: such one-third octave band sound pressure level exceeds the sound pressure level of each adjacent one-third octave band*" criteria is an oversight. Nonetheless, the stimuli in this analysis all complied with this requirement. Thus, the remaining difference in the evaluation of the adjacent bands is either requiring the band of interest:

1) to exceed each (both) of the adjacent bands by the criteria or
2) to exceed the arithmetic average of the two adjacent bands by the criteria.

In this analysis Option 1 is identified as the ISO 1996-2 or ANSI S12.9 Part 4 approach and Option 2 is identified as the ANSI S12.9 Part 3 approach.

---

[1] Unconfirmed at the time of this writing, but it is suspected that the Illinois or Oregon regulations predate the ANSI standard.

[2] Oregon's regulations are similarly clear on this point.

ISO 1996-2 and ANSI S12.9 Part 4 (in both the 2005 and 1997 versions) use the "time-average sound pressure levels of both adjacent one-third-octave bands by some constant level difference" phrasing. The addition of the "time-averaged" phrase in the ANSI S12.9 Part 4 standard may have been intended to avoid confusion with the Illinois or Oregon use of "*arithmetic average of the sound pressure levels*". However, neither the Oregon nor Illinois regulations intended to imply the time varying sound pressure levels were to be arithmetically averaged. Rather in Oregon the "sound pressure level" referred to in the regulations are specified as an hourly L_{50} and in Illinois as L_{eq}. The result is that a strict reading of ISO 1996-2 or ANSI S12.9 Part 4 indicates that each (both) side band need to exceed the criteria, rather than the average of the sidebands per ANSI S12.9 Part 3.

3.2 Investigated methods for narrowband analysis

The narrowband method analysis utilized in this study was the ISO PAS 20065. A comparison of this method to other narrowband methods can be found in reference [6].

ISO PAS 20065

The ISO PAS 20065 method is the preferred engineering method in the informative annex J of ISO 1996-2:2017. The ISO PAS 20065 methods is based on the German DIN 45681 method and is a narrowband method.

ISO 1996-2:2017 suggests that an appropriate tonal adjustment, K_T, by using ISO PAS 20065 could be:

$\Delta L \leq 2$ dB:	$K_T = 0$ dB
2 dB $< \Delta L \leq 9$ dB:	$K_T = 3$ dB
$\Delta L > 9$ dB:	$K_T = 6$ dB

4. Data Collection and Processing

This study reanalysed previously collected data including the listening test results and narrowband analysis presented in reference [6] and [14]. These prior analyses are briefly summarized below. The additional 1/3 octave band analysis is also briefly described.

4.1 Stimuli

A very large library of original recordings of wind turbines was reviewed and 24 samples were specifically selected given the presence of tones. Some recordings were collected close to a turbine as part of a sound power level measurements (following either the IEC 61400-11 standard or the Danish noise regulation for wind turbines, which in terms of measurement positions are identical to IEC 61400-11). Other recordings were collected at receiver/residential locations or a distance from the turbine(s) identical to nearest receiver. If several equally prominent tones in different critical bands were present in a recording, the least prominent were attenuated to facilitate an unambiguous and comparable assessment in the listening test and tone analysis. In addition, seven samples of stationary industrial noise with tones were included, yielding a total of 31 stimuli. These stimuli were specifically developed to have a large range of tonal audibility and tone frequency. One should not misinterpret these stimuli as indicating that tones are always present. Each stimuli consisted of a mono recording with a duration of 20 seconds. The stimuli were level aligned so that the resulting A-weighted levels were the same for all samples. The intended listening level was 50 dB(A).

4.2 Subjective data

Formal listening tests were performed to assess the subjective prominence which was compared to the tonal audibility (ΔL). The listening tests was performed in 2012-2013 and is reported in [14]. The participant was presented for each stimuli in randomized order, and asked to evaluate how prominent the tones were, on an adjustable scale by using a slider as shown in Figure 1 (from the listening test). The position on the scale is then converted to a number in mm ranging from 0 to 150 mm. For this study will the tonal prominence value of 64.4 mm be used to divide the stimuli into either "non-prominent tones" or "prominent tones"[3]. As can be seen in Figure 1 the prominence value of 64.4 mm corresponds to the response "Prominent".

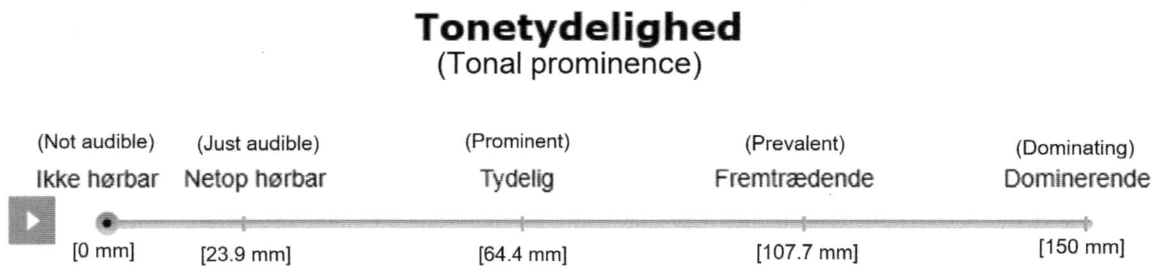

Figure 1 Prominence scale from the listening test. Both the original Danish wording and their equivalent English translation is shown, together with the words position on the scale.

4.3 Narrowband data

For this analysis the found tone frequency (f_T) and tonal audibility (ΔL) by use of ISO PAS 20065 will be used for comparison. The stimuli used had a found tone frequency (f_T) ranging from 70 to 7,000 Hz and tonality audibility (ΔL) per ISO PAS 20065 of -10 dB to 15 dB. The data is taken from Table 2 in [6].

4.4 1/3 octave band data

The 31 stimuli have been reanalysed by 1/3-octave filters conforming to IEC 61260using the program noiseLAB (from FORCE Technology). The analysis was performed both unweighted (lin) and A-weighted.

4.5 Summary

ISO 1996-2:2017 suggests that an appropriate tonal adjustment (penalty), K_T, could be 3 dB for medium tonal audibility (2 dB < ΔL ≤ 9 dB) and 6 dB for high tonal audibility (ΔL > 9 dB).
In this study it will be assumed that the stimuli are prominent if the tonal prominence are above 64.4 mm. As noted previously, when compared to prominence in a subjective listening test the percentage of explained variance of the linear regression (83 %) was good for the narrowband method [6].

[3] Naturally, an exact threshold/value cannot be set for which the stimuli are prominent or not, so care should be taken in the area close to 64.4 mm

5. Results

Relevant results from the narrowband analysis, listening test and 1/3 octave band analysis has been combined in Table 1.

No#	f_T	ΔL	Subjective prominence	On Boundary/ No tone	1/3 Octave Criteria	Tone detected by ANSI S12.9 Part 3		Tone detected by ISO 1996-2 + ANSI S12.9 Part 4	
	[Hz]	[dB]	[mm]		[dB]	(A)	(Z/lin)	(A)	(Z/lin)
O01	1930	14,7	101,2	No	5	Yes	Yes	No	No
O09	1776	14,6	103,2	Yes	5	No	No	No	No
M10	380	14,5	100,1	No	8	Yes	Yes	Yes	Yes
M03	70	12,7	99,6	Yes	15	No	No	No	No
M06	6986	12,7	93,9	()	5	Yes	Yes	Yes	Yes
M08	888	12,3	93,6	Yes	5	No	No	No	No
O08	1776	12,1	79,1	Yes	5	No	No	No	No
M01	1776	11,2	85,2	Yes	5	No	No	No	No
O20	698	10,7	76,5	Yes	5	No	No	No	No
O13	2096	9,7	97,6	No	5	Yes	Yes	No	No
O19	104	9,7	52,0	No	15	No	No	No	No
O10	380	9,6	79,9	No	8	Yes	Yes	Yes	No
O05	136	9,4	57,0	()	15	No	No	No	No
M05	748	9,1	67,5	No	5	Yes	Yes	Yes	Yes
O02	740	8,2	67,1	No	5	Yes	Yes	Yes	No
M04	1344	7,8	93,0	No	5	Yes	Yes	Yes	Yes
O11	224	7,5	77,1	Yes	8	No	No	No	No
O12	1776	7,5	78,4	Yes	5	No	No	No	No
O17	1172	6,4	77,4	No	5	Yes	Yes	Yes	Yes
O14	152	5,6	68,6	No	8	Yes	Yes	Yes	No
O04	6984	5,3	53,3	()	5	Yes	Yes	No	No
M07	98	4,9	58,1	No	15	No	No	No	No
O15	1750	4,9	54,2	Yes	5	No	No	No	No
O16	380	4,7	53,2	No	8	No	No	No	No
O06	126	4,4	50,0	No	15	No	No	No	No
O18	1776	3,6	55,4	Yes	5	No	No	No	No
M09	162	2,0	61,4	No	8	No	No	No	No
O07	380	-0,1	29,3	No	8	No	No	No	No
M02	380	-5,0	16,1	No	8	No	No	No	No
O03	142	-9,3	13,7	No tone	8	No	No	No	No
O21	4340	-9,9	13,4	No tone	5	No	No	No	No

Table 1 Overview of the 31 stimuli and the analysed data. f_T and ΔL is found by ISO PAS 20065. The prominence score is from the listening test. The () indicate the tone was near the 1/3 octave boundary. For stimuli O03 and O21 there is no tone in the signal. The 1/3 octave criteria threshold is determined from f_T by ISO 1996-2 and ANSI S12.9. The last four columns are the results by using the 1/3 octave band methods. The blue colours in column 3 marks either medium tonal audibility (2 dB < ΔL ≤ 9 dB) or high tonal audibility (ΔL > 9 dB). The yellow colours in column 4 marks whether the tonal prominence is above or below 64.4 mm.

The information summarized in Table 1 is depicted in more detail graphically in Figure 2.

Figure 2 shows in the top plot the narrowband analysis, marks the tone frequency, f_T, and displays the tonal audibility, ΔL. In the middle plot is shown the results of the A-wighted 1/3 octave band analysis. The applicable one-third octave band tonal threshold (5, 8 or 15 dB) is displayed. The 1/3 octave band in question is shown in the middle of the plot surrounded by the adjecent 1/3 octave bands. On each adjecent 1/3 octave band is displayed the difference between the level of the center 1/3 octave band to the level of the adjecent 1/3 octave band. On the center 1/3 octave band is displayed the arithmetic average difference between the levels of the adjecent bands and the level of the center band. If the text on the center 1/3 octave bands is red a tone is detected by ANSI S12.9 Part 3 (if the text is black a tone is not detected). If the text on both the adjacent bands is red, a tone was identified by ISO 1996-2 and ANSI S12.9 Part 4 (note that both side bands must exceed the threshold for a tone to be found and the text to be red). For the bottom plot is shown the results of the non-weighted (linear/Z-weighted) 1/3 octave band analysis. Otherwise the displayed information is the same as the middle plot. For the right plot is shown the subjective prominence based on the listening test results. Corresponding plots for the other 30 stimuli is shown in Figure 6 to Figure 35 in the Appendix.

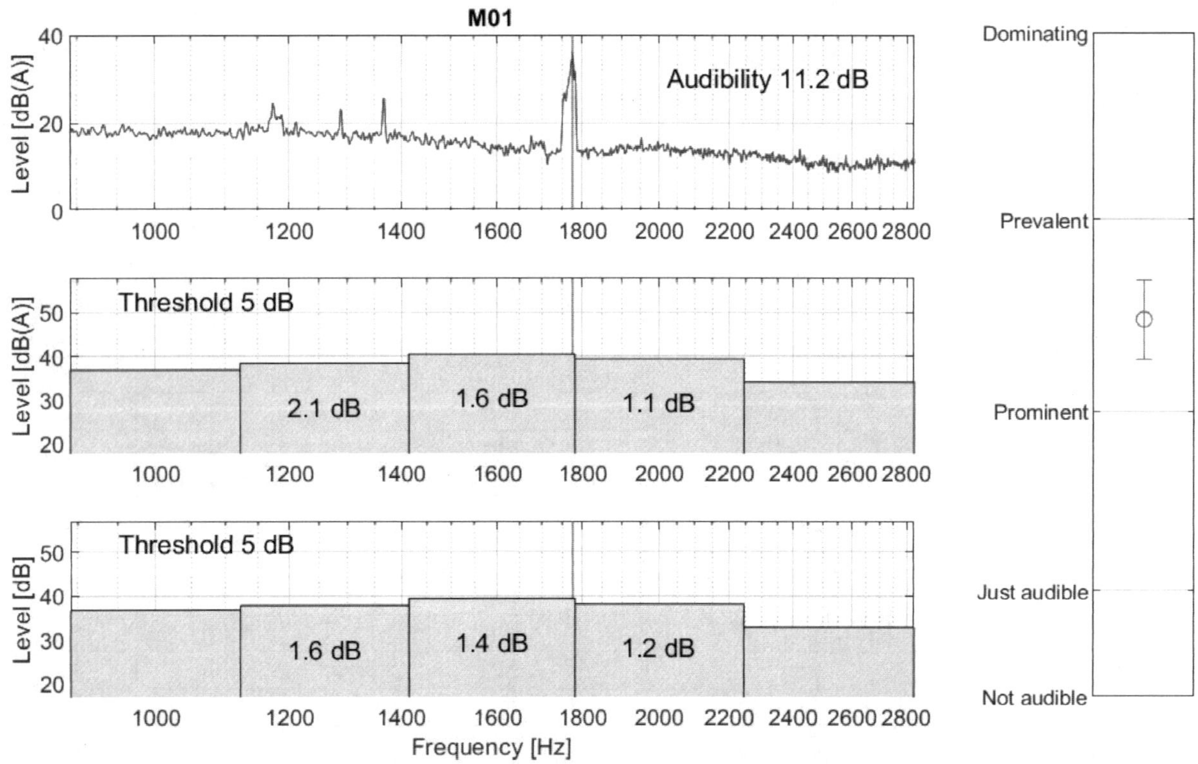

Figure 2 Top left: Narrowband analysis and narrowband results. Middle left: A-weighted 1/3 octave band analysis and results. Bottom left: Unweighted 1/3 octave band analysis and results. Right: Subjective prominence.

Figure 3 to Figure 5 show the subjective prominence (perceptual assessment) as a function of the calculated tonal audibility for the 31 stimuli, together with their 95% confidence intervals. Note that the data in each of these figures is the same, as the tonal audibility and prominence of the 31 stimuli are unchanged. What changes is the colour-coding of the symbols. The colour coding depicts the detection or non-detection of the 1/3 octave band analysis. Red indicates that tones were detected, and black indicates that tones were not detected. Additionally, the shape of the markers divides tonal stimuli into subcategories – triangles and filled circles respectively indicate the tones were on or near the 1/3 octave band boundary. Two vertical blue lines divides the figures according to the suggested tonal adjustment, K_T, as described in ISO 1996-2:2017, and a horizontal yellow line divides the figures into prominent and non-prominent tones.

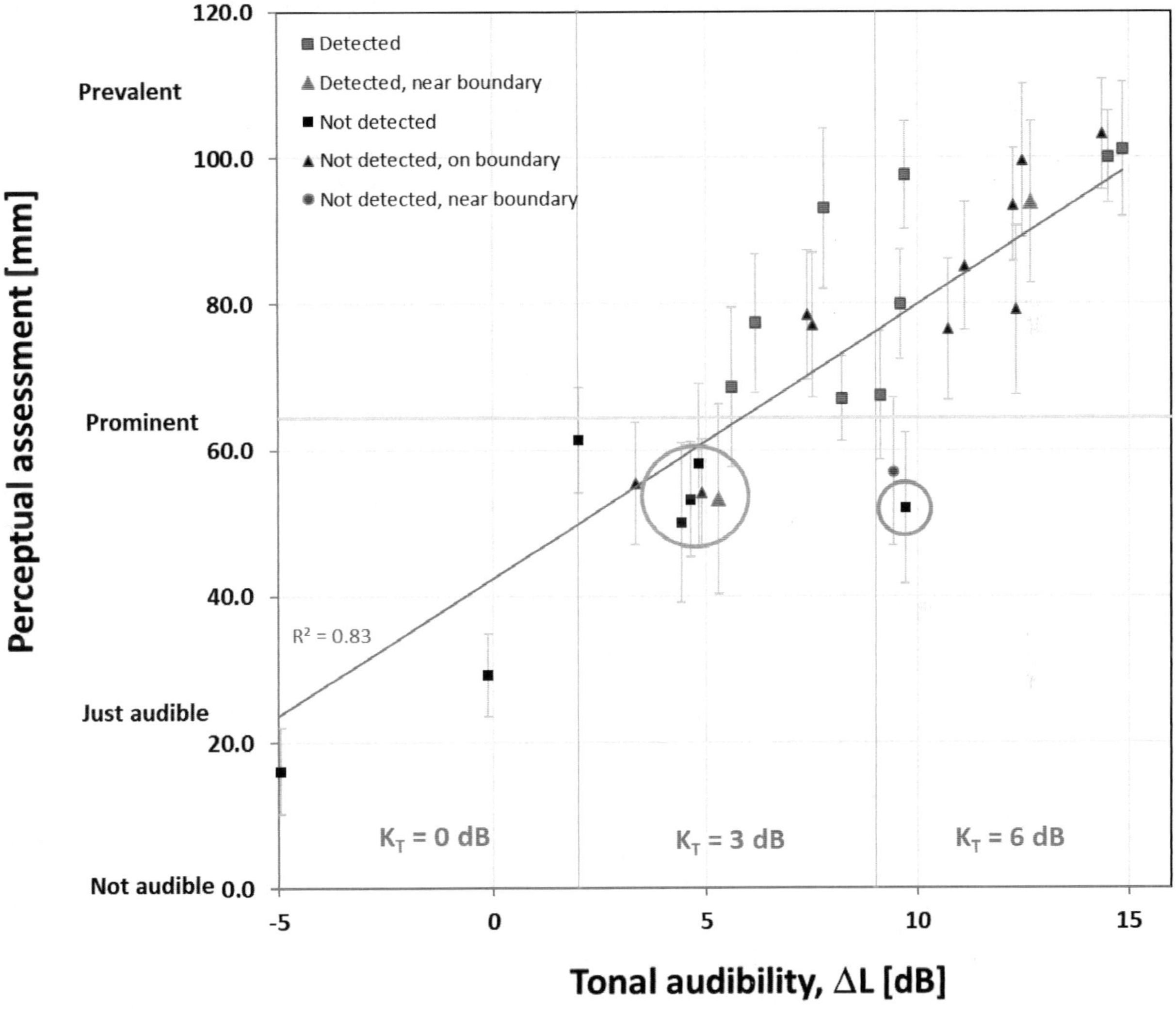

Figure 3 Perceptual assessment, ISO/PAS tonal audibility and ANSI S12.9 Part 3 1/3 octave band analysis

The further into the top right corner the markers are placed (high reported perceptual assessments and higher calculated tonal audibility) the more important it is for the 1/3 octave band methods to detect a tone.[4] If the 1/3 octave band methods are effective, the top right portion should contain only red coloured symbols). As can be seen in Figure 3 there are both black and red markers in the top right of the figure (the part of the figure with both a suggested tonal adjustment and

[4] False detections would also be problematic and would be indicated by red symbols in the bottom left of which there are none.

perceptually identified as “prominent”, above the yellow line and in the K_T =3 or 6 region). Closer scrutiny reveals that all the black markers are triangle shaped, indicating that the tone frequency is just on the border between two adjacent 1/3 octave bands. Thus, it can be concluded that the ANSI S12.9 Part 3 method which evaluates tones based on the averaging of the two adjacent bands works as intended except when the tone frequency falls near the boundary between two adjacent 1/3 octave bands. This is a recognized shortcoming of 1/3 octave band tonality methods that users of such methods should be aware of. As shown in Table 1, there was no difference in detections between the A-weighted and Z-weighted/linear spectra for these stimuli, thus Figure 3 presents both findings.

For Figure 4 the color-coding shows the detection of the A-weighted ISO 1996-2 1/3 octave band method. Compared to Figure 3 it can be seen that two tones with both high tonal audibility and high subjective prominence (in the upper right) has not been detected and their tone frequency is not in the boundary region between 2 adjacent 1/3 octave band (black square). As such it can be concluded that for these stimuli the A-weighted ISO 1996-2 (ANSI S12.9 Part 4) 1/3 octave band method has a poorer detection than the ANSI S12.9 Part 3 method (averaging of sidebands).

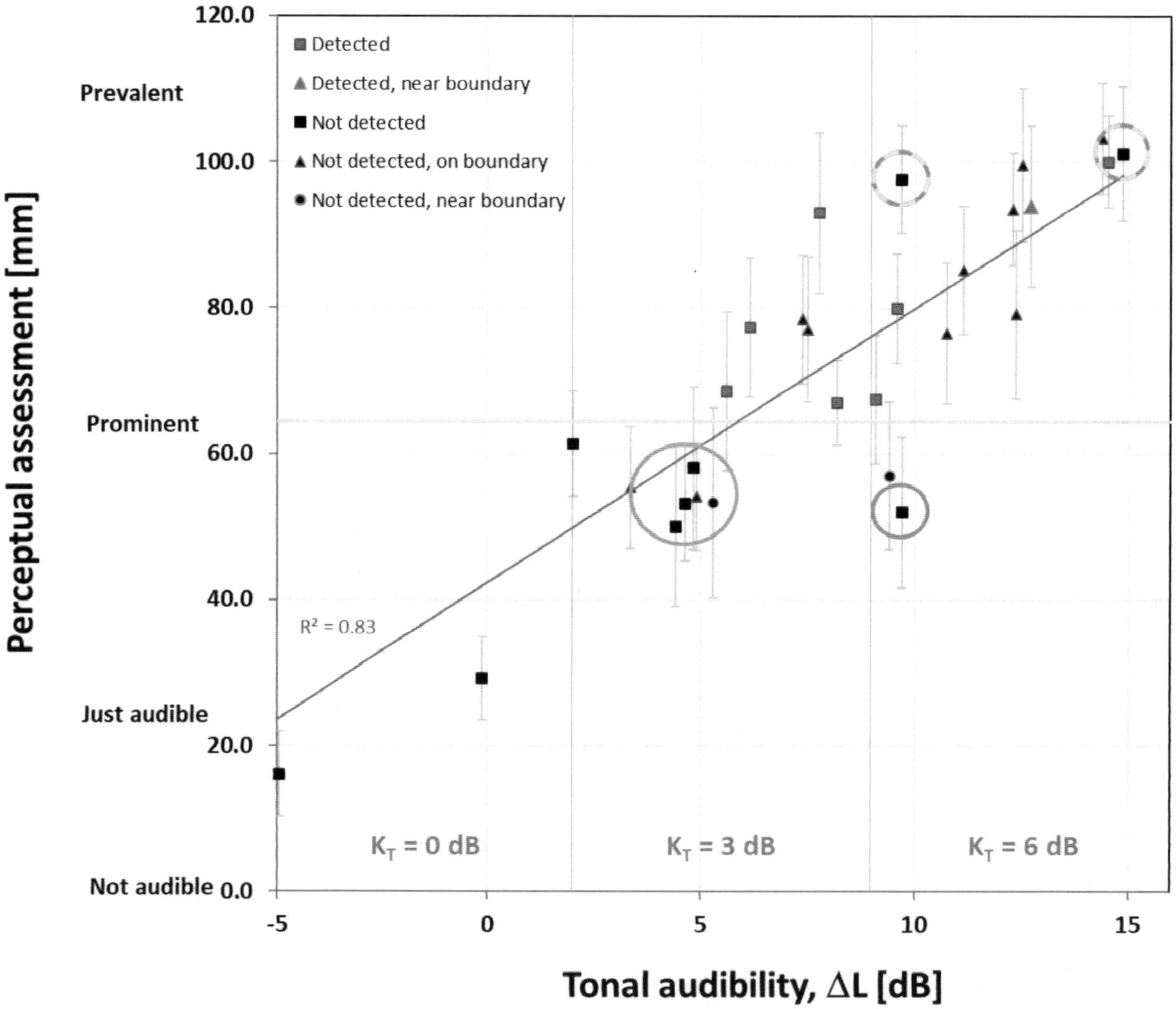

Figure 4 Perceptual assessment, ISO/PAS audibility and ISO 1/3 octave analysis (A)

For Figure 5 the color-coding shows the detection of the unweighted ISO 1996-2 (ANSI S12.9 Part 4) 1/3 octave band method. Compared to Figure 4 it can be seen that even more tones with both high tonal audibility and high subjective prominence has not been detected and these undetected tones were not in the boundary region between 2 adjacent 1/3 octave band. As such it can be concluded that the unweighted ISO 1996-2 1/3 octave band method has a poorer detection than both the ANSI S12.9 Part 3 method (averaging of sidebands) and the A-weighted implementation of the ISO 1996-3 1/3 octave band method for these stimuli.

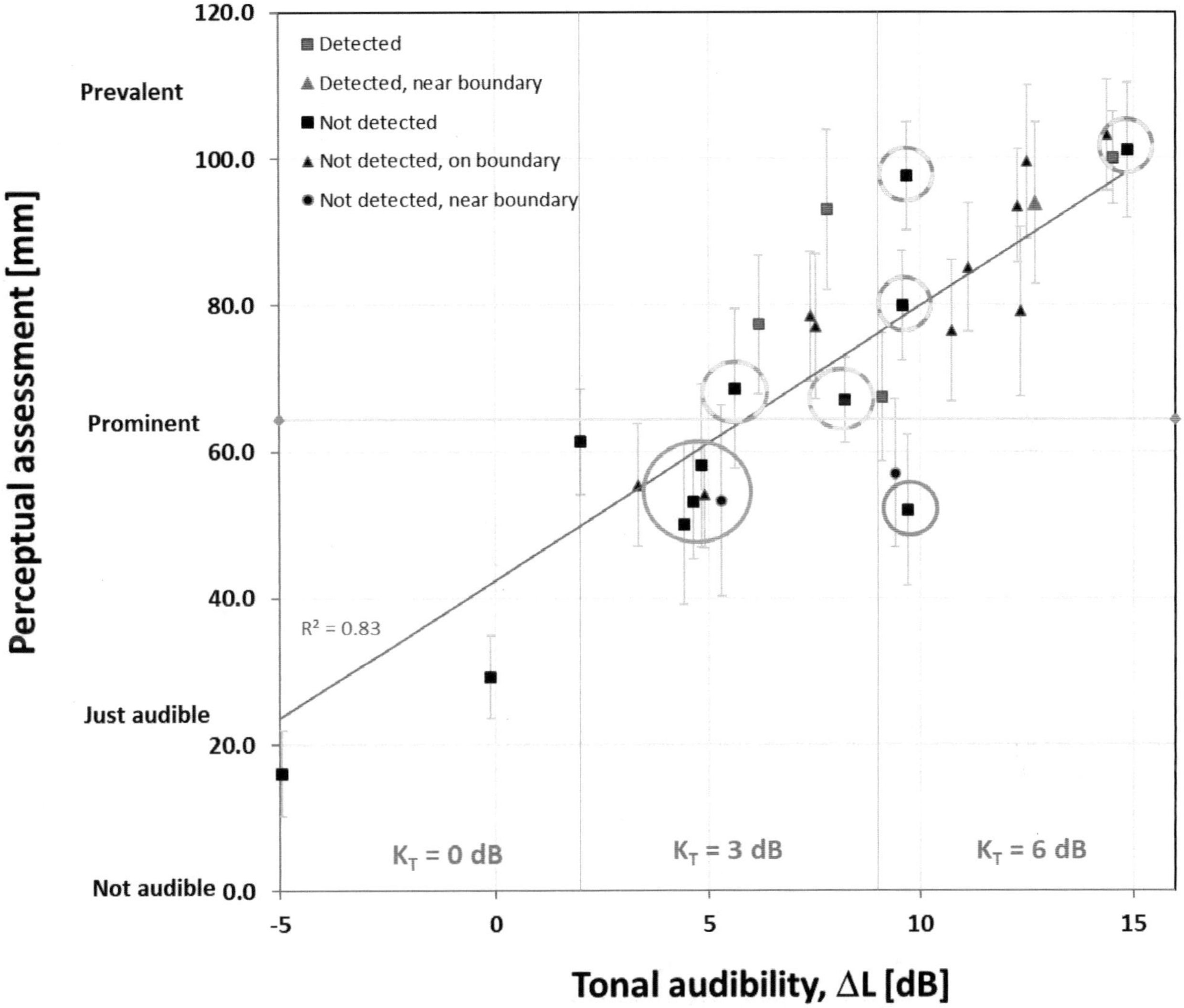

Figure 5 Perceptual assessment, ISO/PAS audibility and ISO 1/3 octave analysis (lin)

6. Discussion

The 31 stimuli in Table 1 can be divided into 4 groups, where the divided stimuli and results are shown in Table 2 to Table 5. The tables are sorted by calculated tonal audibility (ΔL).

The first group (Table 2) is the 16 stimuli which contains a tone and where the tone frequency is not close to the boundary between adjacent 1/3 octave bands. As can be seen the ANSI S12.9-Part 3 method (average of the side bands) in general detects the stimuli for tonal audibility (ΔL) higher than 5 dB. The only exception is stimuli O19 (ΔL of 9.7) which is a low frequency tone (f_T of 104), a frequency for which the ANSI/ISO criteria is 15 dB. While Stimuli O19 has a tonal audibility of 9.7 dB, the subjective prominence of 52 mm is quite low. For ISO 1996-2 there are 4 or 6 (A or Z weighted respectively) false negatives in the same tonal audibility range (ΔL greater than 5) and a subjective prominence higher than 64 mm.

No#	f_T	ΔL	Promi nence	On Boundary	Criteria	ANSI S12.9 Part 3		ISO 1996-2 + ANSI S12.9 Part 4	
	[Hz]	[dB]	[mm]		[dB]	(A)	(Z/lin)	(A)	(Z/lin)
O01	1930	14,7	101,2	No	5	Yes	Yes	No	No
M10	380	14,5	100,1	No	8	Yes	Yes	Yes	Yes
O13	2096	9,7	97,6	No	5	Yes	Yes	No	No
O19	104	9,7	52,0	No	15	No	No	No	No
O10	380	9,6	79,9	No	8	Yes	Yes	Yes	No
M05	748	9,1	67,5	No	5	Yes	Yes	Yes	Yes
O02	740	8,2	67,1	No	5	Yes	Yes	Yes	No
M04	1344	7,8	93,0	No	5	Yes	Yes	Yes	Yes
O17	1172	6,4	77,4	No	5	Yes	Yes	Yes	Yes
O14	152	5,6	68,6	No	8	Yes	Yes	Yes	No
M07	98	4,9	58,1	No	15	No	No	No	No
O16	380	4,7	53,2	No	8	No	No	No	No
O06	126	4,4	50,0	No	15	No	No	No	No
M09	162	2,0	61,4	No	8	No	No	No	No
O07	380	-0,1	29,3	No	8	No	No	No	No
M02	380	-5,0	16,1	No	8	No	No	No	No

Table 2 Overview of the 16 stimuli which contains a tone and with the tone frequency not close to the boundary between adjecent 1/3 octave bands. The stimuli are sorted according to tonal audibility, ΔL. See further information in Table 1.

The second group is shown in Table 3, where 3 stimuli have a tone frequency near the boundary between adjacent 1/3 octave bands. As can be seen sometime tones are detected and sometimes not confirming the 1/3 octave band methods have challenges with tones near the boundary.

No#	f_T	ΔL	Promin ence	On Boundary	Criteria	ANSI S12.9 Part 3		ISO 1996-2 + ANSI S12.9 Part 4	
	[Hz]	[dB]	[mm]		[dB]	(A)	(Z/lin)	(A)	(Z/lin)
M06	6986	12,7	93,9	()	5	Yes	Yes	Yes	Yes
O05	136	9,4	57,0	()	15	No	No	No	No
O04	6984	5,3	53,3	()	5	Yes	Yes	No	No

Table 3 Overview of the 3 stimuli which contains a tone and with the tone frequency near the boundary between adjecent 1/3 octave bands. The stimuli are sorted according to tonal audibility, ΔL. See further information in Table 1.

The third group is shown in Table 4, which is the 10 stimuli which contains a tone and with the tone frequency in the boundary between adjacent 1/3 octave bands. As can be seen none of the 1/3 octave tonality methods detect a tone. Since this is a recognised shortcoming with the 1/3 octave band methods this is as expected.

No#	f_T	ΔL	Prominence	On Boundary	Criteria	ANSI S12.9 Part 3		ISO 1996-2 + ANSI S12.9 Part 4	
	[Hz]	[dB]	[mm]		[dB]	(A)	(Z/lin)	(A)	(Z/lin)
O09	1776	14,6	103,2	Yes	5	No	No	No	No
M03	70	12,7	99,6	Yes	15	No	No	No	No
M08	888	12,3	93,6	Yes	5	No	No	No	No
O08	1776	12,1	79,1	Yes	5	No	No	No	No
M01	1776	11,2	85,2	Yes	5	No	No	No	No
O20	698	10,7	76,5	Yes	5	No	No	No	No
O11	224	7,5	77,1	Yes	8	No	No	No	No
O12	1776	7,5	78,4	Yes	5	No	No	No	No
O15	1750	4,9	54,2	Yes	5	No	No	No	No
O18	1776	3,6	55,4	Yes	5	No	No	No	No

Table 4 Overview of the 10 stimuli which contains a tone and with the tone frequency in the boundary between adjecent 1/3 octave bands. The stimuli are sorted according to tonal audibility, ΔL. See further information in Table 1.

The last group is shown in Table 5, which is the 2 stimuli which do not contains a tone. As can be seen both cases are correctly not detected as a tone.

No#	f_T	ΔL	Prominence	No tone	Criteria	ANSI S12.9 Part 3		ISO 1996-2 + ANSI S12.9 Part 4	
	[Hz]	[dB]	[mm]		[dB]	(A)	(Z/lin)	(A)	(Z/lin)
O03	142	-9,3	13,7	No tone	8	No	No	No	No
O21	4340	-9,9	13,4	No tone	5	No	No	No	No

Table 5 Overview of the 2 stimuli which does no contain a tone. The stimuli are sorted according to tonal audibility, ΔL. See further information in Table 1.

7. Conclusions

31 stimuli have been analysed with 1/3 octave band tonality methods and these findings were compared those of the ISO PAS 20065 narrowband method and to subjective prominence scores (listening test results). Most of the stimuli were based on modified wind turbine noise recordings and 29 of the 31 stimuli contain one or more tones.

The 31 stimuli were reused from previous investigations. Coincidently for 13 out of the 29 stimuli with tones, their tone frequency is in or near the boundary area between two adjacent 1/3 octave bands. Tones in this area is a known shortcoming of the 1/3 octave band tonality methods. Not surprisingly the 1/3 octave band methods do not detect a tone for the 10 cases with tone frequency in the boundary region and were inconsistent in identifying the 3 tones near the boundary.

Boundary and near boundary areas set aside the ANSI S12.9 Part 3 method which is based on evaluation of the average of the side bands, detects all tones in this study which both have a either medium ($\Delta L > 2$ dB) or high ($\Delta L > 9$ dB) tonal audibility (calculated with the ISO PAS 20065 method) and a prominent (tonal prominence > 64.4 mm) tone (from the listening test). In the United States, this 1/3 octave band approach is used in the States of Oregon and Illinois.

The 1/3 octave band method described in ISO 1996-2 and ANSI S12.9 Part 4 are identical and may require both side bands, rather than the average of the sidebands, to exceed the criteria (5, 8, 15 dB). In this study, this method did not detect all tones which have both an either medium or high tonal audibility and a prominent tone. In this study, the tonal detection rate was improved slightly if the 1/3 octave band analysis is performed on A-weighted rather than unweighted spectra.

Those relying on 1/3 octave band methods may need to investigate the presence of tones near the boundary. If present, narrow band methods may be helpful and while not investigated in this study, one might consider altering the 1/3 octave band center frequencies to avoid tones near the boundary.

Lastly, this study evaluated 31 stimuli with a wide range of tonal audibility and tone frequency but does not necessarily represent the complete range of possible tones or tonal combinations which can found when a subjective complaint or concern is assessed. As always, professional judgement and experience is required. All methods have their advantages and disadvantages and this paper has highlighted some of the factors to be aware of.

References

[1] T. H. Pedersen, "The 'Genlyd' Noise Annoyance Model," 2007.

[2] *Bekendtgørelse om støj fra vindmøller*, no. 135. Denmark: Miljø- og Fødevareministeriet, 2019, pp. 1–24.

[3] IEC, "IEC 61400-11 Wind turbines - Part 11: Acoustic noise measurement techniques." 2012.

[4] ISO, "ISO 1996-2:2007 Acoustics — Description, measurement and assessment of environmental noise — Part 2: Determination of environmental noise levels." 2007.

[5] ISO, "ISO/PAS 20065:2016 Acoustics – Objective method for assessing the audibility of tones in noise – Engineering method." p. 33, 2020.

[6] L. S. Søndergaard, C. Thomsen, and T. H. Pedersen, "Prominent tones in wind turbine noise - Round-robin test of the IEC 61400-11 and ISO/PAS 20065 methods for analysing tonality content," in *8th International Conference on Wind Turbine Noise*, 2019, no. June, pp. 1–12.

[7] M. Bastasch, "Regulation of Wind Turbine Noise in the Western United States," in *First International Meeting on Wind Turbine Noise : Perspectives for Control*, 2005, no. October.

[8] J. D. Barnes and R. A. Gomez, "A Variety of Wind Turbine Noise Regulations in the United States –2007," in *Second International Meeting on Wind Turbine Noise*, 2007.

[9] Alberta Utilities Commission, *Rule 012 Noise Control*. Alberta, 2021.

[10] IEC, "IEC/TC 88/PT 61400-11-2 membership and scope." [Online]. Available: https://www.iec.ch/dyn/www/f?p=103:14:11826201584989::::FSP_ORG_ID:22106. [Accessed: 20-Mar-2021].

[11] ISO, "ISO 1996-2:2017 Acoustics – Description, measurement and assessment of environmental noise – Part 2: Determination of sound pressure levels." 2017.

[12] ANSI/ASA, "ANSI/ASA S12.9-2005/Part 4 Quantities And Procedures For Description And Measurement Of Environmental Sound - Part 4: Noise Assessment And Prediction Of Long-Term Community Response." 2005.

[13] ANSI/ASA, "ANSI/ASA S12.9-2013/Part 3 Quantities And Procedures For Description And Measurement Of Environmental Sound - Part 3: Short-Term Measurements With An Observer Present." 2013.

[14] L. S. Søndergaard and T. H. Pedersen, "Tonality in wind turbine noise . IEC 61400-11 ver . 2 . 1 and 3 . 0 and the Danish / Joint Nordic method compared with listening tests," in *5th International Meeting on Wind Turbine Noise*, 2013, no. August, pp. 2–10.

Appendix

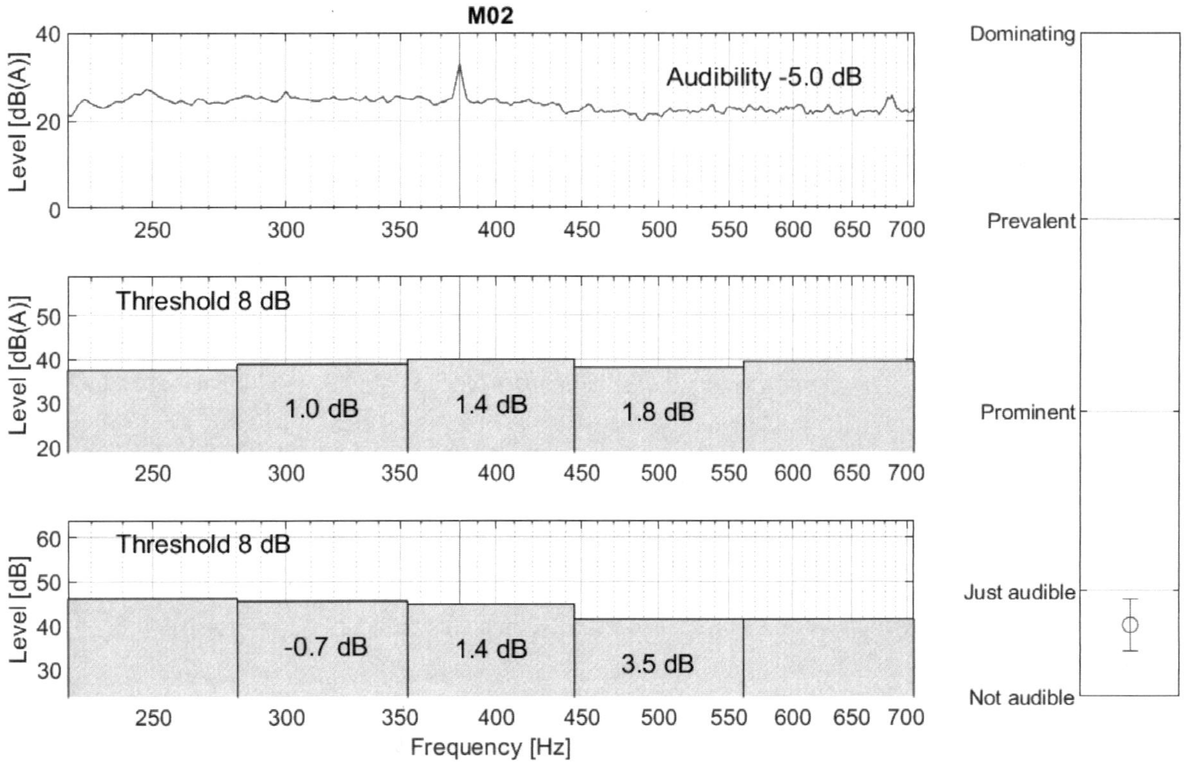

Figure 6 Detailed result for stimuli M02. See further information in Figure 2 at page 8 and in section 5.

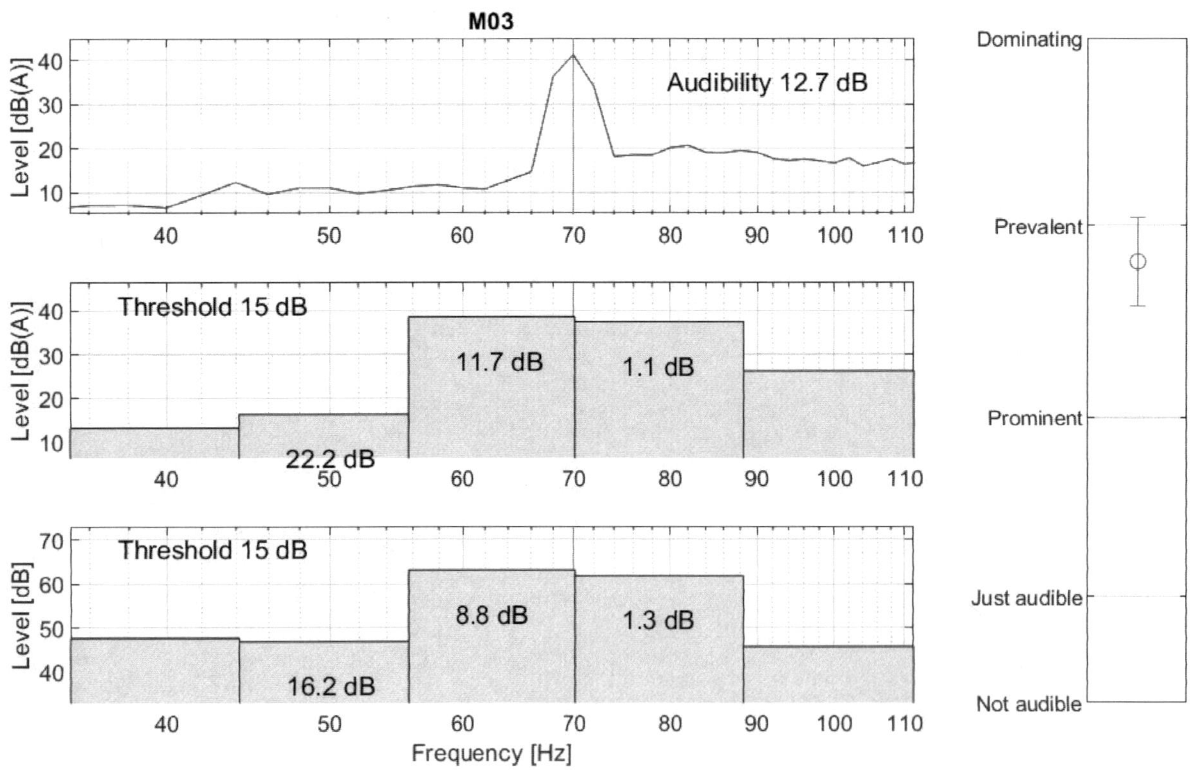

Figure 7 Detailed result for stimuli M03. See further information in Figure 2 at page 8 and in section 5.

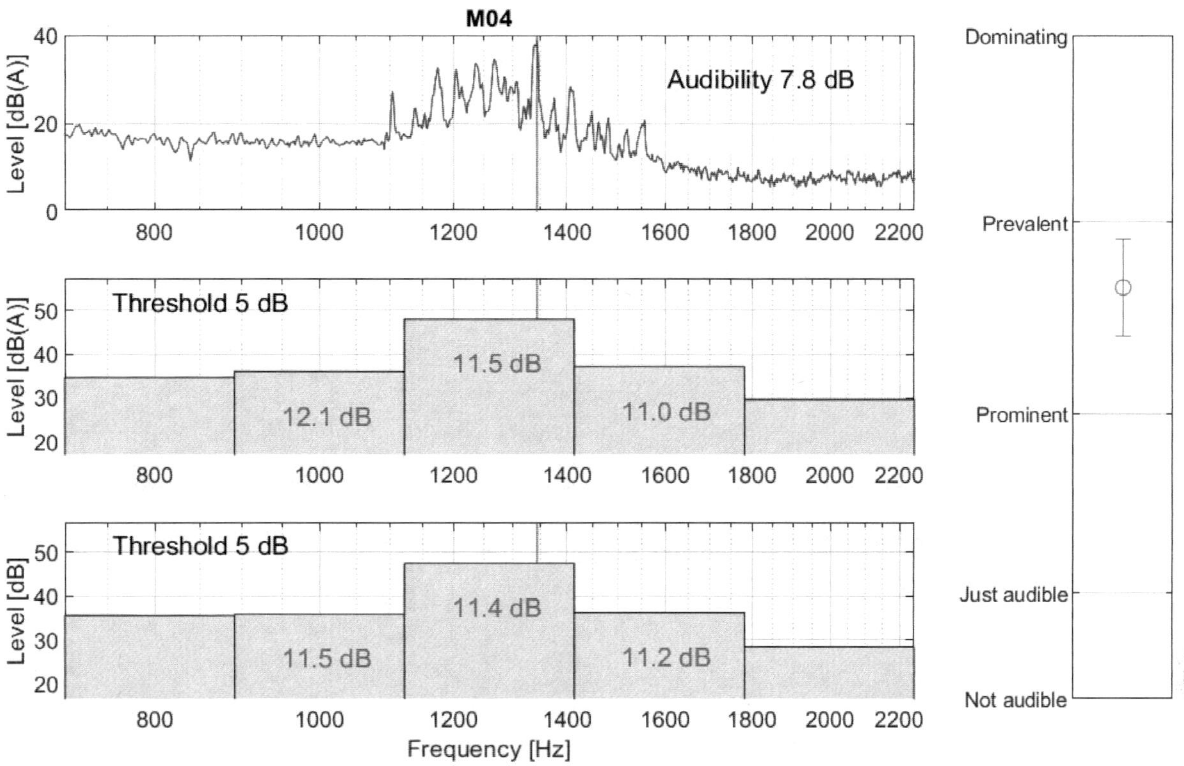

Figure 8 Detailed result for stimuli M04. See further information in Figure 2 at page 8 and in section 5.

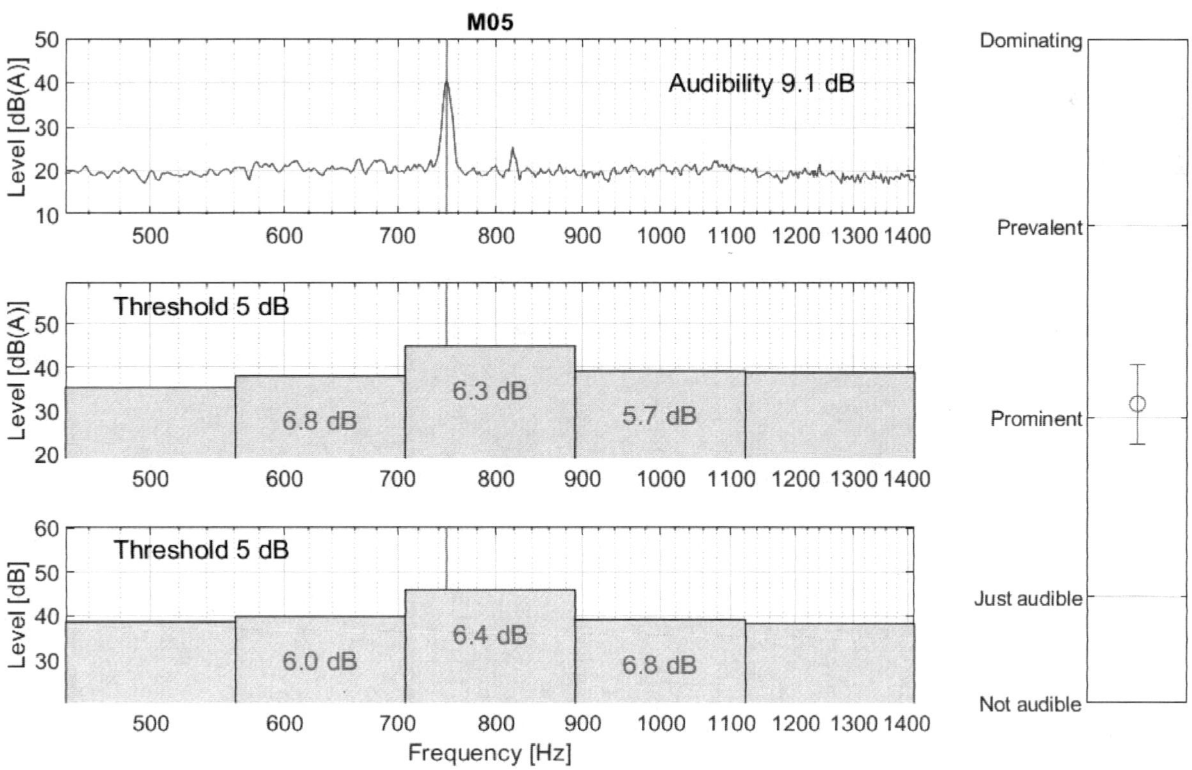

Figure 9 Detailed result for stimuli M05. See further information in Figure 2 at page 8 and in section 5.

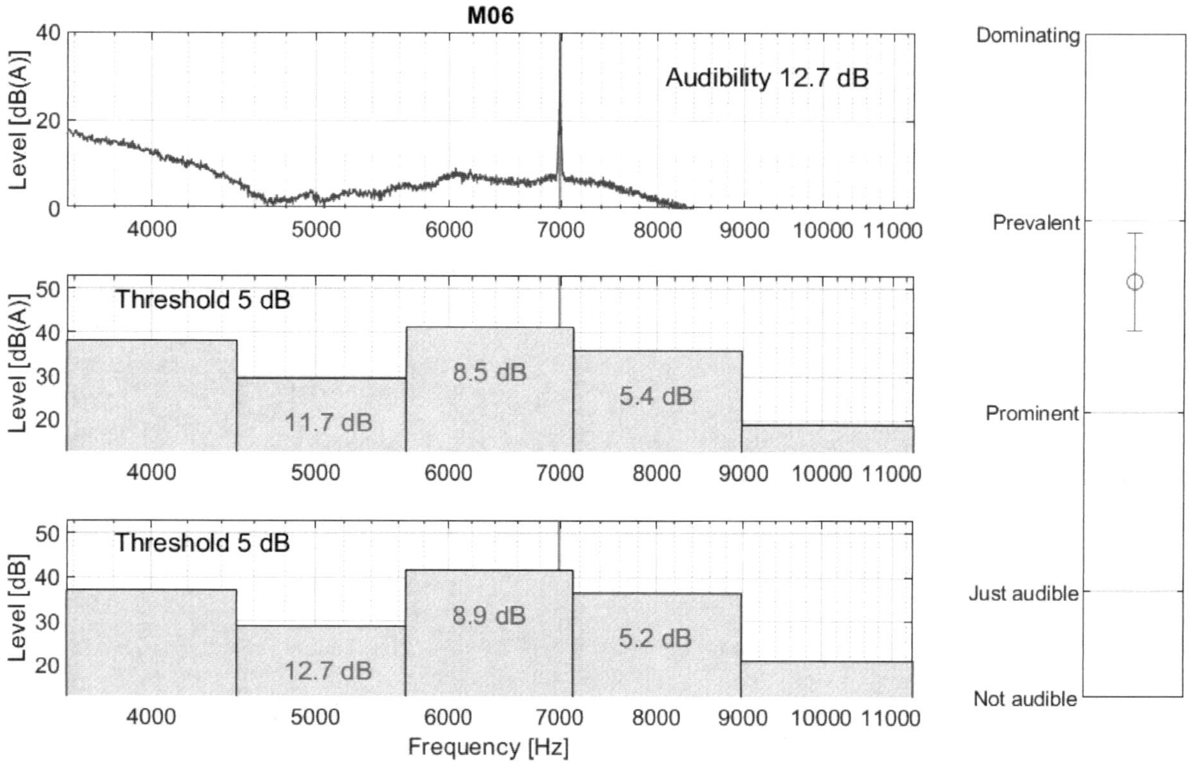

Figure 10 Detailed result for stimuli M06. See further information in Figure 2 at page 8 and in section 5.

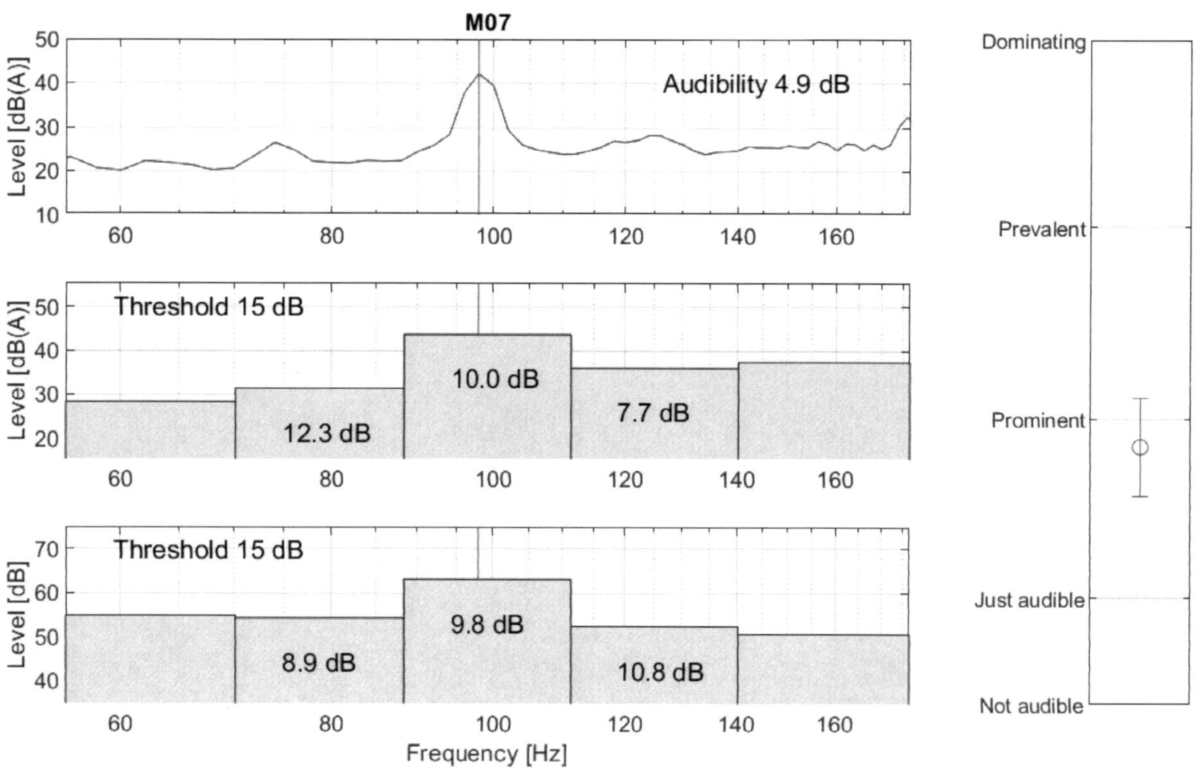

Figure 11 Detailed result for stimuli M07. See further information in Figure 2 at page 8 and in section 5.

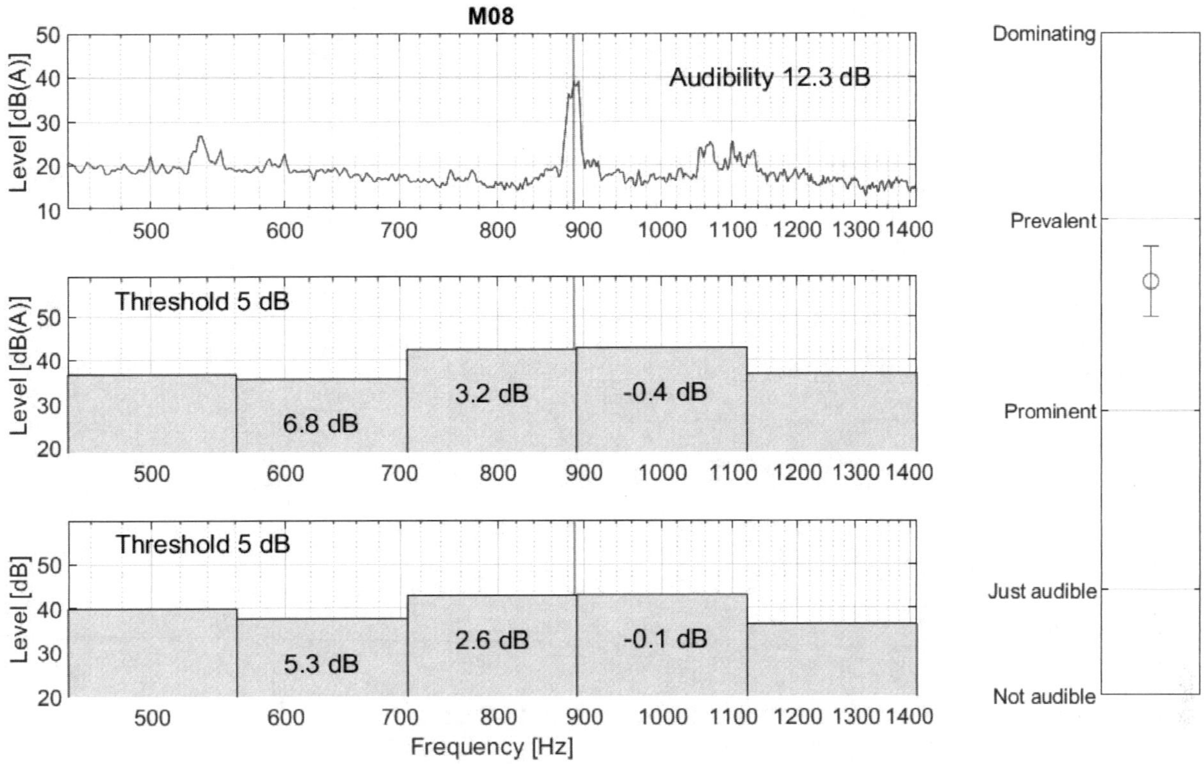

Figure 12 Detailed result for stimuli M08. See further information in Figure 2 at page 8 and in section 5.

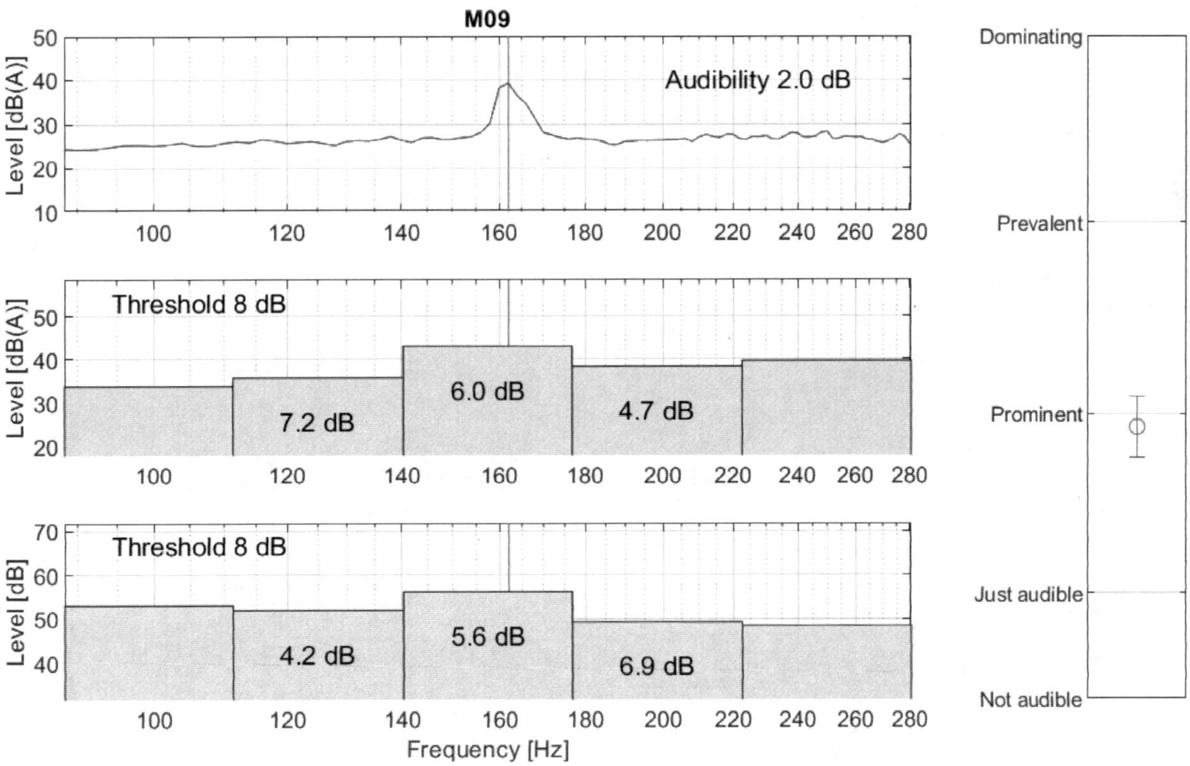

Figure 13 Detailed result for stimuli M09. See further information in Figure 2 at page 8 and in section 5.

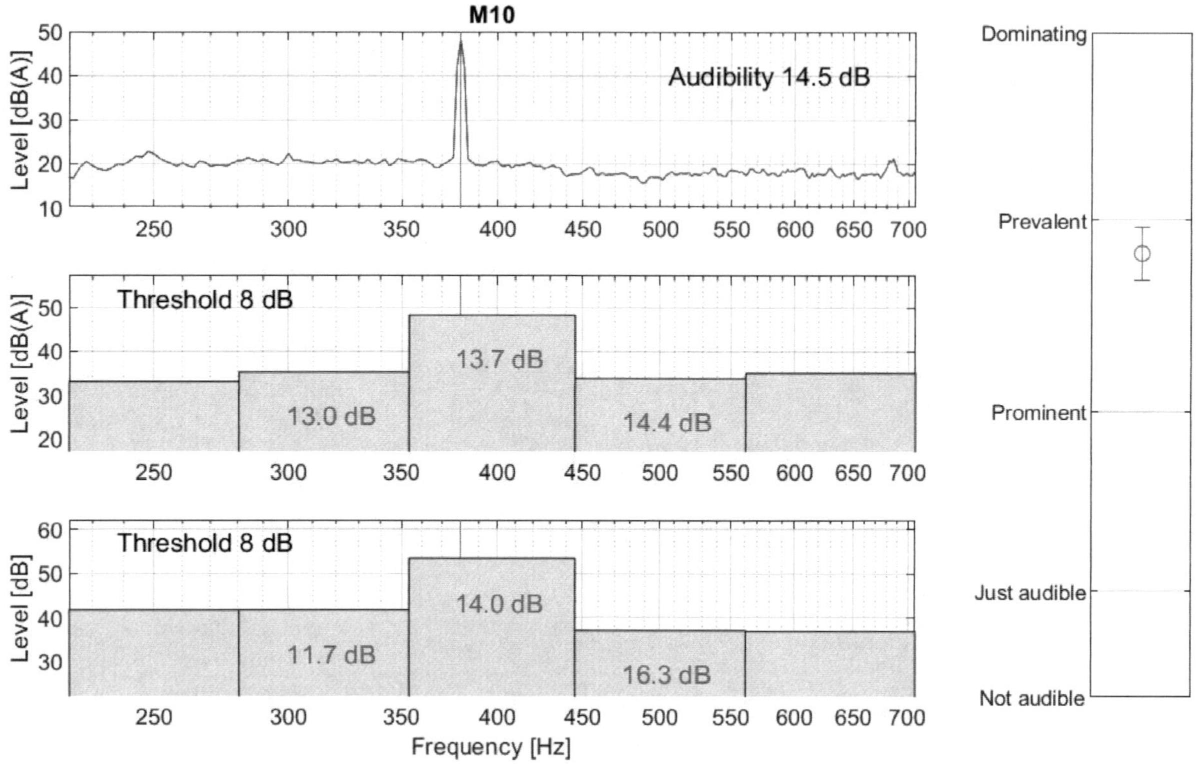

Figure 14 Detailed result for stimuli M10. See further information in Figure 2 at page 8 and in section 5.

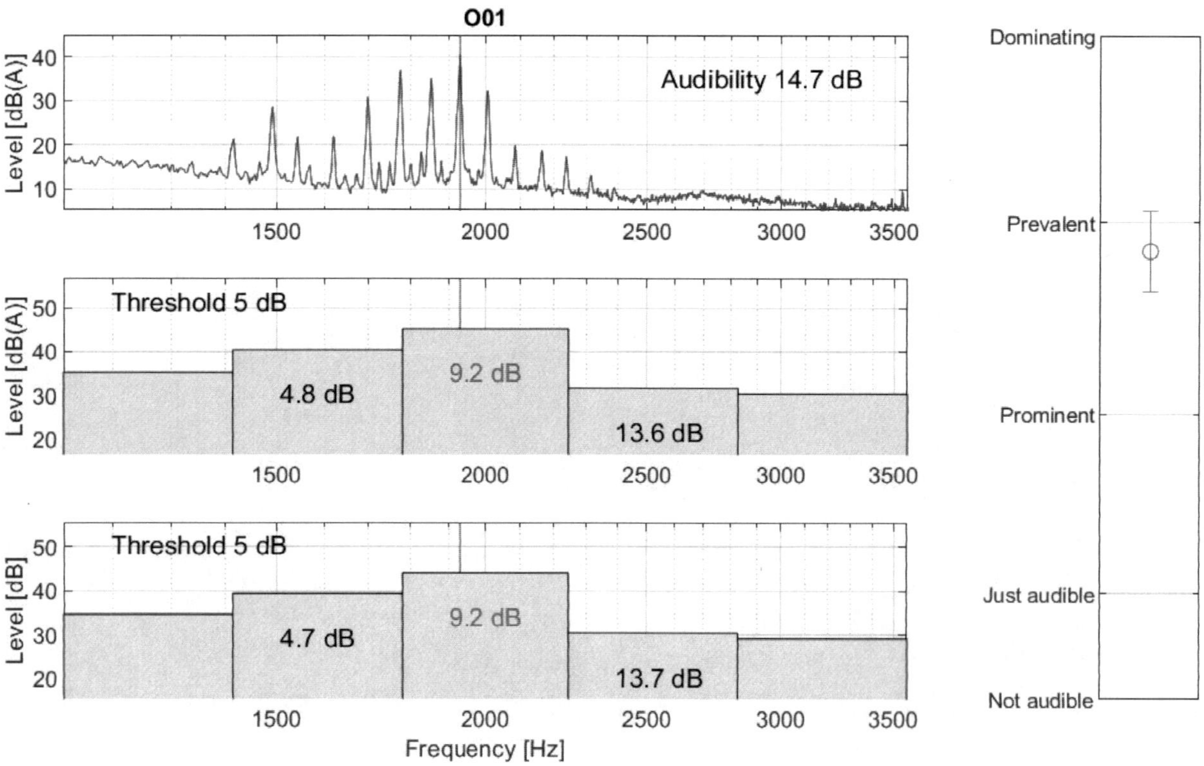

Figure 15 Detailed result for stimuli O01. See further information in Figure 2 at page 8 and in section 5.

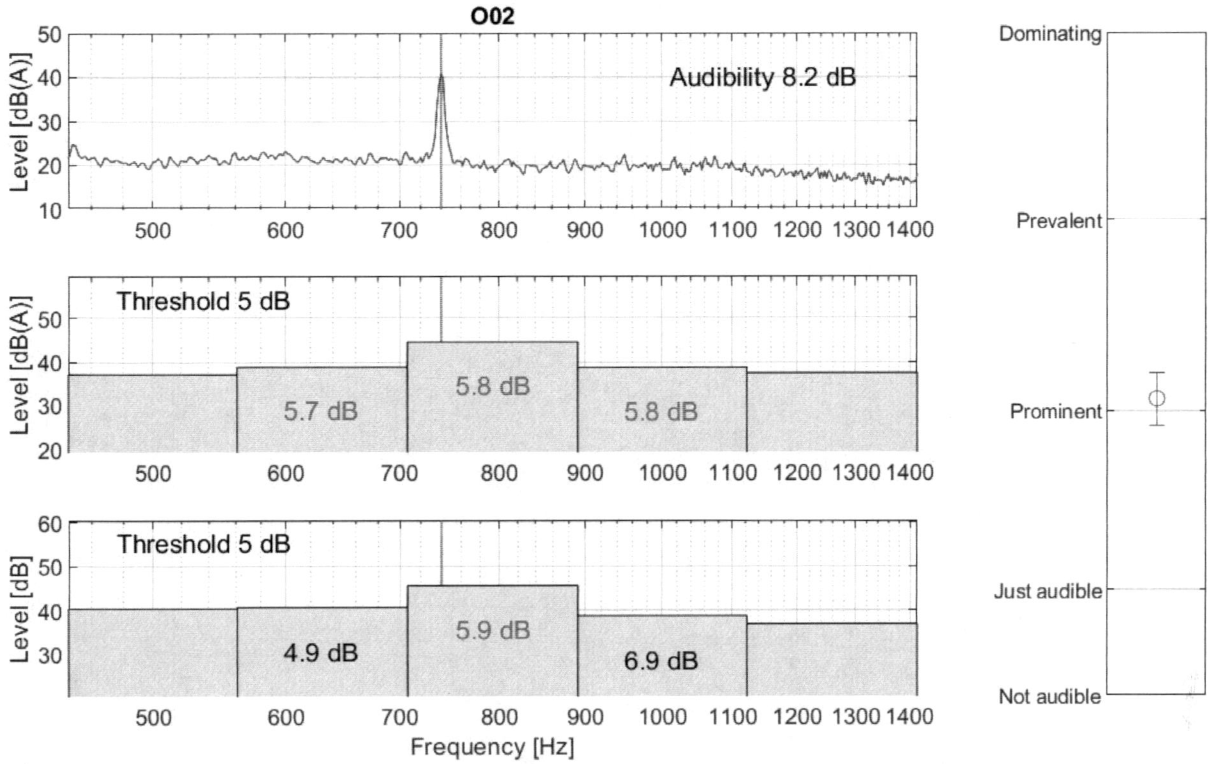

Figure 16 Detailed result for stimuli O02. See further information in Figure 2 at page 8 and in section 5.

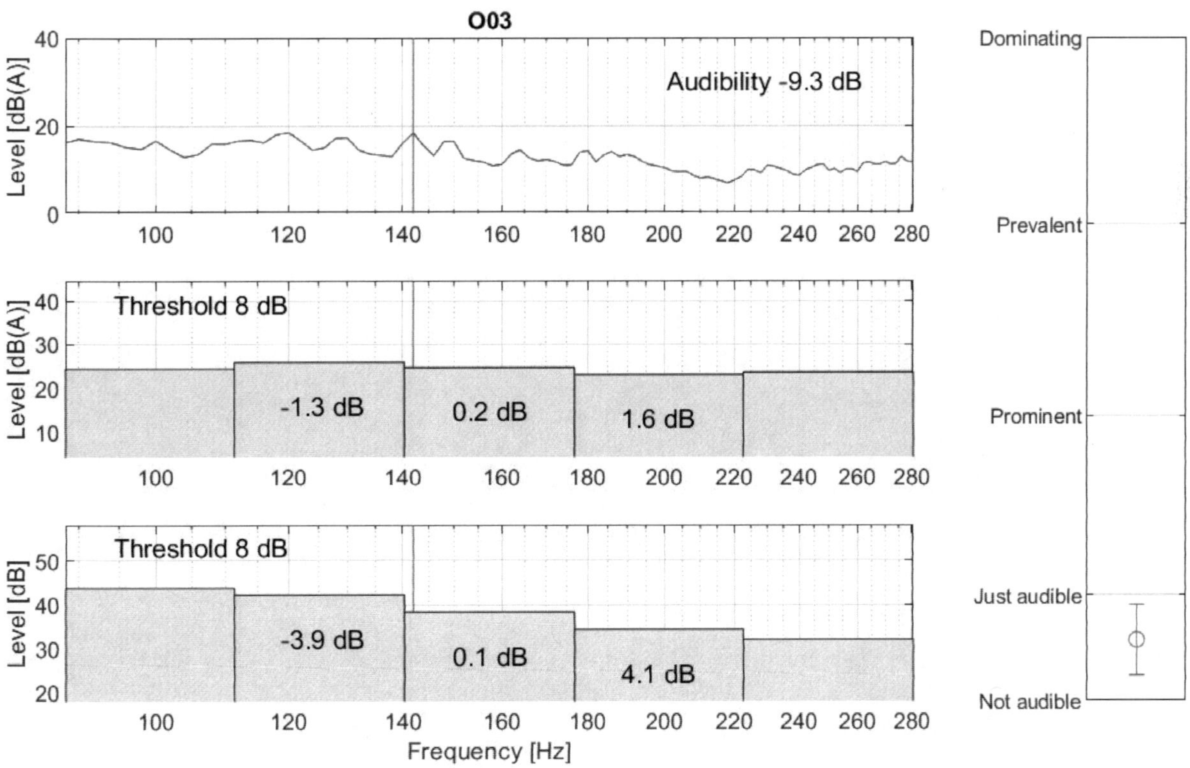

Figure 17 Detailed result for stimuli O03. See further information in Figure 2 at page 8 and in section 5.

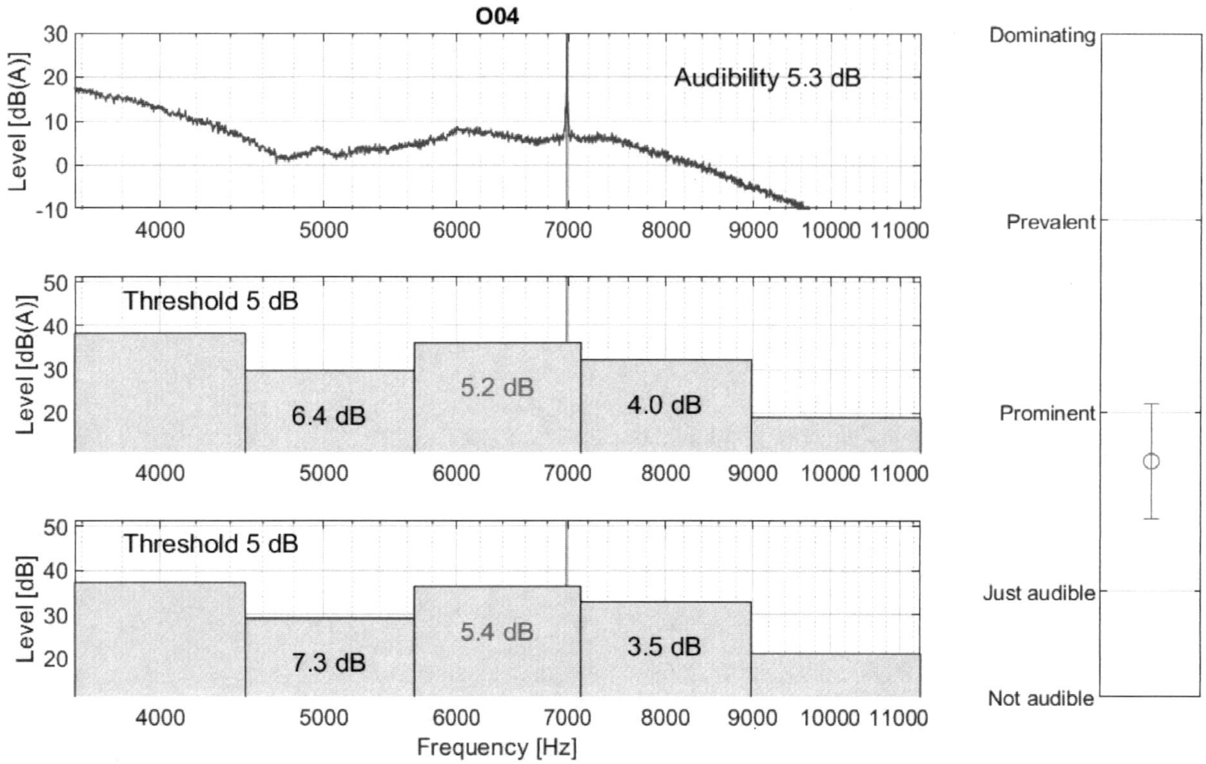

Figure 18 Detailed result for stimuli O04. See further information in Figure 2 at page 8 and in section 5.

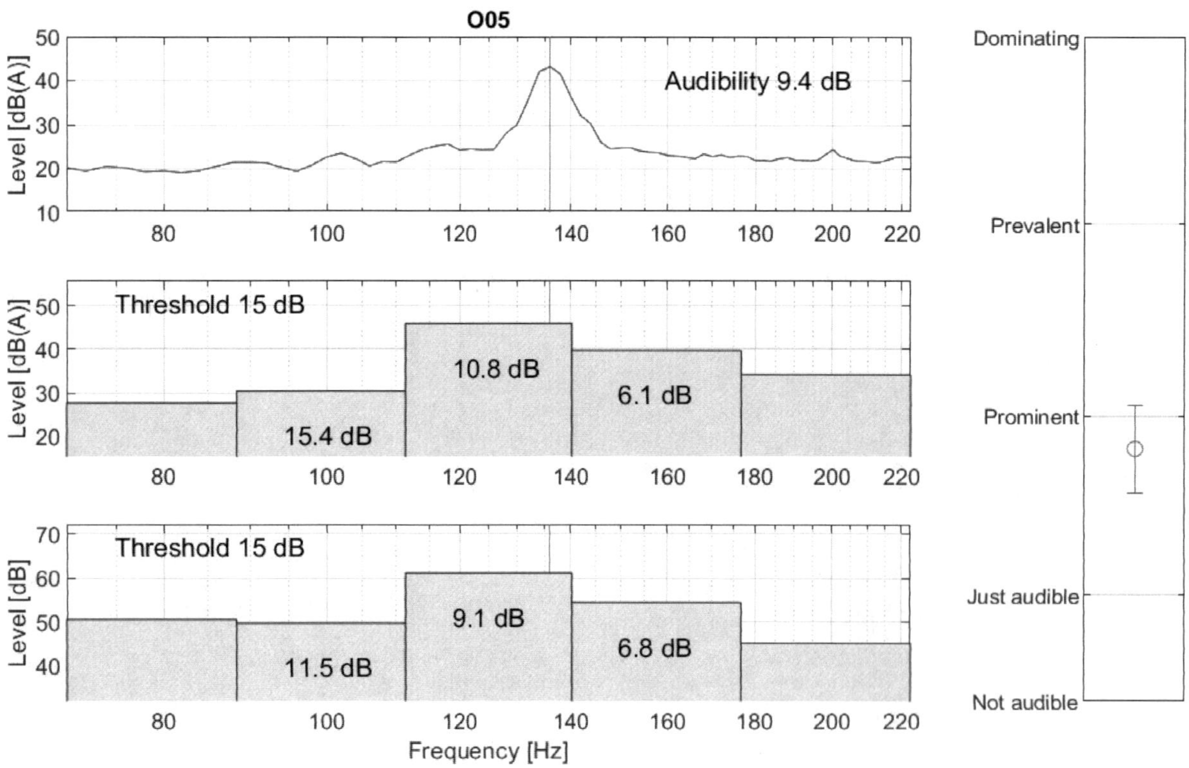

Figure 19 Detailed result for stimuli O05. See further information in Figure 2 at page 8 and in section 5.

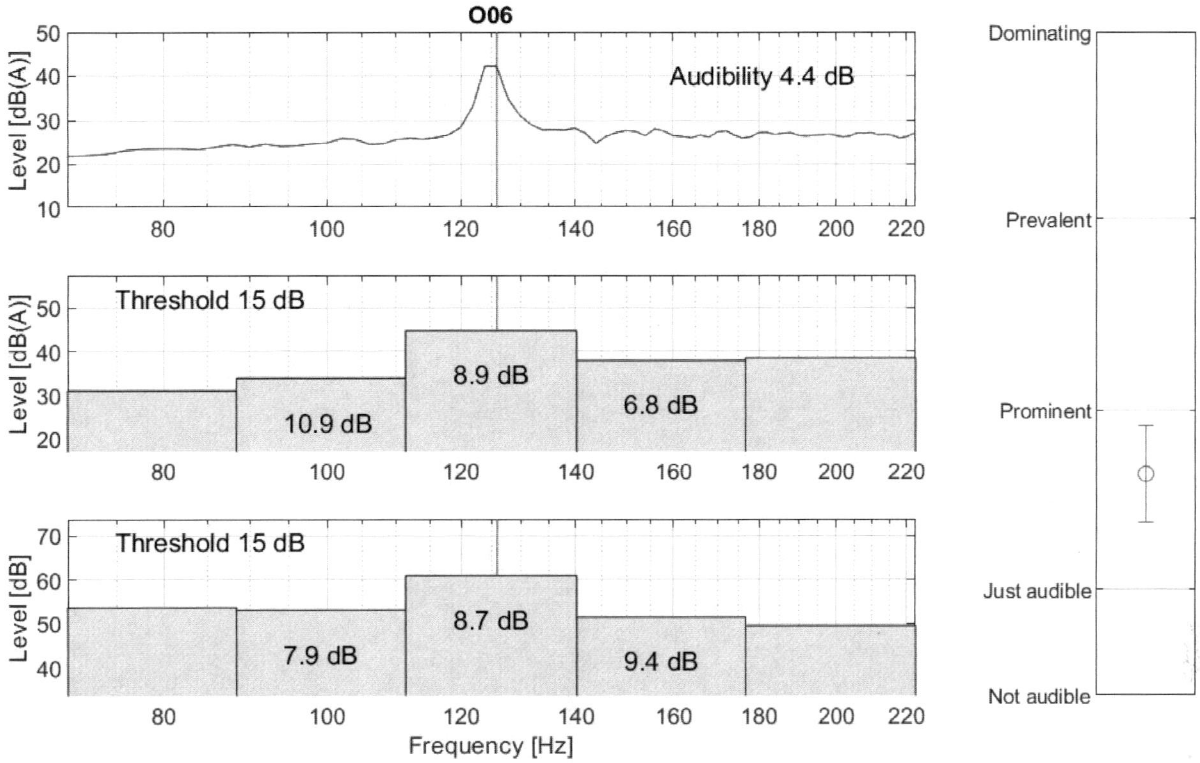

Figure 20 Detailed result for stimuli O06. See further information in Figure 2 at page 8 and in section 5.

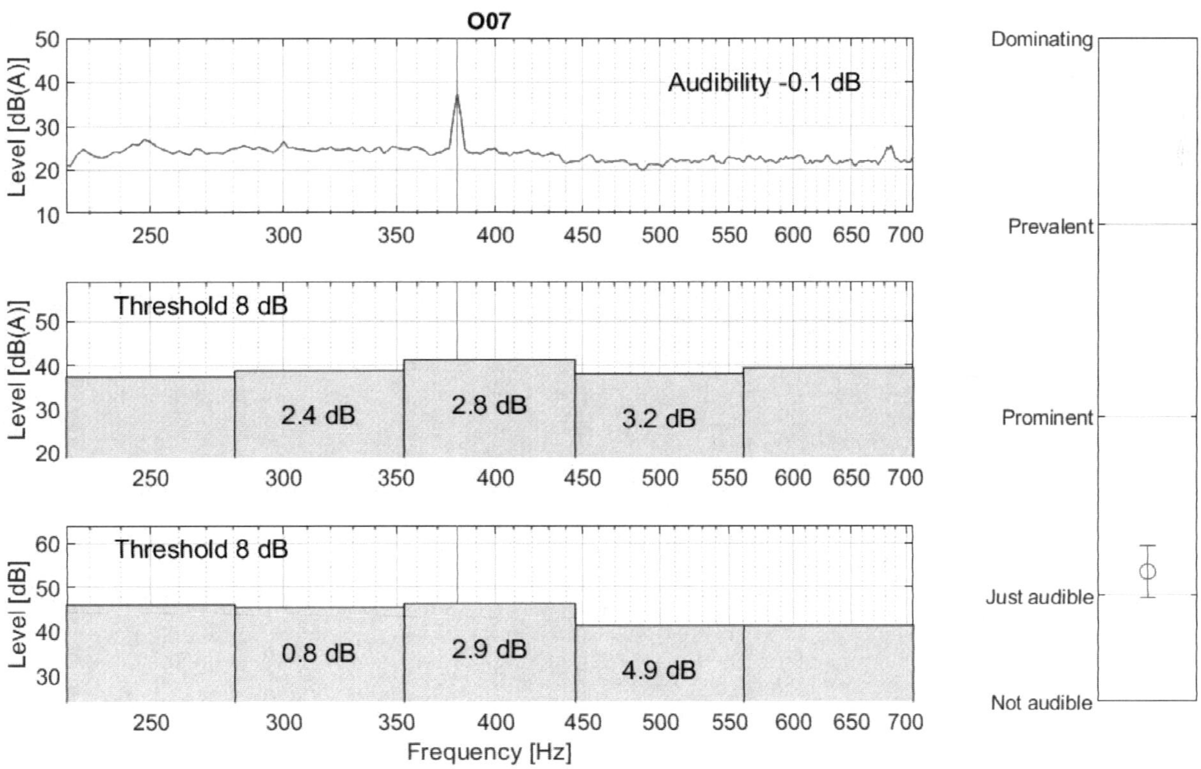

Figure 21 Detailed result for stimuli O07. See further information in Figure 2 at page 8 and in section 5.

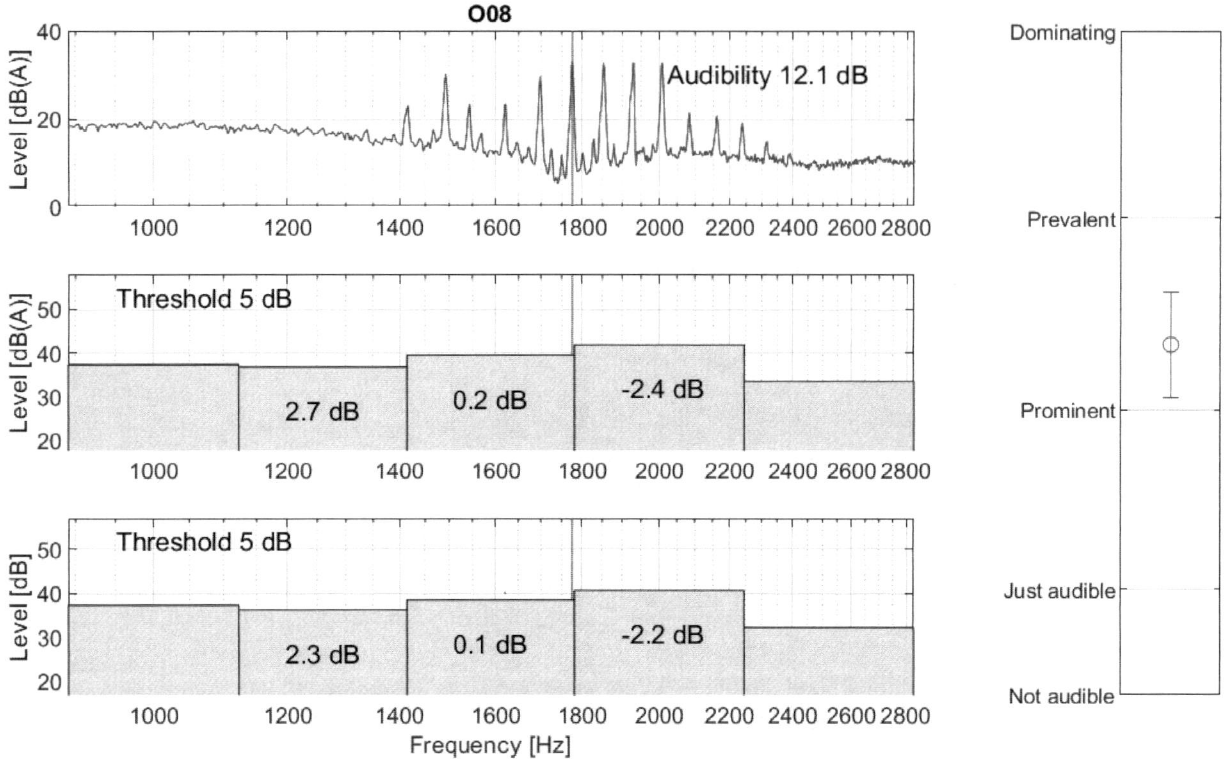

Figure 22 Detailed result for stimuli O08. See further information in Figure 2 at page 8 and in section 5.

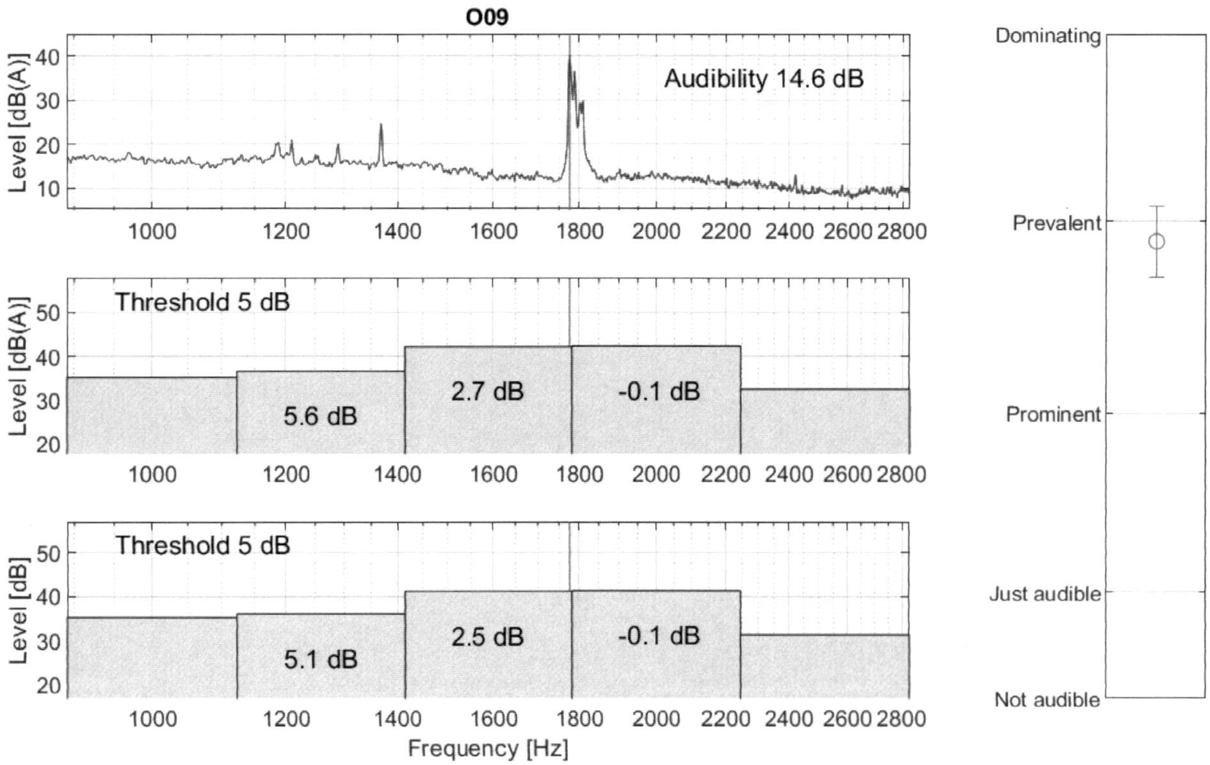

Figure 23 Detailed result for stimuli O09. See further information in Figure 2 at page 8 and in section 5.

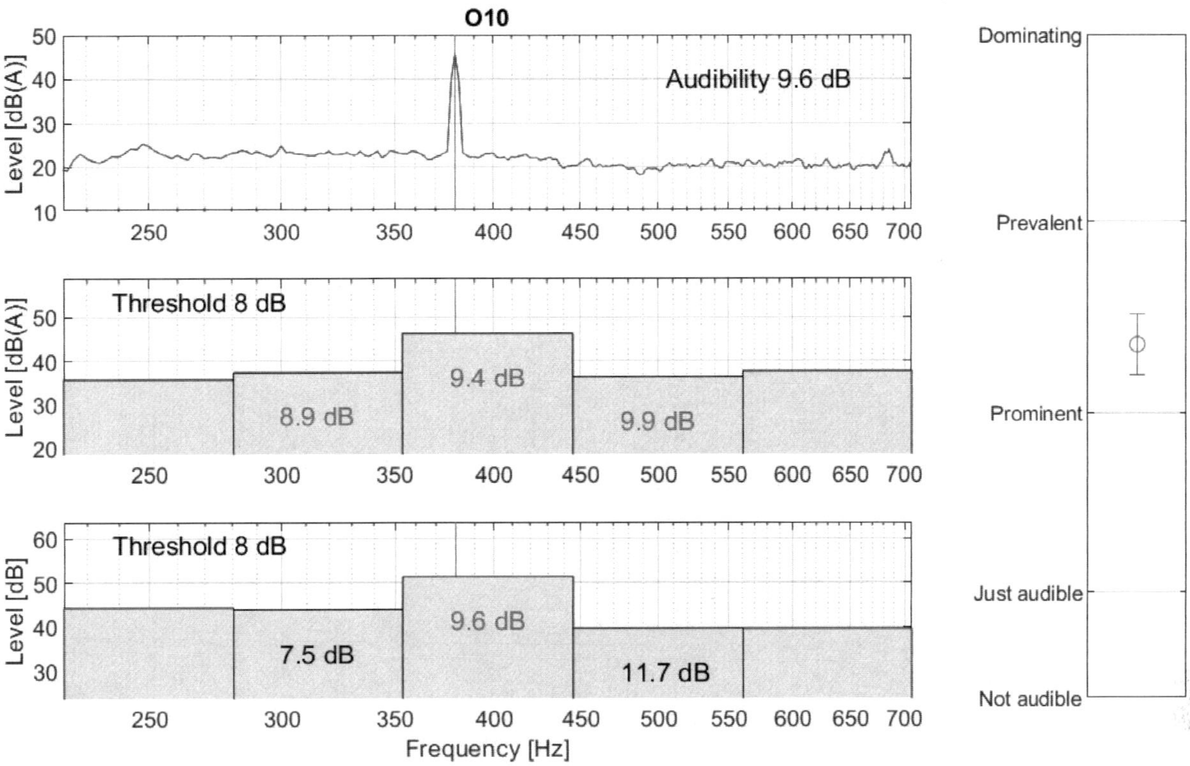

Figure 24 Detailed result for stimuli O10. See further information in Figure 2 at page 8 and in section 5.

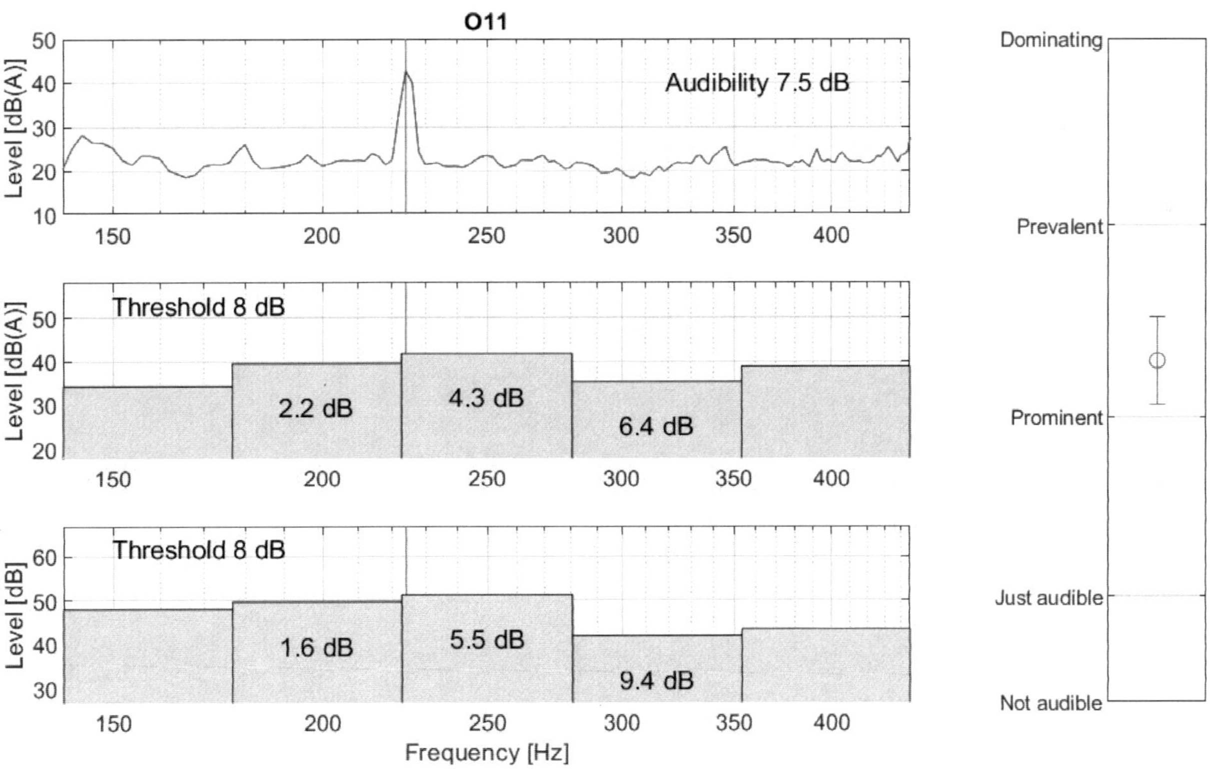

Figure 25 Detailed result for stimuli O11. See further information in Figure 2 at page 8 and in section 5.

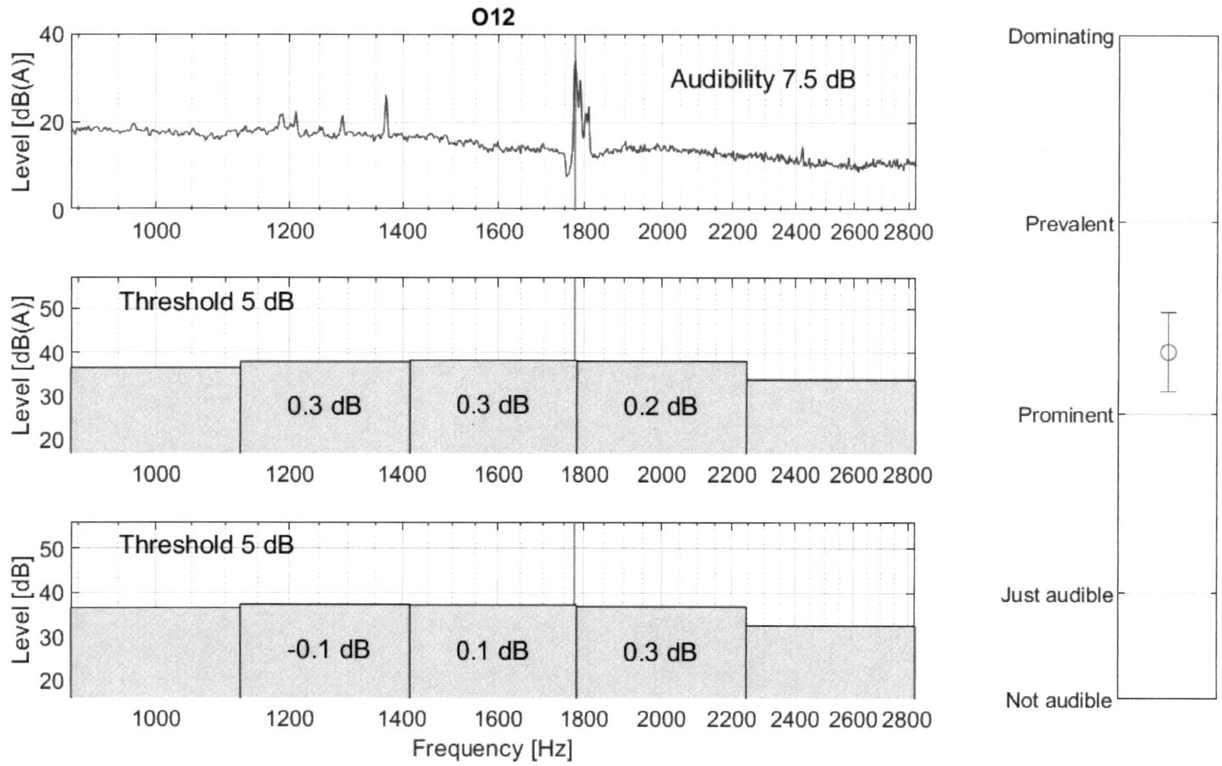

Figure 26 Detailed result for stimuli O12. See further information in Figure 2 at page 8 and in section 5.

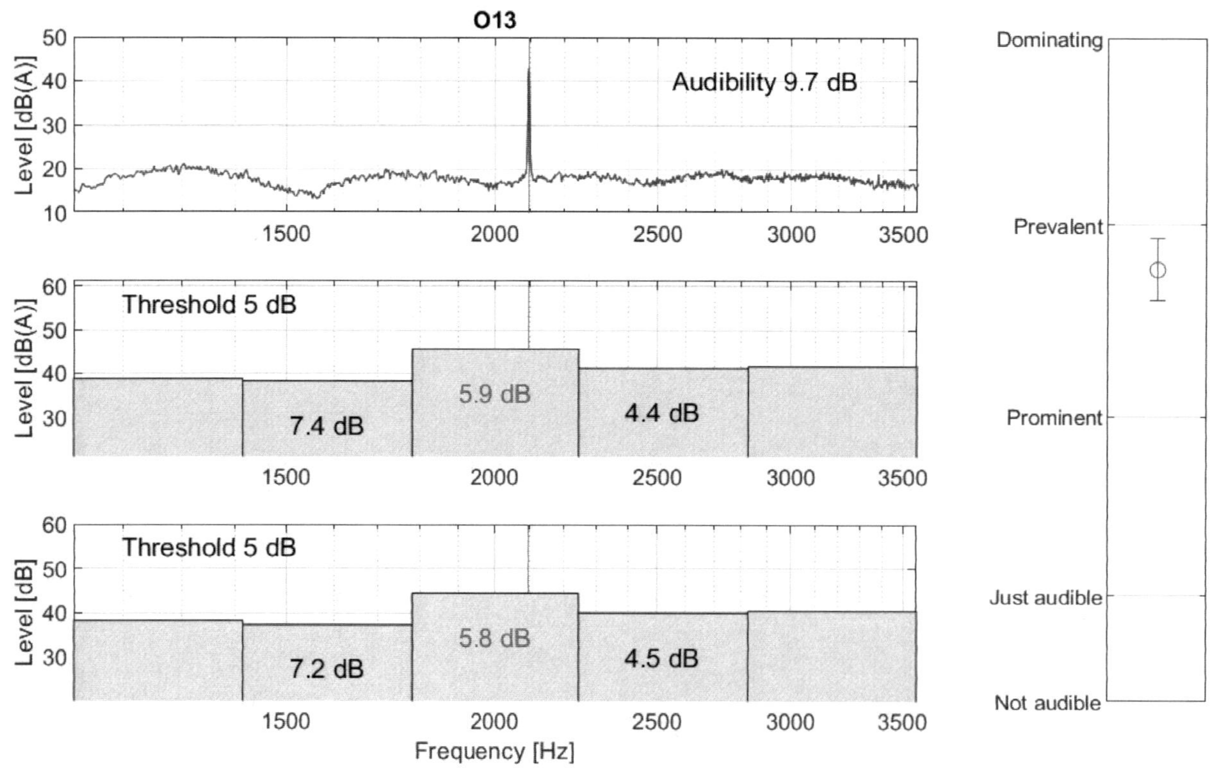

Figure 27 Detailed result for stimuli O13. See further information in Figure 2 at page 8 and in section 5.

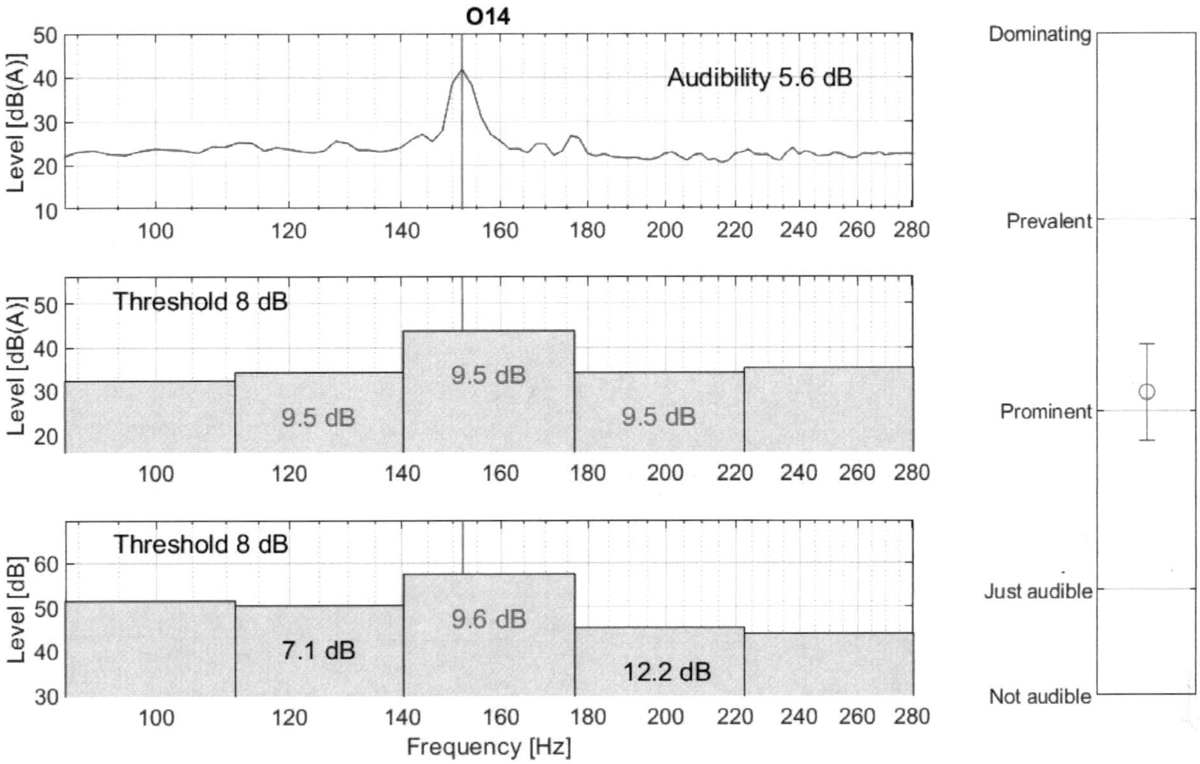

Figure 28 Detailed result for stimuli O14. See further information in Figure 2 at page 8 and in section 5.

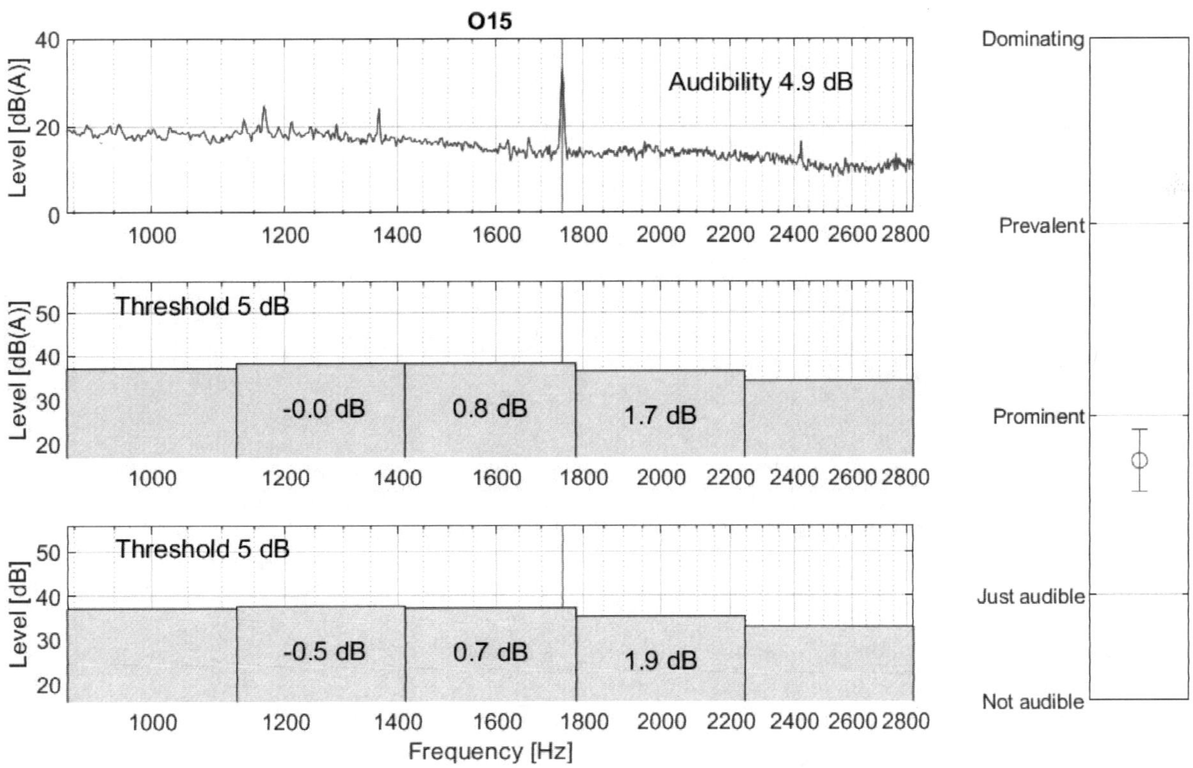

Figure 29 Detailed result for stimuli O15. See further information in Figure 2 at page 8 and in section 5.

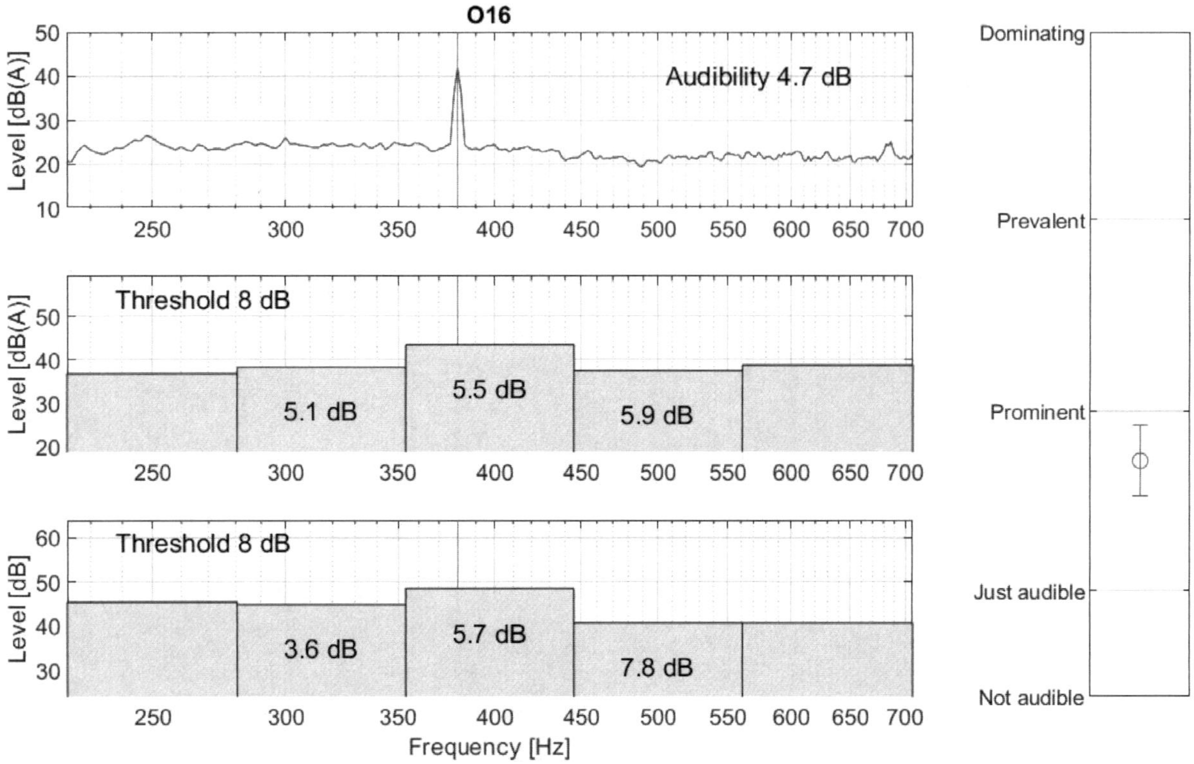

Figure 30 Detailed result for stimuli O16. See further information in Figure 2 at page 8 and in section 5.

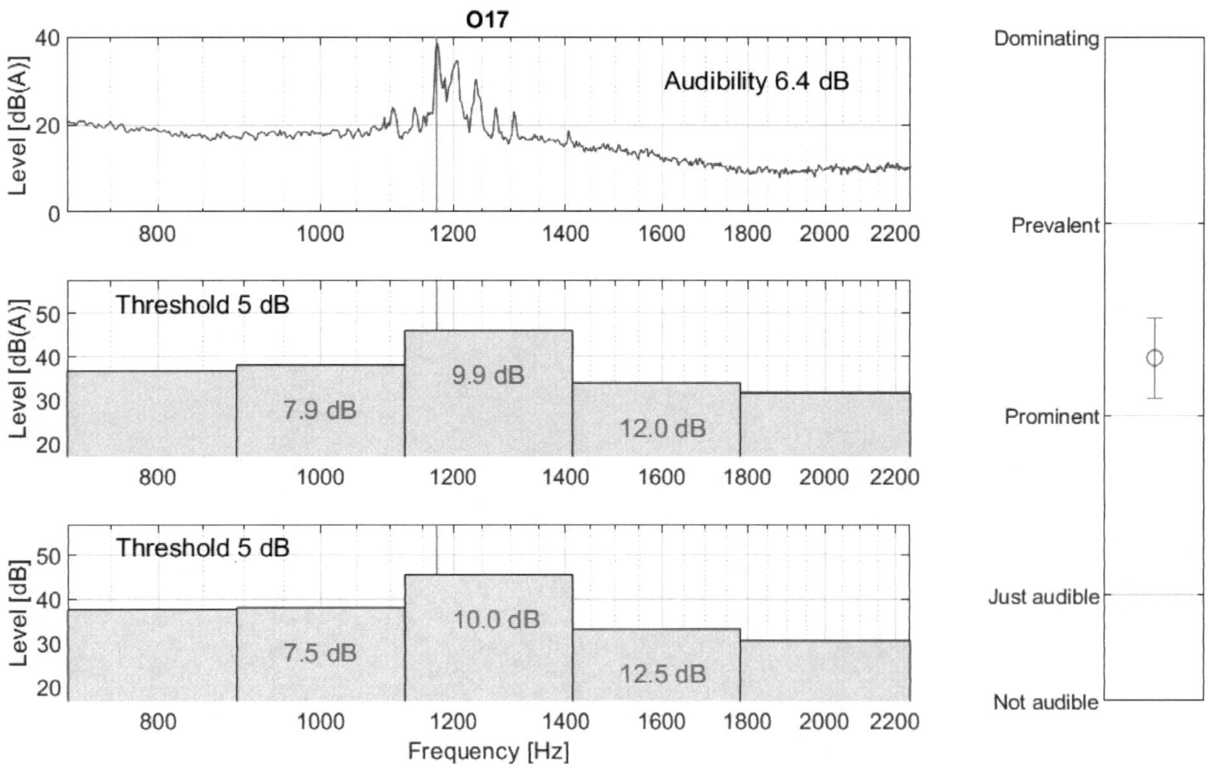

Figure 31 Detailed result for stimuli O17. See further information in Figure 2 at page 8 and in section 5.

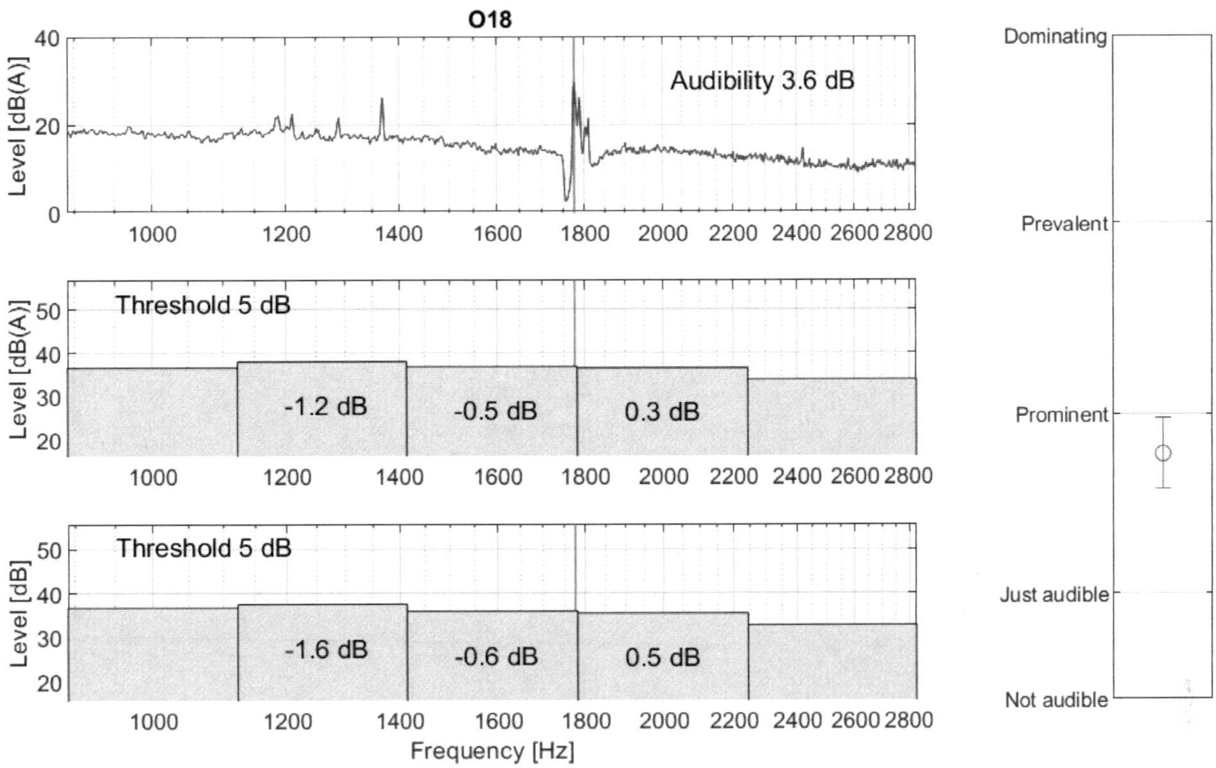

Figure 32 Detailed result for stimuli O18. See further information in Figure 2 at page 8 and in section 5.

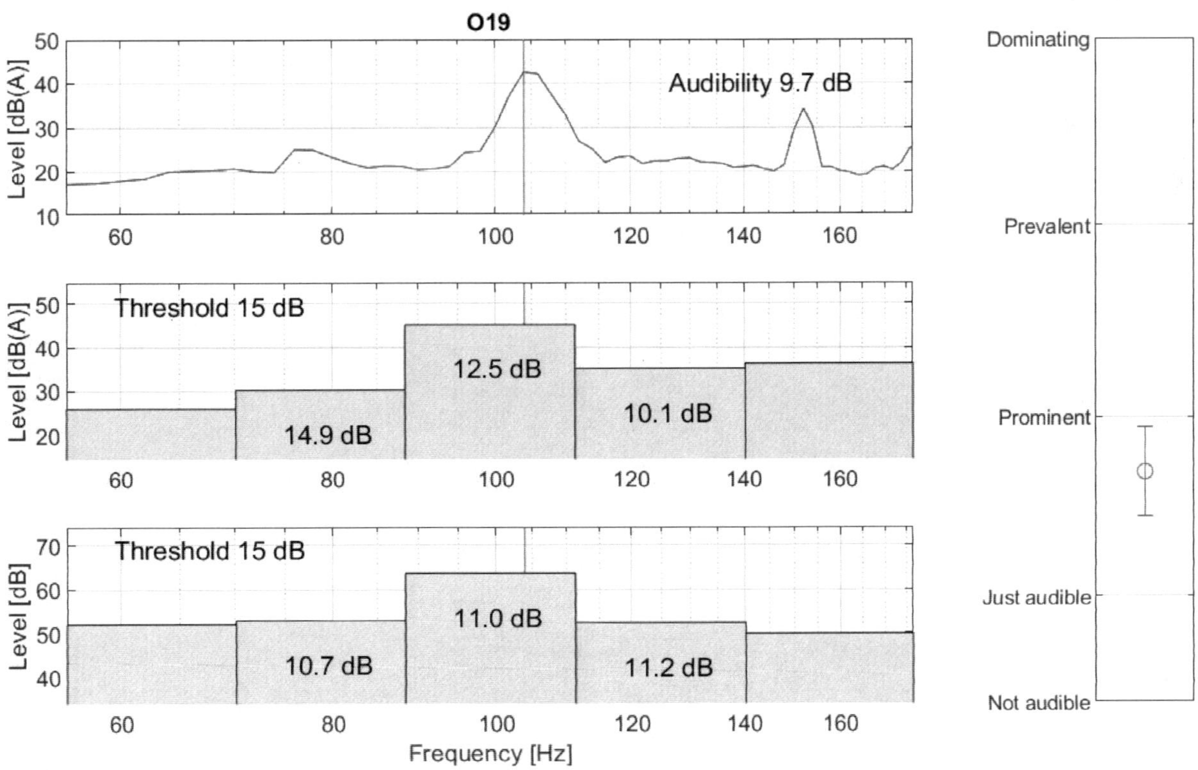

Figure 33 Detailed result for stimuli O19. See further information in Figure 2 at page 8 and in section 5.

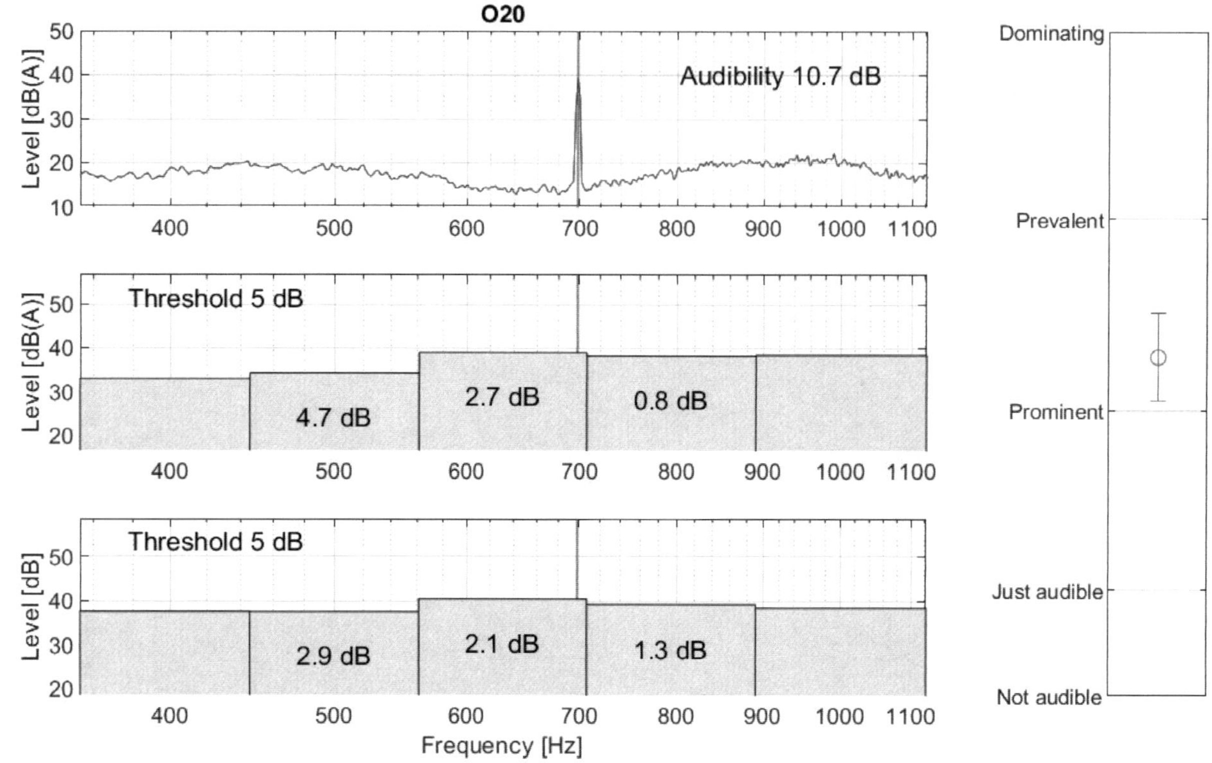

Figure 34 Detailed result for stimuli O20. See further information in Figure 2 at page 8 and in section 5.

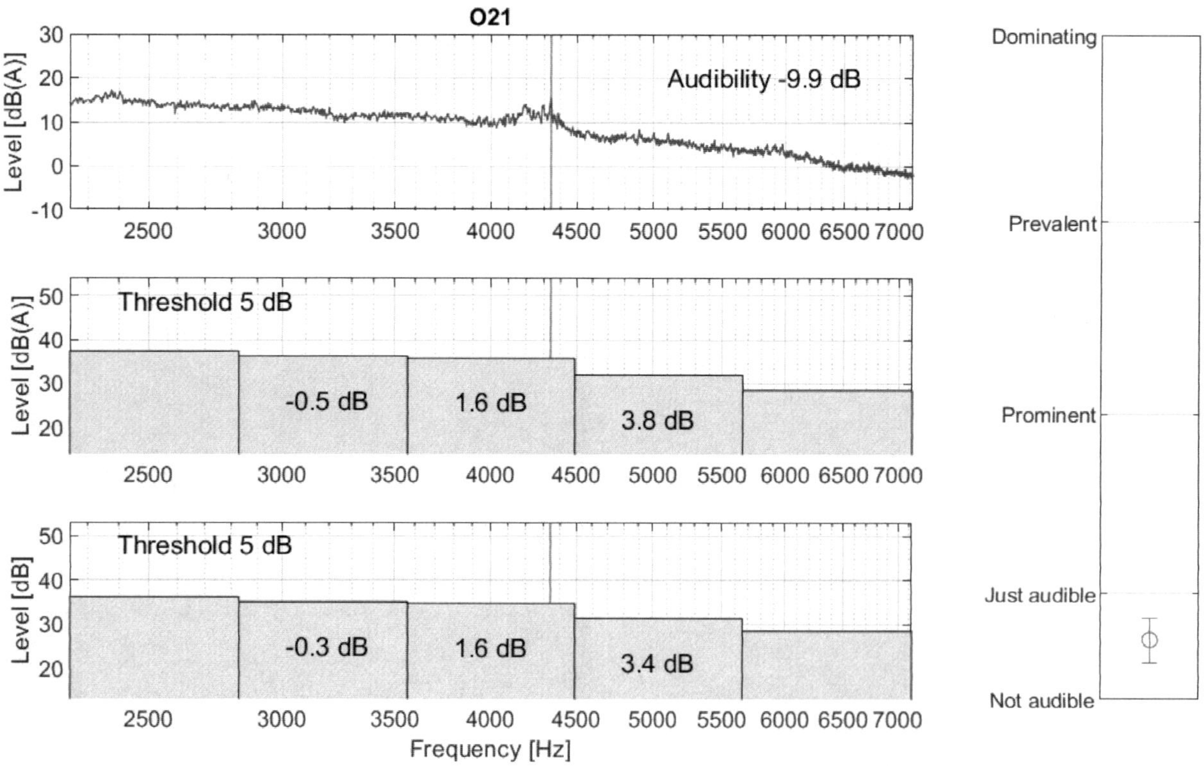

Figure 35 Detailed result for stimuli O21. See further information in Figure 2 at page 8 and in section 5.

INCEEUROPE

9th International Conference
on
Wind Turbine Noise
Remote from Europe – 18th to 21st May 2021

Long distance noise propagation over water for an elevated height-adjustable sound source

Lars Sommer Søndergaard, FORCE Technology, lss@force.dk
Erik Thysell, FORCE Technology, erth@force.dk
Christian Claumarch, formerly FORCE Technology
Andrea Vignaroli, DTU Wind Energy, andv@dtu.dk
Carsten Thomsen, FORCE Technology, ctho@force.dk
Kurt S. Hansen, DTU Wind Energy, kuhan@dtu.dk

Summary

This paper presents the results of measurement campaigns for downwind noise propagation over water for elevated sound sources. This is relevant for offshore wind turbines, near shore wind turbines or wind turbines on land close to large water bodies. Under certain meteorological conditions, in theory noise spreading over an acoustical hard surface can be approximated by cylindrical spreading as a result of multiple reflections. This theory is evaluated in a measurement setup with a height-adjustable sound source (81 m, 50 m and 30 m above ground) and microphones positioned downwind (at shore and ~100 m inland) of the sound source (~3 km, ~5 km and ~7 km distance from the source) where the sound only propagates over water between the sound source and the microphones placed at the shore.

The meteorological conditions (wind speed, wind direction, atmospheric stability, temperature, humidity, etc.) were monitored continuously at both ends of the setup, utilizing both a tall met mast, a wind profiler and sonic anemometers at multiple heights.

The results have been compared with both the current and the previous propagation models described in Danish regulation for wind turbines for noise propagation over water. The results indicate that the current model better captures the effect of possible multiple reflections at the shore.

1. Introduction

Denmark has the longest history in the world with offshore wind turbines (since 1991). Current offshore capacity is 1.7 GW, and approx. 5-6 GW is in planning. Some of the offshore sites in planning are so-called ”nearshore” turbines, with a distance to population where the turbines might be audible This has led to concern, and in 2019 new regulations were introduced to address this [1]. In 2005 noise from offshore turbines was investigated in Denmark [2], but with primary focus on how to measure the sound power level from turbines. A study of propagation models in Danish legislation for noise from Wind Turbines indicated that correction for multiple reflections is necessary in order to protect neighbours.

To our knowledge, only one measurement validation campaign for long distance noise propagation over water has been published [3] The paper includes data for one source and receiver combination, where the source height was 30 m and the receiver was positioned 9750 km away (total distance) of which the first 9 km was over water and the last 750 meters were over land.

Therefore, new measurement campaigns have been performed to obtain more detailed information about the effect of source height and distance to the source.

1.1 DecoWind (Development of low-noise and cost-efficient wind farm technology)

The aim of DecoWind is to address a fundamental gap in our knowledge about noise from wind turbines and to use the acquired insight to increase Annual Energy Production (AEP) by 5 % without increasing noise annoyance at the receiver. Achieving this will improve public acceptance of wind turbines and increase on-shore deployment of wind energy.
The project consists of different work packages, one of which is focused on noise propagation over water.
The DecoWind project has 4 partners: DTU Wind Energy, Siemens Gamesa Renewable Energy, EMD International A/S and FORCE Technology.

1.2 Multiple reflections

Multiple reflections (sometimes also called “cylindrical spreading”) [4] is an effect that can happens when sound propagates over relatively large distances, especially over acoustically hard reflecting surfaces such as water. This effect is illustrated in Figure 1.

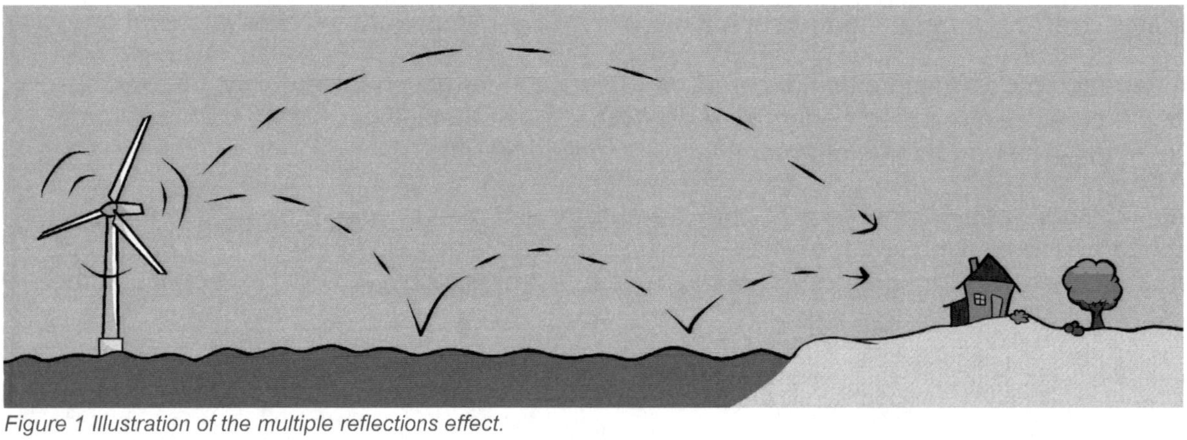

Figure 1 Illustration of the multiple reflections effect.

It is caused by downward refraction up to a sufficiently high altitude that causes the sound field to bend towards and be reflected by the terrain (water) surface. Over longer distances the sound field can bend down and be reflected multiple times. Unlike a situation without downward refraction, the sound cannot “escape”.

The acoustically hard surface contributes a minimum of phase change in the reflection making the sound field more coherent than for reflections from acoustically softer surfaces. The longer the propagation occurs with downward refraction towards the ground, the less is the effect of spherical spreading (hence the name "cylindrical spreading").

The effect varies depending on wind shear, temperature gradient, terrain surface impedance and source height. Increasing wind shear in the sound propagation direction, temperature gradient, and terrain surface impedance all increase the effect by decreasing the distance from the sound source to where the multiple reflections start.
The longer the effect can build up, the larger the effect can become which leads to higher sound levels compared to a situation without multiple reflections. The higher a sound source is located, the greater distance is required, before the sound field is refracted to the ground. Since multiple reflections occur a long distance from the source, less energy can build up resulting in lower noise levels.
The effect therefore varies considerably with these parameters but can lead to significantly higher sound levels (sometimes >10 dB) compared to situations without multiple reflections. With the increasing interest in near-shore/offshore wind turbines, it is therefore very relevant to be able to adequately include this effect in noise calculations/predictions.

2. Choice of location

Previous validation of the Nord2000 method was carried out using an elevated loudspeaker in a certain terrain under certain meteorological conditions. This previous work can be found in a report covering two measurement campaigns in two inland areas [5] and has inspired this measurement campaign to document how sound propagates from an elevated source over water to a distant receiver.

Since the modelling of elevated sources is used for wind turbines, it is most relevant to make the measurement on a windy day in a downwind direction over an appropriate area of water. The combination of downwind and/or temperature gradient that results in downward sound refraction is required to investigate downward reflections.
To minimize the noise from wind-induced waves, inland stretches of water such as fjords, lakes, coves and lagoons were considered to be better candidates than coastal areas. Larger waves would also affect the sound reflection to some extent. This choice introduces another practicality in not having to transport and elevate a large loudspeaker at sea.
The measurement location should also:

- Allow for multiple microphones to be placed on the coast at different distances from the source.
- Preferably be suitable under the most dominant wind direction in Denmark, westerly wind.
- Located in a quiet area where no significant constant noise sources are present (i.e. not close to a highway, airport or similar.)

2.1 First campaign – Dragstrup Vig

A suitable candidate was found named Dragstrup Vig (bay) which is a widening of Limfjorden. On the westerly side is the mainland Jutland, and on the easterly side is the island Mors. Overview of the area is shown in Figure 2. The site gave the possibility for different microphone distances from the source where the distance is only over water. Important as well is that western wind can be used which is the most common wind direction in Denmark.
Microphones was set up at 2 km, 5 km, 6 km and 8 km distances with meteorology logging at both the source and at the 8 km distance.

Figure 2 Layout of the measurement campaign area Dragstrup Vig. The yellow pins show the microphone positions. The blue arrow shows the primary wind direction during the measurement.

The measurement campaign was performed in august 2020 during daytime. The wind direction was southwest. Unfortunately, it turned out that the signals were only audible (and with a usable signal-to-noise) at the 2 km position. The high background noise was primarily due to wave noise. It is likely due to the wind direction, which allowed the large water body to "collect" wind energy, and thereby build up large waves at the shore-side microphone locations.

The first campaign clearly illustrates that background noise from waves at the coastline where multiple reflections are relevant may be significant compared to typical levels from wind turbines.

2.2 Second campaign – Roskilde Fjord

Therefore, a second measurement campaign was considered at a different location at the southern end of Roskilde Fjord next to Risø campus of the Danish Technical University (DTU), see Figure 3.
The following reasons could result in a more successful campaign:

- For the relevant wind direction, the water body "begins" at Risø campus meaning that the water would not be able to collect as much wind energy giving smaller waves.
- The large met mast at the DTU Risø site could be utilized.
- Based on the first measurement campaign it was decided that lower wind speeds (than at the first campaign) were desirable, as long as the temperature gradient was not too negative (preferably positive).
- Chirp signals were added which should improve signal to noise ratio by up to 10 dB at higher frequencies.

Background noise measurements were performed at relevant wind conditions which together with calculations confirmed that it should be possible to obtain a usable signal-to-noise ratio at all measurement location and frequencies. Additionally, measurements were made in the evening/night to give the best possible signal-to-noise ratio.

3. Choice of sound source and signals

3.1 Test signals

The chosen test signals were band-pass filtered pink noise and rapidly repeated exponential chirps. For the first campaign, only the band-pass filtered pink noise was used, as it allows a constant amount of energy to be put into the loudspeakers per octave band. This requires less thermal energy to be dissipated in the loudspeakers compared to broadband noise in order to achieve the same signal-to-noise ratio in a band. Four centre frequencies were selected for the 1/1-octave band-pass filters, namely 63 Hz, 125 Hz, 250 Hz and 500 Hz. Lower frequencies were omitted, since high output levels at low frequencies would require large subwoofers. Output at higher frequencies would be lost in air-absorption over the large propagation distances.

For the second campaign chirp signals were added: Repeated up-down exponential sweeps with a repetition rate of 2 Hz from 1 Hz to 1 kHz. Each chirp sequence lasts 10 seconds and is repeated 30 times with 3 second pauses between sequencies to reduce the risk of damage to the loudspeaker. The test sequence for the second measurement campaign is shown in Table 1.

Signal no.	Signal type	Frequency range	Duration [s]
1	Pink noise	63 Hz 1/1-octave	120
2	Pause	-	20
3	Pink noise	125 Hz 1/1-octave	120
4	Pause	-	20
5	Pink noise	250 Hz 1/1-octave	120
6	Pause	-	20
7	Pink noise	500 Hz 1/1-octave	120
8	Pause	-	40
9	Chirp	0-1kHz up/down exponential Chirp (repeated 30 times)	360
10	Pause	-	10

Table 1 Test signals for the second measurement at Roskilde/Risø. The shown sequence was repeated 12 times: 4 repetitions per source height

3.2 Loudspeaker

The main objective of the loudspeaker is to deliver a high and stable output for a long period of time. For this reason, the demand on the directivity of the source was seen as secondary to the stability and sound power level. Instead it was decided to examine the directivity to allow for angular corrections of source strength in post processing.

The chosen loudspeaker was the Meyer Sound Leo-M, with two 15" woofers and two 4" compression drivers. The crossover point is at about 500 Hz, so most of the frequency range of interest is covered by the woofers.

Two Meyer Sound Leo-M units were stacked and mounted directly on top of one another and driven in phase, i.e. not as a line array.

3.3 Test of Meyer Sound Leo-M at Aalborg University anechoic chamber

In order to determine the sound power level of the two Leo-M loudspeakers in a number of angles, a test was performed in the large anechoic chamber at Aalborg University.

The directivity of two Meyer Sound Leo-M units was measured under free-field conditions in 15° intervals. When the angles changed to 15°, the level decreased by up to 0.5 dB in the 63-250 Hz bands and up to 2.5 dB in the 500 Hz band (in the elevation direction).

At 30° the directivity is significant from 125 Hz about 0.6 dB, and critical at 250 Hz with a difference of about 2.5 dB.
Therefore, it was concluded that the loudspeaker combination was most suitable in a 15° window where it can be assumed to have the same source strength. For angles greater than 15°, the decrease in source strength should be accounted for. The angle between the 3 km position and the 7 km position seen from the source is approx. 20 degrees. By pointing the loudspeaker to the midpoint of the 20 degrees, the maximum of axis angle is less than 15°.
The sound level 2 m in front of the loudspeaker was monitored during the measurements, and combined with the anechoic chamber measurements, the source strength in the on-axis direction was calculated to:

F [Hz]	50	63	80	100	125	160	200	250	315	400	500	630
L_W [dB]	113.9	126.2	130.2	128.9	130.6	130.4	128.1	129.6	129.9	126.4	128.3	129.5

Table 2 Determined equivalent on-axis source strength of the loudspeaker pr 1/3 octave band

Note that the sound levels emitted from the loudspeaker were significantly (15-25 dB) higher than levels emitted from at modern offshore wind turbine in the 10 MW class.

4. Second measurement campaign

The project goal was to perform the measurements over as calm water as possible with several measurement positions at different distances west of the source position. This reduced the number of usable days for the measurements, since westerly wind is the most common wind direction in Denmark.
The measurements were carried out on 30th September 2020, between approx. 20:50-00:30 local time.

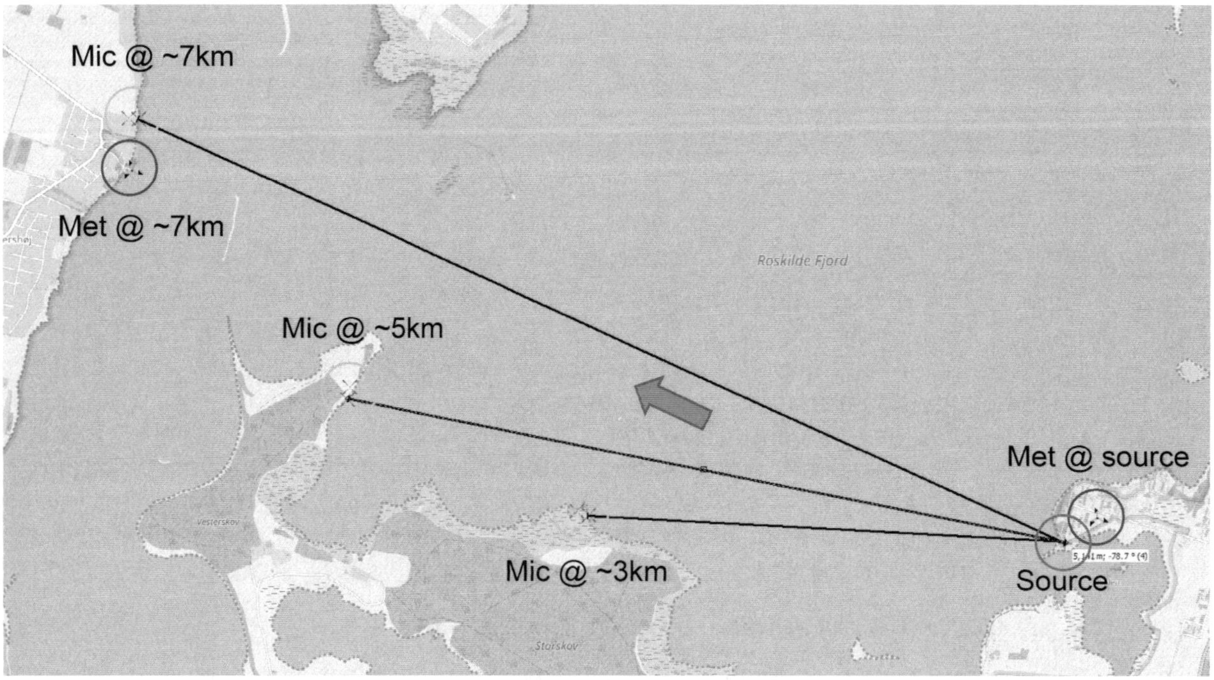

Figure 3 Layout of the measurement campaign area at the southern end of Roskilde Fjord. The approximate general wind direction during the measurements is shown with a large blue arrow.

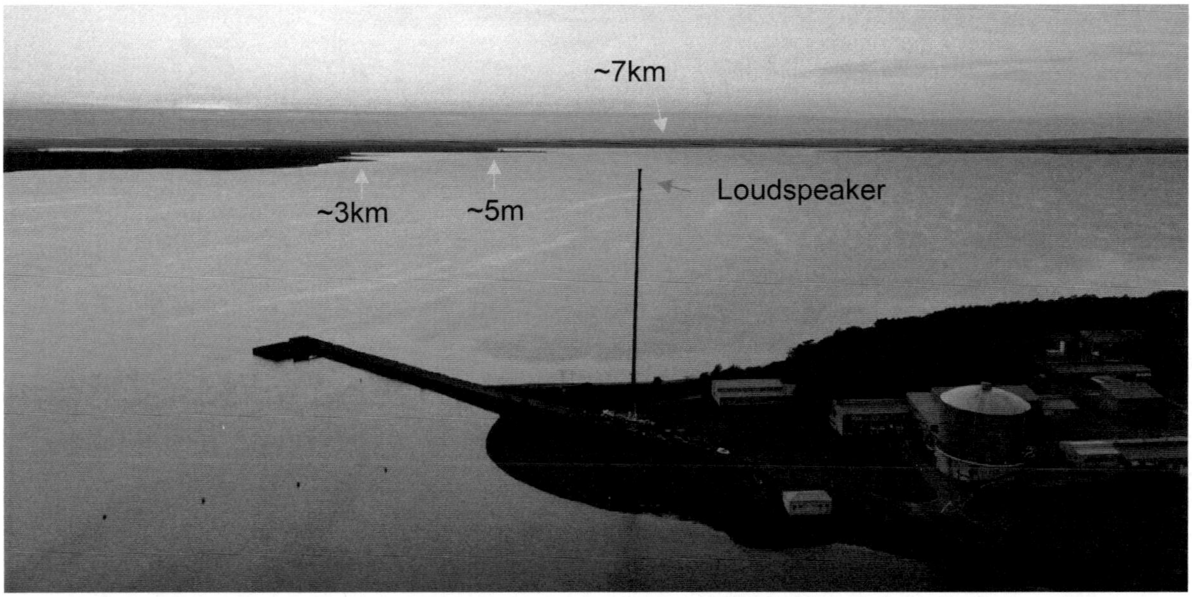

Figure 4 Drone photo showing the area of the measurement campaign looking WNW relative to the loudspeaker.

4.1 Source position

Most of the area around the source position is water. Northeast of the source position is the Risø campus with buildings, facilities and vegetation, see Figure 3 and Figure 4.
The crane with loudspeaker is positioned as close as possible to the water, with a horizontal distance between source and water in the direction of the microphones of ~53 m. The terrain elevation above the water surface is ~2.5 m. The tall met mast at Risø is placed ~290 m northeast of the source (outside the drone photo).
A microphone is mounted in front of the the bottom edge of the loudspeaker and adjusted to match one of the measurement positions from the anechoic lab with a distance of about 2 m from the source centre and an elevation angle of -15°. This microphone is used to verify that the system gives stable sound level during the entire measurement campaign. The loudspeakers were mounted below the crane block. The orientation of the loudspeakers was controlled by two steering ropes attached to the loudspeaker as shown in Figure 5.

Figure 5 Photo of the 240 kg loudspeaker and its mounting below the crane. The microphone is mounted at a distance of 2 m in front of and at the bottom of the loudspeaker.

4.2 Microphone positions

Microphones were set up at ~3 km (3335 m and 3425 m), ~5 km (5145 m and 5150 m) and ~7 km (7160 m and 7225 m) horizontal distances from the source. For each position two microphones were set up: One microphone next to the water referred to in plots as 'shore', the other referred to in plots as 'inland'.

The horizontal (in the direction of the source) distances to water for the three positions were respectively ~80 m, 40 m and 70 m and 1 m, 6 m, and 7 m relative to the water surface. The compass angles to the loudspeaker were 90° (3 km), 100° (5 km) and 110° (7 km).

The microphone on its tripod was 1.5 m above the ground. The microphone was fitted with both a small primary windscreen as well as a large secondary windscreen. The insertion loss of the secondary wind screen has been measured and is negligible at the frequencies of interest in this project.

For each position the microphone signal is sampled at 51.2 kHz using a 24-bit analog-to-digital converter (National Instruments USB 9250) and recorded using noiseLAB 4.04 (from FORCE Technology) running on a Windows 10 laptop.

All equipment in the measurement chain conforms to ISO 61672-1 Class 1 standards for sound level meters and was calibrated both before and after the measurement.

Figure 6 Example of microphone positions both from the ~3 km position. The red light is the Risø met mast close to the source. Left: Microphone at the shore. Right: Microphone on land.

5. Meteorology

The source (crane) was setup close to the tall Risø met mast. Additionally, DTU set up meteorology instrumentation at the harbour of Gershøj (370 m south of the ~7 km microphone position). The instrumentation of these two positions is listed in Table 3 and Table 4..

Measurement	Sensor	Height a.g.l. (boom direction)
Horizontal wind speed	Risø P2546A Cup anemometer	125.2 m (top), 118.0 m (225°), 94.0 m (225°) 92.5 m (45°), 76.7 m (225°), 44.1 m (225°)
Wind direction	Vector W200P F2919A Wind vane	122.5 m (45°) 94.0 m (225°) 76.7 m (225°)
Relative Humidity and absolute temperature	Vaisala HMP155	117.0 m (225°) 2.0 m (225°)
3D wind vector	METEK USA-1 Ultrasonic anemometer	122.5 m (225°) 60.0 m (225°)
Temperature gradient 118 m – 44 m	Risø P2642A temperature sensor with P2029 radiation shield	118.0 m (225°) 44.0 m (225°)
Absolute Temperature	Risø P2449A temperature sensor with P2029 radiation shield	2.5 m (225°)
Pressure	Vaisala PTB110 Air pressure sensor	1.0 m (enclosure)
Solar radiation	Kipp & Zonene pyranometer	120.0 (-)
Rain intensity	F3329A Pronamic rain gauge	Lab roof

Table 3 Risø mast instrumentation.

Measurement	Sensor	Height a.g.l. (location)
3D wind vector	METEK USA-1 Ultrasonic anemometer	5.0 (tripod)
Relative Humidity and absolute temperature	Vaisala HMP155	2.0 m (tripod)
Pressure	Vaisala PTB110 Air pressure sensor	1.0 m (DTU van)
Horizontal wind speed, wind direction and vertical component	Lidar – WindCube V2	40 m, 60 m, 80 m, 100 m, 120 m, 140 m, 160 m, 180 m, 200 m, 230 m, 260 m, 290 m

Table 4 Instrumentation at Gershøj.

6. Measured data

Detailed plots for meteorology are shown in section 6.1.
Section 6.2 describes the measured sound levels.

6.1 Weather

The wind direction was very close to ideal (90-130°) for multiple reflections, as the wind blew from the loudspeaker towards the microphone positions as described in section 4.2. The wind speed profile was generally with a large shear, favourable for multiple reflections, with a little variation during the measurements. The temperature gradient was between -0.4 – +0.1 °C / 100 m which is not ideal but should not prevent the multiple reflections from occurring given the relatively strong wind shear. The estimated shear exponent (power law – here called gamma) to describe the wind speed profile measured with the lidar at the 7 km position was about 0.3. The fact that the wind speed and temperature profile is unknown for higher altitudes presents a limited uncertainty for the effect of multiple reflections. If the wind shear (in propagation direction) and temperature profile both are negative at higher altitudes – that would make the sound field bend upwards at that height, thus limiting the amount of sound energy that is refracted downwards. However, for a situation with neutral atmosphere, zero temperature gradient, a roughness length of 0.001, wind speed of about 6 m/s at 10 m height with a Businger-Dyer profile [6] (up to a height of at least 10 % of propagation distance) in the propagation direction and a sound source at 30 m height, the point where the multiple reflections start to have an effect is at about 2 km from the sound source, see [4]. If the sound source is at a height of about 100 m, the distance is approx. 6.7 km with the same meteorology. The corresponding shear exponent (often called "gamma", γ) for that wind speed profile is about 0.09. This is considerably less than the shear exponent that was derived from the windspeeds observed during the measurement (0.3).

The various meteorology parameters from the 2nd campaign are shown in the following figures (Figure 7, Figure 8 and Figure 9).

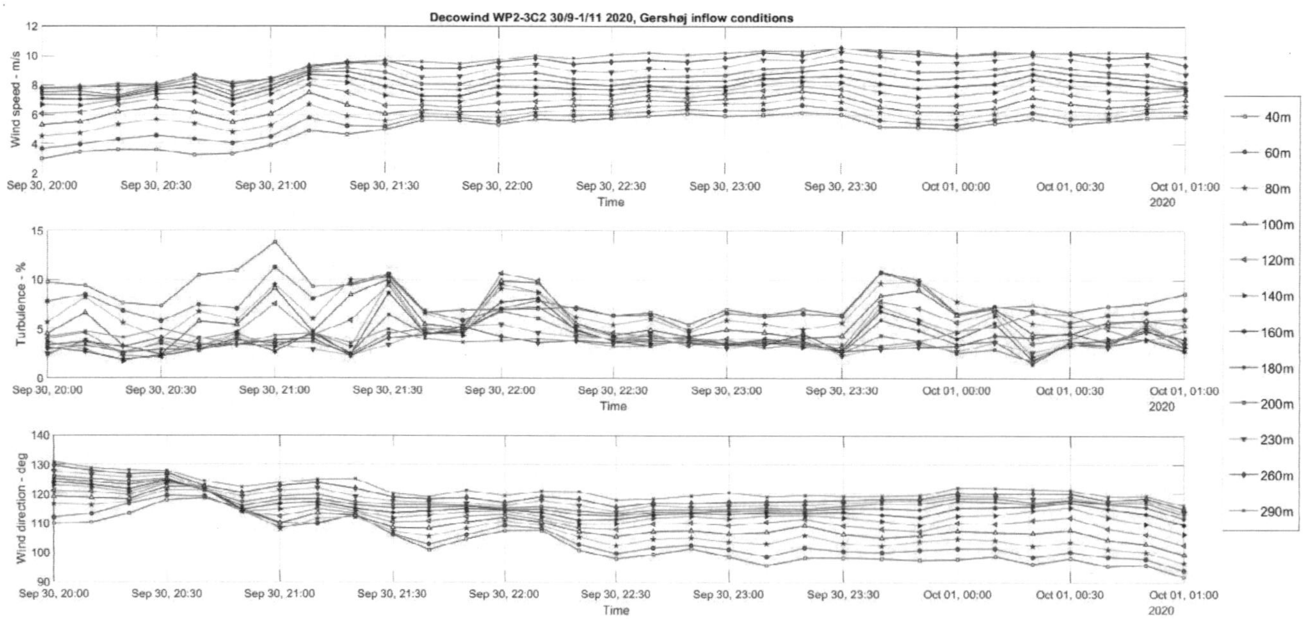

Figure 7 Wind speeds, turbulence intensity and direction for the timespan of the 2nd campaign from the Lidar.

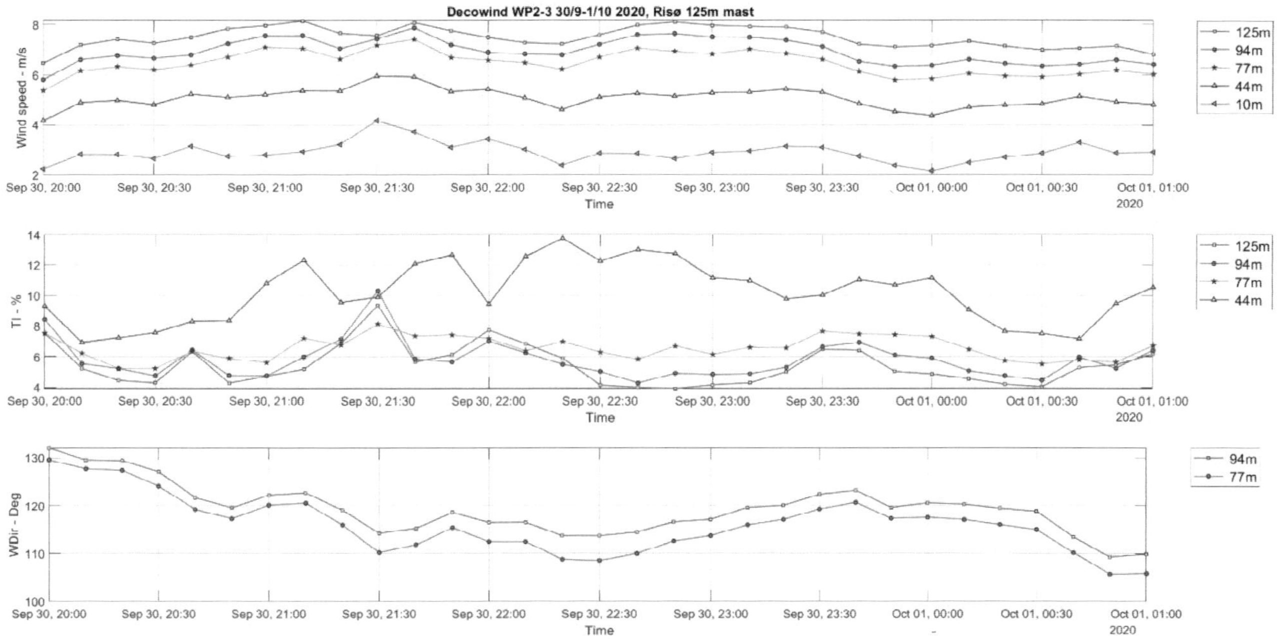

Figure 8 Wind speeds, turbulence intensity and direction for the timespan of the 2nd campaign from DTU met mast at Risø.

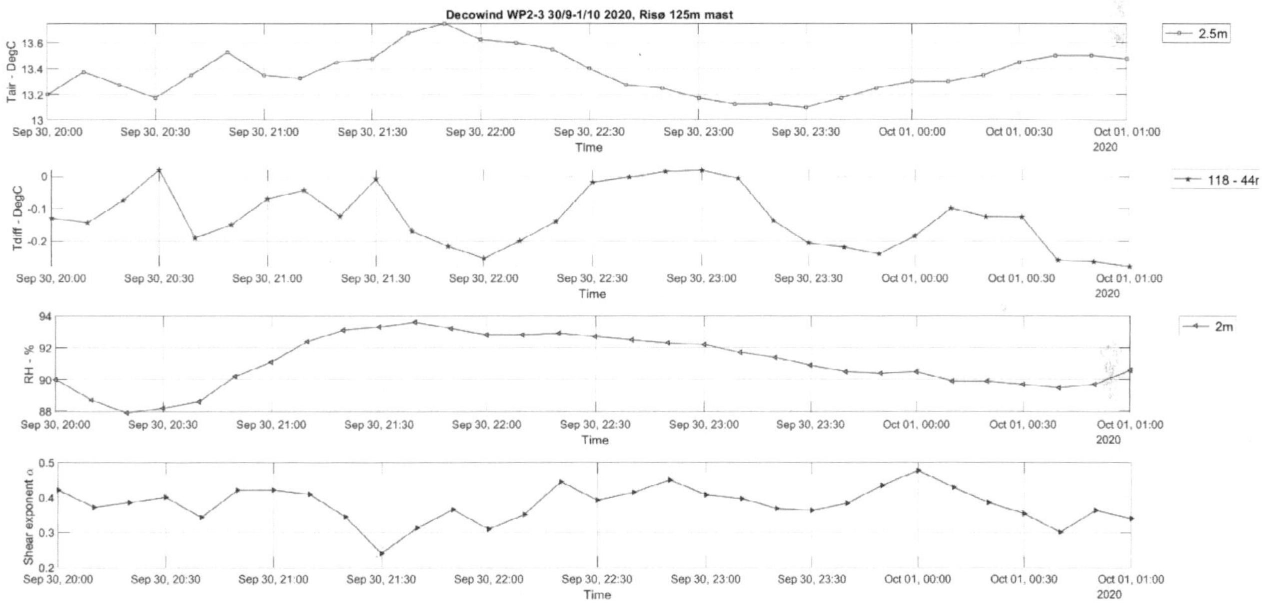

Figure 9 Temperatures, Relative humidity and shear exponent for the wind profile for the 2nd campaign of the measurement from DTU met mast at Risø.

6.2 Noise

Clips of the recorded signals were created corresponding to relevant types of signal as described in Table 1. These clips were analysed by 1/3-octave filters conforming to IEC 61260 in the frequency range 50 Hz – 630 Hz. Recordings from different measurement stations were aligned to better than 100 ms resolution using the sharp pulse from the generated chirp signal. All shown data in this paper are unweighted, i.e. not A-weighted.

The measured noise, when the signal was on, is referred to as “total noise”. Between each signal was a noise pause which is referred to as “background noise”. For each ~16-minute sequence the background noise was analysed and evaluated manually, and an average background noise was calculated for each sequence.

This influence of background noise was used to calculate the corrected level of the generated signal by following the methodology prescribed when handling environmental noise in Denmark [7]:

- Total noise is corrected by the background noise, if the difference is lower than 10 dB.
- Above 10 dB difference the total noise is used directly.
- Below 3 dB difference the data is evaluated as not usable in the further analysis.

Four examples are shown in Figure 10 i.e. for two different positions (~3 km and ~7 km) both at shore and inland, and for two different source heights.

66 % of the datasets have a signal-to-noise ratio of at least 10 dB (total noise minus background noise). 14 % of the datasets have a signal-to-noise ratio below 3 dB, where 5 % are in the 50 Hz 1/3-octave band, and the rest are spread but primarily at the 7 km position at the shore and for the highest frequencies.

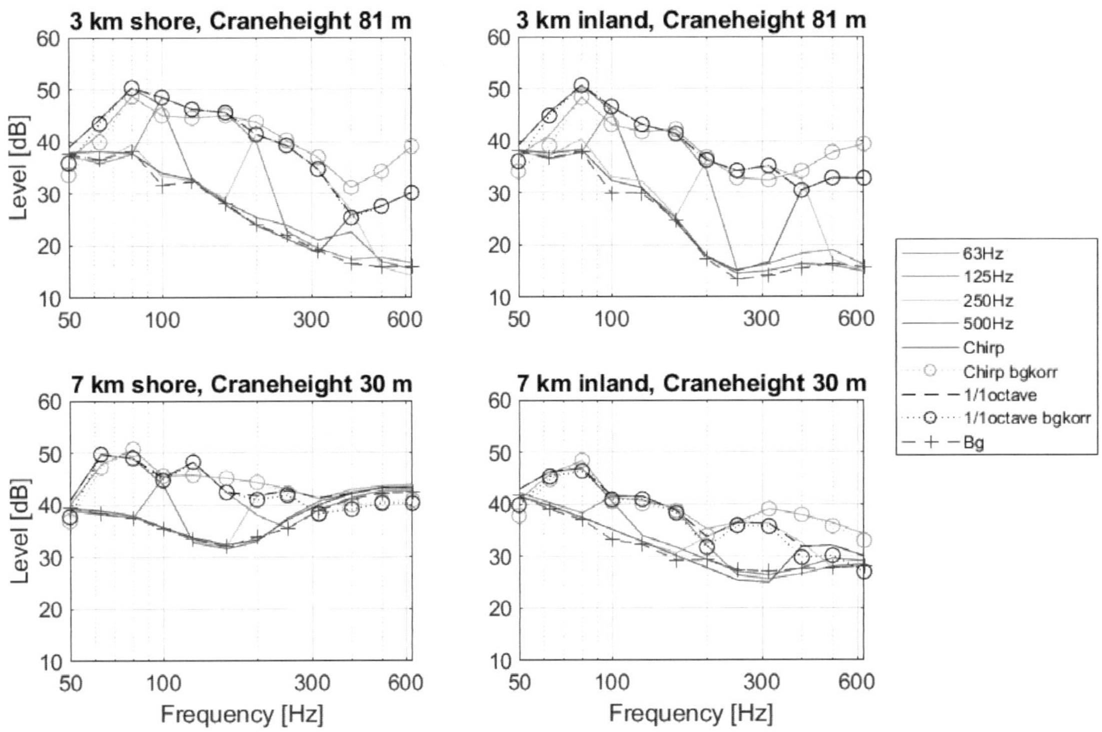

Figure 10 Top: Measured noise levels per 1/3-octave band shown for one sequence for the ~3 km distance position both at the shore and at land with a source height of 81 m. Bottom: Measured noise levels per 1/3-octave band shown for one sequence for the ~7 km distance position both at the shore and at land with a source height of 30 m. The plots both show the measured levels together with the analysed average background noise (red dashed line), chirp levels (green line), chirp levels corrected for background noise (dashed green line), 1/1-octave band levels (dashed black line) and 1/1-octave band levels corrected for background noise (dotted black line).

The measured sound level in front of the loudspeaker did not vary much during the measurement. Figure 11 shows the variation between the arithmetic average for the band-pass filtered signal (on the plot referred to as 1/1-octave). The third octave levels of the chirp signals are also shown relative to the arithmetic average of the band-pass filtered noise. This shows the difference of third octave spectra for the two signals where it can be seen that the bandpass filtered signal has higher relative levels in the low frequency range (50 Hz – 160 Hz) and the chirp signal has higher levels about 200 Hz and up. It can also be seen that the source variation is generally within ~1dB.

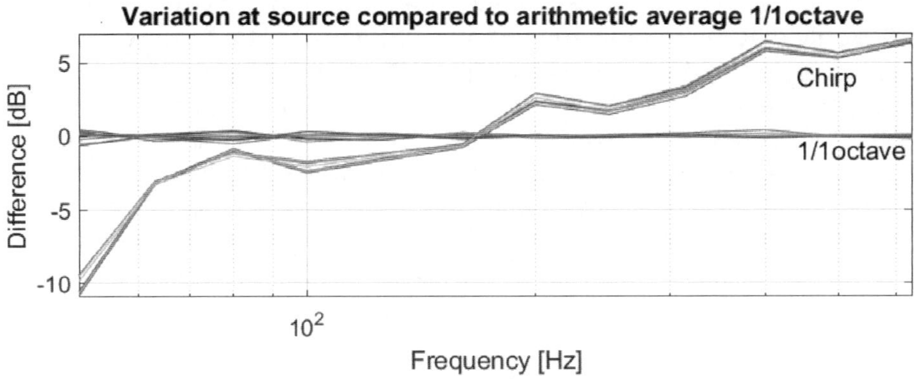

Figure 11 Variation of the source third octave sound levels relative to the arithmetic average of the bandpass signals (1/1-octave).

As shown in section 5, the wind speed at the end of the measurement campaign was higher than in the beginning, resulting in more wave noise at the end (measurement with a source height of 30 m). This can be seen in Figure 10, but more clearly in Figure 12 which shows the background noise as a function of wind speed – both for the microphone close to the water (shore) and the microphone at land (inland). As expected, the background noise (primarily wave noise) primarily affects the higher frequencies.

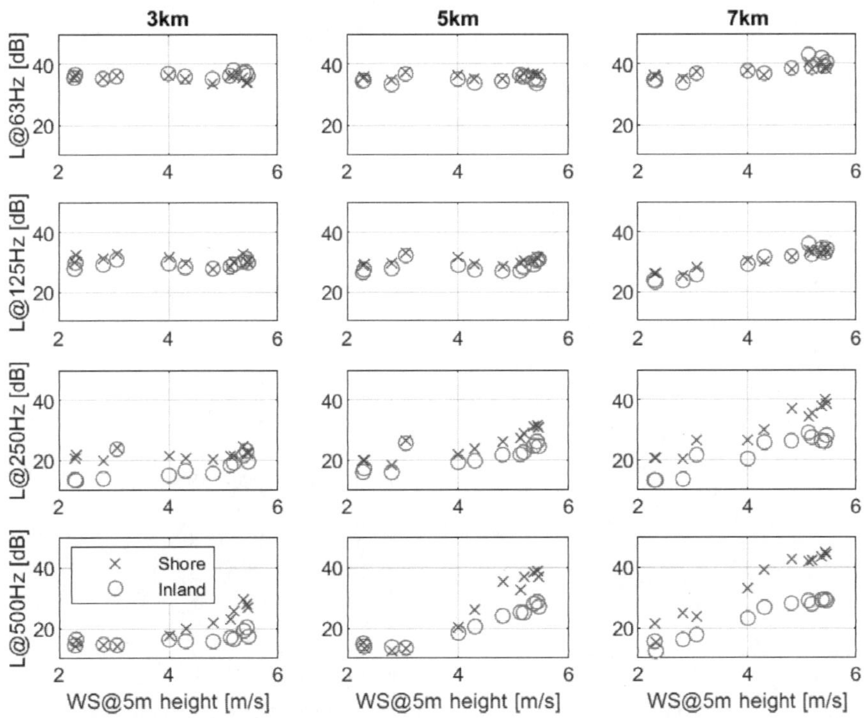

Figure 12 Variation of background noise per position for different wind speeds for the 1/3-octave bands 63 Hz, 125 Hz, 250 Hz and 500 Hz for windspeeds measured with sonic at 5 m height at Gershøj (7 km).

7. Comparison with Danish offshore noise regulations for wind turbines

In this section the signal levels are compared to the noise propagation calculations based on the Danish noise regulation for offshore turbines.

Noise from wind turbines in Denmark has been regulated by statutory orders (DK: Bekendtgørelse) since 1991 [8]. The Danish noise regulation has regularly been updated, latest in 2019 referred to as BEK135 [1]. One of the primary updates in BEK135 is the introduction of multiple reflections for noise propagation over water which is based on [4].

For comparison reasons the measurement results will be compared with the previous statutory order BEK 1736 [9], since the only difference between the two statutory orders for noise propagation over water is the inclusion of the effect of multiple reflections. This effect is calculated for wind speeds of 6 m/s and 8 m/s at 10 m height, even though the windspeeds at 10 m height never exceeded 6 m/s during the measurement. During the measurements, the windspeed at 10 m height was between 2 and 4 m/s at Risø. However, as mentioned in chapter 5 the estimated shear exponent for the BEK 1736 model is about 0.09 whereas in these measurements the shear exponent was estimated around 0.3 for the height span between 40 – 290 m which is considerably higher.

7.1 Modelling

The sound power levels shown in Table 2 are used as input for the noise propagation calculation The measured levels with chirp signals are normalised by using the difference shown in Figure 11.

7.2 Results

In Figure 13 to Figure 15 the average background-noise corrected results are compared with calculations following the methods in BEK135 and BEK 1736 for the 1/3 octave bands 63 Hz, 125 Hz, 250 Hz and 500 Hz. Only results with a signal-to-noise ratio of at least 3 dB are included. However, the calculated arithmetic average is calculated on the basis of both the 1/1 octave bandpass filtered signals and the chirp signals. For this calculation the results with using chirp signals has been normalised corresponding to the information shown in Figure 11. The error bars show the expanded uncertainty from Table 5.

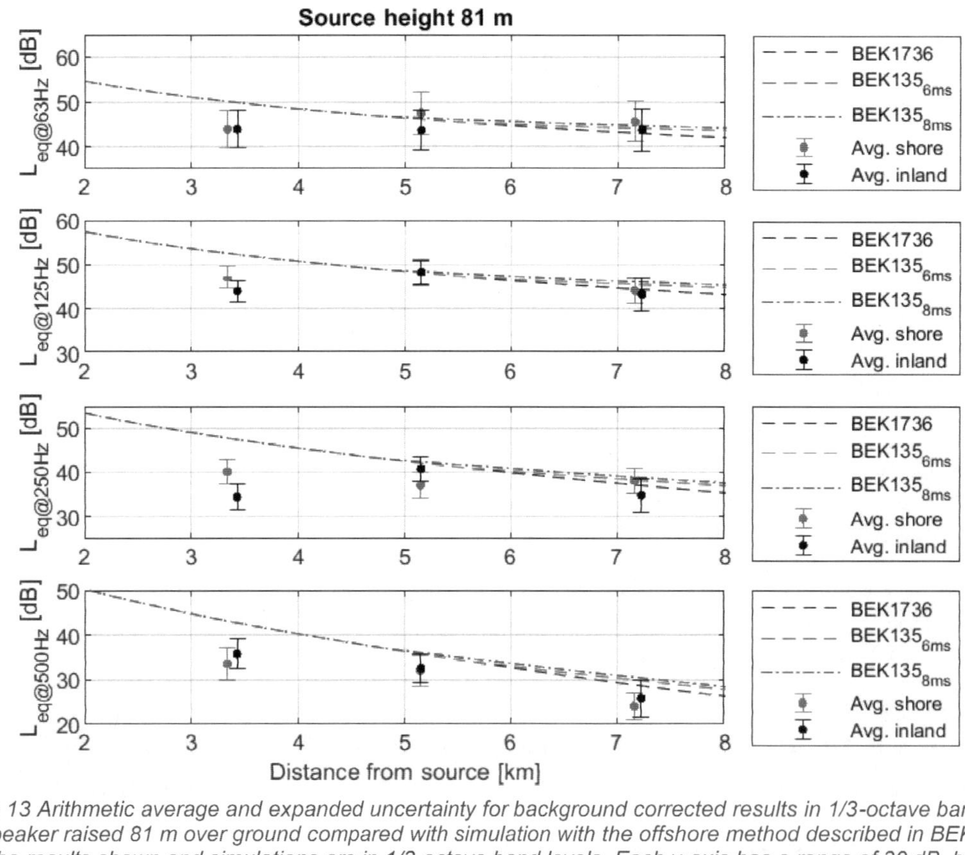

Figure 13 Arithmetic average and expanded uncertainty for background corrected results in 1/3-octave bands with the loudspeaker raised 81 m over ground compared with simulation with the offshore method described in BEK 135 and BEK 1736. Both the results shown and simulations are in 1/3-octave band levels. Each y-axis has a range of 30 dB, but the max and min y-axis have been otherwise scaled to show the results best possible.

There is a reasonable fit between measured and predicted results, but naturally there are levels that are both a little higher as well as lower than the predicted levels using BEK 135. In that sense it is important to note that the BEK 135 predictions are an engineering model and very generalised for situations where wind speed is 6 and 8 m/s at 10 m height while that the measurements in the paper only represent one specific meteorological situation. The wind speed profile in these measurements had a shear exponent that was higher than the shear exponent that is a basis for the BEK 135.

Furthermore, the effect of multiple reflections in BEK 135 describes a sort of "average" effect. Along the propagation, the effect of multiple reflections can be quite chaotic with many peaks and dips in the amplitude response, see [4]. These variations are not possible to predict precisely as they vary substantially with the meteorological situation.

Comparing the plots for different source heights, it can be seen in Figure 15 that the effect of multiple reflections is larger (30 m source height) than in the other two plots (Figure 13 – 81 m source height and Figure 14 – 50 m source height). For the source heights at 50 m and higher, the effect of multiple reflections predicted by BEK 135 is rather limited for measurement distances up to 7 km. Focusing on the 63 Hz 1/3-octave band it is quite clear that there could be an effect of multiple reflections indicated by the approximate average 7 dB difference between BEK 1736 predictions (without effect of multiple reflections) and measurement results at 7 km distance from the source. This is further supported by the BEK 135 predictions (that include the effect of multiple reflections). For higher frequencies this is not clear. It is the assumption that this is due turbulence which reduces the coherence of the reflected sound more for higher frequencies than for lower frequencies.

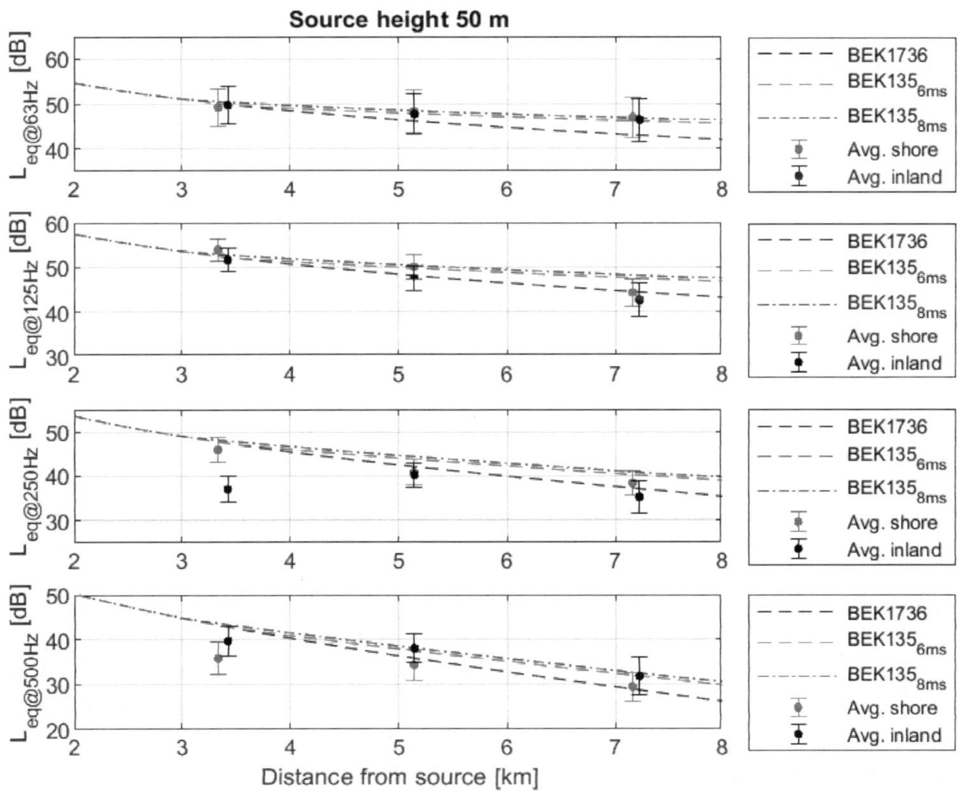

Figure 14 Arithmetic average and expanded uncertainty for background corrected results in 1/3-octave bands with the loudspeaker raised 50 m over ground compared with simulation with the offshore method described in BEK 135 and BEK 1736. See further info at Figure 13.

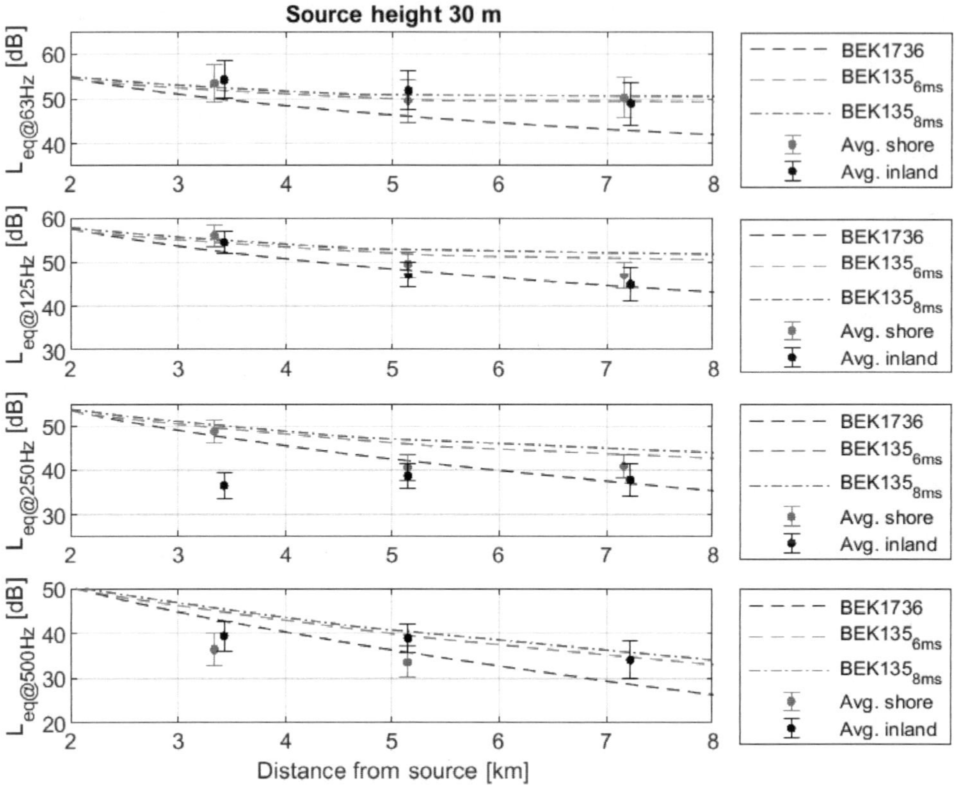

Figure 15 Arithmetic average and expanded uncertainty for background corrected results in 1/3-octave bands with the loudspeaker raised 30 m over ground compared with simulation with the offshore method described in BEK 135 and BEK 1736. See further info at Figure 13.

7.3 Uncertainties

Measurement uncertainty includes the entire measurement chain from the position of the loudspeaker and microphones to meteorological observations to analysis uncertainties. The expanded uncertainty of the measured sound pressure levels with 95 % two-sided confidence interval are:

$$U_{SPL} = 1.96 * \sqrt{\sigma_{instruments}{}^2 + \sigma_{measured\ spl}{}^2 + \sigma_{measurement\ setup}{}^2}$$

where the standard deviations (σ) are:

$\sigma_{instrument} = 0.25\ dB$
$\sigma_{measured\ spl} = 0.65 - 1.56\ dB\ depending\ on\ measurement\ postition\ and\ frequency$
$\sigma_{measurement\ setup} = 0.1\ dB\ (63\ Hz), 0.2\ dB\ (125\ Hz), 0,5\ dB\ (250\ Hz)\ and\ 1,0\ dB\ (500\ Hz)$

(estimated values for variance of the measurement setup).

When comparing measurements to calculation models and methods, the uncertainty of the offset of the noise source power must also be included. The estimated standard deviation for the 63 Hz band is 2 dB and 1.6 dB for the other octave bands based on recommendations of standard deviations from [10]. This total uncertainty is hereafter called $U_{Transfer\ function}$. The relevant expanded uncertainties for this document are listed in Table 5.

Uncertainty	Octave band			
	63 Hz	125 Hz	250 Hz	500 Hz
$U_{SPL\ 3\ km\ shore}$	1.2 dB	1.3 dB	1.5 dB	2.8 dB
$U_{SPL\ 3\ km\ land}$	1.1 dB	1.2 dB	1.9 dB	2.6 dB
$U_{SPL\ 5\ km\ shore}$	2.5 dB	1.8 dB	1.8 dB	2.7 dB
$U_{SPL\ 5\ km\ land}$	1.8 dB	1.6 dB	1.7 dB	2.3 dB
$U_{SPL\ 7\ km\ shore}$	1.9 dB	2.0 dB	1.6 dB	2.2 dB
$U_{SPL\ 7\ km\ land}$	2.5 dB	3.1 dB	3.0 dB	3.5 dB
$U_{Transfer\ function\ 3\ km\ shore}$	4.2 dB	2.5 dB	2.7 dB	3.6 dB
$U_{Transfer\ function\ 3\ km\ land}$	4.2 dB	2.5 dB	2.9 dB	3.4 dB
$U_{Transfer\ function\ 5\ km\ shore}$	4.7 dB	2.8 dB	2.8 dB	3.5 dB
$U_{Transfer\ function\ 5\ km\ land}$	4.4 dB	2.7 dB	2.8 dB	3.1 dB
$U_{Transfer\ function\ 7\ km\ shore}$	4.4 dB	2.9 dB	2.7 dB	3.1 dB
$U_{Transfer\ function\ 7\ km\ land}$	4.8 dB	3.8 dB	3.7 dB	4.2 dB

Table 5. Uncertainties for the measured sound pressure levels and for the transfer functions.

8. Conclusions

Noise propagation over water has been investigated in two measurement campaigns where both utilized a setup with a height-adjustable sound source and microphones positioned far downwind from the sound source. Meteorological conditions were monitored at both ends of the setup.

Even with a sound power level of 130-135 dB, which in the chosen frequency range, is 15-25 dB higher than the sound power level from at modern offshore wind turbine in the 10 MW class, in the first measurement campaign it was not possible to hear or measure the transmitted signals at measurement positions further away than 2 km from the sound source, mainly due to excessive wave noise. Based on the first measurement campaign a new measurement campaign was carefully planned, and with several checks to ensure a successful campaign at another site. The 2nd measurement campaign proved successful and its results can be used for further development and evaluation of methods and models that aim to include the effect of multiple reflections.

Different heights for the sound source were used (81 m, 50 m and 30 m above ground) with microphones positioned downwind (both at the shore and ~100 m inland) of the sound source (~3 km, ~5 km and ~7 km distances to source). The sound only propagated over water between the sound source and the microphones placed at the shore.
The meteorological conditions (wind speed, wind direction, stability, temperature, humidity, etc.) were monitored at both ends of the setup, utilizing both a tall met mast, a wind profiler and sonic anemometers at multiple heights.

For most of the measurement positions and frequency ranges the signal-to-noise was usable with 86 % of the data had a better signal-to-noise ratio than 3 dB, and for 66 % of the data the signal-to-noise ratio was at least 10 dB.

The meteorological conditions were almost ideal for multiple reflections to occur with the only exception being the temperature gradient. The temperature gradient that was approx. zero (0) degrees / m should however not prevent multiple reflections from occurring, even though a positive temperature gradient would have increased the possibility and effect.

When the background-noise corrected results were compared to predictions following the Danish regulation for wind turbine noise over water, a general good agreement was seen.

When comparing the background-noise corrected measured results with predictions following BEK 135 and BEK 1736 it is seen that the measured results for the 63 Hz octave band fit better with BEK 135 than BEK 1736. For the higher frequencies it is more difficult to make any general conclusions. Since the only difference between BEK 135 and BEK 1736 (in this situation) is that multiple reflections are accounted for in BEK 135 it indicates that multiple reflections did occur during the measurement campaign and had an effect on the 63 Hz band. The effect is however often overestimated with BEK 135 for the higher frequencies.

In order to perform simulations of the propagation data for meteorological conditions up to higher altitudes than what was possible to obtain in this measurement is needed. Ideally this is up to about 10 % of the propagation distance, in this campaign layout up to 700 m. There are wind speed measurements up to 290 m height and temperature information up to 118 m.

Observing the variation in measured noise levels, it would be beneficial to have statistics over the meteorological situations that could lead to multiple reflections. I.e. when and how often they occur with what "strength" etc.

Finally, it can be concluded that even low wind speeds at low altitude can have a significant influence on the background noise.

Acknowledgements

Financial support for this work was partly provided by DecoWind project (Development of low-noise and cost-efficient wind farm technology), File number: 8055-00041B under Innovation Fund Denmark. Supplementary financial support was provided by Danish Agency for Institutions and Educational Grants.
We acknowledge gratefully the following colleagues from FORCE Technology for their invaluable assistance: Morten Bording Hansen, Rasmus L. Christensen, Jens Oddershede and Lukas N. Byrne for assisting with the measurements, Jens Oddershede for drone footage, Michelle Herlufsen for your graphical skills, Birger Plovsing for your invaluable knowledge of noise propagation, and finally Inge Lis Kjær for proofreading.
We acknowledge gratefully the following colleagues from DTU Wind for their invaluable assistance: Per Hansen and Lars Christensen for assisting with the measurements.
We also owe great thanks to SGRE for funding crane and loudspeaker rental for the 2nd measurement campaign.
We could not have done it without access, so we would also very much like to thank DTU for granting us access to DTU Risø campus, and to the responsible people at "Bogenæs Skov", and Steffen Sylvest Hansen in Gershøj who let us use their grounds to place microphones.
Finally, we are grateful for the patient and forgiving neighbours that could hear the noise during the measurements.

References

[1] *Bekendtgørelse om støj fra vindmøller*, no. 135. Denmark: Miljø- og Fødevareministeriet, 2019, pp. 1–24.

[2] B. Søndergaard and B. Plovsing, "Noise from offshore wind turbines," 2005.

[3] K. Bolin and M. Boué, "Long range sound propagation over a sea surface," *J. Acoust. Soc. Am.*, vol. 126, no. 5, pp. 2191–2197, 2009.

[4] B. Plovsing and E. Thysell, "PE-beregninger af cylindrisk lydudbredelse," 2017.

[5] B. Søndergaard, B. Plovsing, and T. Sørensen, "Noise and energy optimization of wind farms Validation of the Nord2000 propagation model for use on wind turbine noise," 2009.

[6] BUSINGER JA, WYNGAARD JC, IZUMI Y, and BRADLEY EF, "Flux- profile relationships in the atmospheric surface layer," *J. Atmos. Sci.*, vol. 28, no. 2, pp. 181–189, 1971.

[7] Miljøstyrelsen, "Måling af ekstern støj fra virksomheder, Vejledning nr. 6/1984." 1984.

[8] *Bekendtgørelse om støj fra vindmøller*. Denmark: Miljø- og Fødevareministeriet, 1991.

[9] *Bekendtgørelse om støj fra vindmøller*, no. 1736. Denmark: Miljø- og Fødevareministeriet, 2017.

[10] ISO, "ISO 3745 - Acoustics – Determination of sound power levels and sound energy levels of noise sources using sound pressure – Precision methods for anechoic rooms and hemi-anechoic rooms." 2012.

INCEEUROPE

9th International Conference
on
Wind Turbine Noise
Remote from Europe – 18th to 21st May 2021

Wind farm neighbourship investigated by a daily app questionnaire combining weather, noise, and annoyance

Lars S. Søndergaard, M.Sc., FORCE Technology: lss@force.dk
Christer P. Volk, Ph.D., FORCE Technology
Tomas R. Hansen, M.Sc., Siemens Gamesa
Lars Enggaard, M.Sc., Siemens Gamesa
Thomas Sørensen, M.Sc., EMD International A/S
Alfredo Peña, Ph.D., Technical University of Denmark

Summary

A socio-acoustic study of wind farm annoyance was conducted with 68 participants across 14 locations in Denmark. All participants lived within a 2-km radius of one or more >2MW wind turbines and reported their daily wind turbine noise annoyance for five consecutive weeks, both in terms of current annoyance and annoyance within the last 24 hours. Additional questionnaires provided information on dependent factors, such as age, sleep patterns, other noise sources, noise sensitivity, etc. All data were collected using the ExpiWell smartphone app which allowed for collection of valuable metadata, such as tracking of participant GPS position while answering and time of day. The questionnaire data were enriched with Nord2000 noise predictions based on hourly simulated meteorological data and production data from neighbouring wind turbines in 10-minute means. Collected data included wind turbine and neighbour locations, turbine data and settings (produced power, noise-mode, power-curves, noise-curves, nacelle wind speed and calibrated yaw position) and simulated data (wind speed, wind direction, temperature, relative humidity and Obukhov length). The study focused on establishing knowledge of how annoyance is influenced by factors that could affect the daily operation of wind turbines to the benefit of both owners and neighbours.

1. Introduction

Denmark is densely populated, and even in rural areas dwellings have relatively short distances to each other. Similarly, both existing wind turbines and planned wind turbines are also relatively close to dwellings, and since Denmark has been a pioneer in setting up wind turbines since 1980, there are currently approx. 5700 onshore wind turbines so densely positioned across the country that its land and coastline – including its larger islands – can be sketched solely from their positions (See Figure 1).

While noise is an important concern related to wind farms, the sound pressure level alone however explains only a minor part of annoyance reported by neighbours [1]. It is likely that the background noise level from other sources has an influence, as could the sound insulation of the dwellings, the routines of the families (how much time they spent outdoor) etc. Furthermore, some characteristics of the noise have been reported which may also affect the degree of annoyance (see e.g. [2]). Additionally, wind farms may also displease neighbours by e.g. the visual presence being alien in the green areas in which they are normally set up (e.g. [2]). Furthermore, the value of the property and house may devaluate and be a cause for concerns. Many acoustical and non-acoustical factors have been described in the literature, but differences in culture, weather etc. likely affect the importance of these between regions, making it of interest to study effects reported internationally and understand the significance and interactions of these in Denmark.

Traditionally, wind turbine annoyance has been studied by asking neighbours to report their annoyance over the last year (in compliance with the ISO 15666 standard [3]) in a one-time questionnaire, such as the large study by Pedersen [4] in 1994. These studies usually include a large number of participants which leads to data with large variations in context (geographical, social, personal etc.) to account for in the analyses. While this enables estimates of the *level* of annoyance in studies of e.g. infrastructure and industry noise, it is less suited for understanding noise level moderators depending on weather conditions which does not follow daily or weekly patterns. In this study we were interested in understanding the influence of variations in weather conditions on annoyance and thus decided to ask neighbours daily for five weeks about their current annoyance level. This led to a collection of data on annoyance influences for a large variation in e.g. wind speed and wind direction being documented allowing an in-depth understanding of not just general wind farm annoyance, but conditions in which annoyance was lower and higher.

The factors of most interest in this study were the ones with the potential for wind turbines to produce more power and cause less annoyance. This paper describes the regulatory background for suspecting such factors may exist, which methods were used – including experiences with using an app for questionnaires - and some preliminary results to gauge this data collection methodology.

1.1 Noise legislation

Wind turbine noise in Denmark is currently regulated by Statutory Order 135 dated 2019 [5]. In short, the method is based on declared and confirmed sound power levels, whereafter the sound level at dwellings is calculated by a propagation model which is calibrated with the Nord2000 model [18][19]. For modern pitch-controlled wind turbines set up in Denmark, the maximum (max) production is reached at approximately 7 m/s referenced to 10 m height. Similarly, the max noise level is typically reached at approximately 7 m/s referenced to 10 m height [6].
The Statutory Order specifies noise limits at wind speeds of 6 m/s and 8 m/s respectively. These are listed in Table 1.

	6 m/s	*8 m/s*
Neighbouring dwelling in open countryside	42 dB(A)	44 dB(A)
Areas for noise-sensitive land	35 dB(A)	37 dB(A)

Table 1 Danish noise limits for wind speeds 6 and 8 m/s referenced to 10 m height.

The noise limits are for the cumulative noise from all wind turbines (in practice only the relevant turbines are included).

For the noise propagation calculation, the Danish regulation assumes all neighbours to be downwind from all wind turbines. Since all neighbours are not always downwind from the turbines and since – not even in Denmark – the wind speed is not always above 7 m/s (relative to 10 m height), the noise level from pitch-regulated wind turbines in DK will at neighbour dwellings at the highest be approximately 44 dB(A). For wind speeds lower than 7 m/s and wind directions other than those downwind from the turbines, the noise level will be lower than 44 dB (or 37 dB for noise-sensitive areas). Knowing that sound propagation is affected by wind direction, and assuming that some wind directions are more common than others at wind turbine farm, some neighbours will on average be exposed to higher noise levels than others [7].

1.2 DecoWind (Development of low-noise and cost-efficient wind farm technology)

The aim of the DecoWind project is to address a fundamental gap in our knowledge about noise from wind turbines and to use the acquired insight to increase green energy production without increasing neighbour annoyance. The project consists of different work packages where one of them focuses on improved noise regulation of which this socio-acoustic study is part of.

The DecoWind project has four partners: DTU Wind Energy, Siemens Gamesa Renewable Energy, EMD International A/S and FORCE Technology.

2. Socio-acoustic study

This socio-acoustic study consisted of four main elements: Recruitment by an independent partner, information material to participants including small videos, daily questionnaires answered over an app and continuously monitoring of activity level by each participant with follow-up in case of drops in activity. Even with a modest aim of 70 participants the effort was considerable and required over-recruitment of more than 20 %.

2.1 The choice of using an app

Three purposes were important regarding the choice of a smartphone app as medium for the questionnaires. Firstly, to simulate the noise level at exactly the time and location of each participant which made an app with GPS-tracking of respondents' positions and automatic timestamps ideal. Secondly, to control the time of day at which the questionnaire could be filled out to spread out responses across the day. Thirdly, participants got automatic reminders, which could be crucial for such a long period and finally, we could actively monitor participants' response rate and contact participants (through a third party, see sec. 2.4) falling behind to assist with technical issues or providing renewed motivation. This also allow us to require participants to answer at least a certain amount of days of the total days included in the period to get paid (which they were of course informed of during the recruitment interview).

A smartphone app might, however, also prove a challenge for participants being less comfortable using smartphones (e.g. the older generation). This concern was handled by providing good technical support, and driven by the experience of two smaller pilot studies to get participants started with the app and answering general questions a week prior to the start-up of the daily questionnaire to allow time for support and getting comfortable with the app.

For this purpose, the app ExpiWell [8] was chosen, an app designed specifically for ecological momentary assessments (EMA) which is available for both Android and iOS (iPhone). The app is developed by researchers at Purdue University, US, and supported by universities such as Harvard University and Yale University. The app also provided the opportunity to get images of respondents view from the position they stood in daily while responding, as well as submitting recordings at times with momentarily increased annoyance.

2.2 Study design

The study was divided into four questionnaires: 1) Introduction, 2) Daily questionnaire, 3) Final, and 4) a voluntary momentarily annoyance[1]. The Introduction and Final questionnaire both concerned background information about age, dwelling information, daily noise exposure, etc. The purpose of splitting into two separate questionnaires was to use the Intro questionnaire as app familiarisation with a limited number of questions and wait with questions that was considered more personal/private. Consequently, all questions in the Intro questionnaire were mandatory, while all questions were optional in the Final questionnaire. Furthermore, we did not want to make participants aware of all the factors which may influence annoyance beforehand. A total of 22 background questions and seven daily questions were included in the study.

Background questions were based on a literature study of factors found to significantly influence annoyance as well as questions needed to investigate our hypotheses. Furthermore, a reference question from ISO 15666 [3] was included: *"Thinking about the last (12 months or so), when you are here at home, how much does noise from the nearest wind turbine bother, disturb or annoy you?"*. With these five response choices: Not at all, Slightly, Moderately, Very, Extremely (in Danish translations). The ISO standard requires a response on a 10-point numerical scale as well, but this was found impractical on a small smartphone screen (with the chosen app).

Category	**Questions**
Noise sensitivity and awareness	Noise sensitive today + current noise sensitivity, peak period of annoyance
Noise annoyance and sources	Total current annoyance, current wind turbine noise annoyance, 24-hour wind turbine noise annoyance
Personal routines	Home periods
Covid-19 changes	Expecting days working from home + days extra / days fewer, change of work noise environment (better/worse), change in noise sensitivity.

Table 2 Type of questions included in the daily questionnaire. The Covid-19 questions only appeared on Sundays.

Category	**Questions**
Noise sensitivity and awareness	Frequency of noticing wind turbine noise, experiencing season dependent audibility of wind turbine noise, situational awareness of wind turbine audibility variations
General noise annoyance and sources	Noise acceptable at home, general noise annoyance, 1-year wind turbine noise annoyance, dominating background sources (traffic, wind turbines, etc.)
Personal information	Age
Personal routines	Hours home and awake, noise (level) at home, noise (level) at work, noise (level) from own household (children, etc.), sleeping with open windows + reason if not
Residence	Wind turbine audibility indoor, construction year of residence (or large renovation), window orientation (primary rooms), large windows (primary rooms)
Visual	Visual of nearest wind turbine, image of daily position
Timing	Wind turbine build before/after moving in.
Economic interest	Economic interest (share in ownership, compensated, etc.)

Table 3 Type of questions included in the two questionnaires Introduction and Final.

[1] The voluntary momentarily annoyance questionnaire is not described further in this paper. Few participants used it and of the 14 responses received half was from one person.

To investigate the influence of the time of day, the daily questionnaire was scheduled to be filled out within a fixed time interval. For the first half of the study period, participants could answer between 16-22 local time (hereafter used), and in the second period participants could answer between 06-12. This could have been more controlled and individualised, but the required functionally was under development in the app during the study period. Therefore, the period between 12-16 was avoided where most participants might be at work or in transit, and participants were required to be at home for the evaluation. Specifically, the daily instruction was: *"For this daily questionnaire, you should be listening outside. Position yourself 3-5 metres from your house and on the side where the turbines usually are most audible. It shall be the same position each day. Begin by listening for 10-20 seconds. Notice the background sound and think about the composition of sounds."*

Due to the low wind farm noise levels permitted at neighbours property (see sec. 1.1), it was decided to plan the study for a time of the year where more Danes are outdoor, as the weather becomes warmer and wind farm neighbours more exposed. This was supported by a previous Danish study [4] where the interviewed wind turbine neighbours marked the summer months (May-August) as the period where they were most annoyed. June was chosen, since it is the only one of these months without school vacations. The annoyance levels in this study may consequently be viewed as of interest in relative terms, but not necessarily representative in absolute terms of a yearly average.

Due to the unfortunate situation of dealing with Covid-19 restrictions within the study period (May-July 2020), participants' daily routines may have shifted, and their noise environment and noise sensitivity may also have been influenced by these changes. Consequently, a few extra questions were added to the daily questionnaire on Sundays, asking about their expectation for the coming week.

2.3 Selecting wind farm sites

It was decided to focus on modern pitch-regulated onshore turbines from SGRE, since they are one of the partners of the DecoWind project. On the website of the Danish Energy Agency [9] an Excel list can be downloaded with UTM coordinates for all Danish turbines >6 kW, both existing and decommissioned. The list also contains information about manufacturer, size, rated power, commission date, etc. The positions of the turbines are plotted in Figure 1 which shows all offshore and onshore turbines, and all onshore SGRE turbines. As it can be seen the SGRE turbines are positioned across most of Denmark. All onshore SGRE turbines were divided into wind farm sites and plotted graphically together with eventual neighbour non-SGRE turbines. For all sites addresses of neighbours within a 1.6 km distance of at least one wind turbine was collected. The distance of 1.6 km was empirically found based on learnings from a pilot study.

All possible sites with onshore turbines were visually examined and only sites and neighbours where wind turbines from other manufactures had negligible influences were selected for the further process to ensure easy production data retrieval. A list was constructed based on the above dividing the list of SGRE turbines into 25 sites.

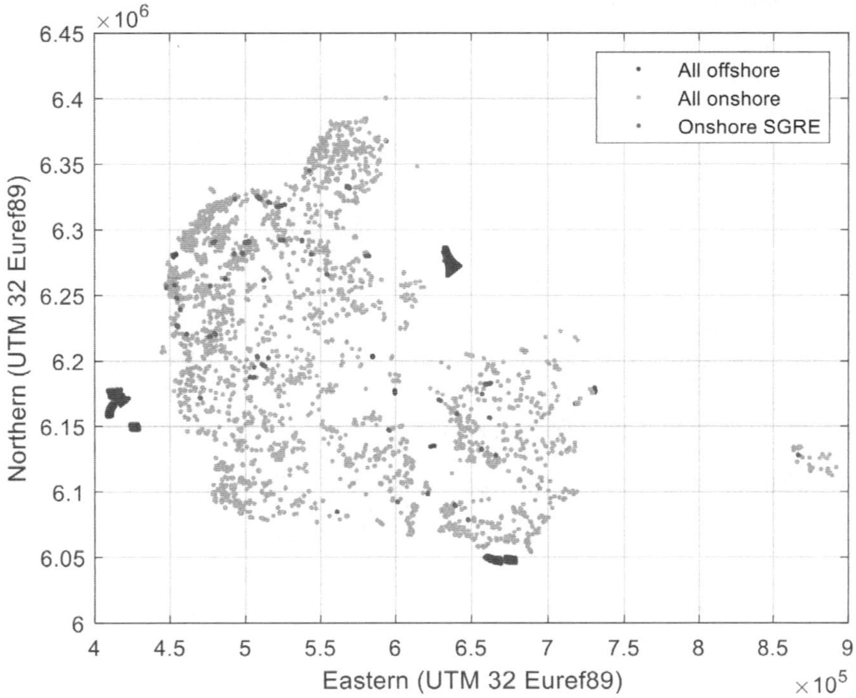

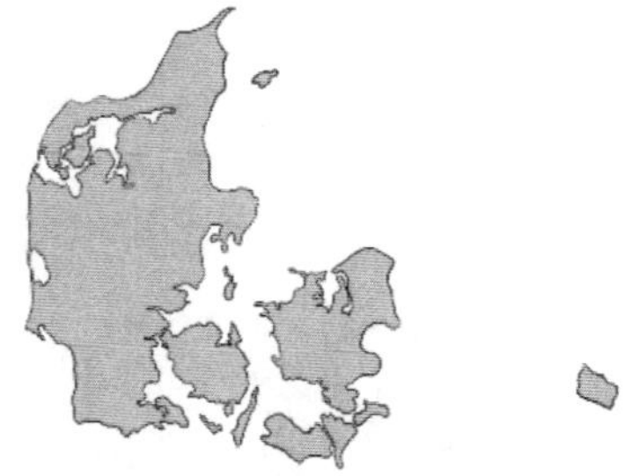

Figure 1 Left: Danish onshore wind turbines ultimo 2019. Dark blue: Offshore turbines, Cyan: Onshore turbines. Red: All turbines marked as "SGRE". Right: Contour of the coastline of Denmark.

2.4 Participant recruitment

All participants were recruited by an external independent partner, Epinion, to avoid direct contact with participants and ensure unbiased selection of wind farm sites and participants. A list was provided with all SGRE wind turbine sites found suited based on the criteria described in the previous section.
Furthermore, it was specified that it was preferred to have multiple participants from each site included in the study, and a total of 70 participants was needed. On this basis (1595 addresses across 25 wind turbine sites) Epinion randomly selected a number of the sites and then from these recruited among the neighbours at the selected sites on the basis of their own randomisation strategy. Epinion enriched neighbour addresses with names and contact information themselves.

In collaboration with Epinion a few screening criteria were specified which were used to vet participants interested in participation: 1) Must be able to operate a smartphone, 2) Must report hearing the wind turbines at least once in a while, 3) Must be able to participate for one month and response to a minimum of 21 of the 30 days (e.g. participants without longer vacation periods planned within the period), 4) Must understand and read Danish fluently. All participants were economically compensated for their participation if participating at least 21 of the 30 days. A few were also compensated for trying but having technical trouble that could not be solved over remote support.

The final list of participants comprised 68 adults from 14 wind farm sites distributed across most of Denmark. On average each site had 4.8 participants with a minimum of 1 and a maximum of 11. Coincidently the total number of turbines was also 68.

3. Meteorological, wind turbine and noise data

In order to simulate the noise levels from the wind turbines at the participants as accurately as possible, it is necessary to obtain information about the meteorology, the turbine production and the produced noise under different wind speeds. Figure 2 gives an overview of the different processes and kind of data achieved which has feed into the socio-acoustic study.

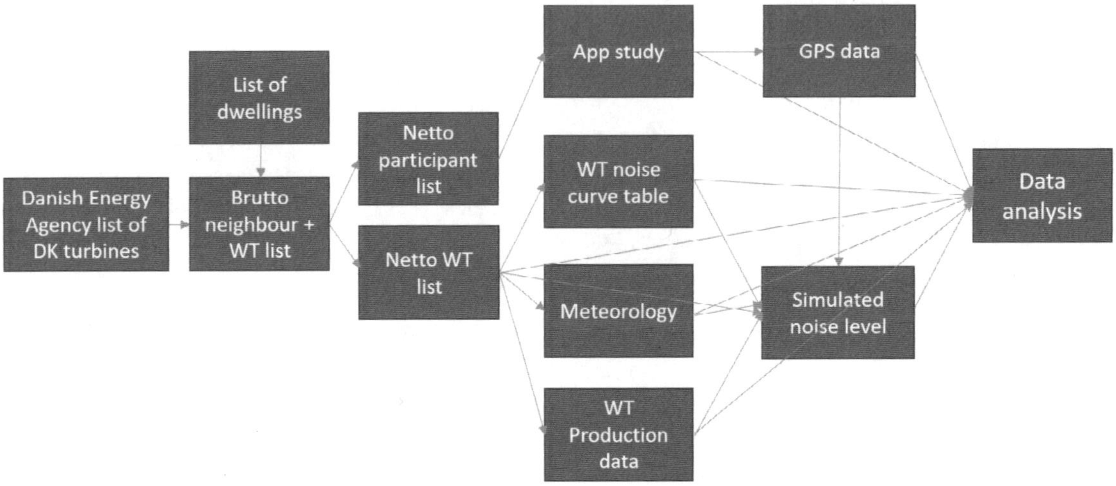

Figure 2 Flowchart of data collection in the socio-acoustic study.

3.1 Connection between wind turbine noise and production data

The relationship between emitted noise and wind turbine production is the key to the international standard for measuring sound power levels for wind turbines [10]. The standard describes the primarily downwind measurement of sound power level per wind speed bin where the normalized wind speed is based on a combination of produced power and nacelle wind speed by a power curve. SGRE typically log this information in 10-minutes intervals.

It was decided that neither SGRE nor the wind turbine owners should be informed on which sites and turbines were used until the study has finished, based on experience in previous projects that some wind turbine neighbours do not trust wind turbine manufacturers and owners to sustain normal operation within measurement periods. Consequently, it could not be confirmed that SGRE had access to production data beforehand. This also removes the option of getting high-resolution data logging, limiting the resolution to 10-minute intervals.
After the study was finished, SGRE was informed which turbines had been included in the study and assisted in compiling a list of owners of the wind turbines.

3.2 Simulated meteorological data

Meteorological data at the nacelle positions for each turbine in the study was simulated, following a similar methodology as in [11].

Model description

The simulated meteorological data come from a real-time weather system, based on the Weather Research and Forecasting (WRF) model [12] which forecasts meteorological conditions over Denmark. The WRF model is a numerical weather prediction and atmospheric simulation system. The real-time weather system was established in 2009 at DTU Wind Energy and uses the version 3.0 of the WRF model.

The setup uses three domains with the innermost covering nearly all Denmark with a spatial resolution of 2 km. It uses 37 vertical levels from the surface to the top of the model at 100 HPa. Within the first 1000 m from the surface, we find 12 of these levels. The setup also uses standard parametrizations including the Yonsei University (YSU) PBL scheme [13]. Other details can be found in [14].

Simulated meteorological data for 68 turbine locations were extracted from the real-time forecasts. The simulated data are instantaneous hourly data covering the full period of the study and correspond to the grid point closest to that of the turbine location. The data correspond to wind speeds, wind direction, both linearly interpolated to the turbine hub height using the two closest vertical levels, 2-m temperature, 2-m relative humidity (computed from the 2-m specific humidity), and the Obukhov length which is computed from the surface-layer scheme that uses Monin-Obukhov similarity theory [15].

3.3 Wind Turbine production data

Wind turbines are regularly stopped due to maintenance, shadow flicker and power grid overproduction. The extent to which this occurred was however higher within the study period than expected. The power grid in Northern Germany has in periods problems with overproduction and has in 2020 hugely utilised the flexibility in the Danish power grid for wind turbines, resulting in 1.46 TWh where Danish turbines were paused [16]. This phenomenon has grown in recent years. For the most affected period of the app study a total of ~81 % of the turbines in the study was stopped at the same time (here defined as nacelle wind speed larger than 5 m/s and a production of less than 20 kW).

Following the completion of the monitored period, the wind turbine owners were contacted and asked for written permission for achieving production data for the relevant time period. The wind turbine owners were promised that the location of the wind farms would not be disclosed. Only one owner did not want to be part of the study.

Data was obtained for 62 turbines out of 68 within the period 30th May – 15th July 2020, and included information about produced power, nacelle wind speed and yaw.
For two turbines it was assessed that data from neighbour turbines could be used instead, resulting in that only data missing from four turbines (1 site). Wind speed and wind direction for all sites were qualified against simulated meteorological data. The yaw direction at SGRE winds turbines is not calibrated to geographical north direction by default, but this was corrected manually for all wind turbines involved.

3.4 Neighbour noise level simulations

A noise curve table was setup with information about sound power level in the frequency area 10 Hz – 10 kHz for each turbine type and noise setting for each integer wind speed from cut in wind speed up to at least the wind speed where rated power is reached. The associated power curve was also added to the table. The 68 turbines in the study were distributed over 18 different turbine types and noise settings.

The normalized wind speed was calculated for each 10-minute period for each turbine following the method described in [10], i.e., combining produced power, nacelle wind speed and power curve.

By using the UTM coordinates for each turbine and participant, all participant-turbine combinations were identified (a total of 307) and the terrain between participant and turbine was extracted by using the Danish Elevation Model [17]. No buildings were considered, as the effect on the noise propagation of these (if any) were assumed to be negligible in the given context. The flow resistivity of the ground was set to 200,000 Ns/m^4 corresponding to normal uncompacted ground (forest floors, pasture field, etc.).
This together with observed meteorological data where available and elsewhere simulated meteorological data was used as input to the noise predictions. Where no information about temperature, relative humidity or temperature gradient/stability was given, the temperature was set to the average temperature in Denmark (i.e. 9 °C), relative humidity was set to 80 %, and Monin-Obukhov length was set to infinity (neutral atmosphere). The relevant noise source power was interpolated from the noise curve table using the normalized wind speed. The wind turbine sound emission was considered to be omnidirectional. This is a simplification that is assumed to be of minor importance. The Nord2000 method [18][19] was used to predict the transfer function between source and receiver. The wake from the turbines was considered to be of limited importance and was not included in the predictions.
A total of 18.434.736 10-minute noise levels was predicted; However, this number also include periods where the turbine was paused, and a real prediction was substituted with a -999 dB value.

4. Data quality

Consistency of the received data was checked, for example by comparing the logged wind turbine production data with the simulated meteorological data, and the correlation between wind speed and wind direction was surprisingly good.
As previously mentioned, one of the benefits of using an app is that GPS position of the response can be checked. This was primarily thought as a way to determine the preferred position for each participant. It however turned out that the participants did not necessarily fill out the app questions from the same daily position as instructed, but some even travelled to other parts of the country (and stayed there for days). 85 % of the total answers were made within 100 m of the instructed position, 90 % of the total answers were made within 1000 of the instructed position, and 95 % of the total answers were made within 10,000 m of the instructed position.

5. Collected questionnaire data

The premise of this study is that while noise from wind turbines contribute to noise annoyance, it alone cannot explain the level of annoyance. This is apparent from the boxplots in Figure 3. The figure is based on filtered responses, as GPS data presented in the previous section indicated that not all participants adhered to the instruction of responding from the same position.
Out of 1711 daily responses (25.2 responses in average per participant), 1453 (85 %) were deemed “valid” responses, e.g. GPS location data within 100 metres of their estimated daily listening position just outside their house when considering use for correlation with current annoyance responses. For 24-hour annoyance, maybe larger distances can be accepted.
An upwards trend is clearly suggesting that annoyance increases with noise level, but it is also obvious that for a given annoyance level, large differences in noise level exist. The differences vary in 41 dB, 34 dB, 29 dB, 22 dB, and 20 dB ranges, respectively.
Notice that the range of noise variation at low wind turbine annoyance levels may be influenced by other sources of background noise. While participants were asked specifically about the noise annoyance from wind turbines, other sources such as traffic noise may have partially or fully masked wind turbine noise at low levels, and even at clearly audible levels it may not be easy to separate contributions from individual noise sources.

At high annoyance levels, the number of responses is much smaller compared to the number of responses at low annoyance levels (only 17 responses at Extremely). Furthermore, a regulatory upper bound on the allowable noise at neighbours may also limit both the noise range and the frequency of Extremely annoyed responses. Even this taken into account, an annoyance level may still span an estimated 15-25 dB range. The range may be accounted for by three main factors: 1) The discrete 5-step response scale, 2) Differences in context (type of house, noise from own household, etc.), and 3) Individual influences (stress level, time spent outdoor, etc.).

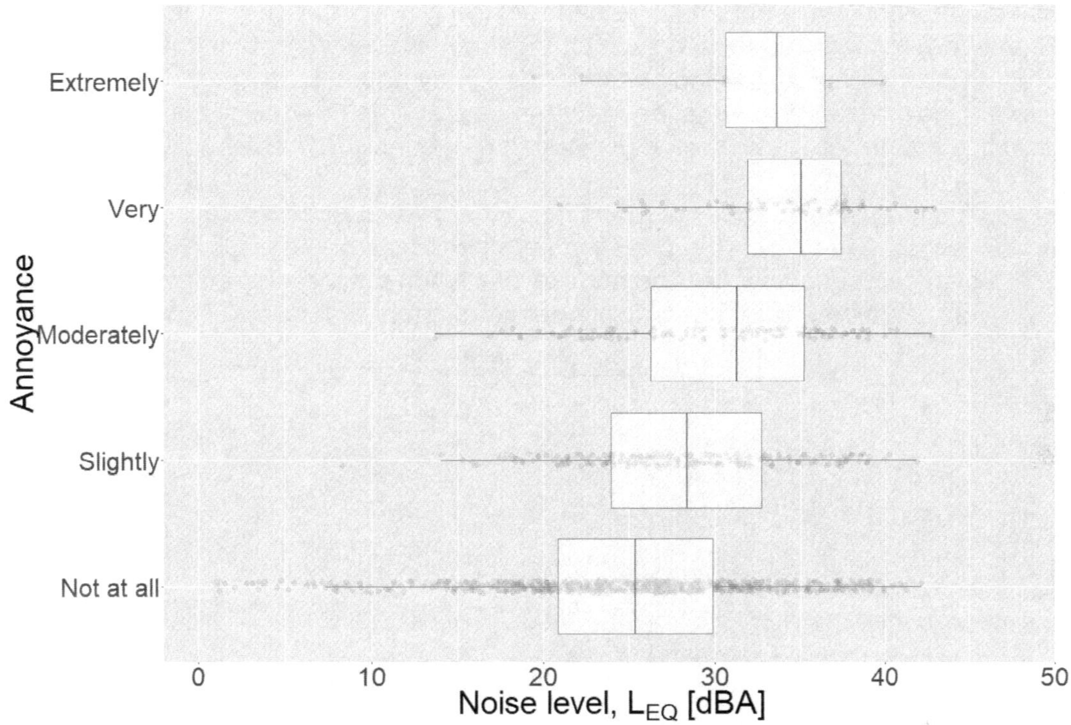

Figure 3 Boxplots of current wind turbine annoyance (at time of responding) versus the simulated combined noise level of all wind turbines at neighbours' daily response positions. Each box encloses the 25 % to 75 % quartile of data. The whiskers mark minimum and maximum values unless these expand beyond 1.5 x the interquartile range. Values outside the whiskers can be considered outliers.

Since the noise levels are modest relative to the potential noise of other noise sources, this constitutes a potential big uncertainty/weakness in this study. To understand the severity of this uncertainty, the daily questionnaire included both a question about the total noise annoyance and the specific noise annoyance contribution from wind turbines. Of the 1483 “valid” responses, 1202 responses (81 %) had total annoyance level equal to the specific wind turbine annoyance level. Consequently, it can be safely assumed that wind turbine noise was perceived as the dominant background noise source in most cases.

5.1 Context variation statistics

Three specific hypotheses of interest in this study were 1) the influence of wind speed, 2) wind direction and 3) time of day on *current annoyance* responses. For gauging whether the study methodology and the duration of the study period were sufficient to obtain the data needed to properly investigate those questions, some preliminarily data analyses were made. The results are depicted in Figure 4 – Figure 6 respectively.

The wind speed data in Figure 4 show that responses were made at wind speeds from 0 m/s to 17 m/s at nacelle height. Each gradient of blue represents a site showing the distributions of wind speeds across all sites. Although that range of wind speeds is satisfactory, the mean at 4-5 m/s is low. This may limit that range of annoyance levels that can be modelled which is also indicated by the low number of “Extremely” and “Very” annoyed in Figure 3.

The plot of observed wind direction at the time of responding depicted in Figure 5 shows that wind from either west or east was dominating in the period while wind from north or south was less likely. This may indicate that a period of 5 weeks cannot be assumed to be sufficient to obtain all wind directions. Consequently, neighbours living north or south of the wind turbines, at their site may not have experienced much change in noise levels and neither of the extremes while others – living west or east of their wind farms – may have experienced larger variations in noise levels. This difference in variation between participants may be of interest to investigate in the exploratory statistical investigation of moderators influencing annoyance.

Finally, in Figure 6 the time of day for responding to the daily questionnaire is depicted. It is distinct that the two times at which the questionnaire became available (06 and 16) are also the most popular times of responding, but the number of responses throughout the periods 06-11 and 16-21 may all contain sufficient data for understanding the influence of time of day on annoyance.

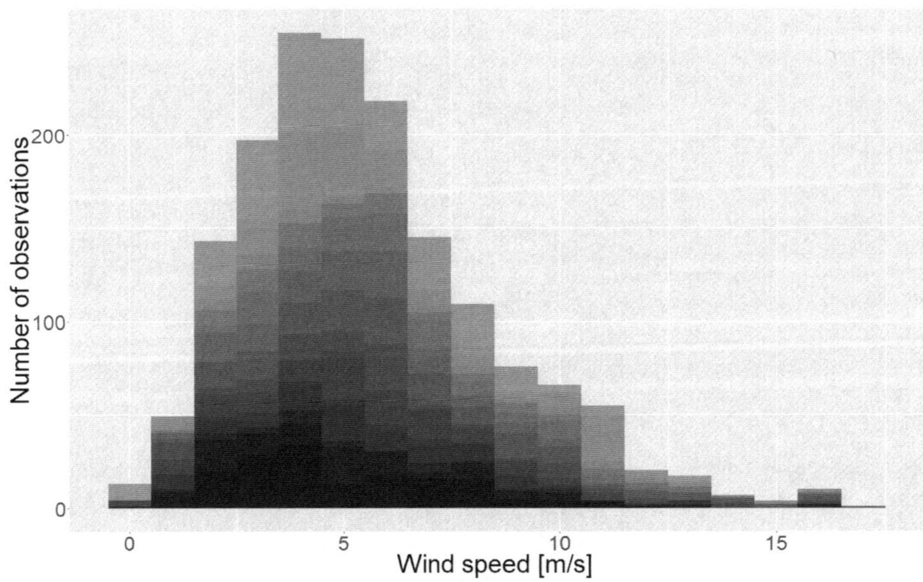

Figure 4 Wind speed variation at nacelle height at the time of responding. Each of the 14 gradients of blue denotes a site.

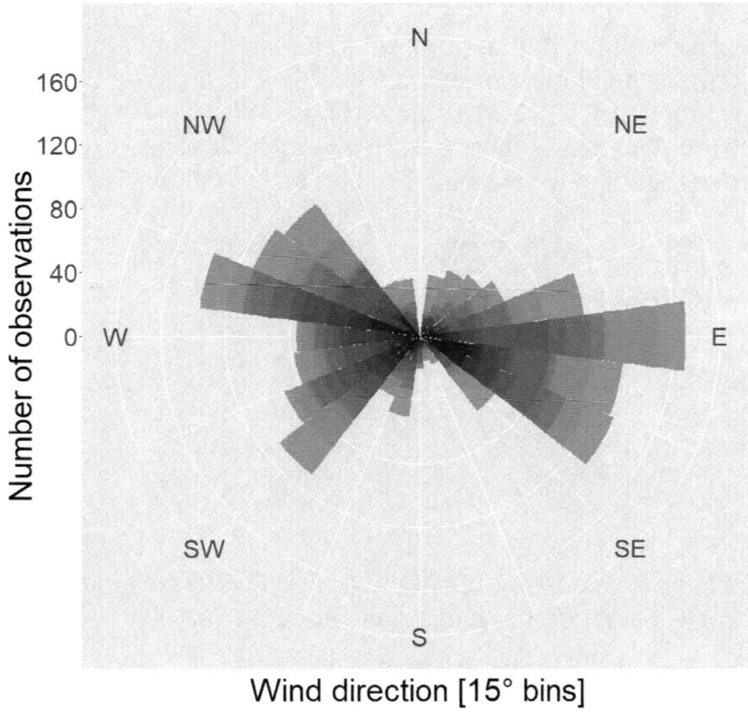

Figure 5 Wind direction variation at the time of responding. Each of the 14 gradients of blue denotes a site.

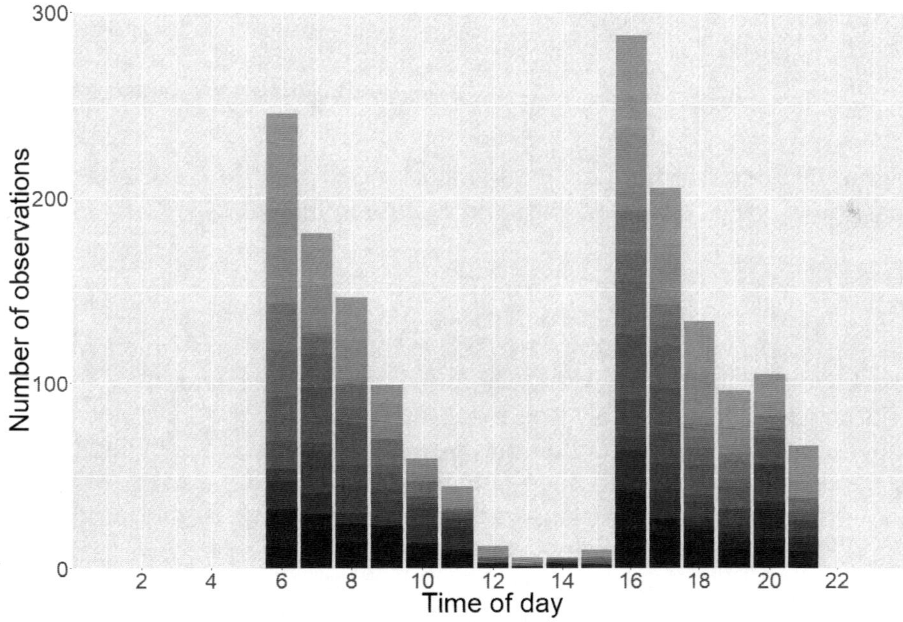

Figure 6 Variation at time of day of responding. Each of the 14 gradients of blue denotes a site.

5.2 Background variable correlated with annoyance

While the in-depth analyses of this dataset are yet to be concluded, a preliminary analysis of the background questions' individual correlation with 24-hour annoyance responses have been investigated and a few stand out, listed in Table 4. Others also showed statistically significant correlation with annoyance but were influenced highly by factor levels (at the extremes of the scale) with few responses leading to uncertainty of their true influence.

Moderator	***Correlation***	***Significance p-value***
Annoyance 1-year (ISO)	0.82	$2.2e^{-16}$
Frequency of noticing WT	0.86	$2.2e^{-16}$
Audibility of WT indoor	0.75	$5.89e^{-12}$

Table 4 Correlation between individual background variables and participants' 24-hour annoyance from the time of responding.

For the questions with binary response options responses were split into "Yes" and "No" subsets and a two-sample t-test used to test significance as listed in Table 5.

Moderator	***Significance p-value***
Sleeping with open window	$2.2e^{-16}$
Economic interest	$4.0e^{-07}$

Table 5 Two-sample t-tests of significance with participants' 24-hour annoyance from the time of responding.

Surprisingly, neither the visual factor nor the timing of moving in before or after turbine construction was significantly correlated with annoyance in this study.

6. Discussion

A concern in this study was the lack of wind speed in the study period which was unusual for the season. Consequently, the period with daily questions was expanded for all participants who agreed to participate. Unfortunately, the extra week was a period with lots of wind in Germany as well, leading to overproduction and powered down wind turbines at the Danish wind farm sites. While annoyance data have sufficient data within the range "Not annoyed at all" to "Very annoyed" to make valuable analyses of noise annoyance moderators, the absolute level of annoyance is unlikely to be representative.

One concern with regards to having daily questions for longer periods was whether repeated and conscious evaluations of annoyance would lead to increase in annoyance. This is a concern both in terms of data analysis, but also long-term effects for participants. To minimize this risk questions and information material were carefully formulated and the daily questionnaire kept short. While the collected data show no tendency of increased annoyance over time, the risk might be further reduced by splitting longer periods into e.g. 1-week periods with breaks in-between which might also improve the planning and timing of periods in terms of desired weather conditions.

7. Conclusions

This socio-acoustic study investigated the influence of weather conditions and non-acoustical moderators on noise annoyance of wind farm neighbours. It utilised a novel methodology of asking participants daily for five consecutive weeks about their current annoyance as well as their annoyance within the last 24 hours from the time of responding. Questionnaires were distributed and responses collected using an app designed specifically for ecological momentary assessment (EMA). Response data were enriched with a wide range of geographic-, weather-, and wind turbine production data (see the overview in Figure 2, p.7) for investigating context influences on short-term annoyance moderators. Compared to traditional ISO-15666 [3] studies, this methodology requires fewer participants to obtain the same amount of data and fixed variables such as wind farm layout, residency variables and as well as participant variations while keeping or increasing similar variation in weather conditions (temperature, wind speed, etc.). The paper investigates the necessary steps involved, the pitfalls and learning points, as well as preliminary findings of significant wind turbine noise annoyance moderators for citizens of Denmark.

8. Acknowledgement

Financial support for this work was partly provided by DecoWind project (Development of low-noise and cost-efficient wind farm technology), File number: 8055-00041B under Innovation Fund Denmark. Supplementary financial support was provided by Danish Agency for Institutions and Educational Grants.
We acknowledge gratefully colleagues from FORCE Technology for assistance in the work; especially Erik Thysell for handling the Nord2000 noise simulations, Torben Holm Pedersen for lending us his long experience in the field of noise annoyance, Michelle Herlufsen for your graphical skills and finally Inge Lis Kjær for proofreading. We also owe great thanks to the Decowind project partners for valuable input and delivery of wind and production data. We would also like to thank all the wind turbine owners, who agreed to provide us with production data. Finally, we would like to thank Epinion for a smooth collaboration and a massive effort to keeping participants going for the full 5-weeks, and of course thanks to all participants who generously lent us their time and effort.

References

[1] R. F. S. Job, "Community response to noise: A review of factors influencing the relationship between noise exposure and reaction," *J. Acoust. Soc. Am.*, vol. 83, no. 3, pp. 991–1001, 1988, doi: 10.1121/1.396524.

[2] E. Pedersen, "Human response to wind turbine noise," The Sahlgrenska Academy, Gothenburg, 2007.

[3] ISO, "ISO/TS 15666:2003(E) Acoustics – Assessment of noise annoyance by means of social and socio-acoustic surveys Deskriptorer :," p. 26, 2003.

[4] T. H. Pedersen, "Genevirkning af støj fra vindmøller. Rapport-150-1994.pdf," 1994.

[5] *Bekendtgørelse om støj fra vindmøller*, no. 135. Denmark: Miljø- og Fødevareministeriet, 2019, pp. 1–24.

[6] C. Søndergaard, Lars Sommer; Backalarz, "Støj fra store , nyere danske vindmøller som funktion af vindhastigheden," Miljø- og Fødevareministeriet, 2016.

[7] Miljøstyrelsen, "Støj fra vindmøller - Vejledning fra Miljøstyrelsen," 2021.

[8] ExpiWell, "Expiwell software." ExpiWell, West Lafayette, IN, USA, 2020, [Online]. Available: https://www.expiwell.com.

[9] Energistyrelsen, "Stamdataregister for vindkraftanlæg," *Energi-, Forsynings- og Klimaministeriet*, 2021.

[10] IEC, "IEC 61400-11 Wind turbines - Part 11: Acoustic noise measurement techniques." 2012.

[11] C. Backalarz, L. S. Søndergaard, and J. E. Laursen, "'Big Noise Data' for wind turbines," 2016.

[12] et al. Skamarock WC, "A description of the advanced research WRF version 3, NCAR Tech. Note, NCAR/TN-468+STR," *Natl. Cent. Atmos. Res. Boulder, Color.*, 2008.

[13] S. Y. Hong, Y. Noh, and J. Dudhia, "A new vertical diffusion package with an explicit treatment of entrainment processes," *Mon. Weather Rev.*, 2006, doi: 10.1175/MWR3199.1.

[14] A. N. Hahmann and A. Peña, "Validation of boundary-layer winds from WRF mesoscale forecasts over Denmark," 2010.

[15] A. S. Monin and A. M. Obukhov, "Basic laws of turbulent mixing in the atmosphere near the ground," *Tr. Akad. Nauk SSSR Geofiz. Inst*, 1954.

[16] T. K. Torp, "Rekordhøj betaling for nedregulering af danske vindmøller i 2020." Naturlig Energi, pp. 22–23, 2021.

[17] S. for D. og Effektivisering, "Danmarks Højdemodel (Danish Height Model)." .

[18] B. Plovsing, "Proposal for Nordtest Method: Nord2000 – Prediction of Outdoor Sound Propagation," 2014. doi: AV1106/07.

[19] B. Plovsing and E. Thysell, "Nord2000 – Prediction of Outdoor Sound Propagation. Amendments to Report AV1106/07 revised 2014 Performed for Vejdirektoratet," 2019.

9th International Conference
on
Wind Turbine Noise
Remote from Europe – 18th to 21st May 2021

Further experience of reviewing noise assessments for wind farms in Scotland and the implementation of the IOA Good Practice Guide to the Application of ETSU-R-97 for the Assessment and Rating of Wind Turbine Noise

Steve Summers, ACCON UK Ltd, steve.summers@accon-uk.com
Graham Parry, ACCON UK Ltd, graham.parry@accon-uk.com

Summary

ACCON UK has been carrying out reviews of noise assessments for onshore wind farms submitted in support of planning applications to two Scottish local authorities. This work has enabled us to gain an overview of the similarities and differences between the approaches of the noise assessments for different projects. Following on from earlier work, this paper draws on assessments carried out over the last three years in addition to previous research based on assessments carried out since 2014. The paper considers emerging trends and complexities arising in the noise assessments, many of which relate to the consideration of cumulative noise in the context of UK's Institute of Acoustics Good Practice Guide (GPG) and ETSU-R-97. Consideration is given to whether parts of the GPG might benefit from revision or the publication of supplementary guidance.

1. Introduction

The 1996 report 'The Assessment and Rating of Noise from Wind Farms', ETSU-R-97[1], sets out the methodology for assessing noise from wind turbines that is approved by the UK Government and the Scottish Government. Following concerns that there was a lack of consistency in how wind farm noise assessments were being carried out, the UK's Institute of Acoustics (IOA) published 'A Good Practice Guidance to the Application of ETSU-R-97 for the Assessment and Rating of Wind Turbine Noise' (the GPG)[2] in 2013. The GPG provides a comprehensive guide to the implementation of the measurement, prediction and assessment procedures given in ETSU-R-97. A framework for setting noise limits is given in ETSU-R-97 and the GPG clarifies this advice. The noise limits recommended by ETSU-R-97 are intended to apply to the noise from all wind turbines affecting each noise sensitive receptor rather than only the turbines considered in a particular consent application. The GPG recognises the importance of this and provides several pages of advice on various approaches to setting noise limits in scenarios where cumulative noise from a number of developments applies.

The IOA GPG has become an established reference document in the United Kingdom. As such the adoption of its recommendations is the generally accepted method of demonstrating that a noise assessment for a new wind farm has followed best practice. ACCON on behalf of a number of Scottish Local Authorities has carried out reviews of noise assessments submitted in support of planning applications for approximately thirty five wind farms in Scotland. This work started in for East Ayrshire Council and has subsequently included wind farms in the South Ayrshire Council area. This work has enabled us to review and compare the approaches of different acoustic consultants. A previous paper[3] reported the results of a review that assessed the similarities and differences between the approaches to the noise assessments submitted for different wind farms. Specifically, the implementation of the various recommendations of the GPG were investigated by completing a summary matrix considering the key requirements of a wind farm noise assessment. That paper considered the approaches used for background noise surveys, use of turbine sound power data, noise prediction modelling, the selection of noise criteria and limits and the assessment of cumulative noise.

The study reported in this paper involved extending the summary matrix to cover the noise assessments reviewed since 2017. This study examines whether or not the guidance provided in the GPG is continuing to be widely applied. Our research also considers the ongoing and expanding complications in the consideration of cumulative noise and issues that affect noise assessments arising from the trend for the use of taller wind turbines.

2. Review of Implementation of Good Practice Guide

The further analysis of wind farm noise assessments following on from the 2017 research has revealed that all assessments are largely implementing the GPG recommended approaches for the following:

- Background noise level surveys
- Turbine sound power levels
- Noise prediction methodology
- Setting standalone noise limits

These aspects are discussed in the following sections.

2.1 Background Noise Surveys

The key elements of background noise surveys required by the GPG include: agreeing monitoring locations with the local authority; carrying out measurements over a sufficient duration; concurrent measurements of wind speed and direction to obtain hub height data; and appropriately excluding rain affected and other anomalous data, including explaining within the reporting the approach to exclusions. Our previous paper identified that the good practice recommendations were being followed in virtually all of the noise assessments reviewed and this has continued to be the case for the more recent cases examined.

2.2 Turbine Sound Power Level Data

Of the sixteen noise assessments rereviewed from 2017 onwards, all except one has adopted a single candidate turbine model for the noise predictions to represent the turbines likely to be installed. One assessment adopted an “envelope” approach which entails considering a range of turbine models and using the highest sound power level at each wind speed from the manufacturers’ data for each model.

All of the assessments included an allowance for uncertainty within the turbine sound power input data used for the noise predictions. In the vast majority of cases a correction of 2.0 dB was utilised. A small number of other assessments used corrections in the range 1.0 dB to 2.6 dB.

All corrections adopted appeared to follow the guidance of paragraph 4.3.6 of the GPG. All the assessments reviewed identified octave band sound power level data applying to the candidate turbine and the turbines used in the cumulative noise predictions for other wind farms.

Most wind turbines assumed in the noise assessments can be supplied in two versions. One version includes serrated trailing edges (STE) incorporated in the turbine blades which typically provides a reduction of around 2 dB compared with the corresponding non-STE model. In a small number of instances, examination of the sound power data used in the noise assessments showed that STE turbine models had been adopted as the candidate turbine but this had not been explicitly stated in the report. In other cases, the octave band data appeared to apply to the non-STE turbine model, although the STE model had been identified as the candidate turbine.

2.3 Noise Prediction Methodology

All of the noise assessments reviewed from 2017 adopted the ISO 9613-2 model input parameters specified in section 4.3 of the GPG. This included using a ground factor setting of G= 0.5, which requires the use of octave band sound power level data. Our review also checked that the noise assessments indicated they had limited topographic screening to 2 dB and allowed for the +3 dB correction for propagation across concave ground profiles. All the assessments indicated that these model adjustments had been made, although only a small minority of the assessment reports set out detailed breakdowns that show how these adjustments have been applied to the predictions for each receptor. In several cases the landforms were such that screening for topographic features was excluded from the noise predictions.

2.4 Standalone Noise Limits

Where noise from other wind farms does not affect the receptors considered for a new application, a cumulative noise assessment is not required. In these cases, noise limits can be set in a straightforward way according to ETSU-R-97 and apply to the wind farm under consideration on a standalone basis. Two of the wind farm noise assessments reviewed fell into this category and both followed ETSU-R-97 and the GPG guidance correctly, including providing a justification of the limits adopted. Some of the noise assessments reviewed set standalone noise limits for the proposed wind farm and assessed the predicted noise limits of the development in isolation against these limits. However, for some of the wind farm assessments reviewed, cumulative noise assessments were necessary that made the initial assessment against standalone limits largely irrelevant.

3. Emerging Issues

Our previous paper identified the increasing complexity relating to the assessment of cumulative noise due to the density of wind turbine developments in certain parts of East and South Ayrshire. Figure 1 presents a map of part of East Ayrshire identifying such an area. The complexities of cumulative noise have continued to apply to many new planning applications with the degree of complexities tending in general to increase because of the growing concentration of wind farm developments. Another emerging issue arises because of the trend for developers to apply to install taller turbines. Several planning applications involve seeking approval for higher turbines of around 200 m blade tip heights on sites where turbines have previously received planning consent for turbines with blade tip heights in the range 120 – 150 m. Where turbine hub heights are significantly higher than those already consented, complications relating to wind shear can arise when relying on previously measured background noise data or ETSU-R-97 derived noise limits. A graph is provided in Figure 2 which displays the average wind turbine blade tip heights for wind farm developments that have been the subject of ACCON's reviews. The more recent trend in turbine heights is evident from the graph.

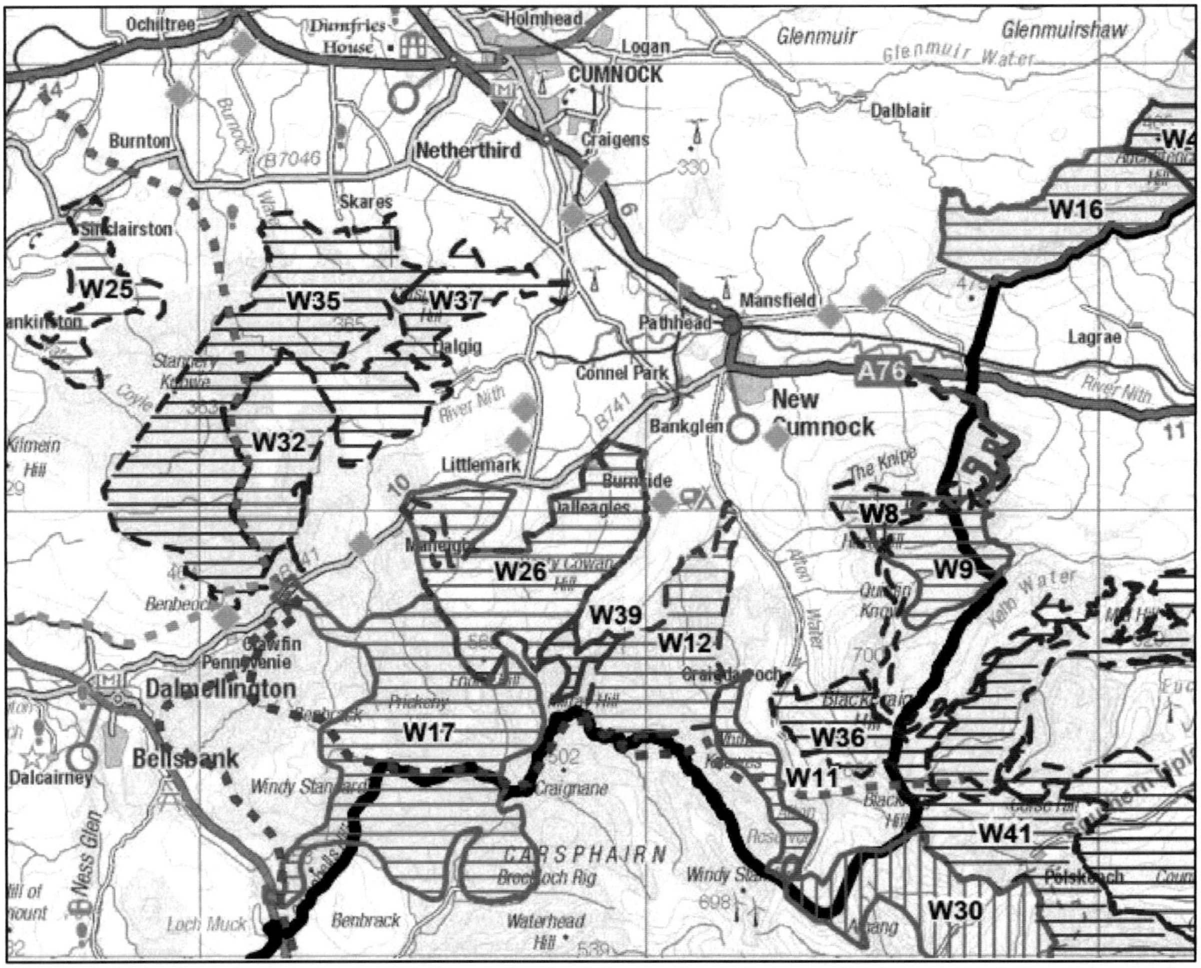

Fig 1: Plan showing concentration of wind farm developments in part of East Ayrshire

3.1 Cumulative Assessment

The ETSU-R-97 noise limits apply to the noise from all wind turbines at the receptors that are affected by the new wind farm, not just the noise immissions from the proposed development. The GPG advises that a cumulative assessment is required where predicted noise levels are within 10 dB of the noise from existing wind turbines at the receptor locations. One approach that may be taken is to set noise limits for the new wind farm that are 10 dB lower than the noise from existing wind farms. This avoids the need to carry out cumulative noise predictions.

Where full cumulative noise predictions are required two methods are commonly used[4]. The first method is to calculate the cumulative noise level from the application site and the existing wind farms and compare this with the ETSU-R-97 noise limits. The second method is often called the 'remaining noise budget' (RNB) method, which involves calculating the noise levels from the existing wind farms excluding the proposed development and subtracting these values logarithmically from the ETSU-R-97 derived cumulative limits. This provides the RNB against which the predicted noise levels from the proposed wind farm can be compared. The two methods are effectively equivalent but the RNB provides a clearer indication of the available noise limits for the proposed wind turbines.

Approximately half of the applications reviewed by ACCON from 2017 adopted the RNB method where a cumulative noise assessment was needed. In several of these cases the noise assessment proceeded straight to a cumulative noise assessment without first carrying out a comparison of the predicted noise from the application wind farm alone against ETSU-R-97 noise limits. The developments using the RNB method were generally those where a large number of other wind farms needed to be included in the assessment, typically a minimum of five wind farms.

For the noise assessment reviews carried out from 2017, one assessment relied on demonstrating that the predicted turbine noise levels from the proposed extension to an existing wind farm would be at least 10 dB below the consented noise limits set for the existing wind farms. The other applications often carried out a two stage noise assessment of comparing the predicted noise levels from the proposed wind farm with ETSU-R-97 standalone derived noise limits and then assessed predicted cumulative noise levels against ETSU-R-97 derived noise limits applicable to the cumulative assessment. In many cases, higher limits are adopted for the cumulative assessment because these limits apply to a greater number of turbines. ETSU-R-97 allows the generating capacity to be considered in setting the lower fixed limit for daytime within the range 35 to 40 dB L_{A90}.

3.2 Planning Conditions relating to Cumulative Noise

Local authorities in Ayrshire are increasingly concerned about how noise from wind farms may be controlled or noise complaints investigated when there are several neighbouring wind farms with different operators and separate planning consents. For some of the wind farm applications considered in the research for this paper, the local planning authority would have preferred a cumulative noise condition to be included in the planning conditions. Such conditions have been set only in a few cases across the United Kingdom. Cumulative conditions potentially hold the operator of the consented wind farm responsible for the noise emitted by the other nearby wind farms, should an investigation demonstrate that the cumulative noise limits are being exceeded. Consequently, cumulative conditions are not popular with wind farm developers. The reason that the local planning authority may seek such conditions is that if a noise complaint is investigated under the consent requirements for a given wind farm, a situation may arise where ETSU-R-97 limits are in breach but it has been demonstrated that the suspected wind farm is actually operating within the noise limits applying to its turbines alone. Without a condition addressing cumulative noise in place, the planning authority cannot require the operator of the first wind farm to provide any further assistance in identifying which other wind farm is responsible for the exceedance of the limits.

Pencloe wind farm in East Ayrshire was consented by the Scottish Government in 2018 with a form of cumulative noise condition. That consent[5] includes two sets of tables of noise limits, one set applying to the Pencloe turbines alone and the other applying to Pencloe turbines in combination with the other wind farms included in the cumulative noise assessment that supported the consent application. In the event that the cumulative noise limits are shown to have been exceeded, the operator of Pencloe wind farm is required to apply mitigation to reduce turbine noise immissions such that either the cumulative limits are met, or the standalone limits for Pencloe are met.

For other wind farm developments in East Ayrshire, planning consents have required the operators to provide a noise complaints assessment protocol to be submitted and approved by the planning authority before the wind farm becomes operational. In these cases, the Council requires a protocol that prevents the operator from avoiding responsibility for seeking to identify the wind turbines likely to be responsible for the noise complaint should the wind farm which is initially considered responsible prove that their turbines are not the cause of the issue.

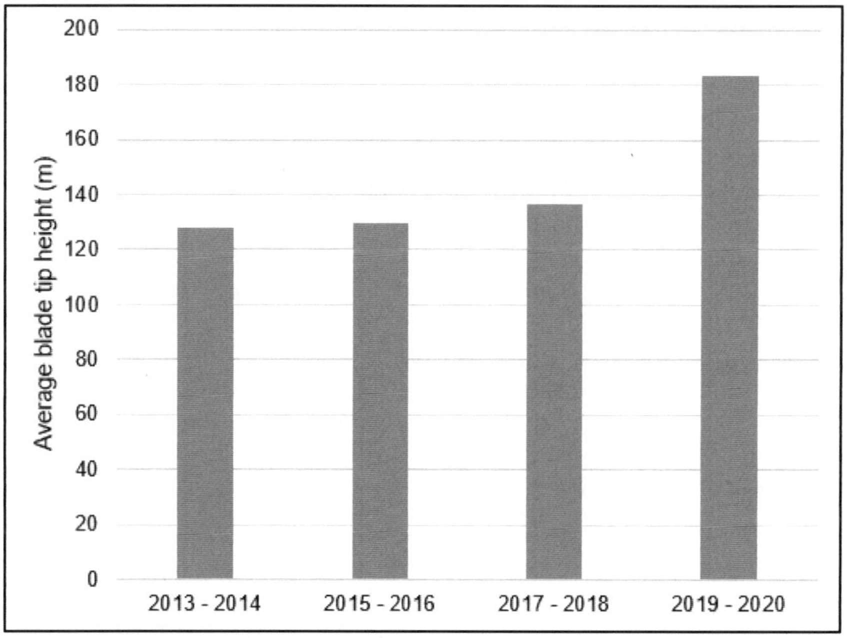

Fig 2: Graph showing average turbine blade tip heights

3.3 Wind Shear Issues

The trend for the use of taller turbines has brought various aspects of the noise assessment process into focus in relation to wind shear. The first of these relates to where a background noise survey has been carried out for a particular wind farm site in the past and a new development is now proposed with significantly taller turbines and utilises the previously existing background noise survey data. Depending on the height of the anemometers used to obtain wind speed data during the noise survey, it may be readily possible to follow the GPG methodology to obtain wind speed data corresponding to the higher turbine hub heights to re-analyse the background noise data referenced to hub height wind speeds, standardised to 10 metre height. If this is not possible, there may be an option to utilise other sources of wind speed data to derive suitable adjustments to the original wind speed data set to derive the required data for the new hub heights. This is illustrated by Example A below.

The guidance concerning the heights at which wind speeds should be measured in relation to the proposed turbine hub heights when the measurement is not taken directly at that height is given in paragraph 2.6.3 of the GPG. The method relies on the use of anemometers at two heights so that an exponential profile can be used to calculate the wind speed at the desired hub height. The GPG advises that the upper anemometer should be '*at a height not less than 60% of the hub height of the proposed turbine*'.

Example B illustrates how the complexities of fully considering all the implications of the use of taller wind turbines can occasionally be missed, especially where a consent application is seeking to encompass a range of options for turbine height configurations.

The other issue relating to increasing hub heights, is the potential need to consider wind shear differences as they affect noise limits derived from background noise measurements applying to cumulative noise from nearby wind farms with significantly different turbine hub heights. This issue is illustrated by Example C. The need to consider wind shear differences has always arisen where large differences in ground elevation (or other geographical features) between wind farms occur. However, such issues are more likely to need to be taken into account given the increased likelihood of large turbine hub height differences as hub heights increase.

Example A
The application proposed turbines of hub heights of up to approximately 160 m. In accordance with good practice, it was necessary to derive hub height wind speeds, for these taller turbines, standardised to 10 m. The acoustic consultants carrying out the assessment decided to utilise the original noise survey data. However, the highest anemometer used for concurrent wind speed measurements was at a height less than 60% of the newly proposed hub height.

An approach was therefore adopted to synthesise wind speed data at 100 m height from the wind data measured concurrently with the background noise survey. This was based on an analysis by a wind resource consultant of wind speed data from a 100 m high mast which was subsequently installed on site, correlating the data with that from the original 80 m mast that remained at the site for a period of approximately 12 months when both meteorological masts were in place at the wind farm site. On this basis, the wind speed data originally measured at 80 m height was corrected to a height of 100 m, facilitating calculation of hub height wind speeds at 160 m height in accordance with the GPG methodology. This data set was then standardised to 10 m height enabling the re-analysis of the background noise measurements.

Example B
The application proposed to install turbines with a range of blade tip heights of 140 m – 180 m at a site previously consented with tip heights in the range 110 m – 140m. The corresponding hub heights for the new development were between approximately 80 m and 120 m, compared with between 65 m and 90 m for the approved wind farm. The noise assessment for the approved development had determined standardised 10 m wind speeds based on wind speeds derived for a 90 m hub height. The resulting noise limits derived from the background noise levels provided a conservative approach for the 65 m hub height turbines and clearly matched the tallest hub heights of 90 m.

The noise assessment for the new application presented in the environmental impact assessment report was based on turbines with three hub heights in the range of 80m – 100 m. The report text did not include any specific discussion of hub height differences between the new and consented wind farm and any consequent wind shear effects on the background noise levels or derived noise limits. The assessment simply adopted the noise limits from the original assessment. We note that any required adjustments to the limits would likely be insignificant between 100 m and 90 m hub heights. However, the project description of the proposed development included in the environmental assessment report explained that a range of turbine configurations were covered by the application. These included turbines with hub heights of up to approximately 120 m. For such hub heights a noise assessment following the requirements of ETSU-R-97 and the IOA GPG would involve re-analysis of the background noise data based on measured wind speeds applicable to the 120 m hub height. Whilst the presented noise assessment considered in isolation did not appear to be ignoring wind shear effects, there was clearly a failure in the EIAR to ensure the noise assessment correctly represented the highest hub heights proposed in the planning application and any consequent implications of wind shear differences.

Example C
The noise assessment for the application for wind farm C acknowledged potential wind shear differences between the application site with wind speeds evaluated for turbine hub heights of 125 m and a closely neighbouring operational wind farm (Site D) with hub heights just below 70 m. For a specific standardised 10 m height wind speed, as determined for a turbine with a specific hub height, the difference in wind speed at the hub height for a different turbine height, depends on the degree of wind shear. In this instance, there may be a discrepancy between the

hub height wind speeds at the two sites, and therefore a consequent difference in the predicted noise levels for a given standardised 10 m wind speed. In effect, this meant that the predicted noise levels for wind farm C should be shifted to the left by 1 to 2 m/s in order to maintain consistency with the difference in wind speed references. The noise report for this application argued that in reality, this effect would not be expected to occur because the degree of wind shear was not expected to be particularly large. The assessment authors considered that the likely difference in wind shear was expected to be less than 1 m/s.

Example C highlights that other noise assessments where cumulative noise limits are used to consider the potential cumulative noise impacts from multiple wind farm sites may be tending to over-simplify wind shear considerations.

4. Conclusions

This study has shown that the recent wind farm noise assessments continue to follow all the principal recommendations in the IOA Good Practice Guide for implementing ETSU-R-97.

Cumulative noise considerations are often complex both in terms of technical aspects of noise assessment and the implications for planning and consent conditions. These complexities have been exacerbated since 2017 due the increasing density of wind farm development in certain parts of East and South Ayrshire. Some facets of the assessment of cumulative noise are not fully addressed by the GPG and this indicates that there may be a need for supplementary good practice guidance on cumulative noise.

There is an increasing trend for wind farm developers to seek to install taller wind turbines. This means that there is a corresponding need for more detailed consideration of wind shear and how this relates to background noise measurements and the resulting noise limits derived in accordance with ETSU-R-97. This aspect of the noise assessment may also benefit from further good practice guidance.

References

1. The Working Group on Noise from Wind Turbines (1996) ETSU-R-97, *The Assessment and Rating of Noise from Wind Farms*, ETSU for the DTI (Department of Trade and Industry).

2. Institute of Acoustics (2013), *A Good Practice Guidance to the Application of ETSU-R-97 for the Assessment and Rating of Wind Turbine Noise.*

3. Summers S and Parry G (2017), *Experience of reviewing wind farm noise assessments for Scottish local authorities and the implementation of the IOA Good Practice Guide to the Application of ETSU-R-97 for the Assessment and Rating of Wind Turbine Noise, 7th International Conference on Wind Turbine Noise.*

4. Wind farms cumulative impact assessment, Acoustics Bulletin, Institute of Acoustics, January/February 2016.

5. Pencloe Wind Farm(2018), Ministerial Determination – Decision Letter, The Scottish Government, Energy Consents Unit, Ref EC00003106.

9th International Conference
on
Wind Turbine Noise
Remote from Europe — 18th to 21st May 2021

Identifying The Flap Side-Edge Noise Contribution Of A Wind Turbine Blade Section With An Adaptive Trailing-Edge

A. Suryadi[1], C. Jätz[2], J.R. Seume[2] and M. Herr[1]
[1] German Aerospace Center, Institute of Aerodynamics and Flow Technology. Lilienthalplatz 7, 38108 Braunschweig, Germany
[2] ForWind and Leibniz University Hannover, Institute of Turbomachinery and Fluid Dynamics. An der Universität 1, 30823 Garbsen, Germany

Summary
Active trailing-edge technology is a promising application for localized load alleviation of large-diameter wind turbine rotors, accomplished using one or more control surfaces in the rotor blade's outer region. This work focuses on identifying noise contributions from the flap side-edge and the trailing edge in a laboratory condition. Measurements were conducted in the Acoustic Wind tunnel Braunschweig (AWB) at the German Aerospace Center's (DLR) Braunschweig site. The small-scale model has a span of 1200 mm and a chord of 300 mm. The control surface, a plain flap, has a span of 400 mm and a chord length of 90 mm. Far-field noise was measured using a phased-microphone array for various flow speeds, angles of attack and flap deflection angles. For sound source identification, two noise reduction *addons* were installed interchangeably: trailing-edge brush and flap side-edge porous foam. Analysis of the far-field noise reveals that, while changes to the flap deflection angle alter the far-field noise spectra, the trailing-edge noise remains the predominant noise source at deflection angles $-5°$ and $5°$. No additional noise level was observed from the flap side-edge within the measurable frequency range at these angles. The flap side-edge noise has an increased role for frequency larger than 2 kHz for the larger flap deflection angles of $-10°$ and $10°$. Furthermore, numerical reproduction of the results will also be presented using the FMCAS (Fast Multipole Code for Acoustic Shielding) toolchain developed at DLR.

1. Introduction
The trend in the wind energy industry is towards larger rotor blades to produce more electricity from a single turbine. Larger rotors will experience stronger and more dynamic loads due to the fluctuating and heterogeneous wind field. Hence, there is an interest in locally distributed aerodynamic control systems. Pitch control is one strategy of load alleviation to maintain the blade's angle of attack, so that its load can be kept, on average, constant. With a larger blade, local changes to the load are preferable. One strategy uses active trailing-edges, such devices can be a flap or an aileron. Either way, it modifies the lift coefficient by changing the camber of the profile.

Load alleviation is studied in the framework of project *SmartBlades 2.0*, which has the overall goal of demonstrating the practical applications of technologies for rotor blade control and validating suitable innovative and manufacturing methods. One of the technologies studied in the project is load alleviation using an active trailing-edge. This is presently accomplished using a plain flap on the outboard section ($0.825 \leq R/\mathbb{R} \leq 0.9$) of the rotor blade from the conceptual wind turbine IWES IWT-7.5-164 [1] which has a blade length of $\mathbb{R}$=80 m. The present plain flap installation means that when active, the trailing edge is no longer a continuous line and a new edge is exposed from the flap's side. The newly exposed edges have the potential to be a noise radiator.

Flap side-edge noise has been covered extensively in the aeronautical industry [2, 3, 4, 5, 6, 7, 8, 9, 10, 11, 12], whereas in the wind energy industry, the study focuses on the implementation and the aerodynamic effect of flap on wind turbine rotor blades [13, 14, 15, 16, 17]. In aeronautics the flap is implemented differently than in the wind energy industry. Namely, in the types of flap used, Fowler flap in aeronautics and plain flap in wind energy, and in the degree of flap deflection, $\delta_f > 20°$ in aeronautics [9, 18] and $|\delta_f| \leq 10°$ in wind energy [15]. To the authors' knowledge the noise radiation from the flap side-edge of a rotor blade has not been extensively explored. A few reasons for this state-of-the-art are: (i) the desired implementation is one that maintains a continuous trailing edge and (ii) the flap deflection angle is small enough that flap side-edge noise is assumed negligible.

In this study, a small-scale test was conducted in the Acoustic Wind tunnel Braunschweig located at the German Aerospace Center's (DLR) Braunschweig site to measure and rank the flap side-edge noise compared to the trailing-edge noise. A segment of the rotor blade, which has the DU08-W-180 profile shape, is scaled down as a 1200 mm span and a 300 mm chord wind tunnel model. The model's shape and chord length are constant. The plain flap has a 400 mm span and a 90 mm chord, capable of being deflected negatively (upwards, towards the suction side) and positively (downwards, towards the pressure side). As a comparison, the small scale wind tunnel model has a chord-based Reynolds number of 0.95×10^6, whereas a similar section in the full-scale rotor has 12.6×10^6.

The paper is structured as follows: first, the experimental setup is detailed, followed by an explanation of the identification of sound sources using noise reduction materials. The results are presented and discussed in the two following sections, and finally, a conclusion from the study is drawn.

2. Experimental Setup

2.1. Wind Tunnel Facility

The Acoustic Wind tunnel Braunschweig (AWB) is a Göttingen type, open section, acoustically insulated wind tunnel. It is capable of producing a maximum wind speed of 65 m/s with 0.3% turbulence intensity [19]. The freestream is introduced into the open section from a 0.8 m wide and 1.2 m high nozzle.

The wind tunnel's test section and the blade model installed within are shown in Figure 1. A phased-microphone array with 96, 1/2-inch LinearX microphones was used to measure the noise radiated from the blade model. The measurement has a sampling rate 65 kHz and a sampling time of 20 s. Noise mapping was performed using the CLEAN-SC algorithm, which allows for separating different noise sources by defining specific regions of interest on the model [20]. The algorithm provides a margin of approximately ± 2 dB within a 95% confidence interval. Measurements were conducted with the phased-microphone array alternately facing the blade model's suction and pressure sides. Because there is no unexpected difference from both sides, the upcoming analysis will present only the suction side.

2.2. Blade Section Model

The blade section model with DU08-W-180 profile was installed vertically in the wind tunnel section, as shown in Figure 1. The model has a span of 1200 mm and a chord length of 300 mm. The plain flap was installed in the mid-span section of the model. The flap's span is 400 mm, and its chord length is 90 mm, see also figure 2. A motor is attached to the outside of the model to drive the flap. During the measurements, the flap angle is fixed, and the flap is only driven

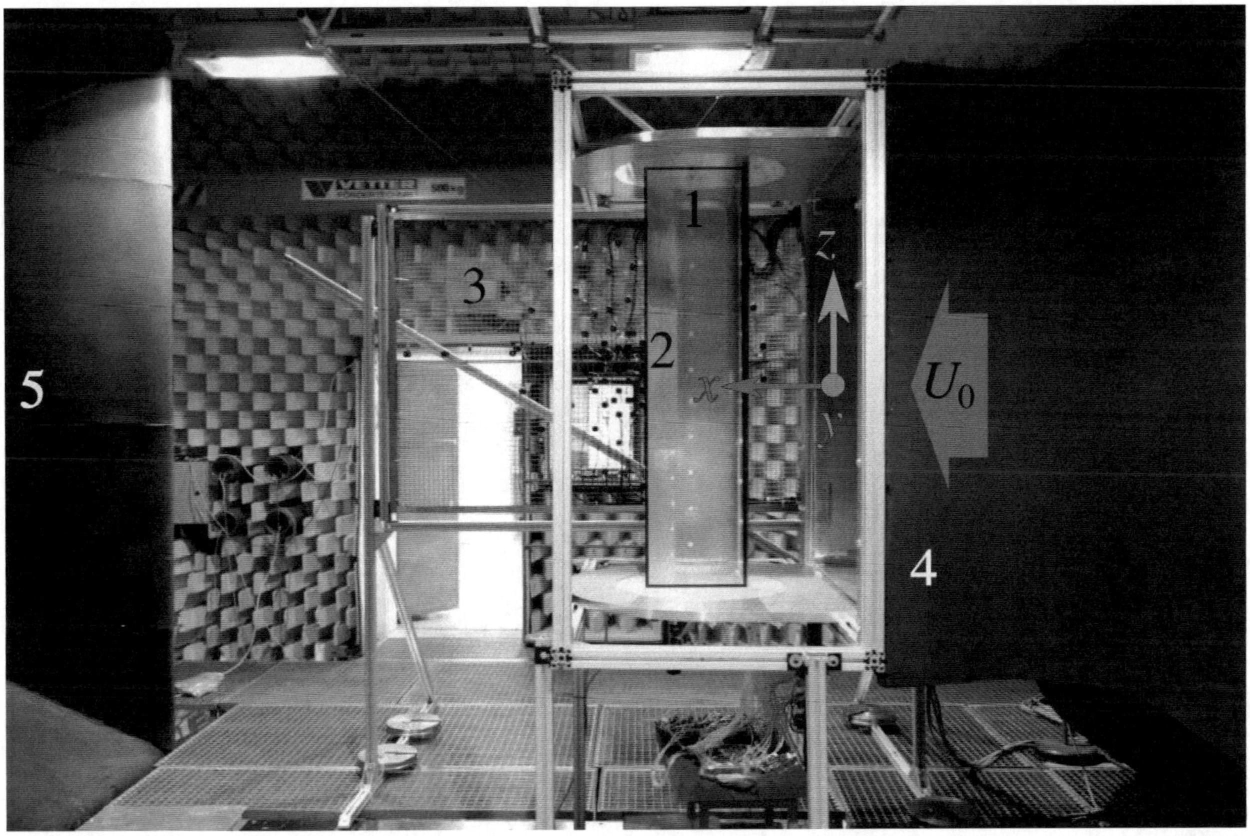

Figure 1: The blade section model, (1), with a plain flap, (2), and the phased-microphone array, (3), inside the AWB test section. Also shown are the wind tunnel nozzle, (4), and collector, (5).

in-between measurements.

By design, wind turbine rotors have a thick blunt trailing edge. The chord-length based scaling down of the model results in a trailing edge thickness of 2 mm leading to bluntness noise in the laboratory scale. Hence, the trailing edge is further reduced to 0.3 mm, akin to a sharp trailing edge.

For motion with minimal friction, the gap between the flap side-edge and the model's side-edge is approximately 0.2 mm. This gap is left untreated. The gap in the span-wise direction between of the flap and the model is sealed using aluminum tape, creating a smooth and continuous transition between both parts and prevents a cross-flow from the pressure side to the suction side, eliminating potential acoustic noise source.

The blade section model is equipped with two sets of static pressure ports. The first set is along the mid-span for monitoring the pressure coefficients when the flap is deflected. The second set is 400 mm away from the mid-span for monitoring the reference, zero-flap-deflection pressure coefficients. The boundary layer was tripped using a 0.205 mm high zig-zag trip on the suction side at the $x/c = 0.05$ and on the pressure side at the $x/c = 0.10$ to emulate high Reynolds number transition location.

2.3. Experimental Parameters

The angular parameters shown in Table 1 were measured in the wind tunnel at windspeeds of $U_0 = 40$ m/s, 50 m/s, and 60 m/s. The angle of attack α_g signifies the geometric angle as set in the wind tunnel. The equivalent aerodynamic angle of attack is α_a. For $\alpha_g = 13°$, the flap deflection angles δ_f is limited because the combination of the large angle of attack and positive flap deflection the wake is deflected outside of the flow collector region.

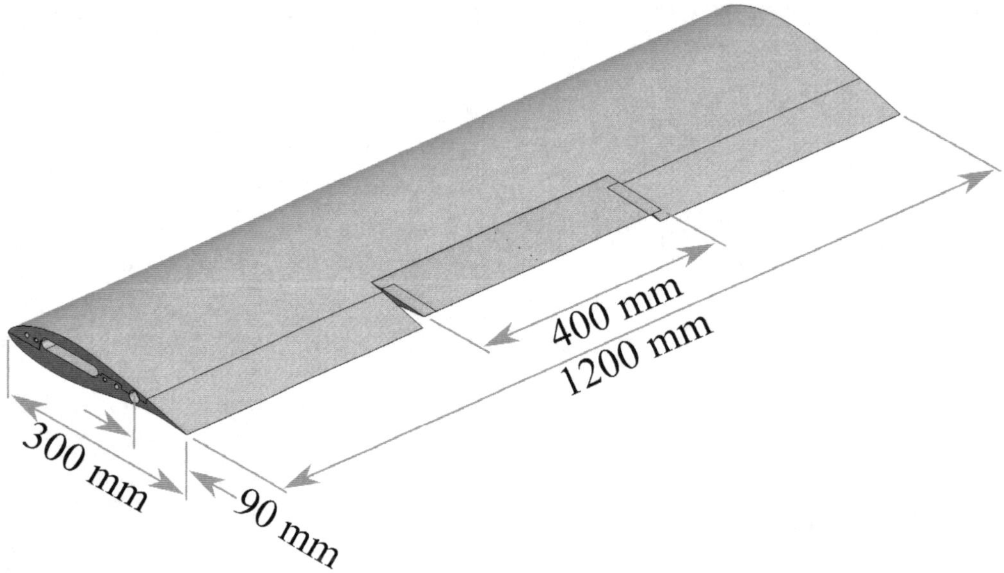

Figure 2: The blade section model DU-08-W-180 with flap set at $\delta_f = -10°$.

3. Numerical Setup

In addition to the experiments in the wind tunnel, numerical simulations were conducted. Steady RANS simulations provide a detailed insight into the flow field and are the input for the numerical aeroacoustic analysis. Due to symmetry, the size of the model and the required high mesh resolution, the numerical domain comprises only a side-edge and extends 150 mm span-wise in both directions.The far field extends to 100 chord lengths.

The mesh is block structured. The topology was carefully designed so that the cells between all blocks in the side-edge region match conformally. This eliminates the need for an interpolating interface, eliminating spurious errors in the region of interest. The initial cell height at the blade surface was chosen to achieve a y^+-value of less than 1. This allows the application of low-Reynolds number turbulence models for a more accurate representation of the boundary layer. The growth ratio in wall-normal direction is 1.1. The cell resolution in the side-edge region was increased according to the requirements for the aeroacoustic analysis. Overall, the mesh contains about $30 \cdot 10^6$ cells.

The simulations were carried out using the coupled implicit solver with a second-order discretization in Star CCM+. For closure of the RANS equations, the k-ω SST turbulence model was selected as this model provides the required flow field quantities for the subsequent aeroacoustic analysis. In order to capture the forced transition by the applied tripping in the wind tunnel, a turbulent suppression approach was selected. Here, the turbulence model is deactivated upstream of the tripping location, resulting in a laminar flow. Downstream of the tripping location, a turbulent boundary layer develops.

The convergence of the simulations was verified by the decrease of the average and maximum residuals. Additionally, lift and drag coefficients were monitored as they converged to constant values.

Table 1: Angle of attack α_g, α_a and flap deflection angle δ_f.

α_g, °	α_a, °	δ_f, °
5	0	-10, -5, -4, ... 4, 5, 10
7	3	-10, -5, -4, ... 4, 5, 10
13	6	-10, -5, -4, ... 0

4. Noise Reduction Technologies

Noise mitigating addons were used to inhibit selective noise radiation and to identify the noise source from a blade section of the present scale. For example, to identify the contribution of the flap side-edge noise, noise-reducing brush was added to the trailing edge. Conversely, to identify the contribution of the trailing-edge noise, the flap side-edge is substituted with a porous copy.

4.1. Trailing-Edge Brush

The trailing-edge brush [21] consists of elastic needles that are 0.4 mm in diameter. The brush is one layer thick and with a density of 250 needles/m along the span and a length of 60 mm. It was installed on the trailing edge's pressure side using a double-sided adhesive tape. Further structural support is provided by an aluminum tape, which also allows for a smooth transition of the boundary layer from the model's surface to the brush. The installed trailing-edge brush is shown in Figure 3*(a)*.

4.2. Porous Side-Edges

Two porous copies of the flap side-edges were manufactured using an electrical discharge machining process. This process allows for shaping the porous foam to the shape of the flap side-edge. These substitutes, when installed, will provide a porous side-edge with a depth of 20 mm. The porous materials are PA 80-110 and PA 120-150, which were supplied by Exxentis AG. The properties of the porous materials are provided in Table 2 and the two materials are shown in Figure 3*(b)* and *(c)*. The porosity, the ratio of open volume to the total volume, and the specific resistance, the pressure loss of a flow passing the material per unit velocity were measured by the Institute for Materials, TU Braunschweig. These materials were extensively investigated for trailing-edge noise reduction [22, 23].

Table 2: Porous material properties

Specimen	Nominal pore size [μm]	Porosity [%]	Specific resistance [Ns/m^3]
PA 80-110	80–110	56	540
PA 120-150	120–150	57	245

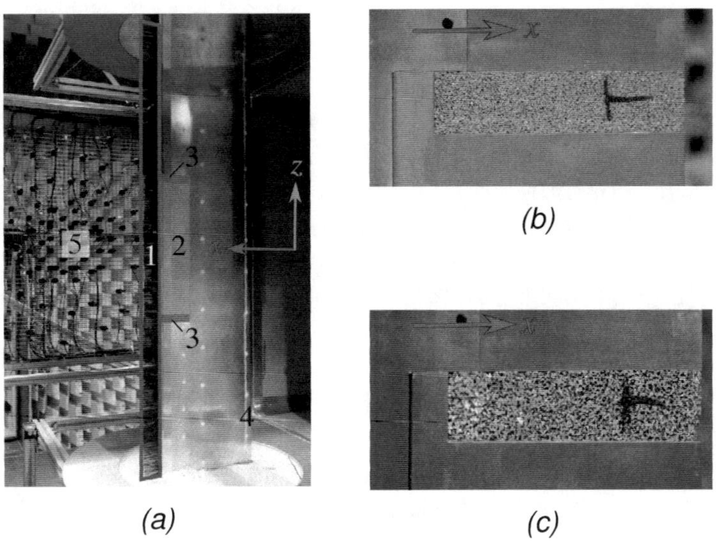

Figure 3: *(a)* The blade section model viewed from the pressure side with the trailing-edge brush (1), the flap (2), the exchangeable flap side-edges (3), the boundary layer tripping (4) and the phased-microphone array (5). Porous flap side-edges, *(b)* PA80-110 and *(c)* PA120-150, viewed from the suction side of the blade section model.

5. Results

5.1. Pressure Coefficents

The measured pressure coefficients are shown in Figure 4 for all measured angles of attack. For brevity only $\delta_f = -10°, -5°, 0°, 5°$ and $10°$ are shown for the pressure coefficients and the sound pressure levels. The margin of error for the measured pressure coefficient is ± 0.026 within a 95% confidence interval.

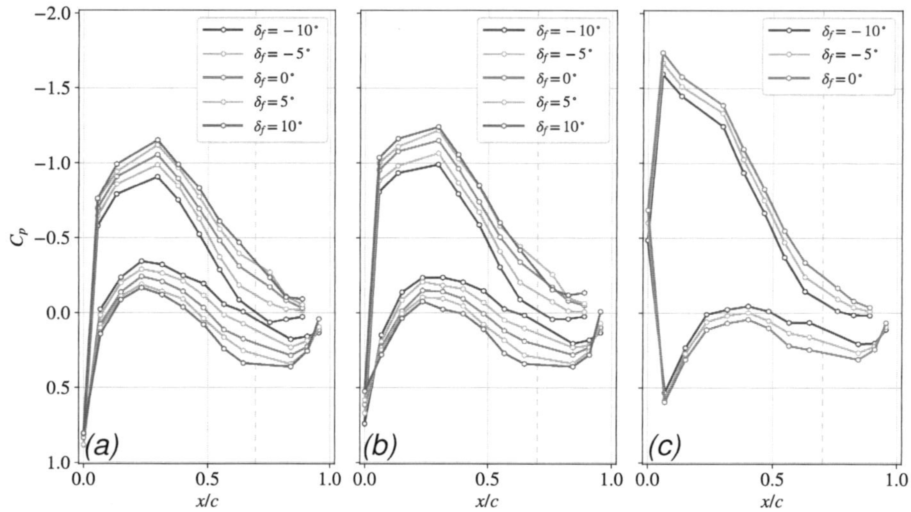

Figure 4: Experimental pressure coefficients at $\delta_f = 0°$ and $|\delta_f| = 5°$ and $10°$, *(a)* $\alpha_g = 5°$, *(b)* $\alpha_g = 7°$, and *(c)* $\alpha_g = 13°$.

5.2. Relevant Frequency Range

The *exterior* noise, anything other than the model's self-noise, was measured with the same tools and relevant measurement parameters without the blade section model installed to identify the meaningful frequency range. For this purpose, the region of interest defined in CLEAN-SC for the analysis of trailing-edge noise is an area with a span width of $b_g = 0.7$ m and streamwise length of $c_g = 0.2$ m. For the analysis of exterior noise, the region of interest encompasses the full-span of the blade section model, $b_g = 1.2$ m and from the AWB nozzle to 1.2 m downstream.

The sound maps of selected one-third octave center frequencies are shown in figure 5(*a*) the exterior noise and in figure 5(*b*) for the trailing-edge noise. The sound maps' red line highlights the sound contour where $\max(L_{p,1/3}) - 3$ dB. The trailing edge is located at $x = 0.55$ m, and at $x = 0.8$ m is the location of the edge of the top and bottom support walls (see figure 1). The sound maps in figure 5(*a*) show that the support walls' edges radiate sound that is louder than the trailing-edge noise at $f_c = 5000$ Hz, and it remains louder for center frequencies higher than 5000 Hz. Despite this, the sound pressure levels at $f_c > 5000$ Hz scale well with the U_0^5 trailing-edge noise, as shown in the next section, because of the decreasing acoustic wavelength. Hence, due to excess noise from the edges of the support walls, the figures involving sound pressure level will exclude the spectral level at $f_c = 5000$ Hz.

5.3. Reference Trailing-Edge Noise

The one-third octave sound pressure level is shown in figure 6 for $\delta_f = 0°$ with and without the trailing-edge noise U_0^5 scaling law. To focus on the trailing-edge noise, the region of interest encompasses the trailing edge with $b_g = 0.7$ m and $c_g = 0.2$ m. The characteristic lengthscale chosen for the scaling law is the suction side boundary layer displacement thickness δ_1 at the trailing edge, and the characteristic velocity is the freestream velocity U_0. All characteristic values are given for $\delta_f = 0°$. Together, they scale the low to medium frequency well [24]. The displacement thicknesses were calculated using XFOIL [25] and are presented in Table 3.

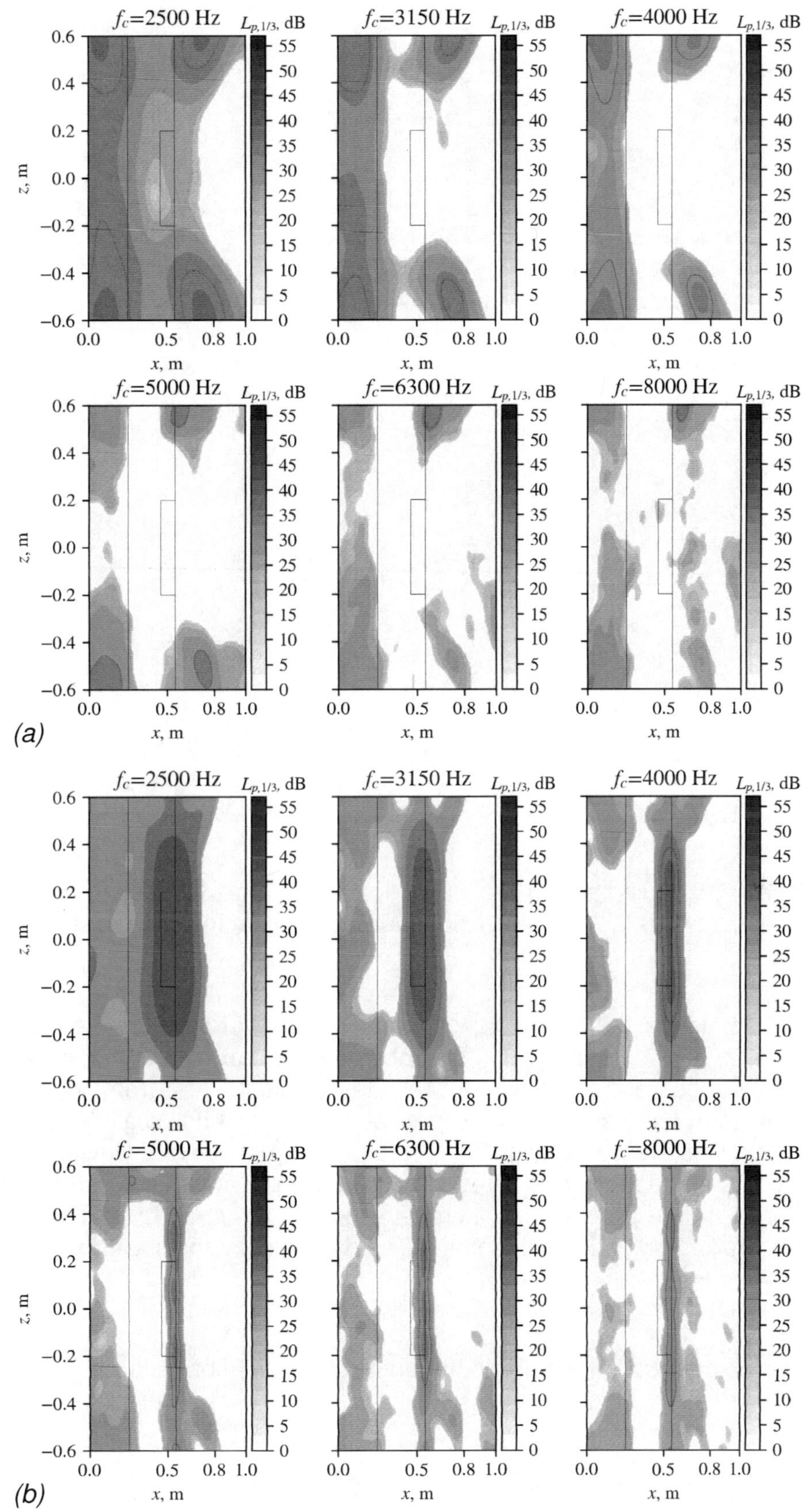

Figure 5: Sound maps with $U_0 = 50$ m/s *(a)* exterior noise and *(b)* trailing-edge noise $\alpha_g = 5°$, and $\delta_f = 0°$. The sound contour where $\max(L_{p,1/3}) - 3$ dB is highlighted by the red line.

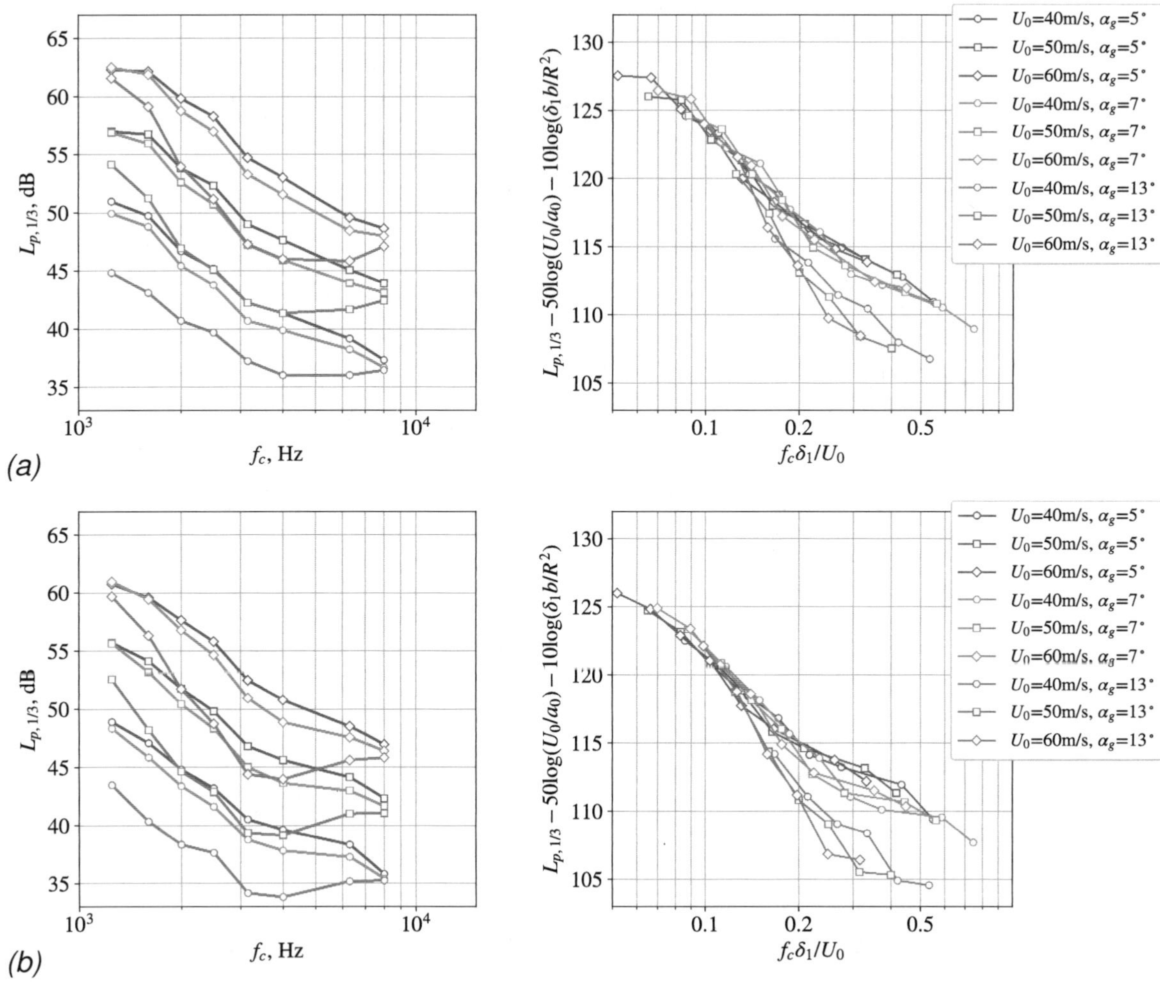

Figure 6: Unscaled and scaled sound pressure level of trailing-edge noise measured with the phased-microphone array *(a)* facing the suction-side and *(b)* facing the pressure-side of blade section model.

It is apparent, from figure 6, the sound pressure level is lower over the measured frequency range for the same U_0 and increasing α_g. This observation can be explained by the increase of boundary-layer thickness and the adverse pressure gradient. As a reminder, $f_c = 5000$ Hz is dropped in the spectra as described in the previous section. After scaling, it was found that the sound pressure levels are self-similar for $f_c < 2000$ Hz, but the self-similar spectra diverge according to α_g at higher frequency as a result of increasingly adverse pressure gradient [26]. Furthermore, at $\alpha_g = 13°$, the spectral level increases for $f_c > 5000$ Hz for the non-scaled spectra, which is atypical for trailing-edge noise. Hence, to avoid confusion, the scaled spectra of $\alpha_g = 13°$ are limited to $f_c \leq 4000$ Hz.

Table 3: Suction side boundary layer displacement thickness δ_1 computed at the trailing edge.

U_0, m/s	δ_1/c		
	$\alpha_g = 5°$	$\alpha_g = 7°$	$\alpha_g = 13°$
40	0.009169	0.012441	0.017808
50	0.008675	0.011728	0.016660
60	0.008299	0.011188	0.015802
150	0.006014	0.008434	0.012626

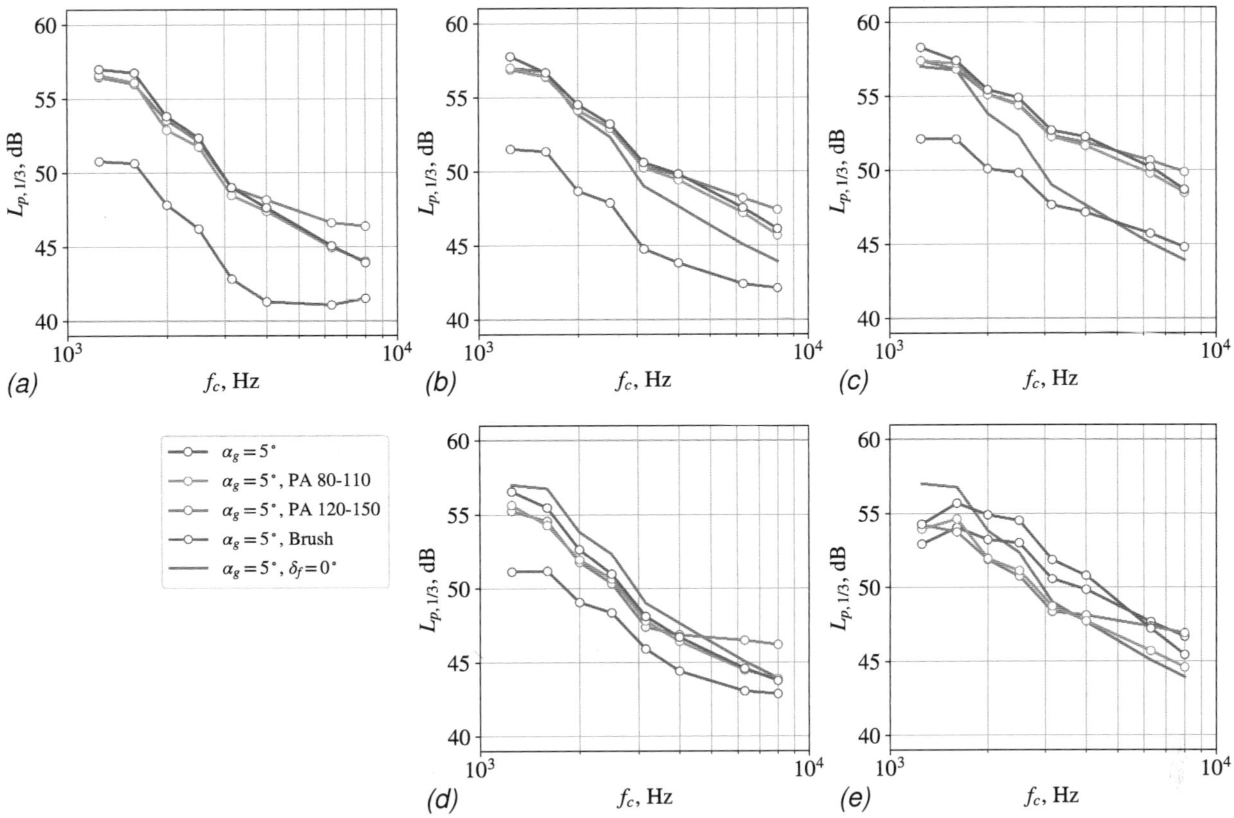

Figure 7: The farfield sound pressure level, $b_g = 0.7$ m. *(a)* $\delta_f = 0°$, *(b)* $\delta_f = -5°$, *(c)* $\delta_f = -10°$, *(d)* $\delta_f = 5°$, *(e)* $\delta_f = 10°$.

5.4. Effect of Edge Noise Reduction Technologies

The flap side-edge porous substitutes were installed and measured in the same way as a solid flap side-edge to localize the noise source. The region of interest is the same one used when investigating trailing edge scaling. The resulting sound pressure levels for porous flap side-edge, trailing-edge brush, and solid flap side-edge are shown in figure 7.

5.4.1 Trailing-Edge Noise

The $\delta_f = 0°$ configuration in figure 7*(a)* has the porous material affecting only 1/30 of the span length of the model's trailing edge. Hence, the porous material has little effect on the trailing-edge noise reduction. Noise increase was measured for PA120-150 side-edge at $f_c > 4000$ Hz because of the increased surface roughness [27]. In contrast, the trailing-edge brush was applied to the full span of the trailing edge. Hence, the porous material's effectiveness for reducing the trailing-edge noise is considerably lower than that of the brush.

The flap-up configuration of $\delta_f = -5°$ in figure 7*(b)* gradually increases the sound pressure level at higher frequencies compared to the non-deflected flap configuration. Whereas the flap-down configuration results in a broadband reduction in sound pressure level. This alteration of the sound pressure level can be attributed to the local boundary layer's change above the flap's surface. For the $\pm5°$ change of flap angle, the sound pressure levels for both porous flap side-edges are similar to those of the solid flap side-edge. Similar to the non-deflected flap case, the trailing-edge brush produces a 6 dB reduction in sound pressure level. Hence, for the $\pm5°$ flap angle deflection, the flap side-edge noise is not as apparent as the trailing-edge noise in the relevant frequency range.

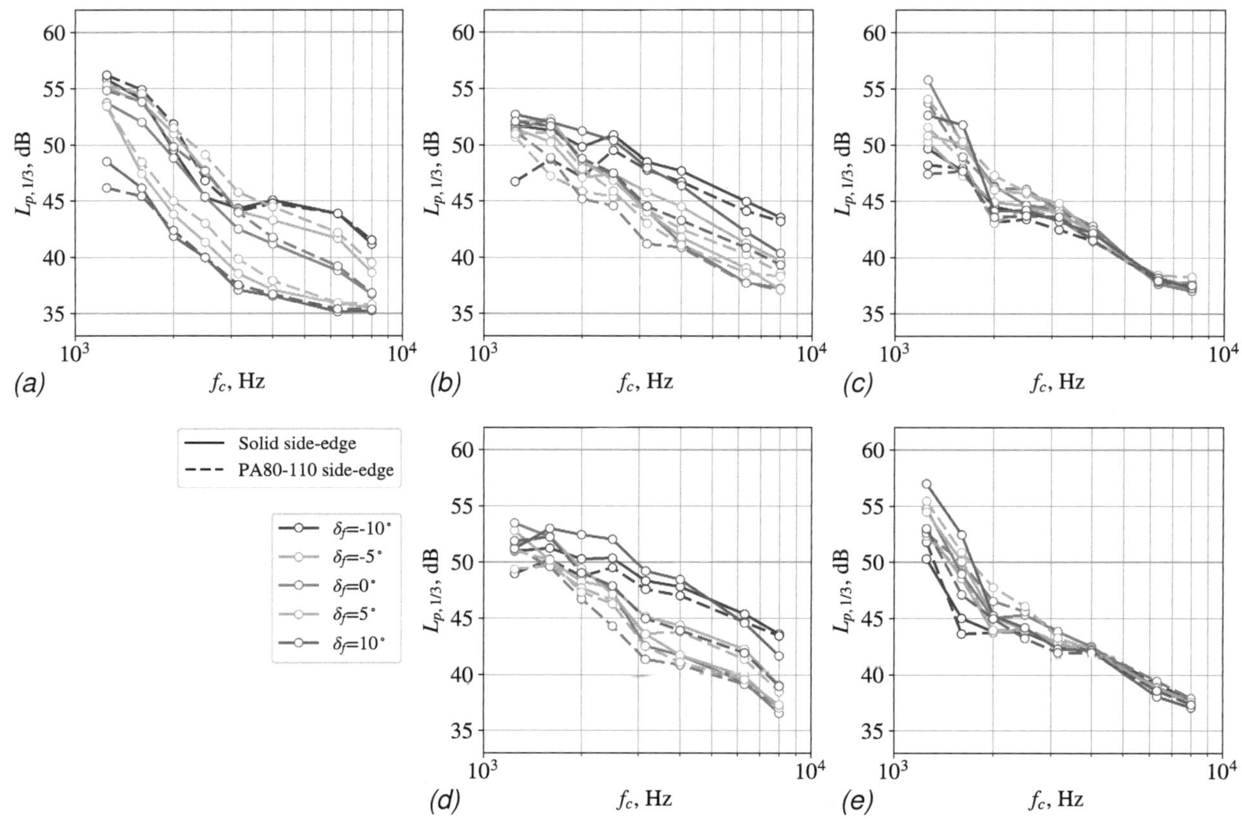

Figure 8: Contributions to the farfield sound pressure level according to the narrow spanwise sections centered at z_c with $b_g = 0.2$ m and $\alpha_g = 5°$. *(a)* $z_c = 0$ m, *(b)* $z_c = -0.2$ m, *(c)* $z_c = -0.4$ m, *(d)* $z_c = 0.2$ m, *(e)* $z_c = 0.4$ m.

5.4.2 Flap Side-Edge Noise

The flap-up configuration $\delta_f = -10°$ in figure 7(*c*) is fundamentally different than the flap-down one $\delta_f = 10°$ in figure 7*(e)*. For $\delta_f = -10°$, the porous flap side-edge does not alter the sound pressure level, whereas for $\delta_f = 10°$ both porous flap side-edges and trailing-edge brush alter the sound pressure level compared to the solid flap side-edge.

In a first look, it appears that the flap side-edge noise is not relevant for $\delta_f = -10°$. However, this outlook could result from the porous material only applied on the flap's side and not on the model's side. The spectra contributed by the narrow strips of the model's trailing edge, $b_g = 0.2$ m, centered at z_c are shown in figure 8, which for brevity shows only the solid flap side-edge and the PA 80-110 flap side-edge. It was confirmed that the sum of sound pressure levels of the narrow sections is equal to the sound pressure level of a region with a span equal to the sum of the span of the narrow sections. Figure 8 classifies 3 general regions of the model as detailed below:

1. The first region is around the midspan of the model, figure 8*(a)*, where the porous flap side-edge shows little relevance. Changes to the sound pressure level in this figure are related to the value of δ_f, which changes the local boundary layer dynamics.

2. The second region encompasses the flap side-edge and shown in figure 8*(b)* and *(d)*. The effect of the porous flap-side edge is most notable for $\delta_f = 10°$ as depicted in figure 7. The porous flap side-edge significantly affects the sound pressure level even for $\delta_f = 0°$, because the region processed by CLEAN-SC is a narrow strip focused around the flap side-edge.

3. The final region is the outboard region shown in figure 8*(c)* and *(e)*. The effect of δ_f to the sound pressure level is limited to $f_c < 2000$ Hz. The effect of porous side-edge is shown only for $\delta_f = 10°$. The sound pressure levels at $f_c > 2000$ Hz is only dependent to α_g.

Comparison of the flap-side edge region, figure 8*(b)* and *(d)*, and the outboard region, figure 8*(c)* and *(e)*, shows that the solid flap side-edge increases the localized sound pressure level for $f_c > 2000$ Hz and it is observed for $\delta_f = -10^\circ$ and 10°. However, the porous side-edge is effective in reducing flap side-edge noise only at $\delta_f = 10^\circ$. This behavior can be explained by figure 9 that shows the flap side-edge vortex for *(a)* $\delta_f = -5^\circ$ and *(b)* $\delta_f = 5^\circ$. Figure 9*(a)* shows that the predominant side-edge vortex develops from the side-edge of the static model, whereas figure 9*(b)* shows that it develops from the flap side-edge. Because the porous material is only installed on the flap side-edge, figure 9 is a compelling argument that the porous side-edge is ineffective in reducing flap side-edge noise in the flap-up configuration. The flap deflection angles of the simulated flow field are smaller than those observed in the measurement, however, we believe that the physical phenomenon explained here is retained.

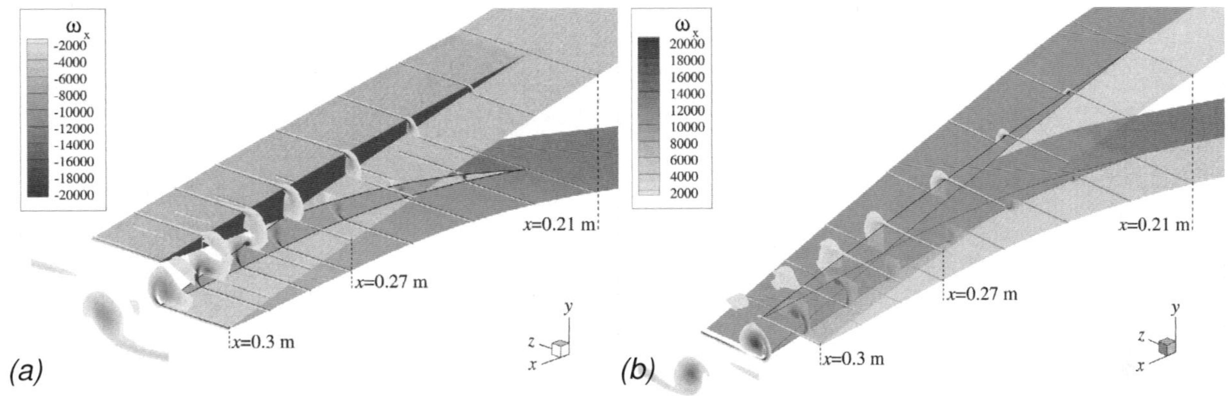

Figure 9: k-ω SST simulations of the flap side-edge vortex with $U_0 = 50$ m/s and $\alpha_a = 0^\circ$ *(a)* $\delta_f = -5^\circ$, *(b)* $\delta_f = 5^\circ$. The flap side-edge is demarked by a red line and the color contour depicts the strength of the chordwise vorticity.

6. Discussion

6.1. Noise of Active Trailing-Edge

The IWT-7.5-164 wind turbine's design angle at $R = 76$ m is 6°, which is achieved in the wind tunnel measurement at approximately $\alpha_g = 13^\circ$. However, the blade section model at the flap-down configuration at this angle of attack deflects the flow greatly. Under such conditions the wake would directly impinge on the flow-collector wall, causing blockage effects and potentially structural damage to the wind tunnel. Therefore, the reference angle of attack in the wind tunnel measurement is $\alpha_g = 7^\circ$ or equivalent to the aerodynamic angle of attack α_a of 3°.

For the present blade section model with flap, a flap deflection of $+5^\circ$ results in a change of lift equivalent to -3° change in the effective angle of attack and vice versa. The pressure coefficients and sound pressure levels of the angle of attack and flap deflection angle combination are shown in figure 10. The pressure coefficient distribution of $(\alpha_g = 5^\circ(0^\circ), \delta_f = 0^\circ)$ in figure 10*(a)* is altered because of the flap deflection of $\delta_f = 5^\circ$ and it is now approximately similar to pressure coefficient of the design point in figure 10*(b)*. At $\alpha_g = 13^\circ$, the flap deflection does not alter the pressure coefficient back to the design point, notably because the pressure minimum at $\alpha_g = 13^\circ$ has moved closer to the leading edge.

The corresponding sound pressure levels for these configurations are shown in figure 10*(b)* without flap deflection and figure 10*(c)* with flap deflection. The active trailing-edge equalizes the blade model's self-noise.

Therefore, in maintaining constant lift using a plain flap, within the envelope of $\pm 3^\circ$ effective angle of attack change from the reference angle of attack, there is no noise increase, within the measured frequency range, related to the flap side-edge noise from the flap deflection of $\pm 5^\circ$.

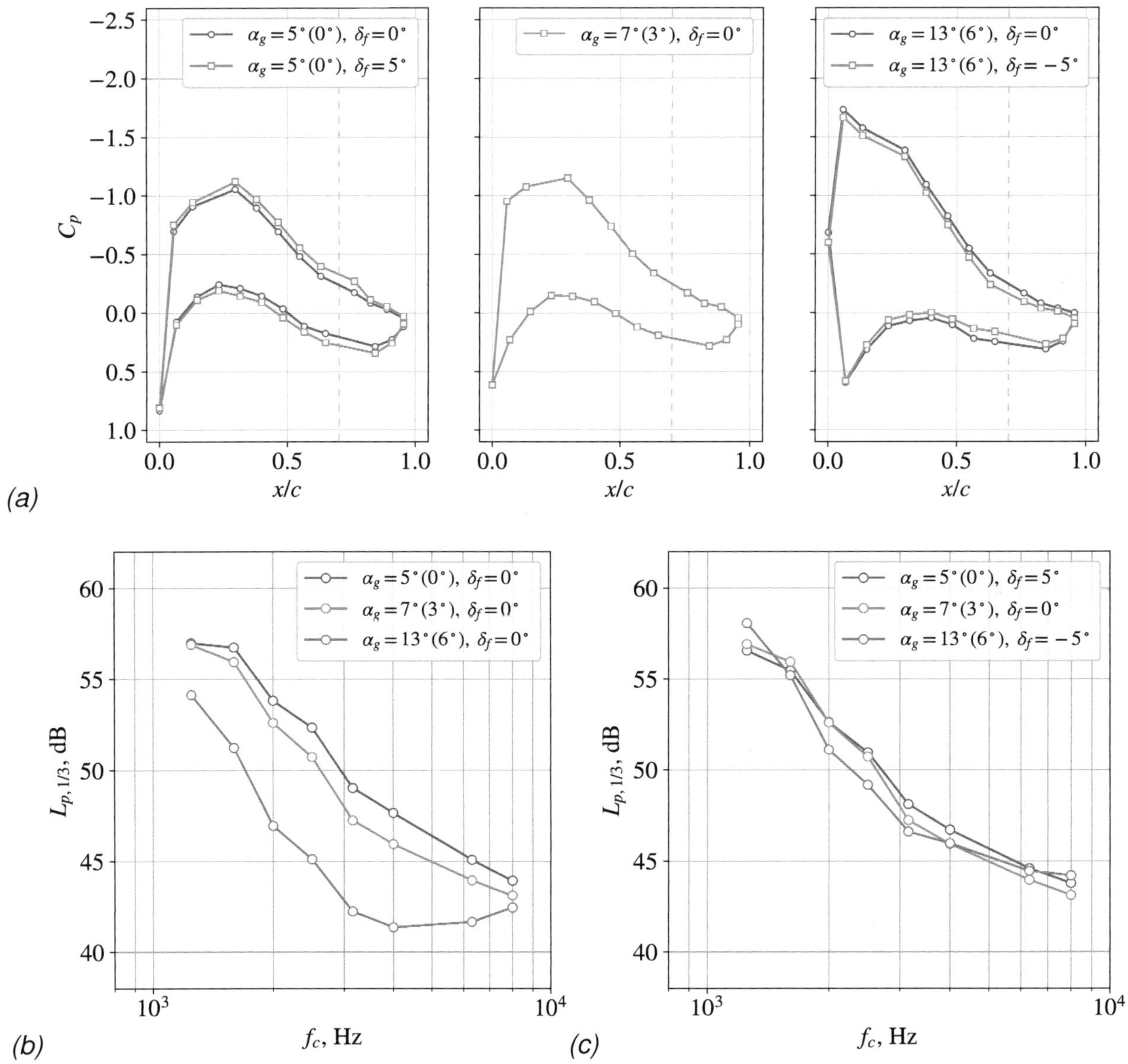

Figure 10: Pressure coefficents *(a)* without and with flap deflection and the one-third octave sound pressure levels *(b)* without flap deflection and *(c)* with flap deflection. The value in parentheses indicates the equivalent aerodynamic angle of attack.

6.2. FMCAS: Fast Multipole Code for Acoustic Shielding

FMCAS is a tool designed to investigate acoustic shielding but is also capable of trailing-edge noise prediction [28, 29]. FMCAS uses the reconstructed turbulent flow field of FRPM (Fast Random Particle Mesh), which is generated with input from RANS, to calculate sound radiation. The details of FMCAS is beyond the scope of this study and will be addressed in a different paper. Figure 11 shows the spectral comparison between FMCAS and the measurements. A small flow patch was used as input for producing the numerical spectrum. This patch contains the flow information around the flap side-edge and is only 64 mm wide to limit the trailing-edge noise's contribution. Hence, the numerical spectrum approximates the flap side-edge noise. The grey lines indicate the upper and lower 95% confidence levels of the simulation due to the stochastic nature of the turbulence flow field reconstruction of FRPM.

The measured spectrum shown in figure 11 is the sound pressure level of the solid flap side-edge, $L_{p,1/3,\text{solid}}$, for no-flap $\delta_f = 0°$, medium-flap $|\delta_f| = 5°$, and high-flap $|\delta_f| = 10°$. These spectra were calculated with CLEAN-SC from a region with a span width of $b_g = 0.1$ m surround-

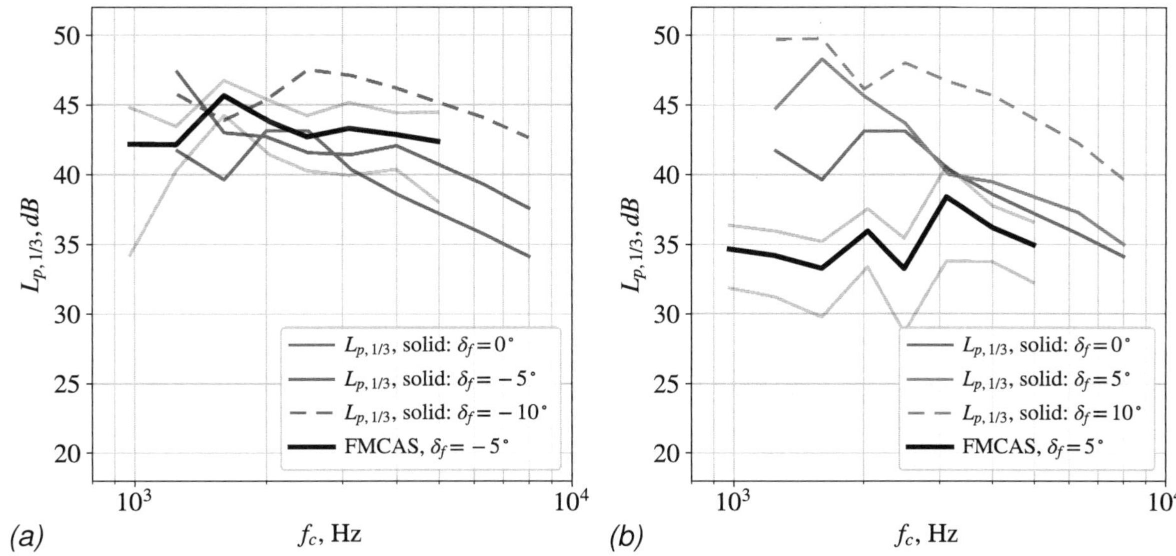

Figure 11: Comparison of the spectra from FMCAS simulation and from the experiment $U_0 = 50$ m/s and $\alpha_a = 0°$ (a) flap-up and (b) flap-down.

ing the flap side-edge. The result shows that FMCAS can reproduce the flap side-edge noise spectrum for $\delta_f = -5°$ in figure 11*(a)*, but less reproducible is the spectrum for $\delta_f = 5°$ in figure 11*(b)*. The reason for the discrepancy could be due to the boundary layer thickening and also the adverse pressure gradient introduced by the flap and not captured in the numerical simulation. For example, The change of δ_f from no-flap to medium-flap in figure 11*(b)* shows a level increase at $1250 \leq f_c \leq 2000$ Hz. This level increase is not observed in figure 11*(a)*.

Another feature of the measured sound pressure levels in figure 11 is that the flap side-edge noise is relevant for $f_c > 2000$ Hz at $\delta_f = \pm 10°$. Whereas for $\delta_f = \pm 5°$ the relevant frequency range is $f_c > 3000$ Hz. This feature is in agreement with the flap side-edge scaling in Ref. [30], which proposed that the flap side-edge noise scales with the diameter of the cross-section of the flap side-edge vortex and the local crossflow velocity.

FMCAS shows promise in predicting flap side-edge noise. The sound pressure level of $\delta_f = -5°$ is well predicted, but less well for $\delta_f = 5°$. Presumably, the localized flow domain used to resolve the numerical spectrum is too small to capture the acoustic effect of the flap-down configuration's thick boundary layer.

7. Conclusions

A small-scale blade section was measured in the Acoustic Wind tunnel Braunschweig to investigate the acoustic effect of implementing an active trailing-edge. The blade section has a 400 mm span wide and 90 mm chord long flap. Three freestream velocities, three angles of attack, and flap deflection angles between $-10°$ to $10°$ were measured using a phased-microphone array. Because of the physical model scale, noise reduction technologies were used to minimize individual sound sources. A trailing-edge brush was implemented at the static and active trailing-edge. A narrow part of the flap side-edge is substituted with a porous material to reduce flap side-edge noise and identify the trailing-edge noise.

The measured far-field noise spectra show that the trailing-edge brush performs best at all flap deflection angles, except for the flap deflection angle of $\delta_f = 10°$, where porous side-edges are more effective for noise reduction. The sound pressure level of a narrow region around the flap side-edge at $\delta_f = -10°$ is higher than the rest of the trailing edge for $f_c > 2000$ Hz. However, at $\delta_f = -10°$, the porous side-edge material does not reduce noise as optimally as at $\delta_f = 10°$. Hence, it can be concluded that flap side-edge noise is not a relevant noise source for the small flap deflection angles between $-5°$ and $5°$.

The active trailing-edge is designed to achieve $5°$ flap deflection to compensate for $-3°$ angle-of-attack change and vice-versa. Both positive and negative trailing edge deflections produce sound pressure levels comparable to the reference angle of attack within the measured frequency range. Hence, the flap side-edge noise is not a significant contributor to an active trailing-edge for the flap deflection angles in the $-5° \leq \delta_f \leq 5°$

Flap side-edge noise prediction was attempted using FMCAS, which shows promise in localizing the contribution of the flap side-edge noise. However, the comparison with measurements has to be undertaken carefully, as the measured spectrum combines other, more predominant noise sources.

Acknowledgement

The authors' are grateful to Jörn Tychsen (TU Braunschweig) for providing the porous foam materials. This study, part of the project SmartBlades 2.0, was funded by the German Ministry for Economic Affairs and Energy (BMWi).

References

[1] IWES Wind Turbine IWT-7.5-164. Rev 4. Technical report, Fraunhofer IWES.

[2] Jay C. Hardin. Noise radiation from the side edges of flaps. *AIAA Journal*, 18(5):549–552, May 1980.

[3] M.S. Howe. On the generation of side-edge flap noise. *Journal of Sound and Vibration*, 80(4):555–573, February 1982.

[4] R. Sen. Local dynamics and acoustics in a simple 2D model of airfoil lateral-edge noise. In *Aeroacoustics Conference*, State College,PA,U.S.A., May 1996. American Institute of Aeronautics and Astronautics.

[5] Jay C. Hardin and James E. Martin. Flap Side-Edge Noise: Acoustic Analysis of Sen's Model. *AIAA Journal*, 35(5):810–815, May 1997.

[6] C. Streett. Numerical simulation of fluctuations leading to noise in a flap-edge flowfield. In *36th AIAA Aerospace Sciences Meeting and Exhibit*, Reno,NV,U.S.A., January 1998. American Institute of Aeronautics and Astronautics.

[7] Y. Guo. Prediction of flap side edge noise. In *5th AIAA/CEAS Aeroacoustics Conference and Exhibit*, Bellevue,WA,U.S.A., May 1999. American Institute of Aeronautics and Astronautics.

[8] Mehdi R. Khorrami, Bart A. Singer, and Ronald H. Radeztsky. Reynolds-Averaged Navier-Stokes Computations of a Flap-Side-Edge Flowfield. *AIAA Journal*, 37(1):14–22, January 1999.

[9] Thomas F Brooks and William M. Humphreys Jr. Flap-edge aeroacoustic measurements and predictions. *Journal of Sound and Vibration*, page 44, 2003.

[10] Yueping Guo. On noise reduction by flap side edge fences. *Journal of Sound and Vibration*, 277(1-2):369–390, October 2004.

[11] D. Angland, X. Zhang, and N. Molin. Measurements of Flow Around a Flap Side Edge with Porous Edge Treatment. *AIAA Journal*, 47(7):1660–1671, July 2009.

[12] Mitsuhiro Murayama, Kazuomi Yamamoto, Takehisa Takaishi, Yasushi Ito, Hiroki Ura, Yuzuru Yokokawa, Kentaro Tanaka, and Tohru Hirai. Airframe Noise Reduction of Flap Side-edge Using Vortex Generators. In *23rd AIAA/CEAS Aeroacoustics Conference*, Denver, Colorado, June 2017. American Institute of Aeronautics and Astronautics.

[13] Torben Juul Larsen, Helge A. Madsen, and Kenneth Thomsen. Active load reduction using individual pitch, based on local blade flow measurements. *Wind Energy*, 8(1):67–80, January 2005.

[14] C.P. van Dam, Dale E. Berg, and Scott J. Johnson. Active load control techniques for wind turbines. Technical Report SAND2008-4809, 943932, July 2008.

[15] T.K. Barlas and G.A.M. van Kuik. Review of state of the art in smart rotor control research for wind turbines. *Progress in Aerospace Sciences*, 46(1):1–27, January 2010.

[16] Damien Castaignet, Thanasis Barlas, Thomas Buhl, Niels K. Poulsen, Jens Jakob Wedel-Heinen, Niels A. Olesen, Christian Bak, and Taeseong Kim. Full-scale test of trailing edge flaps on a Vestas V27 wind turbine: Active load reduction and system identification: Full-scale test of trailing edge flaps on a Vestas V27 wind turbine. *Wind Energy*, 17(4):549–564, April 2014.

[17] Alejandro Gomez Gonzalez, Peder B. Enevoldsen, Athanasios Barlas, and Helge A. Madsen. Field test of an active flap system on a full scale wind turbine. Preprint, Aerodynamics and hydrodynamics, March 2020.

[18] Karl-Stéphane Rossignol. Development of an Empirical Prediction Model for Flap Side-Edge Noise. In *16th AIAA/CEAS Aeroacoustics Conference*, Stockholm, Sweden, June 2010. American Institute of Aeronautics and Astronautics.

[19] Michael Pott-Pollenske and Jan Delfs. Enhanced Capabilities of the Aeroacoustic Wind Tunnel Braunschweig. In *14th AIAA/CEAS Aeroacoustics Conference*, Vancouver, British Columbia, Canada, May 2008. AIAA.

[20] Pieter Sijtsma. CLEAN Based on Spatial Source Coherence. *International Journal of Aeroacoustics*, 6(4):357–374, December 2007.

[21] M. Herr. Experimental Study on Noise Reduction through Trailing Edge Brushes. In Ernst Heinrich Hirschel, W. Schröder, Kozo Fujii, Werner Haase, Bram Leer, Michael A. Leschziner, Maurizio Pandolfi, Jacques Periaux, Arthur Rizzi, Bernard Roux, Hans-Josef Rath, Carsten Holze, Hans-Joachim Heinemann, Rolf Henke, and Heinz Hönlinger, editors, *New Results in Numerical and Experimental Fluid Mechanics V*, volume 92, pages 365–372. Springer Berlin Heidelberg, Berlin, Heidelberg, 2006.

[22] Michaela Herr, Karl-Stephane Rossignol, Jan Delfs, Nicolas Lippitz, and Michael Mößner. Specification of Porous Materials for Low-Noise Trailing-Edge Applications. In *20th AIAA/CEAS Aeroacoustics Conference*, Atlanta, GA, June 2014. American Institute of Aeronautics and Astronautics.

[23] Karl-Stephane Rossignol, Lennart Rossian, Alexandre Suryadi, Michaela Herr, Roland Ewert, and Jan Delfs. Effect of Porous Materials on the Acoustic Emissions of a Circulation-Control High-Lift Airfoil. page 4, 2020.

[24] Michael Goody. Empirical Spectral Model of Surface Pressure Fluctuations. *AIAA Journal*, 42(9):1788–1794, September 2004.

[25] Mark Drela. XFOIL: An Analysis and Design System for Low Reynolds Number Airfoils. In Thomas J. Mueller, editor, *Low Reynolds Number Aerodynamics*, Lecture Notes in Engineering, pages 1–12, Berlin, Heidelberg, 1989. Springer.

[26] Nan Hu. Empirical Model of Wall Pressure Spectra in Adverse Pressure Gradients. *AIAA Journal*, 56(9):3491–3506, September 2018.

[27] Karl-Stéphane Rossignol, Alexandre Suryadi, Michaela Herr, Johannes Schmidt, and Jörn Tychsen. Experimental Investigation of Porous Materials for Trailing-Edge Noise Reduction. *International Journal of Aeroacoustics*, 19(6-8):365–384, 2020.

[28] Markus Lummer, Rinie A. Akkermans, Christoph Richter, Carsten Pröber, and Jan Delfs. Validation of a Model for Open Rotor Noise Predictions and Calculation of Shielding Effects using a Fast BEM. In *19th AIAA/CEAS Aeroacoustics Conference*, Berlin, Germany, May 2013. American Institute of Aeronautics and Astronautics.

[29] Nils Reiche, Roland Ewert, Markus Lummer, and Jan Delfs. Fast Multipole Boundary Element Method with Stochastic Sources for Broadband Noise Simulation. In *23rd AIAA/CEAS Aeroacoustics Conference*, Denver, Colorado, June 2017. American Institute of Aeronautics and Astronautics.

[30] Karl-Stephane S. Rossignol. Flow Field Measurements to Characterize Flap Side-Edge Noise Generation. In *19th AIAA/CEAS Aeroacoustics Conference*, Berlin, Germany, May 2013. American Institute of Aeronautics and Astronautics.

9th International Conference
on
Wind Turbine Noise
Remote from Europe – 18th to 21st May 2021

A Characterization of Wind Turbine and Background Noise Distributions in Far-Field Receptor Testing of Wind Turbine Facilities

Nicholas Tam, Aercoustics Engineering Limited: NicholasT@Aercoustics.com
Duncan Halstead, Aercoustics Engineering Limited: DuncanH@Aercoustics.com

Summary

This study expands on the previous work from the authors regarding the investigation of the variation of measured background sound levels in rural areas. Determinations of the Turbine-Only sound level in far-field sound testing are often challenging, limited by the quality of the measurement data. This study examines statistical factors of a far-field measurement dataset which may impact the analysis. The implications from data skewness, high standard deviation values and uncertainty are discussed.

1. Introduction

Far-field sound testing is a common method of assessing wind farm acoustic compliance [1] [2] [3]. This method of testing is often relied upon by regulators because it provides a determination of sound level at a noise receptor (typically a residence) without the need for sound propagation analysis. However, determinations of sound level using receptor testing are often limited by the quality of the measurement data, as low signal to noise levels between the turbine noise and ambient noise are much more common. The IEC 61400-11 test standard provides a benchmark of 6 dB for a desired signal-to-noise [4]. This benchmark is much more difficult to achieve in receptor testing, given the increased distance from the sound source, long term (and sometimes remote) testing, and higher prevalence of vegetation or other noisy features that are typical around residential locations.

This study builds on the results of the previous paper [1], where it was found that background sound levels approached, and even exceeded, the expected noise from the nearby wind turbines when common foliage elements (trees, vegetation, crops, etc.) were present in the vicinity of the measurement location. This allowed for some insights into factors which may need to be considered when conducting a good measurement campaign. The current study aims to expands on the previous study by examining statistical factors of a far-field measurement dataset and how they may impact the analysis. Other potential factors which may have an impact on the conclusion of the assessment are discussed as well.

2. Background

The measurement equipment, data acquisition parameters, and analysis methodologies are discussed in this section.

2.1 Measurement Methodology

In this study, the audit data used for discussion was collected using the far-field testing methodology established by the Ontario Ministry of the Environment [2]; published in 2017. Aercoustics has conducted numerous measurement campaigns using this protocol at variety of locations across wind farms in southern Ontario, Canada.

Measurement equipment is installed at one or more sensitive receptors to the wind facility under study, typically 500 to 800 metres from the nearest wind turbines. Measurements are conducted remotely every night over a long-term period (typically 4 to 16 weeks). The source emissions and sound propagation are controlled by looking only at periods of high turbine output (typical above 85% of rated power) and under downwind propagation. Assessing the sound levels at night-time also minimizes ambient noise and ensures a more stable atmosphere compared to daytime conditions.

Measurement data is collected in in 1-minute intervals and binned according to the measured 10-m wind speed of the interval. Acoustic and weather/meteorological ("MET") data are acquired synchronously by the monitor at the receptor location. The overall LA_{eq}[1], 1/3rd octave band level, and L90 was recorded for each interval. Weather data recorded includes wind speed, wind direction, temperature, pressure, humidity, and precipitation. Turbine operational information are obtained separately from the wind facility SCADA system; typically in 10-minute average intervals.

The equipment synchronously captures acoustic and MET data at heights of 4.5 meters and 10 metres, respectively. The microphone is fitted with a primary and secondary windscreen to reduce the effect of wind generated self-noise. The 10-metre[2] weather station measures a range of weather parameters for each interval. A picture of a typical measurement apparatus is shown in Figure 1.

Figure 1: Typical measurement apparatus

During the measurement campaign, the nearby turbines are periodically parked to collected ambient noise. For regulatory requirements, a minimum of 60 valid data points when the turbines are operating (Total Noise, or "TON") and 30 valid data points when the turbines are parked (Background, or "BG") are required in each wind bin prior to making an assessment. The "Turbine-Only" sound level for a given wind bin is computed by logarithmically subtracting the average Background level from the average Total-Noise level.

---

[1] Energy-equivalent sound level, A-weighted

[2] Wind speed at 10-metres is referred to as "ground level wind speed" in this paper.

2.2 Statistical Considerations of Measurement Data

In addition to the methodologies outlined in Section 2.1 above, it is important to highlight some additional key concepts that used in this study. Readers may be aware of these as common considerations when designing a measurement plan. Here are some definitions of keys concepts:

Standard Deviation – a measure of the variation of the samples in your dataset. Higher standard deviations indicate a greater variation (or spread) in the samples.

Signal-to-noise – the difference between the desired measurement signal and the background. Higher signal-to-noise levels indicate that the background noise has a lower impact on the overall measured level.

Standard Error (Type A uncertainty) – this is the uncertainty associated with the variation in measured levels. It represents the range about the measured mean that that "true" mean is expected.

Equipment Error (Type B uncertainty) – this is the uncertainty associated with the tolerance of the measurement equipment. It is typically fixed for a given measurement apparatus.

Combined Standard Uncertainty – This is the combination of the Type A and Type B uncertainty and the signal-to-noise of total noise and background measurement datasets.

Confidence Interval – In the context of the mean value, this is interval in which the true mean of the population is expected. The confidence interval is expressed as a percentage. The confidence interval of the mean can be estimated using Standard Error.

Normal Distribution – a dataset where the distribution of samples is centred around the mean, with no bias on the left or right side of the mean. Also known as the "bell curve".

Skewness – a measure of the asymmetry in the distribution of a dataset.

Central Limit Theorem – refers to the tendency for the sample average to be normally distributed from repeated sampling.

Additional considerations may be needed for acoustical data given the logarithmic nature of the collected data. Those wish to learn more on this subject can see Peters [5] for more details.

Many of these factors (including Type A, Type B, and Combined Standard Uncertainty) are considered in the IEC 61400-11 standard. These parameters are considered for logarithmic averages. For far-field receptor measurements, an approach to the statistical evaluation is currently not standardized[3]. Some jurisdictions utilize standard deviation as a measure of data quality; rarely is the combined standard uncertainty evaluated for receptor testing datasets.

2.3 Characterization of Previous Study

In the previous study [6], some characterization of the background sound levels and its variation at different receptors was achieved based on the physical features present in the vicinity of the measurement location. The intent of that study was to provide insights into some factors which may need to be considered when conducting a good measurement campaign. Examinations of the background data found that large variations in samples sound level are

[3] It is noted that there is an IEC standard currently under development for receptor testing (IEC 61400-11-2)

possible in some sites; variations up to 20 dB are possible if there are extraneous noise sources present in the environment (roadways, trees, machinery, etc.).

Data was also examined for different hub height windspeeds with the same ground level wind speed condition. The result showed the ambient sound levels can differ with different windshear conditions. Even when controlling these factors, the ambient background level often approached or exceeded the expected Turbine-Only contribution, making a high signal-to-noise ratio difficult to achieve.

2.4 Studied Locations

To aid in the discussion of the following, this study will cite measurement data examples used in the previous study (Site A – D) as well an addition of four other locations. A brief summary of each site's environment can be found below in Table 1. The locations of the features presented in this table are with reference to the "front" direction pointing towards the nearest turbine.

Table 1 - Description of Monitoring Locations and Surrounding Environment

Location ID	Immediate Surroundings	Distance to Nearest Forest	Distance to Nearest Roadways (and traffic level)
Site A	Front: open field Side: trees, 25m Back: open field	100m	340m (Infrequent traffic)
Site B	Front: open field Side: trees, 15m Back: house, 75m	130m	15m (infrequent traffic)
Site C	Front: open field Side: open field Back: forest, 50m	50m	40m (infrequent traffic)
Site D	Front: open field Side: house, 50m Back: open field	500m	15m (moderate traffic)
Site E	Front: Open field Side Open field Back: building, 100 m	500m	400m (infrequent traffic)
Site F	Front: Open field Side: Open field Back: Open field	150m	40m (infrequent traffic), 3 km (high traffic, major highway)
Site G	Front: Open field Side: Open field Back: Open field	250m	20 m (infrequent traffic)
Site H	Front: forest, 20m Side: building/trees, 30m Back: Open	30m	25 m (infrequent traffic)

In general, the eight locations examined in the current study compose of a selection of environmental conditions which could be observed in Southern Ontario. Several examples, such as Site A, E, and G, are considered ideal with very few contamination sources in the environment. Site B, C and H are hampered by tree noise or vegetation contamination to some degree while Site D, F have influence from traffic noise of nearby roadways.

3. Discussion

3.1 General Observations

To help with later discussion, it is worthwhile to discuss some the typical behaviours of several key parameters in the data from a far-field measurement dataset. The following parameters are of interest: standard deviation and signal-to-noise ratio.

Standard deviation is a measure of the sampling variation observed in the dataset. As found in the previous investigation of background noise [6], the site environmental conditions have a large influence on the variation observed. Figure 2 and Figure 3 summarize the standard deviations observed in from the eight locations of this study.

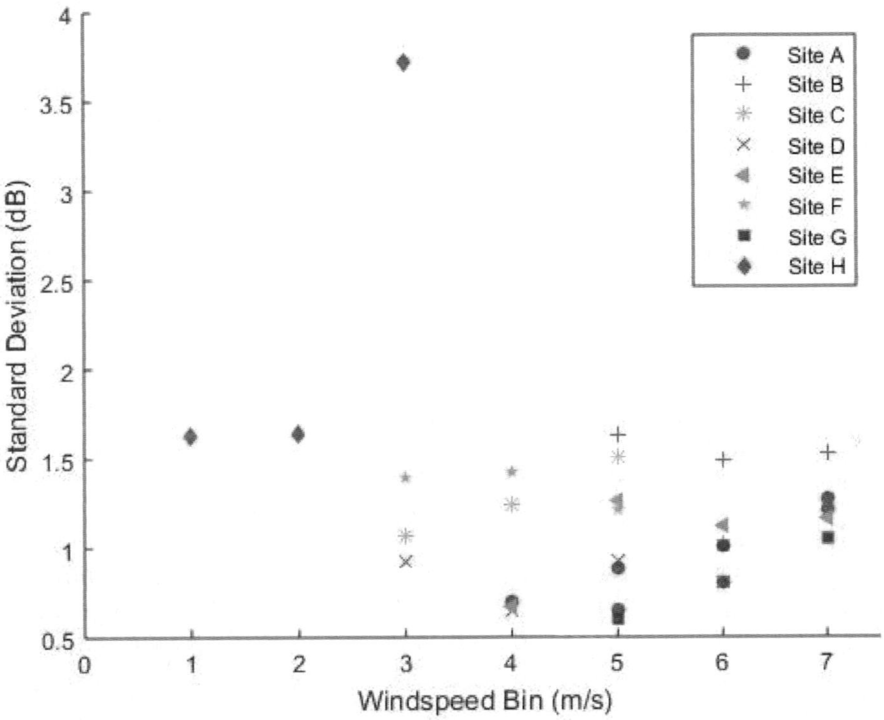

Figure 2- Standard Deviation of TON sound levels with respect to Ground Level Windspeed Bin

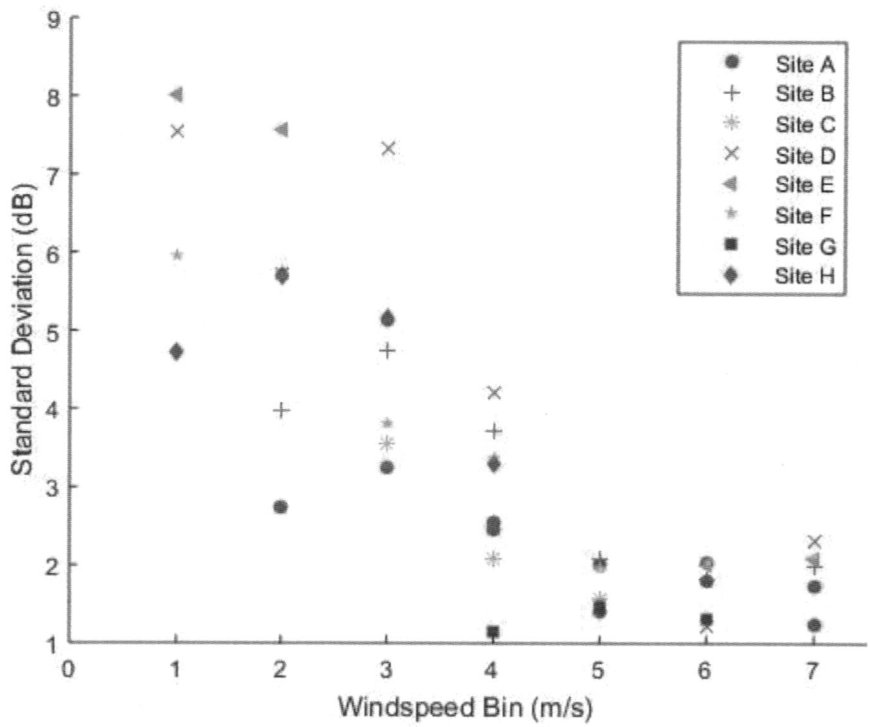

Figure 3- Standard Deviation of Background sound levels with respect to Ground Level Windspeed Bin

As shown in the previous two figures, the standard deviation tends to be higher at the lower wind speed bins, especially for background noise. This is not surprising, as the impact of any extraneous noise is more noticeable at lower wind speeds where the wind-induced noise is the lowest, leading to greater variation in the data range. This trend is much less pronounced with the TON datasets, as the turbine noise itself dominates at low wind speed bins, reducing the influence of extraneous noise on the overall measured level.

It is noted that much of the significant extraneous contamination is already removed from these datasets using Aercoustics' automatic transient noise filtering. With this data reduction, the standard deviation most of the datasets is reduced to below 2 dB for TON data and 4 dB for background data.

From the datasets examined, 25 wind bins can be extracted that have sufficient data for further study. The average signal to noise across all wind bins is 4 dB, although individual wind bins ranged from signal to noise ratios of nearly 0 dBA to almost 12 dBA. The signal to noise ratio tends to decrease with increasing ground level wind speed, as shown in Figure 4.

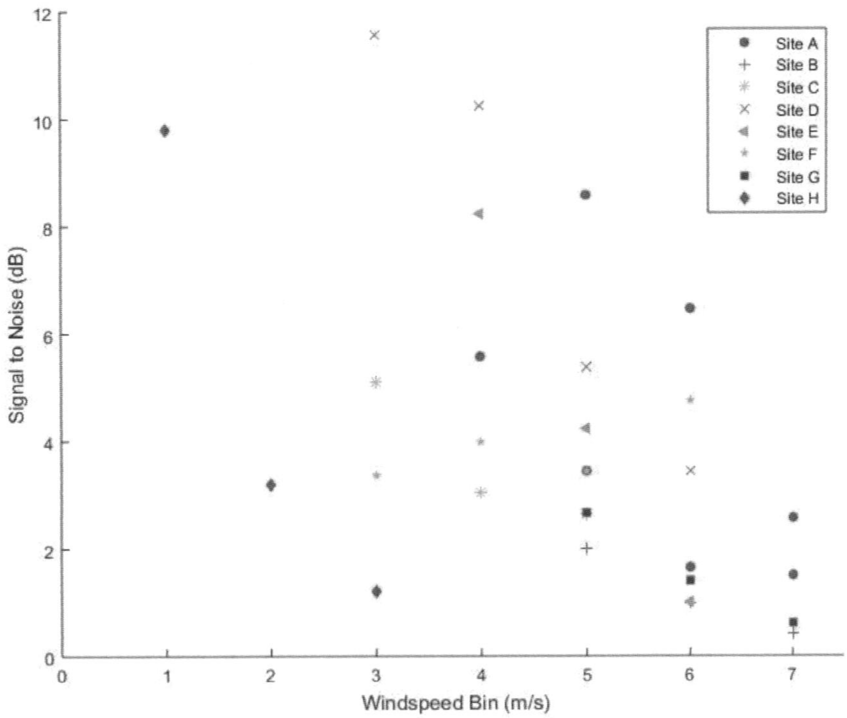

Figure 4- Signal-to-Noise with respect to Ground Level Windspeed Bin

3.2 Effects of Skewed Distribution in the Collected Data

One observation of interest with the dataset is the distribution of the collected data. On some occasions, especially with the presence of a contaminating noise source, the "Raw" dataset (before filtering) may not follow a Normal/Gaussian distribution; tending to skew the distribution left or right of centre.

There may be different explanations for skewed data. Skewness can be caused by ambient noise contamination or inclusion of different environmental conditions not well controlled by the experiment protocol (ex. different wind shear condition, as discussed in the previous study). An example of this is illustrated with the data from Site D, where a notable quantity of data collected was influenced by the presence of traffic noise. This can be seen in the histogram of the "Raw" dataset in Figure 5.

For the datasets examined in this study, the final assessed data, after data reduction, tends to be symmetrical. This can be seen from the histogram of the "Data Reduced" dataset in Figure 5. Data filtering techniques played a large part in removing or reduce sources of ambient contamination.

However, when the skewness cannot be removed, or is inherent to the conditions observed, what would be the implication to the analysis?

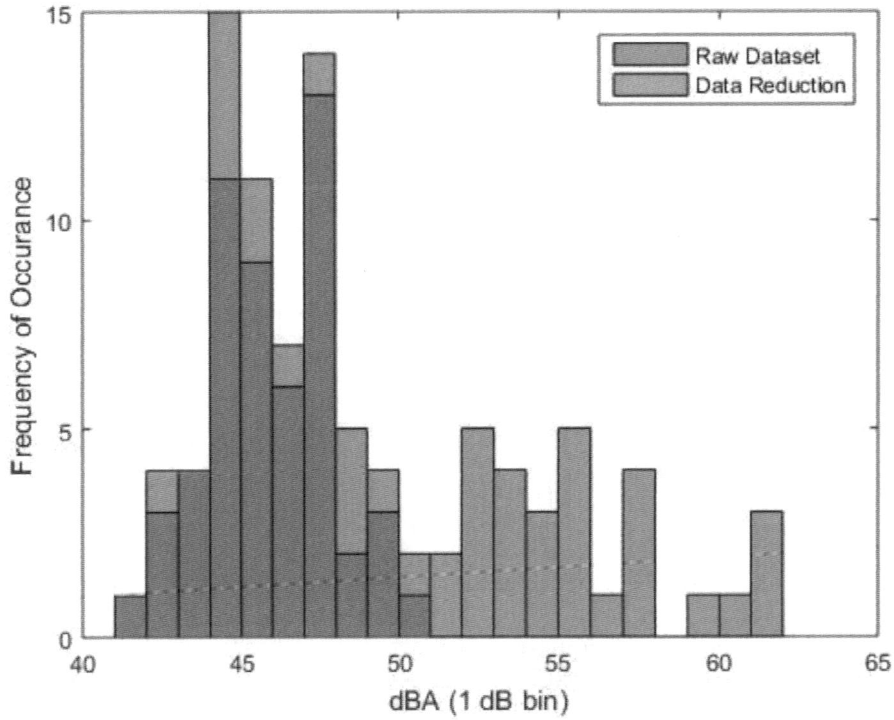

Figure 5- "Raw" and "Data Reduced" Background Dataset for 7 m/s Ground Level Windspeed Bin at Site D

If the quality of interest is the average of the sound pressure level, an inference of the mean value can be made with a sufficiently large sample size (Central Limit Theorem, it is commonly accepted that a minimum of 30 samples would be needed). The mean value is expected to converge to true mean of the population, regardless of the shape of the underlying distribution of the population.

However, if the underlying population is indeed skewed, it begs the question if the mean value would be the best metric to quantify the observed environment. With a skewed distribution, the mean value is not be representative of the most frequent observations. Examining the median or percentile values is a typical approach when evaluating skewed. For comparison, a summary of different metrics calculated from one dataset for both the "Raw" and "Data Reduced" datasets in Figure 5 is summarized in Table 2. It is also worth noting that some of the metrics other than the mean appears to be more resilient to the effects of outliers and contamination.

Table 2- Metrics of the Site D Background Datasets at 7 m/s Ground Level Windspeed Bin

Metric	RAW	With Data Reduction
Linear Average	48 dBA	45 dBA
Logarithmic Average	52 dBA	46 dBA
Standard Deviation	7.4 dBA	2.3 dBA
Median	47 dBA	45 dBA
Mode (1 dB bin)	44-45 dBA	44-45 dBA

3.3 Logarithmic Averaging of Sampled Data

Seen in the Table 2, there is a large difference between the linear average values and the logarithmic average values for the raw dataset, significantly more than the dataset after extraneous data was removed. Indeed, this example demonstrates an effect of the logarithmic averaging tending to deliver higher values compared to the linear average. This difference in the average increases with increasing variation of the sampled dataset.

This is best illustrated in Figure 6; where the difference in the two types of averaging of the example audit datasets are plotted again the dataset's standard deviation values.

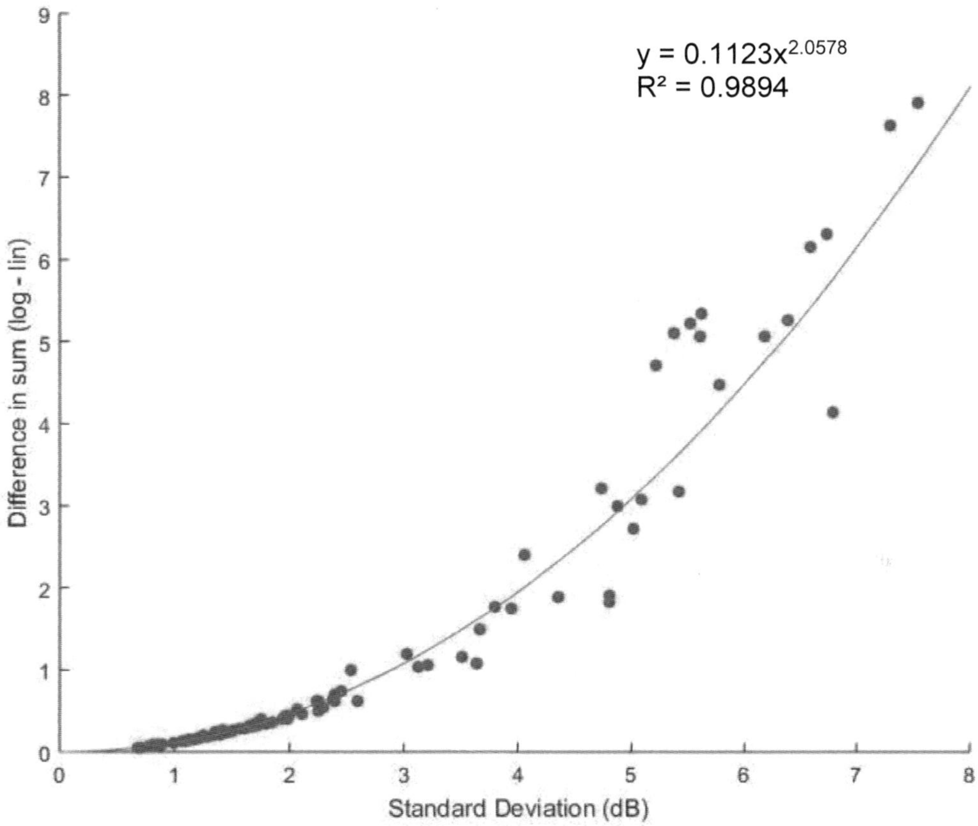

Figure 6 – Difference between Logarithmic Average and Arithmetic Average, as it relates to Standard Deviation

This phenomenon occurs whether the collected dataset is skewed or normally distributed. As the standard deviation increases, the logarithmic average deviates farther above the linear average value.

While Figure 6 does show significant differences, up to 5 dB when measured data is a standard deviation of 6 dB, it is important to remember several mitigating factors:

1.) With proper data reduction, much lower standard deviation can be achieved, in the range of 2 – 4 dB as shown in Section 3.1. The bias of the logarithmic average can be kept low.

2.) High standard deviations are more commonly seen for Background data at low ground level wind speeds. If the signal to noise ratio is sufficiently high (as in the case in many low wind speed situations), the overestimation of the average level may have minor effect on the resultant Turbine-Only value, despite higher standard deviations of the Background dataset.

3.4 Uncertainty

Lastly, a discussion of the uncertainty associated with a far-field measurements. As the reader may know, measurements would ordinarily consider two main type of errors: Type A (Statistical) and Type B (Equipment) uncertainty. Using the methods outlined in the IEC 61400-11 Standard [3], the combined uncertainty of the Turbine-Only Sound Level for the eight sites are estimated.

Selected results are included in the discussion, presented in Table 3. The uncertainty is dependent on three factors from the TON and Background datasets: standard deviation, sample size, and signal-to-noise ratio. These values are included in the table.

Table 3 – Selected Examples of Combined Uncertainties for Far-Field Audit

	Signal-to-Noise Ratio (TON/BG)	Sample Size (TON, BG)	Standard Deviations (TON, BG)	Combined Standard Uncertainty* (dB)
Site A, 4 m/s	6	~200,~100	0.7,2.5	1.0
Site C, 5 m/s	3	~70, ~40	1.5,1.6	1.1
Site D, 3m/s	12	~70,~50	0.9,7.9	1.5
Site H, 2 m/s	3	~500, ~400	1.6, 5.7	1.1

* Type B uncertainty of 1 dB

The cases presented above illustrate the influence on uncertainty of the three different factors. While standard deviation of the measured data may be high for some of the examples shown, the overall Type A uncertainty of the mean is quickly brought low with increasing number of samples, proportionally to root of the sample size. This is illustrated by Site D and Site H: Site H has a lower combined uncertainty than Site D, despite having a similar standard deviation values. In this case, the higher sample size contributed to a lower uncertainty.

Note that there is a diminishing return on the benefits on increased sample size due to the limitations of the measurement equipment (Type B uncertainty). With a sufficiently large sample size, Type A uncertainty is reduced below typical values for Type B uncertainty. Further increasing the sample size thereafter would not result in reducing overall uncertainty of resultant Turbine-Only Levels. This is illustrated by Site A and Site C: Site A has more samples than Site C, however the combined uncertainty is the similar between the two sites.

4. Conclusion

In the current study, several statistical factors involving far-field measurement dataset are examined to understand their impact to analysis. Combined with the Author's earlier study [6], this study provides some statistical insights that are important for conducting a good far-field sound measurement campaign.

References

[1] Ministry of the Environment and Climate Change, "Compliance Protocol for Wind Turbine Noise," Government of Ontario, Toronto, 2017.

[2] "Rule 12: Noise Control," Alberta Utilities Commision, 2021.

[3] "ETSU-R-97- The Assessment & Rating of Noise from Wind Farms," Energy Technology Support Unit, 1997.

[4] "Ed 3.1 IEC 61400-11: Wind Turbines- Part 11: Acoustic noise measurement techniques," International Electrotechnical Commision , 2018.

[5] R. Peters, "Uncertainty in Acoustics: Measurement, Prediction and Assessment," CRC Press, New York, 2020.

[6] D. Halstead and N. Tam, "A Study of Background Noise Levels Measured During Far Field Receptor Testing of Wind Turbine Facilities," in *8th International Conference on WInd Turbine Noise*, Lisbon, Portugal, 2019.

9th International Conference
on
Wind Turbine Noise
Remote from Europe – 18th to 21st May 2021

Audibility and health effects of infrasound

Frits van den Berg, Mundonovo sound research: fvdberg@mundonovo.nl
Irene van Kamp, RIVM (National Institute for Public Health and the Environment, Netherlands)

Summary

A range of residential health effects are attributed to the presence of wind turbines. Infrasound is sometimes mentioned as a possible cause of these effects, also when the infrasound levels are very low or unknown. Acoustically, infrasound is different from sound at higher frequencies: it is attenuated less over larger distances and through building façades. But does infrasound have effects on people that are different from effects of normal sound? Does infrasound sound deserve special consideration with respect to the effects of wind turbine sound?
Infrasound hearing thresholds are known to be very high and at 20 Hz joins the higher frequency thresholds. The audible range of infrasound is known to correspond to a small range of sound levels when compared to higher frequency sound. In a 2017 review report we concluded there were no clear indications that infrasound has health effects that are essentially different from effects of 'normal' sound. Since then a number of studies have shed some light on the relation between infrasound and brain activity and the sense of balance. This included sound above and below the hearing threshold. The recent studies give more insight in how the brain processes infrasound. As yet, the studies largely confirm conclusions based on earlier studies.

1. Introduction

Residential health effects in relation to wind turbines are attributed to infrasound and/or low-frequency sound from wind turbines. Acoustically low-frequency sound and infrasound are different from sound at higher frequencies. When compared to sound at higher frequencies low-frequency sound and infrasound is attenuated less over larger distances and through building façades. But can infrasound or low-frequency sound have effects on people that are different from effects of normal sound? This was investigated in our recent review report (van Kamp and van den Berg, 2020) and this paper provides an overview of the results of that report concerning infra/low-frequency sound, extended with more information and emphasis on the evidence pro and contra sub-audible and/or 'other' effects of infra/low-frequency sound.

Infrasound is defined as acoustic pressure waves in air with frequencies up to 20 Hz. There is no generally accepted definition of low-frequency sound, but it usually involves sound at frequencies up to 100 - 200 Hz and may or may not include infrasound. We will use the term audio sound for sounds at intermediate and higher frequencies, that is: at frequencies above about 100 Hz to 200 Hz. Thus audio sound is similar to what we mean by 'normal sound' except for an overlap where low-frequency sound merges with audio sound –with no clear boundary between the two sound ranges.

Low-frequency sound can be heard daily from road and air traffic and many other sources. It is usually included as part of the normal frequency range of everyday sounds. Less is known about infrasound and the perception of infrasound is not as common as it is for low-frequency or 'normal' sound. However, infrasound is not unique for wind turbines; it is produced by natural sources, big animals and transport and man-made machinery at levels comparable to what wind farms produce. Due to the high threshold of hearing, we are usually not aware of most of this infrasound.
Small differences in the infrasound hearing threshold can result in large differences in sensation because of the steep increase of loudness at infrasound frequencies. This is why in some studies infrasound is not presented at a specific level, but at similar loudness as perceived by the test persons.

2. Perception of infrasound and low-frequency sound

Behler and Uhlenkamp (2020) measured brain activity in 19 young, normal hearing persons exposed to sounds in an MRI-scanner. Individual hearing thresholds at 8 and 32 Hz were comparable to thresholds known from literature. At low loudness for both tones there was some activity in the auditory cortex, at medium loudness there was significant activity at the same brain locations. The auditory cortex is the location in the brain were sound is known to be processed and this study shows that this is also true for infrasound.

Burke et al (2019) investigated whether the ability to hear a sound was influenced by the presence of another sound. This was tested using two infrasound tones (5 Hz and 12 Hz), two audio-sound tones (100 Hz and 1000 Hz), and pink noise between 250 and 4000 Hz. Again, hearing thresholds of each of the 13 young, normal hearing participants were in the range of thresholds known from literature. Adding a soft audio-sound (5 dB above the threshold of that sound) to the 5 or 12 Hz infrasound did not significantly influence the detection threshold for the infrasound. However, the presence of a louder audio-sound (50 dB above threshold) on average did lead to a raise of the detection threshold of the infrasound. Adding sub-audible or barely audible infrasound did not change the detection threshold of the audio-sound. Thus, a medium loud audio-sound raises the detection threshold for infrasound. But a (very) soft audio-sound has no effect on the audibility of infrasound and vice versa.

Krahé et al (2020) did an experimental study with 39 participants exposed to four different infrasounds and complete silence, each for half an hour, in a very quiet, home-like room in a remote building. The infrasound was presented at levels close to a standard threshold used in Germany or at 10 dB above this threshold. It was expected that some participants would be able to hear the sound and some not. The results showed that, on average, the participants perceived the silence period as not annoying, the period with lower frequencies (3 and 5 Hz) as somewhat annoying, with higher frequencies as moderately annoying. However, for most sounds (3, 10, 18 Hz) individual scores covered the entire scale from not annoying to very annoying. For the 5 Hz condition and the silent period scores ranged from not to rather annoying. Participants perceived the 18 Hz sound as rumbling and humming, the others sounds as rumbling and pulsating. 'Predisposed' participants (with an earlier infrasound problem) were not found to react differently from the other participants. The authors concluded that 'essentially' perception is sensed by the ears, even when there is not always a hearing sensation.

Jurado and Gordillo (2019) investigated if fluctuations in the level of a low-frequency sound influenced the perceived loudness of that sound. This was tested with 24 young, normal hearing participants who matched the loudness of three simple low-frequency tones (40, 63 or 80 Hz) and one 1000 Hz tone with a number of tone combinations. Each combination consisted of two tones close in frequency to one of the three simple tones. The combination produces fluctuations (variations in amplitude) at a frequency equal to the difference in the frequencies of both tones. The differences in frequency were 1, 2, 5 and 12 Hz. The results show that the effect of fluctuation at the lower frequencies on loudness was modest and corresponded to 2 dB or less. The results were in agreement with what was already known in literature and loudness models.

Marquardt and Jurado (2018) investigated the perception of amplitude modulation in wind turbine sound: if we hear swishing or beating in wind turbine sound, is that just the sound level variation in the WT audio-sound with the frequency of blades passing the tower, or is there an (added) effect of the infrasound peak at that same frequency? The perception of the (simplified) phenomenon was investigated for two sounds at discrete frequencies and either a modulation of the tone at 8 Hz or an 8 Hz infrasound tone added to the unmodulated sound. It was concluded that a combination of a tone together with a constant 8 Hz infrasound is similar to our ears to the tone amplitude modulated at 8 Hz (without the infrasound): we cannot hear a difference between both.

3. Vestibular effects and infra/low-frequency sound

Jurado and Marquardt (2020b) investigated the effect of airborne infrasound on the vestibular system. They used a clinical method to assess the functioning of the vestibular system by measuring the electric potential related to muscle contraction (EMG). Earlier research has shown that the vestibular system can be activated by a loud mid- to high-frequency sound. In response to this a muscle in the neck and a muscle attached to the eye contract and this can be measured by EMG. In clinical practice loud clicks are used, either from 6 millisecond sound bursts every 0.2 second (= a repetition rate of 5 Hz) or a continuous loud tone modulated at 40 Hz. To these clinically used stimuli three low-frequency stimuli were added: a continuous sound over 120 seconds with a frequency of either 5, 16 or 40 Hz. All these sounds were presented to 15 normal-hearing participants and to each ear separately, all at levels corresponding to loud sounds. Only the electromyogenic (EMG) reaction to vertical acceleration of the head was measured, not a reaction to horizontal acceleration or rotation of the head. The results showed that the 500 Hz sounds (as used in clinical tests) were significantly related to an EMG response for most participants. In contrast, at the low frequencies the response was predominantly not significant. At 4 Hz there was no significant response at all, at 16 and 40 Hz in four of the 15 participants (of which one with both ears). The authors doubt that infrasound can produce accelerations of the head at lower sound levels, such as occurring near wind turbines.

In the report of Krahé et al (2020), mentioned in section 2, participants were also submitted to several neurological tests during exposure. All tests concerned the sense of balance and included keeping balance, performing targeted movements, the occurrence of nystagmus (repetitive, uncontrolled eye movement) and eye fixation. Tests showed no differences between the different sound exposure scenarios and no differences between predisposed participants and others. Participants were also asked to rate their perception of vibration, pressure and unease when exposed to each one of four infrasounds or silence (see description in section 2). All were perceived mainly in the head area (head, brain, ears), much less in other body parts. This applied to every sound scenario, including silence.

Takahashi (2017) investigated if very low-frequency sound could be experienced as a vibration of the head or body. He exposed four normal-hearing participants to six infrasound and low-frequency tones from 16 to 50 Hz in an office type setting. By varying the sound level for each frequency, the hearing threshold was determined, as well as the levels where the sound started to be 'slightly annoying', 'very annoying' or 'too loud to work' and the levels where the sound became unpleasant (unpleasant threshold) and participants felt a 'vibration in the head' (vibration threshold). The results show that the level where participants felt a vibration in the head was on average about 6 dB (at 16 Hz) to 15 dB (40 Hz) above their average hearing threshold. This vibration threshold almost coincided with levels at which the sound started to be slightly annoying. The threshold above which the sound was rated as unpleasant was still higher and was close to levels at which the sound started to be 'too loud to work'. In an earlier study Takahashi (2013) investigated the perception of vibration in the body and head when exposed to low-frequency sound. This also showed that the threshold for vibration in the head was higher than the threshold of hearing when the 14 participants were exposed to sounds at 16 to 50 Hz. Because this threshold was the same as the threshold for 'vibration in the body', he concluded that the head was the most sensitive part of the body to feel vibrations from infrasound.

4. (Alleged) effects of infra/low-frequency sound

Three authors (and their colleagues) have written about adverse effects of the (very) low frequency content of wind turbine sound. They claim that the infra/low-frequency sound could lead to a disturbance of the balance system (Pierpont), a brain reaction (Salt et al) or degradation of body tissue (Alves-Pereira et al).

Pierpont (2006) coined the term Wind Turbine Syndrome (WTS) for "a complex of symptoms which start when local turbines go into operation and resolve when the turbines are off or when the person is out of the area." The symptoms observed in affected persons were: sleep problems; headaches (more frequent or severe); dizziness, unsteadiness and/or nausea; exhaustion, anxiety, anger, irritability, and depression; problems with concentration and learning; tinnitus (Pierpont, 2006). Perhaps the dizziness led to the assumption that the balance system was involved. According to Pierpont wind turbines have an effect on the balance system (eyes, receptors in joints and muscles, balance organ) via visual disturbance (moving blades and shadows) and, she hypothesized, low-frequency sound waves having an effect on the vestibular system (balance organs at inner ear). This was the basis of the assumed affliction VVVD (visceral vibratory vestibular disturbance) causing the WTS.
It is not clear why Pierpont did not consider stress ("a state of mental or emotional strain caused by adverse circumstance", Healthline, 2021a) to be a causal condition for the health complaints. After all, it is well known that chronic annoyance can lead to stress and the reported symptoms fully correspond to known stress symptoms. Sleeping problems, headaches, decreased energy, depression and irritability are some of the symptoms related to stress –from whatever cause. Dizziness and tinnitus are not in themselves causes of stress, but stress and anxiety can aggravate them. In a positive feedback loop, the stress can worsen the dizziness and tinnitus (Hear-it, 2021; Healthline, 2021b) and thus may lead to perceptions that earlier were not problematic.

Salt and Hullar (2010) concluded that human hearing perception, as mediated by the inner hair cells of the cochlea, is "remarkably insensitive to infrasound". But the outer hair cells are more sensitive to infrasound and can be stimulated by low frequency sounds at sub-audible levels. Also, under some clinical conditions (e.g. Meniere's disease, superior canal dehiscence, asymptomatic cases of endolymphatic hydrops) a person may be very sensitive to infrasound.
One year later, Salt and Kaltenbach (2011) stress and stretch the possibilities of harm from subaudible infrasound. They remark that "dismissive statements such as 'there is no significant infrasound from current designs of wind turbines' are undoubtedly false", although in our opinion the falseness can depend on the interpretation of the term 'significant'. They suggest that "the possibility that low-frequency components of the sound could contribute both to high annoyance levels and possibly to other problems that people report as a result of exposure to wind turbine

noise cannot therefore be dismissed out of hand." But, in contrast to high level sounds," there is no evidence that low-level infrasound causes this type of direct damage to the ear. So infrasound from wind turbines is unlikely to be harmful in the same way as high-level audible sounds." With no medical evidence of such harm, evidence is sought in "numerous reports (e.g., Pierpont, 2009; Punch, James, & Pabst, 2010) that are highly suggestive that individuals living near wind turbines are made ill, with a plethora of symptoms that commonly include chronic sleep disturbance." We agree that the absence of research linking possible subaudible effects of infrasound to health complaints does not prove such a link does not exist. But we disagree with the statement there is "ample evidence to support the view that infrasound could affect people" (Salt and Kaltenbach, 2011). In fact, there is only an indication that subaudible sound may lead to a nerve signal, but no evidence at all of any effect of this.

A third claim to detrimental health effects from infra/low-frequency sound from wind turbines is from a Portuguese group. They (and only they) report on non-auditory effects of high-level and low-frequency sound, collectively named vibroacoustic disease (VAD). Only one peer-reviewed experimental study has been published and this was on workers in an aircraft industry with subjects exposed to different sound levels (from <70 dB to >90 dB; sources not stated) (Marciniak et al, 1999). With echography it was found that tissues in the heart were statistically significantly different in the high exposure vs. the low exposure group and this was thought to be a symptom of VAD. A similar study was done by Jensen et al (2009) with military crew chiefs exposed to high-level jet fighter sound (up to 124 dB Leq) while working or having worked close to aircraft with running engines. Health records of 42 crew chiefs were compared to health records of 42 aircraft mechanics not subject to this high sound exposure. The periodic health checks were on a number of items, though not including pericardial thickening. There were slightly more disease cases and less ear diseases amongst the mechanics, but none of the health differences between both groups were significant.

In a conference paper members of the same Portuguese group and local researchers reported significant pericardial thickening in a study of residents of the Caribbean island Vieques (Torres et al, 2001). Residents were exposed to military aircraft sonic booms at levels up to 120 dB Lmax). The results from this study were scrutinized by a panel of experts and they found no clinically relevant difference pericardial thickness between case and control subjects. In fact, the measurements showed any subject's pericardial thickness to be normal, based on published literature (ATSDR, 2001).

In conference papers presented by persons of the Portuguese group (e.g. Alves-Pereira and Castelo Branco, 2007), low frequency sound from wind turbines was mentioned as the cause of health complaints and vibroacoustic disease, but no experimental data were shown to support this.

Most recently Weichenberger et al (2017) found brain activity related to 'inaudible' infrasound. They first determined hearing thresholds and loudness estimates of 12 Hz sound (and other frequencies) with 14 young, normal hearing participants. In an MRI-scanner participants were exposed to three different conditions: the 12 Hz tone was presented to the right ear at either the medium-loud level or 2 dB below the individual threshold, or no sound was presented. Each condition lasted 200 seconds, which is relatively long. When exposed to the medium loud infrasound or to no sound no corresponding brain activity showed up. The authors speculate that this may be due to adaptation of the neurons: with a constant stimulus the activation decreases over time, and averaged over the 200 seconds of exposure the brain activation was not strong enough to show up in the measurements. In an earlier study with almost the same test persons and the same instrumentation the participants were exposed to short bursts (3 seconds) of 12 Hz and medium loudness infrasound, and this did result in significant brain activity in the auditory cortex. In the ;later study exposure to the 200 second 12 Hz infrasound at a level just beneath the individual hearing threshold elicited brain activity not found in the other two conditions. The authors take a big leap when speculating that this could be linked to physiological as well as psychological health effects. In our opinion we first need to be sure this is a true effect of an *in*audible sound. The stimulus was so close to audibility that it probably could stimulate the brain: if the average value (Leq) is 2 dB below the threshold, then the peaks (Lmax) are 1 dB above. Also, the hearing threshold is not –at least in measurements- a precise and definite value: Burke et al (2019) measured for each of 13 young, normal hearing participants the hearing threshold three times and generally for each participant the three outcomes varied over a 5 dB range. Weichenberger et al (2017) suggest that participants were "constantly left guessing, whether stimulation actually occurred or not when near-threshold infrasound was presented", which may support the notion of a signal 'close to' and not entirely 'below' the hearing threshold. Whatever the explanation, the brain activity occurred near the audibility threshold and not at lower levels further away from the threshold. This would be necessary for much lower wind farm infrasound levels to have an effect.

5. Infra/low-frequency sound in the environment

Infrasound as well as low-frequency sound are normal features of the everyday natural, rural and urban environment. The interaction of wind with vegetation and buildings is an important natural source of infrasound, but we usually do not hear it because of the high hearing threshold for infrasound. Traffic is an important artificial source of infrasound. Several measurement campaigns have shown that environmental infrasound levels are comparable to residential infrasound levels from wind turbines. A German measurement survey (Hermann et al, 2016) found infrasound levels at 150 m from a wind turbine ranging from 55 to 80 dB(G). At 650-700 m the level was 55-75 dB(G), irrespective of the wind turbines being operational or not. The infrasound level of city traffic was 55-80 dB(G), corresponding to published infrasound levels of sea surf. Background sound levels at quiet urban locations were mostly between 45 and 65 dB(G). Similar results were found in Australia (Evans et al, 2013): in urban areas infrasound levels between 60 and 70 dB(G) were common and typically 5 to 10 dB(G) lower at night time. Air conditioning systems of office buildings were identified as significant sources of infrasound. In rural areas infrasound levels could be as low as 40 dB(G), but 50-70 dB(G) at higher wind speeds. Again, there was no noticeable difference between locations at or far from wind farm sites. Similar results were found in Poland and it was concluded that "infrasound noise levels emitted by wind turbines do not reach levels causing a threat to people, and are comparable to levels of natural acoustic background common in the environment" (Ingielewicz and Zagubién, 2014).

Infrasound is audible at levels above about 96 dB(G) (Jakobsen, 2001), so urban and rural infrasound will usually not be audible. But audible levels of infrasound do occur in everyday life. E.g. Hermann et al (2016) found that washing machines, while spinning, produce infrasound reaching the audible threshold level. They also showed that the interior noise in a car can be well above threshold: with closed windows a level of 105 dB(G) was measured with closed windows and 139 dB(G) with a rear window open.

In our 2017 review (Van Kamp and van den Berg, 2018) we made reference to papers from Leventhall (2013) and Stead et al (2014). Leventhall (2013) notes that infrasound at low level is not known to have an effect. Normal pressure variations inside the body (from heart beat and breathing) cause infrasound levels in the inner ear that are greater than the levels from wind turbines. From exposure to high levels of infrasound, such as in rocket launches and associated laboratory studies or from natural infrasound sources, there is no evidence that infrasound at levels of 120 – 130 dB causes physical damage to humans, although such exposure may be unpleasant.
Stead et al (2014) came to a similar conclusion when considering the regular pressure changes at the ear when a person is walking at a steady pace. The up and down movement of the head implies a slight change in atmospheric pressure that corresponds to pressure levels in the order

of 75 dB. The pressure changes in the rhythm of the walking frequency are similar in frequency (close to 1 Hz) and in level to pressure changes from infrasound at turbine rotation frequencies measured at houses near wind farms. The eardrum moves in reaction to pressure changes and thus walking would cause an infrasound sensation if we could hear such low levels. But 75 dB at 1 Hz is far below the human perception threshold.

6. Conclusions

Research has shown that the perception of infrasound and low-frequency sound is processed in the same brain areas as 'normal' sound. The threshold of detection of such sound as determined by brain research is in agreement with thresholds determined 'classically' by listening tests.
Further research shows that sub-audible infrasound and very low-frequency sound do not lead to a reaction of the balance system or a sensation of vibration. Such a sensation only occurs at audible levels.
There is no evidence that diseases or afflictions can result from infrasound or low-frequency sound that are different from such effects of 'normal' sound. Any sound, including wind turbine sound, can be annoying and in some people lead to stress. There is no need for a Wind Turbine Syndrome based on erroneous ideas of the impact of sound, just as there is no need for a Road Traffic or Aircraft Syndrome (even though these noise sources can also lead to stress).

The ubiquitous presence of inaudible Infrasound at levels close to or above wind turbine infrasound levels demonstrates that such levels cause no harm. If so, walking would make us ill because of the air pressure variations at the ears during walking.

References

Alves-Pereira, M and Castelo Branco NAA (2007) *In-Home Wind Turbine Noise Is Conducive to Vibroacoustic Disease* Proc. Second International Meeting on Wind Turbine Noise, Lyon

Alves-Pereira, M and Castelo Branco NAA (2014) *Letter to the Editor re: 'How the factoid of wind turbines causing 'vibroacoustic disease' came to be 'irrefutably demonstrated'* Aust N Z J Public Health 38(2), 191-192

ATSDR (2001) *Expert Review of the Vieques Heart Study Summary Report for the Vieques* Heart Study Expert Panel Review (retrieved from https://www.atsdr.cdc.gov/sites/vieques/heart_study_summary.html, 16-2-2021)

Behler, O. and S. Uppenkamp (2020) *Activation in human auditory cortex in relation to the loudness and unpleasantness of low-frequency and infrasound stimuli* PLoS ONE 15(2)

Burke, E., Hensel, J., Fedtke, T., Uppenkamp, S., & Koch, C. (2019). *Detection Thresholds for Combined Infrasound and Audio-Frequency Stimuli* Acta Acustica united with Acustica, 105(6), 1173-1182

Evans, T, Cooper J, Lenchine V (2013) *Infrasound levels near windfarms and in other environments* Report of the South Australian Environment Protection Authority and Resonate Acoustics, Adelaide

Healthline (2021a) *11 Signs and symptoms of too much stress* (retrieved from www.healthline.com/nutrition/symptoms-of-stress, 16-2-2021)

Healthline (2021b) *What's the Link Between Stress, Anxiety, and Vertigo?* (retrieved from www.healthline.com/health/can-stress-cause-vertigo, 16-2-2021)

Hear-it (2021) *Tinnitus and stress* (retrieved from https://www.hear-it.org/Tinnitus-and-stress, 16-2-2021)

Herrmann L, Ratzel U, Bayer O, Krapf K, Hoffmann M, Blaul J, Mehnert C (2016) *Low-frequency noise incl. infrasound from wind turbines and other sources* Report from the State Office for the Environment, Measurement and Nature Conservation of the Federal State of Baden-Wuerttemberg

Ingielewicz R, Zagubien A (2014) *Infrasound noise of natural sources in the environment and infrasound noise of wind turbines* Polish Journal of Environmental Studies 23(4), 1323-1327

Jakobsen J (2001) *Danish guidelines on environmental low frequency noise, infrasound and vibration* J Low Freq Noise, Vib and Active Control 20(3), 141 - 148

Jensen, A, Lund S, Lucke T, Clausen O, Svendsen J (2009) *Non-auditory health effects among air force crew chiefs exposed to high level sound* Noise and Health 11(44)

Jurado C and Marquardt T (2020b) *On the Effectiveness of airborne infrasound in eliciting vestibular-evoked myogenic responses* Journal of Low Frequency Noise, Vibration and Active Control, 39(1) 3-16

Jurado C, D Gordillo, BCJ Moore (2019) *On the loudness of low-frequency sounds with fluctuating amplitudes* Journal of the Acoustical Society of America, v146 n2 1142-1149.

Krahé D, Alaimo Di Loro A, Müller U, Elmenhorst E, De Gioannis R, Schmitt S, Belke C, Benz S, Großarth S, Schreckenberg D, Eulitz C, Wiercinski B, Möhler U (2020) *Lärmwirkungen von Infraschallimmissionen (Noise effects from infrasound immissions)* Report Umweltbundesamt (German Environmental Agency) available at https://www.umweltbundesamt.de/publikationen/laermwirkungen-von-infraschallimmissionen (report in German with English summary)

Leventhall, G (2013) *Infrasound and the ear* Proc. 5th International. Conference on Wind Turbine Noise, Denver

Marciniak, W, Rodriguez E, Olszowska K, Atkov O, Botvin I, Araujo A, Pais F, Soares Ribeiro C, Bordalo A, Loureiro J, Prazeres De Sá E, Ferreira D, Castelo Branco MS, Castelo Branco NA (1999) *Echocardiographic evaluation in 485 aeronautical workers exposed to different noise environments* Aviation Space and Environmental Medicine 70(3), 46-53

Marquardt and Jurado C (2018): Amplitude Modulation May Be Confused with Infrasound. Marquardt T, C Jurado. Acta Acustica united with Acustica, Vol. 104 (2018) 825 - 829

Pierpont, N (2006) *Wind Turbine Syndrome: The Health Effects of Wind Turbine Noise and Shadow Flicker* (retrieved from http://static1.1.sqspcdn.com/static/f/168835/2455987/1233614126047/Nina+Pierpont+-+Wind_turbine_syndrome_with_refs.pdf?token=eaP%2Fo JCfbIArJtkpImmbqpfU6vs%3D, 16-2-2021)

Salt, AN and Hullar TE (2010) *Responses of the ear to low frequency sounds, infrasound and wind turbines* Hear Res. 268(1-2), 12-21.

Salt, AN and Kaltenbach JA (2011) *Infrasound from Wind Turbines Could Affect Humans* Bulletin of Science, Technology & Society 31(4) 296-302

Stead, M, J. Cooper J, Evans T (2014) *Comparison of infrasound measured at people's ears when walking to that measured near wind farms* Acoustics Australia 42(3), 197-203.

Takahashi Y (2013) *Vibratory sensation induced by low-frequency noise: The threshold for 'vibration perceived in the head' in normal-hearing subjects* Journal of Low Frequency Noise Vibration and Active Control 32(1/2), 1-10

Takahashi Y (2017) *Study on the relationship between unpleasantness and perception of vibration in the head of subjects exposed to low-frequency noise* Proc. 46th International Congress and Exposition on Noise Control Engineering

Torres R and many others (2001) *Vibroacoustic Disease Induced by Long-Term Exposure to Sonic Booms* Proc. International Congress and Exhibition on Noise Control Engineering, The Hague, The Netherlands

van Kamp I, van den Berg F (2018) *Health effects related to wind turbine sound, including low-frequency sound and infrasound* Acoustics Australia, 46(1), 31-57.

van Kamp I, van den Berg GP (2020) *Health effects related to wind turbine sound: an update* RIVM report 2020-0150

Weichenberger M, Bauer M, Kühler R, Hensel J, Forlim CG, Ihlenfeld A (2017) *Altered cortical and subcortical connectivity due to infrasound administered near the hearing threshold- Evidence from fMRI* PLoS ONE 12(4)

9th International Conference

on

Wind Turbine Noise

Remote from Europe – 18th to 21st May 2021

An Implementation of ISO/PAS 20065:2016 for the Analysis of Wind Turbine Sound at Receptor Distances

Robin Woodward, Hayes McKenzie Partnership Ltd, on behalf of IEC TC88/PT 61400-11-2: robin@hayesmckenzie.co.uk

Summary

This paper gives an overview of an implementation of the tonal analysis methodology found in ISO/PAS 20065:2016 for use on wind turbine sound at receptor distances. This implementation has been developed by an expert group on tonality formed by the IEC TC88 subcommittee PT61400-11-2, made up of some of the members of said committee, along with other experts in the field of (wind farm) tonal analysis. This implementation includes clarifications of aspects of the method which are not clear, or open to interpretation, in the PAS 20065 as applied to wind turbine sound; specification of input parameters left either to the practitioners discretion, or as a range of options; some deviations from the PAS 20065 method; specifics on how to combine data and calculate results in relation to wind speed; and specifics of what should be reported and how. In addition some aspects of the method were found to require further investigation to be implemented satisfactorily, and these are described.

1. Introduction

Following an initially rejected proposal for a new international standard on the measurement of wind turbine noise characteristics at receptor positions in 2017, the subsequently submitted revised 2nd proposal for a technical specification (TS) received approval by the IEC/TC88 member committees in April 2018. The International Electrotechnical Commission (IEC) Technical Committee TC88 tasked Mr. Bo Søndergaard from Denmark with organizing and convening the new project team, PT61400-11-2 (PT11-2), and the first meeting of the group took place in Hamburg in June 2018. Further meetings were held on a 3 monthly basis through the remainder of 2018, 2019 and up to March 2020. Since March 2020 online meetings have been conducted at least once per month in an effort to try and reach a Committee Draft of the proposed technical specification, and this is expected to be submitted to the IEC by the summer of 2021.

The TS comprises measurement and evaluation methods relating to topics such as sound pressure level, amplitude modulation, tonality, impulsivity, low frequency, and the determination of wind speed and other non-acoustic measures describing the details of the implementation of such measurements under conditions relevant to wind turbine noise. The project team currently includes experts from at least 11 national committees, traditionally strongly represented from Europe and North America, but also a growing number of experts from Asia are interested in collaborating within the project team.

In September 2020 a 'Tonality Expert Group' was formed in an attempt to allow a wider group of experts to dedicate time specifically to this aspect of the TS. Prior to this time, PAS 20065 had been selected by the main PT11-2 committee as the tonality analysis method to be implemented, however few decisions had been made at that time regarding the specifics of its implementation. This expert group consists of members of the PT11-2 committee with experience of tonality analysis of wind turbine noise at receptor distances, but also sought contributions from a wider scope of practitioners, researchers, developers of previous standards, and stakeholders with experience and expertise. The call for experts was sent out both via the national committees for TC88, but also through invitation and peer recommendation. In addition liaison with ISO working group ISO/TC 43/SC 1/WG 45 was instigated and maintained throughout the process. Meetings have been held remotely approximately once a month at a time suitable for European and American time zones, with 4 additional meetings scheduled at times more suitable for Australasian time zones to ensure as wide an input as possible.

Particular thanks should go to contributions from Australia, Canada, Denmark, Germany, Japan and the UK.

2. ISO/PAS 20065:2016

PAS 20065 is the tonality method referenced in ISO 1996-2:2017 and, as such, is a logical choice for implementation for tonal analysis of sound from industrial plant measured as part of environmental noise surveys, such as those carried out for wind turbines. The PAS 20065 method is almost identical to the German DIN 45671:2005-03 standard. Although the DIN 45671 standard has been required for use on wind turbines in Germany since it was published in 2005, this was used primarily as a reference tool to quantify the tonality of individual samples, or short term data-sets, to confirm aural assessments of tonality at receptor locations, and as such is not regularly used in Germany either as a primary assessment tool, or on long-term (larger) data-sets.

Despite the PAS 20065 being published in 2016, and referenced in 1996-2 the following year, neither the PT11-2 committee, nor the tonality expert group were aware of any consultants using the method routinely for analysing tonality of long term measurements of wind turbines at receptor distances.

3. Consistency of Implementation

The tonality expert group initially had three organisations in a position to analyse larger data-sets, and who were doing so as part of consultancy work prior to the forming of the expert group. A reference data-set from four different measurements of three different sites comprising different wind turbine models, totalling 30 minutes of audio recordings, was compiled to allow comparison of the existing and future implementations of the PAS 20065 method. This data-set included wind turbine tone frequencies in the range of 40 to 200 Hz, and each audio recording was at least 2 minutes in length. The participants were asked to provide results based on spectrum averaging times of 3, 10 and 60 seconds, as the choice of averaging time was a specific question raised by the main PT11-2 committee. A results template was provided asking for various statistics of the tonal audibility and tone frequency for the three spectrum averaging times, and a 1 minute result based on each of these.

Within 3 months of the forming of the expert group a total of seven organisations had provided some results for the reference samples. These results varied, in some cases significantly, from organisation to organisation, highlighting a lack of consistency of implementation. Identified sources of variation included:

- Search range for tones (whether an upper limit of tone frequency had been selected and at what frequency)
- Whether an adaption of the PAS method had been implemented to account for tones below 50 Hz
- The frequency grouping of multiple tones in a single critical band
- The definition of the bounds of critical bands
- The narrowband parameters
- The assigning of -10 dB to tones with audibilities below 0 dB

Some of these sources of variation are due to a range of, or no, specifics being detailed for some input parameters (such as narrowband line spacing, window overlap, critical band edges), some are due to lack of clarity in the PAS 20065 text or layout (such as the frequency grouping), and some are due to organisations making practical decisions to account for common issues found when carrying out tonal analysis (such as the limiting of search frequencies, and the adaption to account for tones below 50 Hz).

The former and latter sources of variation identified above are relatively easily addressed by suggesting specific input parameter values, and allowing for routine deviations from the PAS 20065 method. The issue of clarity of the PAS 20065 text or layout was much more difficult to address as the 'accurate' interpretation (as written) had to be ascertained before a discussion of whether it was 'correct' and 'appropriate' or not.

To assist in this process several contributors to the PAS 20065 and DIN 45681 methods were consultedt to ascertain the intention behind some aspects of the text.

It should be noted that it was found to be particularly difficult to interpret aspects of the method (such as frequency grouping) which required reference to both the main PAS 20065 text and the diagrams in its Annex D to understand the intention. Additionally, in some cases, the nomenclature between these two information sources was not consistent.

4. ISO/PAS 20065:2016 Implementation Specifics in IEC TS 61400-11-2

The resulting implementation of PAS 20065 for the analysis of wind turbine sound at receptor distances detailed in the future IEC TS 61400-11-2 (at the time of writing, pre-committee draft) is as detailed below. The implementation states that, where turbine shut downs have been carried out during the measurements, the analysis should be carried out for turbine on and turbine off periods separated, and both data-set results reported. This is to allow the identification of tones emanating from sources other than the turbines, or indeed turbines on other sites, so that they can be excluded from consideration.

4.1 Instrumentation and Narrowband Spectra

Audio recordings are identified as the likely data source. Recordings should be made using a Class 1 sound level meter and should be uncompressed with a sampling rate of at least 12 kHz and a bit depth of at least 24. Note that if tones are suspected above 5 kHz the sampling rate would need to be higher. However, if narrowband spectra are created 'live', audio recording is still required as a reference source for auditioning potential tones, so similar quality requirements are still suggested.

With respect to the narrowband spectra, a fixed line spacing of 2 Hz is specified over a bandwidth of at least 20 Hz to 6 kHz. A hanning window is specified to be used, and an overlap of 75 % is specified for the DFT calculation. With an audio recording sampling rate of 12 kHz and a window size of 6000 samples, 25 individual basic spectra will be merged to create each 3 second average spectrum.

The further specification of narrowband spectra parameters was judged to be useful to try and standardise this aspect of the process to reduce variations between different practitioners. It was found within the expert group that, for the majority of cases, the narrowband spectra parameters were not varied from use to use, and often the implications of the various choices allowable by the PAS 20065 were not readily understood by practitioners. Where a practitioner was both suitably knowledgeable regarding narrowband spectra derivation, and believed that a different set of parameters should be used for analysing a specific data-set, this could be easily detailed as a deviation and justified accordingly.

4.2 Search Frequency Range and Critical Band Width

As wind turbine tones are known to occur below 50 Hz, the implementation allows for this by using a fixed critical band between 20 and 120 Hz for tones between 30 and 50 Hz. Below 30 Hz it is suggested the analysis becomes more problematic both in terms of measuring equipment and artefacts in the analysis. Even at frequencies between 30 and 50 Hz the possibility of false tones being identified due to the steepness of the A-weighting curve, mandated for use in the PAS 20065, should be borne in mind when audible tones are identified in this frequency range, and as such additional diligence is required to interrogate tones in this range. In addition tone levels should be compared to the relevant hearing threshold to confirm the likelihood of audibility (e.g. using ISO 226:2003 and ISO 28961:2012).

The choice of 30 Hz as the lower value was based in part on this accommodating most lower frequency tones measured previously by the expert group, and also that it was very close to the 32 Hz at which the *Perception of Noise from Large Wind Turbines* study from Salford University showed a good correlation of masking threshold between theoretical assumptions and listening test results.

When calculating a critical band around a tone (above 50 Hz) the edge frequencies should be rounded down to the next 2 Hz line for the purposes of allocating spectral lines for inclusion or exclusion within the critical band.

In addition to extending the method to include tones below 50 Hz due to their known occurrence for some wind turbines, it is often prudent to limit the upper search frequencies for tones from wind turbines where these are unlikely to occur, to reduce the amount of tones from other sources needing to be scrutinized and excluded. To this end the implementation includes guidance on limiting the upper search range for tones based on the distance from the nearest turbine and the atmospheric absorption likely to occur at the location based on the equations in ISO 9613-1:1993. However, where tones are reported or expected at higher frequencies than these equations might imply, an extension of this limited search range would be required, and/or the spectral content of the data scrutinised to ascertain the presence of potential tonal energy at higher frequencies.

4.3 Criteria for Tone Being Present

It is stated in Step 2 of 5.3.8 of the PAS that 'if $\Delta L_k>0$, then a tone is present.'. This statement may imply that potential tones where $\Delta L_k<0$ are in fact not tones. For the purposes of this implementation, tones with calculated audibilities below 0 dB should not be excluded from the further calculation steps based on this criteria alone, and therefore should still be included in frequency grouping calculations (see below).

The reasoning behind this divergence from the PAS method is that, although a potential tone with a $\Delta L_k<0$ may not be audible by itself, the perceptual concept of a critical band is that all tone energy within a critical band is perceived as a single tone, and therefore all tone energy is summed to account for this. Whether the energy of a potential tone on its own would cause a perceptible tone is irrelevant to the underlying critical band concept.

4.4 Multiple Tones within A Single Critical Band

Frequency grouping is the process of combining the energy of tones where multiples tones appear within a single critical band. This process is carried out after calculating the tonal audibility (from the combined energy of tone lines around the local maxima, and a subsequent correction for a masking and distinctness check – see Step 1 and Step 2 of 5.3.8 of the PAS 20065) for each individual tone found to be present. This process is detailed in part within the Step 3 of 5.3.8 of PAS 20065, and partly within Figure D.4 - *Detailed diagram 3* within Annex D of the PAS 20065. This process was found to be commonly interpreted in different way by practitioners, and therefore the clarifications below have been specified in the implementation with the aim to avoid these variations from occurring.

This process should be carried out as follows for each audible tone ('origin tone') found in Step 1 and Step 2 (of 5.3.8 of the PAS 20065) in turn:

- A critical band should be centred on the line of the origin tone with the maximum tone level
- The critical band should be searched either side of the origin tone for other previously identified tones
- Where another tone is found whose maximum tone level line falls within the critical band, all lines associated with this tone should be 'flagged' for inclusion within the frequency grouping (whether the individual tone lines, other than the maximum tone level line, fall within or without the critical band)
- Having searched the entire critical band, the origin tone and any other tones flagged should be compared to determine which tone has the maximum audibility calculated in Step 2, and this tone assigned as the 'dominant' tone
- The tone lines of the origin tone and all other flagged tone lines should have their energy summed (ensuring no line is summed multiple times for adjacent tones), and then the tonal audibility of the frequency group calculated using this summed tone level and the masking corrections calculated in Step 2 from the dominant tone
- The frequency assigned to this frequency grouping's tonal audibility should be that of the dominant tone

1)

4.5 Determination of Mean Audibility in each 1-minute Period

The time reference for the tonality output to be aligned with wind speed and wind direction is 1-minute (although other time references could be used if appropriate). The mean audibility of 20 x 3-second averaged spectra decisive audibilities should be calculated, corresponding to a 1-minute mean audibility. This should be done for each 1-minute period (synchronised so that the start of a 1-minute period lines up on the hour precisely).

This calculation should be carried out according to 5.3.9 of PAS 20065, however the adjustment of 3-second decisive audibilities, ΔL_j, to -10 dB where the values are below 0 dB should not be carried out, with these audibilities left with their original negative value for the purposes of energy averaging.

This divergence from the method within PAS 20065 has been made as, in this implementation, the use of 3-second averaged spectra for analysis purposes, which is then combined, via energy averaging, to come to a 1-minute mean audibility, has been chosen over the use of 60-second averaged spectra to allow for the variations within a period to be accounted for. Using the -10 dB audibility adjustment for 3-second spectra where decisive audibility is below 0 dB results in the advantage of following the peaks of audibility within a 1-minute period being off-set somewhat by small dips in audibility bringing down the energy average disproportionately. Where strong tonalities are present this makes little difference to the 1-minute result, but where tones fluctuate around the assumed threshold of audibility this can have more significant effect on results.

4.6 Data Binning

The PAS 20065 method does not have a procedure for correlating tonality with other parameters. However, this is a key factor in determining tonality from wind turbines, where wind speed and wind direction can both have a significant influence on tonalities measured. Therefore, in this implementation, following the calculation of 1-minute mean audibilities for the data-set, these audibilities should be aligned with 1 minute average (or longer time reference if necessary) wind speed and wind direction. These 1-minute data points should then be binned for wind speed and direction with a resolution of 1 m/s and 30 degrees (integer/north centred respectively). The audibilities within each bin should be have an arithmetic mean calculated to come to a bin audibility.

If additional wind direction bins are required for the purpose of the analysis (such as downwind sector), the above procedure should be followed with the alternative direction bin.

Where a bin includes between 1 and 6 datapoints, the values for this bin should be marked with an '*' to indicate that the average values presented are less reliable due to the low number of datapoints upon which they're based.

4.7 Reporting

The reporting required in the implementation includes the following:

- Specifics of the measurement instrumentation/position
- Description and justification of methods used to identify false-positives and false-negatives within the data and to remove extraneous noise sources (including details of the data excluded)
- Tables of sectoral binned results for average tonal audibility; non-excluded data count; and percentage of 1-minute datapoints above 0, 4 and 9 dB audibility.
- Charts of frequency corresponding to maximum decisive audibility for each 1-minute vs wind speed; mean audibility for each 1-minute vs wind speed; and wind speed vs wind direction with tonal audibility shown with colour coding for each 1-minute.
- An optional chart of sound pressure level vs wind speed with tonal audibility shown with colour coding for each 1-minute.

The aim of these reporting requirements is to ensure that the context of the tonality being reported, as well as clear and consistent tabulated values is presented. This should both give improved information from which any required judgements can be made, as well as providing a check on the robustness and relevance of the results.

5. Areas Where Further Investigation is Required

During the discussions and investigation carried out by the expert group, several areas were identified where the implementation of PAS 20065 for wind turbine sound at receptor distances was deemed to have likely scope for improvement, but for which the investigations to inform said improvements were too involved to be completed within the timescales prescribed for the committee draft of the technical specification. These areas, and the suggested investigations are detailed below. It is the intention that the tonality expert group will continue working after the committee draft of the technical specification to try and investigate these areas, with the possibility of including further refinements in future drafts or revisions of the technical specification.

5.1 Combining 3-second Tonal Audibilities

The standard approach to combine 3-second audibilities to create a 1-minute result (or any other large timescale audibility) in line with the PAS 20065 method would be the determination of mean audibility via energy averaging. However, at the time of writing, the project team believe this may not be appropriate, and indeed may not reflect subjective perception. It is thought that perception is likely to correlate more to peak audibility within a 1-minute period, or possibly some statistic of the 3-second audibilities biasing towards the peak audibility.

Although a wide array of literature was reviewed as part of the tonality expert group's efforts, it was not possible to find any examples of the assessment of varying tonality, and listening tests to inform tonality methods and research was found to generally be based on audio samples of at most 20 seconds in length.

To address this, a listening test is under development by members of the expert group to investigate the perception of tonality samples of 1 minute in length. The proposed listening tests will be carried out over headphones with a specified digital audio converter and headphone setup to ensure reproducible playback in a variety of testing locations. The primary tasks of the listening test will be comparative ranking of three sets of five audio samples, with each set being a different wind turbine neighbour receptor location. These samples will be monophonic recordings of actual wind turbine sound, selected so that the background noise throughout each set is consistent, but the level and occurrence of tonality throughout the 1-minute samples varies. The perception results will then be compared with the PAS 20065 and various other tonality methods, as well as several statistical implementations of the PAS 20065 method - likely to include peak 3-second decisive audibility, and various percentiles of mean audibility during a 1-minute period.

5.2 Audibility Threshold

There is anecdotal evidence, when comparing analysis results of wind turbines tones using the PAS 20065 method with both other established methods, and audition of recordings, that tones with calculated audibilities below 0 dB may still be audible. At present this has not been investigated to an extent which allows a recommendation to account for this in this implementation, but, if found to be a common occurrence with marginal tones, this would require further consideration and appropriate guidance or adjustment to account for this.

It is proposed that this also be investigated via the use of listening tests. Initially this will be combined with the listening tests detailed at 5.1 above, with a task purely determining whether or not a range of example tones between -3 and +3 dB audibility are in fact audible to the participant. If this confirms that some, or all, tones below 0 dB are in fact audible, a further listening test specifically investigating this, both in terms of tonal characteristics (frequency of tone and occurrence of tone in sample for example), and in an attempt to define an average level of audibility threshold across a number of samples/tones, will be required.

5.3 Critical Bands Not Centred on Tones

In the Joint Nordic Method 2 (the previous tonality method referenced by ISO 1996-2: 2007) the searching for maximum tonality within a critical band (by combining multiple tones where tones are grouped – referred to in PAS 20065 as frequency grouping) was not limited to critical bands centred on identified tones, as it is in PAS 20065, but included critical bands centred around any line in the spectrum, be it tone, masking or neither. There have been representations that this is in line with the psychoacoustic nature of tone perception. However, this is not the case for the PAS 20065 method, so it requires further investigation as to the most appropriate implementation. It is possible that this choice was made in the PAS 20065 in order to reduce computation time, but this is yet to be ascertained. If no clear reason can be found for this, it requires further investigation to ensure that the method remains robust and appropriate given this simplification.

References

ISO/PAS:20065: 2016 (2016) Acoustics – Objective method for assessing the audibility of tones in noise – Engineering method, ISO/TC 43/SC 1

ISO 1996-2:2017 (2017) Acoustics — Description, measurement and assessment of environmental noise — Part 2: Determination of sound pressure levels, ISO/TC 43/SC 1

DIN 45681:2005-03 (2005) Acoustics - Detection of tonal components of noise and determination of tone adjustment for the assessment of noise immissions, Beuth Verlag, Berlin, Germany. German industry standard.

Perception of Noise from Large Wind Turbines (EFP-06 Project) (2010), Sabine von Hünerbein, Andrew King, Jonathan Hargreaves, Andrew Moorhouse, Chris Plack, University of Salford

ISO 226:2003 (2003) Acoustics — Normal equal-loudness-level contours, ISO/TC 43

ISO 28961:2012 (2012) Acoustics — Statistical distribution of hearing thresholds of otologically normal persons in the age range from 18 years to 25 years under free-field listening conditions, ISO/TC 43

ISO 9613-1:1993 (1993) Acoustics — Attenuation of sound during propagation outdoors — Part 1: Calculation of the absorption of sound by the atmosphere, ISO/TC 43/SC 1

ISO 1996-2: 2007 (2007), Acoustics — Description, measurement and assessment of environmental noise — Part 2: Determination of environmental noise levels, ISO/TC 43/SC 1

AUTHOR INDEX